Lecture Notes in Computer Science 11539

More information about this series at http://www.springer.com/series/7407

João M. F. Rodrigues · Pedro J. S. Cardoso ·
Jânio Monteiro · Roberto Lam ·
Valeria V. Krzhizhanovskaya ·
Michael H. Lees · Jack J. Dongarra ·
Peter M. A. Sloot (Eds.)

Computational Science – ICCS 2019

19th International Conference
Faro, Portugal, June 12–14, 2019
Proceedings, Part IV

Springer

Editors
João M. F. Rodrigues
University of Algarve
Faro, Portugal

Pedro J. S. Cardoso
University of Algarve
Faro, Portugal

Jânio Monteiro
University of Algarve
Faro, Portugal

Roberto Lam
University of Algarve
Faro, Portugal

Valeria V. Krzhizhanovskaya
University of Amsterdam
Amsterdam, The Netherlands

Michael H. Lees
University of Amsterdam
Amsterdam, The Netherlands

Jack J. Dongarra
University of Tennessee at Knoxville
Knoxville, TN, USA

Peter M. A. Sloot
University of Amsterdam
Amsterdam, The Netherlands

ISSN 0302-9743 ISSN 1611-3349 (electronic)
Lecture Notes in Computer Science
ISBN 978-3-030-22746-3 ISBN 978-3-030-22747-0 (eBook)
https://doi.org/10.1007/978-3-030-22747-0

LNCS Sublibrary: SL1 – Theoretical Computer Science and General Issues

This Springer imprint is published by the registered company Springer Nature Switzerland AG
The registered company address is: Gewerbestrasse 11, 6330 Cham, Switzerland

Preface

Welcome to the 19th Annual International Conference on Computational Science (ICCS - https://www.iccs-meeting.org/iccs2019/), held during June 12–14, 2019, in Faro, Algarve, Portugal. Located at the southern end of Portugal, Algarve is a well-known touristic haven. Besides some of the best and most beautiful beaches in the entire world, with fine sand and crystal-clear water, Algarve also offers amazing natural landscapes, a rich folk heritage, and a healthy gastronomy that can be enjoyed throughout the whole year, attracting millions of foreign and national tourists. ICCS 2019 was jointly organized by the University of Algarve, the University of Amsterdam, NTU Singapore, and the University of Tennessee.

The International Conference on Computational Science is an annual conference that brings together researchers and scientists from mathematics and computer science as basic computing disciplines, as well as researchers from various application areas who are pioneering computational methods in sciences such as physics, chemistry, life sciences, engineering, arts, and humanitarian fields, to discuss problems and solutions in the area, to identify new issues, and to shape future directions for research.

Since its inception in 2001, ICCS has attracted an increasingly higher quality and numbers of attendees and papers, and this year was no exception, with over 350 participants. The proceedings series have become a major intellectual resource for computational science researchers, defining and advancing the state of the art in this field.

ICCS 2019 in Faro was the 19th in this series of highly successful conferences. For the previous 18 meetings, see: http://www.iccs-meeting.org/iccs2019/previous-iccs/.

The theme for ICCS 2019 was "Computational Science in the Interconnected World," to highlight the role of computational science in an increasingly interconnected world. This conference was a unique event focusing on recent developments in: scalable scientific algorithms; advanced software tools; computational grids; advanced numerical methods; and novel application areas. These innovative novel models, algorithms, and tools drive new science through efficient application in areas such as physical systems, computational and systems biology, environmental systems, finance, and others.

ICCS is well known for its excellent line-up of keynote speakers. The keynotes for 2019 were:

- Tiziana Di Matteo, King's College London, UK
- Teresa Galvão, University of Porto/INESC TEC, Portugal
- Douglas Kothe, Exascale Computing Project, USA
- James Moore, Imperial College London, UK
- Robert Panoff, The Shodor Education Foundation, USA
- Xiaoxiang Zhu, Technical University of Munich, Germany

This year we had 573 submissions (228 submissions to the main track and 345 to the workshops). In the main track, 65 full papers were accepted (28%); in the workshops, 168 full papers (49%). The high acceptance rate in the workshops is explained by the nature of these thematic sessions, where many experts in a particular field are personally invited by workshop organizers to participate in their sessions.

ICCS relies strongly on the vital contributions of our workshop organizers to attract high-quality papers in many subject areas. We would like to thank all committee members for the main track and workshops for their contribution to ensure a high standard for the accepted papers. We would also like to thank Springer, Elsevier, and Intellegibilis for their support. Finally, we very much appreciate all the local Organizing Committee members for their hard work to prepare this conference.

We are proud to note that ICCS is an A-rank conference in the CORE classification.

June 2019

João M. F. Rodrigues
Pedro J. S. Cardoso
Jânio Monteiro
Roberto Lam
Valeria V. Krzhizhanovskaya
Michael Lees
Jack J. Dongarra
Peter M. A. Sloot

Organization

Workshops and Organizers

Advanced Modelling Techniques for Environmental Sciences – AMES

Jens Weismüller
Dieter Kranzlmüller
Maximilian Hoeb
Jan Schmidt

Advances in High-Performance Computational Earth Sciences: Applications and Frameworks – IHPCES

Takashi Shimokawabe
Kohei Fujita
Dominik Bartuschat

Agent-Based Simulations, Adaptive Algorithms, and Solvers – ABS-AAS

Maciej Paszynski
Quanling Deng
David Pardo
Robert Schaefer
Victor Calo

Applications of Matrix Methods in Artificial Intelligence and Machine Learning – AMAIML

Kourosh Modarresi

Architecture, Languages, Compilation, and Hardware Support for Emerging and Heterogeneous Systems – ALCHEMY

Stéphane Louise
Löic Cudennec
Camille Coti
Vianney Lapotre
José Flich Cardo
Henri-Pierre Charles

Biomedical and Bioinformatics Challenges for Computer Science – BBC

Mario Cannataro
Giuseppe Agapito
Mauro Castelli

Riccardo Dondi
Rodrigo Weber dos Santos
Italo Zoppis

Classifier Learning from Difficult Data – CLD²

Michał Woźniak
Bartosz Krawczyk
Paweł Ksieniewicz

Computational Finance and Business Intelligence – CFBI

Yong Shi
Yingjie Tian

Computational Methods in Smart Agriculture – CMSA

Andrew Lewis

Computational Optimization, Modelling, and Simulation – COMS

Xin-She Yang
Slawomir Koziel
Leifur Leifsson

Computational Science in IoT and Smart Systems – IoTSS

Vaidy Sunderam

Data-Driven Computational Sciences – DDCS

Craig Douglas

Machine Learning and Data Assimilation for Dynamical Systems – MLDADS

Rossella Arcucci
Boumediene Hamzi
Yi-Ke Guo

Marine Computing in the Interconnected World for the Benefit of Society – MarineComp

Flávio Martins
Ioana Popescu
João Janeiro
Ramiro Neves
Marcos Mateus

Multiscale Modelling and Simulation – MMS

Derek Groen
Lin Gan
Stefano Casarin
Alfons Hoekstra
Bartosz Bosak

Simulations of Flow and Transport: Modeling, Algorithms, and Computation – SOFTMAC

Shuyu Sun
Jingfa Li
James Liu

Smart Systems: Bringing Together Computer Vision, Sensor Networks, and Machine Learning – SmartSys

João M. F. Rodrigues
Pedro J. S. Cardoso
Jânio Monteiro
Roberto Lam

Solving Problems with Uncertainties – SPU

Vassil Alexandrov

Teaching Computational Science – WTCS

Angela Shiflet
Evguenia Alexandrova
Alfredo Tirado-Ramos

Tools for Program Development and Analysis in Computational Science – TOOLS

Andreas Knüpfer
Karl Fürlinger

Programme Committee and Reviewers

Ahmad Abdelfattah
Eyad Abed
Markus Abel
Laith Abualigah
Giuseppe Agapito
Giovanni Agosta
Ram Akella
Elisabete Alberdi
Marco Aldinucci
Luis Alexandre
Vassil Alexandrov
Evguenia Alexandrova
Victor Allombert
Saad Alowayyed
Stanislaw Ambroszkiewicz
Ioannis Anagnostou
Philipp Andelfinger
Michael Antolovich
Hartwig Anzt
Hideo Aochi

Rossella Arcucci
Tomasz Arodz
Kamesh Arumugam
Luiz Assad
Victor Azizi Tarksalooyeh
Bartosz Balis
Krzysztof Banas
João Barroso
Dominik Bartuschat
Daniel Becker
Jörn Behrens
Adrian Bekasiewicz
Gebrail Bekdas
Stefano Beretta
Daniel Berrar
John Betts
Sanjukta Bhowmick
Bartosz Bosak
Isabel Sofia Brito
Kris Bubendorfer
Jérémy Buisson
Aleksander Byrski
Cristiano Cabrita
Xing Cai
Barbara Calabrese
Carlos Calafate
Carlos Cambra
Mario Cannataro
Alberto Cano
Paul M. Carpenter
Stefano Casarin
Manuel Castañón-Puga
Mauro Castelli
Jeronimo Castrillon
Eduardo Cesar
Patrikakis Charalampos
Henri-Pierre Charles
Zhensong Chen
Siew Ann Cheong
Andrei Chernykh
Lock-Yue Chew
Su Fong Chien
Sung-Bae Cho
Bastien Chopard
Stephane Chretien
Svetlana Chuprina
Florina M. Ciorba
Noelia Correia
Adriano Cortes
Ana Cortes
Jose Alfredo F. Costa
Enrique Costa-Montenegro
David Coster
Camille Coti
Carlos Cotta
Helene Coullon
Daan Crommelin
Attila Csikasz-Nagy
Loïc Cudennec
Javier Cuenca
Yifeng Cui
António Cunha
Ben Czaja
Pawel Czarnul
Bhaskar Dasgupta
Susumu Date
Quanling Deng
Nilanjan Dey
Ergin Dinc
Minh Ngoc Dinh
Sam Dobbs
Riccardo Dondi
Ruggero Donida Labati
Goncalo dos-Reis
Craig Douglas
Aleksandar Dragojevic
Rafal Drezewski
Niels Drost
Hans du Buf
Vitor Duarte
Richard Duro
Pritha Dutta
Sean Elliot
Nahid Emad
Christian Engelmann
Qinwei Fan
Fangxin Fang
Antonino Fiannaca
Christos Filelis-Papadopoulos
José Flich Cardo
Yves Fomekong Nanfack
Vincent Fortuin
Ruy Freitas Reis
Karl Frinkle
Karl Fuerlinger
Kohei Fujita
Wlodzimierz Funika
Takashi Furumura
Mohamed Medhat Gaber
Jan Gairing
David Gal
Marco Gallieri
Teresa Galvão
Lin Gan
Luis Garcia-Castillo
Delia Garijo
Frédéric Gava
Don Gaydon
Zong-Woo Geem
Alex Gerbessiotis
Konstantinos Giannoutakis
Judit Gimenez
Domingo Gimenez
Guy Gogniat
Ivo Gonçalves
Yuriy Gorbachev
Pawel Gorecki
Michael Gowanlock
Manuel Graña
George Gravvanis
Marilaure Gregoire
Derek Groen
Lutz Gross
Sophia Grundner-Culemann
Pedro Guerreiro
Kun Guo
Xiaohu Guo
Piotr Gurgul
Pietro Hiram Guzzi
Panagiotis Hadjidoukas
Mohamed Hamada
Boumediene Hamzi
Masatoshi Hanai
Quillon Harpham

William Haslett
Yiwei He
Alexander Heinecke
Jurjen Rienk Helmus
Alvaro Herrero
Bogumila Hnatkowska
Maximilian Hoeb
Paul Hofmann
Sascha Hunold
Juan Carlos Infante
Hideya Iwasaki
Takeshi Iwashita
Alfredo Izquierdo
Heike Jagode
Vytautas Jancauskas
Joao Janeiro
Jiří Jaroš
Shantenu Jha
Shalu Jhanwar
Chao Jin
Hai Jin
Zhong Jin
David Johnson
Anshul Joshi
Manuela Juliano
George Kallos
George Kampis
Drona Kandhai
Aneta Karaivanova
Takahiro Katagiri
Ergina Kavallieratou
Wayne Kelly
Christoph Kessler
Dhou Khaldoon
Andreas Knuepfer
Harald Koestler
Dimitrios Kogias
Ivana Kolingerova
Vladimir Korkhov
Ilias Kotsireas
Ioannis Koutis
Sergey Kovalchuk
Michał Koziarski
Slawomir Koziel
Jarosław Koźlak
Dieter Kranzlmüller
Bartosz Krawczyk
Valeria Krzhizhanovskaya
Paweł Ksieniewicz
Michael Kuhn
Jaeyoung Kwak
Massimo La Rosa
Roberto Lam
Anna-Lena Lamprecht
Johannes Langguth
Vianney Lapotre
Jysoo Lee
Michael Lees
Leifur Leifsson
Kenneth Leiter
Roy Lettieri
Andrew Lewis
Jingfa Li
Yanfang Li
James Liu
Hong Liu
Hui Liu
Zhao Liu
Weiguo Liu
Weifeng Liu
Marcelo Lobosco
Veronika Locherer
Robert Lodder
Stephane Louise
Frederic Loulergue
Huimin Lu
Paul Lu
Stefan Luding
Scott MacLachlan
Luca Magri
Maciej Malawski
Livia Marcellino
Tomas Margalef
Tiziana Margaria
Svetozar Margenov
Osni Marques
Alberto Marquez
Paula Martins
Flavio Martins
Jaime A. Martins
Marcos Mateus
Marco Mattavelli
Pawel Matuszyk
Valerie Maxville
Roderick Melnik
Valentin Melnikov
Ivan Merelli
Jianyu Miao
Kourosh Modarresi
Miguel Molina-Solana
Fernando Monteiro
Jânio Monteiro
Pedro Montero
James Montgomery
Andrew Moore
Irene Moser
Paulo Moura Oliveira
Ignacio Muga
Philip Nadler
Hiromichi Nagao
Kengo Nakajima
Raymond Namyst
Philippe Navaux
Michael Navon
Philipp Neumann
Ramiro Neves
Mai Nguyen
Hoang Nguyen
Nancy Nichols
Sinan Melih Nigdeli
Anna Nikishova
Kenji Ono
Juan-Pablo Ortega
Raymond Padmos
J. P. Papa
Marcin Paprzycki
David Pardo
Héctor Quintián Pardo
Panos Parpas
Anna Paszynska
Maciej Paszynski
Jaideep Pathak
Abani Patra
Pedro J. S. Cardoso
Dana Petcu
Eric Petit
Serge Petiton
Bernhard Pfahringer

Daniela Piccioni
Juan C. Pichel
Anna Pietrenko-Dabrowska
Laércio L. Pilla
Armando Pinto
Tomasz Piontek
Erwan Piriou
Yuri Pirola
Nadia Pisanti
Antoniu Pop
Ioana Popescu
Mario Porrmann
Cristina Portales
Roland Potthast
Ela Pustulka-Hunt
Vladimir Puzyrev
Alexander Pyayt
Zhiquan Qi
Rick Quax
Barbara Quintela
Waldemar Rachowicz
Célia Ramos
Marcus Randall
Lukasz Rauch
Andrea Reimuth
Alistair Rendell
Pedro Ribeiro
Bernardete Ribeiro
Robin Richardson
Jason Riedy
Celine Robardet
Sophie Robert
João M. F. Rodrigues
Daniel Rodriguez
Albert Romkes
Debraj Roy
Philip Rutten
Katarzyna Rycerz
Augusto S. Neves
Apaar Sadhwani
Alberto Sanchez
Gabriele Santin
Robert Schaefer
Olaf Schenk
Ulf Schiller
Bertil Schmidt
Jan Schmidt
Martin Schreiber
Martin Schulz
Marinella Sciortino
Johanna Sepulveda
Ovidiu Serban
Vivek Sheraton
Yong Shi
Angela Shiflet
Takashi Shimokawabe
Tan Singyee
Robert Sinkovits
Vishnu Sivadasan
Peter Sloot
Renata Slota
Grażyna Ślusarczyk
Sucha Smanchat
Maciej Smołka
Bartlomiej Sniezynski
Sumit Sourabh
Hoda Soussa
Steve Stevenson
Achim Streit
Barbara Strug
E. Dante Suarez
Bongwon Suh
Shuyu Sun
Vaidy Sunderam
James Suter
Martin Swain
Grzegorz Swisrcz
Ryszard Tadeusiewicz
Lotfi Tadj
Daniele Tafani
Daisuke Takahashi
Jingjing Tang
Osamu Tatebe
Cedric Tedeschi
Kasim Tersic
Yonatan Afework Tesfahunegn
Jannis Teunissen
Andrew Thelen
Yingjie Tian
Nestor Tiglao
Francis Ting
Alfredo Tirado-Ramos
Arkadiusz Tomczyk
Stanimire Tomov
Marko Tosic
Jan Treibig
Leonardo Trujillo
Benjamin Uekermann
Pierangelo Veltri
Raja Velu
Alexander von Ramm
David Walker
Peng Wang
Lizhe Wang
Jianwu Wang
Gregory Watson
Rodrigo Weber dos Santos
Kevin Webster
Josef Weidendorfer
Josef Weinbub
Tobias Weinzierl
Jens Weismüller
Lars Wienbrandt
Mark Wijzenbroek
Roland Wismüller
Eric Wolanski
Michał Woźniak
Maciej Woźniak
Qing Wu
Bo Wu
Guoqiang Wu
Dunhui Xiao
Huilin Xing
Miguel Xochicale
Wei Xue
Xin-She Yang

Dongwei Ye
Jon Yosi
Ce Yu
Xiaodan Yu
Reza Zafarani
Gábor Závodszky
H. Zhang
Zepu Zhang
Jingqing Zhang
Yi-Fan Zhang
Yao Zhang
Wenlai Zhao
Jinghui Zhong
Xiaofei Zhou
Sotirios Ziavras
Peter Zinterhof
Italo Zoppis
Chiara Zucco

Contents – Part IV

Track of Data-Driven Computational Sciences

Track of Machine Learning and Data Assimilation for Dynamical Systems

Track of Data-Driven Computational Sciences

Nonparametric Approach to Weak Signal Detection in the Search for Extraterrestrial Intelligence (SETI)

Anne D. Brooks[1] and Robert A. Lodder[2(✉)]

[1] Stetson University, DeLand, FL 32723, USA
[2] University of Kentucky, Lexington, KY 40536, USA
Lodder@g.uky.edu

Abstract. It might be easier for intelligent extraterrestrial civilizations to be found when they mark their position with a bright laser beacon. Given the possible distances involved, however, it is likely that weak signal detection techniques would still be required to identify even the brightest SETI Beacon. The Bootstrap Error-adjusted Single-sample Technique (BEST) is such a detection method. The BEST has been shown to outperform the more traditional Mahalanobis metric in analysis of SETI data from a Project Argus near infrared telescope. The BEST algorithm is used to identify unusual signals and returns a distance in asymmetric nonparametric multidimensional central 68% confidence intervals (equivalent to standard deviations for 1-D data that are normally distributed, or Mahalanobis distance units for normally distributed data of d dimensions). Calculation of the Mahalanobis metric requires matrix factorization and is order of d^3. Furthermore, the accuracy and precision of the BEST metric are greater than the Mahalanobis metric in realistic data collection scenarios (many more wavelengths available then observations at those wavelengths). An extension of the BEST to examine multiple samples (subclusters of data) simultaneously is explored in this paper.

Keywords: Parallel algorithm · Bootstrap · Supernova · Gamma ray burst · Solar transit

1 Introduction

1.1 Scope of the Problem

SETI is at least a complex 5-dimensional problem. Five dimensions is a lot of space to search. The first three dimensions, length, height, and width, are the (X, Y, Z) spatial coordinates that everyone is used to in daily life. The fourth dimension is frequency or wavelength. The system must be listening at the right optical wavelength or microwave frequency in order to detect a signal. Time is the fifth dimension. In addition to looking in the right place, and listening at the right frequency, the system also must be listening when the signal comes in. Five dimensions is a lot of space to search, and this problem partially explains why finding signals has been so difficult. Moreover, the rotation of planets and the revolution around stars means that transmitting and receiving antennas

J. M. F. Rodrigues et al. (Eds.): ICCS 2019, LNCS 11539, pp. 3–15, 2019.
https://doi.org/10.1007/978-3-030-22747-0_1

rarely line up. In fact, the drift scan transit time of high gain receivers and antennas on earth is usually on the order of seconds. Finally, the Doppler shift from planetary movement complicates signal averaging to increase signal-to-noise ratio. It is well-known that signal adds as n while random noise adds as the square root of n (where n = the number of times the signal is measured, or the signal integration time) [4]. This fact is used to increase the signal-to-noise ratio by a factor of the square root of n by averaging signals over time. However, the Doppler shift imposed on signals by planetary motion is enough to limit the averaging of signals because the signals are moving [7]. Doppler shifts can be compensated for in software, and most SETI systems that work off-line are able to apply dozens or even hundreds of different Doppler shifts in the signal averaging process in order to enhance weak signals.

Detecting a signal is also a somewhat complex process. The statistical hypothesis tested in software tests the hypothesis that no intelligent signal is present against the alternative that a signal from ETI has been detected. Before signal collection can begin, every station must verify proper system operation with a signal generator or another celestial source. This procedure is generally repeated at the end of the data collection run. During a data collection run, an ETI signal should show:

1. Coherence not achievable by known natural emission mechanisms.
2. A signal intensity variation that is consistent with the known antenna pattern and the aiming coordinates (azimuth and elevation). A directional signal should drop in intensity when a directional antenna is moved away from the signal source.
3. A Doppler shift consistent with planetary motion (or the motion of a reasonable object in space, like a spaceship). Satellites can be mistaken for ETI signal sources, but they show a Doppler shift that changes with their angular velocity.
4. Finally, before any signal detection can be announced, there must be a simultaneous detection with proper Doppler shifts at widely separated terrestrial coordinates. Ideally, this detection takes place at a station on the other side of the earth, where the same terrestrial sources of interference would not be present. Even an interfering signal from an airplane or satellite would be unlikely to hit two SETI telescopes on opposite sides of the planet at the same time. If the satellite was far enough out in space to hit two SETI telescopes on opposite sides of the planet at the same time, it would likely still show up at different celestial coordinates (right ascension and declination).

1.2 Reducing the Scope

One way to collapse the five-dimensional problem of signal detection into a more tractable problem is to use time and direction synchronizers in SETI. Reducing the dimensionality of the problem is possible with appropriate synchronizers that attract the attention of scientists. A synchronizer should be a big enough event to attract Galactic attention. Supernovas and gamma ray bursts fit the requirement (see Fig. 1) [3]. For example, type IA supernovas are used as standard candles to measure the expansion of the universe. They are interesting to our scientists and are probably interesting to alien scientists for the same reasons. When one of these events occurs in the Milky Way galaxy or even in another faraway galaxy (for example, Supernova Refsdal is in a galaxy 9.3 billion light-years away from the Milky Way galaxy and earth, making it a good target for measuring the expansion of the universe), the light will eventually reach an alien planet on the way to earth. At that time, ETI will direct its transmitters in a direction roughly collinear to the received supernova light and away from the supernova. ETI will also probably direct its signal in a small cone so that the image of their Beacon appears to the side of the image of the supernova. The light from the alien Beacon and the supernova should arrive at the earth at the same time. In this way, ETI wishing to advertise the presence of a Galactic Internet could take advantage of a high-energy signal source to set four of the five variables in the SETI search space (the x, y, and z coordinates in space as well as the time coordinate), leaving only the frequency or wavelength variable to be determined.

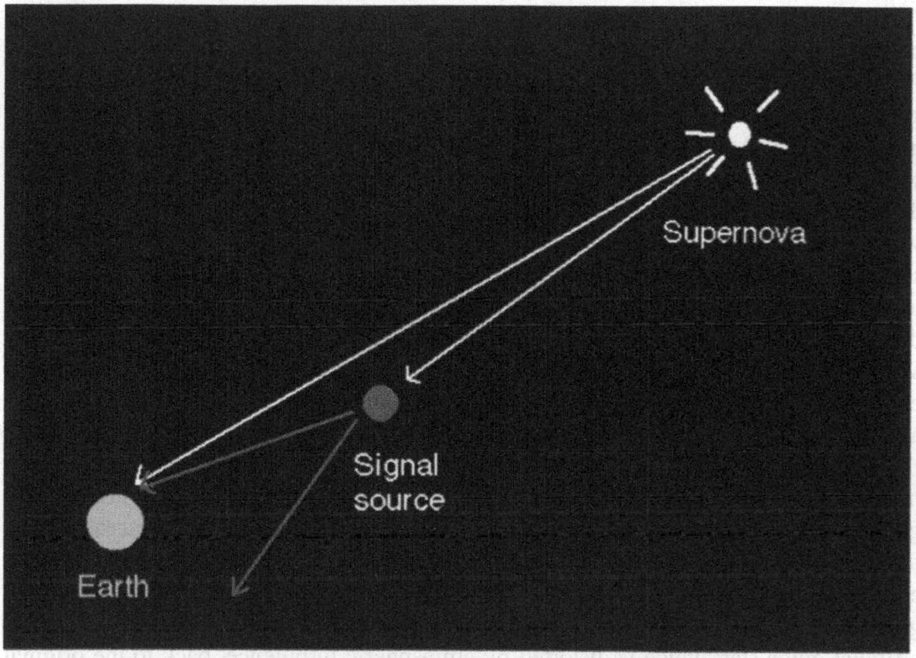

Fig. 1. The supernova SETI synchronizer strategy

Another possible SETI synchronizer is the solar transit. This synchronizer takes advantage of the solar eclipse. When the Earth passes in front of the sun, it blocks a small part of the sun's light. Potential observers outside our solar system might be able to detect the resulting dimming of the sun and study the Earth's atmosphere. This transit method has helped to find most of the thousands of exoplanets known to exist today.

The last variable in the five-dimensional space, the frequency or wavelength, is worth special consideration. Most SETI research has been done in the microwave region since Frank Drake's Project Ozma in the early 1960s [5]. Microwave SETI searches for pulses of electromagnetic radiation within the microwave portions of the spectrum. The beam is wider in the microwave region and thus targeting is not as much of an issue. Historically, microwaves have been viewed as better able to penetrate the atmosphere and thus more likely to be used for interstellar communications. However, developments in the last decade or two in adaptive optics have made arguments for microwave SETI far less convincing. Optical SETI searches for pulses of laser light in or near the portion of the light spectrum that is visible. The beam is narrow, enabling a higher power density to be directed toward a distant target. The use of near-infrared and infrared light in certain bands enables the energy to both escape the atmosphere and avoid being blocked by dust in the galactic disk.

The advent of adaptive optics [1] has given great impetus to near-infrared and optical SETI. Adaptive optics was developed for destroying incoming ICBMs by President Reagan's Star Wars program. Originally telescope mirrors were very thick to keep them from bending and going out of focus when the telescope was angled in a new direction. A 1 to 6 ratio was commonly used in mirror construction (in other words, a 6-inch mirror had to be 1 inch thick, and a six-foot mirror had to be 1 foot thick) in order to be mechanically stable in all directions. A 10 m mirror built in this way would be huge and impractical. Optical telescopes began to approach the 10 m size when the mirrors were instead made very thin and lightweight and were connected to an array of electromechanical actuators to bend in the mirror back into shape when the telescope was moved to a new angle (an electronically deformable mirror). The Star Wars scientists realized that distortion caused by refraction in the Earth's atmosphere could be corrected by these electromechanical actuators if the actuators could be moved at high speed (e.g., 1000 times per second).

Adaptive optics use an artificial guide star directed toward a layer of ions above most of the Earth's atmosphere (and certainly above the turbulent part). This artificial guide star is created by a laser on the surface of the earth that excites fluorescence in the ions. This laser is placed next to the large telescope (i.e., 10 m mirror telescope) on earth and excites ions in the field of view of the telescope. A computer is then programmed to deform the mirror at $\sim$1000 times per second to correct the shape of the artificial guide star, which is distorted by atmospheric turbulence, back to the original shape transmitted by the laser on earth. The distortions required to correct the original shape of the artificial guide star also correct all the other turbulence in the optical path.

This correction does more than take away the twinkle of the stars when light is received. The correction can also be used to take away the twinkle of a transmitted signal. When humans are ready to join the Galactic Internet, a small laser to excite ions above the atmosphere will join a METI (Messaging Extraterrestrial Intelligence) laser on an adaptive optics telescope. The smaller laser will create the artificial guide star, and the larger METI laser will be directed toward the deformable mirror. The mirrors deformations will then cause the refractions in the turbulent atmosphere to rebuild the transmitted light beam in the process of exiting the Earth's atmosphere, leading to a clean signal transmitted to a distant planet or spacecraft.

Infrared light includes wavelengths too long to be visible, from approximately 700 nm to about 1 mm in wavelength. Visible light seen by the human eye ranges over about 400 to 700 nm in wavelength. Light is called near-infrared or near-ultraviolet based upon its proximity to the visible portion of the spectrum. So, near-infrared light is the highest in energy and the shortest in wavelength of the infrared region, while near UV light is the longest in wavelength and the lowest in energy of the ultraviolet region.

Most cosmic dust particles are between a few molecules to 100 nm in size. Near-infrared light penetrates the Milky Way galaxy better than visible light because of reduced scattering. For example, the star-forming region G45.45+0.06 is visible from earth at 2200 nm but obscured by galactic dust at 1250 nm. Light scattering falls off as one over wavelength to the fourth power. In other words, doubling the wavelength reduces light scattering by a factor of 16. Infrared light would be better than near infrared light, except that infrared light is absorbed more by the atmosphere of the earth. In a recent paper in The Astrophysical Journal, two researchers at MIT argue that it might be easier for intelligent extraterrestrial civilizations to be found when they mark their position with a bright laser beacon [2]. Given the possible distances involved, however, it is likely that weak signal detection techniques would still be required to identify even the brightest SETI beacon.

2 Experimental

Modern microwave and near-infrared/optical systems now often incorporate a software-defined radio (SDR). An SDR is a radio communication system where components that have been typically implemented in hardware (e.g., mixers, filters, amplifiers, modulators/demodulators, detectors, etc.) are instead implemented by means of software on a personal computer or embedded system. While the concept of SDR is not new, rapidly evolving digital electronics render practical many processes which used to be only theoretically possible. This approach greatly reduces the cost of instrumentation and is the approach we have adopted for our microwave and infrared SETI telescopes (Project Argus station EM77to). The software defined radio acts as a very sensitive spectrum analyzer, displaying the Fourier transform of the signals present at the InGaAs detector. In Fig. 3, the center detection frequency is set at 147.5 MHz. The SDR# software displays all signals between 146.3 and 148.7 MHz in single spectrum (top) and waterfall (bottom) mode. A Fourier transform converts signals in the time domain to signals in the frequency domain. In other words, a sine wave with amplitude on the Y axis and the time on the X axis appears in a graph as a single spike following Fourier transformation. The single spike appears at the frequency of the sine wave in a graph that still has amplitude on the Y axis, but now frequency on the X axis. An inverse Fourier transform converts signals in the frequency domain back into signals in the time domain. The Fast Fourier Transform (FFT) simply refers to an efficient algorithm for performing a discrete Fourier transform on data.

Our group uses two near-infrared telescopes, a 6-inch Newtonian reflector with all gold first-surface optics, and a one-meter Fresnel refractor with an aluminum compound parabolic concentrator. Both telescopes use Dobsonian az-el mounts. The fully assembled Newtonian reflector telescope is shown in Fig. 2 with the near infrared detector installed in the eyepiece. The handle on the primary mirror is visible in the end of the telescope. The telescope can be programmed using an ordinary laptop computer. The software defined radio attaches to the computer through a USB port. The computer currently runs Microsoft Windows 10.

Fig. 2. This near-infrared telescope for SETI uses all gold first surface optics and a high speed InGaAs detector. An SDR connects the detector to the computer for data monitoring and collection.

The detectors are high speed InGaAs photodiodes and the photodiode signals feed into the SDRs through coaxial cable. SDR#, an open source spectrum analysis program for SDRs, is used as the GUI for the telescope and for data collection.

Figure 3 shows the effect of FM radio interference on signals near 105.18 MHz. Because FM radio signals are modulated, they do not appear as narrow spikes (i.e., a delta function). Because the signals are terrestrial in origin, they also appear brighter and stronger than we would expect a SETI signal to appear. Neither of the FM radio station signals appear at the center frequency, which is shown by the red vertical line in the upper graph. Another clue that these signals are not from deep space is the absence of Doppler shift caused by the motion of the earth. An actual signal from deep space would also include a Doppler shift from motion of the source of the signal.

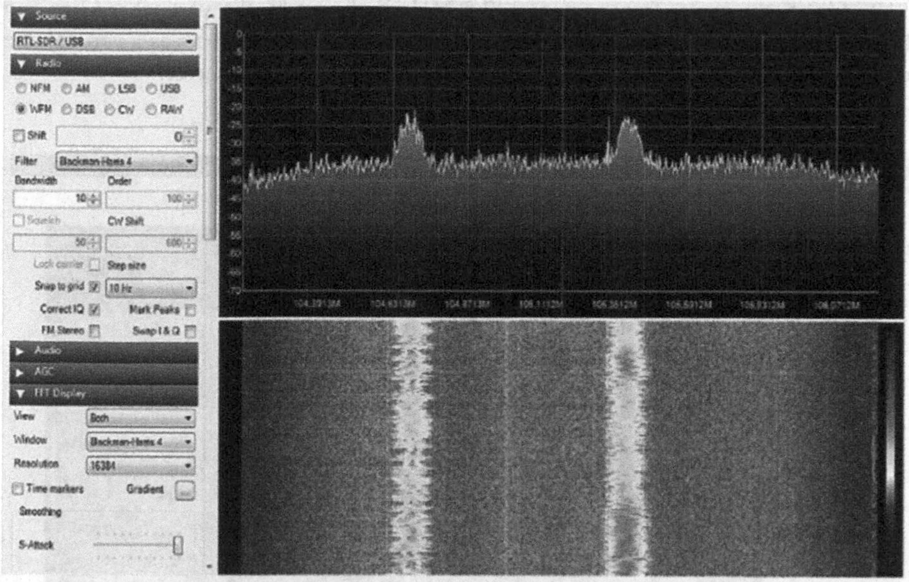

Fig. 3. An example of FM radio interference. The signals have a large bandwidth compared to a beacon. This red signal is not Doppler shifted, suggesting that it is terrestrial in origin. Good shielding will prevent this sort of problem. (Color figure online)

Figure 4 depicts a test using frequency modulated light pulses. The position of the pulses is varied by voice information. Side bands are seen around the central red signal. This red signal is not Doppler shifted, suggesting that it is terrestrial in origin.

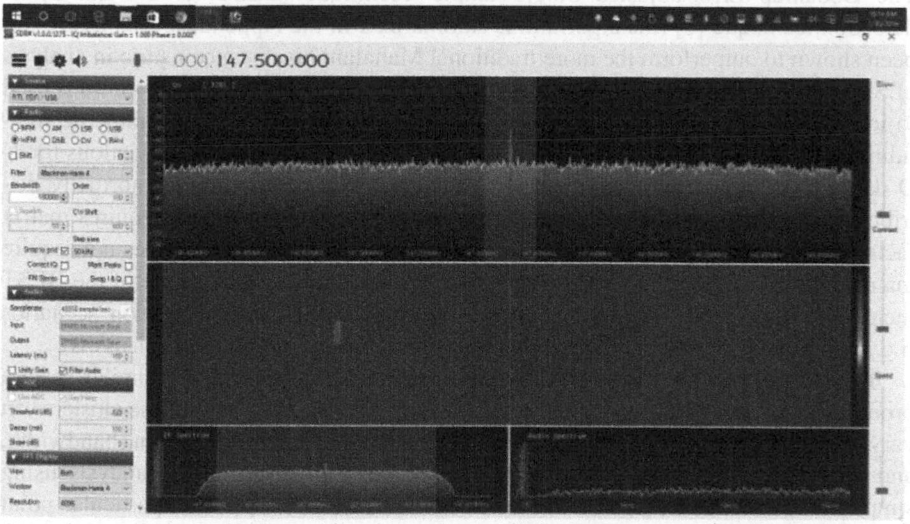

Fig. 4. A system test using frequency modulated light pulses. (Color figure online)

In Fig. 5, the center frequency of 147.5 MHz is shown by the vertical red line and the delta function in the top graph. The weak, SETI-like signal appears as a vertical line to the left of the red line marking the center frequency of the receiver.

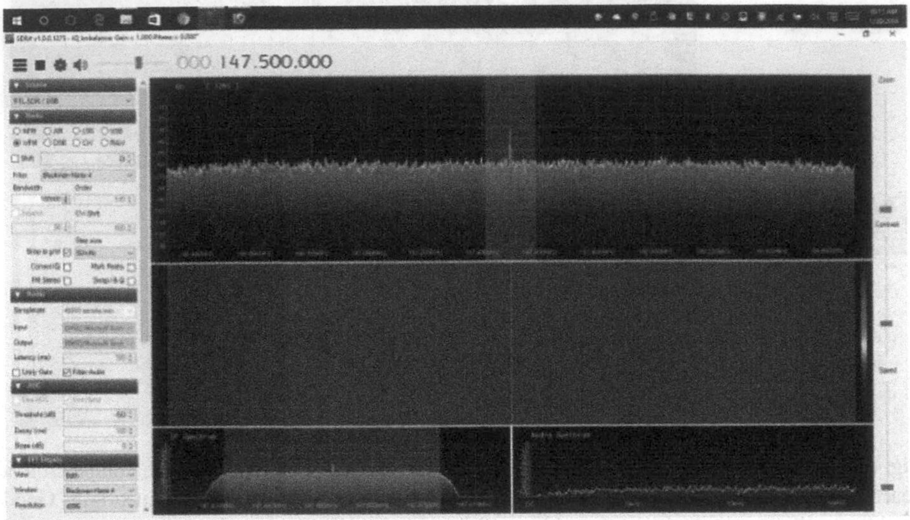

Fig. 5. A beacon-like signal with a more realistic intensity, but still lacking in Doppler shift (Color figure online)

3 The Extended BEST for Subcluster Detection

The Bootstrap Error-adjusted Single-sample Technique (BEST) is a weak signal detection technique [6] (the algorithm is summarized in the Appendix). The BEST has been shown to outperform the more traditional Mahalanobis distance metric in analysis of SETI data from a Project Argus near-infrared telescope. The BEST algorithm is used to identify unusual signals, and returns a distance in asymmetric nonparametric multidimensional central 68% confidence intervals (equivalent to standard deviations for 1-D data that are normally distributed, or Mahalanobis distance units for normally distributed data of d dimensions). Calculation of the Mahalanobis metric requires matrix factorization and is $O(d^3)$. In contrast, calculation of the BEST metric does not require matrix factorization and is $O(d)$. Furthermore, the accuracy and precision of the BEST metric are greater than the Mahalanobis metric in realistic data collection scenarios (i.e., many more wavelengths available than observations at those wavelengths).

In near-infrared multivariate statistical analyses, ETI emitters with similar spectra produce points that cluster in a similar region of spectral hyperspace. These clusters can vary significantly in shape and size due to variation in signal modulation, bandwidth, and Doppler shift. These factors, when combined with discriminant analysis using simple distance metrics, produce a test in which a result that places a particular point inside a particular cluster (the training data are typically noise collected in a specific region of sky) does not necessarily mean that the point is actually a member of the cluster. Weak signal strength may be insufficient to move a data point beyond 3 or 6 SDs of a cluster. Instead, the point may be a member of a new, slightly different cluster that overlaps the first. This happens when the test data contain a weak artificial signal not present in the training noise. A new cluster can be shaped by factors like signal modulation, bandwidth, and Doppler shift. An extension added to part of the BEST can be used to set nonparametric probability-density contours inside spectral clusters as well as outside, and when multiple points begin to appear in a certain region of cluster-hyperspace the perturbation of these density contours can be detected at an assigned significance level. When we have more than a single point sample, it is possible that a larger sample of data points from the test set will produce a new cluster with a different mean and standard deviation that overlaps the training set. If we could collect a sufficiently large sample of these spectra, we might be able to detect a signal even inside the three standard deviation limit on single points from the training cluster center. To do this, the algorithm

- Integrates the training samples from the center of the training set outward, and
- Integrates the test samples AND the training samples combined from the center of the training set outward.

These two integrals are compared in a QQ plot. The detection of candidate ETI signals both within and beyond 3 SDs of the center of the noise training set is possible with this method. Using this technique, distinctive diagnostic patterns form in the QQ plots that are discussed below (see Fig. 6). These patterns have predictable effects on the correlation coefficient calculated from the QQ plots.

A population **P** in a hyperspace **R** represents the universe of possible spectrometric samples (the rows of **P** are the individual samples, and the columns are the independent information vectors such as wavelengths or energies). **P*** is a discrete realization of **P** based on a calibration set **T**, chosen only once from **P** to represent as nearly as possible all the variations in **P**.

P* is calculated using a bootstrap process by an operation k(**T**). **P*** has parameters **B** and **C**, where **C** = E(**P**) (the group mean of **P**) and **B** is the Monte Carlo approximation to the bootstrap distribution [6].

Given two data sets $\mathbf{P}_1^*$ and $\mathbf{P}_2^*$ with an equal number of elements *n*, it is possible to determine whether $\mathbf{P}_1^*$ and $\mathbf{P}_2^*$ are drawn from the same population even if the distance between them is <3 SDs (standard deviations). Quantile-quantile (QQ) plots and a simple correlation test statistic are used [6].

$$\rho\left(\left\{\int_R \mathbf{P}_1^*\right\}, \left\{\int_R \mathbf{P}_1^*\right\} \cup \left\{\int_R \mathbf{P}_2^*\right\}\right)$$

A bootstrap method is employed to set confidence limits on ρ, the correlation coefficient. The central 68% confidence interval on ρ is also used to calculate σ_ρ, a distance in SDs that is sensitive to small differences in location and scale between $\mathbf{P}_1^*$ and $\mathbf{P}_2^*$.

This approach to spectral analysis has significant advantages. More wavelengths can be used in the calibration than there are samples in the calibration set, without degrading the results. Full spectra can be used without down-weighting some of the information at certain wavelengths, reducing the possibility of missing something new that may appear in future samples. Also, the method is completely nonparametric, and the shape, scale, and skew of the spectral sample distributions do not affect the quality of the results.

Figure 6a depicts a pure location difference between the training set (noise) and the test set (noise and a signal). A pure location difference is the situation that might exist when a fairly strong signal with no Doppler shift is detected. In this example, the two populations are identical except for their locations (centers). The shapes of the distributions have been arbitrarily selected to be circles (or hyperspheres in hyperspace of larger dimension) with the same standard deviation in all directions.

Figure 6b illustrates the Cumulative Distribution Functions (CDFs) of the training set (blue) and test set (red) from (a). The x axis values represent the sorted normalized Euclidean distances of each point from the center of the training set.

The QQ plot of the Location Difference Only example in (a) is given in Fig. 6c. A correlation coefficient calculated for the QQ plot gives an indication of how well the two distributions (Training Set and Test Set) match. A correlation coefficient of $r = 1$ indicates perfectly matching distributions, and no SETI signals when the algorithm is trained on galactic background noise. This QQ plot has a break in the line that would indicate the presence of a signal.

Figure 6d illustrates the effect of a Location Difference Only in (a) on the correlation coefficient of the QQ plot as the distance between the centers of the clusters increases. The horizontal line represents a confidence limit on the training set calculated with the use of cross validation samples. When the line drops below the confidence limit a signal has been detected.

A Scale Difference Only, in which the Training Set is larger than the Test Set, is depicted in ref [8]. This illustration shows a training set and a test set in hyperspace, the two population distributions share the same center, and the training set population distribution is larger in scale than the test set distribution. This situation would occur when there was no SETI signal present but the noise level in the receiver dropped.

Ref [8] shows a QQ plot from a Training Set and Test Set that differ only in scale when the Training Set is larger than the Test Set as in (e). There are two bends in the plot that reduce the correlation coefficient calculated from the QQ plot.

The effect of a pure scale difference (when the Test Set is smaller than the Training Set) on the correlation coefficient calculated from the QQ plot as the scaling factor changes is shown in [8]. The x axis values represent the distance factor by which the test set is smaller in scale than the training set.

Ref [8] shows a Scale Difference Only with the Training Set Smaller than the Test Set. Ref [8] illustrates a training set and a test set in spectral hyperspace with the size relationship opposite to that just observed. The two population distributions share the same center, and the training set population distribution is smaller in scale than the test set population distribution. An increase in background noise or a modulated signal could cause this pattern to emerge in the data.

Ref [8] reveals a QQ plot from the subcluster detection method corresponding to the pure scale difference situation in Fig. 6h. The test set is larger in scale than the training set, and a test set forms the lower line with the larger slope in the figure. The bend in the line is slight because the difference between the two set scales is only a factor of 2.5.

The effect of a pure scale difference (training set smaller than the test set) on the correlation coefficient calculated from a QQ plot as the distance scaling factor changes is in [8].

Ref [8] illustrates the situation in which Simultaneous Location and Scale Differences exist and the Training Set is smaller than the Test Set. This is the situation commonly encountered when a signal is detected.

Ref [8] The QQ plot when a training set and a test set exhibit simultaneous location and scale differences, and the test set population distribution is larger in scale than the training set population distribution. There is both a bend and a break in the QQ plot line that lowers the correlation coefficient. The training set forms the lower line (blue) in the figure, and the test set forms the upper line (red).

Ref [8] shows how the correlation coefficient is affected by changes in scaling factor and distance between the clusters when the Training Set is smaller than the Test Set. The highest line represents a test set that is a factor of 2 larger than the training set, the middle line a test set that is a factor of 5 larger than the training set, and the lowest line a test set: that is a factor of 10 larger than the training set. The horizontal line at the top of the graph is a 98% confidence limit. Only one test set crosses the 98% limit (meaning it is considered the same as noise), and that test set is a factor of two larger in scale than the training set, with the two set centers less than 0.5 standard deviation of the training set apart.

Ref [8] shows Simultaneous Location and Scale Differences with the Training Set Larger than the Test Set. A strong terrestrial signal could cause this effect.

Ref [8] shows the effect of simultaneous location and scale differences on the correlation coefficient calculated from a QQ plot when the test set is larger in scale than the training set. Only when the test set is 2x the size of the training set is it ever identified as being the same as the training set, and then only when the two sets are about 0.1 SD of the training set apart. The test sets 5x and 10x the size of the training set are always identified as being different distributions (i.e., a signal is detected).

3.1 November/December 2018 Observations

Near-IR spectra from the vicinity of AT2018ivc, a supernova discovered in M77 on Nov. 24, 2018, were analyzed using the BEST subclustering method to identify unusual signals. Observations were made on Nov. 26, 28, 29, 2018, and on Dec. 2 and 6, 2019 (2 Gb collected each night). All runs were negative. The collected data produced patterns similar to Fig. 6h with the horizontal line at the 99.9999% level. As usual, weather is the major limiting factor on data collection. Cloud cover and precipitation are problematic for optical and near-IR SETI methods.

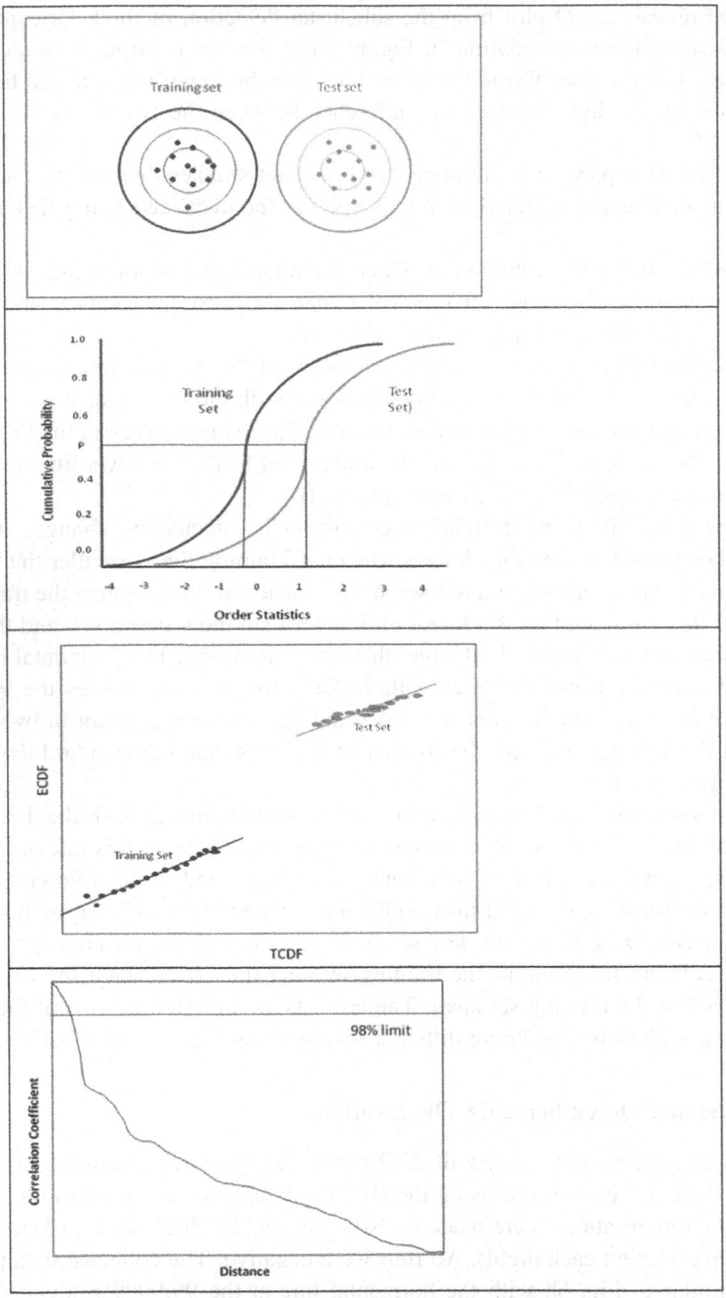

Fig. 6. Different patterns that can emerge in QQ plots as a result of different received signals in the training set and test set, along with the effect on correlation coefficient calculated from the QQ plot as the training set and test set vary in location and scale. Spectra at *d* wavelengths are represented as points in a *d*-dimensional hyperspace. The quantiles are integrals from the center of the Training Set outward in all directions. (Color figure online)

4 Future Work and Conclusions

In the future our research will continue to focus on using the SETI synchronizer strategy based on supernovas and gamma ray bursts, and will introduce some solar transit synchronizer experiments. Planned upgrades to the microwave radio telescope system include the addition of a vacuum-sealed, liquid helium-cooled front end (low noise amplifier, mixer, and antenna probe). For some frequencies, it may be easier to omit the amplifier and send the antenna signal directly into the mixer to down convert it to a lower frequency, where lower noise gain is easier to achieve. An SIS mixer (Superconductor-Insulator-Superconductor) can introduce nonlinearity from quantum tunneling between the two superconductors, achieving low noise in the mixing process.

Collapsing the 5-dimensional SETI problem with synchronizers may never be proven the most fruitful approach to the search for extraterrestrial intelligent (ETI) life. Until we detect the first ETI (and in fact, many more) and know how those detections were made, it will be impossible to say with certainty what is the best approach. Until then, scientists need instrumentation and algorithms capable of collecting and processing increasingly large amounts of Big Data from their searches.

References

1. Beckers, J.M.: Adaptive optics for astronomy - principles, performance, and applications. Ann. Rev. Astron. Astrophys. **31**, 13–62 (1993). (A94-12726 02-90)
2. Clark, J.R., Cahoy, K.: Optical detection of lasers with near-term technology at interstellar distances. Astrophys. J. **867**(2), 97 (2018)
3. Corbet, R.H.: The use of gamma-ray bursts as time and direction markers in SETI strategies. In: IAF, International Astronautical Congress, 51 Rio de Janeiro, Brazil (2000)
4. Downs, R.: Electronic Design. Understand the Tradeoffs of Increasing Resolution by Averaging (2012). https://www.electronicdesign.com/print/51659. Accessed 06 Mar 2019
5. Drake, F.: How can we detect radio transmissions from distant planetary systems, Project Ozma. In: Cameron, A.G.W. (ed.) Interstellar Communication (1963). (Chaps. 16 and 17)
6. Lodder, R.A., Hieftje, G.M.: Detection of subpopulations in near infrared reflectance analysis. Appl. Spectrosc. **42**(8), 1500–1512 (1988)
7. SETI@Home: More about signals (2019). https://setiathome.berkeley.edu/nebula/web/signals.php. Accessed 06 Mar 2019
8. Lodder, R.A.: Weak signal detection in SETI. Contact Context (2018). https://drive.google.com/open?id=1kHGIaFP2J0_7RaGsCHQrE6L56js5OIOB

Parallel Strongly Connected Components Detection with Multi-partition on GPUs

Junteng Hou[1,2(✉)], Shupeng Wang[1], Guangjun Wu[1], Ge Fu[3(✉)], Siyu Jia[1(✉)], Yong Wang[1], Binbin Li[1], and Lei Zhang[1]

[1] Institute of Information Engineering, Chinese Academy of Sciences, Beijing, China
{houjunteng,wangshupeng,wuguangjun,jiasiyu, wangyong,libinbin,zhanglei1}@iie.ac.cn
[2] School of Cyber Security, University of Chinese Academy of Sciences, Beijing, China
[3] National Computer Network Emergency Response Technical Team/Coordination Center of China, Beijing, China
fg@cert.org.cn

Abstract. The graph computing is often used to analyze complex relationships in the interconnected world, and the strongly connected components (SCC) detection in digraphs is a basic problem in graph computing. As graph size increases, many parallel algorithms based on GPUs have been proposed to detect SCC. The state-of-the-art parallel algorithms of SCC detection can accelerate on various graphs, but there is still space for improvement in: (1) Multiple traversals are time-consuming when processing real-world graphs; (2) Pivot selection is less accurate or time-consuming. We proposed an SCC detection method with multi-partition that optimizes the algorithm process and achieves high performance. Unlike existing parallel algorithms, we select a pivot and traverse it forward, and then select a vice pivot and traverse the pivot and the vice pivot backwards simultaneously. After updating the state of each vertex, we can get multiple partitions to parallelly detect SCC. At different phases of our approach, we use a vertex with the largest degree product or a random vertex as the pivot to balance selection accuracy and efficiency. We also implement weakly connected component (WCC) detection and 2-SCC to optimize our algorithm. And the vertices marked by the WCC partition are selected as the pivot to reduce unnecessary operations. We conducted experiments on the NVIDIA K80 with real-world and synthetic graphs. The results show that the proposed algorithm achieves an average detection acceleration of 8.8× and 21× when compared with well-known algorithms, such as Tarjan's algorithm and Barnat's algorithm.

Keywords: Strongly connected components detection · GPU · Multi-partition scheme · Real-world graphs

This work was supported by the National Natural Science Foundation of China (No. 61601458) and the National Key Research and Development Program of China (2016YFB0801305).

J. M. F. Rodrigues et al. (Eds.): ICCS 2019, LNCS 11539, pp. 16–30, 2019.
https://doi.org/10.1007/978-3-030-22747-0_2

1 Introduction

In the interconnected world, many practical applications need to explore large-scale data sets represented by graphs. The strongly connected components (SCC) detection is a fundamental graph computing algorithm that is widely used in many applications, such as network analysis based on web archives, scientific computing [1] and model checking [2].

In a digraph, SCC is the largest subgraph in which any two vertices are mutually reachable. In the early, many efficient algorithms on SCC detection have been proposed including Tarjan's [3], Dijkstra's [4] and Kosaraju's [5]. However, these algorithms are based on depth-first search (DFS) traversal that is difficult to accelerate by parallelization [6]. Barnat et al. [7] introduced three distributed algorithms for detecting SCC: forward and backward (FB) algorithm, Coloring algorithm, and forward and backward with OWCTY elimination (OBF) algorithm. Shrinvas et al. [8] analyzed these algorithms by implementing them on GPU, and summarized that the FB algorithm is better than the other two algorithms. Fleischer et al. first proposed FB algorithm [9], which can obtain one SCC and three subgraphs in each traversal, then continue to iteratively detect SCC on each subgraph. McLendon et al. [10] designed the FB-Trim algorithm that improves the efficiency of SCC detection by trimming the 1-SCCs before forward and backward BFS.

Parallel SCC detection on real-world graphs is difficult because the vertices in real-world graphs obey the power-low distribution [13]. Hong et al. [11] proposed a two-phase algorithm to detect SCCs of different sizes, and they also detect weakly connected component (WCC) and the SCC consisting of two vertices (2-SCC) to optimize the algorithm. Li et al. [12] further propose data-level parallel scheme in the phase of large-scale SCC detection and the task-level parallel scheme in the phase of small-scale SCC detection. Devshatwar et al. [13] proposed an algorithm that selects the vertex with the largest product of in-degree and out-degree as the pivot to ensure the selected pivot is on the largest SCC. Li et al. [14] reduce the traversals by dividing the original graph into multiple partitions and parallel detect on them, but it requires more processing of the vertices on the partition boundary. Aldegheri et al. [15] concluded that no algorithm can beat all other algorithms on various types of graphs. They combine multiple algorithms and use the trained model to adjust their order for different graphs.

We propose a scheme that reduces the traversals by selecting vice pivot to increase the partitions generated on each iteration, and we also optimize some other operations. The contributions are as follows:

(1) We add a vice pivot selection between the pivot's forward and backward traversal. Therefore, the backward traversals of the vice pivot and the pivot are performed simultaneously. In parallel algorithms [9–15], detection on each partition generates one SCC and three partitions. Our approach can generate one SCC and five partitions, which can effectively reduce the number of traversals and significantly accelerate the algorithm.

(2) We balance the accuracy and runtime of the pivot selection. In the phase of large SCC detection, we select the vertex with the maximum product of in-degree and out-degree as the pivot to ensure that the selected pivot is on the largest SCC. In following phase, we use parallel random method to quickly select pivots and vice pivots.
(3) We combine WCC detection and 2-SCC detection with our algorithm to speed up the detection of trivial SCCs. After WCC detection, the vertices marked as WCC labels are directly used as the pivots to reduce a pivot selection without damaging the selection effect.

The rest of this paper is organized as follows: we illustrate the background in Sect. 2. Our proposed design and its operations on different algorithms are introduced in Sect. 3. We present the experimental results in Sect. 4. The conclusion is given in Sect. 5.

2 Background

In this section, we briefly introduce the background of SCC detection on directed graphs.

2.1 Synthetic Graphs and Real-World Graphs

The real-world graphs and synthetic graphs are different in structures, which may cause algorithms that perform well on synthetic graphs inappropriate for real-world graphs.

The most outstanding characteristic of real-world graphs is the power law distribution. Most real-world graphs contain a large SCC and lots of small SCCs. The number of these SCCs generally obey the power law distribution [19]. The small-world property is another feature of real-world graphs. It indicates that the diameter of real-world graph is very small, where the diameter of a graph is the length of the shortest path between the two most distanced vertices.

Synthetic graphs are directly generated by code to simulate real-world graphs. Common synthetic graphs methods include Random, R-MAT [17] and SSCA#2 [18]. The Random scheme directly creates vertices and generates directed edges between any two vertices. R-MAT requires four parameters (a, b, c, d) summed to 1 to respectively represent the probability of selecting four quadrants in the next iteration. Graphs created by SSCA#2 simulate the real-world graphs according to the clique characteristics. It uses an adjacency list to represent vertices and adjacency arrays with auxiliary arrays to store the created edges. It inserts edges between vertices according the clique parameters set by users.

2.2 Sequential Algorithms of SCC Detection

Most of classic sequential algorithms are based on depth-first search(DFS) traversal for DFS only needs to search each edge once. For example, Tarjan's algorithm is inspired by DFS, and it traverses vertices in a way of DFS except for two added mark arrays *Low* recording the minimum connected vertex and *DFN* recording the accessed order of current vertex. The main difference between Tarjan's and DFS is: (1) If the adjacent vertex of the current vertex has not been traversed or has been stored in the stack, then the *Low* value of the adjacent vertex is updated; (2) In the backtracking process, if the *Low* value and *DFN* value of the current vertex are equal, it means that all the vertices whose *Low* value is equal to the current vertices form an SCC, and these vertices are sequentially stored in the stack. Then we can obtain the SCC by popping the stack.

2.3 Parallel Algorithms of SCC Detection

Parallel SCC detection is mostly based on parallel breadth-first search (BFS) traversal algorithms, and GPU can significantly accelerate parallel traversal [20]. Barnat et al. [8] classified the parallel SCC detection algorithms as follows: FB algorithm [9,21], Coloring algorithm, and Recursive OBF algorithm [21], and concluded that the FB algorithm is better than the other two algorithms. The theoretical basis is the following Theorem 1. In Theorem 1, $FW_G(u)$ represents the forward readability closure of vertex u, which is the set of vertices reachable from u. Correspondingly, $BW_G(u)$ represents the backward reachable closure of vertex u, which is the set of vertices reachable to u.

Theorem 1. *Let $G = (V, E)$ be a directed graph with a vertex $u \in V$. Then $FW_G(u) \cap BW_G(u)$ is an SCC containing u. Moreover, every other SCC in G is contained in either $FW_G(u) \setminus BW_G(u)$, $BW_G(u) \setminus FW_G(u)$, or $V \setminus (FW_G(u) \cup BW_G(u))$.*

The main process of the algorithms [11–15] based on FB algorithm is selecting a vertex as the pivot and performing forward BFS and backward BFS from this vertex. According to subgraphs formed by traversal, it can obtain one SCC and three partitions, and the next traversal can be performed in each partition separately until all the SCCs are detected. McLendon et al. [21] proposed the FB-Trim algorithm, which adds a Trim scheme to detect the 1-SCCs separately.

3 Our Approach

In this paper, we propose an effective parallel algorithm of SCC detection, and we introduce it in detail in this section.

3.1 Graph Representation

Aligned memory access and merged memory access are the main characteristics of GPU memory access. And compressed spare row (CSR) is a suitable graph representation for parallel calculation on GPU, which is also a common graph representation for SCC detection on GPU [11,12]. As shown in Fig. 1, the array

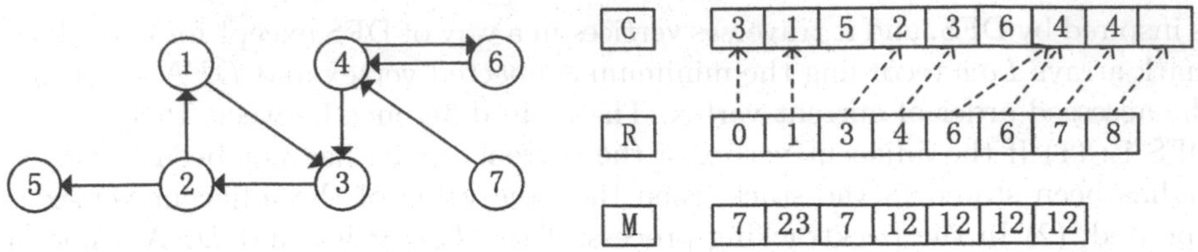

Fig. 1. CSR represent of an example

C saves the ID of vertices linked from the current vertex in order, and the *i-th* element of the array R holds the starting position of all the vertices linked from the *i-th* vertex in array C. Therefore, all the linked vertices of the *i-th* vertex are saved from the $R[i]$-*th* element to the previous of the $R[i+1]$-*th* element in the array C. In our algorithm, we define an integer-type array M to mark the state of each vertex when detecting SCCs.

3.2 SCC Detection with Multi-partition

Many algorithms [11–14] propose improvements based on the FB-Trim algorithm [10] in different aspects, which can accelerate the FB-Trim algorithm, but they do not consider how to improve the detection of medium-sized SCCs, that is, the SCCs expect the largest SCC and 1-SCCs. According to Theorem 1 in Sect. 2.3, the generated partitions of parallel SCC detection algorithm is shown in Fig. 2(a), where p is the selected pivot, vertices traversed by the pivot forward BFS constitute the set $\{A_1, A_2\}$, vertices traversed by the pivot backward BFS constitute the set $\{A_1, A_3\}$. Then A_1 is an SCC, and A_2, A_3, A_4 are three partitions that can respectively perform SCC detection on them. Therefore, it can detect 3^{n-1} SCCs at most on the *n-th* traversal of SCC detection. Assuming that the number of these medium-sized SCCs is x, then the minimal traversal number of the FB-Trim algorithm is $O(\log_3 x)$ under ideal conductions.

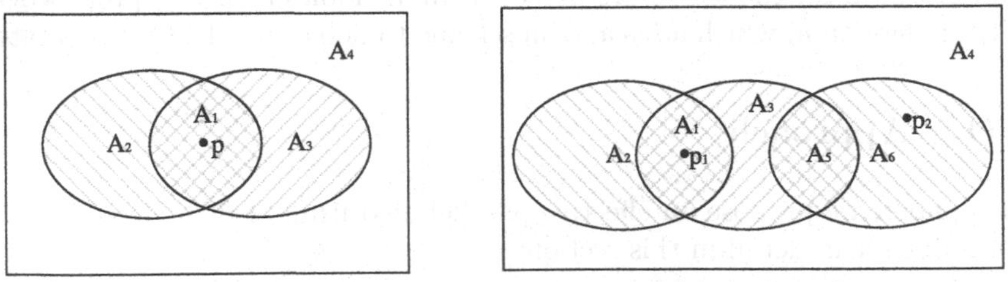

(a) FB-Trim (b) our algorithm

Fig. 2. Partitions in each traversal

We can get the following conclusion from Theorem 1: (1) After forward or backward BFS traversal by an arbitrary vertex, a graph will be divided into two partitions, then each SCC of this graph belongs to one of the two partitions.

There is no SCC cross the partitions, that is, if a part of the SCC belongs to one of these partitions, the rest of the SCC belongs to this partition, too. (2) For the partitions formed by the forward or backward traversals of any two vertices, each partition independently contains some SCCs, and there is no SCC cross the partitions, too. The proof is as follows:

Proof. (1) Let $G = (V, E)$ be a directed graph with an arbitrarily selected vertex $v_0 \epsilon V$. Vertices that are forward traversed by v_0 constitute the partition V_1, and the remaining vertices constitute the partition V_2, so $V_2 = V/V_1$.

Assuming that there is an SCC S crossing the two partitions, and v_1,v_2 are two vertices of S, $v_1 \epsilon V_1$, $v_2 \epsilon V_2$. In partition V_1, v_0 can traverse forward to v_1. In S, v_1 can traverse forward to v_2. Because forward traversal is transferable, v_0 can traverse forward to v_2. Then we can get $v_2 \epsilon V_1$. It is a contradictory to the assumption. Therefore, there is no SCC that can make a crossover in the two partitions formed by traversing.

(2) Let $G = (V, E)$ be a directed graph, and forward or backward traversals of any two vertices v_1, v_2 can obtain four partitions: vertices that can be traversed by v_1 and v_2 constitute partition A_1, vertices that can be traversed by v_1 but can't be traversed by v_2 constitute the partition A_2, the vertices that can be traversed by v_2 but can't traversed by v_1 form partition A_3, and the vertices that can't be traversed by v_1 and v_2 constitute partition A_4.

Suppose that S is an arbitrary SCC in graph G. According to the conclusion obtained in (1), from the perspective of partitions formed by v_1, $S \epsilon \{V_1, V_2\}$ or $S \epsilon \{V_3, V_4\}$, so we only need to prove that S does not have the crossover in V_1 and V_2, or V_3 and V_4. From the perspective of partitions formed by v_2, $S \epsilon \{V_1, V_3\}$ or $S \epsilon \{V_2, V_4\}$. So S does not have the crossover in V_1 and V_2, and S does not have the crossover in V_3 and V_4, either. Therefore, S does not cross between any two of V_1, V_2, V_3, and V_4.

According to the above conclusions, after forward BFS of the pivot, we select another pivot called the vice pivot outside the formed partition. Then the pivot and the vice pivot perform backward BFS simultaneously. The selection of vice pivot has following advantages: (i) The pivot and the vice pivot perform backward BFS traverse with the same method, which can increase the parallelism of BFS to take advantage of the GPUs. (ii) The vice pivot is selected outside of the partition formed by forward traversal of the pivot. So the vice pivot is not on the newly formed SCC. (iii) The partition formed by forward traversal of the pivot does not intersect with the partition formed by backward traversal of the vice pivot, the reason is similar to the conclusion (1). So each partition can intersect with at most one other partition.

Partitions formed by our method are shown in Fig. 2(b), where P_1 is the pivot, and forward BFS traversal of the pivot generates partition $\{A_1, A_2\}$, and then a vice pivot P_2 is selected outside this partition. P_1 and P_2 is simultaneously traversed by backward BFS, where the backward BFS traversal of pivot P_1 generates partitions $\{A_1, A_3, A_5\}$, and the backward BFS traversal of vice pivot P_2 generates partitions $\{A_5, A_6\}$. Among the six partitions, A_1 is an SCC, and the rest partitions can perform SCC detection respectively in the next traversal.

By this way, the minimum traversal of the FB-Trim algorithm based on Theorem 1 can be reduced from $O(\log_3 x)$ to $O(\log_5 x)$, which effectively reduces the number of traversals of the algorithm. Our approach is presented in Algorithm 1.

Algorithm 1. FB-Trim-MP SCC Detection Algorithm

```
procedure FB-TRIM-MP(G(V, E), SCC)
    /* Phase 1 */
    Trim (G, SCC, M, P)
    Pivot-choose1 (G, SCC, M, P)
    do in parallel
        Forward-traverse (G, SCC, M, P)
        Backward-traverse (G, SCC, M, P)
    end do
    Update-state (G, SCC, M, P)
    repeat
        Trim (G, SCC, M, P)
    until no 1-SCC generated
    Trim2 (G, SCC, M, P)
    repeat
        Trim (G, SCC, M, P)
    until no 1-SCC generated
    WCC-detect (G, SCC, M, P)
    /* Phase 2 */
    repeat in parallel
        Forward-traverse (G, SCC, M, P)
        vicePivot-choose (G, SCC, M, P)
        Backward-traverse (G, SCC, M, P)
        Trim (G, SCC, M, P)
        Update-state (G, SCC, M, P)
        Pivot-choose2 (G, SCC, M, P)
    until no SCC generated
end procedure
```

3.3 Selection of Pivot and Vice Pivot

Devshatwar et al. [13] took the vertex with the largest product of in-degree and out-degree as the pivot. They first apply an array *pivots* to store the pivot of the *i-th* partition in *pivots*[*i*], then calculates the product of the in-degree and out-degree of the pivot and the current vertex in parallel. If the product of the current vertex is larger than that of the pivot, they will save current vertex's ID in *pivots*[*i*]. However, all the threads in the GPU calculate products and store pivots in parallel, so in the *i-th* partition, all the vertices whose products of in-degree and out-degree are larger than the product of the pivot will store their ID in *pivots*[*i*] almost at the same time. Therefore, these storage operations are disordered, and they can only guarantee that the product of the vertex stored in *pivots*[*i*] is larger than that of the vertex originally stored in *pivots*[*i*], but it is not the largest. we propose a method by iteratively executing the kernel function until *pivots*[*i*] is no longer changed to guarantee the obtained pivot has the largest product. The algorithm is shown in Algorithm 2.

This scheme can significantly improve the accuracy of locating the maximum SCC, but it is time consuming. So after selecting the first pivot by *Pivot-choose*1, we take another pivot selection method that randomly selects a vertex in each partition as the pivots in the *phase* 2, and it is also used to select the vice pivots. The random selection method is to store all the vertices of the same partition to the same memory location in parallel, and we take a method to prevent the partitions traversed by the pivot and vice pivot from approaching each other. When selecting the pivot, there is an atomic operation to prevent the other vertices from being stored after the first vertex is stored. When selecting the vice pivot, there is no atomic operation to block the storage of any vertices. Therefore, the pivot is the first vertex stored in the fixed position, and the vice pivot is the last stored vertex so that the pivot and the vice pivot will not be close to each other.

Algorithm 2. choose pivot Algorithm

procedure PIVOT-CHOOSE1((G, SCC, M, P))
  repeat
    Pivotchoose-Kenel ($G, SCC, M, P, pivots$)
  until M is not change
    update-pivot ($M, pivots$)
end procedure

procedure PIVOTCHOOSE-KENEL($(G, SCC, M, P, pivots)$)
  $v \leftarrow threadId$
  if $M(v)$ is not marked as trimmed **then**
    $u \leftarrow pivots(P(v))$
    $uDegree \leftarrow inDegree(u) * outDegree(u)$
    $vDegree \leftarrow inDegree(v) * outDegree(v)$
    if $vDegree > uDegree$ **then**
      $pivots(P(v)) \leftarrow v$
    end if
  end if
end procedure

3.4 Improvement Details

In Algorithm 1, we added 2-SCC detection and WCC detection between *phase* 1 and *phase* 2. 2-SCC is a small SCC composed of two vertices, which is also abundant in real-world graphs, and most of them are easy to detect. For each undetected vertex, detect the vertex that it directly links to and directly links from simultaneously. If the two vertices have no in-degree or out-degree, they can form an independent 2-SCC. WCC detection can divide a graph into several disconnected partitions. Firstly, each vertex is initialized to form a WCC by itself, so the WCC ID of a vertex is set to its own ID. Then it is checked whether there is a vertex in the adjacent of current vertex whose WCC ID is smaller than the WCC ID of the current vertex. If it exists, the WCC ID of the current vertex is set to the smaller value, and the above process is iterated until all WCC ID do not change. In *phase* 2, we take the WCC ID as the pivot of this partition directly, which can save the time for selecting the pivots.

3.5 Expend Algorithms Devshatwar-MP and Li-MP

The algorithm proposed by Devshatwar et al. [13] is based on the FB-Trim algorithm, which can be improved by our algorithm. Compared with the FB-Trim algorithm, Devshatwar's algorithm mainly has following improvements: (1) There are two modes to traverse vertices: vertex-centric and virtual warp-centric; (2) It increase 2-SCC detection and WCC detection; (3) When selecting the pivot of each partitions, it uses the vertex with the maximum product of in-degree and out-degree as the pivot of each partition. Besides applying the multi-partition scheme to the above processes, our Devshatwar-MP algorithm improve the Devshatwar's algorithm at the following points: (1) We still adopt vertex-centric and virtual warp-centric modes; (2) After the WCC detection, the ID of WCC region are directly used as the pivot of current region; (3) We use the pivot selection scheme proposed by Devshatwar et al. to select the pivots and vice pivots.

Li et al. [12] also propose improvements on the FB-Trim algorithm, and we can implement our method on the Li's algorithm. Compared with the FB-Trim algorithm, Li's algorithm mainly has the following improvements: (1) There are two traversal modes named data-driven and topology-driven, which is similar to virtual warp-centric and vertex-centric in Devshatwar's algorithm. (2) 2-SCC detection and WCC detection are added; (3) It increases operation of loading balance, and adapts Topo-lb in *phase* 1 and Topo in *phase* 2. Besides applying the multi-partition scheme to the above processes, our Li-MP algorithm improve Li's algorithm at the following points: (1) The algorithm framework using Topo-lb mode in *phase* 1 and Topo mode in *phase* 2 is still adopted; (2) After the WCC detection, the ID of each region marked at the WCC detection is directly used as the pivot of the current region; (3) The selection of the pivots and the vice pivots is optimized by the method mentioned in Sect. 3.3.

4 Experimental Evaluation

Graphs used in our experiment include synthetic graphs and real-world graphs. The synthetic graphs are the following three types of graphs generated by *GeorgiaTech.graphgenerator* (GTgraph) [16]: Random, R-MAT [17] and SSCA#2 [18], as shown in Table 1. Real-world graphs come from two commonly used benchmarks [11–13]: SNAP database [22] and Koblenz Network Collection database [23], as shown in Table 2.

Table 1. The detailed parameters of generated graphs

Type	Name	Vertices	Edges	Parameters
R-MAT	GT-rmata	10,000,000	100,000,000	a = 0.25, b = 0.25, c = 0.25, d = 0.25
R-MAT	GT-rmatb	10,000,000	100,000,000	a = 0.45, b = 0.15, c = 0.15, d = 0.25
Random	GT-randa	10,000,000	100,000,000	p = 0.8
Random	GT-randb	10,000,000	100,000,000	p = 0.6
SSCA#2	GT-sscaa	10,000,000	80,771,507	maxCliqueSize = 10, maxParalEdges = 2
SSCA#2	GT-sscab	10,000,000	95,068,514	maxCliqueSize = 12, maxParalEdges = 2

Table 2. The details of real-world graphs

Name	Vertices	Edges	Average degree	Maximum degree
Amazon0302	400,727	3,200,440	7.99	2,757
Amazon0312	262,111	1,234,877	4.71	425
Amazon0505	410,236	3,356,824	8.18	2,770
Slashdot0811	77,360	905,468	11.70	5,048
NotreDame	325,727	1,497,134	4.60	10,721
Google	875,713	5,105,039	5.83	6,353
pokec-relationships	1,632,803	30,622,564	18.75	20,518
LiveJournal	4,874,571	68,993,773	14.15	22,889

4.1 Experiment Setup

We compare five implementations including: (1) Tarjan: Tarjan's sequential SCC detection algorithm is a classic and representative sequential algorithm [3,8]; (2) Barnat: Barnat's SCC detection method is a classical parallel algorithm, which is compared in many improved algorithms [8,11–13]; (3) FB-Trim-MP: our parallel SCC detection algorithm with multi-partition; (4) Devshatwar-MP: Devshatwar's algorithm [13] improved by our multi-partition algorithm. (5) Li-MP: Li's algorithm [12] improved by our multi-partition algorithm. We use gcc and nvcc with the -O3 optimization option for compilation along with -arch=sm_37 when compiling for the GPU. We execute all the benchmarks 10 times and collect the average execution time to avoid system noise.

4.2 Effectiveness of Multi-partition Scheme

In order to verify the effectiveness of the multi-partition scheme, we improve the Barnat's algorithm by our multi-partition scheme, and the other operations are not changed. The comparison of execution time and the number of traversals between Barnat algorithm and the improved algorithm (Barnat-MP) are summarized in the following Table 3. It can be seen that the Barnat-MP doesn't accelerate at R-MAT and Random graphs. Because these two kinds of graphs only contain one large SCC and lots of 1-SCCs, and they don't need many traversals to process the medium-sized SCCs. However, in real-world graphs and SCCA#2 graphs, there are lots of medium-sized SCCs. It requires multiple traversals to detect these SCCs. For example, the Barnat's algorithm requires 12225 traversals to detect all SCCs in real-world graph *LiveJournal*. Therefore, in real-world graphs, the number of partitions generated by each traversal has a great influence on the number of traversals in SCC detection. As shown in Table 3, the Barnat-MP algorithm can reduce the number of traversals to 50% to 64% of the Barnat algorithm and the speed of Barnat-MP algorithm can also be increased to 1.3× to 13× compared with the Barnat algorithm. Therefore, the multi-partition scheme can significantly accelerate the parallel algorithm of SCC detection.

Table 3. Comparison of execution time (left) and number of iterations (right)

	Comparison of traversal number			Comparison of running time		
Name	Barnat	Barnat-MP	Iteration proportion	Barnat(s)	Barnat-MP(s)	Time proportion
Amazon0302	426	242	56.81%	0.365	0.254	143.70%
Amazon0312	1,250	668	53.44%	0.829	0.588	140.99%
Amazon0505	326	170	52.15%	0.277	0.199	139.20%
Slashdot0811	240	121	50.42%	0.0976	0.059	165.42%
NotreDame	1,180	611	51.78%	0.814	0.641	126.99%
Google	5,371	3,394	63.19%	6.486	4.645	139.63%
pokec-relationships	1,080	684	63.33%	1.317	1.017	129.50%
LiveJournal	12,225	7,539	61.67%	33.286	25.236	131.90%
rmata	1	1	100.00%	0.263	0.293	89.76%
rmatb	1	1	100.00%	0.277	0.312	88.78%
randoma	1	1	100.00%	0.289	0.299	96.66%
randomb	1	1	100.00%	0.289	0.299	96.66%
sscaa	2,236	1,397	62.48%	9.733	0.727	1,338.79%
sscab	1,520	973	64.01%	6.928	0.701	988.30%

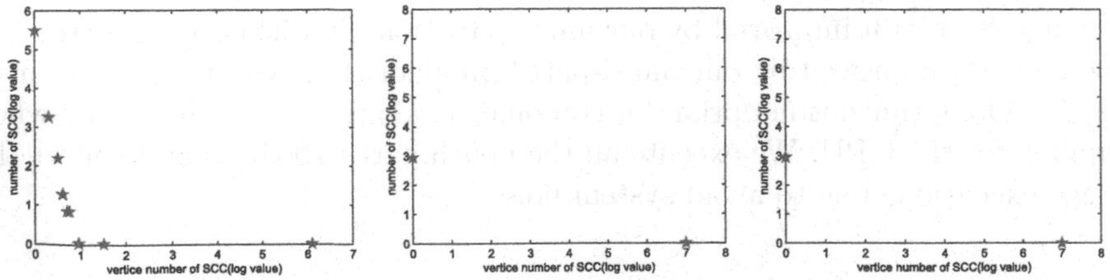

(a) pokec-relationships (b) rmata (c) randoma

Fig. 3. The number distribution of SCCs of different graphs

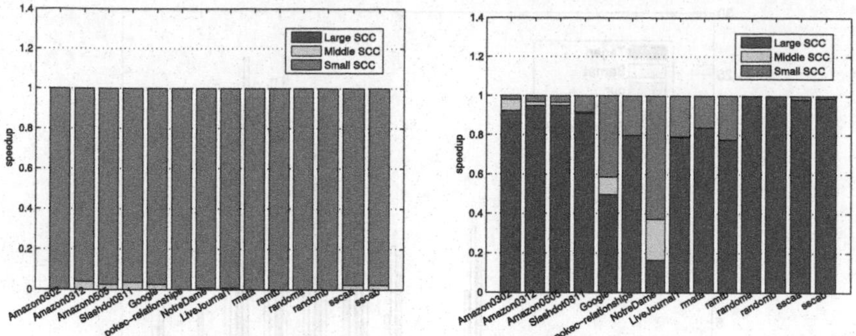

(a) proportion of SCCs (b) proportion of vertices in SCCs

Fig. 4. The proportion of SCCs and their vertices

4.3 Analysis of SCC Distribution

Before the performance analysis, we compare and analyze the SCC distribution of the graphs used in our experiment. Figure 3 shows the distribution of SCCs in synthetic graphs and real-world graphs. In order to clearly show the number of medium-sized SCCs, we use the natural logarithm of the number of SCCs as the ordinate value. Figure 4 compares the number of the largest SCCs, small SCCs including 1-SCCs and 2-SCCs, and remaining medium-sized SCCs, and it also compares the number of vertices contained in these SCCs. It is clear in Fig. 3 that synthetic graphs of R-MAT and Random are different from real-world graphs in SCC distribution. In R-MAT and Random synthetic graphs, there are only two types of SCCs: one large SCC and lots of small 1-SCCs. It also verifies the reason why SCC detection on the R-MAT and Random graphs is decelerated in Sect. 4.2. There are many medium-sized SCCs in real-world graphs. In Fig. 4, we divide SCCs into three categories. It can be found from the R-MAT and Random graphs in Fig. 4 that medium-sized SCCs don't account for any proportion on no matter the number of SCCs or the number of vertices. In real-world graphs and SCCA#2 graphs, the number of medium-sized SCCs and the number of their vertices account for a little proportion on SCCs number and vertices number. For all SCCs, the number of vertices contained in the largest SCC accounts for a large proportion. So it is necessary to be specifically detected. *Phase* 1 in our algorithm mainly deals with the largest SCC. Each small SCC only contains one or two vertices, but the number of small SCCs is large. So *Trim* and *Trim*2 in our algorithm are used to deal with these vertices. The medium-sized SCCs are not large in vertices number and SCCs number, but they need to be traversed many times, which is the main part accelerated by our multi-partition method.

4.4 Performance Analysis

We implement the five algorithms mentioned in Sect. 4.1 on six synthetic graphs and eight real-world graphs. In order to clearly display the experiment results,

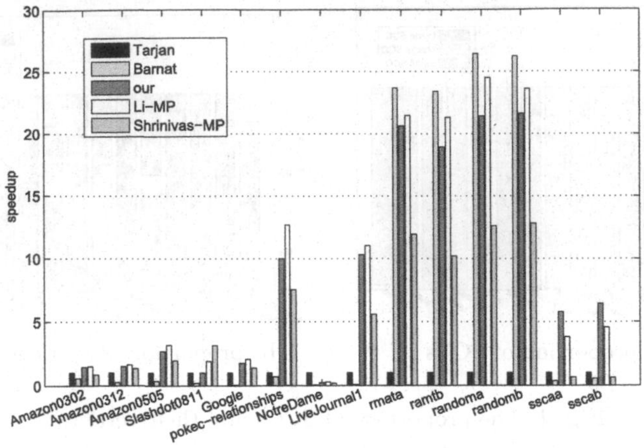

Fig. 5. Acceleration of various algorithms compared with Tarjan's algorithm

we normalize the execution time of all algorithms by that of the Tarjan algorithm and display the speedup in Fig. 5. The result shows that our algorithm achieves an average acceleration of 8.8× and 21× compared to Tarjan's algorithm and Barnat's algorithm. Li-MP, Shrinovas-MP and our algorithm are significantly faster than Tarjan's algorithm in most graphs, which can reach 10× or even 20× speedup. Since there is no medium-sized SCCs, the speed of them is slightly lower than Barnat's algorithm, but when detecting SSCA#2 graphs and real-world graphs, Barnat's algorithm is poor. In these graphs, the speed of Barnat's algorithm is even lower than that of Tarjan's algorithm, while Li-MP, Shrinovas-MP and our algorithm can still maintain a certain acceleration in most cases.

Compared with Li-MP and Shrinovas-MP, our algorithm is not inferior. Li-Mp is consistent with our algorithm except for loading balance. Figure 5 shows that the Li-MP is faster than our algorithm on most graphs, and the average acceleration ratio is 10%. The largest acceleration is on graph *Slashdot*, where the speed of Li-MP is about 1.9× of our algorithm. However, on SSCA#2 graphs, the speed of Li-MP is only 0.66× −0.71× of our algorithm. Therefore, the loading balance can accelerate most graphs but not all. Compared with our algorithm, Shrinivas-Mp adds virtual warp-centric mode in *phase* 1, and uses its own pivot selection scheme. The operation of virtual warp-centric in Shrinivas-MP is similar to the data-driven mode in Li-MP. As shown in Fig. 5, Shrinivas-MP is not as fast as Li-Mp and our algorithm on most graphs. This is mainly because that the pivot selection scheme of Shrinivas-MP takes some time in *phase* 2, so it is necessary to optimize the pivot selection.

It can be concluded in Fig. 5 that parallel SCC detection algorithms show better acceleration on large graphs. The Tarjan's algorithm only processes one edge at a time, so its runtime is positively correlated with the scale of graphs. On real-world graphs *Amazon*0312, *Amazon*0302, *Amazon*0505, and *Google*, whose vertices are not more than 1M, Li-MP, Shrinovas-MP and our algorithm only accelerate several times, and the detection speed on *NotreDame* with only 0.3M vertices is even lower than that of Tarjan algorithm. This is mainly because the scale of graphs is not large, and plenty of threads on GPU are idle. The real-world graphs *LiveJournal* and *pokec-relationships* are commonly used in the experiments of many parallel SCC detection [11–13], and these algorithms can achieve about 10× acceleration on these two graphs. The size of the two graphs is significantly larger than the previous graphs. Therefore,only when the graphs is large enough, can the parallel algorithms make full use of all threads of the GPUs.

5 Conclusion

Graph computing is important to abstract and solve problems in the interconnected world. Parallel SCC detection is an important part of graph computing accelerating algorithms. In this paper, we propose a parallel algorithm of SCC detection with multi-partition to reduce the number of traversals by increasing the number of partitions generated in each traversal. We select a vice pivot after the forward traversal of the pivot and make it traverse backward together with the pivot, which can increase the number of partitions in each traversal. And we also improved the selection method of pivots and vice pivots. We combine 2-SCC detection as well as WCC detection and make the vertices marked at WCC detection as the pivots to accelerate the algorithm. Experimental results demonstrate that our algorithm outperforms existing SCC detection algorithms on synthetic graphs and real-world graphs. The proposed algorithm achieves an average acceleration of 8.8× and 21× over Tarjan's algorithm and Barnat's algorithm. In the future, we will further discuss the structure and proportional of the largest SCC with other SCCs and combine them with various fine-grained acceleration schemes, and adopt different schemes according to the internal features of various graphs to ensure the efficiency and the stability of the algorithm.

References

1. Xie, A., Beerel, P.A.: Implicit enumeration of strongly connected components and an application to formal verification. IEEE Trans. Comput.-Aided Des. Integr. Circ. Syst. **19**(10), 1225–1230 (2000)
2. Simona, O.: On distributed verification and verified distribution. Ph.D. dissertation, Center for Mathematics and Computer Science (CWI) (2004)
3. Tarjan, R.: Depth-first search and linear graph algorithms. SIAM J. Comput. **1**(4), 146–160 (1972)
4. Dijkstra, E.W.: A Discipline of Programming, 1st edn. Prentice Hall, Englewood Cliffs (1976)
5. Cormen, T.H., Leiserson, C.E., Rivest, R.L., Stein, C.: Introduction to Algorithms, 3rd edn. The MIT Press, Cambridge (2009)
6. Reif, J.H.: Depth-first search is inherently sequential. Inf. Process. Lett. **20**(5), 229–234 (1985)

7. Barnat, J., Chaloupka, J., Jaco, V.D.P.: Distributed algorithms for SCC decomposition. J. Logic Comput. **21**(1), 23–44 (2011)
8. Barnat, J., Bauch, P., Brim, L., Ceska, M.: Computing strongly connected components in parallel on CUDA. In: Sussman, A., Mueller, F., Beaumont, O., Kandemir, M.T., Nikolopoulos, D.(eds.) IPDPS 2011, pp. 544–555. IEEE(2011). https://doi.org/10.1109/ipdps.2011.59
9. Fleischer, L.K., Hendrickson, B., Pınar, A.: On identifying strongly connected components in parallel. In: Rolim, J. (ed.) IPDPS 2000. LNCS, vol. 1800, pp. 505–511. Springer, Heidelberg (2000). https://doi.org/10.1007/3-540-45591-4_68
10. McLendon III, W., Hendrickson, B., Plimpton, S.J., Rauchwerger, L.: Finding strongly connected components in parallel in particle transport sweeps. In: SPAA 2001, pp. 328–329. ACM, Crete (2011). https://doi.org/10.1145/378580.378751
11. Hong, S., Rodia, N.C., Olukotun, K.: On fast parallel detection of strongly connected components (SCC) in small-world graphs. In: SC 2013, pp. 1–11. ACM, Denver (2013). https://doi.org/10.1145/2503210.2503246
12. Li, P., Chen, X, Shen, J., Fang, J., Tang, T., Yang, C.: High performance detection of strongly connected components in sparse graphs on GPUs. In: PMAM@PPoPP 2017, pp. 48–57. ACM, Texas (2017). https://doi.org/10.1145/3026937.3026941
13. Devshatwar, S., Amilkanthwar, M., Nasre, R.: GPU centric extensions for parallel strongly connected components computation. In: GPGPU@PPoPP 2016, pp. 2–11. ACM, Barcelona (2016). https://doi.org/10.1145/2884045.2884048
14. Li, G.H., Zhu, Z., Cong, Z., Yang, F.M.: Efficient decomposition of strongly connected components on GPUs. J. Syst. Archit. **60**(1), 1–10 (2014)
15. Aldegheri, S., Barnat, J., Bombieri, N., Busato, F., Češka, M.: Parametric multi-step scheme for GPU-accelerated graph decomposition into strongly connected components. In: Desprez, F., et al. (eds.) Euro-Par 2016. LNCS, vol. 10104, pp. 519–531. Springer, Cham (2017). https://doi.org/10.1007/978-3-319-58943-5_42
16. Madduri, K., Bader, D.A.: GTgraph: a suite of synthetic graph generators. https://github.com/dhruvbird/GTgraph. Accessed 15 Sept 2012
17. Chakrabarti, D., Zhan, Y., Faloutsos, C.: R-MAT: a recursive model for graph mining. In: SDM 2004, pp. 442–446. Society for Industrial and Applied Mathematics, Orlando (2004). https://doi.org/10.1137/1.9781611972740.43
18. Bader, D.A., Madduri, K.: Design and implementation of the HPCS graph analysis benchmark on symmetric multiprocessors. In: Bader, D.A., Parashar, M., Sridhar, V., Prasanna, V.K. (eds.) HiPC 2005. LNCS, vol. 3769, pp. 465–476. Springer, Heidelberg (2005). https://doi.org/10.1007/11602569_48
19. Kumar, R., Novak, J., Tomkins, A.: Structure and evolution of online social networks. In: KDD 2006, pp. 611–617. ACM, New York (2006). https://doi.org/10.1145/1150402.1150476
20. Defour, D., Marin, M.: Regularity versus load-balancing on GPU for treefix computations. Procedia Comput. Sci. **60**, 309–318 (2013)
21. McLendon III, W., Hendrickson, B., Plimpton, S.J., Rauchwerger, L.: Finding strongly connected components in distributed graphs. J. Parallel Distrib. Comput. (JPDC) **65**(8), 901–910 (2005)
22. Leskovec, J., Krevl, A.: SNAP Datasets: Stanford Large Network Dataset Collection. http://snap.stanford.edu/data. Accessed Jun 2014
23. Koblenz network collection. http://konect.uni-koblenz.de/. Accessed 25 Apr 2018

Efficient Parallel Associative Classification Based on Rules Memoization

Michel Pires[1,3](✉), Nicollas Silva[3], Leonardo Rocha[2], Wagner Meira[3], and Renato Ferreira[3]

[1] Centro Federal de Educação Tecnológica de Minas Gerais, Divinópolis, MG, Brazil
michel@cefetmg.br

[2] Universidade Federal de São João del-Rei, São João del-Rei, MG, Brazil
lcrocha@ufsj.edu.br

[3] Universidade Federal de Minas Gerais, Belo Horizonte, MG, Brazil
{michelpires,ncsilvaa,meira,renato}@dcc.ufmg.br

Abstract. Associative classification refers to a class of algorithms that is very efficient in classification problems. Data in such domain are multidimensional, with data instances represented as points of a fixed-length attribute space, and are exploited from two large sets: training and testing datasets. Models, known as classifiers, are mined in the training set by class association rules and are used in eager and lazy strategies for labeling test data instances. Because test data instances are independent and evaluated by sophisticated and high costly computations, an expressive overlap among similar data instances may be introduced. To overcome such drawback, we propose a parallel and high-performance associative classification based on a lazy strategy, which partial computations of similar data instances are cached and shared efficiently. In this sense, a PageRank-driven similarity metric is introduced to reorder computations by affinity, improving frequent-demanded association rules memoization in typical cache strategies. The experiments results show that our similarity-based metric maximizes the reuse of rules cached and, consequently, improve application performance, with gains up to 60% in execution time and 40% higher cache hit rate, mainly in limited cache space conditions.

Keywords: Parallel associative classification · Memoization · Class association rules

1 Introduction

Classification is an important task in data mining, and the associative classification (ACs) its branch that describes a class of algorithms based on two well-known mining paradigms, pattern classification and association rules mining [4]. Pattern classification assigns a class label to a given data input instance by a

J. M. F. Rodrigues et al. (Eds.): ICCS 2019, LNCS 11539, pp. 31–44, 2019.
https://doi.org/10.1007/978-3-030-22747-0_3

specific model named classifier, and association rules mining is the task that discovers correlations or other relationships in large datasets to turned classification process accurate [5].

Several efforts have shown that ACs are capable of building efficient and accurate classification systems, by three typical steps. Firstly, class association rules (CARs) are mined from attribute space through a training dataset. Such CARs are weighted according to a given function, and support and confidence thresholds pruning weak rules to make up a frequent item set. Posteriorly, frequent items are assumed as part of a classifier into eager or lazy approaches to associate classes for unlabeled input data instances of a test data set [15].

In order to achieve accurate classifiers, eager strategies look at frequent items by a global searching into attribute space. An expressive number of useless rules may be introduced due to weak similarity among part of generated rules and unlabeled data attribute space. To overcome such problem, lazy approaches, as the lazy associative classification (LAC) [21], investigates the attribute space by local searches during the classification time, whenever a novel unlabeled data input is provided. Thus, suitable models are afforded for each data instance while a higher similarity between classifiers and data attribute space is ensured [22].

The local models look at into attribute space by a large variability of descriptions or subproblems, which it can afford a better global approximation of target function [21,22], which leads to a specific advantage in relation to traditional classification models, such as decision trees and rule induction [5,16]. Because local models are designed independently for each unlabeled data instance, an expressive computation overlap between similar classifiers may be introduced. This is because different classifiers overlap on the attribute space, which leads to a costly rework under a significant amount of similar CARs. Indeed, an opportunity to employ typical cache strategies to overcome such drawback, since CARs can be efficiently cached. Further, because classifiers are independent, the classification process can be exploited within parallel execution models.

In this paper, we propose a parallel and high-performance LAC execution model based on MapReduce concepts in that a list of unlabeled data instances is orchestrated toward an effective CARs memoization. We propose modeling unlabeled data instances attribute space as a weighted graph in which vertices are drawn as attributes and edges as correlations among them. The weight in each edge exposes the number of times that each relationship between two vertices occurs. Because the local models are designed based on the attribute space, we evaluate the relationship of different data instances by a similarity metric based on PageRank. Basically, we use PageRank to discover relevant attributes and, the eigenvectors and eigenvalues yielded by it for clustering data instances in an attempt to ensure high cohesion and low coupling among CARs demanded by classifiers. PageRank is used to discover correlated Web pages with a high-accuracy, we use such accuracy to identify data instances with high-similar attribute subsets keeping them close to each other, which translates in local models with the high relationship of CARs. In our experiments, the results show that our similarity-based metric maximizes the number of rules reused in the cache and, consequently, improve application performance, with gains up to 60% in execution time and 40% in the cache hit rate, mainly in limited cache space conditions.

To summarize, the main contributions of this paper are: *(i) the proposal of a parallel approach to optimize the data analysis process on LAC for recurrent and high-cost computations; (ii) a powerful cache implementation in which a PageRank-driven similarity metric is employed to deal with computation affinities and CARs demand-relationship; (iii) a thorough evaluation based on different datasets to explain the potential of CAR memoization, as well as execution time improvements and cache hit rate.*

Roadmap. This paper is organized as follows. Section 2 introduces a background of ACs, LAC, MapReduce, and related works. Section 3 describes our approach in details, and as LAC job instance is executed in one of the most important open-source distributed general-purpose engine based on MapReduce, the Spark. Section 4 presents experiments setup and discuss results around the number of generated rules, execution time and cache hit rates. Finally, in Sect. 5, we introduce the conclusions.

2 Background

In this section, we introduce basic definitions that are necessary to understand the associative classification problem, the lazy associative classification approach and MapReduce programming model over the Spark engine. We also show related works to the proposed theme and its benefits for the classification process.

2.1 Associative Classification Problem

AC is a data mining branch in which useful patterns are discovered in large data sets by exploiting class association rules. The first concepts correlated with such domain were introduced in the paper "*Mining association rules between sets of items in large databases*" by [1]. After that, other techniques have been proposed, including emerging patterns methods (CAEP) [8], multiple class association rules (CMAR) [13] and (MCAR) [17], predictive association rules (CPAR) [25], instance centric rule generation (HARMONY) [24]. There are also recent efforts in which parallel and distributed approaches are investigated [5,12,18].

Formally, we explain such approaches denoting training set as Λ and test set as Γ. We address ACs as association rules mining cases in which data instances (also referred to as examples - Λ) are looks at as pairs in the form $\lambda_i = \langle x_i, c_i \rangle$. Each x_i is drawn as a point in a fixed-length attribute space, that is, outlined by an item set $\langle a_1, a_2, \ldots, a_k \rangle$ with a_k as the kth attribute-value mapped in such space. Each c_i is expressed as a value in a discrete and finite set of possibilities $\langle v_1, v_2, \ldots, v_p \rangle$ with $1 \leq i \leq p$ and designates the class to which λ_i belongs. The classification process consists, for the instances of Γ on which c_i is unknown, find a conditional probability distribution $P(c|x)$, mapped of the relationships between points and classes in Λ by a function in a form $F : \chi \rightarrow C$. The performance of a given $f \in F$ is expressed by some accuracy criterion using Γ.

Definition 1: An item in Λ and Γ can be described as a combination of attribute name A_i and value a_i, denoted $\langle(A_i, a_i)\rangle$.

Definition 2: A CAR r, mined in Λ, is denoted as $\chi \rightarrow c_i$ where χ is a itemset (i.e., a set of pairs $\langle(A_1, a_1), \ldots (A_k, a_k)\rangle$) and c_i the ith class in a discrete and finite set of possibilities in C.

Definition 3: The incidence of a r (denoted $I(r)$) is given by the number of data instances in Λ that have as antecedent r, that is:

$$I(r) = \sum_{i=1}^{|\Lambda|} \lambda_i;\ \forall \lambda_i \subseteq \Lambda \mid \lambda_i \in \chi \quad (1)$$

Definition 4: The support threshold of r or $S(r) = S(\chi \rightarrow c_i) = P(c_i|I(r))$ is referenced as:

$$S(r) = \frac{|P|}{|\Lambda|} = \frac{I(r) \rightarrow c_i}{|\Lambda|}. \quad (2)$$

Definition 5: The confidence threshold of r or $E(r) = E(\chi \rightarrow c_i) = P(c_i|I(r))$, is represented as:

$$E(r) = \frac{|P|}{|I(r)|} = \frac{S(\chi \rightarrow c_i)}{S(\chi)} \quad (3)$$

The main task in a CA is to find rules set able of associating classes with unlabeled data instances, that is, for pairs drawn as $\tau_i = \langle x_i, ?\rangle$. In other words, discovering a target function $f \in F$ that express the conditional probability distribution $P(c|x)$ into a higher accuracy function $f(x, c) = \tau$, for each unlabeled data instance $\tau \in \Gamma$.

2.2 Lazy Associative Classification

Introduced by [21], LAC is a demand-driven associative classification that uses local searches in Λ to compose classifiers, whenever an unlabeled data instance $\tau \subseteq \Gamma$ is provided. Therefore, it is assumed that pairs in Λ are in some sense related to pairs in Γ, sampled independently and identically by the same distribution $P(c|x)$. Facts that make LAC capable to afford a better global approximation of the target function F by local models, which leads to greater accuracy than noticed in traditional classification models [19–21]. The classification process performed by LAC is described in the Algorithm 1.

In Algorithm 1, Λ is accommodated into hash tables η_X and η_C. η_X describe the relationship between items and transaction indexes in the dataset. Each item, defined as a key of the hash, is linked to transactions indexes in which the item's pair is addressed, that is, $\langle(A_i, a_i)\rangle \rightarrow \lambda_i\ \forall \lambda_i \subseteq \Lambda \mid \langle(A_i, a_i)\rangle \subseteq \lambda_i$. Similarly, η_C is modeled to express the classes associations.

For each unlabeled data instance τ, a target function $f \in F$ is modeled through a projection of τ in η_X. Only keys in η_X directly related to τ are considered for classification. CARs are mined by the aforementioned Definitions 3, 4 and 5 to extract the relevant rules, named frequent itemset. Thus, each τ is labeled sorting the frequent itemset and finding the higher class threshold in η_C.

Algorithm 1. Lazy Associative Classification [21]

input : η_X hash of items
input : η_C hash of classes
input : Γ set of unlabeled data instances
output: $\tau \rightarrow c_i$ for each $\tau \subseteq \Gamma$
for $\tau \in \Gamma$ **do**
  Given η_τ the projection of τ in η_X
  Given χ_{η_τ} CARs mining in η_τ and submitted to equations 2 and 3
  Sort χ_{η_τ} by thresholds of each $c_i \subseteq \eta_C \wedge \chi_{\eta_\tau}$
  Get better $c_i \in \eta_C$ and use it to classify τ
end

2.3 MapReduce over Spark

MapReduce is a concept for processing a large amount of data in parallel and distributed environments based on a functional programming [7]. Proposed by Google in 2004, its computational flow is based on two core constructions, *map* and *reduce*. The idea behind such construction is providing means to partition data instances of a dataset into independent tasks, while distributing its computations on a cluster, avoiding communications and possible failures, at the same time that ensures an efficient disk usage and partial results arrangements [23].

In a MapReduce execution, a dataset is automatically partitioned into independent subsets and each one processed by an independent machine with a copy of user program. One copy is denoted as a *master* and is used to schedules and handles such subsets for other ones that perform them, named *workers*.

In last years, several efforts have been employed to deal with large datasets over MapReduce concepts. One of such efforts has led to a fast and general engine for large-scale data processing, named Spark [26]. In Spark, the programming model, based on MapReduce, is extended to introduce a data-sharing abstraction denoted as resilient distributed datasets (RDDs), which are fault-tolerant objects distributed across the cluster that can be handled in parallel. In addition, Spark introduces a large range of novel actions and transformation functions, as well as, explicit support for data sharing among computations, such as *accumulator* and *broadcast* variables. Advantages that allows introducing high-performance mining in a wide range of workloads whose composition and size until then required separate engines.

2.4 Related Works

ACs are gaining more and more attention as an ever-increasing amount of data are becoming available about many interesting events of everyday life and, classify such data is an important task of science and engineering. In this sense, the MapReduce is a well-known paradigm addressed for more than one decade that deals with data parallelization in a large number of machines by a series of popular and open-source engines. In recent years, efforts have been directed attention to improving performance for different AC algorithms, many of these based on MapReduce [2,5,9,12,18]. It is also possible to observe a significant trend to improve performance for MapReduce applications by the sophisticated aware-resource usage strategies [11,27] and data schedulers [6,14].

Several researches, e.g., [5,10], show MapReduce applications performance depends on cluster configurations and jobs, as well as, input data. According to [3], an effectively data-scheduling is the major challenges in MapReduce frameworks. Thus, important observations have been comparing First-in-First-Out (FIFO) data scheduler performance with more sophisticated strategies, some of these based on data locality [11,27].

In [11], an extensive investigation of data locality is reported and, data-scheduling issues addressed by a mathematical model and theoretical analysis in which impacts from cluster configurations and tasks are observed. In the same way, [27] investigate a cache strategy and looks at results reuse under a perspective of the Map transformations. Effective solutions for many applications, though is not adequate for situations where results produced by each data instance overlaps in random times during execution, which often happens when a naïve execution order is adopted.

As above mentioned, advances task-scheduler and efficient use of resources in MapReduce have been drawn the attention of different research. In AC, solutions employ such advantages to improve the classification step. In [18], a variation of MCAR (denoted MRMCAR) is presented to search frequent items in the rule discover step under large training data by proposing a new learning method that repeatedly transforms the data spaces. In the work presented by [12], a similar idea is used to develop an online algorithm for performing frequent items discover in a data streaming. In another hand, [2,5,9] use parallelization benefits of MapReduce to introduce classifications strategies that treat with typical methods toward to the big data solutions. In this sense, a fuzzy associative classifier [9], a FP-Growth [5] parallel approach and a new storage format are presented. Solutions that involve several perspectives of performance under different research fields with great results in computational resource and execution time maximization. However, to our best knowledge, such efforts do not consider a pre-analysis of data instances to predict computations and results before executions to achieve high-performance. In addition, many such research fields do not evaluate the overlap among partial computations of different data instances to maximize result reuse as part of their performance goals.

In summary, methods and strategies discussed have shown relevance and great benefits to different research fields. However, such efforts deal with performance through solutions that promote more responsible execution time for applications by an individual analysis of each data instance on execution time, making it the singular point of attention. According to our investigations, different approaches present performance variations if the execution order of data instances changes, in particular, there are significant impacts in the relevance of partial computations under data locality principle when cache strategies are considered. In this sense, we perceived a lack of alternatives that addressed these issues without requiring a costly redefinition of concepts and application under novel paradigms, which led us to discuss our parallel LAC approach, as presented in next section.

3 Parallel Lazy Associative Classification

In the aforementioned section, we discussed the importance of associative classification in current days and some of their problems, concentrating our attention in the lazy associative classification demand driven-basis approach – LAC. We demonstrated the advantages of such an approach when we described the ability of the local models generating well-approximations of the target function F in classification time. We explained MapReduce concepts and their benefits in a recent open-source solution named Spark. Now, we describe an approach that combines LAC and Spark advantages in a parallel AC in which computations of similar data instances can be cached and shared efficiently.

Our insight to propose such a solution is the drawback imposed by sequential LAC execution process, in which similar classifiers introduce an expressive computation overlap in the classification time. A condition caused by similar CARs used to label classes for distinct but related unknown data instances. Because CARs of independent local models are directly related to the attribute space of Γ, we propose a two-step based strategy to improve execution performance during classifications. For explain it, let us consider τ_1 to τ_4 as four data instances of Γ with some similarity and a 5-dimensional attribute space as shown in Fig. 1.

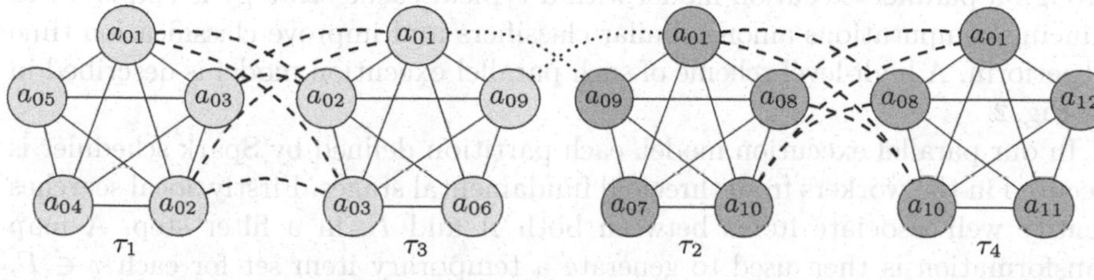

Fig. 1. A toy example of the modeling stage for attribute relationships in Γ using four data instances with some overlap on a 5-dimensional attribute space.

In Fig. 1, attributes are described as vertices and edges are drawn to show the relationship between each attribute pair, in a model designed as undirected and weighted graph $G = \{V, E\}$. Each data instance is reported as a complete graph with a high-correlation among their attributes. In another hand, edges are mapped in an adjacency matrix M, with $M_{n,m} \in \mathbb{R}$ corresponding to the relationship weight between n^{th} and m^{th} attributes, that is, the number of incidences of each attribute pair in the attribute space of Γ.

Considering a naive execution order as τ_1, τ_2, τ_3 and τ_4, CARs yielded by τ_1 will prematurely leave the cache since τ_2 not overlapping them and cache space is naturally limited. As a consequence, an expressive re-work is employed on classification time. However, in G there are some subsets of higher relationship attributes (i.e, linked by dotted edges), that is, τ_1 and τ_3, as well as τ_2 and τ_4 drawn greater overlap than τ_1 and τ_2 or τ_2 and τ_3. Thus, if Γ data instances are reordered by such high-correlation, we believe that it is possible to reduce the re-work on CARs mining process. Consequently, an outperform execution time is achieved. In this sense, we introduce a PageRank-driven similarity metric to compose a pre-analysis that investigates the relevance for each attribute pair in G and, from such relevance, decides the data execution order. As G is a weighted graph with similar representativeness than Web pages relationship description, such a strategy can be used toward an execution order effective.

As incidence values of each attribute pair in G are mapped in M with $M_{i,j}$ showing occurrence value of some pair, we introduce the PageRank to getting an eigenvalue for each attribute in G, which is used as score for mapping data instances in Γ in a pre-analysis stage that uses the following equation:

$$\tau_k = \sum_{a_i \wedge a_j \in \tau_k} G_{v_i,v_j} \tag{4}$$

such that, $\forall i \wedge \forall j \mid M_{i,j} > 0$ with a_i and a_j as i^{th} and j^{th} items of the data instance k.

Data instances are sorted in descending order by representativeness look at attribute-eigenvalues that designed it on the attribute space. Considering the toy example of Fig. 1, data instances are performed as $\tau_1, \tau_3, \tau_2, \tau_4$, ensuring a effective CAR memoization. In addition, a process performed in a second stage through a parallel execution model with a typical cache strategy is employed to reducing computations among similar classifiers and improve classification time outperform. A high-level scheme of such parallel execution model is described in the Fig. 2.

In our parallel execution model, each partition defined by Spark scheduler is executed in the workers from threefold fundamental stages. Firstly, local searches identify well-associate items between both Λ and Γ, in a filter step. A map transformation is then used to generate a temporary item set for each $\tau \in \Gamma$. Then, the generated item sets are performed in a parallel flow in which CARs are mining and shared them in a cache structure. Finally, a filter gets the better c_i for each τ and a reduce action consolidates results on the master, as well as, updates the cache with the frequent CARs. To ensure adequate CARs sharing, the cache structure is designed from accumulators and broadcast variables. Accumulators are employed to storing novel CARs while broadcast variables spread them to the workers. Thus, when each novel task partition is performed, the similar CARs are sharing and reused among classifiers avoiding costly computations on classification time by effective memoization. The performance of our proposal is reported based on observations of different naïve execution performed by the same parallel classification model through a FIFO execution order.

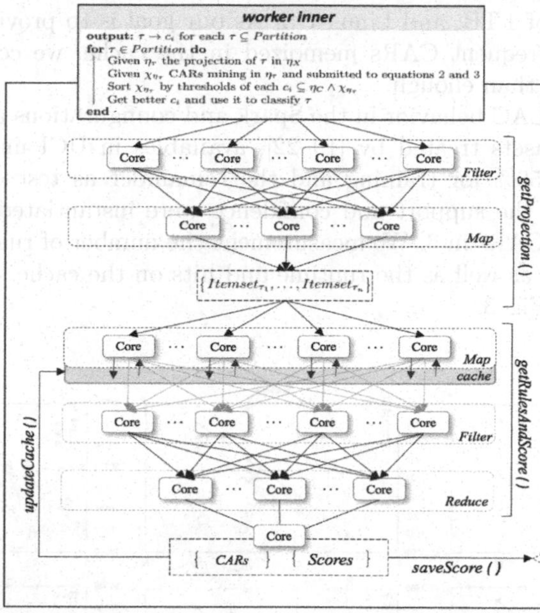

Fig. 2. Execution flow of LAC in Apache Spark programming model in which each unlabeled data partition is performed in threefold core stages: (1) projection, (2) CARs mining and class score, and (3) cache update with frequent CARs.

4 Experimental Results

In order to evaluate the effectiveness of the proposed parallel LAC and measurement the impacts of our similarity metric in memoization of frequent CARs, we investigated the Round-Robin (RR) and Least Recently Used (LRU) cache policies to different CARs storage spaces. For each list of unlabeled data instance evaluated, we performed initial computations, without data collection, to generate training models and cache buffers. Training models, containing data produced in the pre-processing step, were employed in all experiments. The cache buffers, in turn, were used to create a baseline to demonstrate, from 100% advance knowledge of the CARs, how far the executions with caches of smaller size, i.e., 20%, 40%, 60%, and 80%, and without previous initialization are optimal. Initial computations were executed with a thread as master and worker while the parallel evaluation addressed by one master and twelve worker threads.

To baseline for our evaluations, we addressed a First-Come-First-Served (FIFO) with the same cache sizes but without reordering the data instances. We evaluate our proposal (i.e., PageRank), denoted in the graphics as T3, comparing the results achieved by it with the results of FIFO execution with the same data input but randomly organized in three different manners, expressed as T0, T1, and T2. The architecture used to generating the experiments was constituted from four quad-core machines with a 2.7 GHz Intel i5 processor, 16 GB of RAM, an HD of 1 TB, and Linux OS. As our goal is to provide an inherent maximization of frequent CARs memoized in the cache, we considering such architecture more than enough.

We evaluated LAC behavior in the Spark and configurations above proposed using typical datasets treated by [19–22], availables in UCI machine learning repository[1] with 50% for training and the remainder as test data instances. For each dataset, the support and confidence were instantiated as 0, and the maximum size of CARs in 3. We measurement the number of rules produced on classification time as well as the runtime and hits on the cache. The number of rules is shown in Fig. 3.

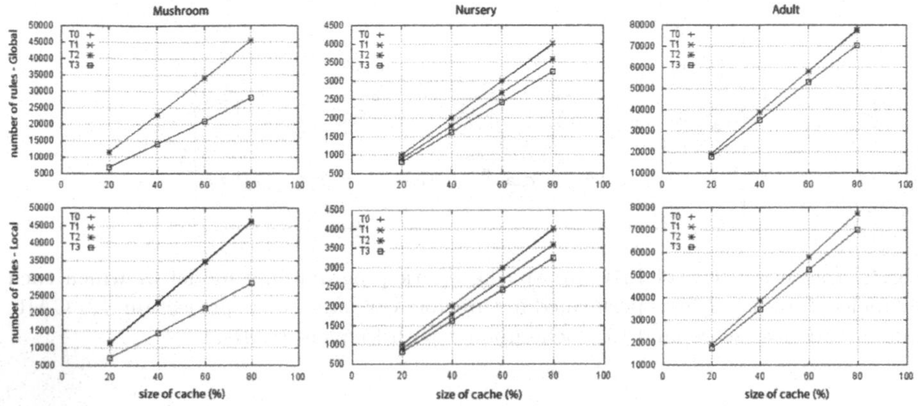

Fig. 3. Experiments overview for the number of rules produced using a typical cache Least Recently Used (LRU) and different data distributions based on FIFO (T0 to T2) and PageRank (T3) execution order.

The results expressed in Fig. 3 make it clear that an expressive reduction in CAR generation is achieved and, as a consequence, relevant gains are acquired to memoization of frequent itemset. Because PageRank produces greater coherence of frequent CARs on the cache, an outperform in the execution time and cache hit rate can be achieved. The Figs. 4 and 5 present such gains.

As can be seen in above results (i.e., Figs. 4 and 5), our PageRank-driven similarity metric can offer significant gains in CAR memoization (i.e., high hit rate), which consequently improves outperform to the total execution time and turned low re-work with costly computations (i.e., Fig. 3). An improvement that occurs because our proposal is able to exploit correlations among items in attribute space in each dataset given as input. Furthermore, the most meaningful results are observed when the cache size is more restrictive, a typical condition in many applications and real computational structures.

[1] https://archive.ics.uci.edu.

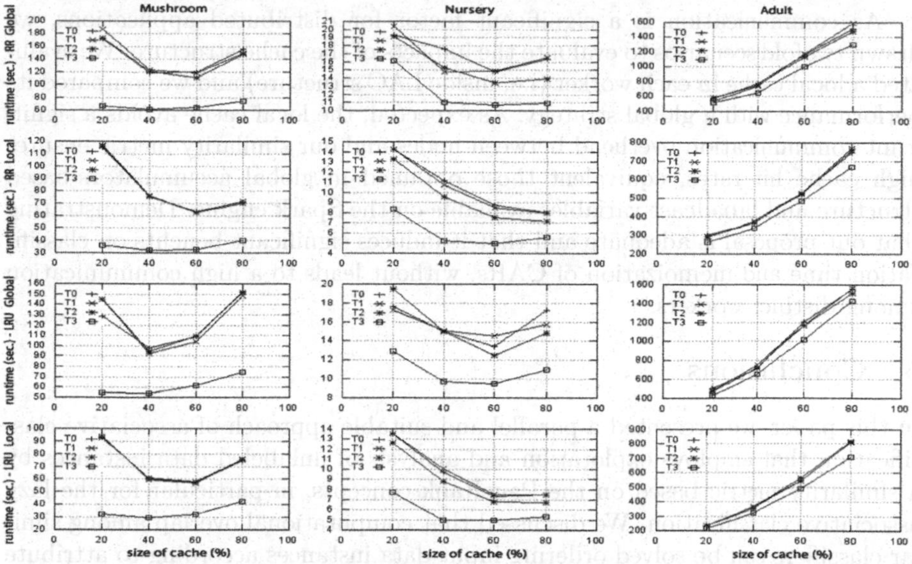

Fig. 4. Experiments overview to LAC execution time on Apache Spark broadcast and accumulators variables, Round-Robin (RR) and Least Recently Used (LRU) typical caches, based on FIFO (T0 to T2) and PageRank (T3) execution orders.

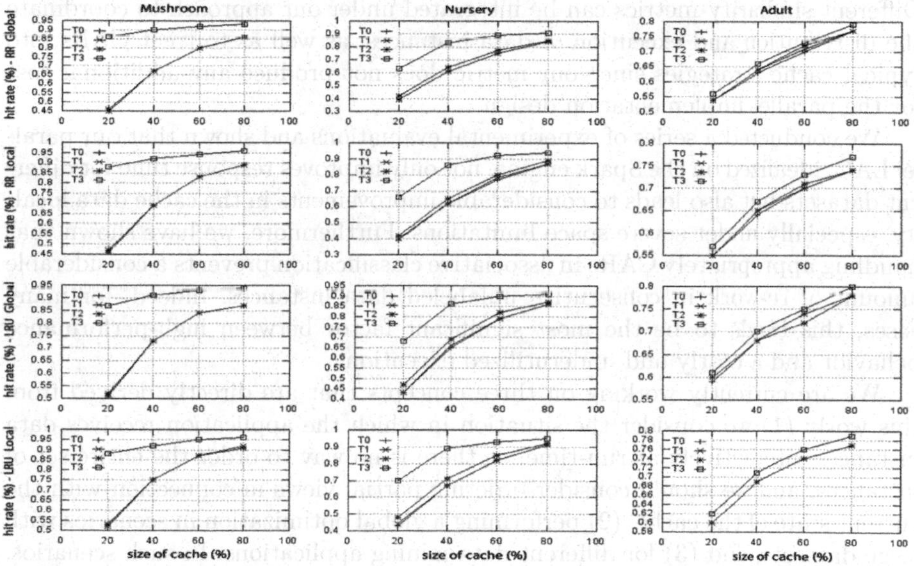

Fig. 5. Experiments overview to LAC cache hit rate on Apache Spark broadcast and accumulators variables, Round-Robin (RR) and Least Recently Used (LRU) typical caches, based on FIFO (T0 to T2) and PageRank (T3) execution orders.

As communication is a significant factor for distributed applications, we drawn twofold scenarios to evaluate the impacts of the cache structure. We evaluated a local cache in each worker (i.e, inside LAC structure) and we compared its performance with a global strategy. As expected, the local cache avoids a significant communication overhead between nodes and our similarity metric ensures high cache hit rates, equivalent those obtained in global accumulators-based structure and broadcast variables available on the Spark engine. Demonstrating that our proposal is adequate and that it induces significant benefits on classification time and memoization of CARs, without leads to a high communication among distinct workers.

5 Conclusions

In this paper, we presented a parallel and suitable approach of associative classification that employs exploration and analysis of unlabeled data instances by a similarity metric based on the PageRank concepts, in particular for the lazy associative classification. We discussed that computational overlap among similar classifiers can be solved ordering input data instances according to attribute space, and demonstrated a parallel implementation of the lazy associative classification can benefits from such order toward to an execution time outperforming. The main contribution of the proposed approach is the embedded similarity metric that can be used to maximize the memoization of computed rules (i.e, CARs). Different similarity metrics can be integrated under our approach to coordinate the distribution and execution of data instances, as well as to treat CARs into typical cache strategies since our metric does not produce any additional cost for the parallel implementation design.

We conducted a series of experimental evaluations and shown that our parallel LAC, idealized on the Spark engine, not only improves response time for different datasets but also leads to considerable improvements in the cache data locality, especially under severe space limitations. Furthermore, we have shown that handling appropriately CARs in associative classification prevents a considerable amount of re-work in consecutive unlabeled data instances. Indeed, for many cases, this tends to be the most significant factor between high-performance behavior and a costly and underutilized execution.

We are currently working on three concepts that are directly derived from this work: (1) we consider the situation in which the application receives data instances interactively at run-time, so there is no way to order the entire set of instances, and we should consider ordering partial views in connection with the current state of the cache, (2) performing a global optimization in scenarios with large datasets, and (3) for different data mining applications. In such scenarios, we have an additional concern about how to distribute the computation between several workers, while we maintaining high cache utilization on each of them. Lastly, we will tackle the problem in a dynamic scenario in which data instances are created at run-time while at the same time workers come and go, with their respective caches. Further, we have also begun conducting evaluations for larger scenarios and massively parallel environments.

Acknowledgements. This work was partially funded by INCT Cyber, MASWeb, CAPES, CNPq, Finep, FAPEMIG, and CEFET-MG, as well as EUBRA-BigSea and Atmosphere projects.

References

1. Agrawal, R., Imieliński, T., Swami, A.: Mining association rules between sets of items in large databases. In: Proceedings of the 1993 ACM SIGMOD International Conference on Management of Data. SIGMOD 1993, pp. 207–216. ACM, New York (1993). https://doi.org/10.1145/170035.170072
2. Almasi, M., Abadeh, M.S.: A new MapReduce associative classifier based on a new storage format for large-scale imbalanced data. Cluster Comput. **21**(4), 1821–1847 (2018). https://doi.org/10.1007/s10586-018-2812-9
3. Althebyan, Q., Jararweh, Y., Yaseen, Q., AlQudah, O., Al-Ayyoub, M.: Evaluating map reduce tasks scheduling algorithms over cloud computing infrastructure. Concurrency Comput.: Pract. Exp. **27**(18), 5686–5699 (2015). https://doi.org/10.1002/cpe.3595
4. Antonelli, M., Ducange, P., Marcelloni, F., Segatori, A.: A novel associative classification model based on a fuzzy frequent pattern mining algorithm. Expert Syst. Appl. **42**(4), 2086–2097 (2015). https://doi.org/10.1016/j.eswa.2014.09.021. http://www.sciencedirect.com/science/article/pii/S0957417414005600
5. Bechini, A., Marcelloni, F., Segatori, A.: A MapReduce solution for associative classification of big data. Inf. Sci. **332**, 33–55 (2016). https://doi.org/10.1016/j.ins.2015.10.041
6. Cheng, D., Rao, J., Guo, Y., Zhou, X.: Improving MapReduce performance in heterogeneous environments with adaptive task tuning. In: Proceedings of the 15th International Middleware Conference. Middleware 2014, pp. 97–108. ACM, New York (2014). https://doi.org/10.1145/2663165.2666089
7. Dean, J., Ghemawat, S.: MapReduce: simplified data processing on large clusters. Commun. ACM **51**(1), 107–113 (2008). https://doi.org/10.1145/1327452.1327492
8. Dong, G., Zhang, X., Wong, L., Li, J.: CAEP: classification by aggregating emerging patterns. In: Arikawa, S., Furukawa, K. (eds.) DS 1999. LNCS (LNAI), vol. 1721, pp. 30–42. Springer, Heidelberg (1999). https://doi.org/10.1007/3-540-46846-3_4
9. Ducange, P., Marcelloni, F., Segatori, A.: A MapReduce-based fuzzy associative classifier for big data. In: 2015 IEEE International Conference on Fuzzy Systems (FUZZ-IEEE), August, pp. 1–8 (2015). https://doi.org/10.1109/FUZZ-IEEE.2015.7337868
10. Gautam, J.V., Prajapati, H.B., Dabhi, V.K., Chaudhary, S.: Empirical study of job scheduling algorithms in hadoop MapReduce. Cybern. Inf. Technol. **17**(1), 146–163 (2017). https://content.sciendo.com/view/journals/cait/17/1/article-p146.xml
11. Guo, Z., Fox, G., Zhou, M.: Investigation of data locality in MapReduce. In: 2012 12th IEEE/ACM International Symposium on Cluster, Cloud and Grid Computing (CCGRID 2012), May, pp. 419–426 (2012). https://doi.org/10.1109/CCGrid.2012.42
12. Lakshmi, K.P., Reddy, C.R.K.: Fast rule-based prediction of data streams using associative classification mining. In: 2015 5th International Conference on IT Convergence and Security (ICITCS), pp. 1–5. IEEE (2015)

13. Li, W., Han, J., Pei, J.: CMAR: accurate and efficient classification based on multiple class-association rules. In: Proceedings of the 2001 IEEE International Conference on Data Mining. ICDM 2001, pp. 369–376. IEEE Computer Society, Washington, DC (2001). http://dl.acm.org/citation.cfm?id=645496.657866
14. Lin, C., Guo, W., Lin, C.: Self-learning MapReduce scheduler in multi-job environment. In: 2013 International Conference on Cloud Computing and Big Data, December, pp. 610–612 (2013). https://doi.org/10.1109/CLOUDCOM-ASIA.2013.95
15. Liu, B., Hsu, W., Ma, Y.: Integrating classification and association rule mining. In: 1998 Knowledge Discovery and Data Mining Conference (KDD), pp. 80–86 (1998)
16. Qureshi, M.N., Aldheleai, H.F.H., Tamandani, Y.K.: An improved documents classification technique using association rules mining. In: 2015 IEEE International Conference on Research in Computational Intelligence and Communication Networks (ICRCICN), November, pp. 460–465 (2015). https://doi.org/10.1109/ICRCICN.2015.7434283
17. Thabtah, F., Cowling, P., Peng, Y.: MCAR: multi-class classification based on association rule. In: The 3rd ACS/IEEE International Conference on Computer Systems and Applications, January (2005). https://doi.org/10.1109/AICCSA.2005.1387030
18. Thabtah, F., Hammoud, S.: Parallel associative classification data mining frameworks based MapReduce. Parallel Process. **25**(02), 1550002 (2015)
19. Veloso, A., Meira, W., Gonçalves, M., Almeida, H.M., Zaki, M.: Calibrated lazy associative classification. Inf. Sci. **181**(13), 2656–2670 (2011). https://doi.org/10.1016/j.ins.2010.03.007. http://www.sciencedirect.com/science/article/pii/S0020025510001192. Including Special Section on Databases and Software Engineering
20. Veloso, A., Meira Jr., W., Gonçalves, M., Almeida, H.M., Zaki, M.: Calibrated lazy associative classification. Inf. Sci. **181**(13), 2656–2670 (2011)
21. Veloso, A., Meira Jr, W., Zaki, M.J.: Lazy associative classification. In: ICDM 2006: Proceedings of the Sixth International Conference on Data Mining, December, pp. 645–654. IEEE Computer Society (2006)
22. Veloso, A.A.: Classificação Associativa sob Demanda. Ph.D. thesis, Universidade Federal de Minas Gerais, March 2009
23. Wang, J., Li, X.: Task scheduling for MapReduce in heterogeneous networks. Cluster Comput. **19**(1), 197–210 (2016). https://doi.org/10.1007/s10586-015-0503-3
24. Wang, J., Karypis, G.: Harmony: efficiently mining the best rules for classification. In: Proceedings of SDM, pp. 205–216 (2005)
25. Yin, X., Han, J.: CPAR: classification based on predictive association rules. In: Proceedings of the International Conference on Data Mining. SIAM (2003)
26. Zaharia, M., et al.: Apache spark: a unified engine for big data processing. Commun. ACM **59**(11), 56–65 (2016). https://doi.org/10.1145/2934664
27. Zhao, Y., Wu, J., Liu, C.: Dache: a data aware caching for big-data applications using the MapReduce framework. Tsinghua Sci. Technol. **19**(1), 39–50 (2014). https://doi.org/10.1109/TST.2014.6733207

Integrated Clustering and Anomaly Detection (INCAD) for Streaming Data

Sreelekha Guggilam[1(✉)], Syed Mohammed Arshad Zaidi[2(✉)], Varun Chandola[1,2(✉)], and Abani K. Patra[1(✉)]

[1] Computational Data Science and Engineering, University at Buffalo, State University of New York (SUNY), Buffalo, USA
{sreelekh,chandola,abani}@buffalo.edu
[2] Computer Science and Engineering, University at Buffalo, State University of New York (SUNY), Buffalo, USA
szaidi2@buffalo.edu

Abstract. Most current clustering based anomaly detection methods use scoring schema and thresholds to classify anomalies. These methods are often tailored to target specific data sets with "known" number of clusters. The paper provides a streaming clustering and anomaly detection algorithm that does not require strict arbitrary thresholds on the anomaly scores or knowledge of the number of clusters while performing probabilistic anomaly detection and clustering simultaneously. This ensures that the cluster formation is not impacted by the presence of anomalous data, thereby leading to more reliable definition of *"normal vs abnormal"* behavior. The motivations behind developing the INCAD model [17] and the path that leads to the streaming model are discussed.

Keywords: Anomaly detection · Bayesian non-parametric models · Extreme value theory · Clustering based anomaly detection

1 Introduction

Anomaly detection heavily depends on the definitions of expected and anomalous behaviors [14,20,22]. In most real systems, observed system behavior typically forms natural clusters whereas anomalous behavior either forms a small cluster or is weakly associated with the natural clusters. Under such assumptions, clustering based anomaly detection methods form a natural choice [7,10,18] but have several limitations.

Firstly, clustering based methods usually require baseline assumptions that are often conjectures and generalizing them is not always trivial. This leads to inaccurate choices for model parameters such as the number of clusters or the thresholds that are required to classify anomalies. Score based models have thresholds that are often based on data/user preference. Such assumptions result in models that are susceptible to modeler's bias and possible over-fitting.

J. M. F. Rodrigues et al. (Eds.): ICCS 2019, LNCS 11539, pp. 45–59, 2019.
https://doi.org/10.1007/978-3-030-22747-0_4

Secondly, setting the number of clusters has additional challenges when dealing with streaming data, where new behavior could emerge and form new clusters. Non-stationarity is inherent as data evolves over time. Moreover, the data distribution of a stream changes over time due to changes in environment, trends or other unforeseen factors [13,21]. This leads to a phenomenon called *concept drift*, due to which an anomaly detection algorithm cannot assume any fixed distribution for data streams. Thus, there arises a need for a definition of an anomaly that is dynamically adapted.

Thirdly, when anomaly detection is performed post clustering [2,12], the presence of anomalies gives a skewed (usually slight) definition of traditional/normal behavior. However, since the existence of anomalies impacts the clustering as well as the definition of the '*normal*'[1] behavior, it seems counter-intuitive to classify anomalies based on such definitions[2]. To avoid this, simultaneous clustering and anomaly detection needs to be performed.

Table 1. Comparison with other anomaly detection methods

	Neural networks	LOF	KNN	Kmeans –	Kernel function based	Gaussian model based	INCAD
Clustering based	✗	✗	✓	✓	✗	✗	✓
Multi-dimension	✓	✗	✓	✓	✓	✓	✓
Unsupervised	✗	✓	✓	✓	✓	✓	✓
Non-parametric	✗	✗	✗	✗	✓	✗	✓
Adaptable to streaming settings	✗	✗	✗	✗	✗	✗	✓
Adaptive thresholds	✗	✗	✗	✗	✗	✗	✓
Probabilistic scoring	✗	✗	✗	✗	✓	✗	✓

In addition to the above challenges, extending these assumptions to the streaming context leads to a whole new set of challenges. Many supervised [8,16] and unsupervised anomaly detection techniques [4,8,15,18] are offline learning methods that require the full data set in advance for data mining which makes them unsuitable for real-time streaming data. Although supervised anomaly detection techniques may be effective in yielding good results, they are typically unsuitable for anomaly detection in streaming data [16]. We propose a method called Integrated Clustering and Anomaly Detection (INCAD), that couples Bayesian non-parametric modeling and extreme value theory to simultaneously perform clustering and anomaly detection. Table 1 summarizes the

[1] Non-anomalous behavior is described as "normal" behavior. Should not be confused with Gaussian/Normal distribution.

[2] Clustering and defining "normal/traditional" behavior in presence of anomalies develop in skewed and inconsistent results.

properties of INCAD vs other strategies for anomaly detection. The primary contributions of the paper are as follows:

1. **Generalized anomaly definition with adaptive interpretation.** The model definition of an anomaly has dynamic interpretation allowing anomalous behaviors to evolve into normal behaviors and vice versa. This definition not only evolves the number of clusters with an incoming stream of data (using non-parametric mixture models) but also helps evolve the classification of anomalies.
2. **Combination of Bayesian non-parametric models and extreme value theory (EVT).** The novelty of the INCAD approach lies in blending extreme value theory and Bayesian non-parametric models. Non-parametric mixture models [19], such as *Dirichlet Process Mixture Models* (DPMM) [3,25,28], allow the number of components to vary and evolve during inference. While there has been limited work that has explored DPMM for the task of anomaly detection [26,29], they have not been shown to operate in a streaming mode or ignore online updates to the DPMM model. On the other hand, EVT gives the probability of a point being anomalous which has a more universal interpretation, in contrast to the scoring schema with user-defined thresholds. Although EVT's definition of anomalies is more adaptable for streaming data sets [1,11,27], fitting an extreme value distribution (EVD) on a mixture of distributions or even multivariate distributions is challenging. This novel combination brings out the much-needed aspects in both the models.
3. **Extension to streaming settings.** The model is non-exchangeable which is well suited to capture the effect of the order of data input and utilize this dependency to develop streaming adaptation.
4. **Ability to handle complex data generative models.** The model can be generalized to multivariate distributions and complex mixture models.

2 Motivation

2.1 Assumptions on *Anomalous* Behavior

One of the key drivers in developing any model are the model assumptions. For INCAD model, we assume that the data has multiple *"normal"* as well as *"anomalous"* behaviors. These behaviors are dynamic with a tendency to evolve from *"anomalous"* to *"normal"* and vice versa. Each such behavior (normal/anomalous) forms a sub-population that can be represented using a cluster. These clusters are assumed to be generated from a family of distributions whose cluster proportions and cluster parameters are generated from a non-parametric distribution.

There are two distinct differences between normal and anomalous data that must be identified: (a) *"Anomalous"* instances are different from tail instances of *"normal"* behavior and need to be distinguished from them. They are assumed to be generated from distributions that are different from *"normal"* data. (b)

The distributions for anomalous data result in relatively fewer instances in the observed set.

As mentioned earlier, clustering based anomaly detection methods could be good candidates for monitoring such systems, but require the ability to allow the clustering to *evolve* with the streaming data, i.e., new clusters form, old clusters grow or split. Furthermore, we need the model to distinguish between anomalies and extremal values of the *"normal"*.

Thus, non-parametric models that can accommodate infinite clusters are integrated with extreme value distributions that distinguish between anomalous and non-anomalous behaviors.

We now describe the two ingredients that go into our model namely, mixture models and extensions of EVT.

2.2 EVT and Generalized Pareto Distribution in Higher Dimensions

Estimation of parameters for extreme value distributions in higher dimensions is complex. To overcome this challenge, Clifton et al. [9] proposed an extended version of the generalized Pareto distribution that is applicable for different multi-model and multivariate distributions. For a given data $X \in \mathbb{R}^n$ distributed as $f_X : X \to Y$ where $Y \in \mathbb{R}$ is the image of the pdf values of X, let $Y \in [0, y_{max}]$ be the range of Y where $y_{max} = sup(f_X)$. Let,

$$G_Y(y) = \int_{f_Y^{-1}([0,y])} f_X(x)dx \tag{1}$$

where $f_Y^{-1} : Y \to X$ is the pre-image of f_X given by, $f_Y^{-1}([0, y]) = \{x | f_X(x) \in [0, y]\}$. Then G_Y is in the domain of attraction of generalized Pareto distribution (GPD) G_Y^e for $y \in [0, u]$ as $u \to 0$ given by,

$$G_Y^e(y) = \begin{cases} 1 - (1 - \xi(\frac{y-\nu}{\beta})^{-1/\xi}, \xi \neq 0 \\ 1 - exp(-\frac{y-\nu}{\beta}), \quad \xi = 0 \end{cases} \tag{2}$$

where, ν, β and ξ are the location, scale and shape parameters of the GPD respectively.

2.3 Mixture Models

Mixture models assume that the data consists of sub-populations each generated from a different distribution. It can be used to study the properties of clusters using mixture distributions. In the classic version, when the number of clusters is known, finite mixture models are used with Dirichlet priors. However, when the number of latent clusters is unknown, one can extend finite mixture models to infinite mixture models like Dirichlet process mixture model (DPMM). In DPMM, the mixture distributions being sampled from a Dirichlet process (DP). DP can be viewed as a distribution over a family of distributions, that constitutes a base distribution G_0 which is a prior over the cluster parameters θ and positive scaling parameter α_{DP}. G is a Dirichlet process (denoted as $G \sim DP(G_0, \alpha_{DP})$) if G is a random distribution with same support as the base distribution G_0 and for any measurable finite partition of the support $A_1 \cup A_2 \cup \ldots \cup A_k$, we have $(G(A_1), G(A_2), \ldots, G(A_k)) \sim \mathrm{Dir}(\alpha_{DP} G_0(A_1), \ldots, \alpha_{DP} G_0(A_k))$.

In order to learn the number of clusters from the data, Bayesian non-parametric (BNP) models are used. BNP models like DPMM assume an infinite number of clusters of which only a finite number are populated. It brings forth a finesse in choosing the number of clusters while assuming a prior on the cluster assignments of the data. The prior is given by the Chinese restaurant process(CRP) which is defined analogously to the seating of N customers who sequentially join tables in a Chinese restaurant. Here, the probability of the nth customer joining an existing table is proportional to the table size while the probability the customer forms a new table is always proportional to parameter α, $\forall n \in 1, 2, .., N$. This results in a distribution over the set of all partitions of integers $1, 2, .., N$. More formally, the distribution can be represented using the following probability function:

$$P(z_n = k | z_{1:n-1}) = \begin{cases} \frac{n_k}{n+\alpha-1} & n_k > 0 \text{ (existing cluster)} \\ \frac{\alpha}{n+\alpha-1} & n_k = 0 \text{ (new cluster)} \end{cases} \quad (3)$$

where z_i is the cluster assignment of the ith data point, n_k is the size of the kth cluster, $\alpha > 0$ is the concentration parameter. Large α values correspond to an increased tendency of data points to form new clusters[3].

3 Integrated Clustering and Anomaly Detection (INCAD)

The proposed INCAD model's prior is essentially a modification of a Chinese Restaurant Process (CRP). The seating of a customer at the Chinese restaurant is dependent on an evaluation by a gatekeeper. The gatekeeper decides the ability of the customer to start a new table based on the customer's features relative to existing patrons. If the gatekeeper believes the customer stands out, the customer is assigned a higher concentration parameter α that bumps up their chances of getting a new table. The model was inspired by the work of Blei and Frazier [5] in their distance dependent CRP models. The integrated INCAD model defines a flexible concentration parameter α for each data point. The probabilities are given by (Fig. 1):

$$P(z_n | z_{1:n-1}, \mathbf{x}) = \begin{cases} \frac{n_k}{n+\alpha_2-1}, & n_k > 0 \ (\textit{existing cluster}) \\ \frac{\alpha_2}{n+\alpha_2-1}, & n_k = 0 \ (\textit{new cluster}) \end{cases}$$

[3] Our modified model targets this aspect of concentration parameter to generate the desired simultaneous clustering and anomaly detection.

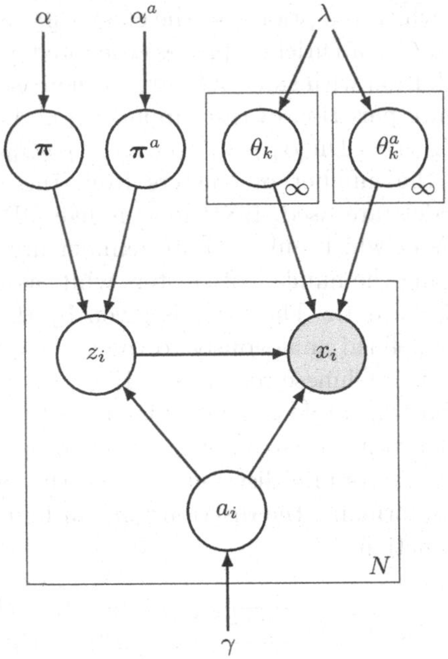

Fig. 1. Graphical representation of the proposed INCAD model.

where z, n, n_k are described earlier and $\alpha_2 = f(\alpha|x_n, \mathbf{x}, \mathbf{z})$ is a function of a base concentration parameter α. The function is chosen such that it is monotonic in $p(x_n)$ where, p is the probability of x_i being in the tail of the mixture distribution. In this paper, f is given by,

$$f(\alpha|x_n, \mathbf{x}, \mathbf{z}) = \begin{cases} \alpha, & if\ not\ in\ tail \\ \alpha^*, & if\ in\ tail \end{cases}$$

where, $\alpha^* = \frac{100}{1-p_n}$ and

$$p_n = p(x_n) = \begin{cases} \text{Probability of } x_n \text{ being anomalous, } x_n \text{ in tail} \\ 0, \quad \text{otherwise} \end{cases} \tag{4}$$

Traditional CRP not only mimics the partition structure of DPMM but also allows flexibility in partitioning differently for different data sets. However, CRP lacks the ability to distinguish anomalies from non-anomalous points. To set differential treatment for anomalous data, the CRP concentration parameter α is modified to be sensitive to anomalous instances and nudge them to form individual clusters. The tail points are clustered using this updated concentration parameter α_2 which is designed to increase with increasing probability of a point being anomalous. This ensures that the probability of tail points forming individual clusters increases as they are further away from the rest of the data.

4 Choice of the Extreme Value Distribution

The choice for G_0^{EV}, the base distribution for the anomalous cluster parameters, is the key for identifying anomalous instances. By choosing G_0^{EV} as the extreme value counterpart of G_0, the model ensures that the anomalous clusters are statistically "far" from the normal clusters. However, as discussed earlier, not all distributions have a well-defined EVD counterpart. We first describe the inference strategy for the case where G_0^{EV} exists. We then adapt this strategy for the scenario where G^{EV} is not available.

4.1 Inference When G_0^{EV} Is Available

Inference for traditional DPMMs is computationally expensive, even when conjugate priors are used. MCMC based algorithms [24] and variational techniques [6] have been typically used for inference. Here we adopt the Gibbs sampling based MCMC method for conjugate priors (Algorithm 1 [24]). This algorithm can be thought of as an extension of a Gibbs sampling-based method for a fixed mixture model, such that the cluster indicator z_i can take values between 1 and $K+1$, where K is the current number of clusters.

Gibbs Sampling. Though INCAD is based on DPMM, the model has an additional anomaly classification variable $a_.$ that determines the estimation of the rest of the parameters. In the Gibbs sampling algorithm for INCAD, the data points $\{x_i\}_{i=1}^N$ are observed and the number of clusters K, cluster indicators $\{z_i\}_{i=1}^N$ and anomaly classifiers $\{a_i\}_{i=1}^N$ are latent. Using Markov property and Bayes rule, we derive the posterior probabilities for z_i $\forall i \in 1, 2, \ldots, N$ as:

$$\begin{aligned}
P(z_i = k|x_., z_{-i}, \alpha, \alpha^*, \pi, \pi^a, \lambda, \{\theta_k\}, \{\theta_k^a\}, a_., \gamma) &= P(z_i = k|x_., z_{-i}, \alpha, \alpha^*, \{\theta_k\}, \{\theta_k^a\}, a_i) \\
&\propto \begin{cases} P(z_i = k|z_{-i}, \alpha, \theta_k) P(x_i|z_i = k, z_{-i}, \theta_k, \alpha) & a_i = 0 \\ P(z_i = k|z_{-i}, \alpha^*, \theta_k^a) P(x_i|z_i = k, z_{-i}, \theta_k^a, \alpha^*) & a_i = 1 \end{cases} \\
&= \begin{cases} \frac{n_k}{K(n+\alpha-1)} F(x_i|\theta_k) & a_i = 0 \\ \frac{n_k}{K(n+\alpha^*-1)} F(x_i|\theta_k^a) & a_i = 1 \end{cases}
\end{aligned} \quad (5)$$

where $\alpha^* = \frac{100}{1-p_i}$, p_i is the probability of x_i being anomalous and K is the number of non-empty clusters. Thus, the posterior probability of forming a new cluster denoted by $K+1$ is given by:

$$\begin{aligned}
P(z_i = K+1|x_., z_{-i}, \alpha, \alpha^*, \pi, \pi^a, \lambda, \{\theta_k\}, \{\theta_k^a\}, a_., \gamma) &= P(z_i = K+1|x_i, z_{-i}, \alpha, \alpha^*, \lambda, a_i) \\
&\propto \begin{cases} P(z_i = K+1|z_{-i}, \alpha, \lambda) P(x_i|z_i = K+1, z_{-i}, \alpha, \lambda, a_i) & a_i = 0 \\ P(z_i = K+1|z_{-i}, \alpha^*, \lambda) P(x_i|z_i = K+1, z_{-i}, \alpha^*, \lambda, a_i) & a_i = 1 \end{cases} \\
&= \begin{cases} \frac{\alpha}{n+\alpha-1} \int F(x_i|\theta) G_0(\theta|\lambda) d\theta & a_i = 0 \\ \frac{\alpha^*}{n+\alpha^*-1} \int F(x_i|\theta^a) G_0^{EV}(\theta^a|\lambda) d\theta^a & a_i = 1 \end{cases}
\end{aligned} \quad (6)$$

Similarly, the parameters for clusters $k \in \{1, 2, \ldots, K\}$ are sampled from:

$$\theta_k \propto G_0(\theta_k|\lambda)\mathcal{L}(\boldsymbol{x}_k|\theta_k) \text{ if cluster is not anomalous} \quad (7)$$

$$\theta_k^a \propto G_0^{EV}(\theta_k^a|\lambda)\mathcal{L}(\boldsymbol{x}_k|\theta_k^a) \quad \text{if cluster is anomalous} \quad (8)$$

where $\boldsymbol{x}_k = \{x_i | z_i = k\}$ is the set of all points in cluster k. Finally, to identify the anomaly classification of the data, the posterior probability of a_i is given by:

$$P(a_i = 1|x_., z_., \alpha, \alpha^*, \boldsymbol{\pi}, \boldsymbol{\pi}^a, \lambda, \{\theta_k\}, \{\theta_k^a\}, \gamma) = P(a_i = 1|x_i, z_., \alpha^*, \lambda, \{\theta_k^a\}, \gamma)$$
$$\propto \sum_{k=1}^{K+1} P(a_i = 1|x_i, z_i = k, z_{-i}, \alpha^*, \lambda, \{\theta_k^a\}, \gamma) * P(z_i = k|x_i, z_{-i}, \alpha^*, \lambda, \{\theta_k^a\}, \gamma)$$
$$= \sum_{k=1}^{K} P(x_i|\theta_k^a)\gamma \frac{n_k}{K(n+\alpha^*-1)} + \left(\int F(x_i|\theta^a) G_0^{EV}(\theta^a|\lambda) d\theta^a \right) \gamma \frac{\alpha^*}{n+\alpha^*-1} \quad (9)$$

Similarly,

$$P(a_i = 0|x_i, z_., \alpha, \lambda, \{\theta_k\}, \gamma)$$
$$\propto \sum_{k=1}^{K} P(x_i|\theta_k)(1-\gamma) \frac{n_k}{K(n+\alpha-1)} + \left(\int F(x_i|\theta) G_0(\theta|\lambda) d\theta \right) (1-\gamma) \frac{\alpha}{n+\alpha-1} \quad (10)$$

4.2 Inference When G_0^{EV} Is Not Available

The estimation of G_0^{EV} is required on two occasions. Firstly, while sampling the parameters of anomalous clusters when generating the data and estimating the posterior distribution. Secondly, to compute the probability of the point being anomalous when estimating the updated concentration parameter α_2. When estimating G_0^{EV} is not feasible, the following two modifications to the original model are proposed:

1. Since an approximate G_0^{EV} distribution need not belong to the family of conjugate priors of F, we need a different approach to sample the parameters for anomalous clusters. Thus, we assume $\theta^a \sim G_0$ for sampling the parameters $\{\theta_k^a\}_{k=1}^{\infty}$ for anomalous clusters.
2. To estimate the probability of a point being anomalous, use the approach described by Clifton et al. [9].

The pseudo-Gibbs sampling algorithm, presented in Algorithm 2, has been designed to address the cases when G_0^{EV} is not available. For such cases, the modified f is given by,

$$f(\alpha|x_n, \mathbf{x}, \mathbf{z}) = \begin{cases} \alpha, & \textit{if not in tail} \\ \alpha * (1 - ev\_prop) + \frac{100}{1-p_n} * ev\_prop, & \textit{if in tail} \end{cases}$$

where *ev_prop* determines the effect of a anomalous behavior on the concentration parameter[4]. *ev_prop* is solely used to speed up the convergence in the Gibbs sampling. Since the estimation of the G^{EV} distribution is not always possible, an alternate/pseudo Gibbs sampling algorithm has been presented in Algorithm 2.

[4] It can be seen that when the data point x_n has extreme or rare features differentiating it from all existing clusters, the corresponding density $f_y(x_n)$ decreases. This makes it a left tail point in the distribution G_y. The farther away it is in the tail, the lower the probability of its δ-nbd being in the tail and hence a higher $f(\alpha|x_n, \mathbf{x}, \mathbf{z})$.

Algorithm 1. Gibbs Sampling Algorithm when G_0^{EV} is available

Given $z_.^{(t-1)}, a_.^{(t-1)}, \{\theta_k^{(t-1)}\}, \{\theta_k^{a(t-1)}\}$ from previous iterations. Let K be the total number of clusters found till last iteration. Sample $z_.^{(t)}, a_.^{(t)}, \{\theta_k^{(t)}\}, \{\theta_k^{a(t)}\}$ as follows

1. Set $z_. = z_.^{(t-1)}$ and $a_. = a_.^{(t-1)}$
2. **for** *each point* $i \rightarrow 1$ **to** N **do**
   (a) Remove x_i from its cluster z_i.
   (b) If x_i is the only point in its cluster, $n_{z_i} = 0$ after step (2)(a). Remove the cluster and update K to K-1.
   (c) Rearrange cluster indices to ensure none are empty.
   (d) Sample z_i from the Multinomial distribution given by Equations 5 and 6
   (e) If $z_i = K + 1$, then sample new cluster parameters from the following distribution (It must be noted that the above posterior distribution was derived under the assumption of independence and exchangeability of priors for mathematical ease.)

$$\theta \left| x_i, z_., \{\theta_k^{(t-1)}\}, \{\theta_k^{a(t-1)}\}, a_.^{(t-1)} \right.$$

$$\propto \begin{cases} \alpha G_0(\theta|\lambda)F(x_i|\theta) + \sum_{j\neq i} F(x_i|\theta_{z_j})\delta(\theta - \theta_{z_j}^{(t-1)})\delta(a_j^{(t-1)}) & , a_i^{(t-1)} = 0 \\ \alpha^* G_0^{EV}(\theta|\lambda)F(x_i|\theta) + \sum_{j\neq i} F(x_i|\theta_{z_j})\delta(\theta - \theta_{z_j}^{(t-1)})\delta(a_j^{(t-1)} - 1) & , a_i^{(t-1)} = 1 \end{cases}$$

   Update K=K+1.
   (f) For each cluster $k \in \{1, 2, \ldots, K\}$, sample cluster parameters θ_k and θ_k^a using Equations 7 and 8.
   (g) Sample the anomaly classification a_i from the Binomial distribution given by Equations 9 and 10.
   (h) Set $z^{(t)} = z.$ and $a^{(t)} = a.$

4.3 Exchangeability

A model is said to be exchangeable when for any permutation S of $\{1, 2, ..., n\}$, $P(x_1, x_2, ...x_n) = P(x_{S(1)}, x_{S(2)}, ...x_{S(n)})$. Looking at the joint probability of the cluster assignments for the integrated model, we know,

$$P(z_1, z_2, ..z_n|\mathbf{x}) = P(z_1|\mathbf{x})P(z_2|z_1, \mathbf{x})..P(z_n|z_{1:n-1}, \mathbf{x})$$

Without loss of generality, let us assume there are K clusters. Let, for any $k < K$, the joint probability of all the points in cluster k be given by

$$\left(\frac{\alpha * p_{k,1}}{I_{k,1} + \alpha - 1} + \frac{\alpha^* * (1 - p_{k,1})}{I_{k,1} + \alpha^* - 1}\right) \prod_{n_k=2}^{N_k} \left(\frac{(n_k - 1) * p_{k,n_k}}{I_{k,n_k} + \alpha - 1} + \frac{(n_k - 1) * (1 - p_{k,n_k})}{I_{k,n_k} + \alpha^* - 1}\right)$$

where N_k is the size of the cluster k, $I_{k,i}$ is the index of the ith instance joining the kth cluster and $p_{k,i} = p_{I_{k,i}}$. Thus, the joint probability for complete data is then given by

$$\frac{\prod_{k=1}^{K} \left[(I_{k,1} - 1)p_{k,1}(\alpha - \alpha^*) + \alpha^*(I_{k,1} + \alpha - 1) \prod_{n_k=2}^{N_k}(n_k - 1)(I_{k,n_k} + \alpha - 1 + p_{k,n_k}(\alpha^* - \alpha)\right]}{\prod_{i=1}^{N} ((i + \alpha - 1)(i + \alpha^* - 1))}$$

which is dependent on the order of the data. This shows that the model is not exchangeable unless $\alpha = \alpha^*$ or $p_{k,n_k} = 0$ or $p_{k,n_k} = 1$. These conditions effectively reduce the prior distribution to a traditional CRP model. Hence, it can be concluded that the INCAD model cannot be modified to be exchangeable.

Algorithm 2. Gibbs Sampling Algorithm when G_0^{EV} is not available

Given $z_.^{(t-1)}, a_.^{(t-1)}, \left\{\theta_k^{(t-1)}\right\}, \left\{\theta_k^{a(t-1)}\right\}$ from previous iterations. Let K be the total number of clusters found till last iteration. Sample $z_.^{(t)}, a_.^{(t)}, \left\{\theta_k^{(t)}\right\}, \left\{\theta_k^{a(t)}\right\}$ as follows

1. Set $z_. = z_.^{(t-1)}$ and $a_. = a_.^{(t-1)}$
2. **for** *each point* $i \rightarrow 1$ **to** N **do**
   - (a) Steps 2a **to** 2d in Algorithm 1
   - (b) If $z_i = K + 1$, then set the cluster distribution to be multivariate normal with the new cluster mean as x_i and cluster variance as Σ which is pre-defined. Update K=K+1.
   - (c) For each cluster $k \in \{1, 2, \ldots, K\}$, sample cluster parameters θ_k and θ_k^a using Equation 7.
   - (d) Sample the anomaly classification a_i from the Binomial(p_i) where p_i is given by Equation 4. If most of the cluster instances are classified as anomalous, classify all of the cluster's instances as anomalies.
   - (e) Set $z_.^{(t)} = z.$ and $a_.^{(t)} = a.$

Non-exchangeable Models in Streaming Settings. Though exchangeability is a reasonable assumption in many situations, the evolution of behavior over time is not captured by traditional exchangeable models. In particular for streaming settings, using non-exchangeable models captures the effect of the order of the data. In such settings, instances that are a result of new evolving behavior should be monitored (as anomalous) until the behavior becomes relatively prevalent. Similarly, relapse of outdated behaviors (either normal or anomalous) should also be subjected to critical evaluation due to extended lag observed between similar instances. Such order driven dependency can be well captured in non-exchangeable models making them ideal for studying streaming data.

4.4 Adaptability to Sequential Data

One of the best outcomes of having a non-exchangeable prior is its ability to capture the *drift or evolution* in the behavior(s) either locally or globally or a mixture of both. INCAD model serves as a perfect platform to detect these changes and delivers an adaptable classification and clustering. The model has a straightforward extension to sequential settings where the model evolves with every incoming instance. Rather than updating the model for entire data with each new update, the streaming INCAD model re-evaluates only the tail instances. This enables the model to identify the following evaluations in the data (Fig. 3):

1. New trends that are classified as anomalous but can eventually grow to become normal.
2. Previously normal behaviors that have vanished over time but have relapsed and hence become anomalous (eg. disease relapse post complete recovery).

The Gibbs sampling algorithm for the streaming INCAD model is given in Algorithm 3.

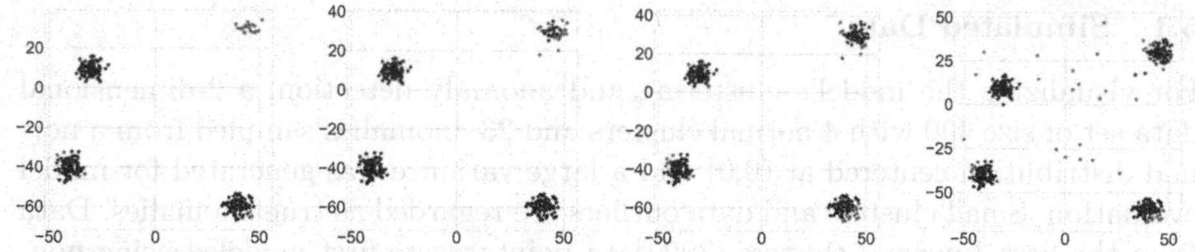

Fig. 2. Evolution of anomaly classification using streaming INCAD: the classification into anomalous and normal instances are represented by ● and ● respectively. Note the evolution of the classification of top right cluster from anomalous to normal with incoming data

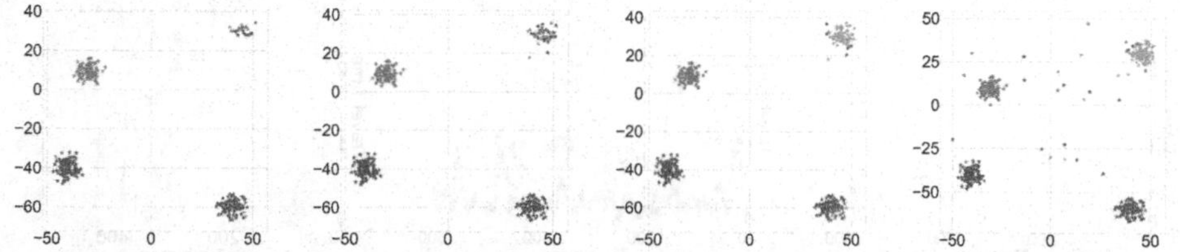

Fig. 3. Evolution of clustering using streaming INCAD: Each cluster is denoted by a different color. Notice the evolution of random points in top right corner into a well formed cluster in the presence of more data (Color figure online)

Algorithm 3. Algorithm for Streaming Extension

Perform clustering on a small portion of the data (10-20%) using non-streaming model
Set $ev_{prop} = exp^{-0.5}$
for *each new data point* x_N **do**

1. Compute the mixture proportions m_para and the mixture density for all the data. Compute $t_1 = q^{th}$ percentile pdf value to identify the tail points
2. For each x_i *s.t.* $f(x_i) < t_1$ repeat steps 2a→2d of Algorithm 2

If cluster size $\leq 0.05 * N$ then, classify all the cluster points as anomalies.

---

5 Results

In this section, we evaluate the proposed model using benchmark streaming datasets from NUMENTA. The streaming INCAD model's anomaly detection is compared with SPOT algorithm developed by Siffer et al. [27]. The evolution of clustering and anomaly classification using the streaming INCAD model is visualized using simulated dataset. In addition, the effect of batch vs stream proportion on quality of performance is presented. For Gibbs sampling initialization, the data was assumed to follow a mixture of MVN distributions. 10 clusters were initially assumed with the same initial parameters. The cluster means were set to the sample mean and the covariance matrix as a multiple of the sample covariance matrix, using a scalar constant. The concentration parameter α was always set to 1.

5.1 Simulated Data

For visualizing the model's clustering and anomaly detection, a 2-dimensional data set of size 400 with 4 normal clusters and 23 anomalies sampled from a normal distribution centered at (0,0) and a large variance was generated for model evaluation. Small clusters and data outliers are regarded as true anomalies. Data from the first 3 normal clusters (300 data points) were first modeled using non-streaming INCAD. The final cluster and the anomalies were then used as updates for the streaming INCAD model. The evolution in the anomaly classification is presented in Fig. 2.

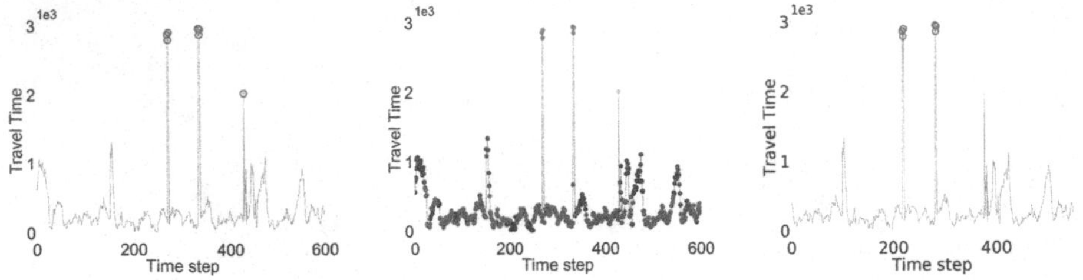

Fig. 4. NUMENTA traffic data: (from left to right) Anomaly detection and clustering using streaming INCAD and anomaly detection using SPOT [27]

5.2 Model Evaluation on NUMENTA Data

Two data sets from NUMENTA [23] namely the real traffic data and the AWS cloud watch data were used as benchmarks for the streaming anomaly detection. The streaming INCAD model was compared with the SPOT algorithm developed by Siffer et al. [27]. Unlike SPOT algorithm, the streaming INCAD is capable of modeling data with more than one feature. Thus, the data instance, as well as the time of the instance, were used to develop the anomaly detection model. Since the true anomaly labels are not available, the model's performance with respect to SPOT algorithm was evaluated based on the ability to identify erratic behaviors. The model results on the datasets using streaming INCAD and SPOT have been presented in Figs. 4 and 5.

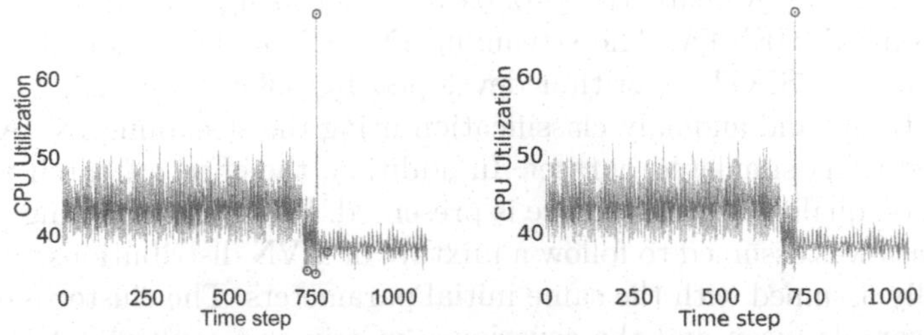

Fig. 5. Anomaly detection on NUMENTA AWS cloud watch data using streaming INCAD (left) and SPOT [27] (right)

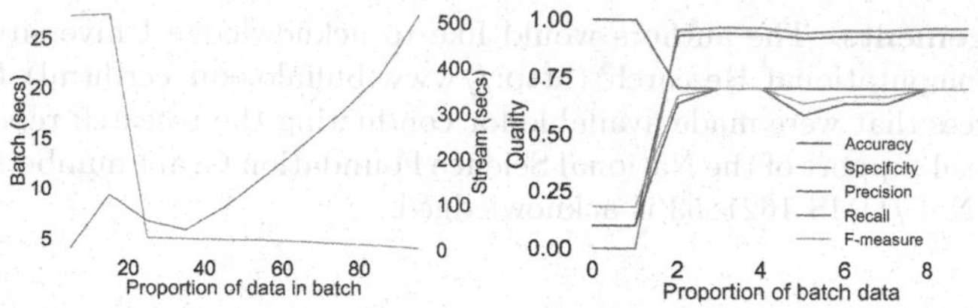

Fig. 6. Proportion of data for batch model vs Quality: Computational time (left) and Proportion of data for batch model (in tens) vs Quality of Anomaly detection (right)

6 Sensitivity to Batch Proportion

Streaming INCAD model re-evaluates the tail data at each update, the dependency of the model's performance on the current state must be evaluated. Thus, various metrics were used to study the model's sensitivity to the initial batch proportion. Figure 6 shows the effect of batch proportion on computational time and performance of anomaly detection. The simulated data defined in Sect. 5.1 was used for the sensitivity analysis. It can be seen that the computational time is optimal for 25% of data used in to run the non-streaming INCAD model. As anticipated, precision, accuracy, specificity, and f-measure for the anomaly detection were observed to plateau after a significant increase.

7 Conclusion and Future Work

A detailed description of the INCAD algorithm and the motivation behind it has been presented in this paper. The model's definition of an anomaly and its adaptable interpretation sets the model apart from the rest of the clustering based anomaly detection algorithms. While past anomaly detection methods lack the ability to simultaneously perform clustering and anomaly detection or to the INCAD model not only defines a new standard for such integrated methods but also breaks into the domain of streaming anomaly detection. The model's ability to identify anomalies and cluster data using a completely data-driven strategy permits it to capture the evolution of multiple behaviors and patterns within the data.

Additionally, the INCAD model can be smoothly transformed into a streaming setting. The model is seen to be robust to the initial proportion of the data subset that was evaluated using the non-streaming INCAD model. Moreover, this sets up the model to be extended to distribution families beyond multivariate normal. Though one of the key shortcomings of the model is its computational complexity in Gibbs sampling in the DPMM clusters, the use of faster methods such as variational inference might prove to be useful.

Acknowledgements. The authors would like to acknowledge University at Buffalo Center for Computational Research (http://www.buffalo.edu/ccr.html) for its computing resources that were made available for conducting the research reported in this paper. Financial support of the National Science Foundation Grant numbers NSF/OAC 1339765 and NSF/DMS 1621853 is acknowledged.

References

1. Al-Behadili, H., Grumpe, A., Migdadi, L., Wöhler, C.: Semi-supervised learning using incremental support vector machine and extreme value theory in gesture data. In: Computer Modelling and Simulation (UKSim), pp. 184–189. IEEE (2016)
2. Amer, M., Goldstein, M.: Nearest-neighbor and clustering based anomaly detection algorithms for rapidminer. In: Proceedings of the 3rd RCOMM 2012, pp. 1–12 (2012)
3. Antoniak, C.E.: Mixtures of Dirichlet processes with applications to Bayesian nonparametric problems. Ann. Stat. **2**(6), 1152–1174 (1974)
4. Bay, S.D., Schwabacher, M.: Mining distance-based outliers in near linear time with randomization and a simple pruning rule. In: Proceedings of the 9th ACM SIGKDD International Conference on Knowledge Discovery and Data Mining, pp. 29–38. ACM (2003)
5. Blei, D.M., Frazier, P.I.: Distance dependent Chinese restaurant processes. In: ICML, pp. 87–94 (2010)
6. Blei, D.M., Jordan, M.I.: Variational methods for the Dirichlet process. In: Proceedings of the Twenty-first International Conference on Machine Learning, p. 12 (2004)
7. Chan, P.K., Mahoney, M.V., Arshad, M.H.: A machine learning approach to anomaly detection. Technical report (2003)
8. Chandola, V., Banerjee, A., Kumar, V.: Anomaly detection: a survey. ACM Comput. Surv. (CSUR) **41**(3), 15 (2009)
9. Clifton, D.A., Clifton, L., Hugueny, S., Tarassenko, L.: Extending the generalised pareto distribution for novelty detection in high-dimensional spaces. J. Sig. Process. Syst. **74**(3), 323–339 (2014)
10. Eskin, E., Arnold, A., Prerau, M., Portnoy, L., Stolfo, S.: A geometric framework for unsupervised anomaly detection. In: Barbará, D., Jajodia, S. (eds.) Applications of Data Mining in Computer Security. ADIS, vol. 6, pp. 77–101. Springer, Heidelberg (2002). https://doi.org/10.1007/978-1-4615-0953-0_4
11. French, J., Kokoszka, P., Stoev, S., Hall, L.: Quantifying the risk of heat waves using extreme value theory and spatio-temporal functional data. Comput. Stat. Data Anal. **131**, 176–193 (2019)
12. Fu, Z., Hu, W., Tan, T.: Similarity based vehicle trajectory clustering and anomaly detection. In: ICIP, vol. 2, p. II–602. IEEE (2005)
13. Gama, J., Žliobaitė, I., Bifet, A., Pechenizkiy, M., Bouchachia, A.: A survey on concept drift adaptation. ACM Comput. Surv. (CSUR) **46**(4), 44 (2014)
14. Garcia-Teodoro, P., Diaz-Verdejo, J., Maciá-Fernández, G., Vázquez, E.: Anomaly-based network intrusion detection: techniques, systems and challenges. Comput. Secur. **28**(1–2), 18–28 (2009)
15. Goldstein, M., Uchida, S.: A comparative evaluation of unsupervised anomaly detection algorithms for multivariate data. PloS One **11**(4), e0152173 (2016)
16. Görnitz, N., Kloft, M., Rieck, K., Brefeld, U.: Toward supervised anomaly detection. J. Artif. Intell. Res. **46**, 235–262 (2013)

17. Guggilam, S., Arshad Zaidi, S.M., Chandola, V., Patra, A.: Bayesian anomaly detection using extreme value theory. arXiv preprint arXiv:1905.12150 (2019)
18. He, Z., Xiaofei, X., Deng, S.: Discovering cluster-based local outliers. Pattern Recogn. Lett. **24**(9–10), 1641–1650 (2003)
19. Hjort, N., Holmes, C., Mueller, P., Walker, S.: Bayesian Nonparametrics: Principles and Practice. Cambridge University Press, Cambridge (2010)
20. Jiang, J., Castner, J., Hewner, S., Chandola, V.: Improving quality of care using data science driven methods. In: UNYTE Scientific Session - Hitting the Accelerator: Health Research Innovation Through Data Science (2015)
21. Jiang, N., Gruenwald, L.: Research issues in data stream association rule mining. ACM Sigmod Rec. **35**(1), 14–19 (2006)
22. Kruegel, C., Vigna, G.: Anomaly detection of web-based attacks. In: Proceedings of the 10th ACM Conference on Computer and Communications Security, pp. 251–261 (2003)
23. Lavin, A., Ahmad, S.: Evaluating real-time anomaly detection algorithms-the numenta anomaly benchmark. In: 2015 IEEE 14th International Conference on Machine Learning and Applications (ICMLA), pp. 38–44. IEEE (2015)
24. Neal, R.M.: Markov chain sampling methods for Dirichlet process mixture models. J. Comput. Graph. Stat. **9**(2), 249–265 (2000)
25. Rasmussen, C.E.: The infinite Gaussian mixture model. In: Advances in Neural Information Processing Systems, vol. 12, pp. 554–560. MIT Press (2000)
26. Shotwell, M.S., Slate, E.H.: Bayesian outlier detection with Dirichlet process mixtures. Bayesian Anal. **6**(4), 665–690 (2011)
27. Siffer, A., Fouque, P.-A., Termier, A., Largouet, C.: Anomaly detection in streams with extreme value theory. In: Proceedings of the 23rd ACM SIGKDD International Conference on Knowledge Discovery and Data Mining, pp. 1067–1075 (2017). ISBN 978-1-4503-4887-4
28. Teh, Y.W., Jordan, M.I., Beal, M.J., Blei, D.M.: Hierarchical Dirichlet processes. J. Am. Stat. Assoc. **101**(476), 1566–1581 (2006)
29. Varadarajan, J., Subramanian, R., Ahuja, N., Moulin, P., Odobez, J.M.: Active online anomaly detection using Dirichlet process mixture model and Gaussian process classification. In: 2017 IEEE WACV, pp. 615–623 (2017)

An Implementation of a Coupled Dual-Porosity-Stokes Model with FEniCS

Xiukun Hu[1] and Craig C. Douglas[2](✉)

[1] Department of Mathematics and Statistics, University of Wyoming, Laramie, WY 82071-3036, USA
xiukun.hu@outlook.com

[2] School of Energy Resources and Department of Mathematics and Statistics, University of Wyoming, Laramie, WY 82071-3036, USA
craig.c.douglas@gmail.com

Abstract. Porous media and conduit coupled systems are heavily used in a variety of areas. A coupled dual-porosity-Stokes model has been proposed to simulate the fluid flow in a dual-porosity media and conduits coupled system. In this paper, we propose an implementation of this multi-physics model. We solve the system with the automated high performance differential equation solving environment FEniCS. Tests of the convergence rate of our implementation in both 2D and 3D are conducted in this paper. We also give tests on performance and scalability of our implementation.

Keywords: Domain decomposition · Finite element method · Multi-physics · Parallel computing · FEniCS

1 Introduction

The coupling of porous media flow and free flow is of importance in multiple areas, including groundwater system, petroleum extraction, and biochemical transport [1–3]. The Stokes-Darcy equation is widely applied in these areas and has been studied thoroughly over the past decade [4–6]. Variants of Stokes-Darcy mode, have also been studied extensively [7–10].

In a traditional Stokes-Darcy system, Darcy's law is applied to the fluid in porous media. Darcy's law, along with its variants, is great in modeling single porosity model with limited Reynolds number, and is widely used in hydrogeology and reservoir engineering. However, for a porous medium with multiple porosities, for example a naturally fractured reservoir, the accuracy of Darcy's law is limited. In contrast, a dual-porosity model assumes two different systems inside a porous media-: the matrix system and the microfracture system. These two systems have significantly different fluid storage and conductivity properties. It gives a better representation of the fractured porous media encountered in hydrology, carbon sequestration, geothermal systems, and petroleum extraction [11–13].

The dual-porosity model itself fails to consider large conduits inside porous media. Thus, the need of coupling both a dual-porosity model with free flow arises [14].

J. M. F. Rodrigues et al. (Eds.): ICCS 2019, LNCS 11539, pp. 60–73, 2019.
https://doi.org/10.1007/978-3-030-22747-0_5

Our paper expands on a coupled Dual-Porosity-Stokes model [15]. Similar to the Stokes-Darcy model, this coupled model contains two nonoverlapping but contiguous regions: one filled with porous media and the other represents conduits. The dual-porosity model describes the porous media and the Stokes equation governs the free flow in the conduits.

In Sect. 2, we define the Dual-Porosity-Stokes model. The equations are presented as well as the variational form. In Sect. 3, we describe the numerical implementation using FEniCS. In Sect. 4, we analyze the accuracy, speed performance, memory usage, and scalability of our implementation. In Sect. 5, we draw some conclusions and discuss future work.

2 Dual-Porosity-Stokes Model

The Dual-Porosity-Stokes model was first presented in [15]. This paper demonstrated the well-posedness of the model and derived a numerical solution in 2D using a finite element method. Several numerical experiments are also in the paper. In this section we will demonstrate this model in detail as well as present the variational form.

To better understand the model, let us first take a look at a simple 2D example presented in Fig. 1. The model consists of a dual-porosity subdomain Ω_d and a conduit subdomain Ω_c, with an interface Γ_{cd} in between. Two subdomains are non-overlapping, and only communicate to each other through the interface Γ_{cd}. Γ_d and Γ_c are boundaries of each subsystem.

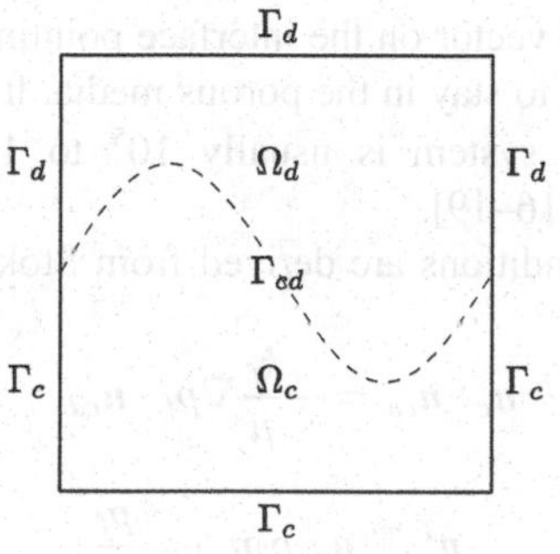

Fig. 1. Coupled model in 2D

Two fluid pressures are presented in Ω_d, p_m and p_f, for fluids in matrix and fractures respectively. The *m* subscript stands for *matrix* and *f* is for *fracture*. We use these two subscripts for other model parameters. The dual-porosity model can be expressed as:

$$\phi_m C_{mt} \frac{\partial p_m}{\partial t} - \nabla \cdot \frac{k_m}{\mu} \nabla p_m = -Q, \tag{1}$$

$$\phi_f C_{ft} \frac{\partial p_f}{\partial t} - \nabla \cdot \frac{k_f}{\mu} \nabla p_f = Q + q_p. \tag{2}$$

The constant σ is a shape factor ranging from 0 to 1. It measures the connectivity between the microfracture and the matrix. μ is the dynamic viscosity. k is the intrinsic permeability. ϕ denotes the porosity. C_{mt} and C_{ft} denote the total compressibility for the two systems respectively. q_p is the sink/source term. Q is the mass exchange between matrix and microfracture systems and can be derived from $Q = \frac{\sigma k_m}{\mu}(p_m - p_f)$.

In the conduit subdomain, we use the linear incompressible Stokes equation to describe the free flow:

$$\frac{\partial \boldsymbol{u}_c}{\partial t} - \nabla \cdot \mathbb{T}(\boldsymbol{u}_c, p) = \boldsymbol{f}, \tag{3}$$

$$\nabla \cdot \boldsymbol{u}_c = 0. \tag{4}$$

The flow velocity $\boldsymbol{u}_c$ and pressure p together describe the free flow. v is the kinematic viscosity. $\boldsymbol{f}$ is a general source term. $\mathbb{T}(\boldsymbol{u}_c, p) := 2v\mathbb{D}(\boldsymbol{u}_c) - p\mathbb{I}$ is the stress tensor, where $\mathbb{D}(\boldsymbol{u}_c) := \frac{1}{2}(\nabla \boldsymbol{u}_c + \nabla^T \boldsymbol{u}_c)$ is the deformation tensor, and $\mathbb{I}$ is the $N \times N$ identity matrix.

On the interface Γ_{cd}, a no-exchange condition between the matrix and the conduit is used,

$$-\frac{k_m}{\mu} \nabla p_m \cdot (-\boldsymbol{n}_{cd}) = 0, \tag{5}$$

where $\boldsymbol{n}_{cd}$ is the unit normal vector on the interface pointing toward Ω_d. This equation forces the fluid in the matrix to stay in the porous media. It is based on the fact that the permeability of the matrix system is usually 10^5 to 10^7 times smaller than the microfracture permeability [16–19].

Three more interface conditions are derived from Stokes-Darcy model:

$$\boldsymbol{u}_c \cdot \boldsymbol{n}_{cd} = -\frac{k_f}{\mu} \nabla p_f \cdot \boldsymbol{n}_{cd}, \tag{6}$$

$$-\boldsymbol{n}_{cd}^T \mathbb{T}(\boldsymbol{u}_c, p) \boldsymbol{n}_{cd} = \frac{p_f}{\rho}, \tag{7}$$

$$-\mathbb{P}_\tau(\mathbb{T}(\boldsymbol{u}_c, p)\boldsymbol{n}_{cd}) = \frac{\alpha v \sqrt{N}}{\sqrt{\text{trace}(\boldsymbol{\Pi})}} \left(\boldsymbol{u}_c + \frac{k_f}{\mu} \nabla p_f \right). \tag{8}$$

ρ is the density of the fluid. $\mathbb{P}_\tau$ is the projection operator onto the local tangent plane of the interface Γ_{cd}. α is a dimensionless parameter which depends on the properties of the fluid and the permeable material, N is the space dimension, and $\boldsymbol{\Pi} = k_f \mathbb{I}$ is the intrinsic permeability of the fracture media. Condition (6) is the conservation of mass on the interface. Equation (7) represents the balance of forces on the interface [20, 21]. Equation (8) is the Beavers-Joseph interface condition [22].

If we introduce a vector valued test function $\vec{v} = [\psi_m, \psi_f, \boldsymbol{v}^T, q]^T$, the variational form for our model can be written as

$$\int_{\Omega_d} \left(\phi_m C_{mt} \frac{\partial p_m}{\partial t} \psi_m + \frac{k_m}{\mu} \nabla p_m \cdot \nabla \psi_m + \frac{\sigma k_m}{\mu} (p_m - p_f) \psi_m \right) d\Omega \tag{9a}$$

$$+ \int_{\Omega_d} \left(\phi_f C_{ft} \frac{\partial p_f}{\partial t} \psi_f + \frac{k_f}{\mu} \nabla p_f \cdot \nabla \psi_f + \frac{\sigma k_m}{\mu} (p_f - p_m) \psi_f \right) d\Omega \tag{9b}$$

$$+ \eta \int_{\Omega_c} \left(\frac{\partial \boldsymbol{u}_c}{\partial t} \cdot \boldsymbol{v} + 2\nu \mathbb{D}(\boldsymbol{u}_c) : \mathbb{D}(\boldsymbol{v}) - p \nabla \cdot \boldsymbol{v} \right) d\Omega \tag{9c}$$

$$+ \eta \int_{\Gamma_{cd}} \left(\frac{1}{\rho} p_f \boldsymbol{v} \cdot \boldsymbol{n}_{cd} + \frac{\alpha \nu \sqrt{N}}{\sqrt{\text{trace}(\boldsymbol{\Pi})}} \mathbb{P}_\tau \left(\boldsymbol{u}_c + \frac{k_f}{\mu} \nabla p_f \right) \cdot \boldsymbol{v} \right) d\Gamma \tag{9d}$$

$$+ \eta \int_{\Omega_c} \nabla \cdot \boldsymbol{u}_c q d\Omega - \int_{\Gamma_{cd}} \boldsymbol{u}_c \cdot \boldsymbol{n}_{cd} \psi_f d\Gamma \tag{9e}$$

$$= \eta \int_{\Omega_c} \boldsymbol{f} \cdot \boldsymbol{v} d\Omega + \int_{\Omega_d} q_p \psi_f d\Omega. \tag{9f}$$

η is a scale factor applied to equations in the conduit subdomain to ensure the whole system is of the same scale.

3 Implementation

Hou et al. [15] numerically solved such a system in 2D using a finite element method with Taylor-Hood elements for the conduit domain and demonstrated the stability and convergence rate of their method.

In this section we describe an implementation of a finite element solver based on FEniCS [23, 24], which allows us to run our model in both 2D and 3D, in parallel, and can be easily modified and extended. FEniCS is a popular open source computing platform for solving partial differential equations (PDEs). The automatic code generation of FEniCS enables people to implement a FEM code using the Unified Form Language (UFL) [25], which is close to a mathematical description of the variational form.

Many multi-physics models have been implemented using FEniCS, e.g., the adaptive continuum mechanics solver Unicorn (Unified Continuum modeling) [26, 27]. It can solve continuum mechanics with moving meshes adaptively. As the coupled systems Unicorn solves always consist of moving meshes, subsystems are solved independently and iteratively until a satisfied error is reached. In our case, we prefer to solve the coupled system as a whole as they together form a linear system and can be solved directly.

A solution to the coupled Navier-Stokes-Darcy model has been implemented with FEniCS by Ida Norderhaug Drøsdal [28]. In the coupled Navier-Stokes-Darcy model, the conduit subdomain and the porosity subdomain contain the same two variables: fluid velocity and pressure. The solver regards the two subsystems as a whole and the variables exist on both subdomains. The interface conditions then are implemented by interior facet integration. However, in our coupled Stokes-Dual-Porosity model, we have two scalar variables p_f and p_m in the dual-porosity domain, but one vector variable $\boldsymbol{u}_c$ and one scalar variable p in the conduit domain.

The disagreement of the variable dimensions on the two subdomains differentiates our model from Navier-Stokes-Darcy model. Here we expand every variable to the whole system and force them to be zero in the opposite subdomain.

3.1 Implementation with FEniCS

Our implementation in Python is described in this section. Since our model involves interior interface integration, adjacent cells need to share information from the common facet. In order for our implementation to run in parallel, the following parameter in FEniCS needs to be set correctly.

```
from fenics import *
parameters['ghost_mode'] = 'shared_facet'
```

For any given mesh object mesh, with any geometric dimension, our function space can be created as:

```
# Given mesh and degree
velem = VectorElement('CG', mesh.ufl_cell(), degree)
selem = FiniteElement('CG', mesh.ufl_cell(), degree)
pelem = FiniteElement('CG', mesh.ufl_cell(), degree-1)

V = FunctionSpace(mesh, MixedElement(selem,selem,
                                      velem,pelem))
```

The four ordered elements `selem`, `selem`, `velem`, `pelem` in the last statement are for p_f, p_m, $\boldsymbol{u}_c$ and p respectively. Note that since the Taylor-Hood method is applied, the degree of p should be less than that of $\boldsymbol{u}_c$.

Initiate constants `phim` $= \varphi_m$, `phif` $= \varphi f$, `km` $= k_m$, `kf` $= k_f$, `mu` $= \mu$, `nu` $= \nu$, `rho` $= \rho$, `sigma` $= \sigma$, `Cmt` $= C_{mt}$, `Cft` $= C_{ft}$, `alpha` $= \alpha$, and `eta` $= \eta$, and function expressions `qp` $= q_p$ and `f` $= \boldsymbol{f}$. Also define initial conditions for all variables, interpolated into our function space `V`, and stored in the variable `x0`. The variational form can be defined in UFL as below. Note the one-to-one correspondence between the variational form below and the one presented in (9a)–(9f).

```
n = FacetNormal(V.mesh()) # n_cd
proj = lambda u: u-dot(u,n('+'))*n('+') # P_τ
pm0, pf0, u0, p0 = x0.split()
avN = alpha*nu/math.sqrt(kf) # αν√N)/√(trace(Π)
pm, pf, u, p = TrialFunctions(V)
psim, psif, v, q = TestFunctions(V)
F = ((phim*Cmt*(pm-pm0)/dt*psim
      + km/mu*dot(grad(pm), grad(psim))
      + sigma*km/mu*(pm-pf)*psim          )*dD # (9a)
   + (phif*Cft*(pf-pf0)/dt*psif
      + kf/mu*dot(grad(pf), grad(psif))
      + sigma*km/mu*(pf-pm)*psif          )*dD # (9b)
   + eta*
     (dot((u-u0)/dt,v)
      + 2*nu*inner(epsilon(u),epsilon(v))
     - p*div(v)                           )*dC # (9c)
   + eta*
     (1/rho*pf('-')*dot(v('+'),n('+'))
      + avN*dot(proj(u('+')+kf/mu*grad(pf('-'))),
               v('+'))                    )*dI # (9d)
   + (eta*(div(u)*q)*dC
      - dot(u('+'),n('+'))*psif('-')*dI)       # (9e)
   - eta*dot(f, v)*dC - qp*psif*dD      )      # (9f)
a, L = lhs(F), rhs(F)
```

Note that the backward Euler scheme can be easily extended to θ method. `dC`, `dD` and `dI` are predefined `Measure` objects in UFL, and represents integrations on Ω_c, Ω_d and Γ_{cd} respectively. Note that the sign in `n('+')` needs to be adjusted for different domain structures. The signs of other variables for interface integration terms are not affecting the result of the model in any of our test cases.

Now we constrain p_m, p_f, $\boldsymbol{u}_c$, p on opposite subdomains by defining the following Dirichlet boundary conditions.

```
fix_pm = DirichletBC(V.sub(0), Constant(0),
                     on_conduit_but_not_interface,
                     method='pointwise')
fix_pf = DirichletBC(V.sub(1), Constant(0),
                     on_conduit_but_not_interface,
                     method='pointwise')
fix_u = DirichletBC(V.sub(2), Constant([0]*N),
                    on_dual_but_not_interface,
                    method='pointwise')
fix_p = DirichletBC(V.sub(3), Constant(0),
                    on_dual_but_not_interface,
                    method='pointwise')
fix_bcs = [fix_pm, fix_pf, fix_u, fix_p]
```

The boundary markers on_conduit/dual_but_not_interface, as their names might suggest, should not include the interface Γ_{cd}. Note that we need to use the method "pointwise" for these special "boundary" conditions.

After creating other Dirichlet boundary conditions bcs, we can solve the PDE as follows.

```
A, b = assemble(a), assemble(L)
for bc in fix_bcs + bcs:
    bc.apply(A, b)
solver = KrylovSolver(A, method='bicgstab',
                      preconditioner='hypre_euclid')
now = 0
while now <= T:  # T is endtime
    now += dt    # dt is length of timestep
    for expr in [pm, pf, u, p, qp, f]:
        expr.t = now
    b = assemble(L)
    for bc in bcs:
        bc.apply(b)
    solver.solve(x.vector(), b)
    x0.assign(x)
```

Due to the interface conditions our matrix A is nonsymmetric. Hence, methods like conjugate gradients and Cholesky decomposition might not work as expected.

4 Result

The implementation works in 2D and 3D with the same code. Here we test our implementation on a unit cubic mesh defined by $\Omega = [0,1] \times [0,1] \times [0,1]$. Let $\Omega_d = \{(x,y,z) \in \Omega | x \leq y\}, \Omega_c = \{(x,y,z) \in \Omega | x \geq y\}$. We simulate our model on the time interval $[0,1]$.

For the constants, we let $k_m = 0.1$ and all the rest be 1. We set up our coefficients and essential boundary conditions so that the following is our solution:

$$p_m = \frac{1}{5}\sin(x+y+z)\cos(\pi t)$$

$$p_f = -2\pi\,\sin(\pi t)\,\sin(x+y+z) + \frac{4}{5}\sin(x+y+z)\cos(\pi t)$$

$$\boldsymbol{u}_c = \begin{bmatrix} 2\pi\,\sin(\pi t)\cos(3x-y+z) - \frac{4}{5}\cos(\pi t)\cos(3y-x+z) \\ 2\pi\,\sin(\pi t)\cos(3y-x+z) - \frac{4}{5}\cos(\pi t)\cos(3x-y+z) \\ 2\pi\,\sin(\pi t)\cos(x+y+z) - \frac{4}{5}\cos(\pi t)\cos(x+y+z) \end{bmatrix}$$

$$p = -8\left(\pi \sin(\pi t) + \frac{2}{5}\cos(\pi t)\right)(\sin(3y - x + z) + \sin(3x - y + z))$$
$$+2\left(\frac{2}{5}\cos(\pi t) - \pi \sin(\pi t)\right)\sin(x + y + z)$$

It is not hard to verify that the above solution satisfies Eq. (1) and (5)–(8). Source terms q_p and f can be calculated from (2) and (3) respectively. However, the divergence of the free flow velocity $\boldsymbol{u}_c$ is not zero, so we need to modify (4) to a more general case $\nabla \cdot \boldsymbol{u}_c = g$, and calculate g from it.

For the boundary conditions, we apply corresponding essential boundaries for all variables except for p.

We tested our implementation on the University of Wyoming's Teton HPC cluster [29].

4.1 Convergence

We examine the convergence rate of our implementation for different timestep length Δt's. To make the result reproducible, all of the experiments are run in a single processor with the direct linear solver MUMPS (Multifrontal Massively Parallel sparse direct Solver) [30, 31]. Piecewise quadratic functions are used for p_m, p_f, and $\boldsymbol{u}_c$. Degree 1 Lagrange elements are used for p.

We examine the convergence rate for both $\Delta t = h$ and $\Delta t = h^2$, where h is the cell size. Recall that our domain is a unit cube. A model with cell size h means our domain is partitioned into $h \times h \times h$ small cubes. Each cube contains 6 tetrahedral cells. The L^2 norm of each variable's error at time $T = 1$ is calculated. The convergence rate is calculated as $\ln(e_i/e_{i-1})/\ln(h_i/h_{i-1})$. The result is shown in Tables 1 and 2.

Table 1. The L^2 error at $T = 1$ with $\Delta t = h$.

h	p_m		p_f		$\boldsymbol{u}_c$		p	
	Error	Rate	Error	Rate	Error	Rate	Error	Rate
1/2	3.56e−2	0.60	1.46e−1	1.12	3.93e−2	3.78	1.50	56.40
1/4	2.01e−2	0.82	5.64e−2	1.37	1.35e−2	1.54	3.48e−1	2.10
-1/8	1.07e−2	0.91	2.17e−2	1.38	4.80e−3	1.49	8.87e−2	1.97
1/16	5.54e−3	0.95	9.04e−3	1.27	1.89e−3	1.34	3.51e−2	1.34

Table 2. The L^2 error at $T = 1$ with $\Delta t = h^2$.

h	p_m		p_f		$\boldsymbol{u}_c$		p	
	Error	Rate	Error	Rate	Error	Rate	Error	Rate
1/2	2.03e−2	1.41	5.59e−2	2.51	1.68e−2	5.01	9.10e−1	57.12
1/4	5.69e−3	1.83	9.38e−3	2.57	2.26e−3	2.90	1.72e−1	2.41
-1/8	1.44e−3	1.97	1.95e−3	2.26	4.08e−4	2.47	3.90e−2	2.14
1/16	3.62e−4	2.00	4.58e−4	2.08	9.98e−5	2.16	9.35e−3	2.06

4.2 Performance

The solver consists of two major parts: linear system assembling and linear system solving. Below we investigate the performance of our implementation in these two parts separately.

Assembling. Despite that the linear form L is assembled at each time step, the assembling of the bilinear form a is usually more time consuming. Figures 2 and 3 show the time spent for assembling the bilinear form under different conditions.

Figure 2 shows the assembling time when using a single CPU versus degrees of freedom (DoF) of our system. We can see that the assembling time is linear to total DoF.

Figure 3 presents the performance of assembling along different number of processors. Each line presents the scalability with fixed problem size. We can see that assembling scales well when the problem size is large enough. However, too many processors may lead to a performance drop.

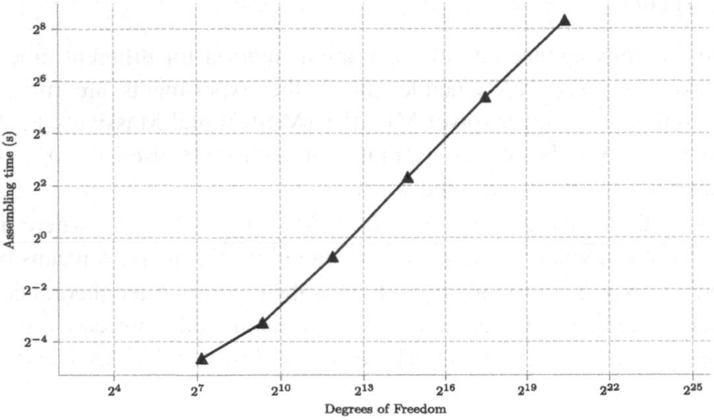

Fig. 2. Assembling time is linear to DoF.

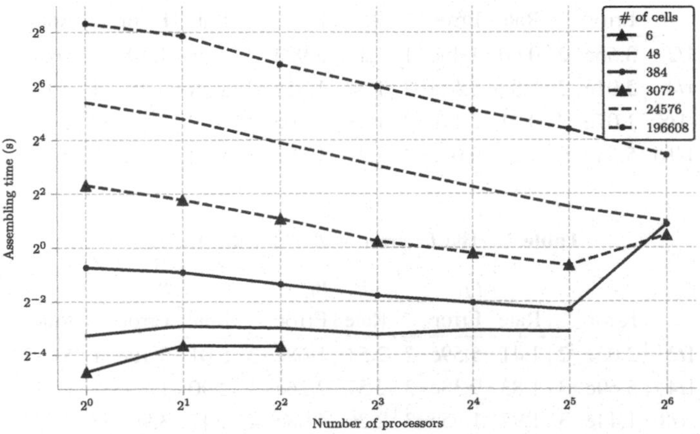

Fig. 3. Assembling scales well in large systems.

Solving. For our time-dependent model, the same linear system is solved at each timestep with varying right hand side. In this case, direct linear solvers can benefit from reuse of decompositions, as we will see in Figs. 4 and 5.

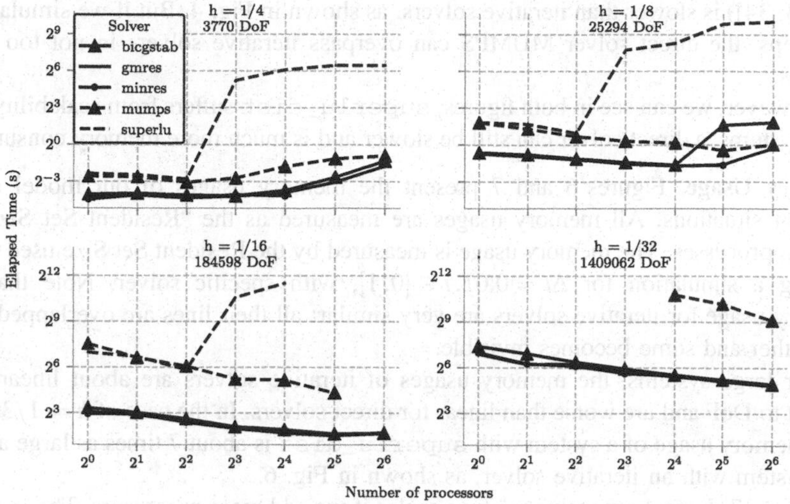

Fig. 4. Time for solving a single step.

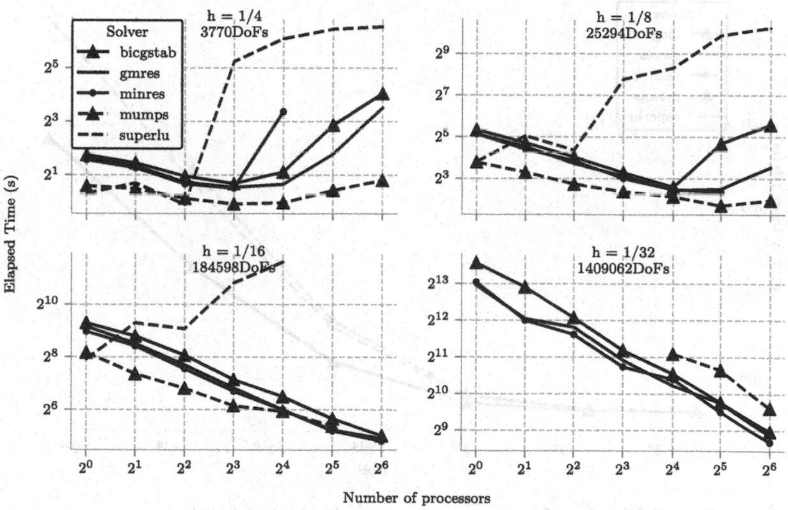

Fig. 5. Time for solving 100 steps.

A collection of high performance solvers and preconditioners are available (callable) from FEniCS, assuming it was built with corresponding packages. To reduce complexity, we choose the ILU preconditioner `hypre_euclid` from Livermore's HYPRE package [32] for all iterative solvers we use.

For a single solve, direct solvers like MUMPS and `superlu_dist` (Supernodal LU [33, 34]) is slower than iterative solvers, as shown in Fig. 4. But if we simulate for 100 steps, the direct solver MUMPS can overpass iterative solvers in not too large systems.

However, we can see in both figures, `superlu_dist` suffers from scalability. For large systems, a direct solver can still be slower and is much more memory consuming.

Memory Usage. Figures 6 and 7 present the memory usages of our model under different situations. All memory usages are measured as the "Resident Set Size" of running processes. The memory usage is measured by the Resident Set Size used when running a simulation for $\Delta t = 0.01, t \in [0, 1]$, with specific solver. Note that the memory usage for iterative solvers are very similar: all their lines are overlapped with each other and some becomes invisible.

For large systems, the memory usages of iterative solvers are about linear with respect to DoF and are worse than linear for direct solvers. In the case of $h = 1/32$, the total memory usage of a system with `superlu_dist` is about 7 times as large as that of a system with an iterative solver, as shown in Fig. 6.

Figure 7 shows how memory usage scales if we add more processors. The memory usage is the memory used by a single processor.

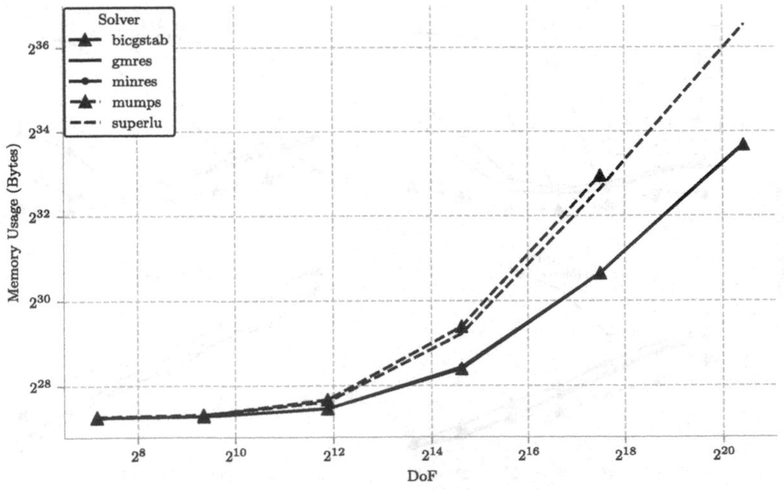

Fig. 6. Memory usage versus degrees of freedom.

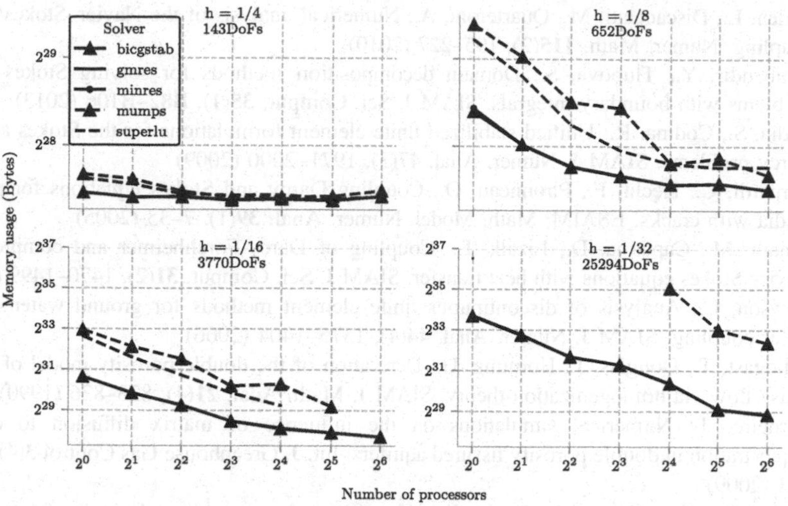

Fig. 7. Memory usage versus number of processors.

5 Conclusions and Future Work

We have presented an implementation of a coupled dual-porosity-Stokes model using the automated FEM solver FEniCS. We proposed a solution to modeling the coupled interface by using FEniCS' built-in interface integration and expanding variables to the whole domain. This approach enables us to simulate both 2D and 3D models in parallel with minimum coding. Future work will include adding data assimilation from active sensors and experimenting with different interface conditions to see better solutions can be computed. Another approach is to implement one of the non-iterative domain decomposition methods that have been developed for Stokes-Darcy systems [35, 36], which can decompose our asymmetric matrix into two small symmetric matrices and reduce communications between two subsystems.

Acknowledgment. This research was supported in part by NSF grant 1722692.

References

1. Çeşmelioğlu, A., Rivière, B.: Primal discontinuous Galerkin methods for time-dependent coupled surface and subsurface flow. J. Sci. Comput. **40**(1–3), 115–140 (2009)
2. Arbogast, T., Brunson, D.: A computational method for approximating a Darcy-Stokes system governing a vuggy porous medium. Comput. Geosci. **11**(3), 207–218 (2007)
3. Cao, S., Pollastrini, J., Jiang, Y.: Separation and characterization of protein aggregates and particles by field flow fractionation. Curr. Pharm. Biotechnol. **10**(4), 382–390 (2009)
4. Babuška, I., Gatica, G.: A residual-based a posteriori error estimator for the Stokes-Darcy coupled problem. SIAM J. Numer. Anal. **48**(2), 498–523 (2010)

5. Badea, L., Discacciati, M., Quarteroni, A.: Numerical analysis of the Navier–Stokes/Darcy coupling. Numer. Math. **115**(2), 195–227 (2010)
6. Boubendir, Y., Tlupova, S.: Domain decomposition methods for solving Stokes-Darcy problems with boundary integrals. SIAM J. Sci. Comput. **35**(1), B82–B106 (2013)
7. Badia, S., Codina, R.: Unified stabilized finite element formulations for the Stokes and the Darcy problems. SIAM J. Numer. Anal. **47**(3), 1971–2000 (2009)
8. Bernardi, C., Hecht, F., Pironneau, O.: Coupling Darcy and Stokes equations for porous media with cracks. ESAIM: Math. Model. Numer. Anal. **39**(1), 7–35 (2005)
9. Amara, M., Capatina, D., Lizaik, L.: Coupling of Darcy-Forchheimer and compressible Navier-Stokes equations with heat transfer. SIAM J. Sci. Comput. **31**(2), 1470–1499 (2009)
10. Dawson, C.: Analysis of discontinuous finite element methods for ground water/surface water coupling. SIAM J. Numer. Anal. **44**(4), 1375–1404 (2006)
11. Arbogast, T., Douglas, J., Hornung, U.: Derivation of the double porosity model of single phase flow via homogenization theory. SIAM J. Math. Anal. **21**(4), 823–836 (1990)
12. Carneiro, J.: Numerical simulations on the influence of matrix diffusion to carbon sequestration in double porosity fissured aquifers. Int. J. Greenhouse Gas Control **3**(4), 431–443 (2009)
13. Gerke, H., Genuchten, M.: Evaluation of a first-order water transfer term for variably saturated dual-porosity flow models. Water Resour. Res. **29**(4), 1225–1238 (1993)
14. Douglas, C., Hu, X., Bai, B., He, X., Wei, M., Hou, J.: A data assimilation enabled model for coupling dual porosity flow with free flow. In: 2018 17th International Symposium on Distributed Computing and Applications for Business Engineering and Science (DCABES), Wuxi, pp. 304–307 (2018)
15. Hou, J., Qiu, M., He, X., Guo, C., Wei, M., Bai, B.: A dual-porosity-stokes model and finite element method for coupling dual-porosity flow and free flow. SIAM J. Sci. Comput. **38**, B710–B739 (2016)
16. Bello, R., Wattenbarger, R.: Multi-stage hydraulically fractured horizontal shale gas well rate transient analysis. In: North Africa Technical Conference and Exhibition (2010)
17. Brohi, I., Pooladi-Darvish, M., Aguilera, R.: Modeling fractured horizontal wells as dual porosity composite reservoirs-application to tight gas, shale gas and tight oil cases. In: SPE Western North American Region Meeting (2011)
18. Carlson, E., Mercer, J.: Devonian shale gas production: mechanisms and simple models. J. Pet. Technol. **43**(04), 476–482 (1991)
19. Guo, C., Wei, M., Chen, H., Xiaoming, H., Bai, B.: Improved numerical simulation for shale gas reservoirs. In: Offshore Technology Conference-Asia (2014)
20. Çeşmelioğlu, A., Rivière, B.: Analysis of time-dependent Navier-Stokes flow coupled with Darcy flow. J. Numer. Math. **16**(4), 249–280 (2008)
21. Chidyagwai, P., Rivière, B.: On the solution of the coupled Navier–Stokes and Darcy equations. Comput. Methods Appl. Mech. Eng. **198**(47–48), 3806–3820 (2009)
22. Beavers, G., Joseph, D.: Boundary conditions at a naturally permeable wall. J. Fluid Mech. **30**(1), 197–207 (1967)
23. Alnæs, M., et al.: The FEniCS project version 1.5. Arch. Numer. Softw. **3**(100), 9–23 (2015)
24. Logg, A., Mardal, K.-A., Wells, G.: Automated Solution of Differential Equations by the Finite Element Method. Springer, Heidelberg (2012). https://doi.org/10.1007/978-3-642-23099-8
25. Alnæs, M.S.: UFL: a finite element form language. In: Logg, A., Mardal, K.-A., Wells, G. (eds.) Automated Solution of Differential Equations by the Finite Element Method. LNCS, vol. 84, pp. 303–338. Springer, Heidelberg (2012). https://doi.org/10.1007/978-3-642-23099-8_17

26. Hoffman, J., Jansson, J., Degirmenci, C., Jansson, N., Nazarov, M.: Unicorn: a unified continuum mechanics solver. In: Logg, A., Mardal, KA., Wells, G. (eds.) Automated Solution of Differential Equations by the Finite Element Method. LNCS, vol. 84, pp. 339–361. Springer, Heidelberg (2012). https://doi.org/10.1007/978-3-642-23099-8_18
27. Hoffman, J., Jansson, J., Jansson, N.: FEniCS-HPC: automated predictive high-performance finite element computing with applications in aerodynamics. In: PPAM 2015, vol. 9573, pp. 356–365. Springer, Cham (2015). https://doi.org/10.1007/978-3-319-32149-3_34
28. Drøsdal, I.: Porous and viscous modeling of cerebrospinal fluid flow in the spinal canal associated with syringomyelia. Master's thesis (2011)
29. Advanced Research Computing Center: Teton Computing Environment, Intel x86_64 cluster. University of Wyoming, Laramie (2018). https://doi.org/10.15786/M2FY47
30. Amestoy, P., Duff, I., Koster, J., L'Excellent, J.-Y.: A fully asynchronous multifrontal solver using distributed dynamic scheduling. SIAM J. Matrix Anal. Appl. **23**(1), 15–41 (2001)
31. Amestoy, P., Cuermouche, A., L'Excellent, J.-Y., Pralet, S.: Hybrid scheduling for the parallel solution of linear systems. Parallel Comput. **32**(2), 136–156 (2006)
32. Falgout, R., Yang, U.: hypre: a library of high performance preconditioners. In: Sloot, P., Hoekstra, A., Tan, C., Dongarra, J. (eds.) ICCS 2002, vol. 2331, pp. 631–641. Springer, Heidelberg (2002). https://doi.org/10.1007/3-540-47789-6_66
33. Li, X., Demmel, J., Gilbert, J., Grigori, I., Yamazaki, M.: SuperLU Users' Guide. Lawrence Berkeley National Laboratory LBNL-44289 (2005). http://crd.lbl.gov/~xiaoye/SuperLU/
34. Li, X., Demmel, J.: SuperLU_DIST: a scalable distributed-memory sparse direct solver for unsymmetric linera systems. ACM Trans. Math. Softw. **29**(2), 110–140 (2003)
35. Cao, Y., Gunzburger, M., He, X., Wang, X.: Parallel, non-iterative, multi-physics domain decomposition methods for time-dependent Stokes-Darcy systems. **83**, 1617–1644 (2014)
36. Feng, W., He, X., Wang, Z., Zhang, X.: Non-iterate domain decomposition methods for a non-stationary Stokes-Darcy model with Beavers-Joseph interface condition. Appl. Math. Comput. **219**, 453–463 (2012)

Anomaly Detection in Social Media Using Recurrent Neural Network

Shamoz Shah and Madhu Goyal(✉)

Centre of Artificial Intelligence, Faculty of Engineering and Information Technology, University of Technology Sydney, PO Box 123, Broadway, NSW 2007, Australia
shamoz.shah@student.uts.edu.au,
madhu.goyal-2@uts.edu.au

Abstract. In today's information environment there is an increasing reliance on online and social media in the acquisition, dissemination and consumption of news. Specifically, the utilization of social media platforms such as Facebook and Twitter has increased as a cutting edge medium for breaking news. On the other hand, the low cost, easy access and rapid propagation of news through social media makes the platform more sensitive to fake and anomalous reporting. The propagation of fake and anomalous news is not some benign exercise. The extensive spread of fake news has the potential to do serious and real damage to individuals and society. As a result, the detection of fake news in social media has become a vibrant and important field of research. In this paper, a novel application of machine learning approaches to the detection and classification of fake and anomalous data are considered. An initial clustering step with the K-Nearest Neighbor (KNN) algorithm is proposed before training the result with a Recurrent Neural Network (RNN). The results of a preliminary application of the KNN phase before the RNN phase produces a quantitative and measureable improvement in the detection of outliers, and as such is more effective in detecting anomalies or outliers against the test dataset of 2016 US Presidential Election predictions.

Keywords: Clustering · Recurrent neural networks · Twitter · Presidential Election

1 Introduction

In today's information environment, data is being collected, stored and analyzed more extensively to make predictions about client behavior, weather patterns and other natural disasters, espionage and many other sequences that would be beneficial to detect and predict ahead of time. For many data mining algorithms to return beneficial information that has some utility in adding predictive value, the underlying dataset needs to be true, accurate and computable. For this reason, among many others, it is important that algorithms are in place which enable the detection of fake or otherwise anomalous data [7].

News is pervasive, and fake new is even more so. Fake news at the very least contributes to the misinformation of society. Taken to extremes, fake news can damage entire cohorts of people, societies, institutions and even whole countries. The most

J. M. F. Rodrigues et al. (Eds.): ICCS 2019, LNCS 11539, pp. 74–83, 2019.
https://doi.org/10.1007/978-3-030-22747-0_6

recent example of the seriousness of this damage can be seen with the 2016 US Presidential Elections. Many accusations have been made that foreign interests and local groups with vested interests have influenced the integrity and progression of the election. The net result of these accusations are the undermining of the confidence in US sovereign processes and a significant blow to the moral of the citizens of the United States in general.

The damaging effects of misinformation is well known and have been studied for some time. There are valid social, ethical and moral reasons to identify, contain and control the creation and dissemination of fake and misleading information. We use this rational to motivate our work on the identification and classification of false and misleading information.

In the second section of this paper, the related work is discussed. The current state of the art is examined and the methodologies, motivations and rational are compared. From these works, a novel methodology is derived, and presented in section three. The results of the methodology are then presented in Sect. 4, followed by a discussion and conclusion in Sect. 5. The paper closes with a discussion on further areas of research.

2 Related Work

Shu et al. [7] outline the human cost of fake, false and misleading information on society as a whole, and we begin there by making a strong case for the need to identify and contain misinformation. Shu makes the point that fake news is very hard to detect from the news item itself, and we need to resort to meta-data such as likes and retweets, friends and followers. Further justification for the utility of meta-data can be found in Akcora [2]. This paper focuses on the clustered behavior of the friend and follower network in Twitter. Users tend to friend and follow other users with the content they are interested in, and in this way Akcora identifies clusters of users who may exhibit the same behavioral patterns.

Telang [8] introduces the use of location data to model the spatio-temporal characteristics of Twitter data, conducting experiments in the identification of spatial patterns in sentiment data and the identification of the temporal aspects of weather data. The innovation in Telang is that they model anomalies as spatial or temporal deviations from the normal distribution. That is to say, if an event occurs out of spatial or temporal locality to the mean, it is deemed to be anomalous. A further development by Oancea (thesis, cannot cite) is the use of statistical noise in combination with a Kalman Filter. The resultant algorithm is called an Extended Kalman Filter, which can not only process linear data, but data that is non-linear and differentiable.

Zhao et al. [10] discusses the use of a novel graphical representation of anomalous data on Twitter. The data is presented as nodes in time, which change color and shape based on the activity surrounding or spawned by that node. In this way, it is possible to track the changing nature of the Twitter data landscape in real-time, with rich visual detail. Rere et al. [6] discusses the use of deep learning to make computing more efficient and to reduce the cost of computing chips, but doesn't remark on the mechanisms by which such occurs. The paper is not available in full, but the discussion centers around improving the training speed via a process called "simulated annealing".

Thom et al. [9] further discusses the utilization of geolocation information from Twitter data. The paper focuses on using Twitter geolocation information to detect anomalies in spatiotemporal data. This approach can be useful for detecting emerging threats and disasters.

Macneil et al. [5] discusses the use of biologically probable weighting algorithms to improve the convergence behavior and stability of recurrent neural networks. Akbari et al. [1] delves into the detection of anomalies using a KNN algorithm, by detecting points that do not conform to a normalized cluster.

3 Methodology

3.1 Definition of Fake News

Fake news has existed since the existence of news itself. Where there has been the dissemination of news for the purposes of knowledge and information, there have been parties which wish to subvert the news for the purposes of propaganda and disinformation. It comes as no surprise then, that the occurrence of fake news increased exponentially with the advent of the printing press and the widespread adoption of the newspaper, printed books and other media. Today, there are many different forms of media being used to create, update and disseminate news. The trend is to move from a digest format to more real-time and close to real-time avenues. One such avenue is Twitter. In fact, Twitter stands today as the most breaking source of news available.

To detect and classify fake and anomalous news, what constitutes such news must be defined and quantified. As there is no standard definition of fake news, a discussion on what qualifies as fake news for the purposes of this paper needs to be defined. A working definition of fake news is that it is intentionally and verifiably false. That is, fake news has the deliberate intent to mislead the consumer and the authenticity of the news itself is deliberately falsified or, at the very least, questionable. This definition implies a two-fold test of accuracy. These points are *authenticity* and *intent*.

For this definition to be applicable, fake news has to be unquestionably verified as false, possibly by a human reader or a machine learning algorithm. The next criterion is that the news has been created to deliberately mislead consumers. Some literature treats satire as fake news, even though satire is deliberately designed to be entertainment oriented and makes its use of deceptiveness clear to the consumer. Other literature treats any deceptive news as fake news, including hoaxes, fabrications and satire. In this paper, we constrain ourselves to a very explicit definition of fake news -

Fake news is a news article that is intentionally and verifiably false [7].

3.2 Proposed Algorithm

A methodology was proposed whereby the Euclidean proximity of data points could be exploited to reveal some hidden underlying features of the data. The spatially enhanced data can then be passed through a further processing phase that can augment the detection of probable anomalies. After some preliminary research on various current algorithms, it was proposed that a clustering algorithm be chained to an artificial neural network and the data evaluated against such a composite classifier.

A few clustering algorithms were considered [3, 4], but the most favorable clustering algorithm for the purposes of this experiment was the K-Nearest Neighbor (KNN) algorithm due to its lack of parametric requirements [11, 12]. As a result, the KNN algorithm makes no assumptions about the probability distributions of the data points being processed [12]. This is a very important feature of KNNs, since it ensures that any pertinent spatial linearity or coherence is preserved for any subsequent processing.

For the second phase of the methodology, it was proposed that the KNN data be passed through a Recurrent Neural Network (RNN). The RNN was trained on the output of a preceding KNN phase and then the results of the classifier chain were tested against a novel dataset.

3.3 Clustered RNN Pseudocode

The pseudocode for the classifier process is as follows:

```
Load(datafile);
Set optimizedRNN = rnn(seed=41, learningrate = 0.5,
epochs = 5000, momentum = 0.1, transferfunction = "Gom-
pertz");
For k  = (1, 10, 20)
  model = KNN (k, datafile);
  trainedModel[k] = optimizedRNN(model);
End For
Return trainedModel[];
```

The data file is first loaded. Next, the RNN training node is created with optimized parameters passed as an argument. The training phase is now ready to begin. For each of the value for k, the KNN is trained and then passed to the optimized RNN node. This is done for three values of k, and the results are returned in a vector.

4 Results

4.1 Dataset

The dataset of 2016 US Presidential Election results is used for prediction and to test the proposed algorithm. The classifier is first trained on the actual election results and then tested against the result predictions to ascertain how effective the classifier optimizations were in predicting voting behavior. To ensure that our algorithm would be applicable to real-world scenarios, the data we obtained was from the 2016 US Presidential Elections. The first set of data that is obtained was the actual (final) result of the

US election vote counting. This data was presented as the number of votes the Democrat Party had won, as a percentage of the total votes. We used this data set as the training input for both the KNN and the RNN phases. The second set of data that was obtained were predictions of the election results before the election had finished and the votes had been counted. There were a number of data points in the data set that was not required, and the data had to be cleaned and formatted for use in the experiments that were being conducted. The cleaned version of this data formed the testing set for both the KNN and RNN algorithms.

5 KNN Clustering Features

It was decided to use standardized algorithms available with R. This would ensure that the results were replicable, accurate and testable. To obtain an accurate idea of the behavior of the KNN algorithm with the dataset in question, a range of k values were used. The k-values used with the KNN algorithm were 1, 10 and 20. Following the training phase, the combined algorithm (Clustered RNN) was tested against a novel dataset to evaluate whether the classifier improved the accuracy of anomaly detection.

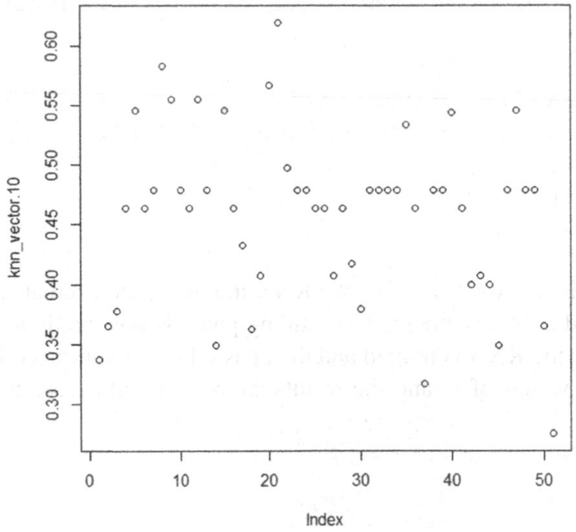

Fig. 1. KNN cluster (mean 10)

As can be seen in Fig. 1, an application of KNN with a mean of 10 does little to cluster the data sufficiently enough to be useful in the prediction of outliers. There are some patterns that are starting to form, and it is advantageous to attempt to increase the expression of these patterns with a further application of the KNN algorithm.

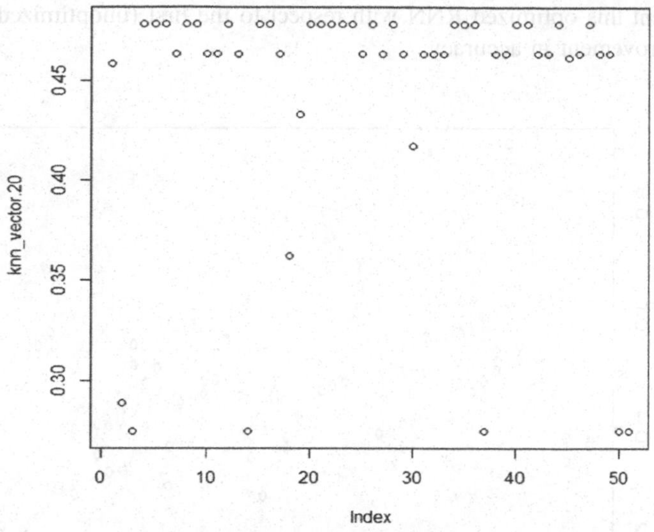

Fig. 2. KNN cluster (mean 20)

In the next iteration of the experiment, we apply the KNN clustering algorithm with a mean of 20. As can be seen from Fig. 2, the adjustment of the mean parameter has started to cluster the data points quite nicely. From this application of the KNN algorithm, it can be seen that there are three points in the middle of the graph that have not conformed to either the top cluster or the bottom cluster. For the purposes of the experiment, we consider these points as anomalous, or outliers to the normal distribution.

The first graph (Fig. 1) shows the result of training an un-optimized RNN against the 2016 US election datasets. The RNN is already somewhat effective in predicting the spatio-temporal distribution of the test dataset (predictions) against the benchmark dataset (winners). However, we wish to delve a little further into the parameters that we train our RNN with, in an attempt to present the most optimal training scenario for our data mining experiment.

To improve the behavior of our prediction model, we next attempted to optimize the RNN portion of the algorithm before adding the clustering step. We assume that the RNN optimization behavior correlates linearly with the application of a clustering algorithm, so it stands to reason that optimizing the RNN would yield more accurate results. We investigated a number of scenarios to optimize the RNN and through experimentation, have come to use the following parameters for the RNN –

- Seed – 41 (for replicability)
- Learning Rate – 0.5
- Epochs – 5000
- Momentum – 0.1
- Transfer Function – Gompertz.

We present this optimized RNN with respect to the first (unoptimized) RNN and note the improvement in accuracy.

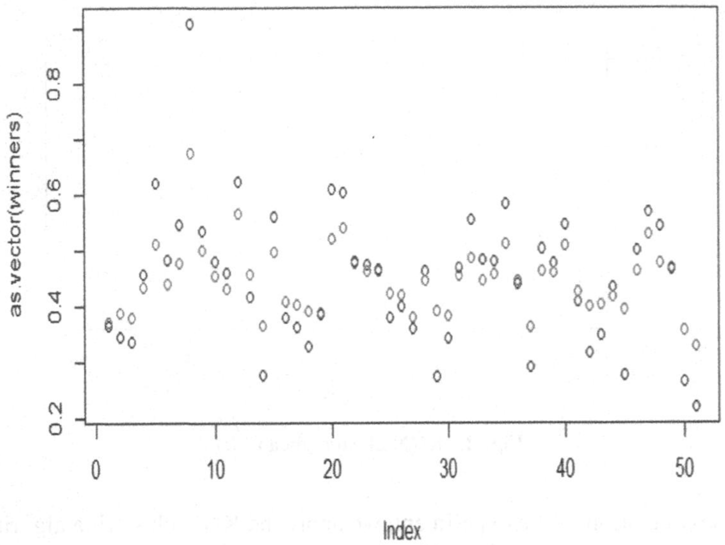

Fig. 3. RNN without optimization

As we can see from the graph (Fig. 2) the optimizations applied to the second RNN enable it to produce results with more spatio-temporal coherence than the first iteration. The prediction accuracy of the optimized RNN has increased significantly. It follows that any clustering algorithm applied to the RNN as a pre-processing step will benefit from this optimization in at least a linear fashion.

For the next phase in the experiment, we aim to combine the optimized RNN with a k nearest neighbor clustering algorithm (KNN). We used a number of different parameters for the KNN, with different distance parameters. What we found was that the final clustered RNN result was highly sensitive to the KNN distance parameter used. For the KNN training phase of the algorithm, we used the following parameters –

- K value – 10, 20

We show the result of running the complete algorithm (KRNN) with a mean value of 1. From what we can see, there is very little difference between the optimized RNN and a KRNN with a mean of 1. The results are shown graphically in Fig. 3.

To examine what significance the clustering parameters have on the final results, we attempt another trial with our experiment, setting the mean value to 10. As can be seen in Fig. 4, the final result is more spatio-temporally coherent than running the algorithm with a mean of 1.

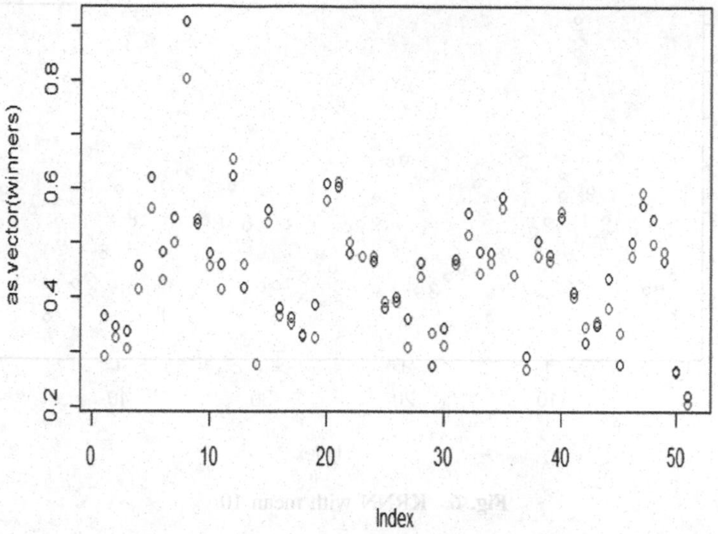

Fig. 4. Optimized RNN

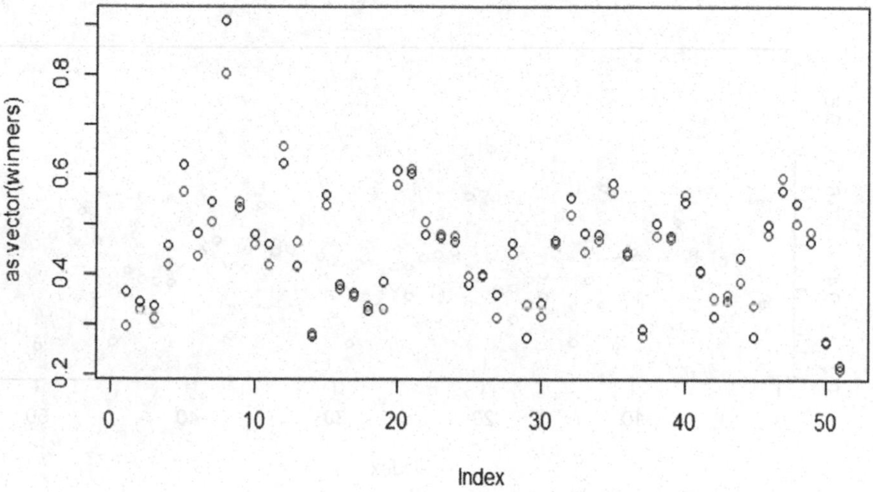

Fig. 5. KRNN with mean 1

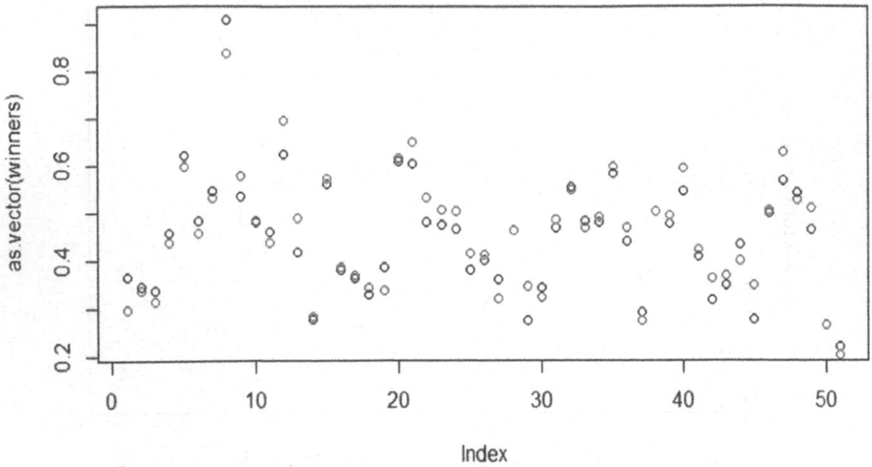

Fig. 6. KRNN with mean 10

For the final test, we change the k parameter to a distance of 20 and re-execute the algorithm. As can be seen in Fig. 5, there is qualitatively an improvement of the spatial behavior of the data, but not a great decrease in the prediction error.

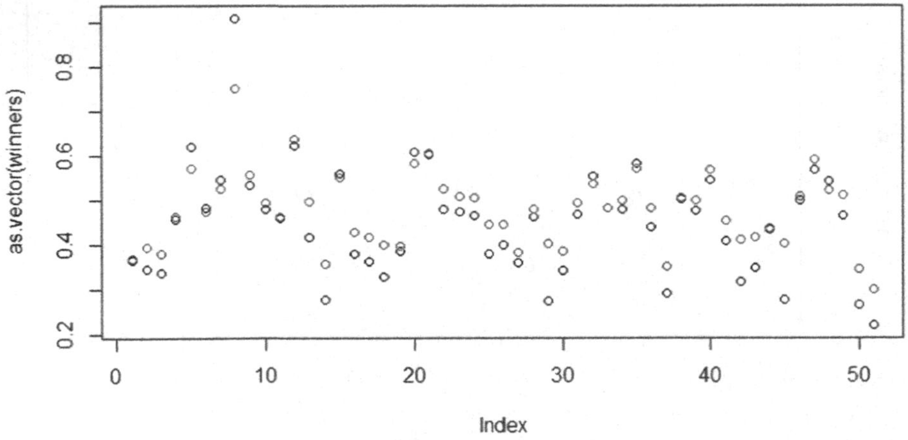

Fig. 7. KRNN with mean 20

From inspection of the KRNN.10 and KRNN.20 results, we find that there is a significant improvement in spatial coherence when we first cluster the data before feeding it through an RNN phase. The predicted data points have better spatial coherence and are aligned to the training set a lot more closely than they were with a single application of RNN alone. From the results of the experiment, it is obvious that applying a clustering phase before the RNN phase results in predictions that are spatially closer to the actual results (Figs. 6 and 7).

6 Conclusions

The experiments and results show that clustered RNNs are highly sensitive to the initial clustering step. The distance chosen affects the distribution of the final points significantly. It was also noted that the distribution of the test dataset does not vary in a linear fashion with respect to the Euclidean distance selected in the KNN step. Further work in this area can focus on generating distribution maps of the KNN stage to further optimize the results that are obtained. Another area that begs examination is the use of different clustering algorithms or even the application of more pre-processing algorithms before the RNN step. Our future work will also focus on comparison with other Fake news detection techniques.

References

1. Akbari, M., Overloop, P.J.V., Afshar, A.: Clustered K nearest neighbor algorithm for daily inflow forecasting. Water Resour. Manag. **25**, 1341–1357 (2010). https://doi.org/10.1007/s11269-010-9748-z
2. Akcora, C.G., Carminati, B., Ferrari, E., Kantarcioglu, M.: Detecting anomalies in social network data consumption. Soc. Netw. Anal. Mining (2014). https://doi.org/10.1007/s13278-014-0231-3
3. Chen, C.-H., Huang, W.-T., Tan, T.-H., et al.: Using K-nearest neighbor classification to diagnose abnormal lung sounds. Sensors **15**, 13132–13158 (2015). https://doi.org/10.3390/s150613132
4. Hu, L.-Y., Huang, M.-W., Ke, S.-W., Tsai, C.-F.: The distance function effect on k-nearest neighbor classification for medical datasets. SpringerPlus (2016). https://doi.org/10.1186/s40064-016-2941-7
5. Macneil, D., Eliasmith, C.: Fine-tuning and the stability of recurrent neural networks. PLoS One (2011). https://doi.org/10.1371/journal.pone.0022885
6. Rere, L.M., Fanany, M.I., Arymurthy, A.M.: Simulated annealing algorithm for deep learning. In: The Third Information Systems International Conference vol. 72, pp. 137–144 (2015). https://doi.org/10.1016/j.procs.2015.12.114
7. Shu, K., Sliva, A., Wang, S., et al.: Fake news detection on social media. ACM SIGKDD Explor. Newslett. **19**, 22–36 (2017). https://doi.org/10.1145/3137597.3137600
8. Telang, A., Deepak, P., Joshi, S., et al.: Detecting localized homogeneous anomalies over spatio-temporal data. Data Mining Knowl. Discov. **28**, 1480–1502 (2014). https://doi.org/10.1007/s10618-014-0366-x
9. Thom, D., Bosch, H., Koch, S., et al.: Spatiotemporal anomaly detection through visual analysis of geolocated Twitter messages. In: 2012 IEEE Pacific Visualization Symposium (2012). https://doi.org/10.1109/pacificvis.2012.6183572
10. Zhao, J., Cao, N., Wen, Z., et al.: #FluxFlow: visual analysis of anomalous information spreading on social media. IEEE Trans. Vis. Comput. Graph. **20**, 1773–1782 (2014). https://doi.org/10.1109/tvcg.2014.2346922
11. Nonparametric statistics (2017). In: Wikipedia. https://en.wikipedia.org/wiki/Nonparametric_statistics. Accessed 11 Dec 2017
12. K-nearest neighbors algorithm (2017). In: Wikipedia. https://en.wikipedia.org/wiki/K-nearest_neighbors_algorithm. Accessed 11 Dec 2017

Conditional BERT Contextual Augmentation

Xing Wu[1,2], Shangwen Lv[1,2], Liangjun Zang[1(✉)], Jizhong Han[1,2], and Songlin Hu[1,2]

[1] Institute of Information Engineering, Chinese Academy of Sciences, Beijing, China
{wuxing,lvshangwen,zangliangjun,hanjizhong,husonglin}@iie.ac.cn
[2] School of Cyber Security, University of Chinese Academy of Sciences, Beijing, China

Abstract. Data augmentation methods are often applied to prevent overfitting and improve generalization of deep neural network models. Recently proposed contextual augmentation augments labeled sentences by randomly replacing words with more varied substitutions predicted by language model. Bidirectional Encoder Representations from Transformers (BERT) demonstrates that a deep bidirectional language model is more powerful than either an unidirectional language model or the shallow concatenation of a forward and backward model. We propose a novel data augmentation method for labeled sentences called conditional BERT contextual augmentation. We retrofit BERT to conditional BERT by introducing a new conditional masked language model (The term "conditional masked language model" appeared once in original BERT paper, which indicates context-conditional, is equivalent to term "masked language model". In our paper, "conditional masked language model" indicates we apply extra label-conditional constraint to the "masked language model".) task. The well trained conditional BERT can be applied to enhance contextual augmentation. Experiments on six various different text classification tasks show that our method can be easily applied to both convolutional or recurrent neural networks classifier to obtain improvement.

1 Introduction

Deep neural network-based models are easy to overfit and result in losing their generalization due to limited size of training data. In order to address the issue, data augmentation methods are often applied to generate more training samples. Recent years have witnessed great success in applying data augmentation in the field of speech area [10,14] and computer vision [17,24,27]. Data augmentation in these areas can be easily performed by transformations like resizing, mirroring,

Supported by the National Key Research and Development Program of China (No. 2017YFB1010000) and the National Natural Science Foundation of China (No. 61702500).

J. M. F. Rodrigues et al. (Eds.): ICCS 2019, LNCS 11539, pp. 84–95, 2019.
https://doi.org/10.1007/978-3-030-22747-0_7

random cropping, and color shifting. However, applying these universal transformations to texts is largely randomized and uncontrollable, which makes it impossible to ensure the semantic invariance and label correctness. For example, given a movie review "The actors is good", by mirroring we get "doog si srotca ehT", or by random cropping we get "actors is", both of which are meaningless.

Existing data augmentation methods for text are often loss of generality, which are developed with handcrafted rules or pipelines for specific domains. A general approach for text data augmentation is replacement-based method, which generates new sentences by replacing the words in the sentences with relevant words (e.g. synonyms). However, words with synonyms from a handcrafted lexical database likes WordNet [19] are very limited, and the replacement-based augmentation with synonyms can only produce limited diverse patterns from the original texts. To address the limitation of replacement-based methods, Kobayashi [15] proposed contextual augmentation for labeled sentences by offering a wide range of substitute words, which are predicted by a label-conditional bidirectional language model according to the context. But contextual augmentation suffers from two shortages: the bidirectional language model is simply shallow concatenation of a forward and backward model, and the usage of LSTM models restricts their prediction ability to a short range.

BERT, which stands for Bidirectional Encoder Representations from Transformers, pre-trained deep bidirectional representations by jointly conditioning on both left and right context in all layers. BERT addressed the unidirectional constraint by proposing a "masked language model" (MLM) objective by masking some percentage of the input tokens at random, and predicting the masked words based on its context. This is very similar to how contextual augmentation predict the replacement words. But BERT was proposed to pre-train text representations, so MLM task is performed in an unsupervised way, taking no label variance into consideration.

This paper focuses on the replacement-based methods, by proposing a novel data augmentation method called conditional BERT contextual augmentation. The method applies contextual augmentation by conditional BERT, which is fine-tuned on BERT. We adopt BERT as our pre-trained language model with two reasons. First, BERT is based on Transformer. Transformer provides us with a more structured memory for handling long-term dependencies in text. Second, BERT, as a deep bidirectional model, is strictly more powerful than the shallow concatenation of a left-to-right and right-to left model. So we apply BERT to contextual augmentation for labeled sentences, by offering a wider range of substitute words predicted by the masked language model task. However, the masked language model predicts the masked word based only on its context, so the predicted word maybe incompatible with the annotated labels of the original sentences. In order to address this issue, we introduce a new fine-tuning objective: the "conditional masked language model" (C-MLM). The conditional masked language model randomly masks some of the tokens from an input, and the objective is to predict a label-compatible word based on both its context and sentence label. Unlike Kobayashi's work [15], the C-MLM objective allows a deep bidirectional representations by jointly conditioning on both left and right context in all layers. In order to evaluate how well our augmentation method improves performance of deep neural network models, following Kobayashi, we experiment it on two most common neural network structures, LSTM-RNN and CNN, on text classification tasks. Through the experiments on six various different text classification tasks, we demonstrate that the proposed conditional BERT model augments sentence better than baselines, and conditional BERT contextual augmentation method can be easily applied to both convolutional or recurrent neural networks classifier. We further explore our conditional MLM task's connection with style transfer task and demonstrate that our conditional BERT can also be applied to style transfer too.

Our contributions are concluded as follows:

- We propose a conditional BERT contextual augmentation method. The method allows BERT to augment sentences without breaking the label-compatibility. Our conditional BERT can further be applied to style transfer task.
- Experimental results show that our approach obviously outperforms existing text data augmentation approaches.

To our best knowledge, this is the first attempt to alter BERT to a conditional BERT or apply BERT on text generation tasks.

2 Related Work

2.1 Fine-Tuning on Pre-trained Language Model

Language model pre-training has attracted wide attention and fine-tuning on pre-trained language model has shown to be effective for improving many downstream natural language processing tasks. Dai [2] pre-trained unlabeled data to improve Sequence Learning with recurrent networks. Howard [8] proposed a general transfer learning method, Universal Language Model Fine-tuning (ULMFiT), with the key techniques for fine-tuning a language model. Radford [23] proposed that by generative pre-training of a language model on a diverse corpus of unlabeled text, large gains on a diverse range of tasks could be realized. Radford [23] achieved large improvements on many sentence-level tasks from the GLUE benchmark [29]. BERT [4] obtained new state-of-the-art results on a broad range of diverse tasks. BERT pre-trained deep bidirectional representations which jointly conditioned on both left and right context in all layers, following by discriminative fine-tuning on each specific task. Unlike previous works fine-tuning pre-trained language model to perform discriminative tasks, we aim to apply pre-trained BERT on generative tasks by perform the masked language model (MLM) task. To generate sentences that are compatible with given labels, we retrofit BERT to conditional BERT, by introducing a conditional masked language model task and fine-tuning BERT on the task.

2.2 Text Data Augmentation

Text data augmentation has been extensively studied in natural language processing. Sample-based methods includes downsampling from the majority classes and oversampling from the minority class, both of which perform weakly in practice. Generation-based methods employ deep generative models such as GANs [7] or VAEs [1,9], trying to generate sentences from a continuous space with desired attributes of sentiment and tense. However, sentences generated in these methods are very hard to guarantee the quality both in label compatibility and sentence readability. In some specific areas [5,11,32]. word replacement augmentation was applied. Wang [30] proposed the use of neighboring words in continuous representations to create new instances for every word in a tweet to augment the training dataset. Zhang [34] extracted all replaceable words from the given text and randomly choose r of them to be replaced, then substituted the replaceable words with synonyms from WordNet [19]. Kolomiyets [16] replaced only the headwords under a task-specific assumption that temporal trigger words usually occur as headwords. Kolomiyets [16] selected substitute words with top-K scores given by the Latent Words LM [3], which is a LM based on fixed length contexts. Fadaee [6] focused on the rare word problem in machine translation, replacing words in a source sentence with only rare words. A word in the translated sentence is also replaced using a word alignment method and a rightward LM. The work most similar to our research is Kobayashi [15]. Kobayashi used a fill-in-the-blank context for data augmentation by replacing every words in the sentence with language model. In order to prevent the generated words from reversing the information related to the labels of the sentences, Kobayashi [15] introduced a conditional constraint to control the replacement of words. Unlike previous works, we adopt a deep bidirectional language model to apply replacement, and the attention mechanism within our model allows a more structured memory for handling long-term dependencies in text, which resulting in more general and robust improvement on various downstream tasks.

3 Conditional BERT Contextual Augmentation

3.1 Preliminary: Masked Language Model Task

Bidirectional Language Model. In general, the language model (LM) models the probability of generating natural language sentences or documents. Given a sequence $\boldsymbol{S}$ of N tokens, $<t_1, t_2, ..., t_N>$, a forward language model allows us to predict the probability of the sequence as:

$$p(t_1, t_2, ..., t_N) = \prod_{i=1}^{N} p(t_i | t_1, t_2, ..., t_{i-1}). \tag{1}$$

Similarly, a backward language model allows us to predict the probability of the sentence as:

$$p(t_1, t_2, ..., t_N) = \prod_{i=1}^{N} p(t_i | t_{i+1}, t_{i+2}, ..., t_N). \tag{2}$$

Traditionally, a bidirectional language model a shallow concatenation of independently trained forward and backward LMs.

Masked Language Model Task. In order to train a deep bidirectional language model, BERT proposed Masked Language Model (MLM) task, which was also referred to Cloze Task [28]. MLM task randomly masks some percentage of the input tokens, and then predicts only those masked tokens according to their context. Given a masked token t_i, the context is the tokens surrounding token t_i in the sequence $\boldsymbol{S}$, i.e., cloze sentence $\boldsymbol{S}\backslash\{t_i\}$. The final hidden vectors corresponding to the mask tokens are fed into an output softmax over the vocabulary to produce words with a probability distribution $p(\cdot|\boldsymbol{S}\backslash\{t_i\})$. MLM task only predicts the masked words rather than reconstructing the entire input, which suggests that more pre-training steps are required for the model to converge. Pre-trained BERT can augment sentences through MLM task, by predicting new words in masked positions according to their context.

3.2 Conditional BERT

As shown in Fig. 1, our conditional BERT shares the same model architecture with the original BERT. The differences are the input representation and training procedure.

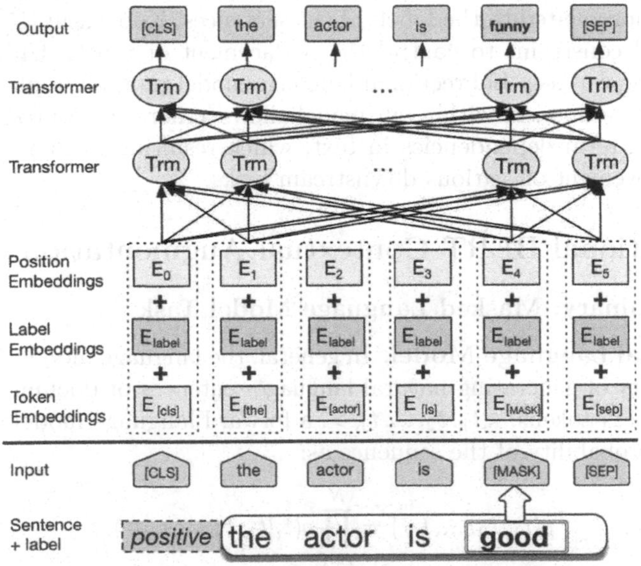

Fig. 1. Model architecture of conditional BERT. The label embeddings in conditional BERT corresponding to segmentation embeddings in BERT, but their functions are different.

The input embeddings of BERT are the sum of the token embeddings, the segmentation embeddings and the position embeddings. For the segmentation embeddings in BERT, a learned sentence A embedding is added to every token of the first sentence, and if a second sentence exists, a sentence B embedding will be added to every token of the second sentence. However, the segmentation embeddings has no connection to the actual annotated labels of a sentence, like sense, sentiment or subjectivity, so predicted word is not always compatible with annotated labels. For example, given a positive movie remark "this actor is good", we have the word"good" masked. Through the Masked Language Model task by BERT, the predicted word in the masked position has potential to be negative word likes "bad" or "boring". Such new generated sentences by substituting masked words are implausible with respect to their original labels, which will be harmful if added to the corpus to apply augmentation. In order to address this issue, we propose a new task: "conditional masked language model".

Conditional Masked Language Model. The conditional masked language model randomly masks some of the tokens from the labeled sentence, and the objective is to predict the original vocabulary index of the masked word based on both its context and its label. Given a masked token t_i, the context $\boldsymbol{S}\backslash\{t_i\}$ and label y are both considered, aiming to calculate $p(\cdot|y, \boldsymbol{S}\backslash\{t_i\})$, instead of calculating $p(\cdot|\boldsymbol{S}\backslash\{t_i\})$. Unlike MLM pre-training, the conditional MLM objective allows the representation to fuse the context information and the label information, which allows us to further train a label-conditional deep bidirectional representations.

To perform conditional MLM task, we fine-tune on pre-trained BERT. We alter the segmentation embeddings to label embeddings, which are learned corresponding to their annotated labels on labeled datasets. Note that the BERT are designed with segmentation embedding being embedding A or embedding B, so when a downstream task dataset with more than two labels, we have to adapt the size of embedding to label size compatible. We train conditional BERT using conditional MLM task on labeled dataset. After the model has converged, it is expected to be able to predict words in masked position both considering the context and the label.

3.3 Conditional BERT Contextual Augmentation

After the conditional BERT is well-trained, we utilize it to augment sentences. Given a labeled sentence from the corpus, we randomly mask a few words in the sentence. Through conditional BERT, various words compatibly with the label of the sentence are predicted by conditional BERT. After substituting the masked words with predicted words, a new sentences is generated, which shares similar context and same label with original sentence. Then new sentences are added to original corpus. We elaborate the entire process in Algorithm 1.

Algorithm 1. Conditional BERT contextual augmentation algorithm. Fine-tuning on the pre-trained BERT, we retrofit BERT to conditional BERT using conditional MLM task on labeled dataset. After the model converged, we utilize it to augment sentences. New sentences are added into dataset to augment the dataset.

1: Alter the segmentation embeddings to label embeddings
2: Fine-tune the pre-trained BERT using conditional MLM task on labeled dataset D until convergence
3: **for** each iteration i=1,2,...,M **do**
4: Sample a sentence s from D
5: Randomly mask k words
6: Using fine-tuned conditional BERT to predict label-compatible words on masked positions to generate a new sentence S'
7: **end for**
8: Add new sentences into dataset D to get augmented dataset D'
9: Perform downstream task on augmented dataset D'

4 Experiment

In this section, we present conditional BERT parameter settings and, following Kobayashi [15], we apply different augmentation methods on two types of neural models through six text classification tasks. The pre-trained BERT model we used in our experiment is BERT_{BASE}, with number of layers (i.e., Transformer blocks) $L = 12$, the hidden size $H = 768$, and the number of self-attention heads $A = 12$, total parameters $= 110M$. Detailed pre-train parameters setting can be found in original paper [4]. For each task, we perform the following steps independently. First, we evaluate the augmentation ability of original BERT model pre-trained on MLM task. We use pre-trained BERT to augment dataset, by predicted masked words only condition on context for each sentence. Second, we fine-tune the original BERT model to a conditional BERT. Well-trained conditional BERT augments each sentence in dataset by predicted masked words condition on both context and label. Third, we compare the performance of the two methods with Kobayashi's [15] contextual augmentation results.

4.1 Datasets

Six benchmark classification datasets are listed in Table 1. Following Kim [12], for a dataset without validation data, we use 10% of its training set for the validation set. Summary statistics of six classification datasets are shown in Table 1.

SST [25] SST (Stanford Sentiment Treebank) is a dataset for sentiment classification on movie reviews, which are annotated with five labels (SST5: very positive, positive, neutral, negative, or very negative) or two labels (SST2: positive or negative).

Table 1. Summary statistics for the datasets after tokenization. c: Number of target classes. l: Average sentence length. N: Dataset size. $|V|$: Vocabulary size. $Test$: Test set size (CV means there was no standard train/test split and thus 10-fold cross-validation was used).

Data	c	l	N	$\|V\|$	$Test$
SST5	5	18	11855	17836	2210
SST2	2	19	9613	16185	1821
Subj	2	23	10000	21323	CV
TREC	6	10	5952	9592	500
MPQA	2	3	10606	6246	CV
RT	2	21	10662	20287	CV

Subj [20] Subj (Subjectivity dataset) is annotated with whether a sentence is subjective or objective.
MPQA [31] MPQA Opinion Corpus is an opinion polarity detection dataset of short phrases rather than sentences, which contains news articles from a wide variety of news sources manually annotated for opinions and other private states (i.e., beliefs, emotions, sentiments, speculations, etc.).
RT [21] RT is another movie review sentiment dataset contains a collection of short review excerpts from Rotten Tomatoes collected by Bo Pang and Lillian Lee.
TREC [18] TREC is a dataset for classification of the six question types (whether the question is about person, location, numeric information, etc.).

4.2 Text Classification

Sentence Classifier Structure. We evaluate the performance improvement brought by conditional BERT contextual augmentation on sentence classification tasks, so we need to prepare two common sentence classifiers beforehand. For comparison, following Kobayashi [15], we adopt two typical classifier architectures: CNN or LSTM-RNN. The CNN-based classifier [12] has convolutional filters of size 3, 4, 5 and word embeddings. All outputs of each filter are concatenated before applied with a max-pooling over time, then fed into a two-layer feed-forward network with ReLU, followed by the softmax function. An RNN-based classifier has a single layer LSTM and word embeddings, whose output is fed into an output affine layer with the softmax function. For both the architectures, dropout [26] and Adam optimization [13] are applied during training. The train process is finish by early stopping with validation at each epoch.

Hyper-parameters Setting. Sentence classifier hyper-parameters including learning rate, embedding dimension, unit or filter size, and dropout ratio, are selected using grid-search for each task-specific dataset. We refer to Kobayashi's implementation in original paper. For BERT, all hyper-parameters are kept the

Table 2. Accuracies of different methods for various benchmarks on two classifier architectures. C-BERT, which represents conditional BERT, performs best on two classifier structures over six datasets. "w/" represents "with", lines marked with "*" are experiments results from Kobayashi [15].

Model	SST5	SST2	Subj	MPQA	RT	TREC	Avg.
CNN*	41.3	79.5	92.4	86.1	75.9	90.0	77.53
w/synonym*	40.7	80.0	92.4	86.3	76.0	89.6	77.50
w/context*	41.9	80.9	92.7	86.7	75.9	90.0	78.02
w/context+label*	42.1	80.8	93.0	86.7	76.1	90.5	78.20
w/BERT	41.5	81.9	92.9	87.7	78.2	91.8	79.00
w/C-BERT	**42.3**	**82.1**	**93.4**	**88.2**	**79.0**	**92.6**	**79.60**
RNN*	40.2	80.3	92.4	86.0	76.7	89.0	77.43
w/synonym*	40.5	80.2	92.8	86.4	76.6	87.9	77.40
w/context*	40.9	79.3	92.8	86.4	77.0	89.3	77.62
w/context+label*	41.1	80.1	92.8	86.4	77.4	89.2	77.83
w/BERT	41.3	81.4	93.5	87.3	78.3	89.8	78.60
w/C-BERT	**42.6**	**81.9**	**93.9**	**88.0**	**78.9**	**91.0**	**79.38**

same as Devlin [4]. The number of conditional BERT training epochs ranges in [1–50] and number of masked words ranges in [1–2].

Baselines. We compare the performance improvements obtained by our proposed method with the following baseline methods, "w/" means "with":

- w/synonym: Words are randomly replaced with synonyms from WordNet [19].
- w/context: Proposed by Kobayashi [15], which used a bidirectional language model to apply contextual augmentation, each word was replaced with a probability.
- w/context+label: Kobayashi's contextual augmentation method [15] in a label-conditional LM architecture.

Experiment Results. Table 2 lists the accuracies of the all methods on two classifier architectures. The results show that, for various datasets on different classifier architectures, our conditional BERT contextual augmentation improves the model performances most. BERT can also augments sentences to some extent, but not as much as conditional BERT does. For we masked words randomly, the masked words may be label-sensitive or label-insensitive. If label-insensitive words are masked, words predicted through BERT may not be compatible with original labels. The improvement over all benchmark datasets also shows that conditional BERT is a general augmentation method for multi-labels sentence classification tasks.

Effect of Number of Fine-Tuning Steps. We also explore the effect of number of training steps to the performance of conditional BERT data augmentation. The fine-tuning epoch setting ranges in [1–50], we list the fine-tuning epoch of conditional BERT to outperform BERT for various benchmarks in Table 3. The results show that our conditional BERT contextual augmentation can achieve obvious performance improvement after only a few fine-tuning epochs, which is very convenient to apply to downstream tasks.

Table 3. Fine-tuning epochs of conditional BERT to outperform BERT for various benchmarks

Model	SST5	SST2	Subj	MPQA	RT	TREC
CNN	4	3	1	2	2	1
RNN	6	2	2	2	1	1

5 Connection to Style Transfer

In this section, we further dip into the connection to style transfer and apply our well trained conditional BERT to style transfer task. Style transfer is defined as the task of rephrasing the text to contain specific stylistic properties without changing the intent or affect within the context [22]. Our conditional MLM task changes words in the text condition on given label without changing the context. View from this point, the two tasks are very close. So in order to apply conditional BERT to style transfer task, given a specific stylistic sentence, we break it into two steps: first, we find the words relevant to the style; second, we mask the style-relevant words, then use conditional BERT to predict new substitutes with sentence context and target style property. In order to find style-relevant words in a sentence, we refer to Xu [33], which proposed an attention-based method to extract the contribution of each word to the sentence sentimental label. For example, given a positive movie remark "This movie is funny and interesting", we filter out the words that contribute largely to the label and mask them. Then

Table 4. Examples generated by conditional BERT on the SST2 dataset. To perform style transfer, we reverse the original label of a sentence, and conditional BERT output a new label compatible sentence.

Original:	there's no disguising this as one of the worst films of the summer
Generated:	there's no disguising this as one of the best films of the summer
Original:	it's probably not easy to make such a worthless film ...
Generated:	it's probably not easy to make such a stunning film ...
Original:	woody allen has really found his groove these days
Generated:	woody allen has really lost his groove these days

through our conditional BERT contextual augmentation method, we fill in the masked positions by predicting words conditioning on opposite label and sentence context, resulting in "This movie is boring and dull". The words "boring" and "dull" contribute to the new sentence being labeled as negative style. We sample some sentences from dataset SST2, transferring them to the opposite label, as listed in Table 4.

6 Conclusions and Future Work

In this paper, we fine-tune BERT to conditional BERT by introducing a novel conditional MLM task. After being well trained, the conditional BERT can be applied to data augmentation for sentence classification tasks. Experiment results show that our model outperforms several baseline methods obviously. Furthermore, we demonstrate that our conditional BERT can also be applied to style transfer task. In the future, (1) We will explore how to perform text data augmentation on imbalanced datasets with pre-trained language model, (2) we believe the idea of conditional BERT contextual augmentation is universal and will be applied to paragraph or document level data augmentation.

References

1. Bowman, S.R., Vilnis, L., Vinyals, O., Dai, A.M., Jozefowicz, R., Bengio, S.: Generating sentences from a continuous space. arXiv preprint arXiv:1511.06349 (2015)
2. Dai, A.M., Le, Q.V.: Semi-supervised sequence learning, pp. 3079–3087 (2015)
3. Deschacht, K., Moens, M.F.: Semi-supervised semantic role labeling using the latent words language model, pp. 21–29 (2009)
4. Devlin, J., Chang, M.W., Lee, K., Toutanova, K.: Bert: Pre-training of deep bidirectional transformers for language understanding. arXiv preprint arXiv:1810.04805 (2018)
5. Ebrahimi, J., Rao, A., Lowd, D., Dou, D.: Hotflip: White-box adversarial examples for NLP. arXiv preprint arXiv:1712.06751 (2017)
6. Fadaee, M., Bisazza, A., Monz, C.: Data augmentation for low-resource neural machine translation. arXiv preprint arXiv:1705.00440 (2017)
7. Goodfellow, I., et al.: Generative adversarial nets, pp. 2672–2680 (2014)
8. Howard, J., Ruder, S.: Universal language model fine-tuning for text classification, vol. 1, pp. 328–339 (2018)
9. Hu, Z., Yang, Z., Liang, X., Salakhutdinov, R., Xing, E.P.: Toward controlled generation of text. arXiv preprint arXiv:1703.00955 (2017)
10. Jaitly, N., Hinton, G.E.: Vocal tract length perturbation (VTLP) improves speech recognition, vol. 117 (2013)
11. Jia, R., Liang, P.: Adversarial examples for evaluating reading comprehension systems. arXiv preprint arXiv:1707.07328 (2017)
12. Kim, Y.: Convolutional neural networks for sentence classification. arXiv preprint arXiv:1408.5882 (2014)
13. Kingma, D.P., Ba, J.: Adam: A method for stochastic optimization. arXiv preprint arXiv:1412.6980 (2014)

14. Ko, T., Peddinti, V., Povey, D., Khudanpur, S.: Audio augmentation for speech recognition (2015)
15. Kobayashi, S.: Contextual augmentation: Data augmentation by words with paradigmatic relations. arXiv preprint arXiv:1805.06201 (2018)
16. Kolomiyets, O., Bethard, S., Moens, M.F.: Model-portability experiments for textual temporal analysis, pp. 271–276 (2011)
17. Krizhevsky, A., Sutskever, I., Hinton, G.E.: Imagenet classification with deep convolutional neural networks, pp. 1097–1105 (2012)
18. Li, X., Roth, D.: Learning question classifiers, pp. 1–7 (2002)
19. Miller, G.A.: Wordnet: a lexical database for English. Commun. ACM **38**(11), 39–41 (1995)
20. Pang, B., Lee, L.: A sentimental education: sentiment analysis using subjectivity summarization based on minimum cuts, p. 271 (2004)
21. Pang, B., Lee, L.: Seeing stars: exploiting class relationships for sentiment categorization with respect to rating scales, pp. 115–124 (2005)
22. Prabhumoye, S., Tsvetkov, Y., Salakhutdinov, R., Black, A.W.: Style transfer through back-translation. arXiv preprint arXiv:1804.09000 (2018)
23. Radford, A., Narasimhan, K., Salimans, T., Sutskever, I.: Improving language understanding by generative pre-training (2018). https://s3-us-west-2.amazonaws.com/openai-assets/research-covers/language-unsupervised/language_understanding_paper.pdf
24. Simard, P.Y., LeCun, Y.A., Denker, J.S., Victorri, B.: Transformation invariance in pattern recognition—tangent distance and tangent propagation. In: Orr, G.B., Müller, K.-R. (eds.) Neural Networks: Tricks of the Trade. LNCS, vol. 1524, pp. 239–274. Springer, Heidelberg (1998). https://doi.org/10.1007/3-540-49430-8_13
25. Socher, R., et al.: Recursive deep models for semantic compositionality over a sentiment treebank, pp. 1631–1642 (2013)
26. Srivastava, N., Hinton, G., Krizhevsky, A., Sutskever, I., Salakhutdinov, R.: Dropout: a simple way to prevent neural networks from overfitting. J. Mach. Learn. Res. **15**(1), 1929–1958 (2014)
27. Szegedy, C., et al.: Going deeper with convolutions, pp. 1–9 (2015)
28. Taylor, W.L.: "cloze procedure": a new tool for measuring readability. Journ. Bull. **30**(4), 415–433 (1953)
29. Wang, A., Singh, A., Michael, J., Hill, F., Levy, O., Bowman, S.R.: Glue: A multi-task benchmark and analysis platform for natural language understanding. arXiv preprint arXiv:1804.07461 (2018)
30. Wang, W.Y., Yang, D.: That's so annoying!!!: a lexical and frame-semantic embedding based data augmentation approach to automatic categorization of annoying behaviors using# petpeeve tweets, pp. 2557–2563 (2015)
31. Wiebe, J., Wilson, T., Cardie, C.: Annotating expressions of opinions and emotions in language. Lang. Resour. Eval. **39**(2–3), 165–210 (2005)
32. Xie, Z., et al.: Data noising as smoothing in neural network language models. arXiv preprint arXiv:1703.02573 (2017)
33. Xu, J., et al.: Unpaired sentiment-to-sentiment translation: A cycled reinforcement learning approach. arXiv preprint arXiv:1805.05181 (2018)
34. Zhang, X., Zhao, J., LeCun, Y.: Character-level convolutional networks for text classification, pp. 649–657 (2015)

An Innovative and Reliable Water Leak Detection Service Supported by Data-Intensive Remote Sensing Processing

Ricardo Martins[1(✉)], Alberto Azevedo[1(✉)], André B. Fortunato[1(✉)], Elsa Alves[1(✉)], Anabela Oliveira[1(✉)], and Alexandra Carvalho[2(✉)]

[1] LNEC – National Laboratory for Civil Engineering/Hydraulics and Environment Department, Lisbon, Portugal
{rjmartins, aazevedo, afortunato, ealves, aoliveira}@lnec.pt

[2] EDIA - Empresa de Desenvolvimento e de Infraestruturas do Alqueva, Beja, Portugal
acarvalho@edia.pt

Abstract. The WADI project (Water-tightness Airborne Detection Implementation), integrated within the H2020 initiative, is developing an airborne water leak detection surveillance service, based on manned and unmanned aerial vehicles. This service aims to provide water utilities with adequate information on leaks in large water distribution infrastructures outside urban areas. Given the high cost associated with water infrastructure networks repairs, a reliability layer is necessary to improve the trustworthiness of the WADI leak identification procedure, based on complementary technologies for leak detection. Herein, a methodology based on the combined use of Sentinel remote sensing data and a water leak pathways model is presented, based on data-intensive computing. The resulting water leak detection reliability service, provided to the users through a web interface, targets prompt and cost-effective infrastructure repairs with the required degree of confidence on the detected leaks. The web platform allows for both data analysis and visualization of Sentinel images and relevant leak indicators at the sites selected by the user. The user can also provide aerial imagery inputs, to be processed together with Sentinel remote sensing data at the satellite acquisition dates identified by the user. The platform provides information about the detected leaks location and time evolution, and will be linked in the future with the outputs from water pathway models.

Keywords: Remote sensing · Water leak service · Data-intensive computing · HPC

1 Introduction

The shortage of freshwater is a growing concern [1]. This shortage has a variety of causes, such as the increase of droughts due to climate change, the population growth, the pollution of water bodies, and the production of crops for biofuel. This problem calls for practices that improve water preservation. In this scope, water losses associated with large distribution networks, frequently used for irrigation purposes, constitute a major concern, given the difficulty to pinpoint the location of leaks over very long

J. M. F. Rodrigues et al. (Eds.): ICCS 2019, LNCS 11539, pp. 96–108, 2019.
https://doi.org/10.1007/978-3-030-22747-0_8

pipes, often of difficult physical access [2]. As a result, considerable water losses occur due to long leak detection times. The costs associated with conventional detection techniques have promoted the development of methodologies for water leak detection based on remote sensing outside of urban areas [3].

Remote sensing imagery data have become an essential tool for the scientific community to study the Earth in many disciplines, such as coastline detection, coastal morphology or marine oil spills follow-up [4–6]. Remote sensing is a data-intensive technology due to the typical resolution and size of the images. With the increase of the computational power and finer and more frequent image gathering, available through public access such as Sentinel satellites outputs, the conditions are met to build information systems to address societal needs for the reduction of water losses. Data-Intensive Scientific Discovery (DISD) tools are one of the avenues to handle the high volumes of remote sensing data, combine them with models to build intelligent services and make them available to society.

Recently, remote sensing has been used to provide water leak detection services at different spatial and temporal scales. Satellite-based leak detection services such as UTILIS [7] have been applied in several types of water leak detection, but their efficiency is limited when the rate of the leak is small (typically about 60% for flow rates under 1 m^3/h). Given the difficulty to reach and repair large water mains, leaks should be detected at their early stages, so a technology that provides better performance for small flows is necessary. A different approach was proposed in the project WADI. In this project, an airborne water leak detection surveillance technology was developed. It consists in coupling and optimizing off-the-shelf optical remote sensing devices (multispectral and infrared cameras) and applying them in two complementary aerial platforms (manned and unmanned). The WADI technology provides the detection of leaks on the acquired multispectral or/and infrared images, a data distribution system for the delivery of the geospatial data to the water utilities, and feedback on service performance. It was validated in an operational environment, to detect small water leaks in the Societé du Canal de Provence infrastructure in France [8]. Being developed with simple and off-the-shelf hardware and fast to operationalize, this technology has the potential to become the leading choice for small water leak detections.

While the performance evaluation of this new service is still underway, its reliability can be strengthened with complementary methodologies that can be used on top of the WADI methodology to reduce false positives and negatives, thus reducing costs and improving efficiency. Herein, a reliability methodology is proposed to be used with two distinct goals: to motivate the use of the WADI service (as a preliminary detection service) and to complement it after the initial leak detection, improving its performance. This new reliability methodology combines global data with local characteristics and merges them in a single evaluation available through a user-friendly web interface. The reliability layer is built by analyzing indexes based on Sentinel images and further complemented with a terrain-following water pathways model. This approach differs from the UTILIS system, based on microwave reflectometry, that works well at any time of the day or night, by being able to travel through atmospheric interferences such as clouds, dust particles, and aerosols.

This paper is organized as follows. Section 2 provides an overview of the WADI detection procedure and highlights its strengths and main characteristics. Section 3 presents the general reliability strategy while Sect. 4 is dedicated to describe in detail the Sentinel remote sensing-based analysis. Section 5 describes the WADI reliability web-portal and its architecture. Finally, Sect. 6 summarizes the conclusions and anticipates the forthcoming work, in particular highlighting how DISD and other approaches can be used in the overall WADI leak detection service.

2 Overview of the WADI Core Water Leak Procedure

One of the main challenges of Horizon 2020 (H2020) is to reduce losses in the water distribution systems [9] and to promote efficiency and resilience on today's society against climate changes. Water availability is already a concern in the context of climate change, affecting most of the continents and in many areas. There is a need to develop better methodologies for easier and faster leak detection in the water systems since the existing ones are unreliable, time-consuming and expensive.

The WADI project, integrated within the H2020 initiative, aims at developing an airborne water leak detection surveillance service to provide water utilities with adequate information on leaks in water infrastructures outside urban areas, thus enabling prompt and cost-effective repairs. WADI's innovative concept consists in coupling and optimizing off-the-shelf optical remote sensing devices (multispectral and infrared cameras) and applying them on two complementary aerial platforms (manned and unmanned) in an operational environment [8]. The WADI service offers several benefits to infrastructure stakeholders: the reduction of water and energy consumption and emissions caused by the leak, a long-term plan for water monitoring in difficult physical access conditions, and the adaptability for accurate and tailored leak detection in water transmission systems.

To provide additional confidence to the WADI service and to support its performance analysis, a reliability layer was included in this service, through the use of complementary methods and data. While aerial images have the advantage of being reliable and precise on quality and image resolution, their application requires dedicated resources and planning for a specific site. Publicly available satellite images have a lower resolution, but provide a continuous stream of data over time that can be used for preliminary identification of leaks. Satellite-based analysis can thus support the application of the WADI service over the areas where leaks are suspected to exist, to direct the airborne leak detection efforts to the most likely leak locations. In addition, satellite data can also be used afterwards to provide added confidence on the aerial detection. This is achieved by taking advantage of the evolution in time of the water and vegetation indexes, combined with water pathways models applied from the network location downhill.

The feasibility of the surveillance service developed by the WADI project was tested in real representative conditions through water leak detection campaigns in the Provence region (France) [8] and will be applied to the Alqueva infrastructure (Portugal) in the coming months.

3 Leak-Detection Reliability Procedure: General Description

The following procedure aims to take advantage of historical, high-frequency satellite images to improve the reliability of WADI's leak detection service. Each set of images (WADI's airborne images and Sentinel 2 images) has its own advantages relative to the other. The former has a higher resolution and is typically obtained under optimal meteorological conditions. In contrast, the latter is free, frequent (every 2–3 days) and continuously available since 2016. These different advantages suggest the possibility of combining the two sources of information to optimize the WADI service.

Information about possible leaks is obtained by analyzing the satellite images in both space and time. The analysis is performed on indexes that describe the humidity of the soil, such as the Normalized Difference Water Index (NDWI), which are computed from the remote sensing images. These indexes are described in the next section.

In the first step, images are analyzed in space to detect possible leaks. The spatial variability is assessed based on the Laplacian (i.e., the sum of the second derivatives along the two horizontal axes) of the water indexes. The signature of a leak is expected to correspond to a local minimum of this Laplacian. The algorithm thus searches for the largest minima. Furthermore, the leaks are expected to produce a signature in the water index very close to their location. Hence, the distance between a local extreme in the water index and the nearest pipe or channel provides the means to eliminate false positives. The next section describes in detail the implementation of this first step. In the future, this algorithm will be improved using the topography data and a water pathway model to further eliminate false negatives.

The second step of the procedure is applied to the points identified in the first step and takes advantage of the wealth of historical Sentinel data. In general, the water index at a given location is expected to follow a seasonal pattern, associated with rainfall or irrigation procedures (Fig. 1). This seasonal pattern is first determined by averaging the yearly signals (climatology) and then removed from the signal. The difference between the initial signal and the climatology, denoted anomaly, is expected to have a smaller temporal variability than the initial signal. Hence, abrupt increases in the water index are easier to identify. These abrupt changes can potentially correspond to the initiation of the leaks.

The outcome of this procedure, which is detailed and illustrated in the following section, is thus a list of the locations most likely to have water leaks. These locations can be ordered based on the likelihood of a leak, measured by the value of the Laplacian of the water indexes. For each point in the list, the date when the potential leak is first detectable can also be identified. This information can be helpful to determine possible reasons to explain the occurrence, whether or not it actually corresponds to a leak. For instance, the date can correspond to an intervention in a pipe which accidentally caused a leak that went unnoticed.

Such a list can be used *a priori*, to help reduce the effort of searching for leaks in the field, by concentrating the efforts in areas where leaks are most likely; or it can be used *a posteriori*, integrated in the workflow of analysis of WADI's images, to provide an additional layer of information for improved reliability. By linking these detections to a water pathway model starting at the network locations, the likelihood of a leak at that location can be corrected. For instance, if the location is far from the network or located uphill from the network, the probability that it corresponds to a leak decreases. This paper focuses on the satellite methodology, supported by Sentinel 2 images, and illustrates it at the second case study of the WADI project - the Monte Novo area in the Alqueva water distribution infrastructure (Portugal).

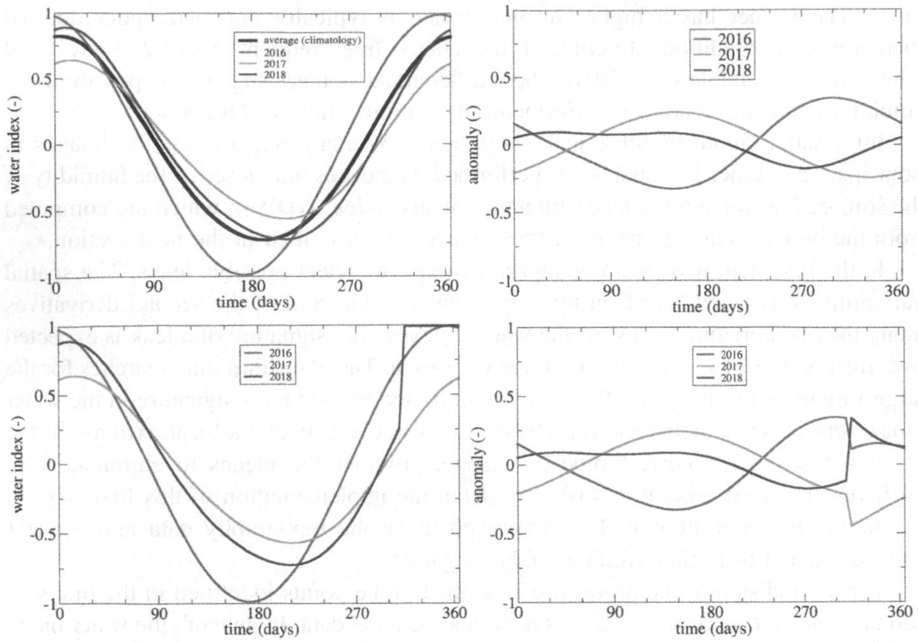

Fig. 1. Theoretical water indexes, climatology (left) and anomalies (right), for a case without (top) and with (bottom) a leak in day 315 in 2018. The leak is identifiable by the strong negative second time derivative of the 2018 anomaly. (Color figure online)

4 Methodology for the Water Leak Detection Based on Sentinel 2 Images

The methodology for processing Sentinel images in order to detect water leaks is described in detail in this section and is based on three parts. In the first part, the water indexes of all the available Sentinel 2 image sets at the area of study are processed. Normalized Difference Water Index (NDWI) and Modified Normalized Difference Water Index (MNDWI) are the two most popular water indexes available for remote sensing methodology processing.

The NDWI was proposed by McFeeters [10] to delineate open water features and enhance their presence in remotely-sensed digital imagery, by using reflected Near Infrared (NIR) radiation and visible green light to enhance the presence of such features while eliminating the presence of soil and terrestrial vegetation features:

$$NDWI = \frac{Green - NIR}{Green + NIR} \tag{1}$$

Xu [11] determined that, while using Landsat imagery, the application of the NDWI in water regions with a built-up land background does not achieve its goal as expected because the extracted water information in those regions is often mixed with built-up land noise. This author proposed the MNDWI, a modified version of the McFeeters' NDWI, that uses Middle Infrared (MIR) instead of Near Infrared to enhance the water detection and even remove built-up land noise as well as vegetation and soil noise. The MNDWI is selected as main water index, since the application detects water leaks on land:

$$MNDWI = \frac{Green - MIR}{Green + MIR} \quad (2)$$

In the second part, the climatology of the indexes is processed, as explained in the previous section. In this case, the climatology consists in a set of images averaging all the images for the same day of the year for all available years (see Fig. 2).

Fig. 2. An example of a climatology image of the MNDWI. The light colors correspond to water bodies

Possible leaks are determined in the third part. The user selects a Sentinel image set for a specific date, and the service starts automatically processing the water index and resamples it to the user-selected region. Because images are not available every day, the anomaly is computed as the difference between the image for the user-selected date and the climatology image for the closest day available in the climatology dataset. The most likely locations for the leaks are identified by applying the Laplacian operator (second derivative) to the anomaly, which is estimated by finite differences as:

$$Q_{i,j} = \frac{\left(P_{i-1,j} - 2P_{i,j} + P_{i+1,j}\right)}{dx^2} + \frac{\left(P_{i,j-1} - 2P_{i,j} + P_{i,j+1}\right)}{dy^2} \tag{3}$$

where P is the anomaly image, Q is the second derivative image, i and j represent the pixel position and dx and dy represent the pixel size, corresponding to 10 m in the Sentinel images. The lowest values of the Laplacian image are marked as possible leaks (Fig. 3).

These locations are then confronted with the water network location and the possibility of water (from a leak) traveling from the network to the hypothetical leak location.

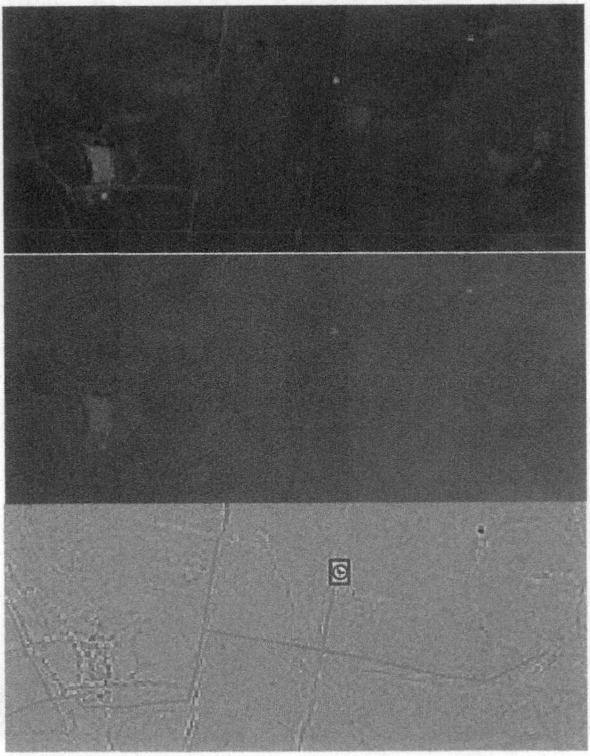

Fig. 3. Example of leak detection: MNDWI (top), Anomaly (middle), Laplacian and possible leaks (bottom) of a small area of the AoS. The network location is shown in orange. (Color figure online)

However, there are some caveats that must be taken into account in this methodology. The first issue is how to process and show water indexes of an area of study where two or more image sets overlap each other in space for the same date. This problem was addressed by merging the water index of the various images into a single image.

A second issue that must be addressed is the different levels of image acquisition in Sentinel 2 images: Level-1C (L1C) and Level-2A (L2A). The algorithm developed herein uses by default the L2A images, which are referenced to the Bottom of Atmosphere through the application of an atmospheric correction. When no L2A images are available in the area of study, the algorithm automatically downloads the L1C images and applies the atmospheric correction of the SNAP toolbox, developed by ESA.

5 A Web Interface for the Reliability Layer in WADI

5.1 Functionalities of the Web Interface

A web-portal interface was developed to facilitate the usage of the reliability methodology. This web-portal is developed in Django, a Python web framework to create web applications with a frontend and a backend. The frontend is the presentation layer of the web application, through which the framework loads the pages for the user to interact with and requests data from them. The backend is the 'engine' running on the back of the portal to attend and process user requests.

The WADI reliability web-portal is structured in "workspaces" and "areas of study" (AoS). Each user can define several workspaces, that typically correspond to a region of interest for water leak analysis. A workspace is defined through a map where users can interact and create/edit/delete areas of study. An AoS is a small area defined by the user on that map to process images and visualize results. In each workspace, there are two panels to interact. The top panel is a menu bar where the user configures the region for leak detection analysis (create/edit/delete/select the AoS) and then proceeds to the functions 'climatology'/'leak detections'. The right panel organizes the analysis performed by a user under a workspace through a list of previously created AoS. Each AoS has some options of actions to be performed by the user. Figure 4 illustrates the typical roadmap for a leak detection study.

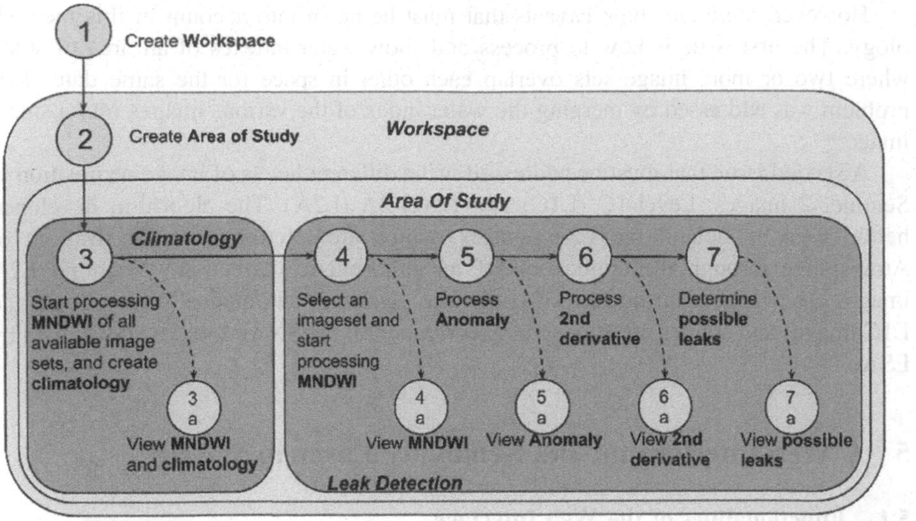

Fig. 4. Sequential steps for creating a leak detection map in an area of study. (Color figure online)

First, the user must create a workspace. Then, in the 'climatology' page, the user selects "Create Area of Study", defines the region to process on the map and names it (see Fig. 5). The new AoS will appear on the right panel.

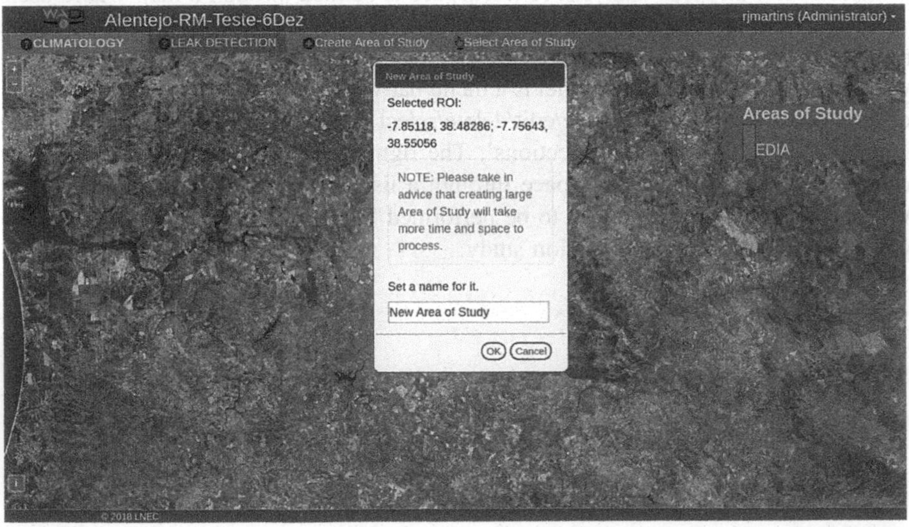

Fig. 5. Climatology overview - create new area of study. (Color figure online)

In order to interact with the AoS and start the image processing, the user selects the AoS on the map and selects “See Climatology Status” on the right panel. A popup is shown to the user with the list of available Sentinel images for processing the climatology (Fig. 6). The user then clicks on “Calculate Climatology”, and the WADI processing service starts processing the images asynchronously. Processing the climatology takes a considerable time, as all the available Sentinel images for that region are processed. The future usage of speed-up strategies like HPC is very important when processing large AoS or many images simultaneously.

On the ‘climatology’ page, the user can see the AoS’ processed indices and climatology and check its status and availability for new Sentinel products available at the ESA Sentinel repository.

After the climatology processing, the user can start validating the data by viewing the processed climatology images and/or checking its pixel value. To check the pixel values, the user selects “Pixel Climatology” and clicks on a point inside the AoS on the map. A balloon pointing to that point is loaded with a chart representing the pixel values of the processed water index images for each day of the year.

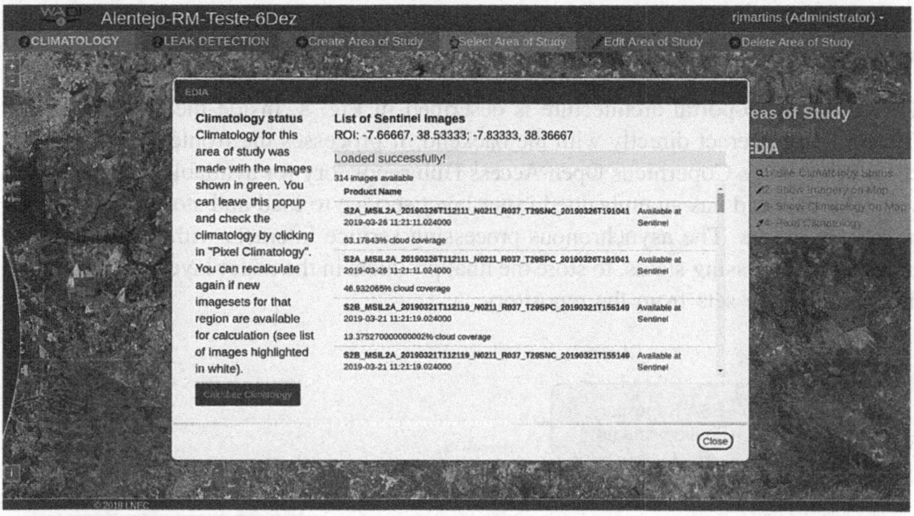

Fig. 6. Climatology overview - list sentinel images available for climatology processing. (Color figure online)

With the processed climatology, the user can start the leak detection for that AoS. In each step, the user checks the results and decides whether the processing should continue. The user selects a Sentinel imageset from the repository, and the service automatically initiates the water indexes processing. The user can then start processing the remaining steps: the anomaly, the Laplacian and the possible leaks identification (Fig. 7).

Fig. 7. Leak detection overview - view possible leaks. (Color figure online)

5.2 Architecture of the Web Interface

The current web-portal architecture is described in Fig. 8. Inside the web-portal, the frontend will interact directly with the backend. It processes the frontend requests and checks the ESA's Copernicus Open Access Hub repository for available sentinel image sets. The backend has an embedded raster layer service to load and store the processed images as rasters. The asynchronous processing service interacts with the backend to provide its processing status, to store the final products in the raster layer service, and to download image sets from the repository.

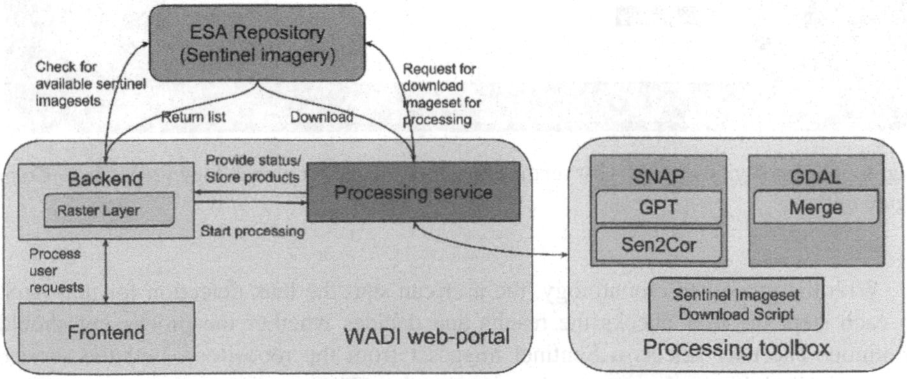

Fig. 8. WADI Architecture diagram. (Color figure online)

The processing service depends on executables from the processing toolbox to perform various tasks: to download/convert/merge image sets, to process the data, to store the processed images and to apply other steps for leak detection. The processing toolbox consists of SNAP, a toolbox for Sentinel image processing with Sen2Cor for conversion, GPT for script processing (rescaling and water index calculation), and GDAL for other tasks, such as image merging. A new approach for processing images needs to be studied and applied in HPC, since SNAP does not presently support parallel processing.

6 Conclusions and Future Work

The reliability strategy proposed herein brings an additional level of confidence to the WADI service, aiming to improve its quality through the usage of data-intensive algorithms applied to public satellite data. Given the low resolution in time of the core WADI service (typically a few plane passages at each location), remote sensing by satellites can work both as a predecessor of the WADI service, to motivate its usage for fine pinpointing of leaks, and also as a post-processing tool, combined with a water pathways model. Both avenues can now be explored through the web portal where this service is available, integrated in the Portuguese National Computing Infrastructure. Herein, we focused on the satellite data component of the methodology, the water pathways model being currently under implementation.

The present methodology has many avenues to evolve to a faster and more detailed service. Future work consists in adapting this methodology into the Data-Intensive Scientific Discovery (DISD) paradigm. Data-Intensive Scientific Discovery focuses on exploring the best strategies that use the available computational resources for processing large datasets, and to provide faster data outputs to analyze. DISD is already being applied for control in critical areas, such as oil industry and airport traffic. Several DISD strategies can now be pursued to improve the performance of the methodology presented here.

High Performance Computing (HPC) is one of them, as it has an important role in the solution of data-intensive and computationally-demanding problems, such as numerical modeling or image processing. HPC takes the advantage of speeding up the computation by efficiently using the resources available on a machine or a cluster of them in cloud/grid environment. Still, many challenges in applying HPC to leak detection remain, such as finding the best strategy for the data decomposition to assure the correct load balancing, node communication and resulting accuracy, and the possible parallel processing limitations caused by hardware bottlenecks.

Future work consists in applying HPC on the reliability WADI service, which is important to provide faster detection and reduce costs and water losses. Our methodology will be tested for speedup the processing by using CPUs and/or GPUs, since the images have a high resolution, and the cost and time to process them sequentially is larger too. One possible way to do that is being studied and consists in 'slicing' equally the images to be processed by each CPU/GPU core, using Python with OpenCV and/or numpy.

Finally, the validation of the methodology in the Alqueva infrastructure for in-situ identified leaks that happened since the Sentinel 2 data became available is underway. This analysis will then be used together with the WADI flights processed images for high accuracy detection results.

Acknowledgements. This work was developed in the scope of the H2020 WADI project (Grant agreement No: 689239) and INCD project funded by Lisboa2020 Operational Program (LISBOA-01-0145-FEDER-022153).

References

1. FAO: Coping with water scarcity. Challenge of the twenty-first century. UN-Water (2007). http://www.fao.org/3/a-aq444e.pdf
2. Carvalho, A., Oliveira, A., Alves, E.: WADI – Detecção inovadora de fugas de água em grandes redes de distribuição a partir da vigilância através de aviões tripulados e drones. In: 14th Water Congress, Évora, Portugal, 7–9 March 2018 (2018)
3. Agapiou, A., Alexakis, D.D., Themistocleous, K., Hadjimitsis, D.G.: Water leakage detection using remote sensing, field spectroscopy and GIS in semiarid areas of Cyprus. Urban Water J. **13**(3), 221–231 (2016). https://doi.org/10.1080/1573062X.2014.975726
4. Azevedo, A., Oliveira, A., Fortunato, A., Bertin, X., Bertin, A.: Application of an Eulerian-Lagrangian oil spill modeling system to the Prestige accident: Trajectory analysis. J. Coast. Res. Special Issue **56**, 777–781 (2009)
5. Evagorou, E., Mettas, C., Agapiou, A., Themistocleous, K., Hadjimitsis, D.: Bathymetric maps from multi-temporal analysis of Sentinel-2 data: the case study of Limassol, Cyprus. Adv. Geosci. **45**, 397–407 (2019). https://doi.org/10.5194/adgeo-45-397-2019
6. Fingas, M., Brown, C.: A review of oil spill remote sensing. Sensors **18**, 91 (2017). https://doi.org/10.3390/s18010091
7. Whitepaper of UTILIS - Satellite Based Leak Detection. https://utiliscorp.com/wp-content/themes/utilis/files/new-files/other/UTILIS_whitepaper_A4_WEB.pdf
8. Chatelard, C., et al.: Multispectral approach assessment for detection of losses in water transmission systems by airborne remote sensing (2018). https://doi.org/10.29007/4xs9
9. EC: H2020-EU.3.5. - SOCIETAL CHALLENGES - Climate action, Environment, Resource Efficiency and Raw Materials (2014). https://cordis.europa.eu/programme/rcn/664389/en
10. McFeeters, S.: The use of normalized difference water index (NDWI) in the delineation of open water features. Int. J. Rem. Sens. **17**, 1425–1432 (1996). https://doi.org/10.1080/01431169608948714
11. Xu, H.: Modification of normalized difference water index (NDWI) to enhance open water features in remotely sensed imagery. Int. J. Rem. Sens. **27**, 3025–3033 (2006). https://doi.org/10.1080/01431160600589179

Track of Machine Learning and Data Assimilation for Dynamical Systems

Scalable Weak Constraint Gaussian Processes

Rossella Arcucci(✉), Douglas McIlwraith, and Yi-Ke Guo

Data Science Institute, Imperial College London, London, UK
r.arcucci@imperial.ac.uk

Abstract. A Weak Constraint Gaussian Process (WCGP) model is presented to integrate noisy inputs into the classical Gaussian Process predictive distribution. This follows a Data Assimilation approach i.e. by considering information provided by observed values of a noisy input in a time window. Due to the increased number of states processed from real applications and the time complexity of GP algorithms, the problem mandates a solution in a high performance computing environment. In this paper, parallelism is explored by defining the parallel WCGP model based on domain decomposition. Both a mathematical formulation of the model and a parallel algorithm are provided. We prove that the parallel implementation preserves the accuracy of the sequential one. The algorithm's scalability is further proved to be $\mathcal{O}(p^2)$ where p is the number of processors.

Keywords: Gaussian processes · Data assimilation · Domain decomposition · Parallel algorithms · Big data

1 Introduction and Motivations

Gaussian processes have been widely used since the 1970's in the fields of geostatistics and meteorology. In geostatistics, prediction with Gaussian processes is termed Kriging, named after the South African mining engineer D. G. Krige by Matheron [10]. Naturally in spatial statistics the inputs to the process are the two or three space dimensions, however, Over the past decade or so, there has been much work on Gaussian processes in the machine learning community, typically over higher dimensionality spaces.

GPs have had substantial impact in technologies including geostatistics [3] and feature reduction [12]. Current applications are in diverse fields such as geophysics, medical imaging, multi-sensor fusion [13,19,21] and sensor placement [8]. The latter of these is well suited application for the GP model we propose in this paper. For example, this could allow the study of optimal sensor placement for collecting air pollution data in big cities.

The main limitations of traditional GP regression are its sensitivity to noisy input and the computational complexity. The contribution of this paper is a new GP algorithm that overcomes these limitations simultaneously. By considering

J. M. F. Rodrigues et al. (Eds.): ICCS 2019, LNCS 11539, pp. 111–125, 2019.
https://doi.org/10.1007/978-3-030-22747-0_9

the inputs to the GP as noisy observations and reformulating the GP, we create a new model in which inputs are assimilated to estimate the true values of inputs. We further demonstrate that our method is highly suitable for parallel processing, both formally and with experimental results.

2 Background and Contribution

In [7], the authors expand the Gaussian process around the input mean (delta method), assuming the random input is normally distributed and they derive a new process whose covariance function accounts for the randomness of the input. In [11] the input noise variances are inferred from the data as extra hyperparameters. Instead, we develop a Weak Constraint Gaussian Process (WCGP) model to be used to improve the accuracy of the classical Gaussian process predictive distribution [15]. Noisy inputs are integrated into the Gaussian process through a data assimilation approach - i.e. by considering information provided by observed values of the noisy input. As the "assimilated observations" are not verified exactly [1], we may consider this a weak constraint over the inputs. The resulting model (which we call WCGP) is still a GP model with modified mean and variance - as we will demonstrate in Sect. 3. The number of processed data points for this model is increased with respect the classical GP and we show the GP time complexity is $\mathcal{O}(N^3)$.

An approach to reduce the time complexity is to introduce approximation methods. In [18] the covariance matrix is approximated by the Nystrom extension of a smaller covariance matrix evaluated on M training observations ($M << N$). Approximation methods help to reduce the computation cost from $\mathcal{O}(N^3)$ to $\mathcal{O}(NM^2)$ and make makes running less expensive, but parameters must still be selected a-priori, and, consequently, important sensitivities may be missed [2].

Due to the large number of states required in real applications plus the time complexity of GP algorithms, the problem mandates the solution in a high performance computing environment. In [22] a scalable Sparse Gaussian process (GP) regression [16] and Bayesian Gaussian process latent variable model (GPLVM) [17] are presented. This work represents the first distributed inference algorithm which is able to process datasets with millions of points. However, even with sparse approximations it is inconceivable to apply GPs to training set sizes of data sets of size more than $\mathcal{O}(10^7)$. In [4] a distributed Gaussian processes model is introduced. Their key idea is to recursively distribute computations to independent computational units and, subsequently, recombine them to form an overall result. Local predictions are recombined by a parent node, which subsequently may play the role of an expert at the next level of the model architecture. Even if this approach allows us to face problems with large data sets, the interaction between the local and the parent node introduces a bottleneck which affects the efficiency of processing for datasets which may be considered "big data".

As claimed in [6], the partitioning problem (i.e, decomposability: to break the problem into small enough independent less complex subproblems) is a universal source of scalable parallelism. In [14] a domain decomposition approach for the classical GP or [20] applied to urban flows simulation are presented based on the definition of boundary conditions for the subproblems. While the introduced approach allows "big data" problems to be tackled, there is a subsequent loss to solution accuracy.

In this paper, we formally address the parallelism problem by defining the parallel Weak Constraint GP (WCGP) model based on the previously introduced domain decomposition approach. The accuracy of the proposed approach is proved by showing that the solution obtained by the parallel algorithm is the same as obtained by the sequential one. In particular, a parallel algorithm to be implemented on a distributed computing architecture is presented. Also, the algorithm's scalability is studied taking into account both the execution time (i.e. the time complexity) and the communication overhead given by the implementation of the algorithm on a parallel and distributed computing architecture. Finally, an upper bound on the achievable performance gain is provided, which turns out to be independent of the computing architecture utilised.

This paper is structured as follows. In Sect. 3 the Weak Constraint Gaussain Process model is described. Then, the Domain Decomposition based Weak Constraint Gaussian Process is introduced in Sect. 4. In this section we investigate the accuracy of the introduced method and a theorem demonstrating the conservation of accuracy is presented. In Sect. 5 the scalability of the resulting algorithm is discussed and experimental results are provided in Sect. 6. Conclusion and future work is summarised in Sect. 7.

3 Weak Constraint Gaussian Process

A spatial noisy input GP regression is formulated as follows: given a training data set $\mathcal{D} = \{(x_i, y_i), i = 1, \ldots, N\}$ of n pairs of noisy inputs x_i and noisy observations y_i, obtain the predictive distribution for the realization of a latent function at a test point x_*, denoted by $f_* = f(x_*)$. We assume that the latent function comes from a zero-mean Gaussian random field with a covariance function $k(\cdot, \cdot)$ on a domain $\Omega \subset \Re^{N\times N}$ and the noisy input x_i and observations y_i are given by

$$y_i = f(x_i + e_{xi}) + e_{yi} \tag{1}$$

where $e_{xi} = \mathcal{N}(0, \sigma_x^2)$ and $e_{yi} = \mathcal{N}(0, \sigma_y^2)$.

Gaussian Process: Denote

$$x = [x_1, x_2, \ldots, x_N]^T$$

and

$$y = [y_1, y_2, \ldots, y_N]^T.$$

The joint distribution of (f_*, y) is

$$(f_*, y) = \mathcal{N}\left(0, \begin{bmatrix} k_{x_* x_*} & k_{xx_*}^T \\ k_{xx_*} & \sigma_y^2 I + K_{xx} \end{bmatrix}\right) \tag{2}$$

where $k_{xx_*} = (k(x_1, x_*), \ldots, k(x_N, x_*))^T$ and K_{xx} is an $N \times N$ matrix $K_{xx} = \{k(x_i, x_j)\}_{i=1,\ldots,N; j=1,\ldots,N}$. By the conditional distribution for Gaussian variables, the predictive distribution of f_* given y is

$$P(f_*|y) = \mathcal{N}\left(k_{xx_*}^T A^{-1} y, k_{x_*x_*} - k_{xx_*}^T A^{-1} k_{xx_*}\right) \tag{3}$$

where

$$A = \sigma_y^2 I + K_{xx} \tag{4}$$

Data Assimilation: At each step i, $i = 1, \ldots, N$ (see Fig. 1), let $o = H(x)$ be the observations vector $o = \{o_i\}_{i=1,\ldots,N}$ where H is a non-linear operator collecting the assimilated observations at each step. The aim of DA problem is to find an optimal tradeoff between the current estimate of the system state (background) defined in (3) and the available assimilated observations. Let R be a covariance matrix whose elements provide the estimate of the errors[1] on o, the assumption of input noise and the assimilation of it by a data assimilation approach (see Chapter 5 of [9]) introduce a corrective term

$$O = H^T R^{-1} H \tag{5}$$

to the output noise. Then the resulting Gaussian process is

$$P(f_*|y, o) = \mathcal{N}\left(k_{xx_*}^T \hat{A}^{-1} y, k_{x_*x_*} - k_{xx_*}^T \hat{A}^{-1} k_{xx_*}\right) \tag{6}$$

where

$$\hat{A} = A + O \tag{7}$$

with A and O defined in (4) and (5) respectively, and where we assume that each input dimension is independently corrupted by noise, thus R is diagonal:

$$R = \sigma_x^2 I \tag{8}$$

The predictive mean $k_{xx_*}^T \hat{A}^{-1} y$ gives the point prediction of $f(x)$ at location x_*, whose uncertainty is measured by the predictive variance $k_{x_*x_*} - k_{xx_*}^T \hat{A}^{-1} k_{xx_*}$. The point prediction given above is the best linear unbiased predictor in the following sense [14]. Consider all linear predictors

$$\mu(x_*) = u(x_*)^t y, \tag{9}$$

satisfying the unbiasedness requirement $E[\mu(x_*)] = 0$. We want to find the vector $u(x_*)$ which minimizes the mean squared prediction error $E[\mu(x_*) - f(x_*)]^2$. Since $E[\mu(x_*)] = 0$ and $E[f(x_*)] = 0$, the mean squared prediction error equals the error variance $var[\mu(x_*) - f(x_*)]$ and can be expressed as

$$\begin{aligned} \sigma(x_*) &= u(x_*)^t E(yy^t) u(x_*) - 2u(x_*)^t E(y f_*) + E(f_*^2) \\ &= u(x_*)^t (\sigma_y^2 I + K_{xx} + H^T R^{-1} H) u(x_*) - 2u(x_*)^t k_{xx_*} + k_{x_*x_*} \end{aligned} \tag{10}$$

[1] i.e., the assimilated observations are not verified exactly, it is a weak constraint over the inputs to the Gaussian Process.

Equation (10) is a quadratic form in $u(x_*)$. It is easy to see $\sigma(x_*)$ is minimized if and only if $u(x_*)$ is chosen to be $(\sigma_y^2 I + K_{xx} + H^T R^{-1} H)^{-1}$. Based on the above discussion, the mean of the predictive distribution in (6) or the best linear unbiased predictor can be obtained by solving the following minimization problem: for $x_* \in \Omega$, compute

$$\bar{u}(x_*) = argmin_{u(x_*) \in \Re^N} J(u(x_*)) \tag{11}$$

with

$$J(u(x_*), \sigma_y, K_{xx}, H, R, \Omega) = u(x_*)^t(\sigma_y^2 I + K_{xx} + H^T R^{-1} H)u(x_*) - 2u(x_*)^t k_{xx_*} \tag{12}$$

To compute the minimum of J in (12), the Jacobian of J has to satisfy the condition $\nabla J = 0$ which implies the solution of a linear system of the matrix $\hat{A}$. Due the ill conditioning of the matrix K_{xx}, the matrix $\hat{A}$ is ill conditioned as well, then is mandatory to introduce a preconditioning [15]. Since K_{xx} is symmetric and positive definite, it is possible to compute its Cholesky factorization $K_{xx} = VV^t$. Let be $\mu(\cdot)$ denote the condition number, by this way it is:

$$\mu(K_{xx}) = \mu(V)^{-\frac{1}{2}}$$

i.e., the Cholesky factorization of the covariance matrix K_{xx} mitigates the ill conditioning of the minimization problem. Let be $v = V^+ u$, where $+$ denotes the generalised inverse of the matrix V. After the preconditioning the minimization problem in (11) and (12) is written as:

$$\bar{v}(x_*) = argmin_{v(x_*) \in \Re^N} J(v(x_*)) \tag{13}$$

with

$$J(v(x_*), \sigma_y, V, H, R, \Omega) = v^t V^t (I + \sigma_y^2 I + H^T R^{-1} H) V v - 2v^t V^t k_{xx_*} \tag{14}$$

The exact treatment of this function would require the consideration of a distribution over Taylor expansions. Due the high computational load, several approximations are usually introduced in order to face this issue [7,11]. Here we face the problem concerning the high computational cost by introducing a domain decomposition approach into the mathematical model. In next sections we provide a proof that the solution obtained by the decomposed problem we are introducing is the same solution of the sequential problem. Also, it is proved that the time complexity is reduced as well.

4 Domain Decomposition Based Weak Constraint Gaussian Process

In this section a synthetic mathematical formalization of the weak constraint GP model based on a domain decomposition approach is presented. Local functions are introduced. These functions are defined on subdomains which constitute a partitioning $DD(\Omega)$ of the domain $\Omega \subset \Re^N$ with overlapping as described in Fig. 2 and such that:

$$DD(\Omega) = \{\Omega_i\}_{i=1,\dots,p} \tag{15}$$

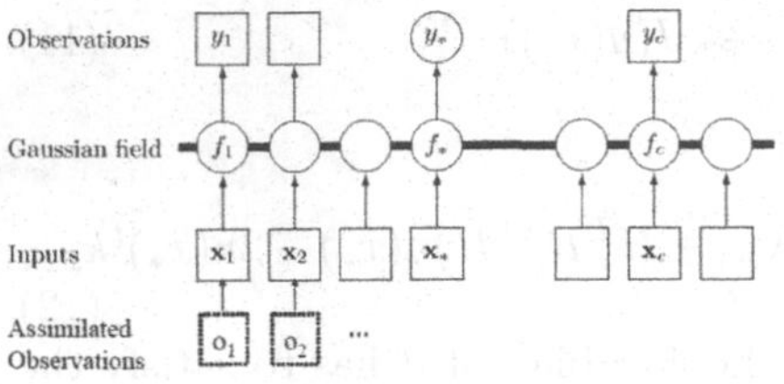

Fig. 1. Graphical model (chain graph) for a WCGP for regression. Squares represent observed variables and circles represent unknowns. Dotted squares represent innovation with respect a classical GP [4].

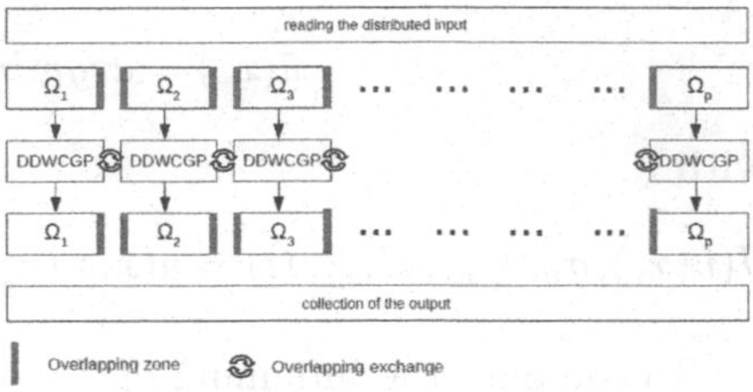

Fig. 2. Domain Decomposition based Weak Constraint Gaussian Process

with $\Omega_i \subset \Re^{r_i}$, $r_i \leq N$ and for $i = 1, \dots, p$, it is such that $\Omega = \bigcup_{i=1}^{p} \Omega_i$ with $\Omega_i \cap \Omega_j = \Omega_{ij} \neq \emptyset$.

The *local* WCGP function describes the local WCGP problem on a sub-domain Ω_i of the domain decomposition. It is obtained restricting the WCGP function J to the sobdomain Ω_i and by adding a *local* constraint onto the overlap region between adjacent domains Ω_i and Ω_j such that:

$$J_{\Omega_i}(v_i) = v_i^t V_i^t (I_i + \sigma_y^2 I_i + H_i^T R_i^{-1} H_i) V_i v_i - 2 v_i^t V_i k_{xx_*} + |v_{ij} - v_{ji}| \tag{16}$$

where v_i, V_i, I_i, R_i and H_i are restrictions on Ω_i of vectors and matrices in (12), and $v_{ij} = v_i/\Omega_{ij}$, $v_{ji} = v_j/\Omega_{ij}$, are restriction on $\Omega_{ij} = \Omega_i \cap \Omega_j$ of v_i and v_j respectively, $\forall j$ such that $\Omega_i \cap \Omega_j \neq \emptyset$.

4.1 Accuracy of the Decomposition Approach

In this section the accuracy of the introduced DDWCGP model is studied. An important point is to ensure that the introduction of a decomposition among the dataset does not introduce errors which affects the accuracy of the GP solution. Let $\bar{v}_i = argmin_{v_i} J_{\Omega_i}(v_i)$ denote the minimum of J_{Ω_i} and let $\widetilde{v}$ denote *extension vector* defined as the sum of the extensions of $\bar{v}_i$ $\forall i$, to the domain Ω:

$$\widetilde{v}_i = \begin{cases} \bar{v}_i & in\ \Omega_i \\ 0 & in\ \Omega - \Omega_i \end{cases} \quad and \quad \widetilde{v} = \sum_{i=1}^{p} \widetilde{v}_i \tag{17}$$

A theorem which ensures that the solution obtained by the parallel algorithm $\widetilde{v}$ is the same as obtained by the sequential one $\bar{v}$ is proved here. First, some preliminary observations are introduced in order to help the mathematical description of the restriction and the extension of the function among the subdomains and the global domain respectively. Let $\widetilde{J}_i$ denote the *extension function* of J_{Ω_i} to Ω:

$$\widetilde{J}_i = \begin{cases} J_{\Omega_i} & in\ \Omega_i \\ 0 & in\ \Omega - \Omega_i \end{cases} \tag{18}$$

Even if the functions $\widetilde{J}_i$ are defined on Ω, we consider here the subscript i to denote the starting subdomain. We can observe that

$$\sum_{i=1}^{p} \widetilde{J}_i = J. \tag{19}$$

Theorem 1. *Let $DD(\Omega)$ be a decomposition of the domain Ω. Also, let $\widetilde{v}$ defined in (17) and $\bar{v} = \bar{v}(x_*)$ defined in (13), it follows that:*

$$\widetilde{v} = \bar{v}.$$

Proof: *Let $\bar{v}_i$ be the minimum of J_{Ω_i} on Ω_i, it is*

$$\nabla J_{\Omega_i}[\bar{v}_i] = 0, \quad \forall i : \Omega_i \in DD(\Omega). \tag{20}$$

From (20) it follows that $\sum_i \nabla J_{\Omega_i}[\bar{v}_i] = 0$ which gives from the property of the gradient:

$$\nabla \sum_i J_{\Omega_i}[\bar{v}_i] = 0 \tag{21}$$

Let consider the extension vector and the extension function as defined in (17) and (18) respectively, from the (21) follows

$$\nabla \sum_i \widetilde{J}_i(\widetilde{v}_i) = 0 \tag{22}$$

which gives from the (19) and the second in (17):

$$\nabla J(\widetilde{v}) = 0 \tag{23}$$

then, from (23) follows that $\widetilde{v}$ is a stationary point for J. As $\bar{v}$ as defined in (13) is the global minimum, it follows that $J(\bar{v}) \leq J(\widetilde{v})$. We prove that $\bar{v} = \widetilde{v}$, then $J(\bar{v}) = J(\widetilde{v})$, by reduction to the absurd. Infact, by assuming

$$J(\bar{v}) < J(\widetilde{v}), \tag{24}$$

on Ω the (24) gives $J(\bar{v}(x_j)) < J(\widetilde{v}(x_j))$ for all $x_j \in \Omega$. In particular, let Ω_i be a subset of Ω, then $J(\bar{v}(x_j)) < J(\widetilde{v}(x_j))$ for all $x_j \in \Omega_i$. Hence, considering the restriction to Ω_i, $\forall i$ and by assuming the (24) we have

$$J_{\Omega_i}(\bar{v}/\Omega_i) < J_{\Omega_i}(\widetilde{v}/\Omega_i), \quad \forall i : \Omega_i \subset \Omega \tag{25}$$

which gives from (17):

$$J_{\Omega_i}(\bar{v}/\Omega_i) < J_{\Omega_i}(\bar{v}_i). \tag{26}$$

Equation (26) is an absurd. In fact, if $\bar{v}/\Omega_i \neq \bar{v}_i$, the (26) says that there exists a point such that the value of the function in that point is smaller than the values of the function in the point of global minimum. Then the theorem is proved.

This result ensures that $\bar{v}$ (the global minimum of J) can be obtained by patching together all the vectors $\bar{v}_i$ (global minimums of the operators J_{Ω_i}), i.e. by using the domain decomposition, the global minimum of the operator J can be obtained by patching together the minimums of the *local* functionals J_{Ω_i}. This result has important implications from the computational viewpoint as it will be explained in the next section.

5 The DD Based WCGP Algorithm (DDWCGP)

An algorithm to implement the minimization problem in (13) and (14) is presented as Algorithm 1. Hereafter it is assumed that the number of computing processors equals the number of subdomains p which constitutes the decomposition as described in (15).

Algorithm 1. $\mathcal{A}(\Omega_i, p)$: The DD based WCGP algorithm for each subdomain Ω_i of a partition of Ω in p subdomains

```
Input: {(x_i, y_i)}_{i=0,...,N}, {o_i}_{i=0,...,N} and x_*
Input: L                      ▷ define the number of points which constitutes the overlapping region
Define σ_y, R_i and K_xx                                          ▷ define the covariances
Compute V ← Cholesky(K_xx)    ▷ compute the Cholesky factorization of the
    covariance matrix K_xx
Compute D_i ← I_i + σ_y^2 I_i
Define H_i                                               ▷ define the interpolation function
Compute D̂_i ← A_i + H_i^T R_i^{-1} H_i
Compute Â_i ← V_i^T D̂_i V_i                                  ▷ compute Â as defined in (7)
Define the initial value of u_i
Compute v_i = V_i^+ u_i
While (convergence on v_i is obtained)                   ▷ start of the minimization steps
    Send and Receive L overlapping values from the adjacent domains
    Compute J_i ← J_i(v_i)                                               ▷ Defined in (16)
    Compute new values for v_i
Compute u_i = V_i v_i
```

5.1 Complexity and Scalability

In this section we provide an estimate of the scalability of our approach through a study of time complexity. We also estimate the performance gain achievable by implementing the algorithm on a parallel and distributed computing architecture. First we provide the following definition of *Scaling factor*, which measures the performance gain in terms of time complexity reduction, as the ratio

$$S_p = \frac{\tau\left(\mathcal{A}(\Omega, 1)\right)}{p\,\tau(\mathcal{A}(\Omega_i, p))}. \tag{27}$$

where $\mathcal{A}$ denotes the parallel algorithm, p is the size of the partition which constitutes the decomposition of Ω and $\tau(\mathcal{A})$ denotes the time complexity of the algorithm.

Theorem 2 (Scaling factor estimation). *The scaling factor of Algorithm 1 is such that*

$$S_p = p^2, \;\forall p \tag{28}$$

Proof. *The time complexity of a classical GP (i.e. when no decompositions are introduced) is $\mathcal{O}(N^3)$. Then it is*

$$\tau(\mathcal{A}(\Omega, 1)) = N^3 \tag{29}$$

By introducing the domain decomposition of Ω as defined in Definition ??, and by considering a partition of size p, we are going to solve, by $\mathcal{A}$, p subproblems of size about N/p, i.e. the time complexity is related to the subproblems and it is

$$\tau(\mathcal{A}(\Omega_i, p)) = \left(\frac{1}{p}N\right)^3 \tag{30}$$

From (27), (29), and (30) we have that

$$S_p = \frac{N^3}{p\left(\frac{1}{p}N\right)^3} = \frac{1}{\left(\frac{1}{p}\right)^2} = p^2$$

then (28) follows.

The parameter p in (27) denotes the rank of the partition which constitutes the decomposition. If we consider the DDWCGP model implemented in an algorithm on a parallel computing architecture of p processors (i.e. such that p denotes the rank of the partition as well as the number of processors involved, one subset Ω_i of the partition for each processor), we may evaluate the execution time of the algorithm running on the computing architecture and then estimate the *Measured Scaling factor*:

$$Measured\ S_p = \frac{T^1_{\mathcal{A}}(N)}{p\,T^p_{\mathcal{A}}(N)}. \tag{31}$$

where p is here the number of computing processors and $T^p_{\mathcal{A}}(N)$ denotes the execution time for solving a problem of size N with p processors.

The execution time $T^p_{\mathcal{A}}(N)$ can be mainly expressed as the sum of the time necessary for computation and the time necessary for communication. Then it is

$$T^p_{\mathcal{A}} = T^p_{flop}(N) + T^p_{comm}(N) \tag{32}$$

Regard $T^p_{flop}(N)$, by denoting with t_{flop} the time required by one floating point operation, it is

$$T^p_{flop}(n) = \tau(\mathcal{A}(\Omega, p))\, t_{flop} \tag{33}$$

In an ideal case, in absence of communication among the processors, by using (32) and (33) in (31) we clearly obtain $Measured\ S_p = S_p$. In reality, the estimation provided by (28) represents an upper bound of the performance gain we may obtain by implementing $\mathcal{A}$ on a parallel computing architecture due an overhead produced by the communication. The *Communication overhead* is the proportion of time the processors spend communicating with each other instead of computing:

$$T^p_{over}(N) = \frac{T^p_{comm}(N)}{T^p_{flop}(N)} \tag{34}$$

where N denotes the size of the problem. The following results hold:

Theorem 3 (Estimation of the Measured Scaling factor).

$$Measured\ S_p = C^p_{over}\ S_p \tag{35}$$

where C^p_{over} is a quantity which depends on the communication overhead defined in (34). Moreover, $C^p_{over} < 1$, which gives

$$Measured\ S_p < S_p. \tag{36}$$

Proof. *From (31) and (32), it is*

$$Measured\ S_p = \frac{T^1_{flop}(N)}{pT^p_{flop}(\frac{N}{p}) + pT^p_{comm}(\frac{N}{p})} \tag{37}$$

dividing by $pT^p_{flop}(\frac{N}{p})$, we have

$$Measured\ S_p = \frac{\frac{T^1_{flop}(N)}{pT^p_{flop}(\frac{N}{p})}}{1 + \frac{pT^p_{comm}(\frac{N}{p})}{pT^p_{flop}(\frac{N}{p})}}$$

also, by using (33), it can be written as

$$Measured\ S_p = \frac{1}{1 + \frac{T^p_{comm}(\frac{N}{p})}{T^p_{flop}(\frac{N}{p})}} \frac{\tau(\mathcal{A}(\Omega, p))\ t_{flop}}{p\ \tau(\mathcal{A}(\Omega, p))\ t_{flop}}$$

which, from (27), gives

$$Measured\ S_p = \frac{1}{1 + \frac{T^p_{comm}(\frac{N}{p})}{T^p_{flop}(\frac{N}{p})}} S_p \tag{38}$$

then, from (34), we have that

$$Measured\ S_p = \frac{1}{1 + T_{over}(\frac{N}{p})} S_p \tag{39}$$

and for

$$C^p_{over} = \frac{1}{1 + T_{over}(\frac{N}{p})} \tag{40}$$

then (35) follows. Moreover, as in (40) $T_{over}(\frac{N}{p}) > 0$, it follows that $C^p_{over} < 1$, which gives (36).

The communication overhead $T_{over}(\frac{N}{p})$ in (39) can be estimated by using the so called surface-to-volume ratio [5] (denoted as S/V) which is a measure of the amount of data exchange (proportional to surface area of domain) per unit operation (proportional to volume of domain). The goal is to minimize this ratio.

Lemma 1. *Concerning Algorithm 1, the surface-to-volume ratio equals to*

$$\frac{S}{V} = \frac{2Lp^3}{(N + 2Lp)^3} \tag{41}$$

Proof. In Algorithm 1 the amount of data exchange (see Step 12) at each iteration is $S = 2L$. *The amount of operations for each processor corresponds to the time complexity of the sub problem solved by the processor. Then it is* $V = \left(\frac{1}{p}N + 2L\right)^3$. *From these estimations of* S *and* V, *we have that (41) holds.*

6 Experimental Results

In this section we provide experimental results that demonstrate the applicability of our approach. In particular, we show experimentally that DDWCGP provides accurate and scalable results as proved in Theorems 1 and 2.

In order to point out the expected performance of the parallel Algorithm 1 in terms of scalability and execution time reduction, test cases for $N = 100$, $N = 1000$ and $N = 10000$ are been studied which correspond to problems of complexity $O(10^6)$, $O(10^9)$ and $O(10^{12})$. We show experimentally that DDWCGP provides accurate and scalable results over a dataset such that the kernel function is the Radial Basis Function kernel $k(x, x') = exp\left(-\frac{1}{2}\left(\frac{x-x'}{\lambda}\right)^2\right)$ with $\lambda = 0.1$, the y values (unbeknownst to our model) from our x are such that $y = sin(x)$ and the input covariance matrix is defined in (8) for $\sigma_x^2 = 0.5$: $R = 0.5\ I$ where I is the identity matrix.

Figure 3 shows results obtained for a number of computing processors (i.e. a number of subdomains which constitutes the decomposition described in (15)) equal to $p = 1$, $p = 2$ and $p = 4$. As shown in Fig. 3, the result obtained by collecting results on subdomains matches perfectly the result obtained by the code for the test case without any decomposition. This constitutes an experimental validation of the accuracy result proved in Theorem 1.

Results in Table 1 show how the performance improves as the size of the problem increasing. This is explicitly confirmed in Fig. 4 shows values of the ratio

$$R_p(N) = \frac{S_p}{Measured\ S_p} \tag{42}$$

this is due to the surface to volume ratio in (41) which is smaller for bigger domains. Namely, the value of S/V is close to zero when the volume is very big with respect to the surface (i.e. the computation is very heavy with respect to

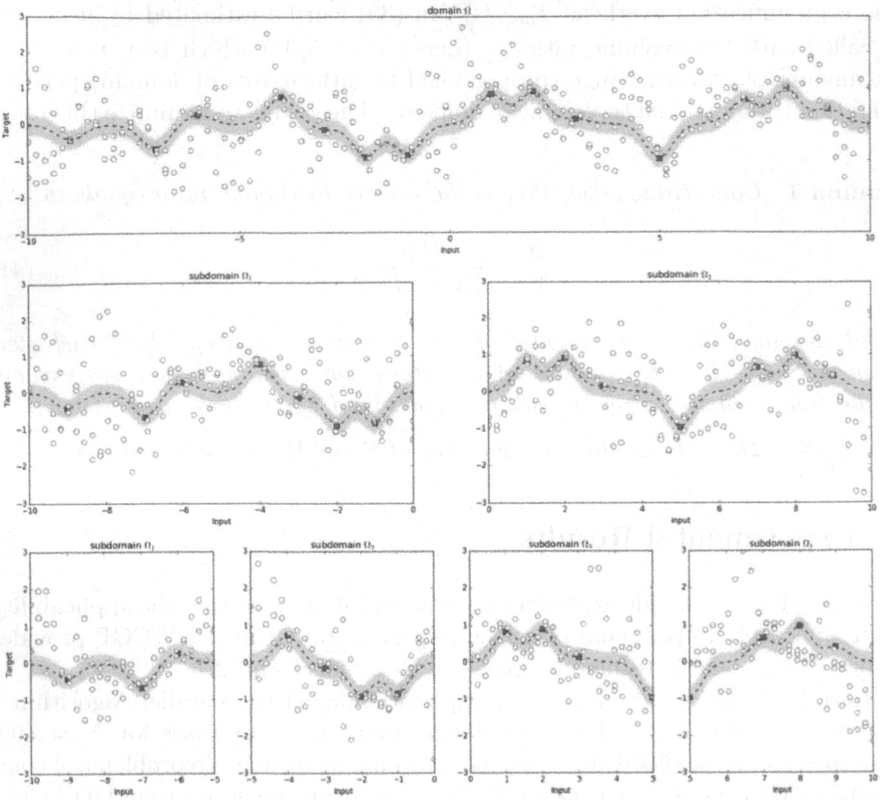

Fig. 3. Results of the test case with $L = 2$ for $p = 1$, $p = 2$, $p = 4$ computing processors (i.e. for $p = 1$, $p = 2$, $p = 4$ subdomains which constitute the decomposition described in (15))

Table 1. Values of Measured Scaling factor ($Measured\ S_p$) for $N = 100$, $N = 1000$ and $N = 10000$ compared with the Scaling Factor (S_p) for a number of processors $2 \leq p \leq 256$

Problem size:		$N = 100$	$N = 1000$	$N = 10000$
p	S_p	$Measured\ S_p$	$Measured\ S_p$	$Measured\ S_p$
2	4	3.9	3.9	4.0
4	16	15.9	15.9	15.9
8	64	63.9	63.9	63.9
16	256	254.1	255.9	255.9
32	1024	994.4	1023.8	1023.9
64	4096	3763.0	4093.0	4095.9
128	16384	13815.3	16314.9	16383.8
256	65536	50694.2	64287.8	65532.2

the communication). For values of S/V close to zero, the quantity C^p_{over} is close to 1. In fact, by the definition of S/V it holds that $C^p_{over} = \frac{1}{1+\frac{S}{V}}$. This is the reason why, in Table 1, by increasing the problem size, the performance of the parallel algorithm improves. As confirmed by results in Table 1, the values of p, up to whom the *Measured* S_p is very close to the value of S_p, depend on the problem size.

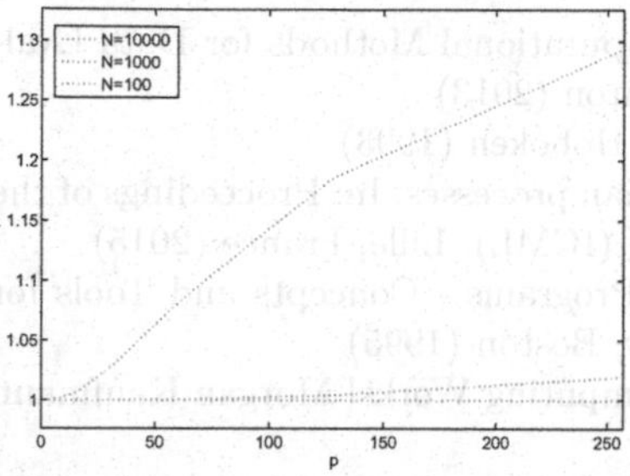

Fig. 4. Values of the ratio $R_p(N)$ in (42) for $N = 100$, $N = 1000$, $N = 10000$ and for a number of processors $1 \leq p \leq 256$

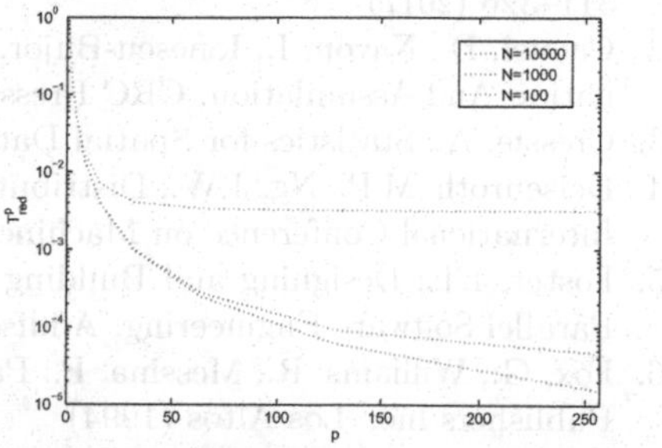

Fig. 5. Values of time reduction T^p_{red} for $N = 100$, $N = 1000$, $N = 10000$ and for a number of processors $1 \leq p \leq 256$ (Color figure online)

Figure 5 shows the time reduction T^p_{red} which is the execution time T^P_A in (32) normalized to 1 s. Also in this case, the obtained results underline how the performance improvement in terms of execution time reduction increases when the volume of the data increases. This is still due to the communication overhead which dominates datasets from smaller domains.

7 Conclusions

In this paper a Weak Constraint Gaussian Process (WCGP) algorithm is presented to integrate noisy inputs into the classical Gaussian Process predictive distribution and we have demonstrate the scalability and the accuracy of our approach. The algorithm developed starts by integrating noisy input into a GP model before to provide a parallel implementation i.e. by decomposing the domain. A mathematical formulation of the model is provided and the mathematical validity of this formulation is proved. Furthermore, the algorithmic scalability has been proved to be $\mathcal{O}(p^2)$ where p is the number of processors. In order to evaluate the performance of the parallel algorithm, a scaling factor (which measures the performance gain in terms of time complexity reduction) is introduced and compared with a measured scaling factor (which measures the performance gain in terms of execution time reduction). The algorithm has been evaluated on data sets for $100 \leq N \leq 10000$. For a fixed number of processors p, results show how the performance improves as the size of the problem size increases due to the communication overhead which decreases.

Acknowledgments. This work is supported by the EPSRC Centre for Mathematics of Precision Healthcare EP/N0145291/1.

References

1. Arcucci, R., D'Amore, L., Pistoia, J., Toumi, R., Murli, A.: On the variational data assimilation problem solving and sensitivity analysis. J. Comput. Phys. **335**, 311–326 (2017)
2. Cacuci, D., Navon, I., Ionescu-Bujor, M.: Computational Methods for Data Evaluation And Assimilation. CRC Press, Boca Raton (2013)
3. Cressie, A.: Statistics for Spatial Data. Wiley, Hoboken (1993)
4. Deisenroth, M.P., Ng, J.W.: Distributed Gaussian processes. In: Proceedings of the International Conference on Machine Learning (ICML), Lille, France (2015)
5. Foster, I.T.: Designing and Building Parallel Programs - Concepts and Tools for Parallel Software Engineering. Addison-Wesley, Boston (1995)
6. Fox, G., Williams, R., Messina, P.: Parallel Computing Works! Morgan Kaufmann Publishers Inc., Los Altos (1994)
7. Girard, A., Murray-Smith, R.: Learning a Gaussian process model with uncertain inputs. Technical report, TR-2003-144, Department of Computing Science, University of Glasgow (2003)
8. Guestrin, C., Krause, A., Singh, A.: Near-optimal sensor placements in Gaussian processes. In: 22nd International Conference on Machine Learning (ICML), pp. 265–272 (2005)
9. Kalnay, E.: Atmospheric Modeling, Data Assimilation and Predictability. Cambridge University Press, Cambridge (2003)
10. Matheron, G.: The intrinsic random functions and their applications. Adv. Appl. Probab. **5**, 439–468 (1973)
11. McHutchon, A., Rasmussen, C.E.: Gaussian process training with input noise. In: Proceedings of the 24th International Conference on Neural Information Processing Systems, NIPS 2011, pp. 1341–1349 (2011)
12. Neil, L.: Probabilistic non-linear principal component analysis with Gaussian process latent variable models. Mach. Learn. Res. **6**, 1783–1816 (2005)
13. Osborne, M., Rogers, A., Ramchurn, A., Roberts, S., Jennings, N.R.: Towards real-time information processing of sensor network data using computationally efficient multi-output Gaussian processes. In: International Conference on Information Processing in Sensor Networks, IPSN 2008 (2008)
14. Park, C., Huang, J.Z., Ding, Y.: Domain decomposition approach for fast Gaussian process regression of large spatial data sets. J. Mach. Learn. Res. **12**, 1697–1728 (2011)
15. Rasmussen, C.E., Williams, C.K.I.: Gaussian Processes for Machine Learning. The MIT Press, Cambridge (2006). ISBN 026218253X
16. Titsias, M., Lawrence, N.: Variational learning of inducing variables in sparse Gaussian processes. In: 12th International Conference on Artificial Intelligence and Statistics (AISTATS) (2009)
17. Titsias, M., Lawrence, N.: Bayesian Gaussian process latent variable model. In: 13th International Conference on Artificial Intelligence and Statistics (AISTATS) (2010)
18. Williams, C.K.I., Seeger, M.: Using the Nystrom method to speed up kernel machines. Adv. Neural Inf. Process. Syst. **12**, 682–688 (2000)

19. Xiao, D., Fang, F., Zheng, J., Pain, C., Navon, I.: Machine learning-based rapid response tools for regional air pollution modelling. Atmos. Environ. **199**, 463–473 (2019)
20. Xiao, D., et al.: A domain decomposition non-intrusive reduced order model for turbulent flows. Comput. Fluids **182**, 15–27 (2019)
21. Xiao, D., et al.: A reduced order model for turbulent flows in the urban environment using machine learning. Build. Environ. **148**, 323–337 (2019)
22. Yarin, G., van der Wilk Mark, E., R.C.: Distributed variational inference in sparse Gaussian process regression and latent variable models. In: Advances in Neural Information Processing Systems (2014)

A Learning-Based Approach for Uncertainty Analysis in Numerical Weather Prediction Models

Azam Moosavi[1], Vishwas Rao[2(✉)], and Adrian Sandu[3]

[1] Biomedical Engineering Department, Case Western Reserve University, Cleveland, USA
azmosavi@vt.edu

[2] Mathematics and Computer Science Division, Argonne National Laboratory, Lemont, USA
vhebbur@anl.gov

[3] Computational Science Laboratory, Department of Computer Science, Virginia Tech, Blacksburg, USA
asandu7@vt.edu

Abstract. This paper demonstrates the use of machine learning techniques to study the uncertainty in numerical weather prediction models due to the interaction of multiple physical processes. We aim to address the following problems: (1) estimation of systematic model errors in output quantities of interest at future times and (2) identification of specific physical processes that contribute most to the forecast uncertainty in the quantity of interest under specified meteorological conditions. To address these problems, we employ simple machine learning algorithms and perform numerical experiments with Weather Research and Forecasting (WRF) model and the results show a reduction of forecast errors by an order of magnitude.

Keywords: Numerical weather prediction · Structural uncertainty · Model errors · Machine learning

1 Introduction

Computer simulation models of the physical world, such as numerical weather prediction (NWP) models, are imperfect and can only approximate the complex evolution of physical reality. Some of the errors are due to the uncertainty in the initial and boundary conditions, forcings, and model parameter values. Other errors, called structural model errors, are due to our incomplete knowledge about the true physical processes; such errors manifest themselves as missing dynamics in the model [11]. Examples of structural errors include the misrepresentation of sea-ice in the spring and fall, errors affecting the stratosphere above polar regions in winter [22], and errors due to the interactions among (approximately represented) physical processes. Data assimilation improves model forecasts by

J. M. F. Rodrigues et al. (Eds.): ICCS 2019, LNCS 11539, pp. 126–140, 2019.
https://doi.org/10.1007/978-3-030-22747-0_10

fusing information from both model outputs and observations of the physical world in a coherent statistical estimation framework [1,15]. While traditional data assimilation reduces the uncertainty in the model state and model parameter values, however, no methodologies to reduce the structural model uncertainty are available to date.

In this study we consider the Weather Research and Forecasting (WRF) model [24], a mesoscale atmospheric modeling system. The WRF model includes multiple physical processes and parametrization schemes, and choosing different model options can lead to significant variability in the model predictions [4, 14]. Among different atmospheric phenomena, the prediction of precipitation is extremely challenging and is obtained by solving the atmospheric dynamic and thermodynamic equations [14]. Model forecasts of precipitation are sensitive to physics options such as the microphysics, cumulus, long-wave, and short-wave radiation [5,9,14].

This paper demonstrates the potential of machine learning techniques to help solve two important problems related to the structural or physical uncertainty in numerical weather prediction models: (1) estimation of systematic model errors in output quantities of interest at future times, and the use of this information to improve the model forecasts, (2) identification of those specific physical processes that contribute most to the forecast uncertainty in the quantity of interest under specified meteorological conditions.

The application of machine learning techniques to problems in environmental science has grown considerably in recent years. In [6] a kernel-based regression method is developed as a forecasting approach with performance close to an ensemble Kalman filter (EnKF). Krasnopol et al. [8] employ an artificial neural network (ANN) technique for developing an ensemble stochastic convection parameterization for climate models.

This study focuses on the uncertainty in forecasts of cumulative precipitation caused by imperfect representations of the physics and their interaction in the WRF model. The total accumulated precipitation includes all phases of convective and non-convective precipitation. Specifically, we seek to use the discrepancies between WRF forecasts and measured precipitation levels in the past in order to estimate the WRF prediction uncertainty in advance. The model-observation differences contain valuable information about the error dynamics and the missing physics of the model. We use this information to construct two probabilistic functions. The first maps the discrepancy data and the physical parameters onto the expected forecast errors. The second maps the forecast error levels onto the set of physical parameters that are consistent with them. Both maps are constructed by supervised machine learning techniques, specifically, using ANN and Random Forests (RF) [13].

The remainder of this study is organized as follows. Section 2 covers the definition of the model errors. Section 3 describes the proposed approach of error modeling using machine learning. Section 4 reports numerical experiments with the WRF model that illustrate the capability of the new approach to answer two important questions regarding model errors. Conclusions are drawn in Sect. 5.

2 Model Errors

First-principles computer models capture our knowledge about the physical laws that govern the evolution of a real physical system. The model evolves an initial state at the initial time to states at future times. All models are imperfect, for example, atmospheric model uncertainties are associated with subgrid modeling, boundary conditions, and forcings. All these modeling uncertainties are aggregated into a component that is generically called *model error* [7,17,18]. In the past decade considerable scientific effort has been spent in incorporating model errors and estimating their impact on the best estimate in both variational and statistical approaches [1,20–22].

Consider the following NWP computer model $\mathcal{M}$ that describes the time-evolution of the state of the atmosphere:

$$\mathbf{x}_t = \mathcal{M}\left(\mathbf{x}_{t-1}, \Theta\right), \quad t = 1, \cdots, T. \tag{1a}$$

The state vector $\mathbf{x}_t \in \mathbb{R}^n$ contains the dynamic variables of the atmosphere such as temperature, pressure, precipitation, and tracer concentrations, at all spatial locations covered by the model and at t. All the physical parameters of the model are lumped into $\Theta \in \mathbb{R}^\ell$.

Formally, the true state of the atmosphere can be described by a physical process $\mathcal{P}$ with internal states υ_t, which are unknown. The atmosphere, as an abstract physical process, evolves in time as follows:

$$\upsilon_t = \mathcal{P}\left(\upsilon_{t-1}\right), \quad t = 1, \cdots, T. \tag{1b}$$

The model state seeks to approximate the physical state:

$$\mathbf{x}_t \approx \psi(\upsilon_t), \quad t = 1, \cdots, T, \tag{1c}$$

where the operator ψ maps the physical space onto the model space, for example, by sampling the continuous meteorological fields onto a finite-dimensional computational grid [11].

Assume that the model state at $t-1$ has the ideal value obtained from the true state via (1c). The model prediction at t will differ from reality:

$$\psi(\upsilon_t) = \mathcal{M}\big(\psi(\upsilon_{t-1}), \Theta\big) + \boldsymbol{\delta}_t(\upsilon_t), \quad t = 1, \cdots, T, \tag{2}$$

where the discrepancy $\boldsymbol{\delta}_t \in \mathbb{R}^n$ between the model prediction and reality is the *structural model error.* This vector lives in the model space.

Although the global physical state υ_t is unknown, we obtain information about it by measuring a finite number of observables $\mathbf{y}_t \in \mathbb{R}^m$ as follows:

$$\mathbf{y}_t = h(\upsilon_t) + \epsilon_t, \quad \epsilon_t \sim \mathcal{N}(0, \mathbf{R}_t), \quad t = 1, \cdots, T. \tag{3}$$

Here h is the observation operator that maps the true state of atmosphere to the observation space, and the observation error ϵ_t is assumed to be normally distributed.

To relate the model state to observations, we also consider the observation operator $\mathcal{H}$ that maps the model state onto the observation space. The model-predicted values $\mathbf{o}_t \in \mathbb{R}^m$ of the observations (3) are

$$\mathbf{o}_t = \mathcal{H}(\mathbf{x}_t), \quad t = 1, \cdots, T. \tag{4}$$

We note that the measurements $\mathbf{y}_t$ and the predictions $\mathbf{o}_t$ live in the same space and therefore can be directly compared. The difference between the observations (3) of the real system and the model predicted values of these observables (4) represent the model error in observation space:

$$\boldsymbol{\Delta}_t = \mathbf{o}_t - \mathbf{y}_t \in \mathbb{R}^m, \quad t = 1, \cdots, T. \tag{5}$$

For clarity, in what follows we make the following simplifying assumptions [11]:

- The physical system is finite dimensional $v_t \in \mathbb{R}^n$.
- The model state lives in the same space as reality; i.e., $\mathbf{x}_t \approx v_t$, and $\psi(\cdot) \equiv id$ is the identity operator in (1c).
- $\mathcal{H}(\cdot) \equiv h(\cdot)$ in (3) and (4).

These assumptions imply that the discretization errors are very small and that the main sources of error are the parameterized physical processes represented by Θ and the interaction among these processes. Uncertainties from other sources, such as boundary conditions, are assumed to be negligible.

With these assumptions, the evolution equations for the physical system (1b) and the physical observations Eq. (3) become, respectively,

$$v_t = \mathcal{M}(v_{t-1}, \Theta) + \boldsymbol{\delta}_t(v_t), \quad t = 1, \cdots, T, \tag{6a}$$
$$\mathbf{y}_t = h(v_t) + \epsilon_t. \tag{6b}$$

The model errors $\boldsymbol{\delta}_t$ (2) are not fully known at any time t, since having the exact errors is akin to having a perfect model. However, the discrepancies between the modeled and measured observable quantities (5) at past times have been computed and are available at the current time t.

Our goal is to use the errors in observable quantities at past times, $\boldsymbol{\Delta}_\tau$ for $\tau = t-1, t-2, \cdots$, in order to estimate the model error $\boldsymbol{\delta}_\tau$ at future times $\tau = t, t+1, \cdots$. This is achieved by unraveling the hidden information in the past $\boldsymbol{\Delta}_\tau$ values. Good estimates of the discrepancy $\boldsymbol{\delta}_t$, when available, could improve model predictions by applying the correction (6a) to model results:

$$\mathbf{v}_t \approx \mathbf{x}_t + \boldsymbol{\delta}_t. \tag{7}$$

3 Approximating Model Errors Using Machine Learning

We propose a multivariate input-output learning model to predict the model errors $\boldsymbol{\delta}$, defined in (2), stemming from the uncertainty in parameters Θ. To this end, we define a probabilistic function ϕ that maps every set of input features $F \in \mathbb{R}^r$ to output target variables $\Lambda \in \mathbb{R}^o$:

$$\phi(F) \approx \Lambda, \tag{8}$$

and approximate the function ϕ using machine learning.

In what follows we explain the problems we wish to address, function mapping ϕ, the input features, and the target variables for each of these problems.

3.1 Problem One: Estimating in Advance Aspects of Interest of the Model Error

Forecasts produced by NWP models are contaminated by model errors. These model errors are highly correlated in time; hence historical information about the model errors can be used as an input to the learning model to gain insight about model errors. We are interested in the uncertainty caused by the interactions between the various components in the physics-based model; these interactions are lumped into the parameter Θ that is supplied as an input to the learning model. We define the following mapping:

$$\phi^{\text{error}}(\Theta, \mathbf{o}_\tau, \boldsymbol{\Delta}_\tau, \mathbf{o}_t) \approx \boldsymbol{\Delta}_t \quad \tau < t. \tag{9}$$

We use a machine learning algorithm to approximate the function ϕ^{error}. Using a supervised learning process, the learning model identifies the effect of physical packages, historical WRF forecast, historical model discrepancy, and the current WRF forecast on the available model discrepancy at the current time. After the model gets trained on the historical data, it yields an approximation to the mapping ϕ^{error}, denoted by $\widehat{\phi}^{\text{error}}$. During the test phase the approximate mapping $\widehat{\phi}^{\text{error}}$ is used to estimate the model discrepancy $\widehat{\boldsymbol{\Delta}}_{t+1}$ in advance. We emphasize that the model prediction (WRF forecast) at the time of interest $t+1$ ($\mathbf{o}_{t+1}$) is available, whereas the model discrepancy $\widehat{\boldsymbol{\Delta}}_{t+1}$ is an unknown quantity. In fact the run time of WRF is much smaller than the time interval between t and $t+1$; in other words, the time interval is large enough to run the WRF model and obtain the forecast for the next time window, estimate the model errors for the next time window and improve the model forecast by combining the model forecast and model errors. At the test time we predict the future model error as follows:

$$\widehat{\boldsymbol{\Delta}}_{t+1} \approx \widehat{\phi}^{\text{error}}(\Theta, \mathbf{o}_\tau, \boldsymbol{\Delta}_\tau, \mathbf{o}_{t+1}), \quad \tau < t+1.$$

As explained in [11], the predicted error $\widehat{\boldsymbol{\Delta}}_{t+1}$ in the observation space can be used to estimate the error $\boldsymbol{\delta}_{t+1}$ in the model space as follows:

$$\boldsymbol{\Delta}_{t+1} \approx \mathbf{H}_t \cdot \boldsymbol{\delta}_{t+1}, \quad \widehat{\boldsymbol{\delta}}_{t+1} \approx \mathbf{H}_t \left(\mathbf{H}_t^T \mathbf{H}_t\right)^{-1} \mathbf{H}_t^T \cdot \widehat{\boldsymbol{\Delta}}_{t+1}, \tag{10a}$$

where we use the linearized observation operator at the current time, $\mathbf{H}_t = h'(\boldsymbol{x}_t)$. A more complex approach is to use a Kalman update formula:

$$\widehat{\boldsymbol{\delta}}_{t+1} \approx \text{cov}(\mathbf{x}_t, \mathbf{o}_t) \left(\text{cov}(\mathbf{o}_t, \mathbf{o}_t) + \mathbf{R}_t\right)^{-1} \widehat{\boldsymbol{\Delta}}_{t+1}, \tag{10b}$$

where $\mathbf{R}_t$ is the covariance of observation errors.

3.2 Problem Two: Identifying the Physical Packages that Contribute Most to the Forecast Uncertainty

Typical NWP models incorporate an array of different physical packages to represent multiple physical phenomena that interact with each other. Each physical package contains several alternative configurations (e.g., parameterizations or numerical solvers) that affect the accuracy of the forecasts produced by the NWP model. A particular scheme in a certain physical package best captures the reality under some specific conditions (for example time of the year, representation of sea-ice). The primary focus of this study is to accurately forecast the cumulative precipitation; therefore we seek to learn the impacts of all the physical packages that affect precipitation. To this end, we define the following mapping:

$$\phi^{\text{physics}}(\boldsymbol{\Delta}_t) \approx \Theta, \tag{11}$$

which estimates the configuration Θ of the physical packages such that the WRF run generates a forecast with an error consistent with the prescribed level $\boldsymbol{\Delta}_t$ (where $\boldsymbol{\Delta}_t$ defined in Eq. (5) is the forecast error in observation space at time t.)

We train the model to learn the effect of the physical schemes on the mismatch between WRF forecasts and reality. The input data required for the training process is obtained by running the model with various physical package configurations Θ_i^{train} and comparing the model forecast against the observations at all past times τ to obtain the corresponding errors $\boldsymbol{\Delta}_{\tau,i}^{\text{train}}$ for $\tau \leq t$ and $i \in \{\textit{training data set}\}$. The output data is the corresponding physical combinations Θ that leads to the input error threshold.

To estimate the physics configuration that contribute most to the uncertainty in predicting precipitation, we take the following approach. The dataset consisting of the observable discrepancies during the current time window $\boldsymbol{\Delta}_t$ is split into a training part and a testing part. In the test phase we use the approximated function $\widehat{\phi}^{\text{physics}}$ to estimate the physical process settings $\widehat{\Theta}_j^1$ that are consistent with the observable errors $\boldsymbol{\Delta}_{t,j}^{\{1\}}$. Here we select $\boldsymbol{\Delta}_{t,j}^{\{1\}} = \boldsymbol{\Delta}_{t,j}^{\text{test}}$ for each $j \in \{\textit{test data set}\}$. Note that in this case, since we know what physics has been used for the current results, we can take $\widehat{\Theta}_j^{\{1\}}$ to be the real parameter values $\Theta_j^{\{1\}}$ used to generate the test data. In general, $\boldsymbol{\Delta}_{t,j}^{\{1\}}$ is chosen appropriately for a given application and the corresponding parameters are estimated.

Next, we reduce the desired forecast error level to $\boldsymbol{\Delta}_{t,j}^{\{2\}} = \boldsymbol{\Delta}_{t,j}^{\{1\}}/2$ and use the approximated function $\widehat{\phi}^{\text{physics}}$ to estimate the physical process setting $\widehat{\Theta}_j^{\{2\}}$ that corresponds to this more accurate forecast. To identify the package setting that has the largest impact on the observable error, we monitor the variability in the predicted parameters $\widehat{\Theta}^{\{2\}} - \widehat{\Theta}^{\{1\}}$. Specifically, the number of times the setting of a physical process in $\widehat{\Theta}_j^2$ is different from its setting in $\widehat{\Theta}_j^1$ is an indicator of the variability in model prediction when that package is changed. A higher variability in predicted physical packages implies a larger contribution to the model errors as estimated by the learning model.

3.3 Machine Learning Algorithms

To approximate the functions (9) and (11), we use regression machine learning methods. Choosing the right learning algorithm is challenging; it largely depends on the problem and the data available [2,3,10,12]. Here, we use RF and ANN as our learning algorithms [13]. Both RF and ANN algorithms can handle nonlinearity in regression and classification. Given that the physical phenomena governing precipitation are highly nonlinear, and atmospheric dynamics is chaotic, we believe that RF and ANN approaches are well suited to capture the associated features. Although there are several other advanced machine learning algorithms that can be deployed here, we note that our aim here is to demonstrate the potential of machine learning approaches and hence the use of simple learning models such as RFs and ANN. Advanced techniques such as long short term memory (LSTM), convolutional neural networks (CNN), and gated recurrent units (GRU) are typically used for handling time series data and we defer such a study with these techniques to our future research. We describe the details regarding the training procedure, selection of the algorithm parameters, obtaining training data, and validation procedure in Sect. 4.

4 Numerical Experiments

We apply the proposed learning models to the WRF model [24] to (1) predict the bias in precipitation forecast caused by structural model errors, (2) predict the statistics associated with the precipitation errors, and (3) identify the specific physics packages that contribute most to precipitation forecast errors for given meteorological conditions.

4.1 WRF Model

In this study we use the non hydrostatic WRF model version 3.3. The simulation domain, shown in Fig. 1, covers the continental United States and has dimensions of 60×73 horizontal grid points in the west-east and south-north directions, respectively, with a horizontal grid spacing of 60 km [23]. The grid has 60 vertical levels to cover the troposphere and lower part of the stratosphere between the surface to approximately 20 km. In all simulations, the six-hourly analysis from the National Centers for Environmental Prediction (NCEP) are used as the initial and boundary conditions of the model [16]. The stage IV estimates are available at an hourly temporal resolution over the continental United States. For experimental purposes, we use the stage IV NCEP analysis as a proxy for the true state of the atmosphere. The simulation window begins at 6AM UTC (Universal Time Coordinated) on May 1, 2017.

The model configuration parameters Θ represent various combinations of microphysics schemes, cumulus parameterizations, short-wave, and long-wave radiation schemes. Specifically, each process is represented by the schema values of each physical parameter it uses, as detailed in WRF model physics options and references [25]. A total of 252 combinations of the four physical modules are used in the simulations. For each of the combinations, the effect of each physics combination on precipitation is investigated. The NCEP analysis grid points are 428 × 614, while the WRF computational model have 60 × 73 grid points. To obtain the discrepancy between the WRF forecast and NCEP analysis, we linearly interpolate the analysis to transfer the physical variables onto the model grid. Figure 1 shows the NCEP analysis at 12PM on May 1, 2017 which is used as "true" (verification) state. Figure 2 shows the forecast at 12PM on May 1, 2017. For the initial conditions, we use the NCEP analysis at 6PM on May 1, 2017. The WRF forecast corresponding to the physics microphysics: Kessler, cumulus physics: Kain-Fritsch, long-wave radiation physics: Cam, shirt-wave radiation physics: Dudhia is illustrated in Fig. 2. Figures 5 and 6 shows contours of discrepancies at 12PM ($\boldsymbol{\Delta}_{t=12\text{PM}}$) discussed in Eq. (5) for two physical combinations, which illustrates the effect that changing the physical schemes has on the forecast.

4.2 Experiments for Problem One: Predicting Pointwise Precipitation Forecast Errors over a Small Geographic Region

We demonstrate our learning algorithms to forecast precipitation in the state of Virginia on May 1, 2017, at 6PM. Our goal is to use the learning algorithms to correct the bias created due to model errors and hence improve the forecast for precipitation. As described in Sect. 3.1, we learn the function ϕ^{error} of Eq. (9) using the training data from the previous forecast window (6AM to 12PM):

$$\phi^{\text{error}}\left(\Theta, \mathbf{o}_\tau, \boldsymbol{\Delta}_\tau, \mathbf{o}_{t=12\text{PM}}\right) \approx \boldsymbol{\Delta}_{t=12\text{PM}}, \quad 7\text{AM} \leq \tau < 12\text{PM}.$$

We use two learning algorithms to approximate the function ϕ^{error}. Specifically, the RF with ten trees and CART learning tree algorithm in the forest and an ANN with four hidden layers and hyperbolic tangent sigmoid activation function in each layer are employed by using Scikit-learn, a machine learning library in Python [19]. For training purposes, we use the NCEP analysis of May 1, 2017 at 6AM as initial conditions for the WRF model. The forecast window is 6 h, and the WRF model forecast final simulation time is 12PM. The input features are as follows

- The physics combinations (Θ).
- The hourly WRF forecasts projected onto observation space o_τ, AM $\leq \tau \leq$ 12PM. The WRF state ($\mathbf{x}_t$) includes all model variables such as temperature, pressure, and precipitation. The observation operator extracts the precipitation portion of the WRF state vector, $\mathbf{o}_t \equiv \mathbf{x}_t^{\text{precipitation}}$. Accordingly, $\boldsymbol{\Delta}_t$ is the discrepancy between WRF precipitation forecast $\mathbf{o}_t$ and the observed precipitation $\mathbf{y}_t$.
- The observed discrepancies at past times ($\boldsymbol{\Delta}_\tau$, 7AM $\leq \tau <$ 12PM).

The output variable is the discrepancy between the NCEP analysis and the WRF forecast at 12PM, that is, the observable discrepancies for the current forecast

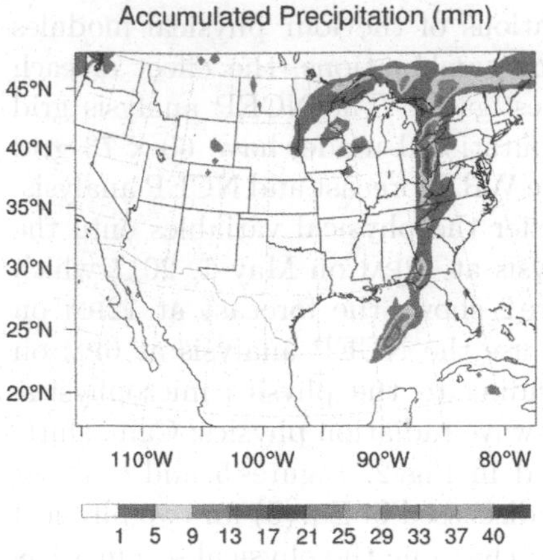

Fig. 1. NCEP analysis at 12PM provides a proxy for the true state of the atmosphere.

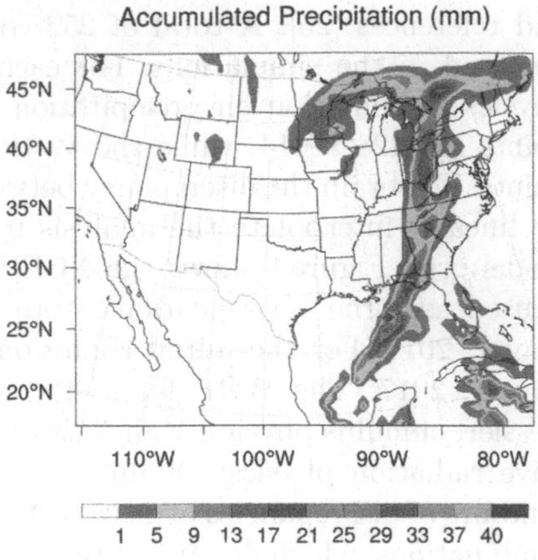

Fig. 2. WRF forecast at 12PM.

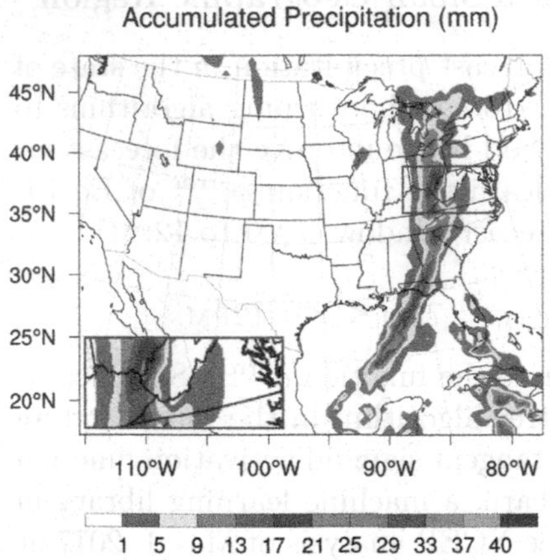

Fig. 3. Original WRF prediction at 6PM on May 1, 2017. Zoom-in panels show the predictions over Virginia.

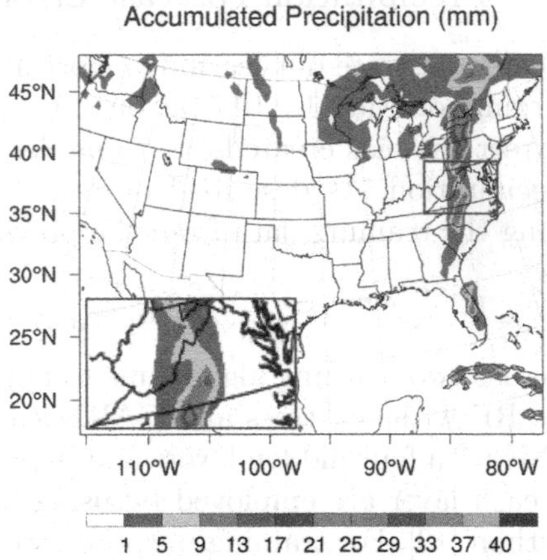

Fig. 4. NCEP analysis at 6PM on May 1, 2017. Zoom-in panels show the predictions over Virginia.

window ($\boldsymbol{\Delta}_{t=12\text{PM}}$). In fact, for each of the 252 different physical configurations, the WRF model forecast and the difference between the WRF forecast and the analysis are provided as input-output combinations for learning the function ϕ^{error}. The number of grid points over the state of Virginia is 14×12. Therefore for each physical combination we have 168 grid points, and the total number of samples in the training data set is $252 \times 168 = 42,336$ with 15 features.

Both ANN and RF are trained with these input-output combinations to obtain an approximation of the function ϕ^{error}, denoted by $\widehat{\phi}^{\text{error}}$, during the

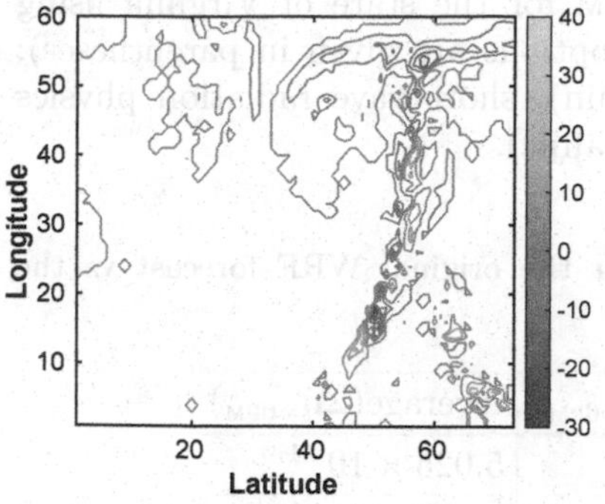

Fig. 5. Microphysics scheme: Kessler; cumulus physics: Kain-Fritsch; short wave radiation: Cam; long wave radiation: Dudhia

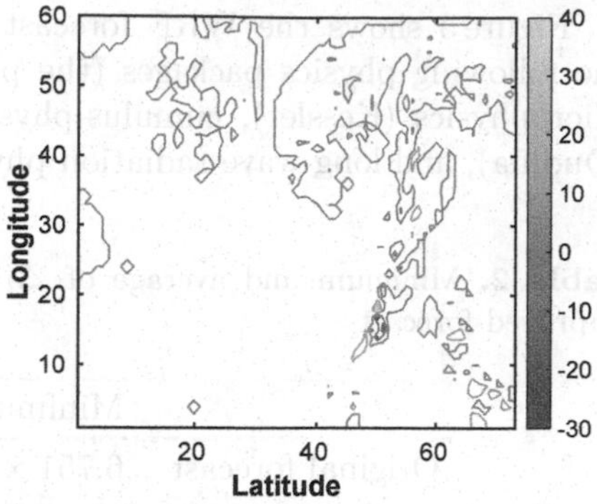

Fig. 6. micro-physics scheme: Lin; cumulus physics: Kain-Fritsch; short wave radiation: RRTM Mlawer; long wave radiation: Cam

training phase. In the testing phase we use the function $\widehat{\phi}^{\text{error}}$ to predict the future forecast error $\widehat{\boldsymbol{\Delta}}_{t=6\text{PM}}$ given the combination of physical parameters as well as the WRF forecast at time 6PM as input features.

$$\boldsymbol{\Delta}_{t=6\text{PM}} \approx \widehat{\phi}^{\text{error}}\left(\Theta, \mathbf{o}_\tau, \boldsymbol{\Delta}_\tau, \mathbf{o}_{t=6\text{PM}}\right), \quad 1\text{PM} \le \tau < 6\text{PM}.$$

To quantify the accuracy of the predicted error we calculate the root mean squared error (RMSE) between the true and predicted discrepancies at 6PM:

$$RMSE = \sqrt{\frac{1}{n}\sum_{i=1}^{n}\left(\widehat{\boldsymbol{\Delta}}^i_{t=6\text{PM}} - \boldsymbol{\Delta}^i_{t=6\text{PM}}\right)^2}, \tag{12}$$

where $n = 168$ is the number of grid points over Virginia. $\widehat{\boldsymbol{\Delta}}^i_{t=6\text{PM}}$ is the predicted discrepancy and $\boldsymbol{\Delta}^i_{t=6\text{PM}}$ is the actual discrepancy at the grid point i. The actual discrepancy is obtained as the difference between the NCEP analysis and the WRF forecast at time t = 6PM. This error metric is computed for each of the 252 physics configurations. The minimum, maximum, and average RMSE over the 252 runs is reported in Table 1.

Table 1. Minimum, average, and maximum RMSE between the predicted $\widehat{\boldsymbol{\Delta}}_{t=6\text{PM}}$ and the true $\boldsymbol{\Delta}_{t=6}$ over 252 physics combinations.

	Minimum $RMSE$	Average $RMSE$	Maximum $RMSE$
ANN	1.264×10^{-3}	1.343×10^{-3}	5.212×10^{-3}
RF	1.841×10^{-3}	1.931×10^{-3}	7.9×10^{-3}

Figure 3 shows the WRF forecast for 6PM for the state of Virginia using the following physics packages (the physics options are given in parentheses): Microphysics (Kessler), cumulus-physics (Kain), short-wave radiation physics (Dudhia), and long-wave radiation physics (Janjic).

Table 2. Minimum and average of $\boldsymbol{\Delta}_{t=6\text{PM}}$ for the original WRF forecast vs the improved forecast

	Minimum($\boldsymbol{\Delta}_{t=6\text{PM}}$)	Average($\boldsymbol{\Delta}_{t=6\text{PM}}$)
Original forecast	6.751×10^{-2}	5.025×10^{-1}
Improved forecast	2.134×10^{-4}	6.352×10^{-2}

Figure 4 shows the NCEP analysis at time 6PM, which is our proxy for the true state of the atmosphere. The discrepancy between the NCEP analysis and the raw WRF forecast is shown in the Fig. 7. Using the model error prediction we can improve the WRF result by adding the predicted bias to the WRF forecast. The discrepancy between the corrected WRF forecast and the NCEP analysis is shown in the Fig. 8. The results show a considerable reduction of model errors when compared with the uncorrected forecast of Fig. 7. Table 2 shows the minimum and average of original model error vs the improved model errors.

4.3 Experiments for Problem Two: Identifying the Physical Processes that Contribute Most to the Forecast Uncertainty

The interaction of different physical processes greatly affects the precipitation forecast, and we are interested in identifying the major sources of model errors in WRF. To this end we construct the physics mapping (11) using the norm and the statistical characteristics of the model-data discrepancy (over the entire U.S.) as input features:

$$\phi^{\text{physics}}\left(\bar{\boldsymbol{\Delta}}_{t=12\text{PM}}, \|\boldsymbol{\Delta}_{t=12\text{PM}}\|_2\right) \approx \Theta.$$

Statistical characteristics include the mean, minimum, maximum, and variance of the field across all grid points over the continental United States. Note that this is slightly different from (11) where the inputs are the raw values of these discrepancies for each grid point. The output variable is the combination of physical processes Θ that leads to model errors consistent with the input pattern $\bar{\boldsymbol{\Delta}}_{t=12\text{PM}}$ and $\|\boldsymbol{\Delta}_{t=12\text{PM}}\|_2$.

In order to build the dataset, the WRF model is simulated for each of the 252 physical configurations, and the mismatches between the WRF forecasts and the NCEP analysis at the end of the current forecast window are obtained. The discrepancy between the NCEP analysis at 12PM and the WRF forecast at 12PM forms the observable discrepancy for the current forecast window $\boldsymbol{\Delta}_{t=12\text{PM}}$. For each of the 252 physical configurations, this process is repeated and statistical characteristics of the WRF forecast model error $\bar{\boldsymbol{\Delta}}_{t=12\text{PM}}$, and the norm of model error $\|\boldsymbol{\Delta}_{t=12\text{PM}}\|_2$ are used as feature values of the function ϕ^{physics}.

Validation of the Learned Physics Mapping. From all the collected data points, 80% (202 samples) are used for training the learning model, and the remaining 20% (50 samples) are used for testing purposes. The learning model uses the training dataset to learn the approximate mapping $\widehat{\phi}^{\text{physics}}$. This function is applied to each of the 50 test samples $\boldsymbol{\Delta}^{\text{test}}_{t=12\text{PM}}$ to obtain the predicted physical combinations $\widehat{\Theta}_1$. To evaluate these predictions, we run the WRF model again with the $\widehat{\Theta}_1$ physical setting and obtain the new forecast $\widehat{\mathbf{o}}_{t=12\text{PM}}$ and the corresponding observable discrepancy $\widehat{\boldsymbol{\Delta}}^{\text{test}}_{t=12\text{PM}}$. The RMSE between the norm of actual observable discrepancies and the norm of predicted discrepancies is shown in Table 3. The small values of the difference demonstrates the performance of the learning algorithm.

Table 3. RMSE between estimated discrepancy using predicted physical combinations $\widehat{\boldsymbol{\Delta}}^{\text{test}}_{t=12\text{PM}}$ and the reference discrepancy $\boldsymbol{\Delta}^{\text{test}}_{t=12\text{PM}}$.

	$RMSE(\|\widehat{\boldsymbol{\Delta}}^{\text{test}}_{t=12\text{PM}}\|_2, \|\boldsymbol{\Delta}^{\text{test}}_{t=12\text{PM}}\|_2)$
ANN	4.1376×10^{-3}
RF	5.8214×10^{-3}

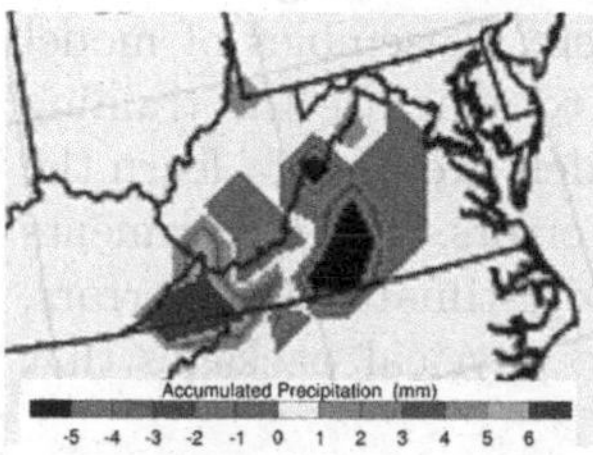

Fig. 7. Discrepancy between original WRF forecast and NCEP analysis

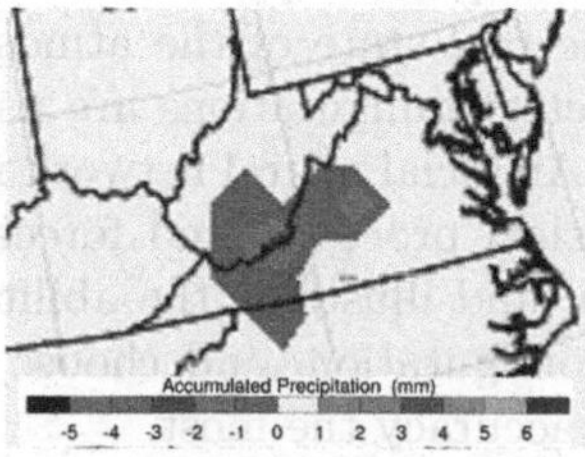

Fig. 8. Discrepancy between the corrected WRF forecast and the NCEP analysis

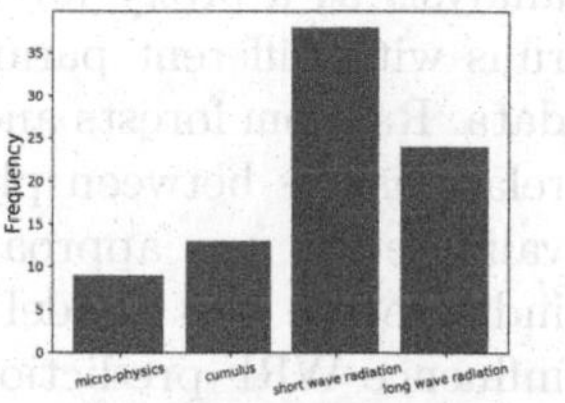

Fig. 9. Frequency of change in the physics with respect to change in the input data from $\boldsymbol{\Delta}^{\text{test}}_{t=12\text{PM}}$ to $\boldsymbol{\Delta}^{\text{test}}_{t=12\text{PM}}/2$. Each data set contains 50 data points, and we report the number of changes of each package.

Analysis of Variability in Physical Settings. We repeat the test phase for each of the 50 test samples with the scaled values of observable discrepancies $\boldsymbol{\Delta}_{t=12\text{PM}}^{\text{test}}/2$ as inputs and obtain the predicted physical combinations $\widehat{\Theta}_2$. The large variability in the predicted physical settings $\widehat{\Theta}$ indicates that the WRF forecast error is sensitive to the corresponding physical packages. We count the number of times the predicted physics $\widehat{\Theta}_2$ is different from $\widehat{\Theta}_1$ when the input data spans the entire test data set.

The results shown in Fig. 9 indicate that microphysics and cumulus physics are not too sensitive to the change of input data, whereas short-wave and long-wave radiation physics are quite sensitive to changes in the input data. Therefore our learning model indicates that having an accurate short-wave and long-wave radiation physics package will aid in greatly reducing the uncertainty in precipitation forecasts due to missing or incorrect physics.

5 Conclusions

This study proposes a novel use of machine learning techniques to understand, predict, and reduce the uncertainty in the WRF model precipitation forecasts. We construct probabilistic approaches to learn the relationships between the configuration of the physical processes used in the simulation and the observed model forecast errors. These relationships are then used to estimate the systematic model error in a quantity of interest at future times, and identify the physical processes that contribute most to the forecast uncertainty in a given quantity of interest under specified conditions.

Numerical experiments are performed with the WRF model using the NCEP analysis as a proxy for the real state of the atmosphere. Ensembles of model runs with different parameter configurations are used to generate the training data. Random forests and Artificial neural network models are used to learn the relationships between physical processes and forecast errors. The experiments validate the new approach, and illustrate the ability to estimate model errors, indicate the best model configuration, and choose the physical packages that influence WRF prediction accuracy the most.

As part of our future work, we will explore other advanced machine learning algorithms that fall under the broad category of recurrent neural nets (such as LSTM, GRU, and CNN) and are known to capture the spatial and temporal correlations well to reduce the uncertainty in medium and long-term forecasts.

Acknowledgments. This work was supported in part by the projects AFOSR DDDAS 15RT1037 and AFOSR Computational Mathematics FA9550-17-1-0205 and by the Computational Science Laboratory at Virginia Tech. The authors would like to thank Dr. Răzvan Ştefănescu for his valuable assistance and suggestions regarding WRF runs and the NCEP dataset.

References

1. Akella, S., Navon, I.M.: Different approaches to model error formulation in 4D-Var: a study with high-resolution advection schemes. Tellus A **61**(1), 112–128 (2009)
2. Asgari, E., Bastani, K.: The utility of Hierarchical Dirichlet Processfor relationship detection of latent constructs. In: Academy of Management Proceedings (2017)
3. Attia, A., Moosavi, A., Sandu, A.: Cluster sampling filters for non-Gaussian data assimilation. arXiv preprint arXiv:1607.03592 (2016)
4. Fovell, R.G.: Impact of microphysics on hurricane track and intensity forecasts. In: Preprints, 7th WRF Users' Workshop, NCAR (2006)
5. Fovell, R.G.: Influence of cloud-radiative feedback on tropical cyclone motion. In: 29th Conference on Hurricanes and Tropical Meteorology (2010)
6. Gilbert, R.C., Richman, M.B., Trafalis, T.B., Leslie, L.M.: Machine learning methods for data assimilation. In: Computational Intelligence in Architecturing Complex Engineering Systems, pp. 105–112 (2010)
7. Glimm, J., Hou, S., Lee, Y., Sharp, D., Ye, K.: Sources of uncertainty and error in the simulation of flow in porous media. Comput. Appl. Math. **23**, 109–120 (2004)
8. Krasnopol sky, V., Fox-Rabinovitz, M., Belochitski, A., Rasch, P.J., Blossey, P., Kogan, Y.: Development of neural network convection parameterizations for climate and NWP models using cloud resolving model simulations. US Department of Commerce, National Oceanic and Atmospheric Administration, National Weather Service, National Centers for Environmental Prediction (2011)
9. Lowrey, M.R.K., Yang, Z.L.: Assessing the capability of a regional-scale weather model to simulate extreme precipitation patterns and flooding in central Texas. Weather Forecast. **23**(6), 1102–1126 (2008)
10. Moosavi, A., Attia, A., Sandu, A.: A machine learning approach to adaptive covariance localization. arXiv preprint arXiv:1801.00548 (2018)
11. Moosavi, A., Sandu, A.: A state-space approach to analyze structural uncertainty in physical models. Metrologia (2017). http://iopscience.iop.org/10.1088/1681-7575/aa8f53
12. Moosavi, A., Stefanescu, R., Sandu, A.: Multivariate predictions of local reduced-order-model errors and dimensions. arXiv preprint arXiv:1701.03720 (2017)
13. Murphy, K.P.: Machine Learning: A Probabilistic Perspective. MIT Press, Cambridge (2012)
14. Nasrollahi, N., AghaKouchak, A., Li, J., Gao, X., Hsu, K., Sorooshian, S.: Assessing the impacts of different WRF precipitation physics in hurricane simulations. Weather Forecast. **27**(4), 1003–1016 (2012)
15. Navon, I.M., Zou, X., Derber, J., Sela, J.: Variational data assimilation with an adiabatic version of the NMC spectral model. Monthly Weather Rev. **120**(7), 1433–1446 (1992)
16. National Oceanic and Atmospheric Administration (NOAA). https://www.ncdc.noaa.gov/data-access/model-data/model-datasets/global-forcast-system-gfs
17. Orrell, D., Smith, L., Barkmeijer, J., Palmer, T.: Model error in weather forecasting. Nonlinear Processes Geophys. **8**, 357–371 (2001)
18. Palmer, T., Shutts, G., Hagedorn, R., Doblas-Reyes, F., Jung, T., Leutbecher, M.: Representing model uncertainty in weather and climate prediction. Annu. Rev. Earth Planet. Sci **33**, 163–93 (2005)
19. Pedregosa, F., et al.: Scikit-learn: machine learning in Python. J. Mach. Learn. Res. **12**, 2825–2830 (2011)

20. Rao, V., Sandu, A.: A posteriori error estimates for the solution of variational inverse problems. SIAM/ASA J. Uncertainty Quantification **3**(1), 737–761 (2015)
21. Tr'emolet, Y.: Accounting for an imperfect model in 4D-Var. Q. J. Roy. Meteorol. Soc. **132**(621), 2483–2504 (2006)
22. Trémolet, Y.: Model-error estimation in 4D-Var. Q. J. Roy. Meteorol. Soc. **133**(626), 1267–1280 (2007)
23. Wang, J., Kotamarthi, V.R.: Downscaling with a nested regional climate model in near-surface fields over the contiguous united states. J. Geophys. Res.: Atmos. **119**(14), 8778–8797 (2014)
24. Weather Research Forecast Model. https://www.mmm.ucar.edu/weather-research-and-forecasting-model
25. WRF Model Physics Options and References. http://www2.mmm.ucar.edu/wrf/users/phys_references.html

Kernel Embedded Nonlinear Observational Mappings in the Variational Mapping Particle Filter

Manuel Pulido[1,2(✉)], Peter Jan vanLeeuwen[1,3], and Derek J. Posselt[4]

[1] Department of Meterology, University of Reading, Reading, UK
m.a.pulido@reading.ac.uk
[2] Department of Physics, Universidad Nacional del Nordeste, Corrientes, Argentina
[3] Department of Atmospheric Science, Colorado State University, Fort Collins, USA
[4] Jet Propulsion Laboratory, California Institute of Technology, Pasadena, CA, USA

Abstract. Recently, some studies have suggested methods to combine variational probabilistic inference with Monte Carlo sampling. One promising approach is via local optimal transport. In this approach, a gradient steepest descent method based on local optimal transport principles is formulated to deterministically transform point samples from an intermediate density to a posterior density. The local mappings that transform the intermediate densities are embedded in a reproducing kernel Hilbert space (RKHS). This variational mapping method requires the evaluation of the log-posterior density gradient and therefore the adjoint of the observational operator. In this work, we evaluate nonlinear observational mappings in the variational mapping method using two approximations that avoid the adjoint, an ensemble based approximation in which the gradient is approximated by the sample cross-covariances between the state and observational spaces the so-called ensemble space and an RKHS approximation in which the observational mapping is embedded in an RKHS and the gradient is derived there. The approximations are evaluated for highly nonlinear observational operators and in a low-dimensional chaotic dynamical system. The RKHS approximation is shown to be highly successful and superior to the ensemble approximation for non-Gaussian posterior densities.

Keywords: Variational inference · Stein discrepancy · Data assimilation

1 Introduction

There is a large number of applications in which the process of interest is not directly measured, a latent process, but it is related through a map to another process which is the one observed. This problem can be framed in the classical Bayesian inference, in which the latent process is inferred from indirect noisy observations [19]. The mapping between the two processes will here be referred

J. M. F. Rodrigues et al. (Eds.): ICCS 2019, LNCS 11539, pp. 141–155, 2019.
https://doi.org/10.1007/978-3-030-22747-0_11

to as the observational mapping. Depending on the application, the observational mapping may be (partially) known through the knowledge of the physical processes involved. An example of particular interest in this work is the inference of atmospheric state variables from satellite measurements of radiation. In other applications, the map is unknown and needs to be estimated. This is one of the central aims in machine learning applications [20].

In modeling and predicting the atmosphere, clouds play a central role. Measurements from spaceborne radars may give information on cloud properties. In this case, the observed variables are radar reflectivity and microwave radiances, while the variables of interest are cloud particle concentrations and distributions. This mapping is represented in models through parameterizations which relate cloud microphysical processes to precipitation and radiative fluxes. In several situations, the joint posterior density of model parameters and the output variables is bimodal [12]. The main factor responsible for the bimodal density is the extremely nonlinear response of model output variables to changes in microphysical parameters. The parameter prior density and observation uncertainty only play a secondary role in the resulting complexity of the posterior density.

If the latent process is governed by a time evolving stochastic dynamical system, the inference is sequential. The time evolution of the latent state is given by a Markov process–the dynamical system– while an observational mapping relates observations with the latent state. These are known as state-space models or hidden Markov models. A rather general method for inference in hidden Markov models is based on Monte Carlo sampling of the prior density, referred to as sequential Monte Carlo or particle filtering [4]. One of the major challenges in high-dimensional particle filtering is to concentrate sample points in the high-probability regions of the posterior density, the so-called typical set. In this case, they produce a non-negligible contribution to expectation estimations. Therefore, sample points are required to be located in the typical set to make the most of them.

Recent works propose to combine variational inference with Monte Carlo sampling [17]. A rigorous well-grounded framework to combine them is via local optimal transport [11,16]. Optimal transport relates a given density with a target density trough a mapping that minimizes a risk. Hence, optimal transport concepts may be used to move sample points to locations where they maximize the amount of Shannon information they can provide. If the mapping function space is constrained to a reproducing kernel Hilbert space, the local direction that minimizes the risk, in terms of the Kullback-Leibler divergence, is well defined. This direction minimizes the Stein discrepancy [10]. An application of the variational mapping using the Stein gradient to sequential Monte Carlo methods in the framework of hidden Markov models was recently developed [16] and has been referred to as variational mapping particle filter (VMPF).

The gradient of the observation likelihood depends on the adjoint of the observational mapping. Thus, most of the approximations used for posterior inference including MAP estimation, the Kalman filter, and the stochastic and square-root ensemble Kalman filters (e.g. [1,7]) require this adjoint of the observational mapping. However, there is a rather large number of complex observational mappings for which the adjoint is not available. In the context of the ensemble Kalman filter, an ensemble approximation of the adjoint of the observational mapping is used [6,7]. However, this approximation may have a detrimental effect in the inference for the highly nonlinear observational mapping of e.g. cloud parameter estimation [13–15] and in other geophysical applications [3].

A description of the VMPF in the context of observational mapping is given in Sect. 2. Two approximations of the adjoint of the observational mapping based on sample points evaluations of the observational operator are derived in Sects. 2.1 and 2.2). Details of the experiments are given in Sect. 3. The VMPF with the exact gradient of the logarithm of the posterior density and the developed approximations is evaluated with nonlinear observational operators in low-dimensional spaces (Sect. 4). The performance of the VMPF in a chaotic dynamical system with a nonlinear observational mapping is also discussed.

2 Observational Function with Variational Mappings

Suppose we want to determine a stochastic latent process $\mathbf{x}$ in $\mathbb{R}^{N_x}$, only sparse observations of another related process $\mathbf{y}$ in $\mathbb{R}^{N_y}$ are available. The relationship between the processes is given through a known nonlinear observational operator $\mathcal{H}$ such that

$$\mathbf{y}_k = \mathcal{H}(\mathbf{x}_k) + \boldsymbol{\eta}_k, \tag{1}$$

where $\boldsymbol{\eta}_k$ is the random observational error which consists of realizations from a density, $p(\boldsymbol{\eta})$, that describes the measurement and representation error, k denotes different realizations of the stochastic process. We assume the observational errors are unbiased, $\mathcal{E}(\boldsymbol{\eta}) = 0$.

Using Bayes rule, the density of the latent process conditioned on the realizations of the observed process is

$$p(\mathbf{x}|\mathbf{y}) \propto p(\mathbf{y}|\mathbf{x})p(\mathbf{x}). \tag{2}$$

Let us assume the prior knowledge of $\mathbf{x}$ is through a sample $\{\mathbf{x}^j, j = 1, \cdots, N_p\} \triangleq \mathbf{x}^{1:N_p}$. A standard importance sampling technique for Bayesian inference assumes that the prior density $p(\mathbf{x})$ is a proposal density of $p(\mathbf{x}|\mathbf{y})$ so that this posterior distributions is written as the sample points of the prior density with weights given by the likelihood of the sample at the points [4]. A better proposal density may be considered assuming knowledge of the observation. In this case, weights are expected to be more equally distributed within sample points so that the variance of the weights is smaller.

Our aim is to find a sequence of mappings, $\mathbf{x}_i = T_i(\mathbf{x}_{i-1})$ that transforms from sample points of $p(\mathbf{x})$ to sample points of $p(\mathbf{x}|\mathbf{y})$. Considering these mappings, the relationship between the transformed density after the mappings and the initial density is

$$q(\mathbf{x}_I(\mathbf{x}_0)) = \prod_{i=1}^{I} |\nabla T_i| \, q(\mathbf{x}_0), \tag{3}$$

where the initial density $q(\mathbf{x}_0)$ is in principle the prior density, while the target density of the transformations is the posterior density $p(\mathbf{x}_0|\mathbf{y})$ and $|\nabla T_i|$ are the Jacobians of the transformations.

Therefore, in order to get equally-weighted sample points that optimally represent the posterior density, we have to find a series of maps T that transforms the prior into the posterior density. In terms of the particles, the goal is to drive them from the prior density to the posterior density. In this work, sample points will also be referred to as particles interchangeably. This process, of driving the particles from one to other density, could be framed as maximizing the likelihood of the particles. Alternatively, it can be formulated as a Kullback-Leibler divergence (KLD) optimization given the well-known equivalence between marginal likelihood maximization and KLD minimization. The KLD between the intermediate density and the target density is

$$\mathcal{D}_{KL}(q_T\|p) = \int q_T(\mathbf{x}) \log \left[\frac{q_T(\mathbf{x})}{p(\mathbf{x}|\mathbf{y})}\right] \mathrm{d}\mathbf{x} \tag{4}$$

The aim is to determine the *local* transformation T that produces the deepest descent in KLD. The derivation of the steepest descent gradient has been already given in previous works [10,16]. Assuming the transformation T is in a reproducing kernel Hilbert space (RKHS), the gradient of the KLD is given by

$$\nabla\mathcal{D}_{KL}(\mathbf{x}) = -\int \left[K(\mathbf{x}',\mathbf{x})\nabla \log p(\mathbf{x}'|\mathbf{y}) + \nabla_{\mathbf{x}'}K(\mathbf{x}',\mathbf{x})\right] \mathrm{d}\mathbf{x}' \tag{5}$$

where $K(\mathbf{x}', \mathbf{x})$ is the reproducing kernel and the gradient is at $\mathbf{x}$ where the local transformation is produced.

Each of the particles is moved along the steepest descent direction $\mathbf{v}(\mathbf{x}) = -\nabla\mathcal{D}_{KL}$,

$$\mathbf{x}_{i+1}^j = T_{i+1}(\mathbf{x}_i^j) = \mathbf{x}_i^j + \epsilon\mathbf{v}(\mathbf{x}_i^j). \tag{6}$$

The particles are tracers in a flow given by the KLD gradient. In essence, the objective is to determine the direction of steepest descent at each sample point and to move them along these directions. The pseudo-time step ϵ in (6) should be small enough so that the particle trajectories do not intersect and therefore the smoothness of the flow is conserved. The overall performance of the variational mapping in a sequential Monte Carlo algorithm is evaluated in [16] and is termed as the variational mapping particle filter (VMPF).

To obtain the gradient of the Kulback-Leibler divergence at a sample point (5), we require an analytical expression of the log-posterior gradient, which can be expressed in terms of the prior density and the observation likelihood using (2),

$$\nabla \log p(\mathbf{x}|\mathbf{y}) = \nabla \log p(\mathbf{x}) + \nabla \log p(\mathbf{y}|\mathbf{x}) \tag{7}$$

Assuming Gaussian observational errors, $p(\boldsymbol{\eta}) \sim \mathcal{N}(\mathbf{0}, \mathbf{R})$, and using (2), the log-observation likelihood gradient is

$$\nabla \log p(\mathbf{y}|\mathbf{x}) = (\nabla\mathcal{H}(\mathbf{x}))^\top \mathbf{R}^{-1}(\mathbf{y} - \mathcal{H}(\mathbf{x})). \tag{8}$$

The observational operator has a major role in (8). For a linear observational mapping, a linear log observation likelihood gradient results. On the other hand, a nonlinear mapping produces a nonlinear likelihood gradient. Therefore, it induces a non-Gaussian posterior distribution. In the case of a non-injective mapping, more than one root of (8) are expected, which results in a multimodal posterior density. This rather common feature in the observational mapping is examined exhaustively in the experiments.

2.1 Observational Mapping in the RKHS

For many applications in geophysical systems, the observational operator is a rather complex mapping that involves physical processes, for instance as mentioned the transformation from cloud properties to observed radar reflectivity. Even when the observational operator is available, the tangent-linear and adjoint operators of the observational mapping are often not available and their development and use could be costly in terms of human resources and computationally demanding in its evaluation. In this work, we derive a Monte Carlo approximation of the term $(\nabla \mathcal{H}(\mathbf{x}))^\top$ in (8). In coherence with the formulation of the variational mapping, we now also assume that the process $\mathcal{H}(\mathbf{x})$ is in the reproducing kernel Hilbert space (RKHS). This assumption is similar to that in support vector machines, where the mapping is also assumed to lie in an RKHS [20]. In that case, we can use the reproducing property for $\mathcal{H}(\mathbf{x})$,

$$\mathcal{H}(\mathbf{x}) = \langle \mathcal{H}(\mathbf{x}')|\, K(\mathbf{x}, \mathbf{x}')\rangle . \tag{9}$$

where $\langle \cdot | \cdot \rangle$ is the RKHS inner product. Using the N_p particles $\mathbf{x}^{1:N_p}$ to generate a finite Hilbert space, the Monte Carlo approximation of the gradient of (9) is

$$\nabla \mathcal{H}(\mathbf{x}) \approx \frac{1}{N_p} \sum_{j=1}^{N_p} \mathcal{H}(\mathbf{x}^j) \nabla_{\mathbf{x}} K(\mathbf{x}, \mathbf{x}^j). \tag{10}$$

We have now an expression of the gradient of the observational operator that only depends on its evaluation at the particle positions. From the RKHS theory, we know that the approximated value in (10) will converge towards the exact one when $N_p \to \infty$ assuming $\mathcal{H}(\mathbf{x})$ is sufficiently smooth. Convergence of the gradient in the RKHS has been examined in [21].

The expression of the gradient of the Kullback-Leibler divergence of the VMPF (5) using a Monte Carlo integration is

$$\nabla \mathcal{D}_{KL}(\mathbf{x}) = -\frac{1}{N_p} \sum_{l=1}^{N_p} \left[K(\mathbf{x}^l, \mathbf{x}) \nabla \log p(\mathbf{x}^l|\mathbf{y}) + \nabla_{\mathbf{x}^l} K(\mathbf{x}^l, \mathbf{x}) \right]. \tag{11}$$

using (7) and (10) in (11), the gradient becomes

$$\nabla \mathcal{D}_{KL}(\mathbf{x}) = -\frac{1}{N_p} \sum_{l=1}^{N_p} \Bigg\{ K(\mathbf{x}^l, \mathbf{x}) \Bigg[\nabla \log p(\mathbf{x}^l) + \left(\frac{1}{N_p} \Sigma_j \mathcal{H}(\mathbf{x}^j) \nabla_{\mathbf{x}^l} K(\mathbf{x}^l, \mathbf{x}^j) \right)^\top \mathbf{R}^{-1}(\mathbf{y} - \mathcal{H}(\mathbf{x})) \Bigg] + \nabla_{\mathbf{x}^l} K(\mathbf{x}^l, \mathbf{x}) \Bigg\}. \tag{12}$$

This expression depends only on the evaluation of the observational operator at the sample points. Therefore, the number of evaluations of the observational operator in (12) is N_p at each mapping iteration. No extra evaluations from the original variational mapping are required. The Gram matrix and the gradient of the kernels are already available since they are required in the second right-hand side term of (11). In conclusion, the main complexity of the algorithm is still of order N_p^2 as in the original VMPF.

There is a problem for the RKHS approximation of the observational mapping (10) in the regions where the sample points are sparse. An experiment to illustrate this drawback is shown in Sect. 3. This problem appears because the kernel values between those sparse points and the rest of the sample points have only few points (the closest ones to the one in consideration) with non-negligible contributions and all the other kernel values are (close to) 0. Note that the square of the bandwidth is chosen to be smaller than the trace of the sample covariance to allow for more detailed structures in the density of $\mathcal{H}(\mathbf{x})$. One way to solve this problem could be using an adaptive kernel bandwidth based on the distance to the k-nearest neighbors. A simpler solution is to normalize the contributions of the kernels

$$\mathcal{H}(\mathbf{x}) \approx \frac{\sum_{j=1}^{N_p} \mathcal{H}(\mathbf{x}^j) K(\mathbf{x}, \mathbf{x}^j)}{\sum_{l=1}^{N_p} K(\mathbf{x}, \mathbf{x}^l)} \tag{13}$$

In this way, the contribution of the kernel functions evaluated at each sample point is normalized. This approximation to the gradient of the observational mapping is evaluated in the experiments.

2.2 Observational Operator in the Ensemble Space

Instead of constraining the observational operator to the RKHS, it can be expressed in the ensemble perturbation space. This type of approximations is common in ensemble Kalman filtering. Indeed, the whole estimation problem may be transformed and determined in the ensemble perturbation space [7]. Here, we derive an approximate expression for the tangent-linear model, i.e. the gradient of the observational mapping, and its adjoint model based on the perturbations of the particles (ensemble members) to the mean.

The increments are approximated with a first-order Taylor series around the mean $\overline{\mathbf{x}}$

$$\mathbf{y} - \mathcal{H}(\mathbf{x}) \approx \mathbf{y} - \mathcal{H}(\overline{\mathbf{x}}) - \mathbf{H}(\mathbf{x} - \overline{\mathbf{x}}), \tag{14}$$

where $\mathbf{H}$ is the tangent-linear operator of $\mathcal{H}$ at $\overline{\mathbf{x}}$. The perturbation matrices in the state and observational spaces are composed by the differences between the ensemble members and the mean, namely

$$\mathbf{X} = \frac{1}{\sqrt{N_p - 1}} \left(\mathbf{x}^{(1)} - \overline{\mathbf{x}}, \mathbf{x}^{(2)} - \overline{\mathbf{x}}, \cdots, \mathbf{x}^{(N_p)} - \overline{\mathbf{x}} \right), \tag{15}$$

$$\mathbf{Y} = \frac{1}{\sqrt{N_p - 1}} \left(\mathcal{H}(\mathbf{x}^{(1)}) - \overline{\mathcal{H}(\mathbf{x})}, \cdots, \mathcal{H}(\mathbf{x}^{(N_p)}) - \overline{\mathcal{H}(\mathbf{x})} \right). \tag{16}$$

where $\mathbf{X}$ is an $N_x \times N_p$ matrix and $\mathbf{Y}$ is $N_y \times N_p$. The included normalization factor is $\sqrt{N_p - 1}$ to avoid bias in the sample covariance. Thus, the prior sample covariance is $\mathbf{P} = \mathbf{X}\mathbf{X}^\top$.

The approximated tangent-linear operator of $\mathcal{H}$ at the ensemble space is then given by

$$\nabla\mathcal{H}(\overline{\mathbf{x}}) = \mathbf{H} \approx \mathbf{Y}\mathbf{X}^\dagger, \tag{17}$$

where $\mathbf{P}_{yx} \triangleq \mathbf{Y}\mathbf{X}^\top$. For the adjoint approximation, the transpose of non-Gaussianity in the posterior density for the inverse model is the nonlinearity in the observational operator. The prior density is N (0.5, 1). The observation corresponds to a true state of 3 with an observational error of R = 0.5. The approximations are evaluated with two nonlinear observation operators. We use a quadratic relationship, H(x) = x2, which is expected to lead to a bimodal posterior density because of its non-injectivity. Non-injective observational operators associate an observation with more than one state. The observation likelihood function then contains several maxima. Therefore, if the prior density is non-null for these states associated to the likelihood maxima, they result in a multimodal posterior density. For a challenging evaluation of the of (17) is used,

$$\nabla\mathcal{H}(\overline{\mathbf{x}})^\top \approx \mathbf{H}^\top = (\mathbf{X}^\dagger)^\top\mathbf{Y}^\top.$$

3 Experiments

In the observational mapping experiments, an iid sample from the prior density is given, which for simplicity is assumed in general to be normally distributed. Furthermore, observational errors are assumed Gaussian. Thus, the only source of non-Gaussianity in the posterior density for the inverse model is the nonlinearity in the observational operator. The prior density is $\mathcal{N}(0.5, 1)$. The observation corresponds to a true state of 3 with an observational error of $R = 0.5$.

The approximations are evaluated with two nonlinear observation operators. We use a quadratic relationship, $\mathcal{H}(x) = x^2$, which is expected to lead to a bimodal posterior density because of its non-injectivity. Non-injective observational operators associate an observation with more than one state. The observation likelihood function then contains several maxima. Therefore, if the prior density is non-null for these states associated to the likelihood maxima, they result in modes of the posterior density. For a challenging evaluation of the gradient approximations of the observational operator, we also use the absolute value $y = |x|$ which contains a discontinuity in its derivative. Representations of the observational mapping through a small number of basis functions is expected to give an inaccurate approximation of this derivative. Although these observational operators are only motivated in evaluating the approximations, they are indeed found in several applications. In particular, the absolute value is a frequent operator that appears when measuring wind and current speeds with instruments that are not able to distinguish flow direction.

Note that the prior density we use is not symmetric around 0, while the chosen observational operators are. Besides, the true state is in a region of very small prior density. These choices have been taken so that the gradient approximation from sample points represents a challenge.

Algorithm 1. Variational mapping algorithm

Input: Given $\mathbf{x}_0^{(1:N_p)}$, $\mathbf{y}$, $\mathcal{H}(\cdot)$, and $p(\boldsymbol{\eta})$
repeat ▷ Mapping iterations
  for $j = 1, N_p$ **do**
    $\mathbf{x}_i^{(j)} \leftarrow \mathbf{x}_{i-1}^{(j)} - \epsilon \nabla \mathcal{D}_{KL}(\mathbf{x}_{i-1}^{(j)})$ ▷ $\nabla \mathcal{D}_{KL}$ using different particle approximations (11), (12). ▷ ϵ obtained with ADAM.
  end for
  i ← i+1
until Stopping criterion met $|\nabla \mathcal{D}_{KL}|/|\nabla \mathcal{D}_{KL0}| < \delta$
Output: $\mathbf{x}_i^{(1:N_p)}$

Algorithm 2. VMPF algorithm

Input: Given $\mathbf{x}_{k-1}^{(1:N_p)}$, $\mathbf{y}_k$, $\mathcal{H}(\cdot)$, $\mathcal{M}(\cdot)$, and $p(\boldsymbol{\eta})$
for $j = 1, N_p$ **do**
  $\mathbf{x}_{k,0}^{(j)} \leftarrow \mathcal{M}(\mathbf{x}_{k-1}^{(j)}, \boldsymbol{\eta}_k)$ ▷ Forecast stage
end for
repeat ▷ Mapping iterations
  for $j = 1, N_p$ **do**
    $\mathbf{x}_{k,i}^{(j)} \leftarrow \mathbf{x}_{k,i-1}^{(j)} - \epsilon \nabla \mathcal{D}_{KL}(\mathbf{x}_{k,i-1}^{(j)})$ ▷ $\nabla \mathcal{D}_{KL}$ using different particle approximations (11), (12). ▷ ϵ obtained with ADAM.
  end for
  i ← i+1
until Stopping criterion met $|\nabla \mathcal{D}_{KL}|/|\nabla \mathcal{D}_{KL0}| < \delta$
Output: $\mathbf{x}_{k,i}^{(1:N_p)}$

In a last set of experiments, we evaluate the use of a nonlinear observation operator in a chaotic dynamical system. The state of the 3-variable Lorenz-63 dynamical system corresponds to the latent process. Observations are obtained with the absolute observational mapping from the latent state and a Gaussian noise. The results from the VMPF using 100 particles are compared with the SIR particle filter [4] using 1000 and 10000 particles.

The pseudo-code of the variational mapping methodology for the observational mapping is shown in Algorithm 1. A single posterior density is estimated through the mapping iterations. The particles are moved along the steepest descent direction as in traditional multidimensional optimization. However, multiple points of the cost function, i.e. KL divergence, are followed at the same time. Furthermore, the distribution of these sample points defines the gradient of the cost function in each iteration. In other words, the particles –sample points– interact during the optimization. The termination criterion of the optimization is based on the mean value of the modulus of the KLD gradient (considering all the points). The pseudo-code of the variational mapping particle filter includes the estimation of a posterior density for each cycle–each time observations are available. The posterior density at one cycle is propagated through the set of particles using the dynamical model to obtain the sample prediction density at the next observation time. Algorithm 2 shows the variational mapping particle filter pseudo-code. A detailed description of the VMPF may be found in [16].

The optimization in the VMPF is conducted through ADAM [9], a second-moment optimization method. The tuning parameters are set to the recommended values, first moment parameter $\beta_1 = 0.9$ and second moment parameter

$\beta_2 = 0.99$. The learning rate was set to 0.03. The maximum number of optimization iterations is set to 500 (this is not reached in any of the experiments), and the criterion for termination is based on the mean value of $|\nabla \mathcal{D}_{KL}|$, the required threshold is $|\nabla \mathcal{D}_{KL}|/|\nabla \mathcal{D}_{KL1}| < 0.01$ where $|\nabla \mathcal{D}_{KL1}|$ corresponds to the first iteration. The required number of iterations under these settings is about 100–150 in the observational mapping experiments and about 50 iterations in the dynamical experiments, however a few cycles may require more than 200 iterations. For more computationally consuming experiments, a higher learning rate may be more efficient. However, we have priorized the smoothness of the mappings in these proof-of-concept experiments.

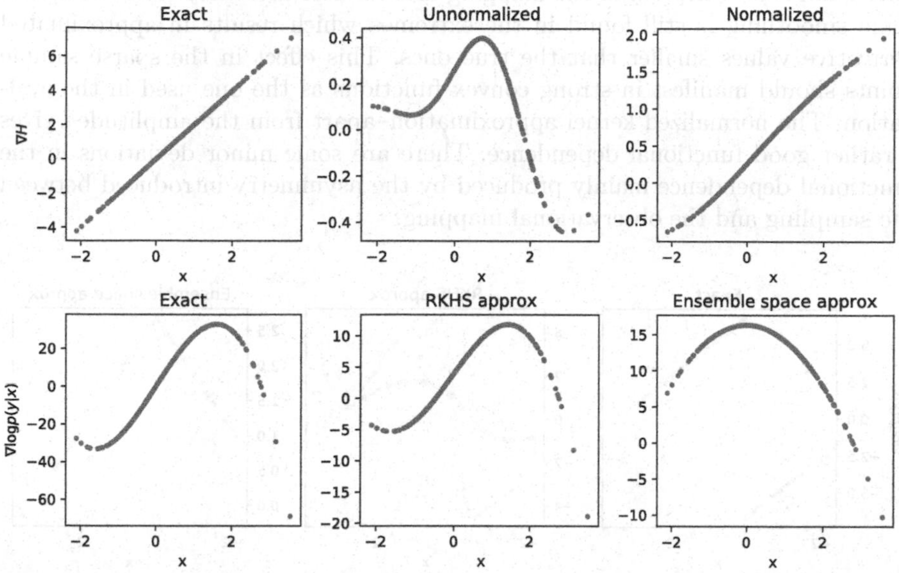

Fig. 1. The gradient of a quadratic observational operator, $\mathcal{H}(x) = x^2$, represented by the samples of the prior density for the exact calculation (left upper panel), the approximation using unnormalized kernels (middle upper panel) and using normalized kernels (right upper panel). The gradient of the log-observation likelihood, (8) with exact gradient (left lower panel), normalized RKHS approximation (middle lower panel) and ensemble approximation (right lower panel).

For all the experiments, radial basis functions are used as kernels. A Mahalanobis distance is taken, $K(\mathbf{x}, \mathbf{x}') = \exp\left(-\|\mathbf{x} - \mathbf{x}'\|^2_{\mathbf{A}}\right)$. The Mahalanobis matrix $\mathbf{A}$, hereinafter referred to as kernel covariance, is chosen proportional to the prior sample covariance in the observational mapping experiments and the model error covariance in the dynamical system experiment. The proportionality factor, which could be interpreted as the bandwidth of an isotropic kernel, is determined with the Scott rule. However, some extra manual tuning of it was required for some of the experiments.

4 Results

Figure 1 shows the results of the derivative of the quadratic observational operator (upper panels) represented by using the sample points of the prior density. The exact calculation is shown in left upper panel of Fig. 1. The approximation in the RHKS using unnormalized kernel functions in the finite space, (10), is in the middle panel and the one using normalized kernels, (13), in the right upper panel. The approximation to the mapping gives small values in isolated sample points because of only a few points contribute to the kernel integration in sparse areas. The normalization factor incorporates weights according to the density of points around the samples, producing a better estimate of the functional representation of the mapping and its derivative. However, note that some smoothing is still found in the extremes which results in approximated derivative values smaller that the true ones. This effect in the sparse sample points should manifest in strong convex functions as the one used in the evaluation. The normalized kernel approximation–apart from the amplitude– gives a rather good functional dependence. There are some minor deviations in the functional dependence mainly produced by the asymmetry introduced between the sampling and the observational mapping.

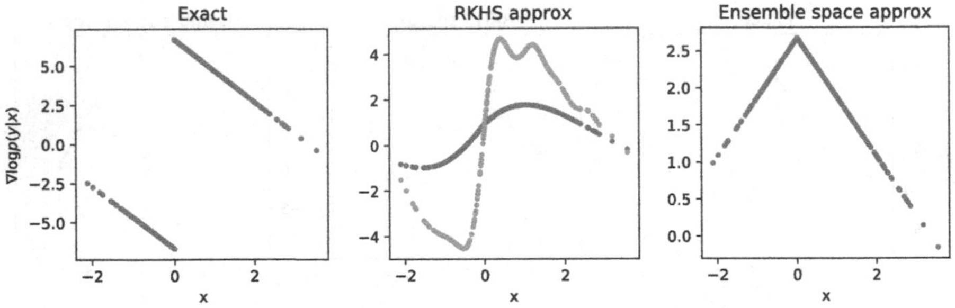

Fig. 2. Gradient of the observation mapping $\mathcal{H}(x) = |x|$ for the exact calculation, the approximation using kernels and the approximation using perturbations in the ensemble space. Two kernel bandwidths, $\gamma = 1$ (blue dots) and $\gamma = 0.3$ (orange dots) are shown for the RKHS approximation. (Color figure online)

The impact of approximating $\nabla\mathcal{H}$ on the gradient of the observation likelihood is shown in the lower panels of Fig. 1. The overall structure using the RKHS approximation is recovered. However, the amplitude of the gradient is underestimated. The ensemble space approximation gives a constant gradient of the observational mapping independent of the sample points, (17), which is expected to give the mean gradient of the mapping. In terms of the gradient of the observation likelihood, it results in a quadratic function, because of the increment term in (8) between observations and the particles. This would only be a good approximation of the true observation likelihood gradient (left panel) close to the observation. For methods that only give the maximum a posteriori solution, a relatively coarse representation of the gradient of the log-likelihood function may be enough to give a good estimation. Thus, they only require a precise gradient of the observational operator close to the observation. On the other hand, an accurate representation in a larger region is required if the inference problem also deals with uncertainty quantification.

Results for the absolute observational operator are shown in Fig. 2. Gaussian kernels act as smoothers (e.g. [20]), so that the approximation with Gaussian kernels to the absolute function is a smooth function and so the derivative is similar to a tanh-function with a smooth transition between the positive and negative values. The transition can be more abrupt if the kernel bandwidth is reduced from $\gamma = 1$ to 0.3 (middle panel in Fig. 2). However, the sampling noise is increased in that case. Note also that the amplitude of the function approximation is closer to the true one for the narrower kernel bandwidth. A narrower kernel bandwidth uses less sample points to approximate the mapping. Hence, it diminishes the smoothing. The ensemble space average produces a correct gradient of the log-likelihood close to the observations in the positive state values, but a wrong one for negative state values (lower right panel). Because the amplitude in $\nabla\mathcal{H}$ results from an average of all the sample points, it is underestimated in the absolute mapping and so in the gradient of the log-likelihood.

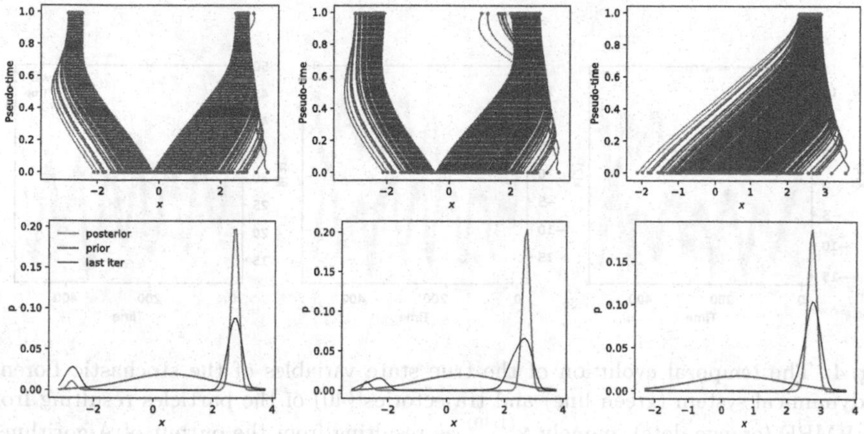

Fig. 3. Evolution in pseudo-time of the sample points for a quadratic observational mapping (upper panels) for the experiment with exact gradient (left panel), RKHS approximation (middle panel) and ensemble approximation (right panel). Posterior density (lower panels) for the exact quadratic observational mapping (red line) and the one obtained with VMPF (black line) (Color figure online)

Figure 3 exhibits the trajectories of the sample points as a function of pseudo-time between the initial iteration of the filter (representing the prior density) and up to the convergence criterion is met which is based on the module of the gradient of the Kullback-Leibler divergence. The experiment corresponds to the quadratic mapping. In both the exact and the RKHS approximation, samples are attracted to two different positive/negative regions which represent a bimodal posterior density. Because of the asymmetry in the prior density (whose the mean is 0.5) more particles are attracted to the positive region. The particles finish more disperse in the RKHS approximation than in the exact gradient calculation. The ensemble space approximation for the gradient of the observational mapping removes the bimodality of the posterior density and the particles are only attracted by the dominant mode (right panel). Lower panels in Fig. 3 compare the analytical posterior density with the one obtained with the VMPF, representing the final sample with kernel density estimation in coherence with the RKHS used in the mappings. The VMPF using the exact observational mapping is shown in the left panel of Fig. 3, the one using the RKHS approximation (middle panel) and the ensemble approximation (right panel). The exact case shows some smoothing of the main mode mainly because the observation is at a low density region of the likelihood. Tests with a narrower kernel bandwidth diminish the effect but it does not disappear. In the case of the RKHS approximation, there is some spread of the sample points toward lower values. This effect may be linked to the lower values of the gradient of the likelihood in this approximation. The ensemble approximation removes the smaller mode and only represents the main one. The slightly wider representation of uncertainty around the main mode is mainly controlled by the kernel bandwidth.

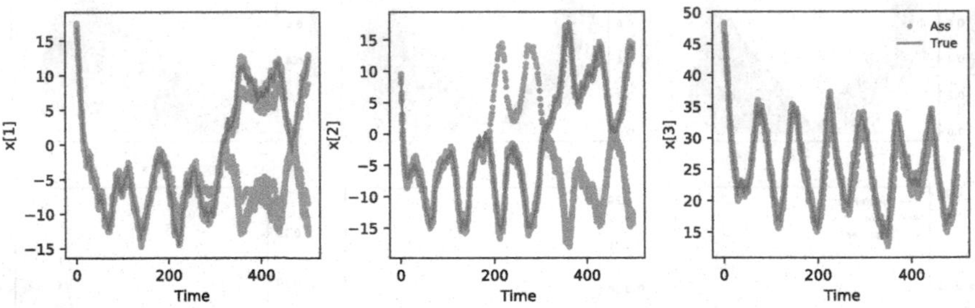

Fig. 4. The temporal evolution of the true state variables of the stochastic Lorenz-63 dynamical system (green line) and trajectories (40) of the particles resulting from the VMPF (orange dots), namely $\mathbf{x}_{1:K}^{(1:40)}$ as resulting from the output of Algorithm 2. Panels show each variable of the Lorenz 63 system. Time units are cycles. (Color figure online)

Figure 4 shows the evolution of the three variables of the Lorenz-63 system for a selected set of particles from VMPF (orange dots) and the true trajectory (green line). Because the apriori density at the initial time is prescribed as Gaussian, the posterior density evolves as unimodal until the true state changes of attractor. This occurs at about the cycle 300. From that time, trajectories of the VMPF particles are located in both attractors because the absolute observations cannot distinguish in which attractor the system is. In other words, the

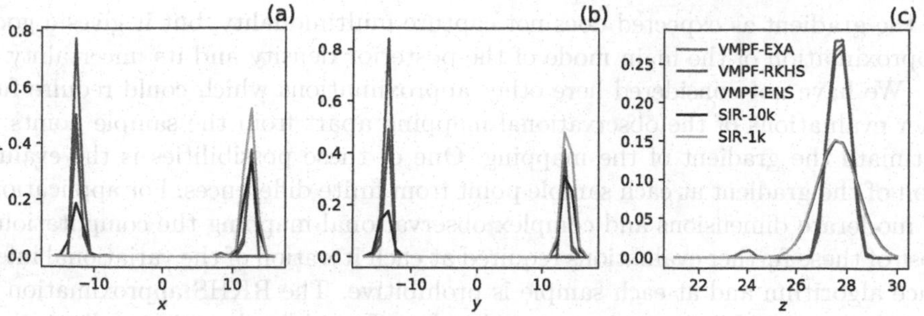

Fig. 5. Marginalized sequential posterior density represented through kernel density estimation, obtained with the VMPF using the exact (VPMF-EXA), RKHS (VMPF-RKHS) and ensemble calculations (VMPF-ENS) of the adjoint. The densities from SIR particle filter with 1000 (SIR-1k) and 10000 particles (SIR-10k) are also shown. Panels show marginalized density as a function of each variable for the Lorenz 63 system at the 500 cycle.

subsequent posterior density evolution from $t = 300$ undergoes a transition to a bimodal density. Figure 5 shows the marginal posterior densities in each variable for the VMPF at cycle 500. Both the exact calculation and the RKHS approximation in the observation likelihood gradient in the VMPF are able to capture the bimodality of the posterior density using 100 particles. On the other hand, the ensemble approximation only gives an unimodal density. For comparison we also show in Fig. 5 the corresponding marginalized posterior density of the SIR particle filter with 1000 and 10,000 particles. The SIR filter requires 10,000 particles to capture the bimodal structure of the posterior density while VMPF only requires 100 particles.

5 Conclusions

This work evaluates the use of a nonlinear observational operator in the variational mapping particle filter. Non-injectivity of the observational mappings leads to multimodal posterior densities which is known to represent a challenge for inference methods. The variational mapping particle filter is able to capture multimodes in the density in offline and online experiments. Particles are attracted to the modes in coherence with the gradient of the posterior density and local optimal transport principles.

Two approximations of the gradient of the observation mapping are evaluated. The representation of the observational mapping in the RKHS which overall exhibits a good performance in non-Gaussian densities. Because of the smoothing associated with this representation, it may slightly shift the modes in multi-modal densities for cases in which observations are in low-density regions of the prior density. The evaluation with the Lorenz-63 shows that the impact of this smoothing in a sequential scheme is negligible even for the absolute value observational mapping–a discontinuous gradient. The ensemble approximation of the gradient as expected does not capture multimodality, but it gives a good approximation of the main mode of the posterior density and its uncertainty.

We have not considered here other approximations which could require further evaluations of the observational mapping apart from the sample points to estimate the gradient of the mapping. One of these possibilities is the evaluation of the gradient at each sample point from finite differences. For applications of moderate dimensions and complex observational mapping the computational cost of these further evaluations required at each iteration of the variational inference algorithm and at each sample is prohibitive. The RKHS approximation of the observational operator is expected to be affected by the curse of dimensionality for high-dimensional states. A potential way to circunvent this limitation could be to divide the state space in the variables which are close to linear dependence from those state variables with a significant nonlinear observational function response. In this case, the partial derivatives of quasi-linear variables may be approximated with the ensemble approximation while the derivatives of highly nonlinear variables may be obtained through the RKHS approximation in the lower-dimensional subspace.

In all the experiments, radial basis functions are used as kernels. We took this choice because the structure of errors was assumed Gaussian. On the other hand, the kernel covariance and in particular the bandwidth are key hyperparameters for a good performance of the inference. In the proof-of-concept experiments an expensive trial-and-error methodology is used to manually tune the hyperparameters. Adaptative estimates of the hyperparameters are highly required. Standard adaptative bandwidth selection [18] does not appear a good option for non-injective observational mappings.

References

1. Burgers, G., Jan van Leeuwen, P., Evensen, G.: Analysis scheme in the ensemble Kalman filter. Monthly Weather Rev. **126**, 1719–1724 (1998)
2. Daum, F., Huang, J.: Nonlinear filters with log-homotopy. In: Signal and Data Processing of Small Targets 2007, vol. 6699, p. 669918 (2007)
3. Evensen, G.: Analysis of iterative ensemble smoothers for solving inverse problems. Comput. Geosci. **22**, 885–908 (2018)
4. Gordon, N.J., Salmond, D.J., Smith, A.F.: Novel approach to nonlinear/non-Gaussian Bayesian state estimation. In: IEE Proceedings F-Radar and Signal Processing, vol. 140, pp. 107–113 (1993)
5. Hoffman, M.D., Blei, D.M., Wang, C., Paisley, J.: Stochastic variational inference. J. Mach. Learn. Res. **14**, 1303–1347 (2013)
6. Houtekamer, P.L., Mitchell, H.L.: A sequential ensemble Kalman filter for atmospheric data assimilation. Mon. Weather Rev. **129**, 123–137 (2001)
7. Hunt, B.R., Kostelich, E.J., Szunyogh, I.: Efficient data assimilation for spatiotemporal chaos: a local ensemble transform Kalman filter. Physica D **230**, 112–126 (2007)
8. Jordan, M.I., Ghahramani, Z., Jaakkola, T.S., Saul, L.K.: An introduction to variational methods for graphical models. Mach. Learn. **37**, 183–233 (1999)

9. Kingma, D., Ba, J.: Adam: a method for stochastic optimization. In: International Conference on Learning Repres (ICLR). arXiv preprint arXiv:1412.6980 (2015)
10. Liu, Q., Wang, D.: Stein variational gradient descent: a general purpose Bayesian inference algorithm. In: Advances in Neural Information Processing Systems, pp. 2378–2386 (2016)
11. Marzouk, Y., Moselhy, T., Parno, M., Spantini, A.: An introduction to sampling via measure transport. In: Ghanem, R., Higdon, D., Owhadi, H. (eds.) To appear in Handbook of Uncertainty Quantification. Springer (2017). arXiv:1602.05023
12. Posselt, D.J., Vukicevic, T.: Robust characterization of model physics uncertainty for simulations of deep moist convection. Mon. Weather Rev. **138**, 1513–1535 (2010)
13. Posselt, D.J., Bishop, C.H.: Nonlinear parameter estimation: comparison of an ensemble Kalman smoother with a Markov chain Monte Carlo algorithm. Mon. Weather Rev. **140**, 1957–1974 (2012)
14. Posselt, D.J., Hodyss, D., Bishop, C.H.: Errors in Ensemble Kalman Smoother Estimates of Cloud Microphysical Parameters. Mon. Wea. Rev. **142**, 1631–1654 (2014)
15. Posselt, D.J.: A Bayesian examination of deep convective squall line sensitivity to changes in cloud microphysical parameters. J. Atmos. Sci. **73**, 637–665 (2016)
16. Pulido M., vanLeeuwen, P.J.: Kernel embedding of maps for Bayesian inference: the variational mapping particle filter. https://arxiv.org/pdf/1805.11380 (2018)
17. Saeedi, A., Kulkarni, T.D., Mansinghka, V.K., Gershman, S.J.: Variational particle approximations. J. Mach. Learn. Res. **18**, 2328–2356 (2017)
18. Scholkopf, B., Smola, A.J.: Learning with Kernels: Support Vector Machines, Regularization, Optimization, and Beyond. MIT Press, Cambridge (2002)
19. Tarantola, A.: Inverse Problem Theory and Methods for Model Parameter Estimation, vol. 89. SIAM (2005)
20. Vapnik, V.: The Nature of Statistical Learning Theory. Springer, Heidelberg (2013)
21. Zhou, D.X.: Derivative reproducing properties for kernel methods in learning theory. J. Comput. Appl. Math. **220**, 456–463 (2008)

Data Assimilation in a Nonlinear Time-Delayed Dynamical System with Lagrangian Optimization

Tullio Traverso[1,2,3] and Luca Magri[3,4(✉)]

[1] LadHyX, Département de Mécanique, Ecole Polytechnique, CNRS, 91128 Palaiseau, France
traverso@ladhyx.polytechnique.fr
[2] Università di Genova, via Montallegro 1, 16145 Genova, Italy
[3] Cambridge University Engineering Department, Trumpington St., Cambridge CB2 1PZ, UK
lm547@cam.ac.uk
[4] Institute for Advanced Study, TU Munich, 85748 Garching, Germany

Abstract. When the heat released by a flame is sufficiently in phase with the acoustic pressure, a self-excited thermoacoustic oscillation can arise. These nonlinear oscillations are one of the biggest challenges faced in the design of safe and reliable gas turbines and rocket motors [7]. In the worst-case scenario, uncontrolled thermoacoustic oscillations can shake an engine apart. Reduced-order thermoacoustic models, which are nonlinear and time-delayed, can only qualitatively predict thermoacoustic oscillations. To make reduced-order models quantitatively predictive, we develop a data assimilation framework for state estimation. We numerically estimate the most likely nonlinear state of a Galerkin-discretized time delayed model of a horizontal Rijke tube, which is a prototypical combustor. Data assimilation is an optimal blending of observations with previous system's state estimates (background) to produce optimal initial conditions. A cost functional is defined to measure (i) the statistical distance between the model output and the measurements from experiments; and (ii) the distance between the model's initial conditions and the background knowledge. Its minimum corresponds to the optimal state, which is computed by Lagrangian optimization with the aid of adjoint equations. We study the influence of the number of Galerkin modes, which are the natural acoustic modes of the duct, with which the model is discretized. We show that decomposing the measured pressure signal in a finite number of modes is an effective way to enhance state estimation, especially when nonlinear modal interactions occur during the assimilation window. This work represents the first application of data

T. Traverso gratefully acknowledges support from the Erasmus+ traineeship grant from the University of Genova to visit the University of Cambridge. L. Magri gratefully acknowledges support from the Royal Academy of Engineering Research Fellowships and the Hans Fischer visiting fellowship of the Technical University of Munich – Institute for Advanced Study, funded by the German Excellence Initiative and the European Union Seventh Framework Programme under grant agreement n. 291763.

J. M. F. Rodrigues et al. (Eds.): ICCS 2019, LNCS 11539, pp. 156–168, 2019.
https://doi.org/10.1007/978-3-030-22747-0_12

assimilation to nonlinear thermoacoustics, which opens up new possibilities for real-time calibration of reduced-order models with experimental measurements.

Keywords: Data assimilation · Nonlinear time-delayed dynamical systems · Thermoacoustics

1 Nonlinear Time-Delayed Thermoacoustic Model

We investigate the acoustics of a resonator excited by a heat source, which is a monopole source of sound. The main assumptions of the reduced-order model are [7]: (i) the acoustics evolve on top of a uniform mean flow; (ii) the mean-flow Mach number is negligible, therefore the acoustics are linear and no flow inhomogeneities are convected; (iii) the flow is isentropic except at the heat-source location; (iv) the length of the duct is sufficiently larger than the diameter, such that the cut-on frequency is high, i.e., only longitudinal acoustics are considered; (v) the heat source is compact, i.e., it excites the acoustics at a specific location, x_f; (vi) the boundary conditions are ideal and open-ended, i.e., the acoustic pressure at the ends is zero. Under these assumptions, the non-dimensional momentum and energy equations read, respectively [5]

$$\frac{\partial u}{\partial t} + \frac{\partial p}{\partial x} = 0, \tag{1}$$

$$\frac{\partial p}{\partial t} + \frac{\partial u}{\partial x} + \zeta p - \dot{\mathcal{Q}}\delta_D(x - x_f) = 0, \tag{2}$$

where u is the acoustic velocity; p is the acoustic pressure; t is the time; x is the axial coordinate of the duct; $\delta_D(x - x_f)$ is the Dirac delta distribution at the heat source location, x_f; ζ is the damping factor, which models the acoustic energy radiation from the boundaries and thermo-viscous losses; and $\dot{\mathcal{Q}}$ is the heat release rate (or, simply, heat release). The heat release, $\dot{\mathcal{Q}}$, is modelled by a nonlinear time delayed law [10]

$$\dot{\mathcal{Q}} \equiv \beta \text{Poly}(u_f(t - \tau)), \tag{3}$$

where τ is the time delay; β is the strength of the heat source; and $\text{Poly}(u(t)) = a_1 u^5(t) + \cdots + a_5 u(t)$. The time delay and strength of the heat source are the two key parameters of a reduced-order model for the flame [3]. Physically, τ is the time that the heat release takes to respond to a velocity perturbation at the flame's base; while β provides the strength of the coupling between the heat source and the acoustics. Velocity, pressure, length and time are nondimensionalized as in [5]. The set of nonlinear time-delayed partial differential Eqs. (1)–(2) provides a physics-based reduced-order model for the nonlinear thermoacoustic dynamics. Owing to the assumptions we made, the model can only qualitatively replicate the nonlinear thermoacoustic behaviour. In this paper, we propose a Lagrangian method to make a qualitative model quantitatively more accurate any time that reference data can be assimilated. Such data can come, for example, from sensors in experiments or time series from high-fidelity simulations.

1.1 Numerical Discretization with Acoustic Modes

We use separation of variables to decouple the time and spatial dependencies of the solution. The spatial basis on to which the solution is projected consists of the natural acoustic modes. When decomposed on the natural acoustic eigenfunctions, the acoustic velocity and pressure read, respectively

$$u(x,t) = \sum_{j=1}^{N_m} \eta_j(t)\cos(j\pi x), \tag{4}$$

$$p(x,t) = \sum_{j=1}^{N_m} \left(\frac{\dot{\eta}_j(t)}{j\pi}\right)\sin(j\pi x), \tag{5}$$

where the relationship between η_j and $\dot{\eta}_j$ has yet to be found and N_m is the number of acoustic modes considered. This discretization is sometimes known as the Galerkin discretization [12]. The state of the system is provided by the amplitudes of the Galerkin modes that represent velocity, η_j, and those that represent pressure, $\dot{\eta}_j/(j\pi)$. The damping has modal components, $\zeta_j = C_1 j^2 + C_2\sqrt{j}$, where C_1 and C_2 are damping coefficients [1,5,6,8,9]. In vector notation, $\boldsymbol{\eta} \equiv (\eta_1, \cdots, \eta_N)^{\mathrm{T}}$ and $\dot{\boldsymbol{\eta}} \equiv (\dot{\eta}_1/\pi, \cdots, \dot{\eta}_N/(N\pi))^{\mathrm{T}}$. The state vector of the discretized system is the column vector $\mathbf{x} \equiv (\boldsymbol{\eta}; \dot{\boldsymbol{\eta}})$. The governing equations of the Galerkin modes are found by substituting (4)–(5) into (1)–(3). Equation (2) is then multiplied by $\sin(k\pi x)$ and integrated over the domain $x = [0, 1]$ (projection). In so doing, the spatial dependency is removed and the Galerkin amplitudes are governed by a set of nonlinear time-delayed differential equations

$$\mathrm{F}_{1j} \equiv \frac{d}{dt}(\eta_j) - j\pi\left(\frac{\dot{\eta}_j}{j\pi}\right) = 0 \qquad t > 0, \tag{6}$$

$$\mathrm{F}_{2j} \equiv \frac{d}{dt}\left(\frac{\dot{\eta}_j}{j\pi}\right) + j\pi\eta_j + \zeta_j\left(\frac{\dot{\eta}_j}{j\pi}\right) = 0 \qquad t \in [0, \tau), \tag{7}$$

$$\mathrm{F}_{2j} \equiv \frac{d}{dt}\left(\frac{\dot{\eta}_j}{j\pi}\right) + j\pi\eta_j + \zeta_j\left(\frac{\dot{\eta}_j}{j\pi}\right) + 2s_j\beta\,\mathrm{Poly}(u_f(t-\tau)) = 0 \quad t \in [\tau, T], \tag{8}$$

where $u_f(t-\tau) = \sum_{j=1}^{N_m} \eta_j(t-\tau)c_j$; $s_j \equiv \sin(j\pi x_f)$ and $c_j \equiv \cos(j\pi x_f)$. The labels $\mathrm{F}_\bullet$ are introduced for the definition of the Lagrangian (Sect. 3.2). Because the Galerkin expansions (4)–(5) are truncated at the N_m-th mode, we obtain a reduced-order model of the original thermoacoustic system (1)–(2) with $2N_m$ degrees of freedom (6)–(8). The reduced-order model is physically a set of $2N_m$ time-delayed oscillators, which are nonlinearly coupled through the heat release term. In the following sections, we employ 4D-Var data assimilation to improve the accuracy of such a reduced-order model[1].

2 Data Assimilation as a Constrained Optimization Problem

The ingredients of data assimilation are (i) a reduced-order model to predict the amplitude of thermoacoustic oscillations, which provides the so-called background state vector $\mathbf{x}^{bg}$ at any time, t (red thick line in Fig. 1); (ii) data from external observations, $\mathbf{y}^i$, which is available from high-fidelity simulations or experiments at times t^i_{obs} (grey diamonds in Fig. 1); and (iii) an educated guess on the parameters of the system, which originates from past experience. The uncertainties on the background solution and observations are here assumed normal and unbiased. $\mathbf{B}$ and $\mathbf{R}$ are the background and observation covariance matrices, respectively, which need to be prescribed. For simplicity, we assume that $\mathbf{B}$ and $\mathbf{R}$ are diagonal with variances B and R (i.e., errors are statistically independent). A cost functional is defined to measure the statistical distance between the background predictions and the evidence from observations. First, we want the state of the system to be as close as possible to the observations. Second, we do not want the improved solution to be "too far" away from the background solution. This is because we trust that our reduced-order model provides a reasonable solution. Mathematically, these two requirements can be met, respectively, by minimizing the following cost functional

$$J = \underbrace{\frac{1}{2}||\mathbf{x}_0 - \mathbf{x}_0^{bg}||^2_{\mathbf{B}}}_{J_{bg}} + \underbrace{\frac{1}{2}\sum_{i=1}^{N_{obs}} ||\mathbf{M}\mathbf{x}^i - \mathbf{y}^i||^2_{\mathbf{R}}}_{J_{obs}} \quad \text{over} \quad [0, T], \tag{9}$$

where N_{obs} is the number of observations over the assimilation window $[0, T]$. $\mathbf{M}$ is a linear measurement operator, which maps the state space to the observable space (see Eqs. (4)–(5)). Moreover, $||\mathbf{x}_0 - \mathbf{x}_0^{bg}||^2_{\mathbf{B}} \equiv \left(\mathbf{x}_0 - \mathbf{x}_0^{bg}\right)^{\mathrm{T}} \mathbf{B}^{-1} \left(\mathbf{x}_0 - \mathbf{x}_0^{bg}\right)$. Likewise, $||\mathbf{M}\mathbf{x}^i - \mathbf{y}^i||^2_{\mathbf{R}} \equiv (\mathbf{M}\mathbf{x}^i - \mathbf{y}^i)^{\mathrm{T}} \mathbf{R}^{-1} (\mathbf{M}\mathbf{x}^i - \mathbf{y}^i)$.

The objective of state estimation is to improve the prediction of the state, $\mathbf{x}$, over the interval $[0, T]$, by reinitializing the background initial conditions, $\mathbf{x}_0^{bg}$, with optimal initial conditions. These optimal initial conditions are called analysis initial conditions, $\mathbf{x}_0^{analysis}$, which are the minimizers of the cost functional (9). We start from a background knowledge of the model's initial conditions, $\mathbf{x}_0^{bg}$, which is provided by the reduced-order model when data is not assimilated. By integrating the system from $\mathbf{x}_0^{bg}$, we obtain the red trajectory in Fig. 1, $\mathbf{x}^{bg}(t)$. The set of observations is assumed to be distributed over an

[1] Although different from our study, it is worth mentioning that a study that combined 4D-Var data assimilation with reduced-order models of the Navier-Stokes equations based on proper orthogonal decomposition can be found in [4].

assimilation window at some time instants. Pictorially, the analysis trajectory corresponds to the green thin line in Fig. 1, which is the minimal statistical distance between the background initial condition (magenta thick arrow) and observations (blue thin arrows). This algorithm is known as 4D-Var in weather forecasting [2]. State estimation enables the adaptive update of reduced-order thermoacoustic models whenever data is available.

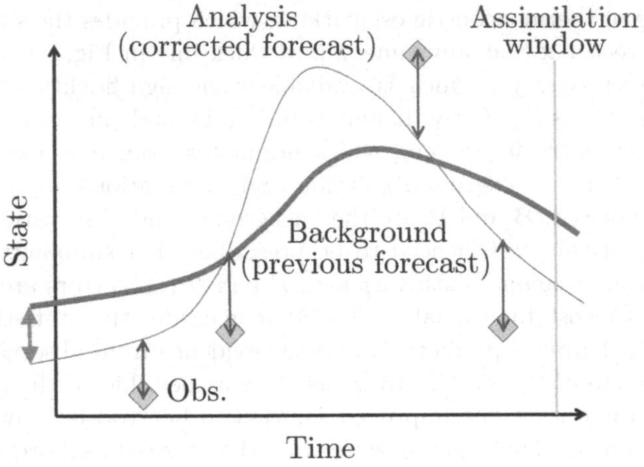

Fig. 1. Pictorial representation of data assimilation. The background error, J_{bg}, is proportional to the length of the magenta thick arrow, while the observation error, J_{obs}, is proportional to the sum of the blue thin arrows. The vertical cyan line marks the end of the assimilation window, after which the forecast begins. (Color figure online)

3 Data Assimilation for Nonlinear Thermoacoustic Dynamics

We propose a set of cost functionals to perform data assimilation with the thermoacoustic model introduced in Sect. 1. We also introduce the formalism to perform Lagrangian optimization, in which adjoint equations enable the efficient calculation of the gradients of the thermoacoustic cost functionals with respect to the initial state.

3.1 Thermoacoustic Cost Functionals

Crucial to data assimilation is the definition of the cost functionals J_{bg} and J_{obs}. Three physical thermoacoustic cost functionals are proposed and compared to reproduce different scenarios. For the background error

$$J_{bg}^{a} = \frac{1}{2B}\left(p(0) - p(0)_{bg}\right)^{2}, \tag{10}$$

$$J_{bg}^{b} = \frac{1}{2B}\sum_{j=1}^{N_m}\left\{\left[\left(\frac{\eta_{j0}}{j\pi}\right) - \left(\frac{\eta_{j0}}{j\pi}\right)_{bg}\right]\sin(j\pi x_m)\right\}^{2}, \tag{11}$$

$$J_{bg}^{c} = \frac{1}{2B}\sum_{j=1}^{N_m}\left[\left(\frac{\dot{\eta}_{j0}}{j\pi}\right) - \left(\frac{\dot{\eta}_{j0}}{j\pi}\right)_{bg}\right]^{2} + \frac{1}{2B}\sum_{j=1}^{N_m}\left[\eta_{j0} - \eta_{j0,bg}\right]^{2}. \tag{12}$$

For the observation error

$$J_{obs}^{a} = \sum_{i=1}^{N_{obs}} J_{obs,i}^{a} = \frac{1}{2R}\sum_{i=1}^{N_{obs}}\left(p(t_{obs}^{i}) - p_{obs}^{(i)}\right)^{2}, \tag{13}$$

$$J_{obs}^{b} = \sum_{i=1}^{N_{obs}} J_{obs,i}^{b} = \frac{1}{2R}\sum_{i=1}^{N_{obs}}\sum_{j=1}^{N_m}\left\{\left[\left(\frac{\dot{\eta}_{j}(t_{obs}^{i})}{j\pi}\right) - \left(\frac{\dot{\eta}_{j}}{j\pi}\right)_{obs}^{(i)}\right]\sin(j\pi x_m)\right\}^{2}, \tag{14}$$

where x_m is the location where the measurement is taken and t_{obs}^{i} is the instant at which the i–th observation is assimilated. On the one hand, by using J_{bg}^{a} and J_{obs}^{a} the analysis solution is optimized against the background pressure at $t = 0$ and the measured pressure at $t = t_{obs}^{i}$, $i = 1, \dots N_{obs}$. Physically, this means that the acoustic pressure is the model output, $p(0)_{bg}$, and the observable from the sensors, p_{obs}^{i}. On the other hand, J_{bg}^{b} and J_{obs}^{b} constrain the pressure modes. Physically, this means that every pressure mode provided by the background solution is a model output, $(\eta_{j0}/(j\pi))_{bg}$, and it is assumed that the modes of the acoustic pressure, $(\dot{\eta}_j/(j\pi))_{obs}^{(i)}$, can be calculated from the sensors on the fly. For the background cost functional, we also defined J_{bg}^{c} as a norm of the modes, which does not have a corresponding observation cost functional because the spatial dependency is not explicit.

To attain a minimum of J, a necessary condition is that the gradient vanishes, i.e.,

$$\nabla_{\mathbf{x}_0}(J) = \nabla_{\mathbf{x}_0}(J_{bg}) + \sum_{i=1}^{N_{obs}} \nabla_{\mathbf{x}_0}(J_{obs,i}) = 0, \tag{15}$$

where $\nabla_{\mathbf{x}_0}$ is the gradient with respect to the initial conditions. There exists $\mathbf{x}_0$ such that $\nabla_{\mathbf{x}_0}(J) = 0$ because of the convexity of the cost functionals in the neighbourhood of a local extremum. To compute $\nabla_{\mathbf{x}_0}(J_{bg})$ and $\nabla_{\mathbf{x}_0}(J_{obs,i})$, we use calculus of variation (Sect. 3.2). The Lagrange multipliers, also known as adjoint, or dual, or co-state variables (Sect. 3.3), provide the gradient information with respect to the initial state.

3.2 Lagrangian of the Thermoacoustic System

The governing equations and their initial conditions are rewritten in the form of constraints, F, which hold over time intervals, while G are the constraints that hold for a specific time only, i.e., $t = t_0$. Together with Eqs. (6)–(8) and by defining the auxiliary variable $\bar{\eta}(t) \equiv u_f(t-\tau)$, they read

$$\mathrm{F}_3 \equiv \bar{\eta}(t) = 0, \qquad t \in [0, \tau) \tag{16}$$

$$\mathrm{F}_3 \equiv \bar{\eta}(t) - u_f(t-\tau) = 0, \qquad t \in [\tau, T]. \tag{17}$$

The constraints for the initial conditions read

$$\mathrm{G}_{1j} \equiv \eta_j(0) - \eta_{j0} = 0, \tag{18}$$

$$\mathrm{G}_{2j} \equiv \left(\frac{\dot{\eta}_j(0)}{j\pi}\right) - \left(\frac{\dot{\eta}_{j0}}{j\pi}\right) = 0. \tag{19}$$

By defining an inner product

$$[a, b] = \frac{1}{T}\int_0^T ab \; dt \tag{20}$$

where a and b are arbitrary functions, the Lagrangian of the nonlinear system can be written as

$$\mathcal{L} \equiv J_{bg} + J_{obs,i} + \sum_{j=1}^{N_m} \mathcal{L}_j - \left[\bar{\xi}(t), \mathrm{F}_3\right], \tag{21}$$

where each $\mathcal{L}_j$ is

$$\mathcal{L}_j \equiv - \left[\frac{\xi_j}{j\pi}, \mathrm{F}_{1j}\right] - [\nu_j, \mathrm{F}_{2j}] - b_{1j}\mathrm{G}_{1j} - b_{2j}\mathrm{G}_{2j}, \tag{22}$$

where $\bar{\xi}$, $\xi_j/j\pi$, ν_j and $b_{\bullet j}$ are the Lagrange multipliers, or adjoint variables, of the corresponding constraints. Because we wish to derive the adjoint equations for the cost functional $J_{obs,i}$, we consider the time window to be $T = t^i_{obs}$.

3.3 Adjoint Equations

We briefly report the steps to derive the evolution equations of the Lagrange multipliers (adjoint equations) [7]. First, the Lagrangian (21) is integrated by parts to make the dependence on the direct variables explicit. Secondly, the first variation is calculated by a Fréchet derivative

$$\left[\frac{\partial \mathcal{L}}{\partial \mathbf{x}}, \delta \mathbf{x}\right] \equiv \lim_{\epsilon \to 0} \frac{\mathcal{L}(\mathbf{x} + \epsilon\delta\mathbf{x}) - \mathcal{L}(\mathbf{x})}{\epsilon}. \tag{23}$$

Thirdly, the derivatives of (21) are taken with respect to the initial condition of each variable of the system, $\partial\mathcal{L}/\partial\mathbf{x}_0$. These expressions will be used later to compute the gradient. Finally, to find the set of Lagrange multipliers that characterizes an extremum of the Lagrangian, $\mathcal{L}$, variations with respect to $\delta\mathbf{x}$ are set to zero. The adjoint equations and their initial conditions are derived by setting variations of $\delta\eta_j$, $\delta\left(\dot{\eta}_j/(j\pi)\right)$ and $\delta\bar{\eta}$ to zero over $[0, T]$.

3.4 Gradient-Based Optimization

The optimization loop consists of the following steps:

(1) Integrate the system forward from $t = 0$ to $t = T$ from an initial state $\mathbf{x}_0$;
(2) Initialize the adjoint variables;
(3) Evolve the adjoint variables backward from $t = T$ to $t = 0$;
(4) Evaluate the gradient using the adjoint variables at $t = 0$.

Once the gradient is numerically computed, the cost functional can be minimized via a gradient based optimization loop. The conjugate gradient [11] is used to update the cost functional until the condition $\nabla_{\mathbf{x}_0}(J) = 0$ is attained to a relative numerical tolerance of 10^{-4}. By using a gradient based approach, we find a local minimum of J. We verify that there is no other local minimum by computing $J = J(\mathbf{x}_0)$ in the vicinity of $\mathbf{x}_0^{analysis}$.

4 Results

We validate the data-assimilation algorithm by twin experiments: The true state solution, $\mathbf{x}^{true}(t)$, is produced by perturbing the unstable fixed point at the origin of the phase space[2]; the background trajectory, $\mathbf{x}^{bg}(t)$, is obtained by perturbing each true mode initial condition with Gaussian error with variance $B = 0.005^2$; the i−th observation is produced by adding Gaussian error with variance $R = 0.005^2$ to $\mathbf{x}^{true}(t^i_{obs})$. The outcome of twin experiments is summarized by the error plots shown in Figs. 3 and 4. The cyan vertical line indicates the end of the assimilation window, the red thick line is the difference between the true pressure and the background pressure, the green thin line is the difference between the true pressure and the analysis pressure. First, it is shown how the number of computed acoustic modes affects the solution of the system. Secondly, we investigate the effects that the different cost functionals have on the analysis solution. Finally, we discuss the effects that the number of observations have on the analysis.

The parameters we use are $\beta = 1$, $\tau = 0.02$, $C_1 = 0.05$, $C_2 = 0.01$ and $(a_1, a_2, a_3, a_4, a_5) = (-0.012, 0.059, -0.044, -0.108, 0.5)$ for the heat release term $\dot{Q}$. The position of the heat source is $x_f = 0.3$ and all the measurements are taken at $x_m = 0.8$.

[2] Strong constraint 4D-Var assumes that the model is perfect and the uncertainty is only in the initial conditions, therefore, the true trajectory can be a model output.

4.1 Remarks on the Thermoacoustic Nonlinear Dynamics

The thermoacoustic model is a set of $2N_m$ nonlinearly coupled oscillators, which we initialize by imposing non-equilibrium initial conditions. We compare two solutions, using $N_m = 3$ and $N_m = 10$ in Fig. 2a and b, respectively. Higher modes are quickly damped out, thus, after a transient where strong nonlinear modal coupling occurs, the solution obtained with $N_m = 10$ is qualitatively similar to the solution obtained with $N_m = 3$. During the transient, if sufficient modes are computed, the dynamics are more unpredictable because of the intricate modal interaction. The twin experiments are performed with $N_m = 10$, which provide a more accurate solution. It is shown that state estimation is markedly affected depending on whether we observe the system during the transient or at regime.

Effect of the Observation Error. We can simulate two main scenarios, depending on the choice of J_{obs} (Sect. 3.1). Figure 3a is obtained using J^a_{obs}, i.e., by modelling observations on the pressure only. The analysis pressure error slightly deviates from zero in the assimilation window. When the forecast window starts, the analysis suddenly approaches the background again. Figure 3b is obtained using J^b_{obs}, therefore the observations contain information about every pressure modes. The forecast quality is considerably enhanced. The assimilation window is $T_{as} = 0.4$, thus, the observations are obtained during the transient, which lasts up to $t \approx 2$, where the dynamics are more unpredictable due to nonlinear interactions between modes. As we will show in the next subsection, increasing N_{obs} is not an effective strategy to improve the forecast during the transient when the pressure is observed.

Effect of the Background Error. From a numerical standpoint, the background error, J_{bg}, acts as an observation at $t = 0$. The analysis trajectory is an optimal blending of the information from the measurements and the previous educated model output. Therefore, we emphasize that the outcome of twin experiments is improved if the cost functionals of the background knowledge and observations are consistent. In the present framework, it means that J^a_{bg} should be used with J^a_{obs} and J^b_{bg} should be used with J^b_{obs}. On the one hand, if the source of assimilated data favours the background knowledge (e.g. using J^b_{bg} or J^c_{bg} together with J^a_{obs}), the analysis trajectory will be closer to the background. On the other hand, if the source of assimilated data favours the observations (e.g. using J^b_{obs} with J^a_{bg}), the analysis trajectory will be closer to the observations.

Effect of the Number of Observations. Generally speaking, the higher the number of observations the more the optimal solution will be similar to the true solution. This can be deduced by inspection of Fig. 4a and b. The value of N_{obs} is increased from 50 to 250, over an assimilation window of 2.5 time units (the observations are not shown in the figures), resulting in a smaller error amplitude when more observations are available.

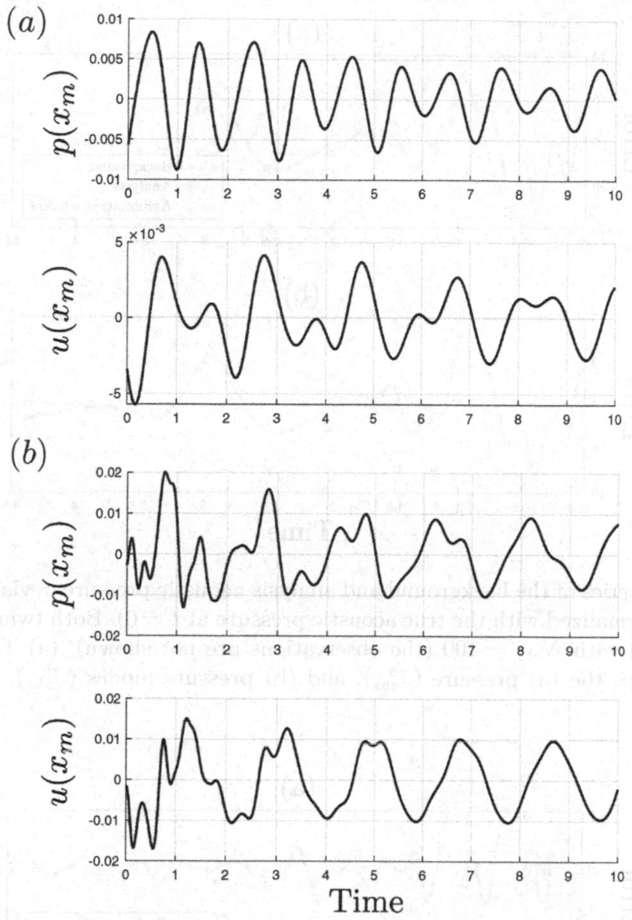

Fig. 2. Time series of the pressure, p and velocity, u evaluated at $x_m = 0.8$ using (a) $N_m = 3$ and (b) $N_m = 10$. When 10 modes are computed, a transient region can be identified for $t \lesssim 2$, which is characterized by irregular fluctuations due to nonlinear modal coupling.

However, it is possible that increasing N_{obs} will not result in a better state estimation. This happens if we measure the pressure during the transient, that is, using J^a_{obs} and $T_{as} < 2$. In the transient, the measured pressure is a combination of all modes. Therefore, the same pressure level is associated with different combinations of modes, hence, no useful information is assimilated to help determine the state of the system by knowing only the pressure (i.e., by using J^a_{obs}). Under these circumstances, the forecast quality will remain poor, as shown in Fig. 3a, regardless of the number of observations. When the assimilation window is extended up to 2.5 time units, as shown in Fig. 4, the pressure is observed also at regime, that is in $t \in [2, 2.5]$, approximately. In this interval, the mea-

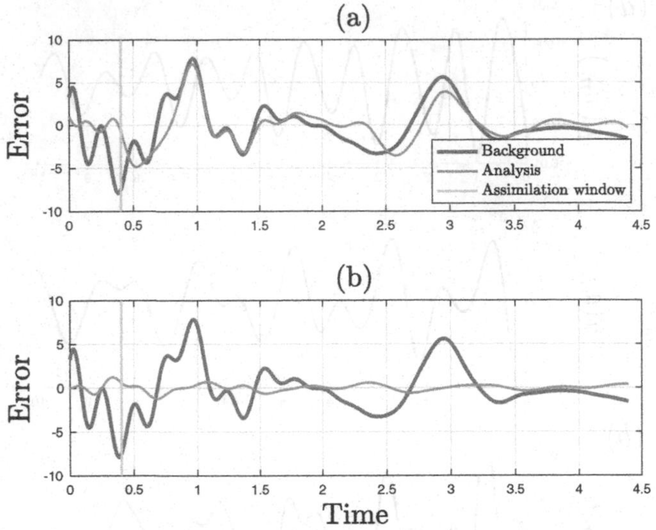

Fig. 3. Time series of the background and analysis acoustic pressure deviation from the true state (normalized with the true acoustic pressure at $t = 0$). Both twin experiments are performed with $N_{obs} = 100$ (the observations are not shown). (a) The cost functional measures the (a) pressure (J^a_{obs}), and (b) pressure modes (J^b_{obs}). (Color figure online)

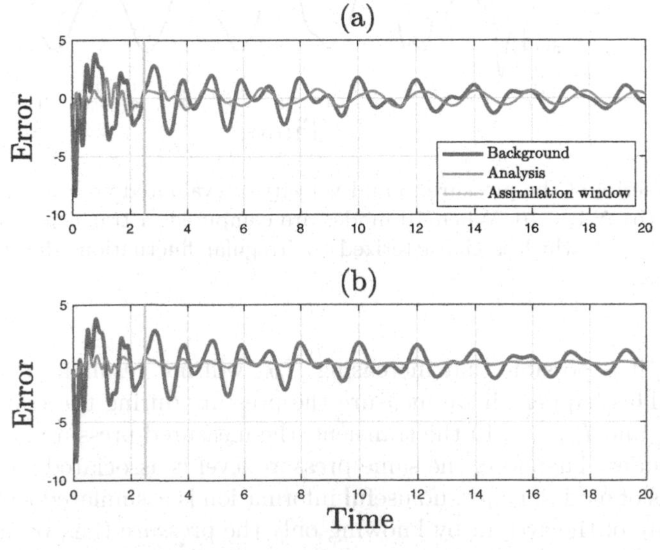

Fig. 4. Error plots of two different twin experiments. The assimilation window is $T_{as} =$ 2.5 time units and the observation error is measured using J^a_{obs} in both cases. The choice of T_{as} implies that the system is observed also at regime. (a) $N_{obs} = 50$ and (b) $N_{obs} = 250$. (Color figure online)

sured pressure signal is produced chiefly by the first three modes because higher modes are damped out. Therefore, the measured pressure becomes more effective information about the state to assimilate. At regime, as intuitively expected, increasing the observation frequency produces a better forecast.

We conclude that having poor information about the system's state cannot be simply balanced by increasing the number of observations. Given a number of observations with their time distribution, it is the synergy between an appropriate cost functional and the recognition of the type of dynamics (transient vs. regime) to be the key for a successful data assimilation.

5 Conclusions and Future Work

Preliminary thermoacoustic design is based on simplified and computationally cheap reduced-order models that capture the inception of thermoacoustic instabilities (linear analysis) and their saturation to finite amplitude oscillations (nonlinear analysis). We propose a Lagrangian method to make a qualitative reduced-order model quantitatively more accurate any time that reference data can be assimilated. Such data can come, for example, from sensors in experiments or time series from high-fidelity simulations. To test the method we perform a series of twin experiments with the thermoacoustic model of a resonator excited by a heat source (horizontal Rijke tube). When sufficient modes are computed, a clear distinction emerges between a transient state and the state at regime. The former is characterized by irregular dynamics due to the interaction between all modes, while at regime the dynamics are chiefly dominated by the first three modes. We find that, at regime, it is possible to enhance the forecast by assimilating data about the pressure. As intuitively expected, the higher the number of observations, the better the forecast accuracy. While testing the effectiveness of data assimilation during the transient, we find that it is not possible to improve the forecast by measuring the pressure only. Moreover, the quality of the forecast remains poor regardless of the number of observations. Therefore, we propose a more effective cost functional, which takes into consideration the spectral content of the measured signal to enable a successful state estimation also during the transient. In state estimation, we implicitly assume that the parameters are correct. However, this is rarely the case in thermoacoustics, where the parameters are uncertain and need to be optimally calibrated. Ongoing work includes simultaneous parameter and state estimation using Lagrangian optimization with state augmentation. This work opens up new possibilities for on-the-fly optimal calibration and state estimation of reduced-order models in thermoacoustics for applications in propulsion and power generation.

References

1. Balasubramanian, K., Sujith, R.I.: Thermoacoustic instability in a Rijke tube: non-normality and nonlinearity. Phys. Fluids **20**(4), 044103 (2008). https://doi.org/10.1063/1.2895634
2. Blayo, É., Bocquet, M., Cosme, E., Cugliandolo, L.F.: Advanced Data Assimilation for Geosciences, 1st edn. Oxford University Press (2015). https://doi.org/10.1093/acprof:oso/9780198723844.001.0001
3. Crocco, L.: Research on combustion instability in liquid propellant rockets. In: Symposium (International) on Combustion, vol. 12, no. 1, pp. 85–99 (1969). https://doi.org/10.1016/S0082-0784(69)80394-2
4. Du, J., Navon, I.M., Zhu, J., Fang, F., Alekseev, A.K.: Reduced order modeling based on POD of a parabolized Navier-Stokes equations model II: trust region POD 4D VAR data assimilation. Comput. Math. Appl. **65**(3), 380–394 (2013). https://doi.org/10.1016/j.camwa.2012.06.001
5. Juniper, M.P.: Triggering in the horizontal Rijke tube: non-normality, transient growth and bypass transition. J. Fluid Mech. **667**, 272–308 (2011). https://doi.org/10.1017/S0022112010004453
6. Landau, L.D., Lifshitz, E.M.: Fluid Mechanics, 2nd edn. Pergamon Press (1987)
7. Magri, L.: Adjoint methods as design tools in thermoacoustics. Appl. Mech. Rev. **71**(2), 020801 (2019). https://doi.org/10.1115/1.4042821
8. Magri, L., Juniper, M.P.: Sensitivity analysis of a time-delayed thermo-acoustic system via an adjoint-based approach. J. Fluid Mech. **719**, 183–202 (2013). https://doi.org/10.1017/jfm.2012.639
9. Matveev, K.I., Culick, F.E.C.: A model for combustion instability involving vortex shedding. Combust. Sci. Technol. **175**, 1059–1083 (2003)
10. Orchini, A., Rigas, G., Juniper, M.P.: Weakly nonlinear analysis of thermoacoustic bifurcations in the Rijke tube. J. Fluid Mech. **805**, 523–550 (2016). https://doi.org/10.1017/jfm.2016.494
11. Press, W.H., Teukolsky, S.A., Vetterling, W.T., Flannery, B.P.: Numerical Recipes, 3rd edn. Cambridge University Press, Cambridge (2007)
12. Zinn, B.T., Lores, M.E.: Application of the Galerkin method in the solution of non-linear axial combustion instability problems in liquid rockets. Combust. Sci. Technol. **4**(1), 269–278 (1971). https://doi.org/10.1080/00102207108952493

Machine Learning to Approximate Solutions of Ordinary Differential Equations: Neural Networks vs. Linear Regressors

Georg Engel(✉)

Christian Doppler Laboratory for Quality Assurance, Methodologies for Autonomous Cyber-Physical Systems, Institute for Software Technology, Graz University of Technology, Graz, Austria
engel@ist.tugraz.at

Abstract. We discuss surrogate models based on machine learning as approximation to the solution of an ordinary differential equation. Neural networks and a multivariate linear regressor are assessed for this application. Both of them show a satisfactory performance for the considered case study of a damped perturbed harmonic oscillator. The interface of the surrogate model is designed to work similar to a solver of an ordinary differential equation, respectively a simulation unit. Computational demand and accuracy in terms of local and global error are discussed. Parameter studies are performed to discuss the sensitivity of the method and to tune the performance.

Keywords: Ordinary differential equations · Machine learning · Surrogate model · Neural network · Multivariate linear regressor

1 Introduction

Machine learning techniques are successful in many fields such as image and pattern recognition. In recent years, interest increases in applying these techniques also to other fields of science, which are conventionally dominated by e.g. physically motivated models. The success of these techniques is usually tied to the amount and quality of data available for training the model. In engineering and science, these data can be generated using measurements or simulations.

The present paper discusses machine learning techniques in the context of simulations based on ordinary differential equations. These are often used to describe physical or engineering systems. Solving these equations can be expensive in practice, in particular for complex systems and when solutions are required repeatedly, like in optimization problems. The computational costs encountered are often too high, e.g. when combining these models with the real world, like in cyber-physical systems, where real-time performance is required.

J. M. F. Rodrigues et al. (Eds.): ICCS 2019, LNCS 11539, pp. 169–177, 2019.
https://doi.org/10.1007/978-3-030-22747-0_13

The purpose of model order reduction or surrogate models is to reduce the computational costs, taking into account some limited reduction of accuracy.

A prominent method for model order reduction is proper orthogonal decomposition, which was successfully applied already many years ago, e.g. to analyze turbulent flows [1]. In the present work, we follow a different approach, adopting the perspective of co-simulation. Co-simulation denotes the dynamic coupling of various simulation units, which exchange information during simulation time, for a review see [2,3]. Several co-simulation interfaces have been defined and implemented, e.g. the Functional Mock-Up Interface standard [4], for a discussion see [5]. A discussion and comparison was performed e.g. in [6]. Typically, in such a setup the computational costs are dominated by only few of the individual simulation units. It is intriguing to replace only the costly simulation units by cheaper surrogate models using machine learning techniques. The framework of co-simulation might be exploited in this context in a way that the master algorithm is left unaltered. However, as models generated by machine learning do not explicitly incorporate a solver, the meaning of co-simulation fades and one could use a notion of model-coupling instead.

Machine learning has been applied to differential equations before, some work dates back more than twenty years, but a considerable boost appeared very recently. Artificial neural networks have been proposed to derive analytical solutions for differential equations by reformulating them as optimization problem [7]. Deep reinforcement learning was suggested for general non-linear differential equations, where the network consists of an actor that outputs solution approximations policy and a critic that outputs the critic of the actor's output solution [8]. An approximation model for real-time prediction of non-uniform steady laminar flow based on convolutional neural networks are proposed in [9]. Considering the simulation of buildings, several efforts have been made to reduce the computation costs, in particular within the IBPSA 1 project [10]. The input and output data of simulators is used to train neural networks in a co-simulation setting (referred to as "intelligent co-simulation"), and the model is compared to model order reduction by proper orthogonal decomposition [11]. A similar approach called "component-based machine learning", for the same field of application, is pursued in [12]. The same setting was investigated using deep-learning neural-networks [13], where a considerable computational speedup over the physical model at similar accuracy is claimed.

The present work focuses on simple case studies where extensive parameter studies are feasible and more insights related to the structure of the method and the case study can be achieved. A first discussion in this respect, considering neural networks for the Dahlquist equation and the Van der Pol oscillator was presented in [14]. The main contribution of this work is as follows:

- The general method for using data models to approximate solutions of ordinary differential equations is discussed.
- A neural network and a multi-variate regressor are introduced as specific examples.
- Accuracy and computational costs are discussed.
- Parameter studies are performed to discuss the sensitivity of the method.

2 Method

Data models shall be trained as surrogate model for a simulation unit. The surrogate model obtains the same input and output variables as the original simulation unit. The input variables are represented as explicit function of time. For simplicity, the output shall be the state variable itself. Output variables as functions of the state variable mean a straightforward generalization of the method. The data model contains short-term memory units to store previous values of the state variable. These serve as inputs for the calculation in addition to the present value and the explicit function of time to predict the next value of the state variable.

This way, the model builds a trajectory, where errors can easily accumulate, which is typical for solutions of differential equations. As training data, we make use of the solution to differential equations generated by a traditional solver. For the latter, we employ the routine *odeint* from the python package *scipy* [15].

For the neural network, the multi-layer perceptron of the package scikit-learn is chosen [16]. Unless stated otherwise, we use one hidden layer with one neuron, a tolerance of 10^{-5} and the solver "lbfgs". Model selection on a test data set is used to apply 100 different random seeds in the optimizer of the neural network, which is common practice and crucial here to obtain a satisfactory validation. While this is satisfactory, note that it can be expected that the true optimum is not found, and hence some "random" additional error is found in the validation when comparing different settings. The multivariate linear regressor is constructed using the Moore-Penrose pseudo-inverse from the python package *numpy*. The regressor is limited to linear functions for the case study considered in the present work. An extension to polynomials is very straightforward but not necessary if the underlying differential equation is linear.

The method is designed to allow for generalization in several respects to be considered in future. In particular, a wrapper for simulation units in a co-simulation can encapsulate the method as a surrogate model. In such applications, it might be better to take more arbitrary snapshots of the data for training, which means a further extension to the consecutive data in simulation time considered here. Finally, for the simple cases considered here, the output coincides with the state variables of the model. In actual applications, only the output as some function of the state variable might be available for training the data model.

3 Model/Case Study

For testing the method, we consider a damped harmonic oscillator amended by an explicit function of time as perturbation,

$$\ddot{y} + 0.1\dot{y} + y + 0.1\sin^2\left(\sqrt{2}t\right) = 0\,, \tag{1}$$

with the initial condition $y(0) = 1$ and $\dot{y}(0) = 0$. The damping and the perturbation are chosen such that the behaviour of $y(t)$ changes significantly from the training to the validation phase, which is supposed to challenge the method. For the same purpose, the frequencies of the harmonic oscillator and the perturbation are chosen differently. Note, however, that the model is chosen such that the solution $y(t)$ is bounded for all t within a range which can easily be covered during the training phase. Hence, even though the $y(t)$ changes qualitatively from training to validation, the data model can be considered to remain within the "interpolation" range, where the training data should provide a good basis for prediction. The perturbation term represents an input for a simulation unit in a co-simulation scenario of an actual application, e.g. weather data in case of a building performance simulation.

4 Results

We choose the interval $t \in [0, 100)$ and 1000 time steps. $t \in [0, 30)$ is used for training, $t \in [30, 50)$ as testing for potential model selection and $t \in [50, 100)$ for validation. The Figures shown in this section always show results for the neural network on the left hand side and results for the linear regressor on the right hand side.

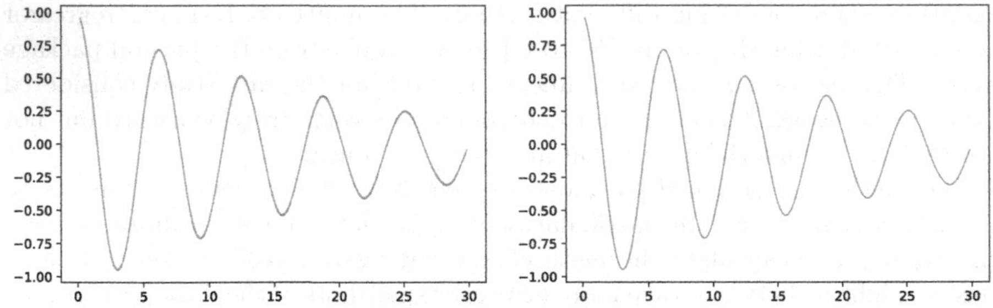

Fig. 1. Training results. The blue curve shows the "exact" solution generated by the traditional solver, the orange curve shows the result for the data model. If only one curve is visible, the predicted one lies one top of the "exact" solution. Left: neural network; right: linear regressor. Neither model shows a significant deviation. (Color figure online)

Results for training, testing and validation are shown in Figs. 1, 2 and 3, respectively. For the chosen parameters, the validation works very well. Note how the behaviour of the function $y(t)$ changes qualitatively from the training to the test and to the validation phase. It is remarkable that the prediction during the validation phase, dominated by the perturbation term, is satisfactory even though the perturbation is barely visible for the eye during the training phase.

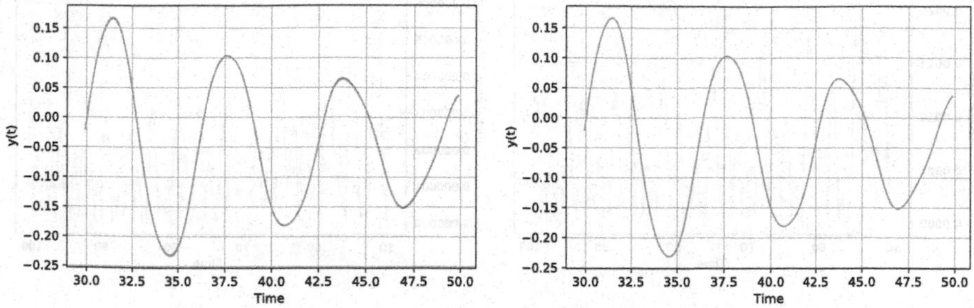

Fig. 2. Like, Fig. 1, but for test results, used for model selection in case of the neural network. Note that the effect of the perturbation becomes visible as distortion of the oscillation.

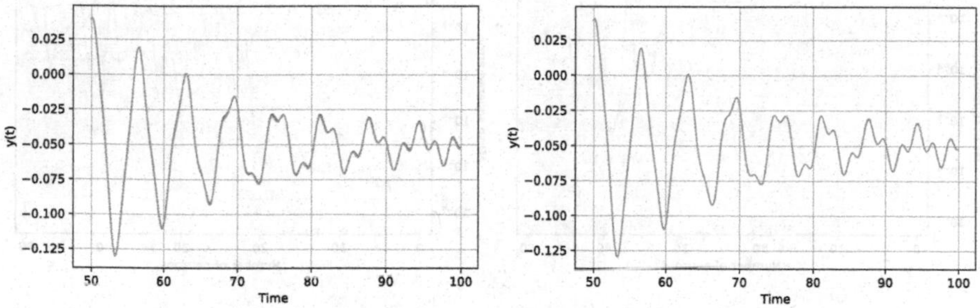

Fig. 3. Like, Fig. 1, but for validation results. Note how the behaviour of the solution $y(t)$ is qualitatively different from the training phase. For the neural network (right hand side), we see some small deviation where the blue curve becomes visible below the orange one. These deviation are better visible in Fig. 5. (Color figure online)

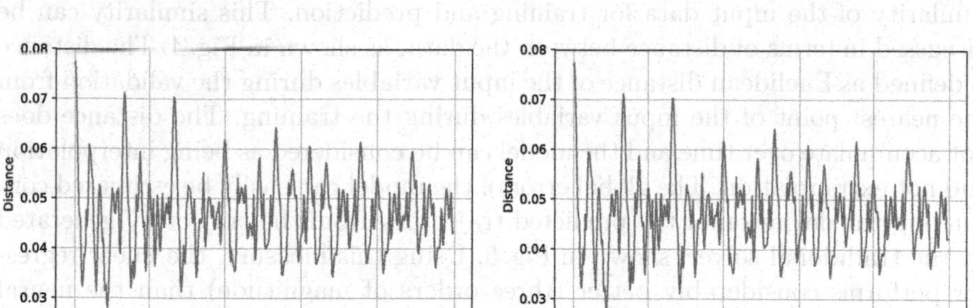

Fig. 4. The distance in theory space from input variables during validation compared to training data is shown. The distance is defined as Euclidean distance from the nearest point from the training data set. Left: Results for the neural network. Right: Same as left, but for the linear regressor.

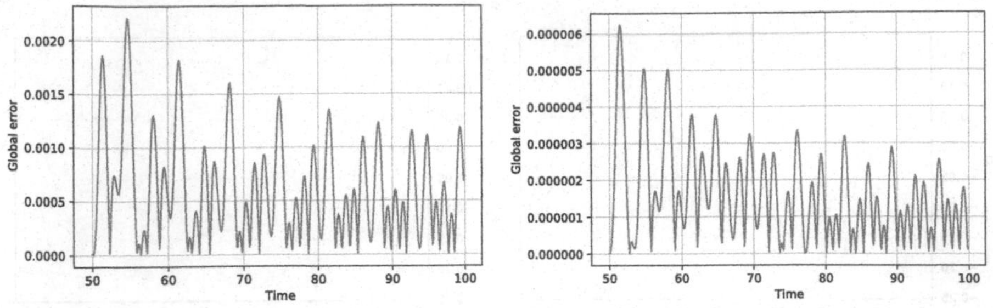

Fig. 5. The global integration error generated by the data model along the trajectory, determined by comparison to a trajectory generated by a traditional solver. Left: Results for the neural network. Right: Same as left, but for the linear regressor. Note the different scale on the y-axis in the two plots.

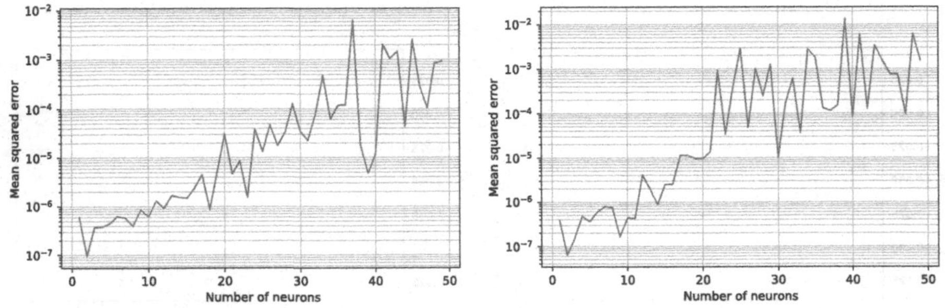

Fig. 6. The mean squared error during the validation, shown versus the number of neurons for one hidden layer. Left: Results for the neural network with one hidden layer. Right: Results for the neural network with two hidden layers.

After successful training, the predictive quality of a data model relates to the similarity of the input data for training and prediction. This similarity can be discussed in terms of distance between the data, as shown in Fig. 4. The distance is defined as Euclidean distance of the input variables during the validation from the nearest point of the input variables during the training. The distance does not accumulate over time and the model can be considered as being interpolation and not extrapolation. The global error of the model can easily be estimated considering the deviation of the predicted trajectory from the trajectory generated by the traditional solver, shown in Fig. 5. Using this measure, the linear regressor performs considerably better (three orders of magnitude) than the neural network. We conjecture that the linear regressor performs so well because the model and the solver are linear except for the perturbation. The non-linearity of the latter is not modelled by the regressor, since it is used as one of the input variables.

Parameter studies have been performed to discuss the sensitivity of the method. The mean square global error during the validation is shown versus the number of neurons in Fig. 6 for one and two hidden layers (each with the

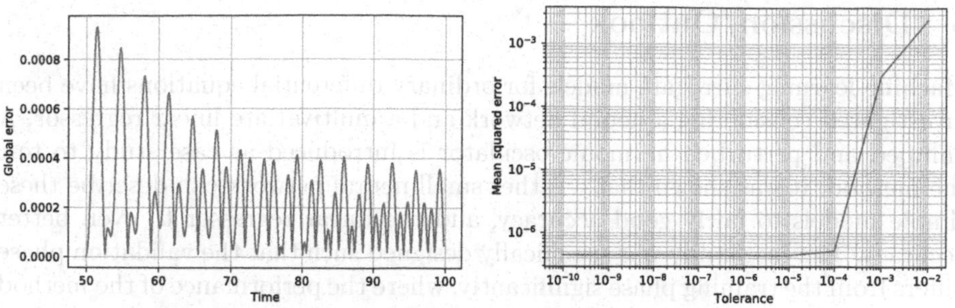

Fig. 7. Left: Global integration error for the validation of the neural network using two neurons in one hidden layer, which is the best neural network found in this study in the sense of least mean squared error, compare Figs. 5 and 6. Right: The mean squared error during the validation, shown versus the required tolerance during the training of the neural network.

same number of neurons). There is a significant amount of noise, which might relate to the difficulty of finding the global optimum in the training phase, as discussed earlier. However, a trend is observed for the method to worsen towards a higher number of neurons, independent of the number of hidden layers. It seems that superfluously increasing of the complexity of the data model is disadvantageous. Different sources of error might accumulate, including the difficulty of finding the optimum. This is true at least at this level of accuracy and for simple models like the considered one here.

The best performing neural network appears to use two neurons in one hidden layer, for which we show results for the global error in Fig. 7, left hand side. Still, the global error is worse than that of the linear regressor by three orders of magnitude. The dependence on the tolerance used in the optimizer of the neural network is shown in Fig. 7, right hand side. We conclude that a further improvement of the performance of the neural network cannot be achieved by such measures for this simple case study. However, for more complex applications, it is expected that larger neural networks will be required.

The computational costs have been measured using the python package *timeit*. The traditional solver took about 7 ms, the neural network 31 ms and the linear regressor 26 ms on a 2.4 GHz Intel Core i5. Note that the computational costs for training the data model are neglected here. This is reasonable, as the training could be done offline or at least easily parallelized. The high computational costs of the data models compared to the traditional solver are not surprising, as no proper model order reduction is applied. Benefits in terms of saved computational costs are however expected for more complicated models where model order reduction is desirable and can be achieved by the proposed method.

5 Discussion/Outlook

Machine learning surrogate models for ordinary differential equations have been investigated considering a neural network and a multivariate linear regressor. A damped and perturbed harmonic oscillator is introduced as case study to test the method. It was shown that a rather small neural network can describe these simple models to fairly good accuracy, and the linear regressor to even better accuracy. The case study was specifically designed such that the validation phase differs from the training phase significantly, where the performance of the method is remarkable. For the neural network, it was crucial to apply various random seeds for the optimizer of the neural network during the training phase to obtain satisfactory performance.

As next step, non-linear and hybrid models (discrete/continuous) shall be discussed, like the Dahlquist's test equation and the bouncing ball equation. The method presented here shall then be tested further for more complicated models, where model order reduction is desired. A realistic case study will be considered in the field of thermal energy engineering. It will be discussed which problems are particularly suited for this kind of surrogate models. The method shall be applied in the context of co-simulation, replacing a simulation unit after sufficient training on the fly.

Acknowledgements. The financial support by the Austrian Federal Ministry for Digital and Economic Affairs and the National Foundation for Research, Technology and Development is gratefully acknowledged. We further acknowledge fruitful discussions with Gerald Schweiger, Claudio Gomes and Philip Ohnewein.

References

1. Berkooz, G., Holmes, P., Lumley, J.L.: The proper orthogonal decomposition in the analysis of turbulent flows. Ann. Rev. Fluid Mech. **25**(1), 539–575 (1993)
2. Gomes, C., Thule, C., Broman, D., Larsen, P.G., Vangheluwe, H.: Co-simulation: state of the art. CoRR, abs/1702.0, February 2017
3. Schweiger, G., Gomes, C., Engel, G., Hafner, I., Schöggl, J., Nouidui, T.S.:. Co-simulation: an empirical survey identifies promising standards, current challenges and research needs (2018). Submitted
4. Blochwitz, T., et al.: The functional mockup interface for tool independent exchange of simulation models. In: 2011 8th International Modelica Conference, pp. 173–184 (2009)
5. Schweiger, G., Gomes, C., Engel, G., Hafner, I., Schoeggl, J.-P., Nouidui, T.S.: Functional mockup-interface : an empirical survey identifies research challenges and current barriers. In: 2018 American Modelica Conference (2018)
6. Engel, G., Chakkaravarthy, A.S., Schweiger, G.: A general method to compare different co-simulation interfaces: demonstration on a case study. In: Obaidat, M.S., Ören, T., Rango, F.D. (eds.) SIMULTECH 2017. AISC, vol. 873, pp. 351–365. Springer, Cham (2019). https://doi.org/10.1007/978-3-030-01470-4_19
7. Lagaris, I.E.E., Likas, A., Fotiadis, D.I.I.: Artificial neural networks for solving ordinary and partial differential equations. IEEE Trans. Neural Netw. **9**(5), 1–26 (1997)

8. Wei, S., Jin, X., Li, H.:. General solutions for nonlinear differential equations: a deep reinforcement learning approach. Technical report (2018)
9. Guo, X., Li, W., Iorio, F.: Convolutional neural networks for steady flow approximation. In: Proceedings of the 22nd ACM SIGKDD International Conference on Knowledge Discovery and Data Mining, San Francisco, CA, USA, 13–17 August 2016, pp. 481–490 (2016)
10. IBPSA. IBPSA Project 1 - https://ibpsa.github.io/project1/
11. Berger, J., Mazuroski, W., Oliveira, R.C.L.F., Mendes, N.: Intelligent co-simulation: neural network vs. proper orthogonal decomposition applied to a 2D diffusive problem. J. Build. Perform. Simul. **11**(5), 568–587 (2018)
12. Geyer, P., Singaravel, S.: Component-based machine learning for performance prediction in building design. Appl. Energy **228**, 1439–1453 (2018)
13. Singaravel, S., Suykens, J., Geyer, P.: Deep-learning neural-network architectures and methods: using component-based models in building-design energy prediction. Adv. Eng. Inform. **38**(May), 81–90 (2018)
14. Engel, G.: Neural networks to approximate solutions of ordinary differential equations. In: Computing Conference 2019. Advances in Intelligent Systems and Computing. Springer, Heidelberg (2019)
15. Jones, E., Oliphant, T., Peterson, P., et al.: SciPy: Open Source Scientific Tools for Python (2001). http://www.scipy.org/. Accessed 22 May 2019
16. Pedregosa, F., et al.: Scikit-learn: machine learning in Python. J. Mach. Learn. Res. **12**, 2825–2830 (2011)

Kernel Methods for Discrete-Time Linear Equations

Boumediene Hamzi[1,2(✉)] and Fritz Colonius[3]

[1] Department of Mathematics, Imperial College London, London, UK
boumediene.hamzi@gmail.com
[2] Department of Mathematics, AlFaisal University, Riyadh, Kingdom of Saudi Arabia
[3] Institut für Mathematik, Universität Augsburg, Augsburg, Germany

Abstract. Methods from learning theory are used in the state space of linear dynamical systems in order to estimate the system matrices and some relevant quantities such as a the topological entropy.

The approach is illustrated via a series of numerical examples.

Keywords: Reproducing Kernel Hilbert spaces · Linear discrete-time equations · Parameter estimation

1 Introduction

This paper discusses several problems in dynamical systems and control, where methods from learning theory are used in the state space of linear systems. This is in contrast to previous approaches in the frequency domain [8,21]. We refer to [8] for a general survey on applications of machine learning to system identification.

Basically, learning theory allows to deal with problems when only data from a given system are given. Reproducing Kernel Hilbert Spaces (RKHS) allow to work in a very large dimensional space in order to simplify the underlying problem. We will discuss this in the simple case when the matrix A describing a linear discrete-time system is unknown, but a time series from the underlying linear dynamical system is given. We propose a method to estimate the underlying matrix using kernel methods. Applications are given in the stable and unstable case and for estimating the topological entropy for a linear map. Furthermore, in the control case, stabilization via linear-quadratic optimal control is discussed.

The emphasis of the present paper is on the formulation of a number of problems in dynamical systems and control and to illustrate the applicability of our approach via a series of numerical examples. This paper should be viewed as a preliminary step to extend these results to nonlinear discrete-time systems

BH thanks the European Commission and the Scientific and the Technological Research Council of Turkey (Tubitak) for financial support received through Marie Curie Fellowships.

J. M. F. Rodrigues et al. (Eds.): ICCS 2019, LNCS 11539, pp. 178–191, 2019.
https://doi.org/10.1007/978-3-030-22747-0_14

within the spirit of [3,4] where the authors showed that RKHSs act as "linearizing spaces" and offers tools for a data-based theory for nonlinear (continuous-time) dynamical systems. The approach used in these papers is based on *embedding a nonlinear system in a high (or infinite) dimensional reproducing kernel Hilbert space (RKHS) where linear theory is applied.* To illustrate this approach, consider a polynomial in $\mathbb{R}$, $p(x) = \alpha + \beta x + \gamma x^2$ where α, β, γ are real numbers. If we consider the map $\phi : \mathbb{R} \to \mathbb{R}^3$ defined as $\phi(x) = [1\,x\,x^2]^T$ then $p(x) = \alpha \cdot [1\,x\,x^2]^T = \alpha \cdot \phi(x)$ is an affine polynomial in the variable $\phi(x)$. Similarly, consider the nonlinear discrete-time system $x(k+1) = x(k)+x^2(k)$. By rewriting it as $x(k+1) = [1\,1]\begin{bmatrix} x(k) \\ x(k)^2 \end{bmatrix}$, the nonlinear system becomes linear in the variable $[x(k)\; x(k)^2]$.

The contents is as follows: In Sect. 2 the problem is stated formally and an algorithm based on kernel methods is given for the stable case. In Sect. 3 the algorithm is extended to the unstable case. In particular, the topological entropy of linear maps is computed (which boils down to computing unstable eigenvalues). Section4 draws some conclusions from the numerical experiments. For the reader's convenience we have collected in the appendix basic concepts from learning theory as well as some hints to the relevant literature.

2 Statement of the Problem

Consider the linear discrete-time system

$$x(k+1) = Ax(k), \tag{1}$$

where $A = [a_{i,j}] \in \mathbb{R}^{n \times n}$. We want to estimate A from the time series $x(1) + \eta_1$, $\cdots$, $x(N) + \eta_N$ where the initial condition $x(0)$ is known and η_i are distributed according to a probability measure ρ_x that satisfies the following condition (this is the *Special Assumption* in [12]).

Assumption. The measure ρ_x is the marginal on $X = \mathbb{R}^n$ of a Borel measure ρ on $X \times \mathbb{R}$ with zero mean supported on $[-M_x, M_x], M_x > 0$.

One obtains from (1) for the components of the time series that

$$x_i(k+1) = \sum_{j=1}^{n} a_{ij} x_j(k). \tag{2}$$

For every i we want to estimate the coefficients $a_{ij}, j = 1, \cdots, n$. They are determined by the linear maps $f_i^* : \mathbb{R}^n \to \mathbb{R}$ given by

$$(x_1, ..., x_n) \mapsto \sum_{j=1}^{n} a_{ij} x_j. \tag{3}$$

This problem can be reformulated as a learning problem as described in the Appendix where f_i^* in (3) plays the role of the unknown function (36) and $(x(k), x_i(k+1) + \eta_i)$ are the samples in (38).

We note that in [12], the authors do not consider time series and that we apply their results to time series.

In order to approximate f_i^*, we minimize the criterion in (41). For a positive definite kernel K let f_i be the kernel expansion of f_i^* in the corresponding RKHS $\mathcal{H}_K$. Then $f_i = \sum_{j=1}^{\infty} c_{i,j}\phi_j$ with certain coefficients $c_{ij} \in \mathbb{R}$ and

$$||f_i||_{\mathcal{H}_K} = \sum_{j=1}^{\infty} \frac{c_{i,j}^2}{\lambda_j}, \tag{4}$$

where (λ_j, ϕ_j) are the eigenvalues and eigenfunctions of the integral operator $L_K : \mathcal{L}_\nu^2(\mathcal{X}) \to \mathcal{C}(\mathcal{X})$ given by $(L_K f)(x) = \int K(x,t)f(t)d\nu(t)$ with a Borel measure ν on $\mathcal{X}$. Thus $L_K\phi_j = \lambda_j\phi_j$ for $j \in \mathbb{N}^*$ and the eigenvalues $\lambda_j \geq 0$.

Then we consider the problem of minimizing over $(c_{i,1}, \cdots, c_{i,N})$ the functional

$$\mathcal{E}_i = \frac{1}{N}\sum_{k=1}^{N}(y_i(k) - f_i(x(k)))^2 + \gamma_i||f_i||^2_{\mathcal{H}_K}, \tag{5}$$

where $y_i(k) := x_i(k+1)+\eta_i = f_i^*(x(k))+\eta_i$ and γ_i is a regularization parameter.

Since we are dealing with a linear problem, it is natural to choose the linear kernel $k(x,y) = \langle x, y\rangle$. Then the solution of the above optimization problem is given by the kernel expansion of $x_i(k+1)$, $i = 1, \cdots, n$,

$$y_i(k) := x_i(k+1) = \sum_{j=1}^{N} c_{ij}\langle x(j), x(k)\rangle, \tag{6}$$

where the c_{ij} satisfy the following set of equations:

$$\begin{bmatrix} x_i(1) \\ \vdots \\ x_i(N) \end{bmatrix} = \left(N\lambda I_d + \mathbb{K}\right)\begin{bmatrix} c_{i1} \\ \vdots \\ c_{iN} \end{bmatrix}, \tag{7}$$

with

$$\mathbb{K} := \begin{bmatrix} \sum_{\ell=1}^{n} x_\ell(1)x_\ell(0) & \cdots & \sum_{\ell=1}^{n} x_\ell(N)x_\ell(0) \\ \vdots & \cdots & \vdots \\ \sum_{\ell=1}^{n} x_\ell(1)x_\ell(N-1) & \cdots & \sum_{\ell=1}^{n} x_\ell(N)x_\ell(N-1) \end{bmatrix}. \tag{8}$$

This is a consequence of Theorem 2.

From (2), we have

$$\begin{aligned} x_i(k+1) &= \sum_{j=1}^{N} c_{ij}\langle x(j), x(k)\rangle = \sum_{j=1}^{N} c_{ij}x(j)^T \cdot x(k) = \sum_{j=1}^{N}\sum_{\ell=1}^{n} c_{ij}x_\ell(j)x_\ell(k) \\ &= \sum_{\ell=1}^{n}\sum_{j=1}^{N} c_{ij}x_\ell(j)x_\ell(k). \end{aligned}$$

Then an estimate of the entries of A is given by

$$\hat{a}_{i\ell} = \sum_{j=1}^{N} c_{i,j} x_\ell(j). \tag{9}$$

This discussion leads us to the following basic algorithm.

Algorithm $\mathcal{A}$: If the eigenvalues of A are all within the unit circle, one proceeds as follows in order to estimate A. Given the time series $x(1), \cdots, x(N)$ solve the system of Eq. (7) to find the numbers c_{ij} and then compute $\hat{a}_{i\ell}$ from (9).

Before we present numerical examples and modifications and applications of this algorithm, it is worthwhile to note the following preliminary remarks indicating what may be expected.

The stability assumption in algorithm $\mathcal{A}$ is imposed, since otherwise the time series will diverge exponentially. Then, already for a moderately sized number of data points ($N \approx 10^2$) Eq. (7) will be ill conditioned. Hence for unstable A, modifications of algorithm $\mathcal{A}$ are required.

While for test examples one can compare the entries of the matrix A and its approximation $\hat{A}$, it may appear more realistic to compare the values $x(1), \cdots, x(N)$ of the data series and the values $\hat{x}(1), \cdots, \hat{x}(N)$ generated by the iteration of the matrix $\hat{A}$.

In general, one should not expect that increasing the number of data points will lead to better approximations of the matrix A. If the matrix A is diagonalizable, for generic initial points $x(0) \in \mathbb{R}^n$ the data points $x(k)$ will approach for $N \to \infty$ the eigenspace for the eigenvalue with maximal modulus. For general A and generic initial points $x(0) \in \mathbb{R}^n$, the data points $x(N)$ will approach for $N \to \infty$ the largest Lyapunov space (i.e., the sum of the real generalized eigenspaces for eigenvalues with maximal modulus). Thus in the limit for $N \to \infty$, only part of the matrix can be approximated. A detailed discussion of this (well known) limit behavior is, e.g., given in Colonius and Kliemann [6]. A consequence is that a medium length of the time series should be adequate.

This problem can be overcome by choosing the regularization parameter γ in (5) and (7) using the method of cross validation described in [10]. Briefly, in order to choose γ, we consider a set of values of regularization parameters: we run the learning algorithm over a subset of the samples for each value of the regularization parameter and choose the one that performs the best on the remaining data set. Cross validation helps also in the presence of noise and to improve the results beyond the training set.

A theoretical justification of our algorithm is guaranteed by the error estimates in Theorem 5. In fact, for the linear dynamical system (1), we have that f^* in (36) is the linear map $f^*(x) = f_i(x)$ in (3) and the samples s in (38) are $(x(k), x_i(k+1) + \eta_i)$. Moreover, by choosing the linear kernel $k(x, y) = \langle x, y \rangle$ we get that $f^* \in \mathcal{H}_K$. In this case, (46) has the form

$$||\hat{x}_i(k+1) - x_i(k+1)||^2 \leq 2C_{\bar{x}}\mathcal{E}_{\text{samp}} + 2||x(k+1)||_K^2(\gamma + 8C_{\bar{x}}\Delta), \tag{10}$$

where $||x_i(k+1)||_{\mathcal{H}_K} = \sum_{j=1}^{\infty} \frac{c_{i,j}^2}{\lambda_j}$.

The first term in the right hand side of inequality (10) represents the error due to the noise (sampling error) and the second term represents the error due to regularization (regularization error) and the finite-number of samples (integration error).

Next we discuss several numerical examples, beginning with the following scalar equation.

Example 1. Consider $x(k+1) = \alpha x(k)$ with $\alpha = 0.5$. With the initial condition $x(0) = -0.5$, we generate the time series $x(1), \cdots, x(100)$. Applying algorithm $\mathcal{A}$ with the regularization parameter $\gamma = 10^{-6}$ we compute $\hat{\alpha} = 0.4997$. Using cross validation, we get that $\hat{\alpha} = 0.5$ with regularization parameter $\gamma = 1.5259 \cdot 10^{-5}$. When we introduce an i.i.d perturbation signal $\eta_i \in [-0.1, 0.1]$, the algorithm does not behave well when we fix the regularization parameter. With cross validation, the algorithm works quite well and the regularization parameter adapts to the realization of the signal η_i. Here, for $e(k) = x(k) - \hat{x}(k)$ with $x(k+1) = \alpha x(k)$ and $\hat{x}(k+1) = \hat{\alpha}\hat{x}(k)$, we get that $||e(300)|| = \sqrt{\sum_{i=1}^{300} e^2(i)} = 0.0914$ and $\sqrt{\sum_{i=100}^{300} e^2(i)} = 1.8218 \cdot 10^{-30}$.

We observe an analogous behavior of the algorithm when the data are generated from $x(k+1) = \alpha x(k) + \varepsilon x(k)^2$ where the algorithm works well in the presence of noise and structural perturbations when using cross validation. When $\varepsilon = 0.1$ and with an i.i.d perturbation signal $\eta_i \in [-0.1, 0.1]$, $\hat{\alpha}$ varies between 0.38 and 0.58 depending on the realization of η_i but $||e(300)|| = \sqrt{\sum_{i=1}^{300} e^2(i)} =$ 0.2290 and $\sqrt{\sum_{i=100}^{300} e^2(i)} = 2.8098 \cdot 10^{-30}$ which shows that the error e decreases exponentially and the generalization properties of the algorithm are quite good.

3 Unstable Case

Consider

$$x(k+1) = Ax(k) \text{ with } A \in \mathbb{R}^{n \times n}, \tag{11}$$

where some of the eigenvalues of A are outside the unit circle. Again, we want to estimate A when the following data are given,

$$x(1), x(2), ..., x(N), \tag{12}$$

which are generated by system (11), thus $x(k) = A^{k-1}x(1)$.

As remarked above, a direct application of the algorithm $\mathcal{A}$ will not work, since the time series diverges fast. Instead we construct a new time series from (12) associated to an auxiliary stable system.

For a constant $\sigma > 0$ we define the auxiliary system by $y(k+1) = \tilde{A}y(k)$ (13) with $\tilde{A} := \frac{1}{\sigma}A$. Thus $y(k) = \left(\frac{A}{\sigma}\right)^{k-1} y(1)$ and with $y(1) = x(1)$ one finds $y(k) = \frac{1}{\sigma^{k-1}}A^{k-1}x(1) = \frac{1}{\sigma^{k-1}}x(k)$. If we choose $\sigma > 0$ such that the eigenvalues of $\frac{A}{\sigma}$ are in the unit circle, we can apply algorithm $\mathcal{A}$ to this stable matrix and hence we would obtain an estimate of $\frac{A}{\sigma}$ and hence of A. However, since the eigenvalues of the matrix A are unknown, we will be content with a somewhat weaker condition than stability of $\frac{A}{\sigma}$.

The data (12) for system (11) yield the following data for system (3): $y(1) := x(1), y(2) := \frac{1}{\sigma}x(2), ..., y(N) := \frac{1}{\sigma^{N-1}}x(N)$. We propose to choose σ as follows: Define

$$\sigma := \max\left\{\frac{\|x(k+1)\|}{\|x(k)\|}, k \in \{0, 1, ..., N\}\right\}. \tag{13}$$

Clearly the inequality $\sigma \leq \|A\|$ holds. We apply algorithm $\mathcal{A}$ to the time series $y(k)$. This yields an estimate of $\frac{A}{\sigma}$ and hence an estimate $\hat{A}$ of A.

For general A, this choice of σ certainly does not guarantee that the eigenvalues of $\frac{A}{\sigma}$ are within the unit circle. However, as mentioned above, a generic data sequence $x(k), k \in \mathbb{N}$, will converge to the eigenspace of the eigenvalue with maximal modulus. Hence $\frac{\|x(k+1)\|}{\|x(k)\|}$ will approach the maximal modulus of an eigenvalue, thus this choice of σ will lead to a matrix $\frac{A}{\sigma}$ which is not "too unstable".

Example 2. Consider $x(k+1) = \alpha x(k)$ with $\alpha = 11.46$. With the initial condition $x(0) = -0.5$, we generate the time series $x(1), \cdots, x(100)$. The algorithm above with the regularization parameter $\gamma = 10^{-6}$ yields the estimate $\hat{\alpha} = 11.4086$. Cross validation leads to the regularization parameter $\gamma = 9.5367 \cdot 10^{-7}$ and the estimate $\hat{\alpha} = 11.4599$. In the presence of a small noise $\eta \in [-0.1, 0.1]$, cross validation yields the regularization parameter $\gamma = 0.002$ and the slightly worse estimate $\hat{\alpha} = 11.1319$.

We observe the same behavior in higher dimensional systems where the eigenvalues are of the same order of magnitude.

The next example is an unstable system with a large gap between the eigenvalues.

Example 3. Consider the system $x(k+1) = Ax(k)$ with $A = \begin{bmatrix} 20 & 0 \\ 0 & -0.1 \end{bmatrix}$. With the initial condition $x(0) = [-1.9, 1]$, we generate the time series $x(1), \cdots, x(100)$. The algorithm above yields the (excellent) estimate $\hat{A} = \begin{bmatrix} 20.0000 & 0.0000 \\ -0.0000 & -0.1000 \end{bmatrix}$. In the presence of noise of maximal amplitude 10^{-4}, the algorithm approximates well only the large entry $a_{11} = 20$: For a first realization of η_i and with cross validation, we get $\hat{A} = \begin{bmatrix} 19.9997 & -0.0111 \\ 0.0000 & -0.1104 \end{bmatrix}$, with $\gamma_1 = 1.5259 \cdot 10^{-5}$ and $\gamma_2 = 2^{20}$. However another realization of η_i leads to $\hat{A} = \begin{bmatrix} 19.9994 & -0.0011 \\ 0.0000 & -0.0000 \end{bmatrix}$, with $\gamma_1 = 3.0518 \cdot 10^{-5}$ and $\gamma_2 = 2.8147 \cdot 10^{14}$. This is due to the fact that the data converge to the eigenspace generated by the largest eigenvalue $\lambda = 20$. However, the eigenvalues of $A - \hat{A}$ are within the unit disk with small amplitude which guarantees that the error dynamics of

$e(k) = x(k) - \hat{x}(k)$ converges to the origin quite quickly. We observe the same phenomenon with

$$A = \begin{bmatrix} -0.5 & 0 \\ 0 & 25 \end{bmatrix}. \tag{14}$$

Here, in the absence of noise, we obtain the estimate

$$\hat{A} = \begin{bmatrix} -0.5000 & 0.0000 \\ -0.0000 & 25.0000 \end{bmatrix}, \tag{15}$$

with $\gamma_1 = \gamma_2 = 0.9313 \cdot 10^{-9}$. In the presence of noise η_i with amplitude 10^{-4}, the data converge to the eigenspace corresponding to the largest eigenvalue $\lambda = 25$: for some realization of η_i one obtains the estimate

$$\hat{A} = \begin{bmatrix} -0.4809 & 0.0008 \\ 0.0164 & 24.9960 \end{bmatrix}, \tag{16}$$

while for another realization of η

$$\hat{A} = \begin{bmatrix} -0.0000 & -0.0000 \\ -1.0067 & 24.8696 \end{bmatrix}. \tag{17}$$

The regularization parameters γ_1 and γ_2 adapt to the realization of the noise.

As already remarked in the end of Sect. 2, we see that "more data" does not always necessarily lead to better results, since the data sequence converges to the eigenspace generated by the largest eigenvalue. However, whether with or without noise, the approximations of A are good enough to reduce the error between $x(k+1) = Ax(k)$ and $\hat{x}(k+1) = \hat{A}\hat{x}(k)$ outside of the training examples, since cross-validation determines a good regularization parameter γ that balances between good fitting and good prediction properties.

The next example has an eigenvalue on the unit circle.

Example 4. Consider $x(k+1) = Ax(k)$ with

$$A = \begin{bmatrix} 2.2500 & -1.2500 & 1.2500 & -49.5500 \\ 3.7500 & -2.7500 & 13.1500 & -20.6500 \\ 0 & 0 & 10.4000 & -32.3000 \\ 0 & 0 & 0 & -21.9000 \end{bmatrix}. \tag{18}$$

The set of eigenvalues of A is $\text{spec}(A) = \{-1.5000, 1.0000, 10.4000, -21.9000\}$. In the absence of noise and initial condition $x = [-0.9, 15, 1.5.2.5]$ with $N = 100$ points, we compute the estimate

$$\hat{A} = \begin{bmatrix} 2.2500 & -1.2500 & 1.2498 & -49.5499 \\ 3.7500 & -2.7500 & 13.1498 & -20.6499 \\ 0.0000 & 0.0000 & 10.3998 & -32.2999 \\ 0.0000 & 0.0000 & -0.0001 & -21.8999 \end{bmatrix}, \tag{19}$$

and regularization parameters $\gamma_1 = \gamma_2 = 0.9313 \cdot 10^{-9}$. In this case, the set of eigenvalues of $\hat{A}$ is

$$\mathrm{spec}(\hat{A}) = \{-21.9000, 10.3999, -1.5000, 1.0000\}. \tag{20}$$

For a given realization of $\eta \in [-10^{-4}, 10^{-4}]$, we obtain the estimate

$$\hat{A} = \begin{bmatrix} 2.2551 & -1.2490 & 1.2187 & -49.5304 \\ 3.7554 & -2.7489 & 13.1175 & -20.6297 \\ 0.0055 & 0.0011 & 10.3669 & -32.2794 \\ 0.0053 & 0.0010 & -0.0325 & -21.8797 \end{bmatrix} \tag{21}$$

with $\gamma_1 = 0.0745 \cdot 10^{-7}$ and $\gamma_2 = 0.1490 \cdot 10^{-7}$. The eigenvalues of $A - \hat{A}$ are of the order of 10^{-4} which guarantees that the error dynamics converges quickly to the origin. However, the set of eigenvalues of $\hat{A}$ is

$$\mathrm{spec}(\hat{A}) = \{-21.8996, 10.3999, -1.5026, 1.0134\}. \tag{22}$$

Hence an additional unstable eigenvalue occurs.

Example 5. Consider $x(k+1) = Ax(k)$ with

$$A = \begin{bmatrix} -0.8500 & 0.4500 & -0.4500 & -77.8500 \\ -1.3500 & 0.9500 & 14.3500 & -11.6500 \\ 0 & 0 & 15.3000 & -55.3000 \\ 0 & 0 & 0 & -40.0000 \end{bmatrix}. \tag{23}$$

The eigenvalues of A are given by

$$\mathrm{spec}(A) = \{-0.4000, 0.5000, 15.3000, -40.0000\}. \tag{24}$$

For an initial condition $x = [-0.9; 15; 1.5; 2.5]$ and with $N = 100$ data points, we get

$$\hat{A} = \begin{bmatrix} -0.8498 & 0.4501 & -0.4499 & -77.8504 \\ -1.3499 & 0.9500 & 14.3501 & -11.6502 \\ 0.0001 & 0.0001 & 15.3001 & -55.3004 \\ -0.0004 & -0.0002 & -0.0004 & -39.9987 \end{bmatrix} \tag{25}$$

with eigenvalues given by

$$\mathrm{spec}(\hat{A}) = \{-40.0000, -0.3974, 0.4982, 15.3008\}. \tag{26}$$

Here we used $\gamma_i = 10^{-12}$, $i = 1, \cdots, 4$. Moreover, the eigenvalues of $A - \hat{A}$ are quite small and such that the error dynamics converges quickly to the origin. In the presence of noise η, the algorithm approximates the largest eigenvalues of A but does not approximate the smaller (stable) ones. For example, for a particular realization of noise with amplitude 10^{-4}, we get the estimate

$$\hat{A} = \begin{bmatrix} -2.1100 & -0.0993 & -1.3259 & -74.4543 \\ -1.7053 & 0.7777 & 13.9397 & -10.5308 \\ -0.8277 & -0.3692 & 14.6466 & -52.9920 \\ -0.8283 & -0.3694 & -0.6539 & -37.6904 \end{bmatrix} \tag{27}$$

and $spec(\hat{A}) = \{-40.0009, 0.1620 \pm 0.8438i, 15.3008\}$.

For another realization of noise with amplitude 10^{-2}, we get the estimate

$$\hat{A} = \begin{bmatrix} -138.0893 & -60.7052 & -105.8111 & 301.5029 \\ -0.2435 & 0.9101 & 12.9638 & -12.6745 \\ -71.1408 & -31.9557 & -40.3842 & 142.3170 \\ -71.1408 & -31.9557 & -55.6843 & 157.6172 \end{bmatrix} \tag{28}$$

and $\text{spec}(\hat{A}) = \{-40.1391, 3.9326, 0.9601, 15.3002\}$.

The algorithm introduced above also allows us to compute the topological entropy of linear systems, since it is determined by the unstable eigenvalues. Recall that the topological entropy of a linear map on $\mathbb{R}^n$ is defined in the following way:

Fix a compact subset $K \subset \mathbb{R}^n$, a time $\tau \in \mathbb{N}$ and a constant $\varepsilon > 0$. Then a set $R \subset \mathbb{R}^n$ is called (τ, ε)-spanning for K if for every $y \in K$ there is $x \in R$ with

$$\left\|A^j y - A^j x\right\| < \varepsilon \text{ for all } j = 0, ..., \tau. \tag{29}$$

By compactness of K, there are finite (τ, ε)-spanning sets. Let R be a (τ, ε)-spanning set of minimal cardinality $\#R = r_{\min}(\tau, \varepsilon, K)$. Then

$$h_{top}(K, A, \varepsilon) := \lim_{\tau \to \infty} \frac{1}{\tau} \log r_{\min}(\tau, \varepsilon, K), h_{top}(K, A) := \lim_{\varepsilon \to 0^+} h_{top}(K, \varepsilon). \tag{30}$$

(the limits exist). Finally, the topological entropy of A is

$$h_{top}(A) := \sup_K h_{top}(K, A), \tag{31}$$

where the supremum is taken over all compact subsets K of $\mathbb{R}^n$.

A classical result due to Bowen (cf. [19, Theorem 8.14]) shows that the topological entropy is determined by the sum of the unstable eigenvalues, i.e.,

$$h_{top}(A) = \sum \max(1, |\lambda|), \tag{32}$$

where summation is over all eigenvalues of A counted according to their algebraic multiplicity.

Hence, when we approximate the unstable eigenvalues of A by those of the matrix $\hat{A}$, we also get an approximation of the topological entropy.

Example 6. For Example 4, we get that $h_{top}(A) = 34.80$ while for the estimate $\hat{A}$ one obtains $h_{top}(\hat{A}) = 34.7999$. For Example 5, we get that $h_{top}(A) = 55.30$ and $h_{top}(\hat{A}) = 55.3008$. These estimates appear reasonably good.

4 Conclusions

This paper has introduced the algorithm $\mathcal{A}$ based on kernel methods to identify a stable linear dynamical system from a time series. The numerical experiments give excellent results in the absence of noise and structural perturbations. In the presence of noise and structural perturbations the algorithm works well in the stable case. In the unstable case, a modified algorithm works quite well in the presence of noise but cannot handle structural perturbations.

Then we have extended algorithm $\mathcal{A}$ to identify linear control systems. In particular, we have used estimates obtained by kernel methods to stabilize linear systems using linear-quadratic control and the algebraic Riccati equation. Here the numerical experiments seem to indicate that the same conclusions on applicability of the algorithm apply.

Extensions of the considered algorithms to nonlinear systems appear feasible and are left to future work.

A Appendix: Elements of Learning Theory

In this section, we give a brief overview of Reproducing Kernel Hilbert Spaces (RKHS) as used in statistical learning theory. The discussion here borrows heavily from Cucker and Smale [7], Wahba [18], and Schölkopf and Smola [17]. Early work developing the theory of RKHS was undertaken by Schoenberg [14–16] and then Aronszajn [2]. Historically, RKHS came from the question, when it is possible to embed a metric space into a Hilbert space.

Definition 1. *Let $\mathcal{H}$ be a Hilbert space of functions on a set $\mathcal{X}$ which is a closed subset of $\mathbb{R}^n$. Denote by $\langle f, g\rangle$ the inner product on $\mathcal{H}$ and let $||f|| = \langle f, f\rangle^{1/2}$ be the norm in $\mathcal{H}$, for f and $g \in \mathcal{H}$. We say that $\mathcal{H}$ is a reproducing kernel Hilbert space (RKHS) if there exists $K : \mathcal{X} \times \mathcal{X} \to \mathbb{R}$ such that*

i. K has the reproducing property, i.e., $f(x) = \langle f(\cdot), K(\cdot, x)\rangle$ for all $f \in \mathcal{H}$.
ii. K spans $\mathcal{H}$, i.e., $\mathcal{H} = \overline{span\{K(x,\cdot)|x \in \mathcal{X}\}}$.

K will be called a reproducing kernel of $\mathcal{H}$ and $\mathcal{H}_K$ will denote the RKHS $\mathcal{H}$ with reproducing kernel K.

Definition 2. *Given a kernel $K : \mathcal{X} \times \mathcal{X} \to \mathbb{R}$ and inputs $x_1, \cdots, x_n \in \mathcal{X}$, the $n \times n$ matrix*

$$k := (K(x_i, x_j))_{ij}, \tag{33}$$

is called the Gram Matrix *of k with respect to $x_1, \cdots, x_n$. If for all $n \in \mathbb{N}$ and distinct $x_i \in \mathcal{X}$ the kernel K gives rise to a strictly positive definite Gram matrix, it is called strictly positive definite.*

Definition 3. *(Mercer kernel map) A function $K : \mathcal{X} \times \mathcal{X} \to \mathbb{R}$ is called a Mercer kernel if it is continuous, symmetric and positive definite.*

The important properties of reproducing kernels are summarized in the following proposition.

Proposition 1. *If K is a reproducing kernel of a Hilbert space $\mathcal{H}$, then*

i. $K(x, y)$ *is unique.*
ii. *For all* $x, y \in \mathcal{X}$, $K(x, y) = K(y, x)$ *(symmetry).*
iii. $\sum_{i,j=1}^{m} \alpha_i \alpha_j K(x_i, x_j) \geq 0$ *for* $\alpha_i \in \mathbb{R}$ *and* $x_i \in \mathcal{X}$ *(positive definiteness).*
iv. $\langle K(x, \cdot), K(y, \cdot) \rangle_{\mathcal{H}} = K(x, y)$.
v. *The following kernels, defined on a compact domain* $\mathcal{X} \subset \mathbb{R}^n$, *are Mercer kernels:* $K(x, y) = x \cdot y^\top$ *(Linear),* $K(x, y) = (1 + x \cdot y^\top)^d$, $d \in \mathbb{N}$ *(Polynomial),* $K(x, y) = e^{-\frac{||x-y||^2}{\sigma^2}}$, $\sigma > 0$ *(Gaussian).*

Theorem 1. *Let* $K : \mathcal{X} \times \mathcal{X} \to \mathbb{R}$ *be a symmetric and positive definite function. Then there exists a Hilbert space of functions* $\mathcal{H}$ *defined on* $\mathcal{X}$ *admitting* K *as a reproducing Kernel. Moreover, there exists a function* $\Phi : X \to \mathcal{H}$ *such that*

$$K(x, y) = \langle \Phi(x), \Phi(y) \rangle_{\mathcal{H}} \quad \text{for} \quad x, y \in \mathcal{X}. \tag{34}$$

Φ *is called a feature map.*

Conversely, let $\mathcal{H}$ *be a Hilbert space of functions* $f : \mathcal{X} \to \mathbb{R}$, *with* $\mathcal{X}$ *compact, satisfying*

$$\text{For all } x \in \mathcal{X} \text{ there is } \kappa_x > 0, \text{ such that } |f(x)| \leq \kappa_x ||f||_{\mathcal{H}}. \tag{35}$$

Then $\mathcal{H}$ *has a reproducing kernel* K.

Remark 1

i. The dimension of the RKHS can be infinite and corresponds to the dimension of the eigenspace of the integral operator $L_K : \mathcal{L}^2_\nu(\mathcal{X}) \to \mathcal{C}(\mathcal{X})$ defined as $(L_K f)(x) = \int K(x, t) f(t) d\nu(t)$ if K is a Mercer kernel, for $f \in \mathcal{L}^2_\nu(\mathcal{X})$ and ν a Borel measure on $\mathcal{X}$.
ii. In Theorem 1, and using property [iv.] in Proposition 1, we can take $\Phi(x) := K_x := K(x, \cdot)$ in which case $\mathcal{F} = \mathcal{H}$ – the "feature space" is the RKHS. This is called the *canonical feature map.*
iii. The fact that Mercer kernels are positive definite and symmetric shows that kernels can be viewed as generalized Gramians and covariance matrices.
iv. In practice, we choose a Mercer kernel, such as the ones in [v.] in Proposition 1, and Theorem 1, that guarantees the existence of a Hilbert space admitting such a function as a reproducing kernel.

RKHS play an important role in learning theory whose objective is to find an unknown function

$$f^* : X \to Y \tag{36}$$

from random samples

$$\mathbf{s} = (x_i, y_i)|_{i=1}^{m}, \tag{37}$$

In the following we review results from [12] (for a more general setting, cf. [7]) in the special case when the data samples $\mathbf{s}$ are such that the following assumption holds.

Assumption 1: The samples in (37) have the special form

$$\mathcal{S}: \quad \mathbf{s} = (x, y_x)|_{x\in\bar{x}}, \tag{38}$$

where $\bar{x} = \{x_i\}|_{i=1}^{d+1}$ and y_x is drawn at random from $f^*(x) + \eta_x$, where η_x is drawn from a probability measure ρ_x.

Here for each $x \in X$, ρ_x is a probability measure with zero mean, and its variance σ_x^2 satisfies $\sigma^2 := \sum_{x\in\bar{x}} \sigma_x^2 < \infty$. Let X be a closed subset of $\mathbb{R}^n$ and $\bar{t} \subset X$ is a discrete subset. Now, consider a kernel $K : X \times X \to \mathbb{R}$ and define a matrix (possibly infinite) $K_{\bar{t},\bar{t}} : \ell^2(\bar{t}) \to \ell^2(\bar{t})$ as

$$(K_{\bar{t},\bar{t}}a)_s = \sum_{t\in\bar{t}} K(s,t)a_t, \quad s \in \bar{t}, a \in \ell^2(\bar{t}), \tag{39}$$

where $\ell^2(\bar{t})$ is the set of sequences $a = (a_t)_{t\in\bar{t}} : \bar{t} \to \mathbb{R}$ with $\langle a, b\rangle = \sum_{t\in\bar{t}} a_t b_t$ defining an inner product. For example, we can take $X = \mathbb{R}$ and $\bar{t} = \{0, 1, \cdots, d\}$.

In the case of a linear dynamical system (1), we are interested in learning the map $x(k) \mapsto x(k+1)$. Here we can apply the following results.

The problem to approximate a function $f^* \in \mathcal{H}_K$ from samples s of the form (37) has been studied in [12,13]. It is reformulated as the minimization problem

$$\bar{f}_{\mathbf{s},\gamma} := \text{argmin}_{f\in\mathcal{H}_{K,\bar{t}}} \left\{ \sum_{x\in\bar{x}} (f(x) - y_x)^2 + \gamma||f||_K^2 \right\}, \tag{40}$$

where $\gamma \geq 0$ is a regularization parameter. Moreover,when $\bar{x}$ is not defined by a uniform grid on X, the authors of [12] introduced a weighting $w := \{w_x\}_{x\in\bar{x}}$ on $\bar{x}$ with $w_x > 0$[1]. Let D_w be the diagonal matrix with diagonal entries $\{w_x\}_{x\in\bar{x}}$. Then, $||D_w|| \leq ||w||_\infty$.

In this case, the regularization scheme (40) becomes

$$\bar{f}_{\mathbf{s},\gamma} := \text{argmin}_{f\in\mathcal{H}_{K,\bar{t}}} \left\{ \sum_{x\in\bar{x}} w_x(f(x) - y_x)^2 + \gamma||f||_K^2 \right\}, \tag{41}$$

Theorem 2. *Assume $f^* \in \mathcal{H}_{K,\bar{t}}$ and the standing hypotheses with X, K, $\bar{t}$, ρ as above, y as in (38). Suppose $K_{\bar{t},\bar{x}} D_w K_{\bar{x},\bar{t}} + \gamma K_{\bar{t},\bar{t}}$ is invertible. Define $\mathcal{L}$ to be the linear operator $\mathcal{L} = (K_{\bar{t},\bar{x}} D_w K_{\bar{x},\bar{t}} + \gamma K_{\bar{t},\bar{t}})^{-1} K_{\bar{t},\bar{x}} D_w$. Then problem (41) has the unique solution*

$$f_{\mathbf{s},\gamma} = \sum_{t\in\bar{t}} (\mathcal{L}y)_t K_t \tag{42}$$

Assumption 2: For each $x \in X$, ρ_x is a probability measure with zero mean supported on $[-M_x, M_x]$ with $\mathcal{B}_w := (\sum_{x\in\bar{x}} w_x M_x^2)^{\frac{1}{2}} < \infty$.

The next theorems give estimates for the different sources of errors.

[1] A suggestion in [12] is to consider the ρ_X−volume of the Voronoi cell associated with $\bar{x}$. Another example is $w = 1$ or if $|\bar{x}| = m < \infty$, $w = \frac{1}{m}$.

Theorem 3. *(Sample Error) [12, Theorem 4, Propositions 2 and 3] Let* Assumptions 1 *and* 2 *be satisfied, suppose that* $K_{\bar{t},\bar{x}} D_w K_{\bar{x},\bar{t}} + \gamma K_{\bar{t},\bar{t}}$ *is invertible and let* $f_{\mathbf{s},\gamma} = \sum_{t \in \bar{t}} c_t K_t$ *be the solution of (41) given in Theorem 2 by* $c = \mathcal{L} y$. *Define*

$$\mathcal{L}_w := (K_{\bar{t},\bar{x}} D_w K_{\bar{x},\bar{t}} + \gamma K_{\bar{t},\bar{t}})^{-1} K_{\bar{t},\bar{x}} D_w^{1/2}$$
$$\kappa := ||K_{\bar{t},\bar{t}}||\ ||(K_{\bar{t},\bar{x}} D_w K_{\bar{x},\bar{t}} + \gamma K_{\bar{t},\bar{t}})^{-1}||^2.$$

Then for every $0 < \delta < 1$, *with probability at least* $1 - \delta$ *we have the sample error estimate*

$$||f_{\mathbf{s},\gamma} - f_{\bar{x},\gamma}||_K^2 \leq \mathcal{E}_{samp} := \kappa \sigma_w^2 \alpha^{-1} \left(\frac{2||K_{\bar{t},\bar{t}} \mathcal{L}_w||\ ||\mathcal{L}_w||\ \mathcal{B}_w^2}{\kappa \sigma_w^2} \log \frac{1}{\delta} \right), \tag{43}$$

where $\alpha(u) := (u-1) \log u$ *for* $u > 1$. *In particular,* $\mathcal{E}_{samp} \to 0$ *when* $\gamma \to \infty$ *or* $\sigma_w^2 \to 0$.

Theorem 4. *(Regularization Error and Integration Error) [12, Proposition 4 and Theorem 5] Let* Assumptions 1 *and* 2 *be satisfied and let* $\bar{X} = (X_x)_{x \in \bar{x}}$ *be the Voronoi cell of* X *associated with* $\bar{x}$ *and* $w_x = \rho_X(X_x)$. *Define the Lipschitz norm on a subset* $X' \subset X$ *as* $||f||_{Lip(X')} := ||f||_{L^\infty(X')} + \sup_{s,u \in X} \frac{|f(s) - f(u)|}{||s-u||_{\ell^\infty(\mathbb{R}^n)}}$ *and assume that the inclusion map of* $\mathcal{H}_{K,\bar{t}}$ *into the Lipschitz space satisfies*[2]

$$C_{\bar{x}} := \sup_{f \in \mathcal{H}_{K,\bar{t}}} \frac{\sum_{x \in \bar{x}} w_x ||f||_{Lip(X_x)}^2}{||f||_K^2} < \infty. \tag{44}$$

Suppose that $\bar{x}$ *is* $\Delta-$*dense in* X, *i.e., for each* $y \in X$ *there is some* $x \in \bar{x}$ *satisfying* $||x - y||_{\ell^\infty(\mathbb{R}^n)} \leq \Delta$.

Then for $f^* \in \mathcal{H}_{K,\bar{t}}$

$$||f_{\bar{x},\gamma} - f^*||^2 \leq ||f^*||_K^2 (\gamma + 8 C_{\bar{x}} \Delta) \tag{45}$$

Theorem 5. *(Sample, Regularization and Integration Errors) [12, Corollary 5] Under the assumptions of Theorems 3 and 4, let* $\bar{X} = (X_x)_{x \in \bar{x}}$ *be the Voronoi cell of* X *associated with* $\bar{x}$ *and* $w_x = \rho_x(X_x)$. *Suppose that* $\bar{x}$ *is* $\Delta-$*dense,* $C_{\bar{x}} < \infty$, *and* $f^* \in \mathcal{H}_{K,\bar{t}}$. *Then, for every* $0 < \delta < 1$, *with probability at least* $1 - \delta$ *there holds*

$$||f_{\mathbf{s},\gamma} - f^*||^2 \leq 2 C_{\bar{x}} \mathcal{E}_{samp} + 2||f^*||_K^2 (\gamma + 8 C_{\bar{x}} \Delta), \tag{46}$$

where $\mathcal{E}_{samp}$ *is given in (43).*

[2] This assumption is true if X is compact and the inclusion map of $\mathcal{H}_{K,\bar{t}}$ into the space of Lipschitz functions on X is bounded which is the case when K is a C^2 Mercer kernel [20]. In fact, if $||f||_{\mathrm{Lip}(X)} \leq C_0 ||f||_K$ for each $f \in \mathcal{H}_{K,\bar{t}}$, then $C_{\bar{x}} \leq C_0^2 \rho_X(X)$.

References

1. Antsaklis, P.J., Michel, A.N.: Linear Systems. Birkhäuser, Boston (2006)
2. Aronszajn, N.: Theory of reproducing kernels. Trans. Am. Math. Soc. **68**, 337–404 (1950)
3. Bouvrie, J., Hamzi, B.: Kernel methods for the approximation of nonlinear systems. SIAM J. Control Optim. **55-4**, 2460–2492 (2017)
4. Bouvrie, J., Hamzi, B.: Kernel methods for the approximation of some key quantities of nonlinear systems. J. Comput. Dyn. **4**(1&2), 1–19 (2017)
5. Cheney, W., Light, W.: A Course in Approximation Theory. Graduate Studies in Mathematics, vol. 101. American Mathematical Society, Providence (2009)
6. Colonius, F., Kliemann, W.: Dynamical Systems and Linear Algebra. Graduate Studies in Mathematics, vol. 158. American Mathematical Society, Providence (2014)
7. Cucker, F., Smale, S.: On the mathematical foundations of learning. Bull. Am. Math. Soc. **39**, 1–49 (2001)
8. Pillonetto, G., Dinuzzo, F., Chen, T., De Nicolao, G., Ljung, L.: Kernel methods in system identification, machine learning and function estimation: a survey. Automatica **50**(3), 657–682 (2014)
9. Evgeniou, T., Pontil, M., Poggio, T.: Regularization networks and support vector machines. Adv. Comput. Math. **13**(1), 1–50 (2000)
10. Rifkin, R.M., Lippert, A.: Notes on regularized least squares. Computer Science and Artificial Intelligence Laboratory Technical repor, MIT, MIT-CSAIL-TR-2007-025, CBCL-268 (2007)
11. Smale, S., Zhou, D.-X.: Estimating the approximation error in learning theory. Anal. Appl. **1**(1), 17–41 (2003)
12. Smale, S., Zhou, D.-X.: Shannon sampling and function reconstruction from point values. Bull. Am. Math. Soc. **41**, 279–305 (2004)
13. Smale, S., Zhou, D.-X.: Shannon sampling II: connections to learning theory. Appl. Comput. Harmonic Anal. **19**(3), 285–302 (2005)
14. Schoenberg, I.J.: Remarks to Maurice Fréchet's article "Sur la définition axiomatique d'une classe d'espace distanciés vectoriellement applicable sur l'espace de Hilbert". Ann. Math. **36**, 724–732 (1935)
15. Schoenberg, I.J.: On certain metric spaces arising from euclidean spaces by a change of metric and their imbedding in Hilbert space. Ann. Math. **38**, 787–793 (1937)
16. Schoenberg, I.J.: Metric spaces and positive definite functions. Trans. Am. Math. Soc. **44**, 522–536 (1938)
17. Schölkopf, B., Smola, A.J.: Learning with Kernels. The MIT Press, Cambridge (2002)
18. Wahba, G.: Spline Models for Observational Data. SIAM CBMS-NSF Regional Conference Series in Applied Mathematics, vol. 59 (1990)
19. Walters, P.: An Introduction to Ergodic Theory. Springer, New York (1982)
20. Zhou, D.-X.: Capacity of reproducing kernel spaces in learning theory. IEEE Trans. Inf. Theory **49**(7), 1743–1752 (2003)
21. Li, L., Zhou, D.-X.: Learning theory approach to a system identification problem involving atomic norm. J. Fourier Anal. Appl. **21**, 734–753 (2015)

Physics-Informed Echo State Networks for Chaotic Systems Forecasting

Nguyen Anh Khoa Doan[1,2(✉)], Wolfgang Polifke[1], and Luca Magri[2,3]

[1] Department of Mechanical Engineering, Technical University of Munich, Garching, Germany
doan@tfd.mw.tum.de

[2] Institute for Advanced Study, Technical University of Munich, Garching, Germany

[3] Department of Engineering, University of Cambridge, Cambridge, UK

Abstract. We propose a physics-informed Echo State Network (ESN) to predict the evolution of chaotic systems. Compared to conventional ESNs, the physics-informed ESNs are trained to solve supervised learning tasks while ensuring that their predictions do not violate physical laws. This is achieved by introducing an additional loss function during the training of the ESNs, which penalizes non-physical predictions without the need of any additional training data. This approach is demonstrated on a chaotic Lorenz system, where the physics-informed ESNs improve the predictability horizon by about two Lyapunov times as compared to conventional ESNs. The proposed framework shows the potential of using machine learning combined with prior physical knowledge to improve the time-accurate prediction of chaotic dynamical systems.

Keywords: Echo State Networks · Physics-Informed Neural Networks · Chaotic dynamical systems

1 Introduction

Over the past few years, there has been a rapid increase in the development of machine learning techniques, which have been applied with success to various disciplines, from image or speech recognition [2,5] to playing Go [14]. However, the application of such methods to the study and forecasting of physical systems has only been recently explored including some applications in the field of fluid dynamics [4,6,13,15]. One of the major challenges for using machine learning algorithms for the study of complex physical systems is the prohibitive cost of data acquisition and generation for training [1,12]. However, in complex physical systems, there exists a large amount of prior knowledge, which

The authors acknowledge the support of the Technical University of Munich - Institute for Advanced Study, funded by the German Excellence Initiative and the European Union Seventh Framework Programme under grant agreement no. 291763. L.M. also acknowledges the Royal Academy of Engineering Research Fellowship Scheme.

J. M. F. Rodrigues et al. (Eds.): ICCS 2019, LNCS 11539, pp. 192–198, 2019.
https://doi.org/10.1007/978-3-030-22747-0_15

can be exploited to improve existing machine learning approaches. These hybrid approaches, called *physics-informed machine learning*, have been applied with some success to flow-structure interaction problems [13], turbulence modelling [6] and the solution of partial differential equations (PDEs) [12].

In this study, we propose an approach to combine physical knowledge with a machine learning algorithm to time-accurately forecast the evolution of a chaotic dynamical system. The machine learning tools we use are based on reservoir computing [9], in particular, Echo State Networks (ESNs). ESNs have shown to predict nonlinear and chaotic dynamics more accurately and for a longer time horizon than other deep learning algorithms [9]. ESNs have also recently been used to predict the evolution of spatiotemporal chaotic systems [10,11]. In the present study, ESNs are augmented by physical constraints to accurately forecast the evolution of a prototypical chaotic system, i.e., the Lorenz system [7].

Section 2 details the method used for the training and for forecasting the dynamical systems, both with conventional ESNs and the newly proposed physics-informed ESNs. Results are presented in Sect. 3 and final comments are summarized in Sect. 4.

2 Methodology

The Echo State Network (ESN) approach presented in [8] is used here. Given a training input signal $\boldsymbol{u}(n)$ of dimension N_u and a desired known target output signal $\boldsymbol{y}(n)$ of dimension N_y, the ESN has to learn a model with output $\widehat{\boldsymbol{y}}(n)$ matching $\boldsymbol{y}(n)$. $n = 1, ..., N_t$ is the discrete time and N_t is the number of data points in the training dataset covering a discrete time from 0 until time $T = (N_t - 1)\Delta t$. In the particular case studied here, where the forecasting of a dynamical system is of interest, the desired output signal is equal to the input signal at the next time step, i.e., $\boldsymbol{y}(n) = \boldsymbol{u}(n+1)$.

The ESN is composed of an artificial high dimensional dynamical system, called a reservoir, whose states at time n are represented by a vector, $\boldsymbol{x}(n)$, of dimension N_x, representing the reservoir neuron activations. This reservoir is coupled to the input signal, $\boldsymbol{u}$, via an input-to-reservoir matrix, $\boldsymbol{W}_{in}$ of dimension $N_x \times N_u$. The output of the reservoir, $\widehat{\boldsymbol{y}}$, is deduced from the states via the reservoir-to-output matrix, $\boldsymbol{W}_{out}$ of dimension $N_y \times N_x$, as a linear combination of the reservoir states:

$$\widehat{\boldsymbol{y}} = \boldsymbol{W}_{out}\boldsymbol{x} \tag{1}$$

In this work, a non-leaky reservoir is used, where the state of the reservoir evolves according to:

$$\boldsymbol{x}(n+1) = \tanh\left(\boldsymbol{W}_{in}\boldsymbol{u}(n) + \boldsymbol{W}\boldsymbol{x}(n)\right) \tag{2}$$

where $\boldsymbol{W}$ is the recurrent weight matrix of dimension $N_x \times N_x$ and the (element-wise) tanh function is used as an activation function for the reservoir neurons.

In the conventional ESN approach, illustrated in Fig. 1a, the input and recurrent matrices, $\boldsymbol{W}_{in}$ and $\boldsymbol{W}$, are randomly initialized once and are not further trained. These are typically sparse matrices constructed so that the reservoir verifies the Echo State Property [3]. Only the output matrix, $\boldsymbol{W}_{out}$, is then trained to minimize the mean-square-error, E_d, between the ESN predictions and the data defined as:

$$E_d = \frac{1}{N_y} \sum_{i=1}^{N_y} \frac{1}{N_t} \sum_{n=1}^{N_t} (\widehat{y}_i(n) - y_i(n))^2 \tag{3}$$

The subscript d is used to indicate the error based on the available data.

In the present implementation, following [11], $\boldsymbol{W}_{in}$ is generated such that each row of the matrix has only one randomly chosen nonzero element, which is independently taken from a uniform distribution in the interval $[-\sigma_{in}, \sigma_{in}]$. $\boldsymbol{W}$ is constructed to have an average connectivity $\langle d \rangle$ and the non-zero elements are taken from a uniform distribution over the interval $[-1, 1]$. All the coefficients of $\boldsymbol{W}$ are then multiplied by a constant coefficient for the largest absolute eigenvalue of $\boldsymbol{W}$ to be equal to a value Λ.

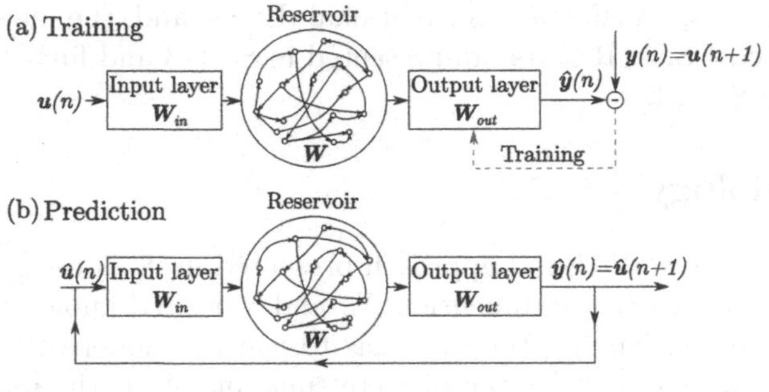

Fig. 1. Schematic of the ESN during (a) training and (b) future prediction. The physical constraints are imposed during the training phase (a).

After training, to obtain predictions from the ESN for future time $t > T$, the output of the ESN is looped back as the input of the ESN and it then evolves autonomously as represented in Fig. 1b.

2.1 Training

As discussed earlier, the training of the ESN consists of the optimization of $\boldsymbol{W}_{out}$. As the outputs of the ESN, $\widehat{\boldsymbol{y}}$, are a linear combination of the states, $\boldsymbol{x}$, $\boldsymbol{W}_{out}$ can be obtained by using a simple Ridge regression:

$$\boldsymbol{W}_{out} = \boldsymbol{Y}\boldsymbol{X}^T \left(\boldsymbol{X}\boldsymbol{X}^T + \gamma \boldsymbol{I}\right)^{-1} \tag{4}$$

where $\boldsymbol{Y}$ and $\boldsymbol{X}$ are respectively the column-concatenation of the various time instants of the output data, $\boldsymbol{y}$, and associated ESN states $\boldsymbol{x}$. γ is the Tikhonov regularization factor, which helps avoid overfitting. Explicitly, the optimization in Eq. (4) reads:

$$\boldsymbol{W}_{out} = \underset{\mathbf{W}_{out}}{\text{argmin}} \frac{1}{N_y} \sum_{i=1}^{N_y} \left(\sum_{n=1}^{N_t} (\widehat{y}_i(n) - y_i(n))^2 + \gamma ||\boldsymbol{w}_{out,i}||^2 \right) \tag{5}$$

where $\boldsymbol{w}_{out,i}$ denotes the i-th row of $\boldsymbol{W}_{out}$. This optimization problem penalizes large values of $\boldsymbol{W}_{out}$, generally improves the feedback stability and avoids overfitting [8].

In this work, following the approach of [12] for artificial deep feedforward neural networks, we propose an alternative approach to training $\boldsymbol{W}_{out}$, which combines the data available with prior physical knowledge of the system under investigation. Let us first assume that the dynamical system is governed by the following nonlinear differential equation:

$$\mathcal{F}(\boldsymbol{y}) \equiv \partial_t \boldsymbol{y} + \mathcal{N}(\boldsymbol{y}) = 0 \tag{6}$$

where $\mathcal{F}$ is a nonlinear operator, ∂_t is the time derivative and $\mathcal{N}$ is a nonlinear differential operator. Equation (6) represents a formal equation describing the dynamics of a generic nonlinear system. The training phase can be reframed to make use of our knowledge of $\mathcal{F}$ by minimizing the mean squared error, E_d, and a physical error, E_p, based on $\mathcal{F}$:

$$E_{tot} = E_d + E_p, \text{ where } E_p = \frac{1}{N_y} \sum_{i=1}^{N_y} \frac{1}{N_p} \sum_{p=1}^{N_p} |\mathcal{F}(\widehat{y}_i(n_p))|^2 \tag{7}$$

Here, the set $\{\widehat{\boldsymbol{y}}(n_p)\}_{p=1}^{N_p}$ denotes "collocation points" for $\mathcal{F}$, which are defined as a prediction horizon of N_p datapoints obtained from the ESN covering the time period $(T + \Delta t) \leq t \leq (T + N_p \Delta t)$. Compared to the conventional approach where the regularization of $\boldsymbol{W}_{out}$ is based on avoiding extreme values of $\boldsymbol{W}_{out}$, our proposed method regularizes $\boldsymbol{W}_{out}$ by using our prior physical knowledge. Equation (7), which is a key equation, shows how to constrain the prior physical knowledge in the loss function. Therefore, this procedure ensures that the ESN becomes predictive because of data training and the ensuing prediction is consistent with the physics. It is motivated by the fact that in many complex physical systems, the cost of data acquisition is prohibitive and thus, there are many instances where only a small amount of data is available for the training of neural networks. In this context, most existing machine learning approaches lack robustness: Our approach better leverages on the information content of the data that the machine learning algorithm uses. Our physics-informed framework is straightforward to implement because it only requires the evaluation of the residual, but it does not require the computation of the exact solution.

3 Results

The approach described in Sect. 2 is applied for forecasting the evolution of the chaotic Lorenz system, which is described by the following equations [7]:

$$\frac{du_1}{dt} = \sigma(u_2 - u_1), \quad \frac{du_2}{dt} = u_1(\rho - u_3) - u_2, \quad \frac{du_3}{dt} = u_1 u_2 - \beta u_3 \tag{8}$$

where $\rho = 28$, $\sigma = 10$ and $\beta = 8/3$. These are the standard values of the Lorenz system that spawn a chaotic solution [7]. The size of the training dataset is $N_t = 1000$ and the timestep between two time instants is $\Delta t = 0.01$.

The parameters of the reservoir both for the conventional and physics-informed ESNs are taken to be: $\sigma_{in} = 0.15$, $\Lambda = 0.4$ and $\langle d \rangle = 3$. In the case of the conventional ESN, the value of $\gamma = 0.0001$ is used for the Tikhonov regularization. These values of the hyperparameters are taken from previous studies [10,11].

For the physics-informed ESN, a prediction horizon of $N_p = 1000$ points is used and the physical error is estimated by discretizing Eq. (8) using an explicit Euler time-integration scheme. The choice of $N_p = 1000$ gives equal importance to the error based on the data and the error based on the physical constraints. The optimization of $\boldsymbol{W}_{out}$ is performed using the L-BFGS-B algorithm with the $\boldsymbol{W}_{out}$ obtained by Ridge regression (Eq. (4)) as the initial guess.

The predictions for the Lorenz system by conventional and physics-informed ESNs are compared with the actual evolution in Fig. 2, where the time is normalized by the largest Lyapunov exponent, $\lambda_{\max} = 0.934$, and the reservoir has 200 units. Figure 2d shows the evolution of the normalized error, which is defined as

$$E(n) = \frac{||\boldsymbol{u}(n) - \widehat{\boldsymbol{u}}(n)||}{\langle ||\boldsymbol{u}||^2 \rangle^{1/2}} \tag{9}$$

where $\langle \cdot \rangle$ denotes the time average. The physics-informed ESN shows a significant improvement of the time over which the predictions are accurate. Indeed, the time for the normalized error to exceed 0.2, which is the threshold used here to define the predictability horizon, improves from 4 Lyapunov times to approximately 5.5 with the physics-informed ESN. The dependence of the predictability

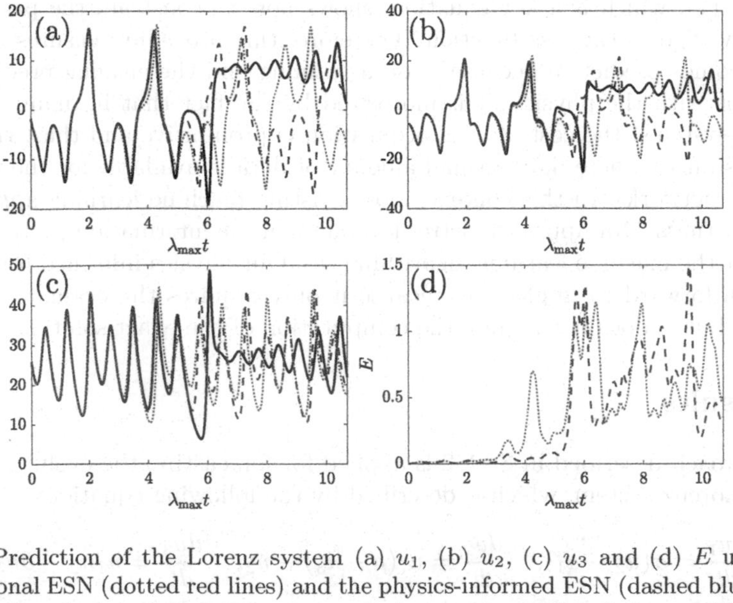

Fig. 2. Prediction of the Lorenz system (a) u_1, (b) u_2, (c) u_3 and (d) E using the conventional ESN (dotted red lines) and the physics-informed ESN (dashed blue lines). The actual evolution of the Lorenz system is shown using full black lines. (Color figure online)

horizon on the reservoir size is estimated as follows (Fig. 3). First, the trained physics-informed and conventional ESNs are run for an ensemble of 100 different initial conditions. Second, for each run, the predictability horizon is calculated. Third, the mean and standard deviation of the predictability horizon are computed from the ensemble. It is observed that the physics-informed approach provides a marked improvement of the predictability horizon over conventional ESNs and, most significantly, for reservoirs of intermediate sizes. The only exception is for the smallest reservoir ($N_x = 50$). In principle, it may be conjectured that a conventional ESN may have a similar performance to that of a physics-informed ESN by ad-hoc optimization of the hyperparameters. However, no efficient methods are available (to date) for hyperparameter optimization [9]. The approach proposed here allows us to improve the performance of the ESN (optimizing $\boldsymbol{W}_{out}$) by adding a constraint on the physics, i.e., the governing equations, and not by ad-hoc tuning of the hyperparameters. This suggests that the physics-informed approach is more robust than the conventional approach.

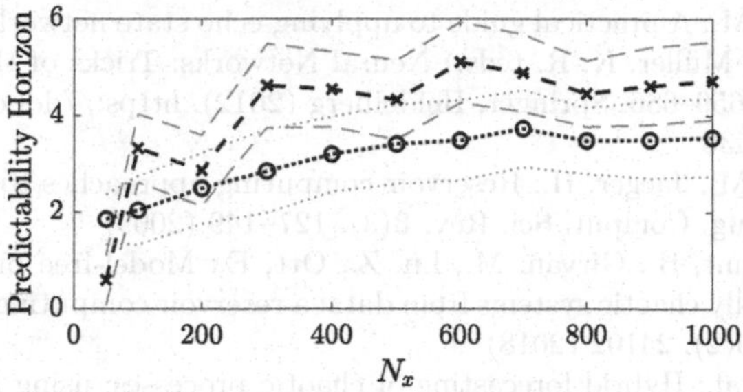

Fig. 3. Mean predictability horizon of the conventional ESN (dotted line with circles) and physics-informed ESN (dashed line with crosses) as a function of the reservoir size (N_x). The associated gray lines indicate the standard deviation from the mean.

4 Conclusions and Future Directions

We propose an approach for training Echo State Networks (ESNs), which constrains the knowledge of the physical equations that govern a dynamical system. This physics-informed Echo State Network approach is shown to be more robust than purely data-trained ESNs: The predictability horizon is markedly increased without requiring additional training data. In ongoing work, (i) the impact that the number of collocation points has on the accuracy of the ESNs is thoroughly assessed; (ii) the reason why the predictability horizon saturates as the reservoirs become larger is investigated; and (iii) the physics-informed ESNs are applied to high dimensional fluid dynamics systems. Importantly, the physics-informed Echo State Networks we propose will be exploited to minimize the data required for training. This work opens up new possibilities for the time-accurate prediction of the dynamics of chaotic systems by constraining the underlying physical laws.

References

1. Duraisamy, K., Iaccarino, G., Xiao, H.: Turbulence modeling in the age of data. Annu. Rev. Fluid Mech. **51**, 357–377 (2019)
2. Hinton, G., et al.: Deep neural networks for acoustic modeling in speech recognition: the shared views of four research groups. IEEE Sig. Process. Mag. **29**(6), 82–97 (2012)
3. Jaeger, H., Haas, H.: Harnessing nonlinearity: predicting chaotic systems and saving energy in wireless communication. Science **304**(5667), 78–80 (2004)
4. Jaensch, S., Polifke, W.: Uncertainty encountered when modelling self-excited thermoacoustic oscillations with artificial neural networks. Int. J. Spray Combust. Dyn. **9**(4), 367–379 (2017)
5. Krizhevsky, A., Sutskever, I., Hinton, G.E.: ImageNet classification with deep convolutional neural networks. Neural Inf. Process. Syst. **25**, 1097–1105 (2012)
6. Ling, J., Kurzawski, A., Templeton, J.: Reynolds averaged turbulence modelling using deep neural networks with embedded invariance. J. Fluid Mech. **807**, 155–166 (2016)
7. Lorenz, E.N.: Deterministic nonperiodic flow. J. Atmos. Sci. **20**(2), 130–141 (1963)
8. Lukoševičius, M.: A practical guide to applying echo state networks. In: Montavon, G., Orr, G.B., Müller, K.-R. (eds.) Neural Networks: Tricks of the Trade. LNCS, vol. 7700, pp. 659–686. Springer, Heidelberg (2012). https://doi.org/10.1007/978-3-642-35289-8_36
9. Lukoševičius, M., Jaeger, H.: Reservoir computing approaches to recurrent neural network training. Comput. Sci. Rev. **3**(3), 127–149 (2009)
10. Pathak, J., Hunt, B., Girvan, M., Lu, Z., Ott, E.: Model-free prediction of large spatiotemporally chaotic systems from data: a reservoir computing approach. Phys. Rev. Lett. **120**(2), 24102 (2018)
11. Pathak, J., et al.: Hybrid forecasting of chaotic processes: using machine learning in conjunction with a knowledge-based model. Chaos **28**(4), 041101 (2018)
12. Raissi, M., Perdikaris, P., Karniadakis, G.: Physics-informed neural networks: a deep learning framework for solving forward and inverse problems involving nonlinear partial differential equations. J. Comput. Phys. **378**, 686–707 (2019)
13. Raissi, M., Wang, Z., Triantafyllou, M.S., Karniadakis, G.: Deep learning of vortex-induced vibrations. J. Fluid Mech. **861**, 119–137 (2019)
14. Silver, D., et al.: Mastering the game of Go with deep neural networks and tree search. Nature **529**(7587), 484–489 (2016)
15. Wu, J.L., Xiao, H., Paterson, E.: Physics-informed machine learning approach for augmenting turbulence models: a comprehensive framework. Phys. Rev. Fluids **3**, 074602 (2018)

Tuning Covariance Localization Using Machine Learning

Azam Moosavi[1], Ahmed Attia[2(✉)], and Adrian Sandu[3]

[1] Biomedical Engineering Department,
Case Western Reserve University, Cleveland, USA
azamosavi@vt.edu

[2] Mathematics and Computer Science Division,
Argonne National Laboratory, Argonne, IL, USA
attia@mcs.anl.gov

[3] Computational Science Laboratory, Department of Computer Science,
Virginia Polytechnic Institute and State University, Blacksburg, USA
asandu@cs.vt.edu

Abstract. Ensemble Kalman filter (EnKF) has proven successful in assimilating observations of large-scale dynamical systems, such as the atmosphere, into computer simulations for better predictability. Due to the fact that a limited-size ensemble of model states is used, sampling errors accumulate, and manifest themselves as long-range spurious correlations, leading to filter divergence. This effect is alleviated in practice by applying covariance localization. This work investigates the possibility of using machine learning algorithms to automatically tune the parameters of the covariance localization step of ensemble filters. Numerical experiments carried out with the Lorenz-96 model reveal the potential of the proposed machine learning approaches.

Keywords: Data assimilation · EnKF · Covariance localization · Machine learning

1 Introduction

Data assimilation (DA) is the set of methodologies that combine multiple sources of information about a physical system, with the goal of producing an accurate description of the state of that system [27]. Statistical DA algorithms apply Bayes' theorem to describe the system state using a probability distribution conditioned by all available sources of information. A typical starting point for most of the algorithms in this approach is the Kalman filter (KF) [26], which assumes that the underlying sources of errors are normally distributed, with known means and covariances. The ensemble Kalman filter (EnKF) [19] follows a Monte-Carlo approach to propagate covariance information, which makes it a practical approach for large-scale settings.

In typical atmospheric applications the model state space has dimension $\sim 10^9$–10^{12}, and a huge ensemble is required to accurately approximate the

J. M. F. Rodrigues et al. (Eds.): ICCS 2019, LNCS 11539, pp. 199–212, 2019.
https://doi.org/10.1007/978-3-030-22747-0_16

corresponding covariance matrices. However, computational resources limit the number of ensemble members to 30–100, leading to "under-sampling" [24] and its consequences: filter divergence, inbreeding, and long-range spurious correlations [4]. Inbreeding and the filter divergence are alleviated by some form of inflation [5]. We focus here only on long-range spurious correlations which are handled in practice by covariance localization [23].

Covariance localization is implemented by multiplying the regression coefficient in the Kalman gain with a decaying distance-dependent function such as a Gaussian [4] or the Gaspari-Cohn fifth order piecewise polynomial [22]. Different localization techniques have been recently considered for different observation types, different type of state variables, or for an observation and a state variable that are separated in time. However, in general, tuning the localization parameter for big atmospheric problems is a very expensive process. Previous efforts for building adaptive algorithms for covariance localization includes the works [3,4,7,10,11,29].

In this study we propose to adapt covariance localization parameters using machine learning algorithms. Two approaches are proposed and discussed. In the *localization-in-time* method the radius of influence is held constant in space, but it changes adaptively from one assimilation cycle to the next. In the *space-time-localization* method, the localization radius is space-dependent and is also adapted for each assimilation time instant. The learning process is conducted off-line based on historical records such as reanalysis data, and the trained model is subsequently used to predict the proper values of localization radii in future assimilation windows.

The paper is organized as follows. Background is given in Sect. 2. Section 3 presents the new adaptive localization algorithms. Experimental setup, and numerical results are reported in Sect. 4. Conclusions and future directions are highlighted in Sect. 5.

2 Background

2.1 Ensemble Kalman Filter (EnKF)

EnKF proceeds in a prediction-correction fashion and carries out two main steps in every assimilation cycle: *forecast* and *analysis*. Assume an analysis ensemble $\{\mathbf{x}^{\mathrm{a}}_{k-1}(e) \mid e = 1, \ldots, \mathrm{N}_{\mathrm{ens}}\}$ is available at a time instance t_{k-1}. In the forecast step, an ensemble of forecasts $\{\mathbf{x}^{\mathrm{f}}_{k}(e) \mid e = 1, \ldots, \mathrm{N}_{\mathrm{ens}}\}$ is generated by running the numerical model forward to the next time instance t_k where observations are available:

$$\mathbf{x}^{\mathrm{f}}_{k}(e) = \mathcal{M}_{t_{k-1}\to t_k}(\mathbf{x}^{\mathrm{a}}_{k-1}(e)) + \eta_k(e), \; e = 1, \ldots, \mathrm{N}_{\mathrm{ens}}, \tag{1a}$$

where $\mathcal{M}$ is a discretization of the model dynamics. To simulate the fact that the model is an imperfect representation of reality, random model error realizations $\eta_k(e)$ are added. Typical assumption is that the model error is a random variable distributed according to a Gaussian distribution $\mathcal{N}(0, \mathbf{Q}_k)$. In this paper we follow a perfect-model approach for simplicity, i.e., we set $\mathbf{Q}_k = \mathbf{0}\ \forall k$.

The generated forecast ensemble provides estimates of the ensemble mean $\overline{\mathbf{x}}_k^{\rm f}$ and the flow-dependent background error covariance matrix $\mathbf{B}_k$ at time instance t_k:

$$\mathbf{B}_k = \frac{1}{\mathrm{N_{ens}} - 1} \mathbf{X}_k^{'} \mathbf{X}_k^{'\,T}; \quad \mathbf{X}_k^{'} = \left[\mathbf{x}_k^{\rm f}(e) - \overline{\mathbf{x}}_k^{\rm f}\right]_{e=1,\dots,\mathrm{N_{ens}}},$$
$$\overline{\mathbf{x}}_k^{\rm f} = \frac{1}{\mathrm{N_{ens}}} \sum_{e=1}^{\mathrm{N_{ens}}} \mathbf{x}_k^{\rm f}(e). \tag{1b}$$

In the analysis step, each member of the forecast is analyzed separately using the Kalman filter formulas [16,19]:

$$\mathbf{x}_k^{\rm a}(e) = \mathbf{x}_k^{\rm f}(e) + \mathbf{K}_k \left([\mathbf{y}_k + \zeta_k(e)] - \mathcal{H}_k(\mathbf{x}_k^{\rm f}(e)) \right), \tag{1c}$$
$$\mathbf{K}_k = \mathbf{B}_k \mathbf{H}_k^T \left(\mathbf{H}_k \mathbf{B}_k \mathbf{H}_k^T + \mathbf{R}_k \right)^{-1}, \tag{1d}$$

where $\mathbf{y}_k$ is the observation collected at time t_k. The relation between a model state $\mathbf{x}_k$ and an observation $\mathbf{y}_k$ is characterized by

$$\mathbf{y}_k = \mathcal{H}_k(\mathbf{k}) + \zeta_k; \quad \zeta_k \sim \mathcal{N}(0, \mathbf{R}_k), \tag{2}$$

with $\mathcal{H}_k$, and $\mathbf{R}_k$ being the observation operator and the observation error covariance matrix, respectively, at time t_k. Here $\mathbf{H}_k = \mathcal{H}_k'(\overline{\mathbf{x}}_k^{\rm f})$ is the linearized observation operator, e.g. the Jacobian, at time instance t_k. Many flavors of EnKF have been developed over time. For a detailed discussion on EnKF and variants, see for example [6,20].

2.2 Covariance Localization

The small number of ensemble members may result in a poor estimation of the true correlations between state components, or between state variables and observations. In particular, spurious correlations might develop between variables that are located at large physical distances, when the true correlation between these variables is negligible. As a result, state variables are artificially affected by observations that are physically remote [2,23]. This generally results in degradation of the quality of the analysis, and eventually leads to filter divergence. Covariance localization seeks to filter out the long range spurious correlations and enhance the estimate of forecast error covariance [23,25]. Standard covariance localization is typically carried out by applying a Schur (Hadamard) product between a correlation matrix ρ with distance-decreasing entries and the ensemble estimated covariance matrix, resulting in the localized Kalman gain:

$$\mathbf{K}_k = (\rho \circ \mathbf{B}_k)\, \mathbf{H}_k^T \left(\mathbf{H}_k \left(\rho \circ \mathbf{B}_k\right) \mathbf{H}_k^T + \mathbf{R}_k \right)^{-1}. \tag{3}$$

Localization can be applied to $\mathbf{H}_k \mathbf{B}_k$, and optionally to the $\mathbf{B}_k$ projected into the observations space, that is, $\mathbf{H}_k \mathbf{B}_k \mathbf{H}_k^T$ [34]. Since the correlation matrix is a covariance matrix, the Schur product of the correlation function and the forecast background error covariance matrix is also a covariance matrix. Covariance localization has the virtue of increasing the rank of the flow-dependent background error covariance matrix $\rho \circ \mathbf{B}_k$, and therefore increasing the effective sample size. A popular choice of the correlation function ρ is a Gaussian function defined by

$$\rho(z, c) = e^{-z^2/2\ell^2}, \tag{4}$$

where $z \equiv z(i, j)$ is a distance function between ith and jth grid points respectively. The value of the correlation coefficient $\rho(z, c)$ is at highest of 1 for a distance $z = 0$, and decreases as the distance increases. Depending on the implementation, z can be either the distance between an observation and grid point or the distance between grid points in the physical space. The radius of influence ℓ must be tuned for each application.

2.3 Machine Learning

Recent studies show that machine learning (ML) algorithms can be helpful in solving computational science problems, including [8,32]. There is a plethora of ML algorithms for regression analysis. In this work, we limit ourselves to the *ensemble* approach [18] which has proven successful in enhancing the performance and results of ML algorithms. Specifically, ensemble methods work by combining several ML models into a single predictive model that can in principle overcome the limitations of the individual ML models. These limitations are generally manifested as bias and/or high-variance. Ensemble ML methods aim to decrease the bias (e.g., boosting) and the variance (e.g., bagging), and hence outperform the individual predictive models. Moreover, ML algorithms work by performing an optimization procedure that my be entrapped in a local optimum. An ensemble ML algorithm enables running the local search, carried out by each individual predictive model, from different starting points and thus enhances the predictive power. Common types of ML ensemble methods include the Bootstrap aggregation – bagging for short – [12], and Boosting [15]. In bagging, the training set is used to train an ensemble of ML models, and all trained models are equally important, i.e. the decisions made by all models are given the same weight. Each of the models is trained using a subset randomly drawn from the training dataset. A widely successful algorithm in this family of methods, is Random Forests (RF) [13]. In the boosting approach, on the other hand, the decisions made by the learners are weighted based on the performance of each model. A widely common algorithm in this approach is Gradient Boosting (GB) [14].

Random Forests. RFs [13] work by constructing an ensemble of decision trees, such that each tree builds a classification or regression model in the form of a tree structure. Instead of using the whole set of features available for the learning algorithm at once, each subtree uses a subset of features. The ensemble of trees is constructed using a variant of the bagging technique, thus yielding a small variance of the learning algorithm [18]. Furthermore, to ensure robustness of the ensemble-based learner, each sub-tree is assigned a subset of features selected randomly in a way that minimizes the correlation between individual learners. Random sampling and bootstrapping [30] can be efficiently applied to RFs to generate a parallel, robust, and very fast learner for high-dimensional data and features.

Gradient Boosting. GB proceeds by incrementally building the prediction model as an ensemble of weak predictors. Specifically, GB algorithm build a sequence of simple regression trees with each constructed over the prediction residual of the preceding trees [21]. This procedure gives a chance to each sub-tree to correct its predecessors, and consequently build an accurate ensemble-based model.

3 Machine Learning Approach for Adaptive Localization

This section develops two machine learning approaches for adaptive covariance localization. Specifically, we let a ML model learn, and consequently predict, the best localization radius to be used in the filtering procedure. Here, we can either allow the localization radius to vary over time only, or both in space and in time. In both cases, RF or GB, or another suitable regression, model is used to construct the learning model that takes the impactful set of features as input, and the localization radius as output.

3.1 Features and Decision Criteria

Under the Gaussianity assumption, the quality of the DA solution is given by the quality of its first two statistical moments. However, including the ensemble mean and correlations as a set of features can be prohibitive in large-scale applications. One idea is to select only model states with negligible correlations among them, e.g., states that are physically located at distances larger than the radius of influence. Another useful strategy to reduce model features is to select descriptive summaries such as the minimum and the maximum magnitude of state components in the ensemble. Similarly, we suggest including blocks of the correlation matrix for variables located nearby in physical space, i.e., for subsets of variables that are highly correlated.

To construct a proper objective function for the ML algorithm to optimize, we need to quantify the accuracy of the mean estimate, and ensemble-based approximation of the covariance matrix generated by the filtering algorithm. To quantify the accuracy of the ensemble mean we use the root mean-squared error (RMSE), defined as follows:

$$RMSE_k = \frac{1}{\sqrt{\mathrm{N_{state}}}} \left\| \mathbf{x}_k - \mathbf{x}^{\mathrm{true}}(t_k) \right\|_2 , \tag{5}$$

where $\mathbf{x}^{\mathrm{true}}$ is the true system state, and $\|\cdot\|_2$ is the Euclidean norm. Since the true state is not known in practice, we also consider the deviation of the state from collected measurements as a useful indication of filter performance. The observation-state $RMSE$ is defined as follows:

$$RMSE_k^{\mathbf{x}|\mathbf{y}} = \frac{1}{\sqrt{\mathrm{N_{obs}}}} \left\| \mathcal{H}(\mathbf{x}_k) - \mathbf{y}_k \right\|_2 . \tag{6}$$

The quality of the analysis state $\mathbf{x} = \mathbf{x}^{\mathrm{a}}$ by either (5) in case of perfect problem settings, or by (6) in case of real applications. *In this work we use the observation-analysis error metric* (6), *denoted by* $RMSE^{\mathbf{x}^{\mathrm{a}}|\mathbf{y}}$, *as the first decision criterion.*

The quality of the ensemble-based covariance can be inspected by investigating the spread of the ensemble around truth (or the observations), using Talagrand diagram (rank histogram) [1,17]. A quality analysis ensemble leads to a rank histogram that is close to a uniform distribution. Conversely, U-shaped and Bell-shaped rank histograms correspond to under-dispersion and over-dispersion of the ensemble, respectively. Ensemble based methods, especially with small ensemble sizes, are generally expected to yield U-shaped rank histograms, unless they are well-designed and well-tuned. In this work we use the uniformity of the analysis rank histogram, in observation space, as the second decision criterion. To quantify the level of uniformity of the rank histogram, we follow the approach proposed in [9]. Specifically, the Kullback-Leibler (KL) divergence [28] between a Beta distribution $Beta(\alpha, \beta)$ fitted to rank histogram of an ensemble, and a uniform distribution. This measure calculated using the forecast ensemble is used as a learning feature, while the one calculated using the analysis ensemble is used as a decision criterion. To account for both accuracy and dispersion, we combine the two metrics into a single criterion, as follows:

$$\mathcal{C}_{\mathbf{r}} = w_1\, RMSE^{\mathbf{x}^{\mathrm{a}}|\mathbf{y}} + w_2\, D_{KL}\big(Beta(\alpha, \beta) \| Beta(1.0, 1.0)\big), \tag{7}$$

where the weighting parameters realize an appropriate scaling of the two metrics. The weights w_1, w_2 can be predefined, or can be learned from the data them as part of the ML procedure. Here, we define the best set of localization radii at every assimilation cycle to be the minimizer of (7).

Adaptive-in-Time Localization. Here, the value of this radius is fixed in space, and only varied from one assimilation cycle to the next. Specifically, at the current cycle we perform the assimilation using all localization radii from a pool of possible value, and for each case compute the cost function (7). The radius associated with the minimum cost function is selected as winner. The analysis of the current assimilation cycle is then computed using the winner radius. During the training phase, at each assimilation cycle, the ML algorithm learns the best localization radius (i.e., winner) corresponding to the selected features. During the test phase, the learned model uses the current features to estimate the proper value of the localization radius.

Space-Time Adaptive Localization. Here, the localization radii vary both in time and in space. In this case, the localization radius is a vector $\mathbf{r}$ containing a scalar localization parameter for each state variable of the system. At each assimilation cycle we collect a sample consisting of the model features as inputs and the winner vector of localization radii as output of the learning model.

Computational Considerations. During the training phase, the proposed methodology requires trying all possible radii from the pool, and re-do the assimilation with the selected radius. This is computationally demanding, but the model can be trained off-line using historical data. The testing phase the learning model predicts a good value of the localization radius, which is then used in the assimilation; no additional costs are incurred except for the (relatively inexpensive) prediction made by the trained model.

4 Numerical Results

In order to study the performance of the proposed adaptive localization algorithm we employ he Lorenz-96 model [31], described by:

$$\frac{dX_k}{dt} = -X_{k-1}\left(X_{k-2}X_{k-1} - X_{k+1}\right) - X_k + F, \quad k = 1, 2, \cdots, K, \tag{8}$$

with $K = 40$ variables, and a forcing term $F = 8$. A vector of equidistant component values ranging from $[-2, 2]$ was integrated forward in time for 1000 steps, each of size 0.005 [units], and the final state was taken as the reference initial condition for the experiments. The background uncertainty is set to 8% of average magnitude of the reference solution. All state vector components are observed, i.e., $\mathcal{H} = \mathbf{I} \in \mathbb{R}^{K \times K}$ with $\mathbf{I}$ the identity operator. To avoid filter collapse, the analysis ensemble is inflated at the end of each assimilation cycle, with the inflation factor set to $\delta = 1.09$.

Assimilation Filter. All experiments are implemented in Python using the DATeS framework [9]. The performance of the proposed methodology is compared against the deterministic implementation of EnKF (DEnKF) with parameters empirically tuned as reported in [35]. The EnKF uses 25 ensemble members, with an inflation factor of 1.09 applied to the analysis ensemble.

Machine Learning Model. Several ML regressors to model and predict the localization radii, for ensemble data assimilation algorithms, have been explored and tested. However, for brevity, we use RF and GB as the main learning tools in the numerical experiments discussed below. We use *Scikit-learn*, the machine learning library in Python [33], to construct the ML models used in this work.

Results with Adaptive-in-Time Localization. This experiment has 100 assimilation cycles, where the first 80% are dedicated to the training phase and the last 20% to the testing phase. The pool of radii for this experiment covers all possible values for the Lorenz model, i.e., $r \in [1, 40]$. We compare the performance of the adaptive localization algorithms against the best hand-tuned fixed localization radius value of 4 which is obtained by letting the localization radius ℓ take all possible integer values in the interval [1, 40]. Figure 1 shows the RMSE results, on a logarithmic scale, of EnKF with a fixed localization radius $r = 4$, and EnKF with adaptive covariance localization with multiple choices of the weights w_1, $w_2 = 1 - w_1$. The RMSE over the training phase is shown in Fig. 1(left), and that of the testing phase is shown in Fig. 1(right). We separate the results into two panels here, to get a closer look at the relative performance between the different experiments during the testing phase, e.g., in Fig. 1(right). The results suggest that increasing the weight of the KL distance measure, that is w_2, enhances the performance of the filter, as long as we don't completely eliminate w_1. For the best choices of the weights, the overall performance of the adaptive localization is slightly better than that of the fixed, hand-tuned radius.

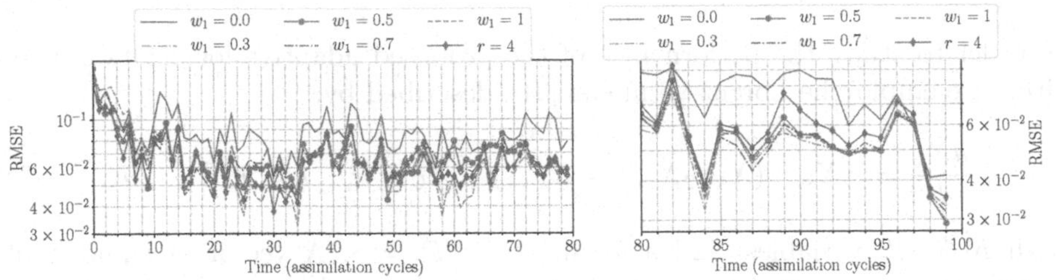

Fig. 1. EnKF results with adaptive-in-time covariance localization, using RF learning model, for different choices of the weighting factors w_1, w_2 of (7), compared to EnKF with fixed localization radius. The training phase consists of 80 assimilation cycles (left panel), followed by the testing phase with 20 assimilation cycles (right panel).

To elaborate more on the results, we pick the weights $w_1 = 0.7$ and $w_2 = 0.3$ of the adaptive localization criterion for this experiment. Figure 2 shows the variability in the tuned localization radius over time for both training and test phase. The adaptive algorithm changes the radius considerably over the simulation.

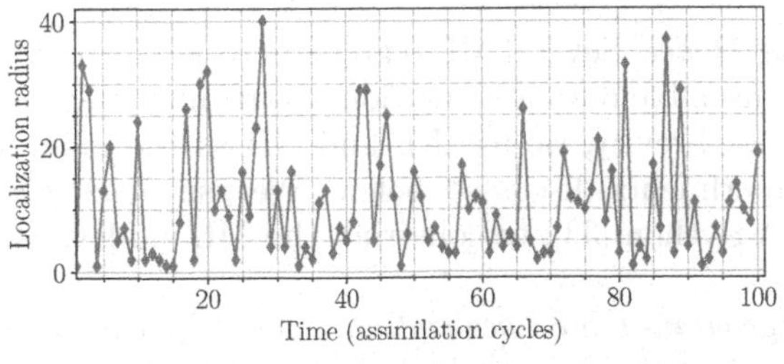

Fig. 2. EnKF results with adaptive-in-time covariance localization, using RF learning model. The evolution of the localization radius in time over all 100 assimilation cycles is shown. The weights of the adaptive localization criterion are $w_1 = 0.7$ and $w_2 = 0.3$.

In decision trees, every node is a condition how to split values in a single feature. The criteria usually is based on Gini impurity, information gain (entropy) or variance. Upon training a tree, it is possible to compute how much each feature contributes to decreasing the weighted impurity. Hence, the RF model helps in recognition and selection of the most important features affecting the target variable prediction. Figure 3 shows the 35 most important features of the Lorenz model which we included in our experiments. These results, as expected, suggest that the information about the first and second order moments are both essential for the learning algorithm.

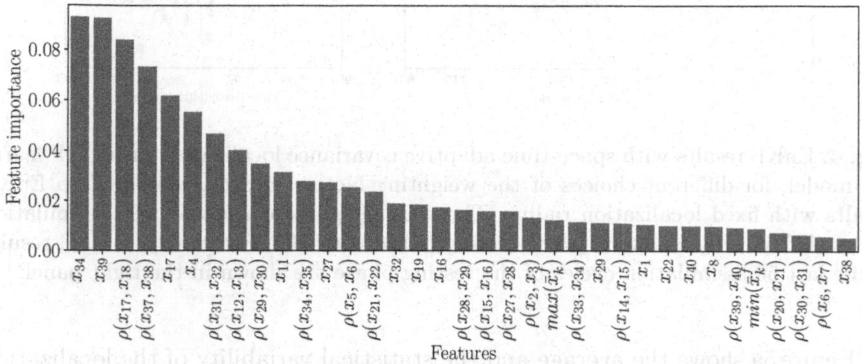

Fig. 3. EnKF results with adaptive-in-time covariance localization, using RF learning model. The plot shows the 35 most important features extracted for the DA experiment with weights $w_1 = 0.7$ and $w_2 = 0.3$.

Results with Space-Time Adaptive Localization. The pool of radii for this experiment consists of vectors of size 40 where each component of the vector can take any value in the interval $[1, 40]$. With the infinite number of possibilities, trying all possible permutations of the localization radii is infeasible. One way to limit the number of trials is to test randomly selected vectors of radii in the pool. For this experiment, we set the number of trials to 30 and at each trial we randomly pick a vector of radii from the pool. The number of target variables to estimate at each assimilation cycle in the test phase is 40 and hence we need more samples for the training phase. The number of assimilation cycles for this experiment is 1000, from which 80% dedicated to the training phase, and 20% to the testing phase.

Figure 4 shows the RMSE results of EnKF with space-time adaptive localization for multiple choices of the weighting parameters w_1, $w_2 = 1 - w_1$. Figure 4(left) shows the results over the training phase, while Fig. 4(right) shows the RMSE results over the last 50 assimilation cycles of the testing phase. The performance of adaptive localization is compared to EnKF with fixed localization radius $r = 4$. The RMSE results of the adaptive localization algorithm are slightly better than those of EnKF with the empirically tuned fixed radius. Of course in practice, the goal is to completely replace the empirical tuning procedure with an automated scheme. These results suggest that the proposed approach, to automatically adjust the space-time covariance localization parameter can produce favorable results without the need for empirical adjustment.

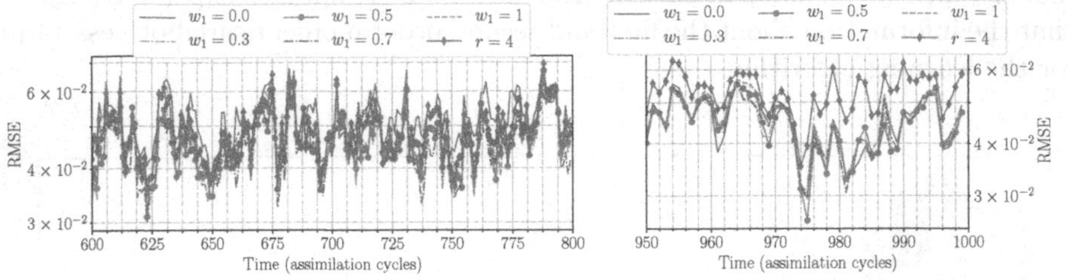

Fig. 4. EnKF results with space-time adaptive covariance localization, using RF learning model, for different choices of the weighting factors w_1, w_2, compared to EnKF results with fixed localization radius. The training phase consists of 800 assimilation cycles (left panel), followed by the testing phase with 200. For clarity, RMSE results of the last 50 assimilation cycles of the testing phase are shown in the right panel.

Figure 5a shows the average and the statistical variability of the localization radii over time, for each state variable of the Lorenz-96 model. The results are found by averaging over all 1000 assimilation cycles, with the weights $w_1 = 0.7$ and $w_2 = 0.3$. From these results, we see that the adaptive values chosen by the algorithm can vary considerably in the temporal domain of the experiment. This variability can be further seen in Fig. 5b, which shows the evolution of localization radii in both time and space, over the last 100 cycles of the testing phase.

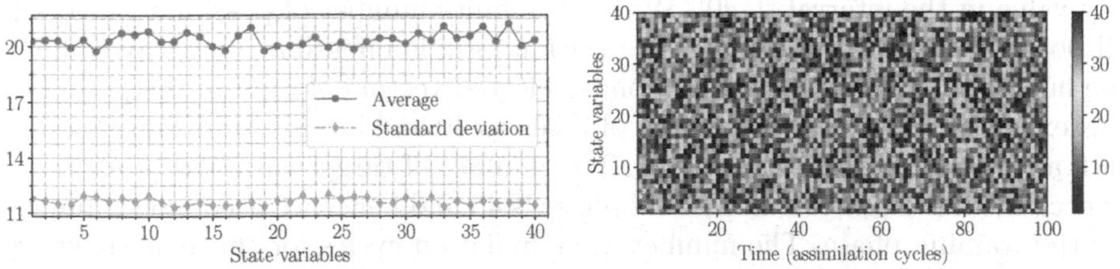

(a) Temporal statistics

(b) Space-time evolution

Fig. 5. EnKF results with space-time adaptive covariance localization, using RF learning model. The weights of the adaptive localization criterion are set to $w_1 = 0.7$ and $w_2 = 0.3$. Panel (a) shows average and standard deviation results of the localization radii for the state variables of the Lorenz-96 model (8). Panel (b) shows the space-time evolution of the localization radii over the last 100 assimilation cycles of the testing phase of the experiment.

On the Choice of the Learning Model. The work in this paper is not aimed to cover or compare all suitable ML algorithms in the context of adaptive covariance localization. In the numerical experiments presented above, we chose the RF as the main learning model, however the method proposed is not limited this choice, and can be easily extended to incorporate other suitable regression model. For example RF could be replaced with GB, however the computational cost of training the regressor, and the performance of the DA algorithm must be both accounted for.

DA Performance. To compare the performance of the DA filter with localization radii predicted by RF against GB, we study the RMSE obtained by incorporating each of these two learning models. Figure 6 shows the average RMSE over the test phase resulting by replacing RF with GB. Here, the RMSE results for both cases, i.e. time-only and space-time adaptivity, resulting by incorporating RF tend to be slightly lower than that resulting when GB is used.

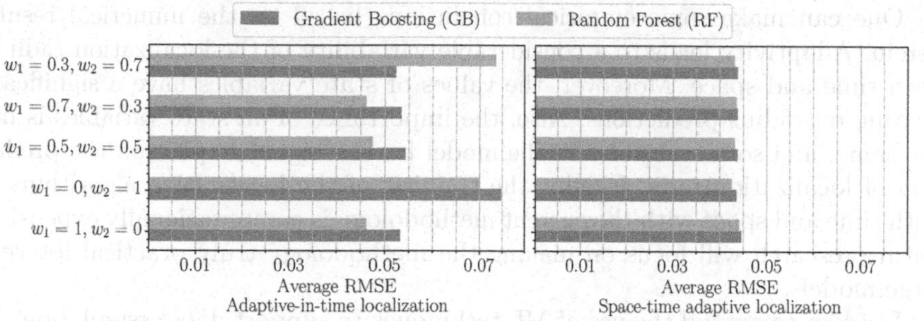

Fig. 6. RMSE results of the adaptive covariance localization approaches are shown for different choices of the weighting factors w_1, w_2. Results are shown for both adaptive-in-time (left), and space-time adaptive localization (right). RMSE is averaged the testing phase of each experiment, obtained by using both RF and GB.

Computational Time. Table 1 shows the CPU-time spent in fitting the training dataset or training the learning model with both RF and GB. Learning RF model is less time consuming than GB, especially in the case of space-time adaptivity.

Table 1. CPU-time of the training time of the two ML algorithms, RF and GB for both time adaptivity and space-time adaptivity approaches.

CPU time (seconds)		Adaptivity type	
		Time	Space-time
ML model	GB	0.0467	16.3485
	RF	0.0308	0.7508

This is mainly because RF, by construction, supports multi-target regression, while GB does not. A simple extension of GB is used for space-time adaptivity, by fitting a regressor to each of the outputs. From both Fig. 6, and Table 1, we can empirically conclude that RF yields a combination of better performance and lower computational time, than GB.

5 Concluding Remarks and Future Work

This study investigates using ML models to adaptively tune the covariance localization radii for EnKF family of data assimilation methods. The learning model can be trained off-line using historical records, e.g., reanalysis data. Once it is successfully trained, the regression model is used to estimate the values of localization radii in future assimilation cycles. Numerical results carried out using two standard ML models, suggest that the proposed automatic approach performs at least as good as the traditional EnKF with empirically hand-tuned localization parameters.

One can make some empirical conclusions based on the numerical results herein. Adaptivity leads to a considerable variability of the localization radii in both time and space. Moreover, the values of state variables have a significant bearing on radius predictions. Also, the importance of all state variables is not the same, and some variables in the model have a higher impact on the prediction of localization radii. Finally, the training of the localization algorithms in both time and space with the current methodology is computationally expensive. Future research will focus on making the methodology truly practical for very large models.

In order to extend the use of ML techniques to support data assimilation, an important question that will be addressed in future research concerns the optimal choice of features in large-scale numerical models. Specifically, one has to select sufficient aspects of the model state to carry the information needed to train a ML model. In the same time, the size of the features vector needs to be relatively small, even when the model state is extremely large. Next, the computational expense of the training phase is due to the fact that the analysis needs to be repeated with multiple localization radii. Future work will seek to considerably reduce the computational effort by intelligently narrowing the pool of possible radii to test, and by devising assimilation algorithms that reuse the bulk of the calculations when computing multiple analyses with multiple localization radii.

Acknowledgments. This work was supported in part by the projects AFOSR DDDAS 15RT1037 and AFOSR Computational Mathematics FA9550-17-1-0205.NSF ACI-1709727, and NSF CCF-1613905.

References

1. Anderson, J.L.: A method for producing and evaluating probabilistic forecasts from ensemble model integrations. J. Clim. **9**(7), 1518–1530 (1996)
2. Anderson, J.L.: An ensemble adjustment Kalman filter for data assimilation. Mon. Weather Rev. **129**(12), 2884–2903 (2001)
3. Anderson, J.L.: An adaptive covariance inflation error correction algorithm for ensemble filters. Tellus A **59**(2), 210–224 (2007)
4. Anderson, J.L.: Localization and sampling error correction in ensemble Kalman filter data assimilation. Mon. Weather Rev. **140**(7), 2359–2371 (2012)
5. Anderson, J.L., Anderson, S.L.: A Monte Carlo implementation of the nonlinear filtering problem to produce ensemble assimilations and forecasts. Mon. Weather Rev. **127**(12), 2741–2758 (1999)
6. Asch, M., Bocquet, M., Nodet, M.: Data Assimilation: Methods, Algorithms, and Applications. SIAM, Philadelphia (2016)
7. Attia, A., Constantinescu, E.: An optimal experimental design framework for adaptive inflation and covariance localization for ensemble filters. arXiv preprint arXiv:1806.10655 (2018)
8. Attia, A., Moosavi, A., Sandu, A.: Cluster sampling filters for non-Gaussian data assimilation. Atmosphere **9**(6) (2018). https://doi.org/10.3390/atmos9060213. http://www.mdpi.com/2073-4433/9/6/213
9. Attia, A., Sandu, A.: DATeS: a highly-extensible data assimilation testing suite v1.0. Geosci. Model Dev. (GMD) **12**, 629–2019 (2019). https://doi.org/10.5194/gmd-12-629-2019. https://www.geosci-model-dev.net/12/629/2019/
10. Bishop, C.H., Hodyss, D.: Flow-adaptive moderation of spurious ensemble correlations and its use in ensemble-based data assimilation. Q. J. Roy. Meteorol. Soc. **133**(629), 2029–2044 (2007)
11. Bishop, C.H., Hodyss, D.: Ensemble covariances adaptively localized with eco-rap. Part 2: a strategy for the atmosphere. Tellus A **61**(1), 97–111 (2009)
12. Breiman, L.: Bagging predictors. Mach. Learn. **24**(2), 123–140 (1996)
13. Breiman, L.: Random forests. Mach. Learn. **45**(1), 5–32 (2001)
14. Breiman, L., et al.: Arcing classifier (with discussion and a rejoinder by the author). Ann. Stat. **26**(3), 801–849 (1998)
15. Bühlmann, P., Hothorn, T.: Boosting algorithms: regularization, prediction and model fitting. Stat. Sci. **22**, 477–505 (2007)
16. Burgers, G., van Leeuwen, P.J., Evensen, G.: Analysis scheme in the ensemble Kalman filter. Mon. Weather Rev. **126**, 1719–1724 (1998)
17. Candille, G., Talagrand, O.: Evaluation of probabilistic prediction systems for a scalar variable. Q. J. Roy. Meteorol. Soc. **131**(609), 2131–2150 (2005)
18. Dietterich, T.G.: Ensemble methods in machine learning. In: Kittler, J., Roli, F. (eds.) MCS 2000. LNCS, vol. 1857, pp. 1–15. Springer, Heidelberg (2000). https://doi.org/10.1007/3-540-45014-9_1
19. Evensen, G.: Sequential data assimilation with a nonlinear quasi-geostrophic model using Monte Carlo methods to forcast error statistics. J. Geophys. Res. **99**(C5), 10143–10162 (1994)
20. Evensen, G.: Data Assimilation: The Ensemble Kalman Filter. Springer, Heidelberg (2009). https://doi.org/10.1007/978-3-642-03711-5
21. Friedman, J.H.: Stochastic gradient boosting. Comput. Stat. Data Anal. **38**(4), 367–378 (2002)

22. Gaspari, G., Cohn, S.E.: Construction of correlation functions in two and three dimensions. Q. J. Roy. Meteorol. Soc. **125**, 723–757 (1999)
23. Hamill, T.M., Whitaker, J.S., Snyder, C.: Distance-dependent filtering of background error covariance estimates in an ensemble Kalman filter. Mon. Weather Rev. **129**(11), 2776–2790 (2001)
24. Houtekamer, P.L., Mitchell, H.L.: Data assimilation using an ensemble Kalman filter technique. Mon. Weather Rev. **126**(3), 796–811 (1998)
25. Houtekamer, P.L., Mitchell, H.L.: A sequential ensemble Kalman filter for atmospheric data assimilation. Mon. Weather Rev. **129**(1), 123–137 (2001)
26. Kalman, R.E., et al.: A new approach to linear filtering and prediction problems. J. Basic Eng. **82**(1), 35–45 (1960)
27. Kalnay, E.: Atmospheric Modeling, Data Assimilation and Predictability. Cambridge University Press, Cambridge (2002)
28. Kullback, S., Leibler, R.A.: On information and sufficiency. Ann. Math. Stat. **22**(1), 79–86 (1951)
29. Lei, L., Anderson, J.L.: Comparisons of empirical localization techniques for serial ensemble Kalman filter in a simple atmospheric general circulation model. Mon. Weather Rev. **142**(2), 739–754 (2014)
30. Liaw, A., Wiener, M., et al.: Classification and regression by randomforest. R News **2**(3), 18–22 (2002)
31. Lorenz, E.N.: Predictability: a problem partly solved. In: Proceedings of Seminar on predictability, vol. 1 (1996)
32. Moosavi, A., Stefanescu, R., Sandu, A.: Multivariate predictions of local reduced-order-model errors and dimensions. Int. J. Numer. Methods Eng. (2017). https://doi.org/10.1002/nme.5624
33. Pedregosa, F., et al.: Scikit-learn: machine learning in Python. J. Mach. Learn. Res. **12**, 2825–2830 (2011)
34. Petrie, R.: Localization in the ensemble Kalman filter. M.Sc. Atmosphere, Ocean and Climate University of Reading (2008)
35. Sakov, P., Oke, P.R.: A deterministic formulation of the ensemble Kalman filter: an alternative to ensemble square root filters. Tellus A **60**(2), 361–371 (2008)

Track of Marine Computing in the Interconnected World for the Benefit of the Society

Marine and Atmospheric Forecast Computational System for Nautical Sports in Guanabara Bay (Brazil)

Rafael Henrique Oliveira Rangel[1], Luiz Paulo de Freitas Assad[2(✉)], Elisa Nóbrega Passos[1], Caio Souza[1], William Cossich[1], Ian Cunha D'Amato Viana Dragaud[1], Raquel Toste[1], Fabio Hochleitner[1], and Luiz Landau[1]

[1] Laboratório de Métodos Computacionais em Engenharia, LAMCE/COPPE/UFRJ, Av. Athos da Silveira Ramos, 149,Centro de Tecnologia – Bloco I – Sala 214 Cidade Universitária, Rio de Janeiro, RJ 21941-996, Brazil

[2] Departamento de Meteorologia, Rua Athos da Silveira Ramos, 274, Bloco G1, Cidade Universitária, Rio de Janeiro, RJ 21941-916, Brazil

lpaulo@lamce.coppe.ufrj.br

Abstract. An atmospheric and marine computational forecasting system for Guanabara Bay (GB) was developed to support the Brazilian Sailing Teams in the 2016 Olympic and Paralympic Games. This system, operational since August 2014, is composed of the Weather Research and Forecasting (WRF) and the Regional Ocean Modeling System (ROMS) models, which are both executed daily, yielding 72-h prognostics. The WRF model uses the Global Forecast System (GFS) as the initial and boundary conditions, configured with a three nested-grid scheme. The ocean model is also configured using three nested grids, obtaining atmospheric fields from the implemented WRF and ocean forecasts from CMEMS and TPXO7.2 as tidal forcing. To evaluate the model performances, the atmospheric results were compared with data from two local airports, and the ocean model results were compared with data collected from an acoustic current profiler and tidal prediction series obtained from harmonic constants at four stations located in GB. According to the results, reasonable model performances were obtained in representing marine currents, sea surface heights and surface winds. The system could represent the most important local atmospheric and oceanic conditions, being suitable for nautical applications.

Keywords: Tidal currents · Coastal winds · Ocean modelling · Atmospheric modelling · Computational operational system

1 Introduction

The application of environmental information to sports has become a common practice in the last two decades, especially regarding the Olympic Games (e.g., Powell and Rinard 1998; Horel et al. 2002; Golding et al. 2014). The sport of sailing is extremely

J. M. F. Rodrigues et al. (Eds.): ICCS 2019, LNCS 11539, pp. 215–228, 2019.
https://doi.org/10.1007/978-3-030-22747-0_17

sensitive to weather conditions, as wind is a limiting factor for the occurrence of yachting events, and a lack of wind can make it impossible to practice the sport. In this sense, forecasts are required to provide information to manage competition schedules and establish strategies for better performance of athletes during nautical competitions. For sailing, oceanic and atmospheric models have been employed to provide wind and sea surface current forecasts (Powell and Rinard 1998; Katzfey and McGregor 2005; Vermeersch and Alcoforado 2013). These predictions have been used to support the Olympic and Paralympic Games, as weather has become an important issue in terms of planning, training and safety (Powell and Rinard 1998; Rothfusz et al. 1998; Spark and Connor 2004; Golding et al. 2014).

The Brazilian Olympic Games were held in the period between August 5 to 21, 2016, and the Paralympic Games were held from September 7 to 18, 2016 in Rio de Janeiro. Nautical sports occurred mainly in Guanabara Bay (GB), located in the metropolitan region. A regional forecast system with ocean and atmospheric models for the GB was developed to support the Brazilian Olympic Sailing Team during the 2016 Olympic Games and during prior training and official test events. This system consisted of two numerical models that yield wind, sea surface height and current forecasts, which then became available to Brazilian coaches and athletes.

The goal of the present paper is to evaluate the modelling system considering atmospheric and oceanic data collected in situ and to provide an overview of the computational modelling forecast system applied to GB.

2 Study Area

GB is one of the most important coastal marine environments in Brazil due to its economic, social and political characteristics. This bay is located near the second largest Brazilian metropolitan region, surrounded by Rio de Janeiro, Niterói, Sao Gonçalo, Magé and Duque de Caxias municipalities. There are two oil refineries, the second largest Brazilian port and two international airports in the metropolitan area around GB. The bay constitutes an important marine traffic area for many commercial and fishery ships and vessels. The importance of the bay was highlighted during the 2016 Olympic and Paralympic Games by its use as the main area for Olympic sailing competitions.

GB is located between 22.68° S and 22.97° S latitude and 43.03° W and 43.30° W longitude (Fig. 1), covering an area of 384 km^2. Its longitudinal length is approximately 30 km, with a zonal length of approximately 28 km. There are several islands located inside the bay, which cover an area of almost 60 km^2. The average depth of the entire bay is approximately 4 m, but in the main navigation channel, depths of approximately 50 m are found (Kjerfve et al. 1997). The shallower regions are in GB's northern area, which is directly influenced by river and sediment discharges.

The river basin that flows into GB contains 45 rivers and streams, corresponding to an annual average flow of 100 m^3/s. Nevertheless, the continental water volume is small compared to the water volume of the bay and to the marine water inputs. The most important rivers that flow into GB are located in the northern part of the bay (Fig. 1). December and July are the months with the highest and lowest average discharge, respectively, which demonstrates the semiannual variability of the river discharge in the region. GB marine hydrodynamics and meteorological characteristics are briefly described in Sects. 2.1 and 2.2, respectively.

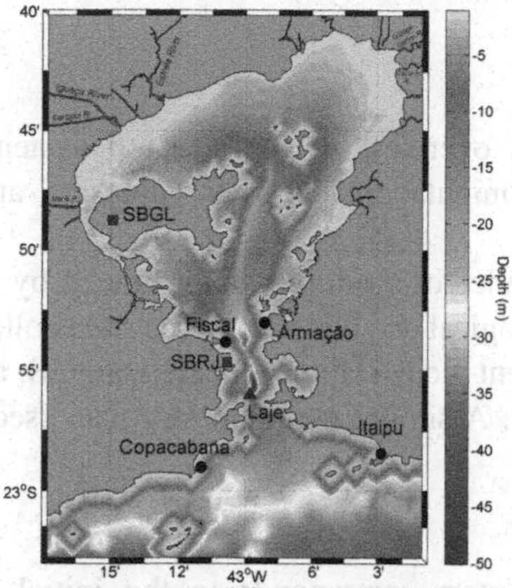

Fig. 1. Bathymetry of the oceanic grid domain G2 (color shading) and position of atmospheric (square dots) and oceanic stations used for the model performance evaluation. The circles represent the tidal stations, and the triangle marks the ADCP location. The G2 location is shown in Fig. 2. (Color figure online)

2.1 Hydrodynamic Conditions

Kjerfve et al. (1997) described the GB water circulation as a combination of gravitational circulation and residual tidal circulation, modified by prevailing net wind effects. The authors classified the GB as an estuary dominated by tidal influence. Tidal currents are important because they are responsible for the process of water transport and the mixture between the estuary and the ocean (Miyao and Harari 1989). This type of current is also influenced by the wind through momentum transfer once sufficient wind speed is attained. GB has a pattern of unequal tidal currents, presenting flood currents that are faster than ebb currents (Kjerfve et al. 1997). The most intense velocities are associated with currents aligned with the main navigation channel. Bérgamo (2006) concluded that the tidal currents dominate marine circulation inside and, to some extent, outside the GB estuary.

2.2 Meteorological Conditions

The GB area features a warm, rainy climate in the summer and a cold, dry climate in the winter (Dereczynski et al. 2013). The wind regime over GB is influenced by meteorological processes at different spatial and temporal scales, such as the South Atlantic Subtropical High, cold fronts, and extratropical cyclones, and the local circulation is modulated by sea/land breeze (Pimentel et al. 2014).

Surface wind data from a weather station located at the Santos Dumont Airport (SBRJ) presents a clear pattern in the north-south direction, whereas data from a weather station located at the Antonio Carlos Jobim International Airport (SBGL) in the middle of the bay shows that the wind has a southeast-east pattern (Pimentel et al. 2014).

3 Methodology

In this section, a general overview of the adopted system is presented, and brief information about the computational models, datasets and the analysis are also provided.

To verify the performance of the forecasts produced by this system, hindcast runs were performed. Meteorological and oceanographic data collected by fixed stations and predicted tides for different tidal stations located at GB are used in this paper to evaluate the model results. A similar methodology was used in Blain et al. (2012).

3.1 System Overview

The first stage of this system corresponds to the initial and boundary conditions downloaded from the Global Forecast System (GFS), which are used as input to the atmospheric model, and from Copernicus Marine Environment Monitoring Service (CMEMS), which are used in the ocean model. After this stage, the acquired initial and boundary conditions are interpolated for use in the models. In the next stage, the atmospheric model is executed, and after the integration, the 10-m wind fields and net heat flux at the surface are used as boundary conditions for the following ocean model integration.

Both models are executed daily to yield 72-h prognostics using a three nested grid scheme. The prognostics of the wind fields, sea surface levels and surface current fields are updated daily and are made available on a website. Additionally, to produce more reliable results from the system, observational data are also used to evaluate the model results.

3.2 Atmospheric Model

The atmospheric model used in this study is the Weather Research and Forecasting (WRF) model, which is widely used by the scientific community; its details can be found in Skamarock et al. (2008). The model was configured with three online nested grids, each using a time step of 120 s and 28 vertical levels. The forecast range is 96 h, and the results are generated at 1-h intervals; however, only 72 h are available on the Nautical Strategy Project website. The coarser grid domain covers the South, Middle East and part of the Northeast Brazilian territory. This domain is configured with 100 × 118 grid cells with a horizontal resolution of 27 km. The intermediary domain covers the entire Rio de Janeiro State and part of the Minas Gerais and São Paulo States. This grid is configured with 61 × 58 cells with a horizontal resolution of 9 km. The finest grid domain is centered on GB. It was configured with 64 × 70 grid cells with a horizontal resolution of 3 km. The hierarchy of the WRF grids is illustrated in Fig. 2.

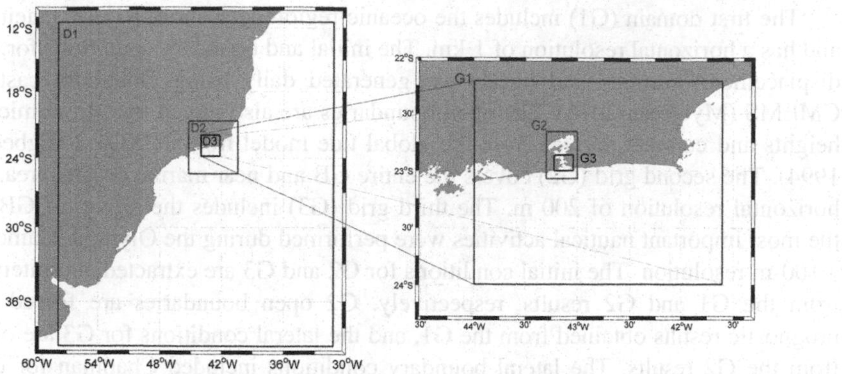

Fig. 2. Atmospheric and oceanic nested grid domains meshed: D1 corresponds to the coarse resolution of the atmospheric grid domain (27 km), D2 corresponds to the intermediary domain (9 km) and D3 represents the finest domain (3 km). G1 corresponds to the coarse resolution of the oceanic domain (1 km), G2 corresponds to the intermediary domain (0.2 km) and G3 represents the finest domain (0.1 km).

The Global Forecast System (GFS) results from the 0000 UTC run, with a 0.5° horizontal resolution and a 3-h temporal resolution obtained from the NCEP/NOAA ftp server. These results are used as the initial and boundary conditions for the WRF coarse domain. The intermediate domain used boundary conditions from the coarse domain, and the smallest nested domain used boundary conditions from the intermediate domain. To verify the performance of the atmospheric system, a hindcast was performed for the period of July 2014 to May 2015, and it was used for the analyses presented in the following sections.

3.3 Ocean Model

The ocean model used to simulate the GB hydrodynamics is the Regional Ocean Modelling System (ROMS) (Shchepetkin and McWilliams 2005). ROMS is a free-surface numerical model widely used by the scientific community and has been applied in several studies using different spatial and temporal scales. The model solves the Navier-Stokes equations with hydrostatic and Boussinesq approximations using finite-difference methods. The primitive equations are solved in the horizontal direction in an orthogonal curvilinear coordinate system discretized with an Arakawa-C grid (Shchepetkin and McWilliams 2005; Haidvogel et al. 2008). In the vertical direction, the model uses a terrain-following coordinate system, denoted as S-coordinates that behave as equally spaced sigma-coordinates in shallow regions and as geopotential coordinates in deep regions (Song and Haidvogel 1994).

ROMS was configured in baroclinic mode with three numerical grids using offline nesting (G1, G2 and G3), all with 20 vertical levels (Fig. 2). The bottom topography used in all of the grids was generated using digitalized nautical charts provided by the Brazilian Navy, merged with the bathymetry from ETOPO 2 (National Geophysical Data Center 2006) used for the oceanic region. It was applied a space filter on the generated bathymetry field in order to avoid hydrodynamics inconsistencies during the numerical integration.

The first domain (G1) includes the oceanic region adjacent to Rio de Janeiro State and has a horizontal resolution of 1 km. The initial and boundary conditions for surface displacement, currents and tracers are generated daily using global forecasts from CMEMS (MyOcean 2016). The open boundaries are also forced by astronomical tidal heights and currents derived from the global tide model from TPXO7.2 (Egbert et al. 1994). The second grid (G2) covers the entire GB and near marine coastal area, with a horizontal resolution of 200 m. The third grid (G3) includes the region of GB where the most important nautical activities were performed during the Olympic Games, with a 100-m resolution. The initial conditions for G2 and G3 are extracted and interpolated from the G1 and G2 results, respectively. G2 open boundaries are forced by the prognostic results obtained from the G1, and the lateral conditions for G3 are obtained from the G2 results. The lateral boundary conditions included Chapman for the free surface (Chapman 1985), Flather for the 2D momentum (Flather 1976), and Radiation for the 3D momentum and tracers (Marchesiello et al. 2001).

At the surface, the three oceanic models are forced by the hourly prognostic fields generated by the WRF (previously described), including the wind stress and the net heat flux at the surface. These atmospheric prognostic fields are extracted from the highest resolution WRF grid as a subset. In the operational system, the ROMS simulation is executed daily with a 24-h spin-up period to provide predictions for the following 72 h. In the present paper, a hindcast was performed to obtain results regarding daily oceanic forecasts for the period of September 15 to October 2, 2014. The results from the atmospheric and oceanic models were evaluated using in situ data, as described in the next section.

3.4 Available Data

To evaluate the WRF forecasts, Meteorological Aerodrome Report (METAR) data were used from SBGL and SBRJ stations from July 2014 to May 2015. These fields were selected due to their location near GB (Fig. 1). The wind direction and intensity were extracted from METAR raw data and were decomposed into zonal (u) and meridional (v) vectoral components to compare with the model results. A total of 5,274 observations were used from SBRJ, and 7,128 observations were used from SBGL. During the whole period, SBRJ data between 0300 UTC and 0800 UTC were not available. Also, there was a total of 6 days of missing data for SBRJ.

The oceanographic data used to evaluate the ocean model forecast skill consisted of current profiles and sea surface height data acquired from an Acoustic Doppler current profiler (ADCP) sensor situated near the GB connection with the open ocean at 22.92° S and 43.15° W (Fig. 1), named Laje Station. The ADCP was deployed at a depth of 27 m. The measured data used in this study were the direction and magnitude of the ocean current and pressure converted into depth. The dataset is composed of time series records for the period between September 14 and October 2, 2014. The data were measured at 10-min intervals; however, the data were subsampled, and a 1-h interval was used in the performed analysis. A 48-h high-pass filter and a 6-h low-pass filter were applied to the time series of ADCP data and the modeled results to maintain the tidal oscillation as the main forcing. The spatial variability of the tidal wave representation by the ocean model was also evaluated using the tidal time series predicted for the same period of analysis using sets of harmonic constants from four tidal stations (Fig. 1).

3.5 Performance Assessments

Comparison of the Meteorological Data and Model Results

To evaluate the WRF predictions, statistical indices were used, including a bias, which is commonly used to evaluate model tendencies to over or underestimate wind speed and the root mean square error (RMSE), which demonstrates numerical accuracy. The advantage of these metrics is given by the presentation of error values in the same dimensions as the variable analyzed. The bias and RMSE were calculated using the difference between the atmospheric model result and the observation (model results minus observation). Several articles can be found in the literature in which these statistical indices were used to perform a comparison between the WRF results and observed wind data (Jiménez and Dudhia 2012; Jiménez et al. 2013).

For RMSE, near-zero values indicate better model performance; however, for the bias, also known as systematic error, null values do not necessarily correspond to greater forecast accuracy, because positive and negative errors cancel each other. The dimensionless Pearson linear correlation (COR), which varies from −1 to 1, was also calculated. If the index is equal to 1 (−1), then a perfect positive (negative) correlation between the two variables exists. If this index is null, then the predicted and observed values have no linear relationship. However, one cannot rule out the existence of nonlinear dependencies between the variables with this metric. Thus, from the statistical metrics, one can determine how many days in advance the best prediction will be obtained. To qualitatively evaluate the model, wind roses were generated for observed values and predicted data. Only wind roses of observed data and WRF results for the first 24 h are shown because the 48-, 72- and 96-h results were very similar.

Comparison of the Oceanographic Data and Model Results

To evaluate the skill of the ocean model, hindcast results of G3 were compared to in situ data collected by the ADCP. The time series average (mean) and standard deviation (SD) were used to perform a global assessment of the errors. The central frequency (CF) was computed as the percentage of errors within the following limits. The acceptable error limits used in the present work for the water level displacement and current speed are 10 cm and 0.03 m/s, respectively, and these values were chosen based on the observed levels and speed. The current direction was not considered in this evaluation.

The RMSE, the refined index of agreement (RIA) (Willmott et al. 2012) and COR between model results and observations were calculated. The RIA varies from −1 to 1 and is based in the central tendency of the model-prediction errors and describes the relative co variability of observations and the prediction errors. Positive RIA indicates small model errors in relation to the variance of the observational data, and when the index is equal to 1, a perfect match is observed between predictions and observations. If this index is null the prediction error is twice the sum of observed deviations. Negative RIA values represent poor estimates according as they get close to −1. Nevertheless, RIA values equal to −1 may also mean that little variability is presented in the observed data (Willmott et al. 2012).

These metrics were applied to the ADCP data and model results obtained for G2 model. The RMSE, RIA and COR indices were also computed for the water level time series of the four tidal stations (described above) using local tidal harmonic constants. The complete set of analysis is presented in the next section.

4 Results and Discussion

4.1 Analysis of the Atmospheric Model Results

An analysis of the Table 1 shows that the RMSE increases and COR decreases with time, except for the meridional component RMSE. Furthermore, bias analysis demonstrates that in forecasts with horizons longer than 24 h, the wind speed is overestimated. These results suggest that the forecasts made for the first day exhibit better results compared to other forecasts.

Table 1. Root Mean Square Error (RMSE) and Correlation (COR) between SBGL observations and WRF results for the zonal (u) and meridional (v) wind vector components and wind speed (WS), and bias for WS, in m/s, for each day of prediction.

SBGL	u		v		WS		
Day	RMSE	COR	RMSE	COR	Bias	RMSE	COR
1	2.5	0.4	3.0	0.3	0.0	2.4	0.3
2	2.8	0.3	3.2	0.2	0.2	2.5	0.2
3	3.0	0.2	3.0	0.2	0.2	2.5	0.2
4	3.1	0.2	3.1	0.2	0.3	2.6	0.2

In the wind roses generated with SBGL data (Fig. 3), northeasterly and easterly winds are shown more frequently in the model results than in the observational data. The opposite pattern occurs with southeasterly winds, which occur more frequently than indicated by the model. Furthermore, the percentage of observed calm wind cases is greater than the simulated percentage, which favours wind speed with a positive bias.

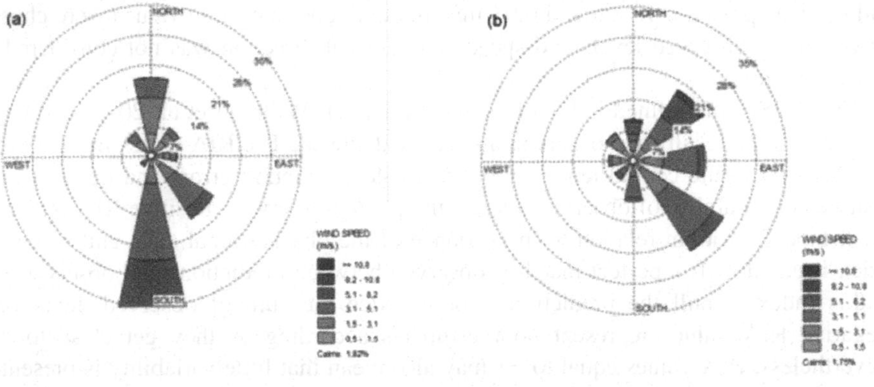

Fig. 3. Wind roses for (a) observed and (b) predicted winds for the first 24 h of each simulation at Antônio Carlos Jobim International Airport (SBGL - ICAO Code). The SBGL location is shown in Fig. 1.

Wind roses generated for SBGL show a model deficiency in representing the wind's north-south pattern, which is often observed at SBRJ (not shown). In addition, because synoptic-scale circulation strongly influences the wind directions represented by the model, this circulation also contributes to the predominance of winds from the east quadrant (between the northeast and southeast) in numerical simulations.

The prevailing wind direction and intensity found for both fields correspond with the results of Pimentel et al. (2014), who characterised the land/sea breeze mechanisms in the GB. However, the WRF model results for both aerodromes indicate the strong influence of the synoptic pattern, as the winds from the northeast on the airfield are directly connected with the Subtropical South Atlantic High. This could also be due to topographic misrepresentation because increasing errors are expected in complex terrain regions, as described in Jiménez and Dudhia (2012).

4.2 Analysis of the Ocean Model Results

Sea Surface Height

The sea surface displacements from the mean sea level computed by the ocean model were compared to the level measured by the ADCP at Laje station and the sea level extracted from the harmonic constituents of Armação, Copacabana, Fiscal and Itaipu stations.

The plot of the filtered sea surface height time series measured by the ADCP and the sea level computed by the ocean model is presented in Fig. 4 for a sampling rate of 1 h. The analysis of the observed data reveals the mixed semi-diurnal characteristics of the tides in GB during two neap tide periods and one spring tide period. The same pattern was observed for the water level computed by the ocean model, although some differences were found. The largest differences occur during the spring periods, as the model computes lower amplitudes than those observed, which leads to a better performance during neap tides, as tide amplitudes are expected to be lower. In Fig. 4, it is noted that the model results underestimate the ebb and flood tide heights. Despite these differences in magnitude, the tidal phase signal strongly agrees with the ocean model results, mainly considering the tidal phase.

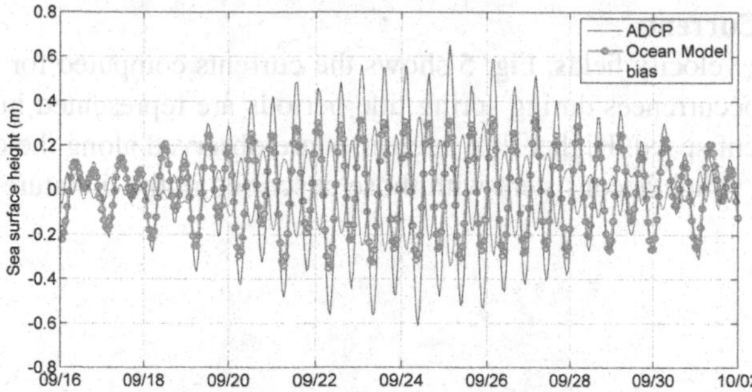

Fig. 4. Observed ADCP data (blue) and modelled (grey) sea surface heights at Laje Station, and the bias (model minus observation) between these two time series (red) in meters. (Color figure online)

To quantify the model performance regarding the representation of the sea surface level, some statistical metrics were calculated and are summarized in Tables 2 and 3. In general, the mean and SD of sea level have similar orders of magnitude in all stations, but different absolute sea levels were measured compared to those computed (Table 2). One possible cause to these differences is the river inputs that are not considered in the ocean model, and an average amount of 100 m^3/s flow to GB per year (Kjerfve et al. 1997). Considering the sea level predicted at the tidal stations, the ocean model presented high positive correlation values and quadratic errors less than 15 cm (Table 3). The same correspondence is observed considering the refined index of agreement proposed by Willmott et al. (2012). RIA evaluates model performances considering the differences between the model and observed deviations, where values of 1.0 indicate a perfect match. Good agreement indices were obtained in the present work, with the worst index obtained at the Itaipu station. At this station, large average errors were observed. Therefore, this behaviour is expected, as the Itaipu station is located in a region susceptible to the effect of coastal currents and local recirculation.

Table 2. Mean sea level and standard deviations, root mean squared errors (RMSE), correlation (COR) and refined index of agreement (RIA) values of the observed/predicted and modelled water levels at the tidal stations and the Laje mooring.

Station\Statistic	Mean ± SD (cm) Observed/predicted	Mean ± SD (cm) Modelled	RMSE (cm)	COR	RIA
Armação	0.10 ± 26.31	0.10 ± 19.95	14.85	0.83	0.72
Copacabana	0.12 ± 24.21	0.23 ± 22.47	5.81	0.97	0.98
Fiscal	−0.31 ± 26.34	0.11 ± 19.88	13.47	0.87	0.76
Itaipu	1.86 ± 24.29	0.23 ± 22.07	14.69	0.80	0.68
Laje (ADCP)	0.06 ± 26.30	0.06 ± 16.96	11.13	0.96	0.81

The next lowest RIA values were observed in Armação and Fiscal Station that also have high RMSE. Both stations are located in the narrow area that borders the main navigation channel at GB (Fig. 1).

Near-Surface Currents

Considering the velocity fields, Fig. 5 shows the currents computed for the G2 model. Ebb and flood occurrences during spring tide periods are represented in this figure, as well as the current speed. Higher flux intensities are observed along the deeper channel in both periods, which is consistent with those described in the literature (Kjerfve et al. 1997).

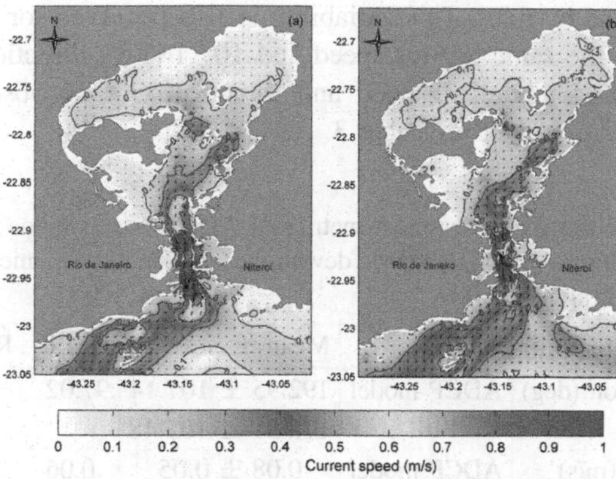

Fig. 5. Ebbing (a) and flooding (b) currents computed by the ocean model during the spring tide period at the surface layer. The arrows represent surface current speed and direction, and the colors represent current speed (m/s) for the grid G2. (Color figure online)

Several metrics were also used to evaluate the model skill in the representation of GB currents and are presented in Table 3. For this analysis the G3 model results were used. This analysis was performed using the velocities extracted from ADCP measurements (the Laje station) to provide information about the model representativeness of near-surface velocities. The time series for the observed and computed velocities for this layer are shown in Fig. 6.

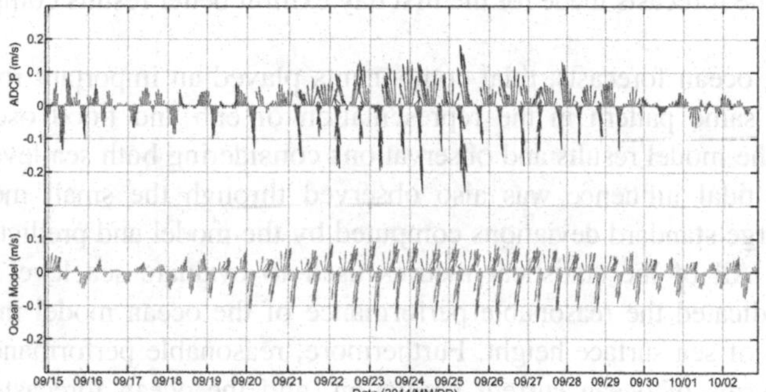

Fig. 6. Hourly series of the observed (blue arrows) and modelled (black arrows) velocities at the Laje station. (Color figure online)

The mean current speed computed by the G3 ocean model is 0.09 m/s, with a mean direction of 178.51°. Considering the observed currents, the mean speed in the near-surface layer is 0.08 m/s with a mean azimuth of 192.95° (Table 3). Considering the G3 results, the standard deviations computed for magnitude and direction are 0.05 m/s and 105.48°, respectively. This indicates the high variability of the mean flux direction along the integration time, which is expected because the model domain represents a bay mainly dominated by tides. This variability is also observed for currents measured by the ADCP, with the same SD for speed and 107.14° for direction. This similarity between observed and modelled speed and direction can be observed using other statistical metrics, as presented in Table 3.

Table 3. Summary of the skill assessment metrics of ADCP observations and modelled ocean currents at Laje station (SD = standard deviation; RMSE = root mean squared errors; RIA = refined index of agreement).

Attribute (unit)		Mean ± SD	RMSE	RIA
Direction (deg)	ADCP model	192.95 ± 107.14	97.02	−0.17
		178.51 ± 105.48		
Speed (m/s)	ADCP model	0.08 ± 0.05	0.06	0.37
		0.09 ± 0.05		

5 Summary and Conclusions

The Marine and Atmospheric Computational Forecast System developed for nautical sports in GB were demonstrated to be capable of representing the main local atmospheric and oceanic conditions. Considering wind directions, at SBRJ, the observed data presented the North-South axis as its preferential direction, whereas the model indicated a deviation toward the southeast direction. At SBGL, differences were observed with deviations from the observed southeasterly winds ranging from the northeast to southeast directions computed by the WRF. The results from statistics suggests that the forecasts made for the first day exhibit better results compared to other forecasts.

Regarding ocean forecasts, tidal components played an important role in GB circulation. The same pattern in the representation of ebb and flood oscillations was observed for the model results and observations considering both sea level and marine currents. The tidal influence was also observed through the small mean sea level heights and large standard deviations computed by the model and predicted at the four tidal stations. All of the statistical metrics used to evaluate sea level displacement predictions indicated the reasonable performance of the ocean model in terms of the representation of sea surface height. Furthermore, reasonable performances were also observed in terms of ocean current prediction, and the ocean forecasts were representative of local currents. To improve the skill of the present forecast system, some additional developments are planned for future applications. These developments include the analysis of a larger amount of oceanographic and meteorological data and the improvement of spatial resolutions for both models. Additionally, for the atmospheric model, other sets of parameterizations should be tested to construct a WRF ensemble system. For the ocean model, different settings should be tested, and the tidal forcing should be adjusted to better represents the main tidal constituents, in addition to the possible use of data assimilation and the inclusion of water discharges from rivers.

Finally, it is important to emphasize the support that the developed system gave to the Brazilian Sailing Team (BST) during the Rio 2016 Olympic games. The BST conquered one of the best results in olympic games including one gold medal in the 49er class. The sailing coaches and athletes also contribute to the development of the forecast system by means of internal communications about the forecast performance.

References

Bérgamo, A.L.: Características Hidrográficas, da Circulação e dos Transportes de Volume e Sal na Baía de Guanabara (RJ): Variações sazonais e Moduladas pela Maré. [Dissertation]. Universidade de São Paulo, São Paulo (2006)

Blain, C.A., Cambazoglu, M.K., Linzell, R.S., Dresback, K.M., Kolar, R.L.: The predictability of near-coastal currents using a baroclinic unstructured grid model. Ocean Dyn. **62**(3), 411–437 (2012)

Chapman, D.C.: Numerical treatment of cross-shelf boundaries in a barotropic coastal ocean model. J. Phys. Oceanogr. **15**, 1060–1075 (1985)

Dereczynski, C.P., Luiz Silva, W., Marengo, J.A.: Detection and projections of climate change in Rio de Janeiro. Braz. Am. J. Clim. Change **2**, 25–33 (2013). https://doi.org/10.4236/ajcc.2013.21003

Egbert, G.D., Bennett, A.F., Foreman, M.G.G.: TOPEX/POSEIDON tides estimated using a global inverse model. J. Geophys. Res. **99**, 24821–24852 (1994)

Flather, R.A.: A tidal model of the north west European continental shelf. Mem. Soc. R. Sci. Liege. **10**, 141–164 (1976)

Golding, B.W., et al.: Forecasting capabilities for the London 2012 Olympics. Bull. Am. Meteorol. Soc. **95**(6), 883–896 (2014)

Haidvogel, D.B., et al.: Ocean forecasting in terrain-following coordinates: formulation and skill assessment of the Regional Ocean Modeling System. J. Comput. Phys. **227**(7), 3595–3624 (2008)

Horel, J., et al.: Weather support for the 2002 winter Olympic and Paralympic Games. Bull. Am. Meteorol. Soc. **83**(2), 227–240 (2002)

Jiménez, P.A., Dudhia, J.: Improving the representation of resolved and unresolved topographic effects on surface wind in the WRF model. J. Appl. Meteorol. Climatol. **51**(2), 300–316 (2012)

Jiménez, P.A., et al.: An evaluation of WRF's ability to reproduce the surface wind over complex terrain based on typical circulation patterns. J. Geophys. Res. Atmos. **118**(14), 7651–7669 (2013)

Katzfey, J.J., McGregor, J.L.: High-resolution weather predictions for the America's cup in Auckland: a blend of model forecasts, observations and interpretation. In: Proceedings of the World Weather Research Program Symposium on Nowcasting and Very Short Range Forecasting, Toulouse, France (2005)

Kjerfve, B., Ribeiro, C.H.A., Dias, G.T.M., Filippo, A.M., Da Silva, Q.V.: Oceanographic characteristics of an impacted coastal bay: Baía de Guanabara, Rio de Janeiro. Braz. Cont. Shelf Res. **17**(13), 1609–1643 (1997)

Marchesiello, P., McWilliams, J.C., Shcheptkin, A.: Open boundary conditions for long-term integration of regional oceanic models. Ocean Model. **3**, 1–20 (2001)

Miyao, S.Y., Harari, J.: Estudo preliminar da maré e das correntes de maré da região estuarina de Cananéia (25S–48°W). Boletim do Instituto Oceanográfico **37**(2), 107–123 (1989)

MyOcean: Operational Mercator Global Ocean Analysis and Forecast System. Global Ocean 1/12° Physics Analysis and Forecast Updated Daily (2016). www.myocean.eu

National Geophysical Data Center: 2-minute gridded global relief data (ETOPO2) v2. National Geophysical Data Center, NOAA (2006)

Pimentel, L.C.G., Marton, E., Da Silva, M.S., Jourdan, P.: Caracterização do regime de vento em superfície na Região Metropolitana do Rio de Janeiro. Engenharia Sanitaria e Ambiental **19** (2), 121–132 (2014)

Powell, M.D., Rinard, S.K.: Marine forecasting at the 1996 centennial Olympic games. Weather Forecast. **13**(3), 764–782 (1998)

Rothfusz, L.P., McLaughlin, M.R., Rinard, S.K.: An overview of NWS weather support for the XXVI Olympiad. Bull. Am. Meteorol. Soc. **79**(5), 845–860 (1998)

Shchepetkin, A.F., McWilliams, J.C.: The regional oceanic modeling system (ROMS): a split-explicit, free-surface, topography-following-coordinate oceanic model. Ocean Model. **9**(4), 347–404 (2005). https://doi.org/10.1016/j.ocemod.2004.08.002

Skamarock, W.C., et al.: A description of the advanced research WRF Version 2 (No. NCAR/TN-468+STR). National Center for Atmospheric Research, Boulder, CO, Mesoscale and Microscale Meteorology Div (2008)

Song, Y., Haidvogel, D.: A semi-implicit ocean circulation model using a generalized topography-following coordinate system. J. Comput. Phys. **115**(1), 228–244 (1994)

Spark, E., Connor, G.J.: Wind forecasting for the sailing events at the Sydney 2000 Olympic and Paralympic games. Weather Forecast. **19**(2), 181–199 (2004)

Vermeersch, W., Alcoforado, M.J.: Wind as a resource for summer nautical recreation. Guincho beach study case. Finisterra **95**, 105–122 (2013)

Willmott, C.J., Robeson, S.M., Matsuura, K.: A refined index of model performance. Int. J. Climatol. **32**, 2088–2094 (2012)

An Integrated Perspective of the Operational Forecasting System in Rías Baixas (Galicia, Spain) with Observational Data and End-Users

Anabela Venâncio[1(✉)], Pedro Montero[2(✉)], Pedro Costa[1], Sabela Regueiro[1], Swen Brands[1], and Juan Taboada[1]

[1] MeteoGalicia, Subdirección Xeral de Meteoroloxía e Cambio Climático Dirección Xeral de Calidade Ambiental e Cambio Climático Consellería de Medio Ambiente, Territorio e Vivenda Xunta de Galicia, Santiago de Compostela, Spain
numerico.meteogalicia@xunta.gal

[2] INTECMAR, Instituto Tecnolóxico para o Control do Medio mariño de Galicia, Vilaxoán, Villagarcía de Arousa, Spain
pmontero@intecmar.gal

Abstract. Rías Baixas is a coastal region located in northwestern Spain (Galicia), between Cape Fisterra and the Portugal-Spain border. Its rich natural resources, which are key for the welfare of the region, are highly vulnerable to natural and anthropogenic stress. In this study, the operational ocean forecasting system developed at the *meteorological agency of the Galician government* (MeteoGalicia) is presented focussing on the Rías Baixas region. This system includes four models providing daily output data: the hydrodynamic models ROMS and MOHID, the atmospheric model WRF and the hydrological model SWAT. Here, MOHID's implementation for the Rías Baixas region is described and the model's performance with respect to observations is shown for those locations where Current-Temperature-Depth (CTD) profiles are obtained weekly by the *Technological Institute for the Monitoring of the Marine Environment* in Galicia (INTECMAR). Although the hydrodynamical conditions of this region are complex, the model skillfully reproduces these CTDs. The model results and derived products are publicly available through MeteoGalicia's web page and data server (www.meteogalicia.gal).

Keywords: Rías Baixas · Modeling · Observational data

1 Introduction

Located in the northwest of the Iberian Peninsula, the region of Galicia is characterised by a singular morphology. Along the coastline, there are deep coastal inlets with SW-NE (southwest-northwest) orientation, called *rías*, which are Pleistocene river valleys flooded by the sea at the end of the Würm Glaciation. They have a typical V form and, as we get closer to the platform, become wider

J. M. F. Rodrigues et al. (Eds.): ICCS 2019, LNCS 11539, pp. 229–239, 2019.
https://doi.org/10.1007/978-3-030-22747-0_18

and deeper. According to their hydrodynamic and sedimentologic characteristics, the *rías* can be subdivided into distinct zones [14] and in each zone there exists a predominating type of water circulation. The inner zone is influenced by river discharges and tides, the outer zone is dominated by coastal winds that promote water exchange with the open ocean and the middle zone is subject to both influences.

From north to south, four *rías* can be found in the region (see Fig. 1): Muros and Noia, Arousa, Pontevedra and Vigo. All of them have similar characteristics, influenced by tides, winds and river plumes. These estuaries are usually classified as partially mixed during the whole year. In winter, stratification is determined by the river freshwater input, while in summer it is caused by solar heating. The *rías* biodiversity is favoured by the meteorological conditions of the Galician coast. The interaction between the ocean and the atmosphere is affected by the presence of the low-pressure system of the North Atlantic [5]. Namely, the movement associated with the anticyclone of Azores has repercussions in the wind field that interferes with the ocean circulation along the Iberian coast. When the wind blows off the coast, it causes a displacement of the surface waters known as Ekman's transport. Perpendicular to the direction of the wind the transport is directed to the right in the Northern Hemisphere due to Coriolis force. If northerly winds prevail, they drive a surface current directed to the open ocean that is compensated by a current in opposite direction generated in depth. Near the coastline, these cold waters emerge in a process in a process known as "upwelling". On the other hand, "downwelling" occurs when southerly winds drive the warm surface waters towards the coast, where they are forced to sink to the bottom. Upwelling waters are usually rich in nutrients and, in combination with a sufficiently intense solar radiation, promote a rapid growth in phytoplankton populations.

In addition, this area is affected by freshwater plumes originating from the discharges of several rivers, as well as by the Western Iberian Circulation, which both exhibit a strong seasonal cycle. During periods of intense precipitation and associated large river discharges, a buoyant plume may form in the ría and propagate towards the open sea altering key features of the marine ecosystem such stratification, nutrients, turbidity and circulation patterns [15,16]. The complex dynamics of the Rías Baixas ecosystem have been the subject of many previous studies [2,4,7,13,16,19,20].

The Galician coast is part of an important upwelling system extended along the east coast of the North Atlantic from approximately 10°N to 44°N. The high productivity of the *rías* is exploited by a very active mussel farming industry using rafts (*bateas*), as well as by the fishing and aquaculture industry. Moreover, a variety of human activities such as harbours, industrial complexes, buildings, agriculture, sewage emissions, maritime traffic and tourism have direct or indirect effects *rías* and their surroundings. Also, most of the Galicia's population lives in the coastal zone which is subject to a considerable anthropogenic pressure and continuous situations of environmental stress.

Over the last years, several research institutes and agencies have developed methodologies for monitoring and forecasting aquatic systems in order to mitigate the negative anthropogenic impacts and to promote the benefits arising from the natural resources. These efforts imply the integration of information technologies using numerical models and sensor devices which increasingly contribute to the sustainable management of the water resources. It is a multidisciplinary field covering three main components: measurements, modelling and data dissemination, [1]. Observational data obtained from satellites, buoys and CTDs directly provide ecosystem information and can be used for the calibration and validation of numerical models, making these more accurate to simulate the behaviour of these systems. It is also necessary to have easy access to this information through tools that allow the analysis and visualisation of the data in an effective way. This integrated concept will serve as a decision support tool for offshore operations, navigation, coastal management, tourism activities and the monitoring of marine pollution episodes, as well as other emergency situations. All of these activities critically dependent on precise predictions of the oceanographic and meteorological conditions.

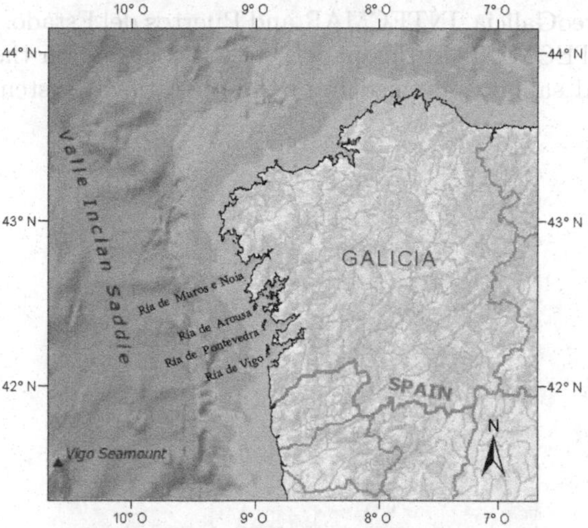

Fig. 1. Geographical overview of the Rías Baixas region, comprising the four coastal inlets/*rías* Muros and Noia, Arousa, Pontevedra and Vigo.

2 Operational Modeling System

At present, the operational scheme involves the coupling of four models, all running daily at MeteoGalicia and each of them covering different scales. At the regional scale, ROMS (Regional Ocean Modeling System) model [17], covers the Northern Iberian Peninsula with a horizontal resolution of 2 km and a vertical discretisation of 41 levels. The model is nested to the MyOcean global model (Copernicus Marine Environment Monitoring Service (CMEMS)) that runs with a horizontal resolution of 1/12° [6]. The numerical output from ROMS provides the initial and boundary conditions for the high resolution near-shore MOHID (Water Modeling System) model developed by MARETEC (Marine and Environmental Technology Research Center) [10,12] that runs at local scale. MOHID is used to simulate the main Galician *rías*: Muros and Noia, Arousa, Pontevedra and Vigo, and runs at a horizontal resolution of 300 m with 29, 34 and 27 vertical levels respectively, (see Fig. 2). The Weather Research and Forecasting Model (WRF), run at 12 and 4 km resolution provides the atmospheric forcing for ROMS and MOHID, respectively. The atmospheric variables ingested by the hydrodynamics models are sea level pressure, winds, surface air temperature, surface specific humidity and radiation on hourly timescale.

To simulate the effect of river discharges, the Soil Water Assessment Tool (SWAT) developed by the Agricultural Research Service and Texas A&M University is applied [18]. This model calculates the daily average flow and temperature of the region's principal rivers and is used to feed both hydrodynamic models.

The models were calibrated and validated using satellite data, buoy data provided by MeteoGalicia, INTECMAR and Puertos del Estado, and CTDs data collected by INTECMAR. In this paper, we are focusing on the comparison of temperature and salinity obtained from the operational system and the CTD

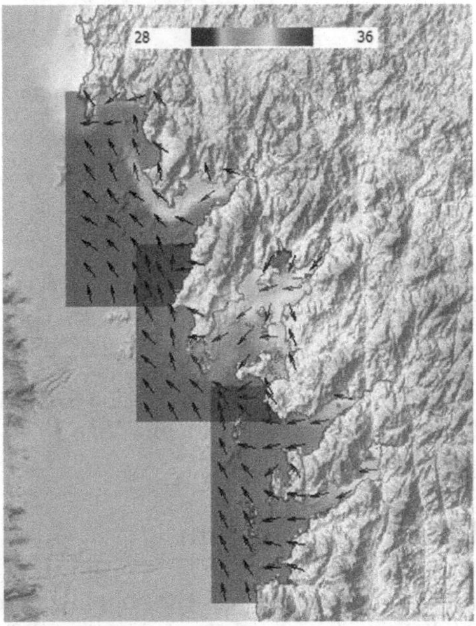

Fig. 2. Salinity and velocity fields from MOHID model for the Rías Baixas (Muros and Noia, Arousa, Pontevedra and Vigo).

profiles. INTECMAR weekly monitors the hydrography of Galician coast since 1992. The current oceanographic network is formed by 43 oceanographic stations distributed along Rías Baixas (Fig. 3) and the Ría de Ares (not shown). All data are downloaded, processed and saved on the INTECMAR data center (and distributed through www.intecmar.gal).

With this operational system, the water level, velocities, temperature, and salinity fields are predicted once a day. The model output is mapped on MeteoGalicia's official web page and the corresponding files are published on a threads server. The data is publicly available and can be used for both research or commercial purposes. Several protocols for remote access via the threads server have been established and some specific products have been created from this system, such forecast reports for associations and harbours, nautical sports, beach activities as well as mobile applications.

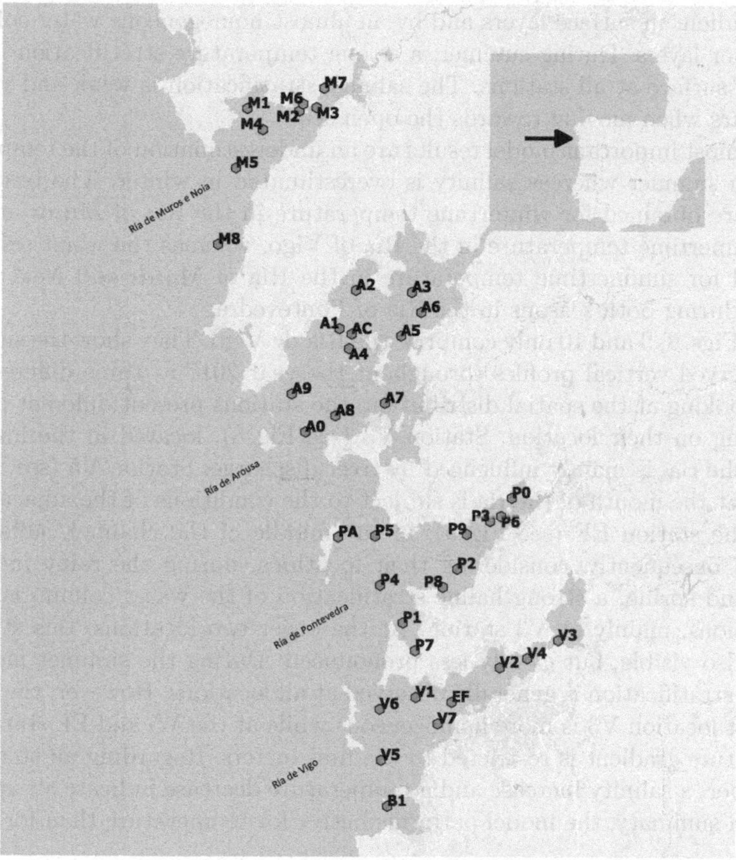

Fig. 3. CTDs stations location map of Rías Baixas (Muros and Noia, Arousa, Pontevedra and Vigo).

3 Results and Discussion

In Figs. 4, 5, 6 and 7, a comparison between the model (dashed lines) results and CTD data (continuous lines), collected at the locations shown in Fig. 3, is provided for the *rías* of Muros and Noia, Arousa, Pontevedra and Vigo, respectively. The chosen days represent known patterns of the vertical profiles, that commonly occur during the winter (blue lines) and summer (red lines) seasons. Temperature profiles are presented in the first column and salinity profiles in the second. The distinct panels in each column represent distinct locations, moving towards the open sea from top to bottom.

In the CTD data, the typical two-layer estuarine circulation of this *rías* appears. During winter events, and particularly at those sites located near the river mouths, the salinity of the surface layers is strongly influenced by freshwater input from river discharges whereas the salinity of the deeper layers is similar to ocean water. The temperature profiles of these sites are characterised by a week gradient in surface layers and by an almost homogeneous water column in the deeper layers. During summer, a strong temperature stratification is found near the surface at all stations. The salinity stratification is weak and virtually disappears when moving towards the open sea.

The most important model results are an underestimation of the temperature values in summer whereas salinity is overestimated in winter. The best model results are obtained for wintertime temperature in the Ría of Muros and Noia and summertime temperature in the Ría of Vigo, whereas the worst results are obtained for summertime temperature in the Ría of Muros and Noia and for salinity during both seasons in the Ría of Pontevedra.

The Figs. 8, 9 and 10 only comprise the Ría de Vigo. They show the simulated and observed vertical profiles throughout the year 2017 at three different locations. Looking at the spatial distribution, the stations present different patterns depending on their location. Station V3 (see Fig. 8), located in the innermost part of the ría, is mainly influenced by river discharge. Station V5 (see Fig. 10), located at the mouth of the ría is subject to the conditions of the adjacent platform. The station EF (see Fig. 9), in the middle of the channel, suffers both effects. Consequently, considering their locations, during the rainy months of winter and spring, a strong haline stratification of the water column is seen in observations, mainly at V3 station. At the other two locations, this stratification is also visible, but clearly less pronounced. During the summer months, a thermal stratification is generally observed at all locations. However, the vertical profile at location V3 is more homogeneous while at the V5 and EF stations the temperature gradient is restricted to the first meters. Regarding all stations, in September, a salinity increase and a temperature decrease indicate an upwelling event. In summary, the model performs better for temperature than for salinity profiles.

Albeit MOHID correctly reproduces the main features of the vertical profiles, predictions are not perfect and the modelling system is subject to continuous improvement. One promising approach is to enhance the performance of the auxiliary models ROMS, WRF and SWAT, providing boundary conditions

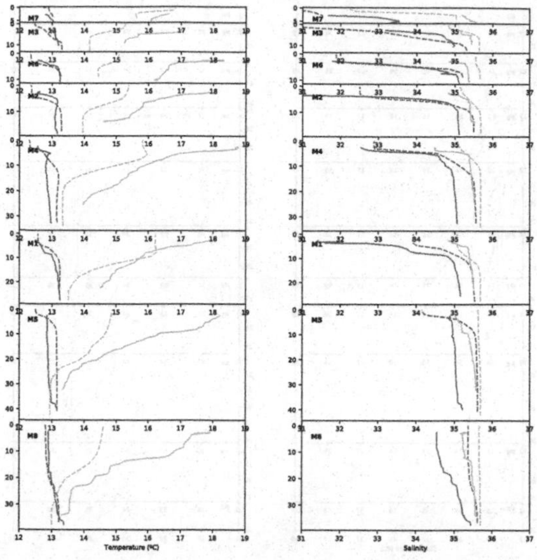

Fig. 4. Vertical profiles for temperature (left panel) and salinity (right panel) at the Ría de Muros and Noia CTDs stations, for the days 2018/01/15 and 2018/07/10 corresponding to a winter and summer situation (blue and red lines, respectively). CTD profile (continuous line) and MOHID model profile (dash line). (Color figure online)

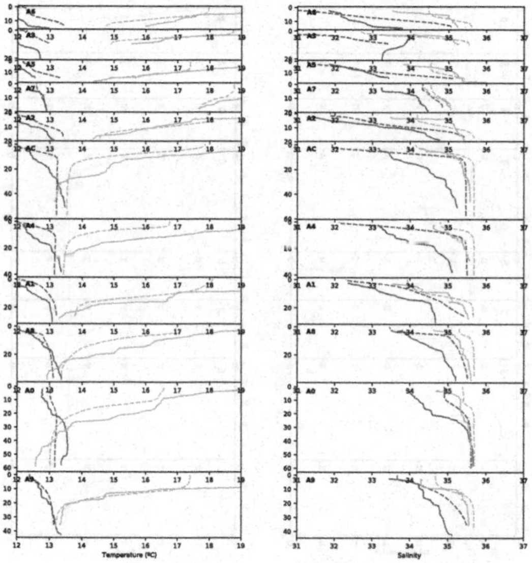

Fig. 5. Vertical profiles for temperature (left panel) and salinity (right panel) at the Ría de Arousa CTDs stations, for the days 2018/01/21 and 2018/07/10 corresponding to a winter and summer situation (blue and red lines, respectively). CTD profile (continuous line) and MOHID model profile (dash line). (Color figure online)

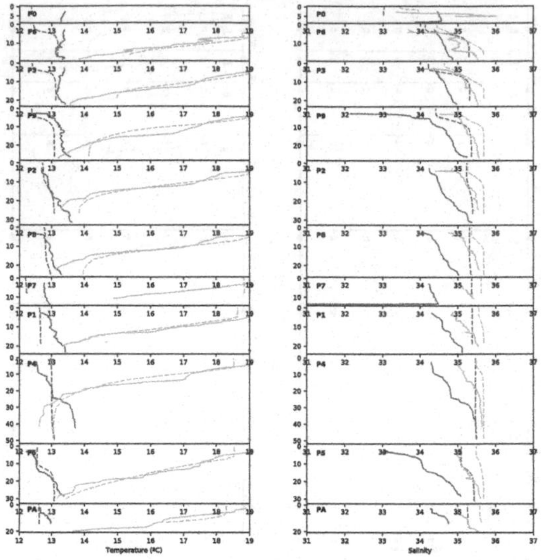

Fig. 6. Vertical profiles for temperature (left panel) and salinity (right panel) at the Ría de Pontevedra CTDs stations, for the days 2018/01/15 and 2018/07/10 corresponding to a winter and summer situation (blue and red lines, respectively). CTD profile (continuous line) and MOHID model profile (dash line). (Color figure online)

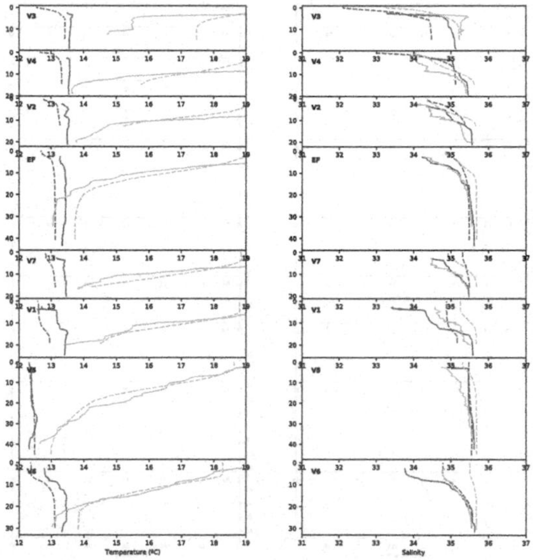

Fig. 7. Vertical profiles for temperature (left panel) and salinity (right panel) at the Ría de Vigo CTDs stations, for the days 2018/01/31 and 2018/07/10 corresponding to a winter and summer situation (blue and red lines, respectively). CTD profile (continuous line) and MOHID model profile (dash line). (Color figure online)

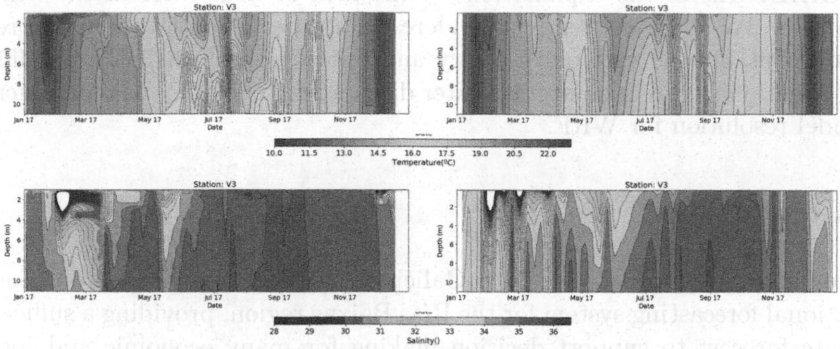

Fig. 8. Vertical profiles at V3 station along the year 2017. Weekly monitoring campaigns. CTD (left), MOHID model (right). Temperature (up), Salinity (down).

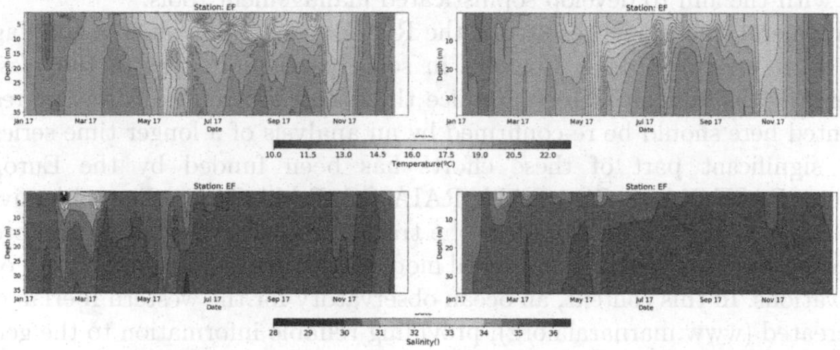

Fig. 9. Vertical profiles at EF station along the year 2017. Weekly monitoring campaigns. CTD (left), MOHID model (right). Temperature (up), Salinity (down).

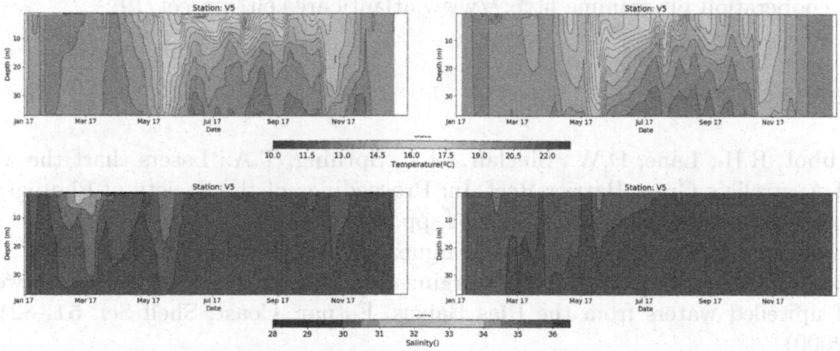

Fig. 10. Vertical profiles at V5 station along the year 2017. Weekly monitoring campaigns. CTD (left), MOHID model (right). Temperature (up), Salinity (down).

to MOHID. One of the explanations of the lack of freshwater in the MOHID domains is SWAT model tends to underestimate peak flows although it reproduces acceptably the variability of dry and wet seasons. Currently, an effort is being spent on a correction of freshwater discharges in SWAT and on an increase in model resolution for WRF.

4 Conclusions

For now more than a decade, MeteoGalicia has been employing an open-access operational forecasting system for the Rías Baixas region, providing a sufficiently accurate forecast to support decision making for many economic and societal activities. This ocean modelling system has been subject to continuous development and improvement, and agile validation tools have been built to visualise the system's behaviour. The main focus of the current efforts is to increase the integration between the operational prediction systems and ocean observation data, with the aim to develop sophisticated management tools.

In general, the model results for the Rías Baixas region show a good agreement with measured data. However in some particular areas of the *rías* the model still has difficulties to reproduce the observed profiles. Also, the results presented here should be re-confirmed by an analysis of a longer time series.

A significant part of these efforts has been funded by the European projects EASY/EASY-CO, RAIA/RAIA-CO/RAIA-TEC. The objective of those projects was the development of a trans-boundary observation and prediction infrastructure based on numerical models and an extensive network of ocean observations. In this context, an ocean observatory for the western Iberian coast was created (www.marnaraia.org), providing reliable information to the general public.

Acknowledgements. Part of this contribution has been funded by the European Union MyCOAST project: Coordinated Atlantic Coastal Operational Oceanographic Observatory (EAPA_285/2016), through INTERREG Atlantic Area European transnational cooperation programme http://www.atlanticarea.eu/project/19.

References

1. Abbot, R.H., Lane, D.W., Sinclair, M.J., Spruing, T.A.: Lasers chart the waters of Australia's Great Barrier Reef. In: Proceedings of the Society of Photographic Instrumentation Engineers, vol. 2964, pp. 72–90 (1996)
2. Alvarez-Salgado, X.A., Gago, J., Miguez, B.M., Gilcoto, M., Pérez, F.F.: Surface waters of the NW Iberian margin: upwelling on the shelf versus outwelling of upwelled waters from the Rías Baixas. Estuar. Coast. Shelf Sci. **51**, 821–837 (2000)
3. Blanton, J.O., Atkinson, L.P., de Castillejo, F.F., Montero, A.L.: Coastal upwelling off the Rias Bajas, Galicia, northwest Spain I: hydrographic studies. Rapp. PV Reun. Cons. Int. Explor. Mer **183**, 79–90 (1984)

4. Blanton, J.O., Tenore, K.R., Castillejo, F., Atkinson, L.P., Schwing, F.B., Lavin, A.: The relationship of upwelling to mussel production in the rias on the western coast of Spain. J. Mar. Res. **45**(2), 497–511 (1987)
5. Carracedo, P., Balseiro, C.F., Penabad, E., Gómez, B., Pérez-Muñuzuri, V.: One year validation of wave forecasting at Galician coast. J. Atmos. Ocean Sci. **10**(4), 407–419 (2005)
6. Costa, P., Gómez, B., Venâncio, A., Pérez, E., Muñuzuri, V.: Using the regional ocean modeling system (ROMS) to improve the sea surface temperature from MERCATOR ocean system. Adv. Span. Phys. Oceanogr. Sci. Mar. **76**, S1 (2012)
7. Fraga, F.: Upwelling off the Galician coast, Northwest Spain. In: Coastal Upwelling, pp. 176–182. American Geophysical Union, Washington DC (1981)
8. Fraga, F., Margalef, R.: Las rías galegas. In: Estudio y Explotación del Mar en Galicia (ed.) Edição Universidad de Santiago de Compostela, pp. 101–121 (1979). ISBN 9788-4719-113-77
9. Fernandes, R.M.: Modelação Operacional no estuário do Tejo. Dissertação para a obtenção do grau de mestre em Engenharia do Ambiente Gestão e Modelação dos Recursos Marinhos, Instituto Superior Técnico, Universidade Técnica de Lisboa, Lisboa, Portugal, p. 95 (2005)
10. Leitão, P.: Integração de Escalas e de Processos na Modelação do Ambiente Marinho. Dissertação para a obtenção do grau de Doutor em Engenharia do Ambiente. Instituto Superior Técnico, Universidade Técnica de Lisboa, Lisboa, p. 279 (2003)
11. Martins, F.: Modelação Matemática Tridimensional de Escoamentos Costeiros e Estuarinos usando uma Abordagem de Coordenada Vertical Genérica. Ph.D. thesis - Universidade Técnica de Lisboa, Instituto Superior Técnico (1999)
12. Martins, F., Leitão, P., Silva, A., Neves, R.: 3D modelling in the Sado estuary using a new generic vertical discretization approach. Oceanologica Acta **24**, 51–62 (2001)
13. McClain, C.R., Chao, S., Atkinson, L.P., Blanton, J.O., de Castillejo, F.: Wind-driven upwelling in the vicinity of Cape Finisterre, Spain. J. Geophys. Res. **91**(C7), 8470–8486 (1986)
14. Mendez, G., Vilas, F.: Geological antecedents of the Rias Baixas (Galicia, Northwest Iberian Peninsula). J. Mar. Syst. **54**(1–4), 195–207 (2005)
15. Otero, P., Ruiz-Villarreal, M., Peliz, A.: River plume fronts off NW Iberia from satellite observations and model data. ICES J. Mar. Sci.: Journal du Conseil **66**(9), 1853–1864 (2009)
16. Peliz, Á., Rosa, T.L., Santos, A.M.P., Pissarra, J.L.: Fronts, jets, and counter-flows in the Western Iberian upwelling system. J. Mar. Syst. **35**(1), 61–77 (2002)
17. Shchepetkin, A.F., McWilliams, J.C.: The regional ocean modeling system: a split-explicit, free-surface, topography following coordinates ocean model. Ocean Model. **9**, 347–404 (2005)
18. Srinivasan, R., Arnold, J.G.: Integration of a basin-scale water quality model with GIS1. J. Am. Water Resour. Assoc. **30**(3), 453–462 (1994)
19. Tilstone, G.H., Figueiras, F.G., Fraga, F.: Upwelling-downwelling sequences in the generation of red tides in a coastal upwelling system. Mar. Ecol. Progr. Ser. **112**, 241–253 (1994)
20. Wooster, W.S., Bakun, A., McLain, D.R.: The seasonal upwelling cycle along the eastern boundary of the North Atlantic. J. Mar. Res. **34**(2), 131–141 (1976)

Climate Evaluation of a High-Resolution Regional Model over the Canary Current Upwelling System

Ruben Vazquez[1(✉)], Ivan Parras-Berrocal[1], William Cabos[2], Dmitry V. Sein[3], Rafael Mañanes[1], Juan I. Perez[2], and Alfredo Izquierdo[1]

[1] University of Cadiz, 11510 Cadiz, Spain
{ruben.vazquez, ivan.parras, rafael.salinas, alfredo.izquierdo}@uca.es

[2] University of Alcala, 28801 Alcala de Henares, Spain
{william.cabos, nacho.perez}@uah.es

[3] Alfred Wegener Institute for Polar and Marine Research, 27570 Bremerhaven, Germany
dmitry.sein@awi.de

Abstract. Coastal upwelling systems are very important from the socio-economic point of view due to their high productivity, but they are also vulnerable under changing climate. The impact of climate change on the Canary Current Upwelling System (CCUS) has been studied in recent years by different authors. However, these studies show contradictory results on the question whether coastal upwelling will be more intense or weak in the next decades. One of the reasons for this uncertainty is the low resolution of climate models, making it difficult to properly resolve coastal zone processes. To solve this issue, we propose the use of a high-resolution regional climate coupled model. In this work we evaluate the performance of the regional climate coupled model ROM (REMO-OASIS-MPIOM) in the influence zone of the CCUS as a first step towards a regional climate change scenario downscaling. The results were compared to the output of the global Max Planck Institute Earth System Model (MPI-ESM) showing a significant improvement.

Keywords: ROM · Canary current · Regional climate modelling

1 Introduction

The Eastern Boundary Upwelling Systems (EBUSs) are highly productive coastal ocean areas where cold water upwells by the action of favourable winds [1]. The upwelling is associated to the along-shore Trade winds dominating these subtropical regions and causing Ekman transport from the coast to the open ocean. There are four EBUSs, being the Canary Current Upwelling System (CCUS) one of the most important fishery grounds in the world [2]. The Canary current is part of the North Atlantic subtropical gyre, extending from the northern tip of the Iberian Peninsula at (43°N) to the south of Senegal at about 10°N (Fig. 1).

J. M. F. Rodrigues et al. (Eds.): ICCS 2019, LNCS 11539, pp. 240–252, 2019.
https://doi.org/10.1007/978-3-030-22747-0_19

The CCUS has a seasonal variability related to that of the Trade winds and the Intertropical Convergence Zone (ITCZ) migration (Fig. 1). At the Iberian coast the CCUS presents a marked seasonal cycle with the upwelling season beginning in spring and extending through summer and early autumn, and downwelling events frequently observed during wind relaxations in winter [3]. The length of the upwelling season increases progressively as latitude decreases, with upwelling becoming mostly a year-round phenomenon at tropical–subtropical latitudes [4].

In the last decades many authors have studied the EBUEs due to their vulnerability under global warming conditions. Bakun [5] hypothesized that the increase in the ocean-land thermal gradient due to greenhouse warming would result in stronger along-shore winds intensifying the upwelling of deeper water to the surface.

Sydeman et al. [6], through a meta-analysis of the existing literature on upwelling-favourable wind intensification, revealed contradictory results between observational data and model-data reanalysis. Their results showed equivocal wind intensification in the Canary upwelling, in agreement with the analysis of Varela et al. [7], which also highlighted the importance of high resolution wind database to properly resolve conditions at the scale of coastal upwelling in intense and localized upwelling zones.

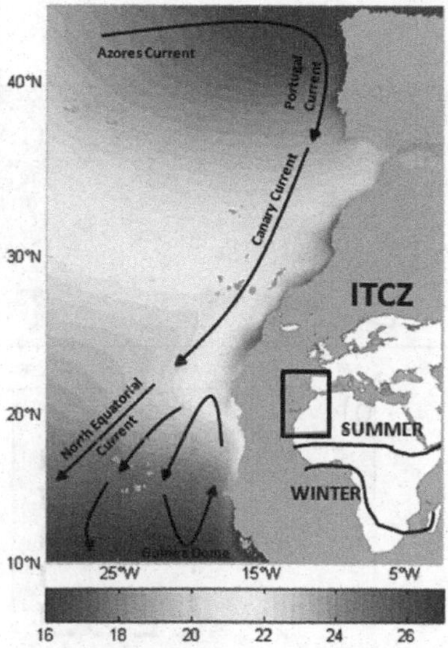

Fig. 1. Sea surface temperature (°C) and upper ocean circulation in the Canary current system. Also it is showed the seasonal migration of the ITCZ (based on Benazzouz et al. [8]).

Thus we present and validate the performance of a high-resolution regionally coupled atmosphere ocean model in the CCUS, so the advantages of this modelling approach can be assessed, and the uncertainty in the assessment of upwelling-favourable wind intensification at the CCUS under global warming reduced.

2 Models and Data Sets

The regionally coupled climate model ROM [9] comprises the REgional atmosphere MOdel (REMO), the Max Planck Institute Ocean Model (MPI-OM), the HAMburg Ocean Carbon Cycle (HAMOCC) model, the Hydrological Discharge (HD) model, the soil model [10] and dynamic/thermodynamic sea ice model [11] which are coupled via OASIS [12] coupler, and was called ROM by the initials REMO-OASIS-MPIOM. The regional downscaling allows the interaction of atmosphere and ocean in the region covered by REMO (Fig. 2a), while the rest of the global ocean is driven by energy, momentum and mass fluxes from global atmospheric data used as external forcing.

The oceanic component of ROM is the Max Planck Institute Ocean Model (MPI-OM) developed at the Max Planck Institute for Meteorology (Hamburg, Germany) [13, 14]. The MPI-OM configuration used for all experiments features the grid poles over North America and Northwestern Africa (Fig. 2a). The horizontal resolution ranges from 5 km (close to the NW African coast) to 100 km in the southern oceans (Fig. 2b). This feature allows a local high resolution in the region of interest allowing the study of local-scale processes while maintaining a global domain (e.g. Izquierdo and Mikolajewicz [15]).

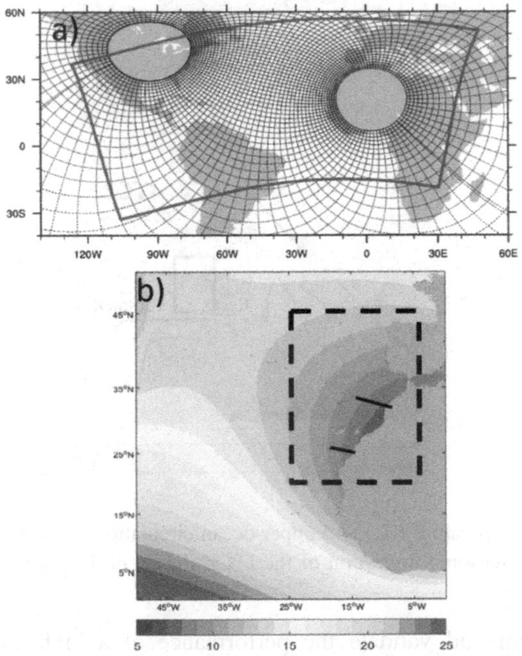

Fig. 2. (a) Atmosphere and ocean ROM grids. MPI-OM variable resolution grid (black lines, drawn every twelfth), REMO domain (red line). (b) MPI-OM grid resolution (km) in the Canary current system. Localization of the study zone (dashed black line) and of the 2 choosen transects (solid black lines). (Color figure online)

The atmospheric component of ROM is the REgional atmosphere MOdel (REMO) [16]. The dynamic core of the model as well as the discretization in space and time are based on the Europa-Model of the Germany Weather service [17]. The physical parameterizations are taken from the global climate model ECHAM versions 4 and 5 [18, 19]. To avoid the largely different extensions of the grid cells close to the poles, REMO uses a rotated grid, with the equator of the rotated system in the middle of the model domain with a constant resolution of 25 km [9].

ROM was compared with the Max Plank Institute for Meteorology – Earth System Model (MPI-ESM) to analyze the differences between a regional and global model. The MPI-ESM has been used in the context of the CMIP5 process (Coupled Models Intercomparison Project Phase 5) and consists of the coupled general circulation models for the atmosphere (ECHAM6) and the ocean (MPI-OM) and the subsystem models for land and vegetation JSBACH [20, 21] and for the marine biogeochemistry HAMOCC5 [22].

The MPI-ESM has been developed for a variety of configurations differing in the resolution of ECHAM6 or MPI-OM (MPI-ESM-LR, -MR). The low resolution (LR) configuration uses for the ocean a bipolar grid with 1.5° resolution and the medium resolution (MR) decreases the horizontal grid spacing of the ocean to 0.4° with a tripolar grid, two poles localized in Siberia and Canada and a third pole at the South Pole [23]. The experiments used for the evaluation and the analyses were the historical run and the RCP4.5 scenario.

Table 1. Observational data products used in the ROM validation.

Product	Description
OSTIA	Global high-resolution (6 km) SST from the Operational Sea Surface Temperature and Sea Ice Analysis (OSTIA) [24]. The temporal resolution is daily from April 2006 to the present
SCOW	Scatterometer Climatology of Ocean Winds (SCOW) based on 8 years (September 1999 to August 2007) of QuikSCAT data. It provides a monthly wind stress climatology with 1/4° resolution [25]
WOA18	The World Ocean Atlas 2018 (WOA18) is a set of objectively analyzed (1° grid) climatological fields of in situ temperature, salinity, dissolved oxygen, nutrients, … [26]
WOD	The World Ocean Database (WOD) includes in situ measurements of temperature, salinity, dissolved oxygen and nutrients from 1773 to the present [27]

Finally, the different datasets used in the present paper to evaluate and analyze ROM are described in the Table 1. OSTIA (the system uses data from a combination of infrared and microwave satellites as well as in situ data) appears to be one of the best options with respect to other available reanalysis data for mesoscale processes [28].

3 Results

3.1 Sea Surface Temperature

ROM Sea Surface Temperature (SST) was evaluated using OSTIA in the period 2008–2012. In addition, ROM output was compared to MPI-ESM-LR and MR (Fig. 3).

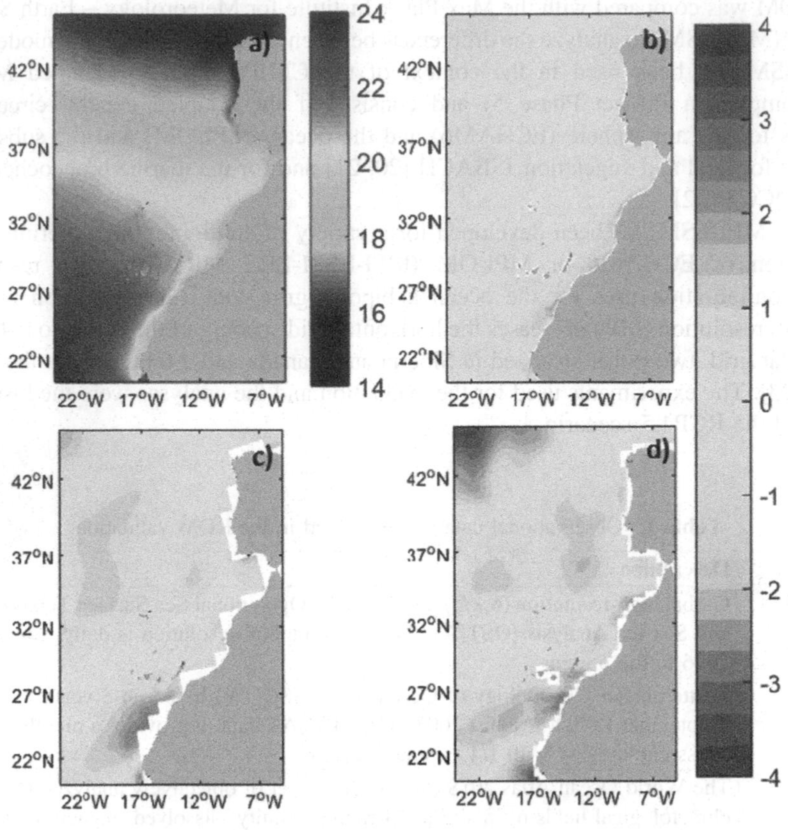

Fig. 3. Mean SST (°C): (a) OSTIA and differences with ROM (b), MPI-ESM-LR (c) and MPI-ESM-MR (d) for 2008–2012.

Mean SST 2008–2012

The time averaged SST in OSTIA (Fig. 3a) shows a clear meridional gradient with lower SST by the coast as a result of the upwelled waters.

Comparison to model SSTs shows that, although all three reproduce the general OSTIA SST pattern reasonably well, ROM clearly outperforms MPI-ESM, attaining smaller differences (Fig. 3b–d). This is more remarkably close to the coast, where ROM higher resolution plays a role.

It allows ROM to properly reproduce some smaller scale features in the field of SST lacking in the MPI-ESMs, notably along the Iberian coast and by the Strait of Gibraltar. However, we can observe a cold bias along the coast in ROM from the Strait of Gibraltar to Cape Blanc (from 35°N to 22°N). This is evident in Fig. 4a, plotting the SST of the grid-point closest to the coast from 20°N to 42°N for all four datasets.

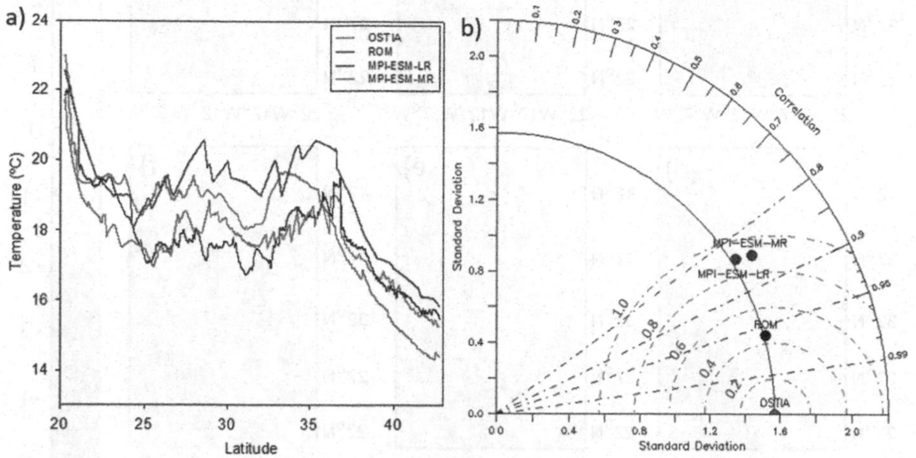

Fig. 4. (a) Meridional distribution of SST in the ocean grid-point closest to the coast. (b) Taylor Diagram for the CCUS region SST during 2008–2012 period. The diagram summarizes the relationship between standard deviation (°C), correlation (r) and RMSE (red lines, °C) among all datasets. (Color figure online)

The model performance in the CCUS was evaluated by means of a Taylor diagram (Fig. 4b) averaging SST over the box enclosed by 22°N to 45°N and 5°W to 25°W. The SST temporal variability, expressed by the standard deviation, is similar in all datasets, with a value around 1.6 °C. However, when comparing the models output to analysis data (OSTIA) ROM clearly improves both MPI-ESM configurations, showing a higher correlation coefficient (0.96 vs 0.85) and a lower RMSE (0.4 °C vs 1.0 °C). Despite differences in horizontal resolution, MPI-ESM-LR and MPI-ESM-MR show a very similar performance.

Seasonal Cycle

The CCUS experiences an important seasonal cycle in SST characterized by the winter (DJF) and summer (JJA) SSTs. Figure 5 shows SST differences between OSTIA and the models output (ROM, MPI-ESM-LR, and MPI-ESM-MR). ROM biases are lower than in any of the MPI-ESM configurations, and generally within ±1 °C, notably improving a JJA warm bias in the SW Iberian margin and a cold bias south of Canary Islands.

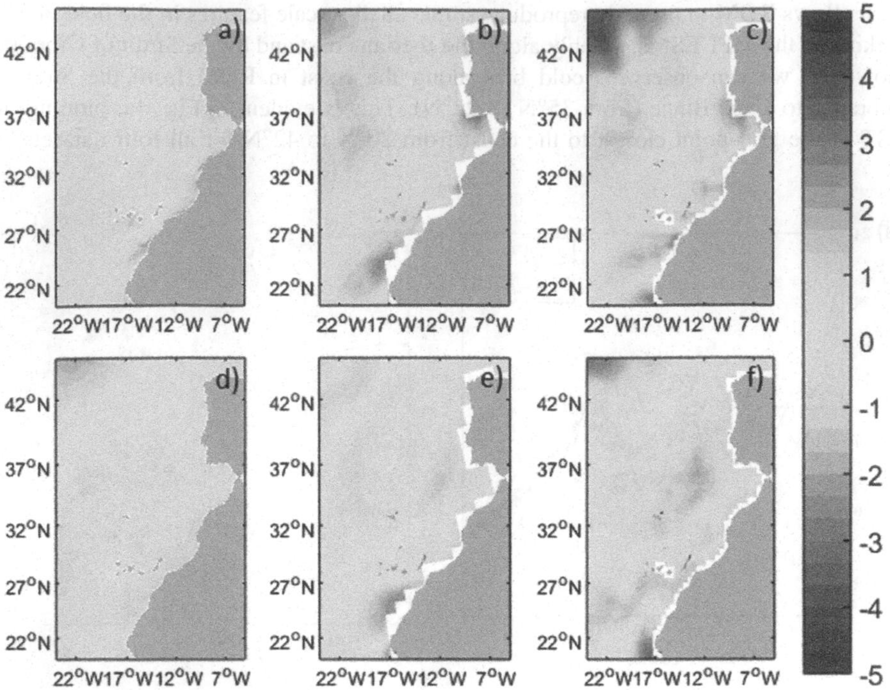

Fig. 5. OSTIA SST (°C) difference with ROM (a), MPI-ESM-LR (b) and MR (c) in summer (JJA); and ROM (d), MPI-ESM-LR (e) and MR in winter (DJF)

Table 2 shows the seasonal statistics comparing model SST values in the defined CCUS box with OSTIA analysis. All models perform better (according to correlation coefficient and RMSE) in winter. However even in summer ROM presents a high correlation of 0.93, while correlation for MPI-ESM-LR and MR drastically drops down to 0.69 and 0.72, respectively. Analogous situation takes place for RMSE, with ROM showing always better results than MPI-ESM configurations.

Table 2. Seasonal statistics comparing OSTIA and model SST

	Winter			Summer		
	r	RMSE (°C)	Sd (°C)	r	RMSE (°C)	Sd (°C)
ROM	0.98	0.38	1.60	0.93	0.56	1.49
MPI-ESM-MR	0.92	0.78	1.99	0.72	0.94	1.25
MPI-ESM-LR	0.91	0.78	1.87	0.69	0.97	1.20
OSTIA			1.79			1.26

3.2 Wind Stress

The ROM surface wind stress was evaluated with the Scatterometer Climatology of Ocean Winds (SCOW) comprising QuikSCAT observations from 1999 to 2007.

Figure 6 presents the time averaged (1999–2007) zonal and meridional components of wind stress corresponding to SCOW, ROM, MPI-ESM-LR and MPI-ESM-MR. The zonal component of wind stress shows a similar spatial pattern for all datasets, with a change of sign around 36°N, but ROM presents smaller biases than the MPI-ESM. SCOW and ROM fields of the meridional component of wind stress are very similar, both showing local maxima south of Cap Ghir and Cape Blanc. MPI-ESM-LR properly reproduces the general pattern, but not the smaller scale details. Remarkably, MPI-ESM-MR shows a different pattern, with northern winds predominating north 37°N.

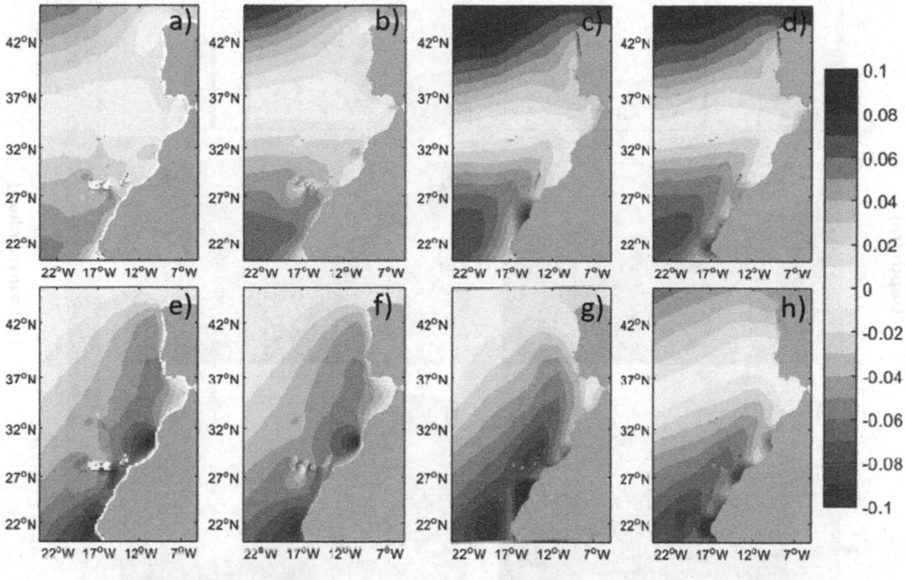

Fig. 6. Zonal component of wind stress (N/m^2) from SCOW (a), ROM (b), MPI-ESM-LR (c) and MPI-ESM-MR (d). Meridional component of the wind stress (N/m^2) from SCOW (e), ROM (f), MPI-ESM-LR (g) and MPI-ESM-MR (h).

3.3 Thermal Vertical Structure

The thermal vertical structure of the upper 200 m from ROM was compared with the climatology World Ocean Atlas 2018 (WOA18) [26] and with the historical in-situ data from the World Ocean Database (WOD) [27]. The analysis was realized through two across-shore transects at Cape Ghir and to the south of the Canary Islands (Cape Bojador). In the comparison we used 12 available CTD/XBT stations for Cape Ghir transect and 21 for Cape Bojador. All the observations were taken in summer periods between 1985 and 2004.

WOD observations show a clear thermal stratification at both transects (Figs. 7a and b) with temperatures ranging from 14 °C at 200 m depth to 24 °C at the surface, and somewhat colder at Cape Ghir than at Cape Bojador. ROM temperature transects properly reproduce this structure (Figs. 7c and d), but limiting the extension of colder waters (≈14 °C) to the coast, and showing a small warm bias at depth. Interestingly, when compared to direct observations WOA18 (Figs. 7e and f) shows a larger warm bias than ROM, which is more evident near the coast. This is a clear evidence that the higher horizontal resolution of ROM allows to reproduce smaller scale processes that are partly masked in the climatology.

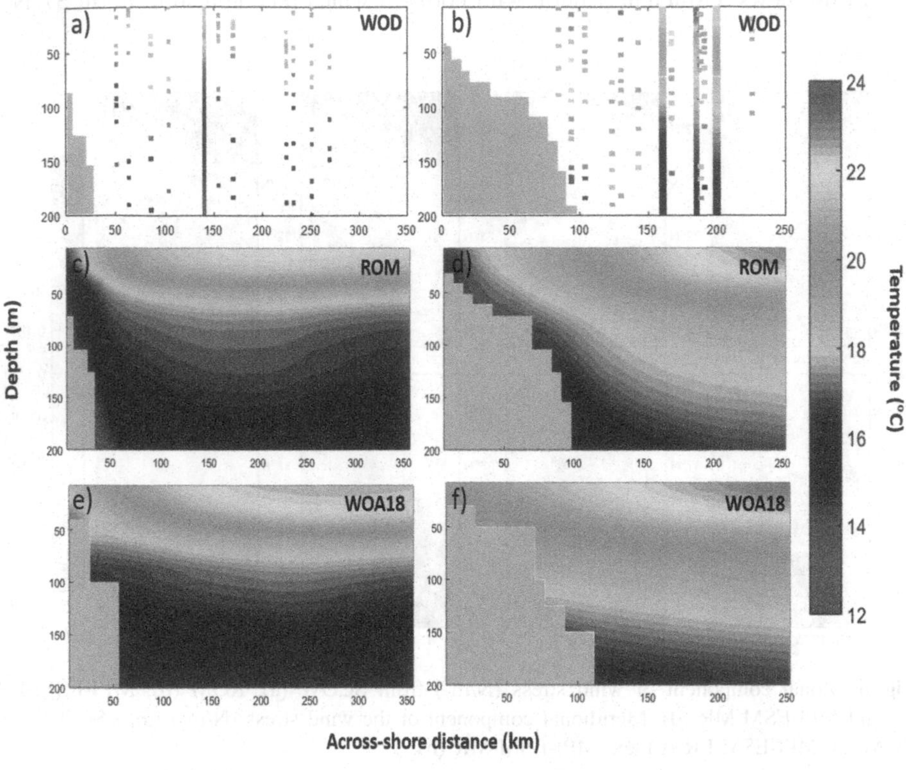

Fig. 7. Temperature (°C) transects at Cape Ghir (a, c, e) and Cape Bojador (b, d, f) for summers periods between 1985 and 2004. Transect referred to in Fig. 2.

To clearly show the differences near coast Fig. 8 plots the temperature profiles corresponding to the location of the nearest to coast observations from WOD. In the first vertical profile localized at Cape Ghir (Fig. 8a), we can observe a similar temperature profiles for the three datasets in the upper 100 m, deeper the temperature profiles start to diverge, being the warm bias larger in WOA18 than in ROM.

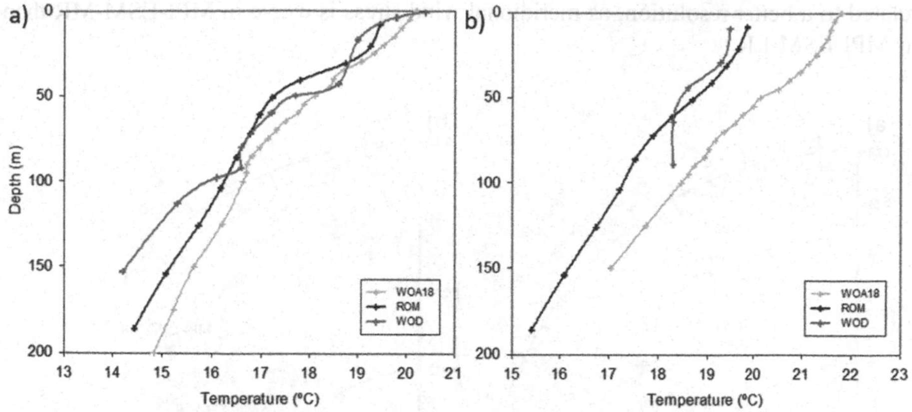

Fig. 8. Temperature (°C) profile at Cape Ghir (a) and Cape Bojador (b) in 1985–2004 summer period.

The WOD Cape Bojador profile (Fig. 9b) is much shallower, however at surface WOA18 is 2 °C warmer than WOD and ROM.

4 Discussion and Conclusions

In this study, the regional atmosphere-ocean model coupled was validated for the CCUS region. The ROM ocean outputs analyzed were SST, surface wind stress and thermal vertical structure.

ROM time mean and seasonal SST was validated against OSTIA data set, showing biases largely bounded to ± 1 °C, correlations coefficients above 0.9 and RMSE below 0.4 °C. All the statistics showed a better performance than MPI-ESM-LR and MPI-ESM-MR. Interestingly, between both MPI-ESM configurations there was almost no improvements, which is an indication that ROM is not only providing a better result due to its higher resolution, but also because it is able of better reproducing mesoscale coastal processes. ROM also presented cold biases along the North African coast stronger in summer periods. Li et al. [29] regional model for the California upwelling showed a similar cold bias when compared to satellite data. Mason et al. [30] reported a similar bias in their ROMS model for the Canary upwelling, and blamed the uncertainty in the nearshore model wind structure. However, that bias can also be a consequence of the analysis system used in OSTIA since it assumes that the observation errors are not locally biased. OSTIA corrects the observation errors in a global way [24], therefore in zones with intense mesoscale dynamics, like the coastal strip along the CCUS, it could generate biases.

Other important variable to evaluate in the CCUS is the surface wind stress. ROM surface wind stress was compared to SCOW, showing small differences. The biases found did not exceed 0.006 N/m^2. Taylor diagrams for SCOW and model wind stress components averaged over the CCUS box (Fig. 9) show the better performance of ROM as compared to MPI-ESM configurations and again, this improvement is not only related to a better resolution, as meridional wind stress is worse in MPI-ESM-MR than in MPI-ESM-LR.

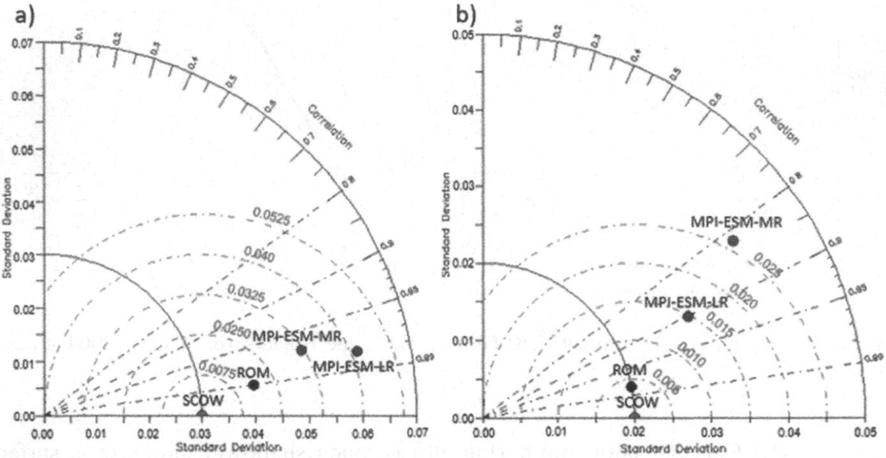

Fig. 9. Taylor diagrams for CCUS zonal (a) and meridional (b) surface wind stress components during 1999–2009 period.

ROM simulated wind stress was clearly better than MPI-ESM, pointing out to the need of regional downscaling to properly simulate the CCUS dynamics.

The analysis of the vertical thermal structure in two across-shore transects also showed that ROM is able to reproduce near coast temperature gradients, likely masked if the resolution is not very high.

In conclusion, ROM shows clear improvements in reproducing the surface wind and ocean temperature fields in the CCUS when compared to global climate models as MPI-ESM. The improvement is related to a much higher horizontal resolution in the atmosphere and in the ocean, which allows to better simulate the dominant mesoscale coastal dynamics at the CCUS. The results here give ground to the future use of ROM to have deeper insight into the state changes expected to happen at CCUS by the end of 21st century.

Acknowledgement. This work has been developed within the framework of the European Cooperation Project Interreg VA España-Portugal "OCASO" (Southwest Coastal Environmental Observatory,0223_OCASO_5_E).

References

1. Tim, N., Zorita, E., Hünicke, B., Yi, X., Emeis, K.-C.: The importance of external climate forcing for the variability and trends of coastal upwelling in past and future climate. Ocean Sci. **12**, 807–823 (2016). https://doi.org/10.5194/os-12-807-2016
2. Hagen, E., Feistel, R., Agenbag, J.J., Ohde, T.: Seasonal and interannual changes in intense Benguela upwelling (1982–1999). Oceanol. Acta **24**(6), 557–568 (2001). https://doi.org/10.1016/S0399-1784(01)01173-2

3. Cordeiro, N., Dubert, J., Nolasco, R., Barton, E.D.: Transient response of the Northwestern Iberian upwelling regime PLoS ONE **13**(5) (2018). https://doi.org/10.1371/journal.pone.0197627
4. Wang, D.W., Gouhier, T.C., Menge, B.A., Ganguly, A.R.: Intensification and spatial homogenization of coastal upwelling under climate change. Nature **518** (2015). https://doi.org/10.1038/nature14235
5. Bakun, A.: Global climate change and intensification of coastal ocean upwelling. Science **247**, 198–201 (1990). https://doi.org/10.1126/science.247.4939.198
6. Sydeman, W.J., et al.: Climate change and wind intensification in coastal upwelling ecosystems. Science **345**, 77–80 (2014). https://doi.org/10.1126/science.1251635
7. Varela, R., Alvarez, I., Santos, F., deCastro, M., Gomez-Gesteira, M.: Has upwelling strengthened along worldwide coasts over 1982–2010? Sci. Rep. **5** (2015). https://doi.org/10.1038/srep10016
8. Benazzouz, A., et al.: An improved coastal upwelling index from sea surface temperature using satellite-based approach – the case of the Canary current upwelling system. Cont. Shelf Res. **81**, 38–54 (2014). https://doi.org/10.1016/j.csr.2014.03.012
9. Sein, D.V., et al.: Regionally coupled atmosphere-ocean-sea icemarine biogeochemistry model ROM: 1. Description and validation. J. Adv. Model. Earth Syst. **7**, 268–304 (2015). https://doi.org/10.1002/2014ms000357
10. Rechid, D., Jacob, D.: Influence of monthly varying vegetation on the simulated climate in Europe. Meteorol. Z. **15**, 99–116 (2006). https://doi.org/10.1127/0941-2948/2006/0091
11. Hibler III, W.D.: A dynamic thermodynamic sea ice model. J. Phys. Oceanogr. **9**, 815–846 (1979). https://doi.org/10.1175/1520-0485(1979)009<0815:adtsim>2.0.co;2
12. Valcke, S.: The OASIS3 coupler: a European climate modelling community software. Geosci. Model Dev. **6**, 373–388 (2013). https://doi.org/10.5194/gmd-6-373-2013
13. Marsland, S.J., Haak, H., Jungclaus, J.H., Latif, M., Roeske, F.: The Max-Planck - Institute global ocean/sea ice model with orthogonal curvilinear coordinates. Ocean Model. **5**(2), 91–127 (2003). https://doi.org/10.1016/S1463-5003(02)00015-X
14. Jungclaus, J.H., et al.: Characteristics of the ocean simulations in MPIOM, the ocean component of the MPI-Earth system model. J. Adv. Model Earth Syst. **5**, 422–446 (2013). https://doi.org/10.1002/jame.20023
15. Izquierdo, A., Mikolajewicz, U.: The role of tides in the spreading of mediterranean outflow waters along the southwestern Iberian margin. Ocean Model. **133**, 27–43 (2019). https://doi.org/10.1016/j.ocemod.2018.08.003
16. Jacob, D.: A note to the simulation of the annual and interannual variability of the water budget over the Baltic Sea drainage basin. Meteorol. Atmos. Phys. **77**(1–4), 61–73 (2001). https://doi.org/10.1007/s007030170017
17. Majewski, D.: The Europa modell of the Deutscher Wetterdienst. In: Seminar Proceedings ECMWF, vols. 2, 5, pp. 147–191. ECMWF, Reading (1991)
18. Roeckner, E., et al.: The atmospheric general circulation model ECHAM-4: model description and simulation of present-day-climate. Report 218, MPI für Meteorol., Hamburg, Germany (1996)
19. Roeckner, E., et al.: The atmospheric general circulation model ECHAM 5. PART I: model description. Report 349, MPI für Meteorol., Hamburg, Germany (2003)
20. Reick, C.H., Raddatz, T., Brovkin, V., Gayler, V.: The representation of natural and anthropogenic land cover change in MPIESM. J. Adv. Model. Earth Syst. **5**, 1–24 (2013). https://doi.org/10.1002/jame.20022
21. Schneck, R., Reick, C.H., Raddatz, T.: The land contribution to natural CO2 variability on time scales of centuries. J. Adv. Model. Earth Syst. **5**, 354–365 (2013). https://doi.org/10.1002/jame.20029

22. Ilyina, T., Six, K., Segschneider, J., Maier-Reimer, E., Li, H., Nunez-Riboni, I.: Global ocean biogeochemistry model HAMOCC: model architecture and performance as component of the MPI-Earth system model in different CMIP5 experimental realizations. J. Adv. Model. Earth Syst. (2013). https://doi.org/10.1029/2012ms000178
23. Giorgetta, M.A., et al.: Climate and carbon cycle changes from 1850 to 2100 in MPI-ESM simulations for the coupled model intercomparison project phase 5. J. Adv. Model. Earth Syst. **5**, 572–597 (2013). https://doi.org/10.1002/jame.20038
24. Stark, J.D., Donlon, C.J., Martin, M.J., McCulloch, M.E.: OSTIA: an operational, high resolution, real time, global sea surface temperature analysis system. In: OCEANS 2007—Europe, pp. 1–4 (2007). https://doi.org/10.1109/oceanse.2007.4302251
25. Risien, C.M., Chelton, D.B.: A global climatology of surface wind and wind stress fields from eight years of QuikSCAT Scatterometer data. J. Phys. Oceanogr. **38**, 2379–2413 (2008). https://doi.org/10.1175/2008JPO3881.1
26. Locarnini, R.A., et al.: World Ocean Atlas 2018: Temperature. A. Mishonov Technical Ed. vol. 1 (2018, in preparation)
27. Boyer, T.P., et al.: World Ocean Database 2018 (2018, in preparation)
28. Desbiolles, F., et al.: Upscaling impact of wind/sea surface temperature mesoscale interactions on Southern Africa austral summer climate. Int. J. Climatol. **38**(12), 4651–4660 (2018). https://doi.org/10.1002/joc.5726
29. Li, H., Kanamitsu, M., Hong, S.Y.: California reanalysis downscaling at 10 km using an ocean-atmosphere coupled regional model system. J. Geophys. Res. Atmos. **117**, D12 (2012). https://doi.org/10.1029/2011jd017372
30. Mason, E., et al.: Seasonal variability of the Canary current: a numerical study. J. Geophys. Res. Oceans **116**(C6) (2011). https://doi.org/10.1029/2010jc006665

Validating Ocean General Circulation Models via Lagrangian Particle Simulation and Data from Drifting Buoys

Karan Bedi[1], David Gómez-Ullate[2,3](✉), Alfredo Izquierdo[4], and Tomás Fernández Montblanc[5]

[1] Department of Mathematics, Indian Institute of Technology Roorkee, Roorkee, India
[2] Higher School of Engineering, University of Cádiz, Puerto Real, Spain
david.gomezullate@uca.es
[3] Departamento de Física Teórica, Universidad Complutense de Madrid, Madrid, Spain
[4] Applied Physics Department, University of Cádiz, 11510 Puerto Real, Spain
[5] Faculty of Marine Science (CACYTMAR), University of Cádiz, Puerto Real, Spain

Abstract. Drifting Fish Aggregating Devices (dFADs) are small drifting platforms with an attached solar powered buoy that report their position with daily frequency via GPS. We use data of 9,440 drifting objects provided by a buoys manufacturing company, to test the predictions of surface current velocity provided by two of the main models: the NEMO model used by Copernicus Marine Environment Monitoring Service (CMEMS) and the HYCOM model used by the Global Ocean Forecast System (GOFS).

Keywords: Copernicus CMEMS · HYCOM model · Lagrangian trajectory simulation · Fish Aggregating Devices

1 Introduction

Drifting Fish Aggregating Devices (dFADs) are man-made objects that consist of a bamboo raft covered by old pieces of purse seine net. Throughout the 2000s, several technological improvements were added to dFADs, including the use of GPS buoys to more accurately locate dFADs and logs, and the introduction of echosounder buoys to monitor the amount of biomass aggregated under them [1]. While the reasons why the fish choose to aggregate below these devices are still under debate [2], there is no doubt that they indeed perform their task, as evidenced by the large scale deployment of dFADs on the ocean, with thousands of new buoys deployed every year.

The main purpose of these devices is naturally to help fishing companies in their task. The buoys can be equipped with other sensors and technologies that allow them to provide useful data for other scientific purposes as a by product.

J. M. F. Rodrigues et al. (Eds.): ICCS 2019, LNCS 11539, pp. 253–264, 2019.
https://doi.org/10.1007/978-3-030-22747-0_20

Data collected from dFADs have given rise to a considerable number of studies, mostly concentrating on the ecological aspects of fisheries management, and also on the echosounder technology and the reliability of their reported measures. For a good review on the subject, see [3] and the references therein.

The purpose of this paper is to highlight the relevance of data gathered from dFADs for other scientific studies. Buoys equipped with an accelerometer can be useful to measure wave motion, their echosounder measures concentration of fish on a certain volume below them, and sensors for salinity, temperature, etc. can provide useful data to feed and validate forecasting models.

In this paper we will only use the daily position of the buoys, registered through a daily GPS report, to draw some information about the ocean currents at those points. In particular, since we have data from ca. 10,000 buoys and each of them has an average lifetime of ca. 100 days, we have ca. 1M daily displacements, which we will test against the prediction of daily displacements obtained by Lagrangian integration of virtual particles propagating with the velocity fields provided by two OGCM models: the NEMO model used by Copernicus Marine Environment Monitoring Service (CMEMS) and the HYCOM model used by the Global Ocean Forecast System (GOFS).

1.1 Ocean General Circulation Models (OGCMs)

For the purpose of this study, we have used the global ocean analyses from two state-of-the-art global ocean monitoring and forecasting systems: the Global Ocean Forecast System (GOFS) and Copernicus Marine Environment Monitoring Service (CMEMS).

GOFS 3.1 analysis is a product developed by the Naval Research Laboratory (NRL), with global coverage. It uses the HYCOM model [4] with 41 vertical levels and uniform 1/12° resolution between 80.48°S and 80.4°N. It runs daily assimilating observed data through NCODA (Navy Coupled Ocean Data Assimilation [5]) system. The ocean analysis expands from July 2014 until present, with the 3D fields available every 3 h.

GLOBAL_ANALYSIS_FORECAST_PHY_001_024 is developed by CMEMS, providing daily mean ocean physical fields with a global coverage at 1/12° resolution. It uses version 3.1 of NEMO ocean model [6] with 50 vertical levels. It runs daily assimilating altimetry data, vertical profiles of salinity and temperature and satellite sea surface temperature (SST). CMEMS global ocean analysis expands from January 2016 until present.

Both of the models cover the region and time span for which we have data of real displacements, allowing us to compare the predictions based on different analyses. For simplicity, in the rest of this paper we will refer to the models as HYCOM and Copernicus.

1.2 Drifting Fish Aggregating Devices (FADs)

Drifting Fish Aggregating Devices (dFADs) are synthetic floating devices specially designed to attract ocean going pelagic fish such as tuna, marlin, etc. They exploit the fact that many fish species naturally assemble near floating objects. dFADs usually consist of a floating raft, synthetic netting submerged in water, and a smart buoy equipped with echo sounder and GPS systems. dFADs carry a heavy suspended weight in order to provide stability. They estimate the fish density beneath it and transmit this information along with its position to satellites.

1.3 Lagrangian Particle Simulation

The validation of OGCMs by comparison with drifting buoys was conducted using a Lagrangian model, allowing us to calculate the predicted trajectories from the velocity field provided by the OGCM. The Lagrangian ocean analysis model Parcels v2.0.0beta was used to simulate the trajectories of the dFADs. Parcels is an open source Python based code that simulates the advection and dispersion of passive particles given a 3D hydrodynamic velocity field. The code enables a potential development in order to account for flow-particle interactions and/or more complex particle behaviour (see further details and model description in [7, 8]). For a recent review of Lagrangian Ocean modelling, the reader is advised to see [9] and the references therein.

A careful treatment of the physical properties of drifting dFADs needs to be performed. The devices have a suspended weight to stabilize them, that usually lies 40–60 m below the surface, so it is unclear how to model their drifting motion: what should be their drag coefficient, at which depth are they being driven, etc. This will be the object of further study. For the preliminary analysis reported in this paper, we chose to model dFADs as virtual particles (which are just driven by the current without interaction) and we have selected current velocity fields corresponding to the surface layer. Numerical tests performed at different depths suggest that surface layer is the best choice to describe their driving force.

2 Simulations and Analysis

2.1 Exploratory Data Analysis

Describing the Dataset. Our dataset includes the daily positions of 9,440 buoys, corresponding to the period ranging from Jan 01 2016 to Nov 30 2016. It consists of 925,187 entries, each entry having a Buoy Id (a unique identifier), Latitude-Longitude (co-ordinates for the position of the Buoy) and a timestamp (date-time when the position was recorded).

Data Cleaning. After eliminating entries with missing data, upon closer study it becomes evident that some of the daily displacements cannot possibly be due to drifting buoys, as these displacements are larger than 200 km in a single day. These anomalous displacements are likely due to the fact that the buoy was being carried by a fishing boat. A filter is applied to eliminate from the study any displacement greater than 120 km in a single day. The resulting trajectories are broken into smaller trajectories if such displacements occur in the middle of the original trajectory (see Fig. 2). Only the new trajectories having 5 or more data-points after filtering are considered for further analysis. This procedure increases the number of trajectories from 9,440 to 29,241, while it decreases the total number of observations from 925,187 to 810,165.

This filtering process to discard non-drifting buoys is a first approach that probably needs to be refined: while all discarded displacements are probably correct, there remains the chance that some buoys carried by boats are still present in the analysis. A finer analysis will be performed, either by considering the whole time series for fixing the criterion, or by considering external information (e.g. echo sound signal) provided by the company.

A histogram with the number of observations for each buoy before and after the filtering process can be seen in Fig. 2, while one trajectory with anomalous jumps has been shown for illustrative purposes in Fig. 1.

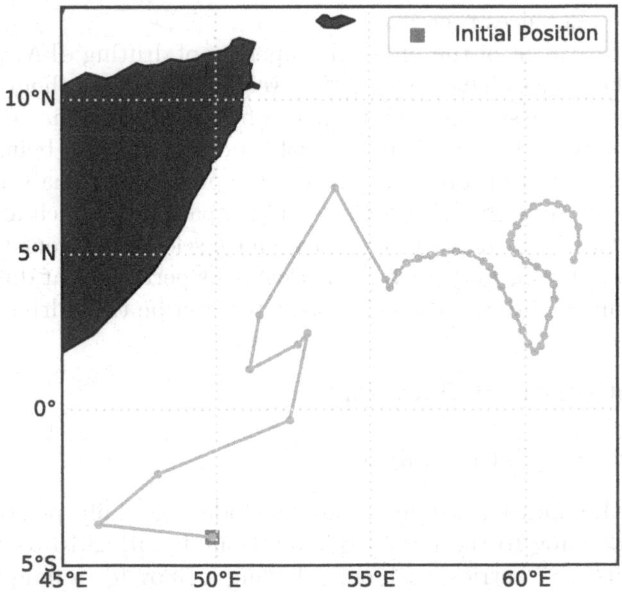

Fig. 1. Trajectory fo 49 days of a buoy with anomalous daily displacements, probably moved by external means during the first 9 days, and drifting for the rest of the days.

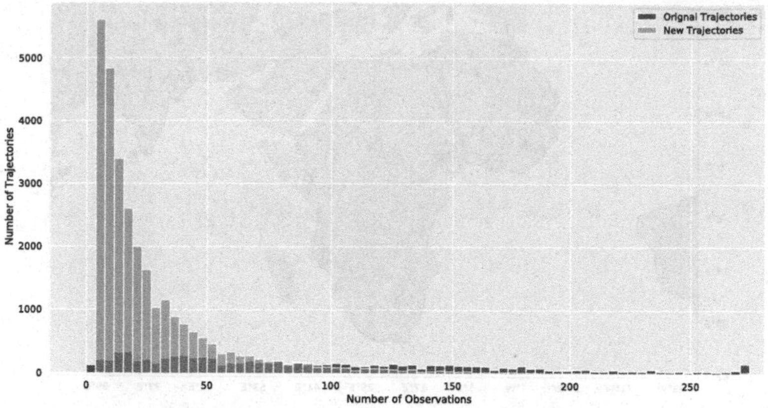

Fig. 2. Histogram of number of observations for new trajectories

Data Exploration. The initial position of the buoys are spread across both the Atlantic and Indian Ocean. The initial positions of the original 9,440 can be seen in Fig. 4, while the initial positions of the 29,241 trajectories obtained after the filtering process explained above, are depicted in Fig. 5. Notice in Fig. 4 that there is a gap with very few initial positions of buoys off the coast of Somalia, which is probably due to security reasons, in order to avoid pirate ships. The geographical distribution of the 810,165 daily displacements can be seen in Fig. 3. The plot corresponds to 180° longitude degrees and has been divided in 300 hexagonal units, so each unit spans an area of roughly 0.6°, i.e. a circle of radius approximately 30 km. The plot shows clearly that the region around the equator in the Indian ocean has a higher concentration of data points.

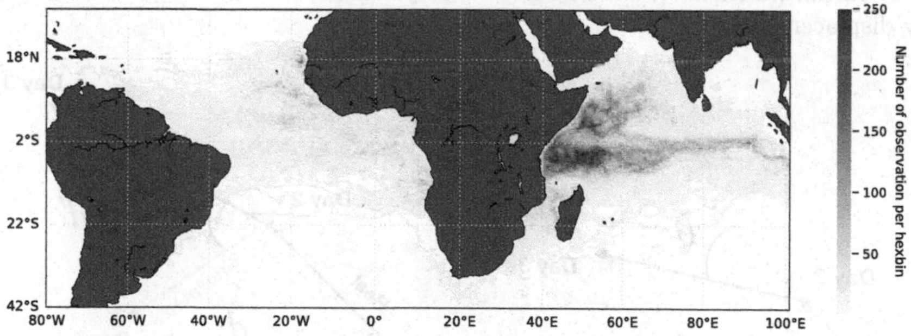

Fig. 3. Geographical distribution of the 810,165 data-points (daily positions) in the dataset

As part of the exploratory analysis, we have studied the distribution of daily displacements and daily changes in direction from the real dataset. The daily displacements of the buoys have a mean value of 28.63 km (which corresponds

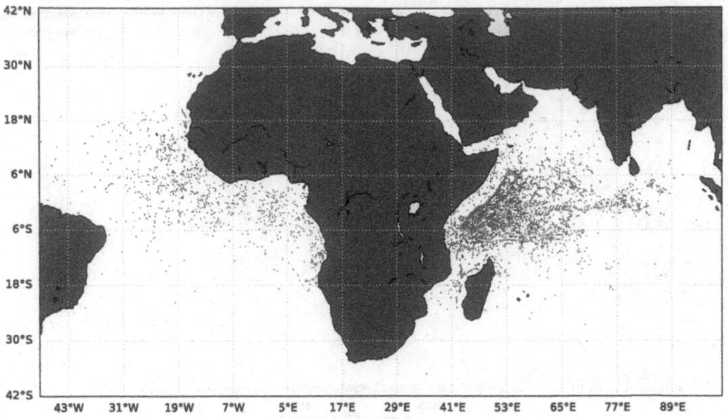

Fig. 4. Initial positions of the 9,440 original trajectories

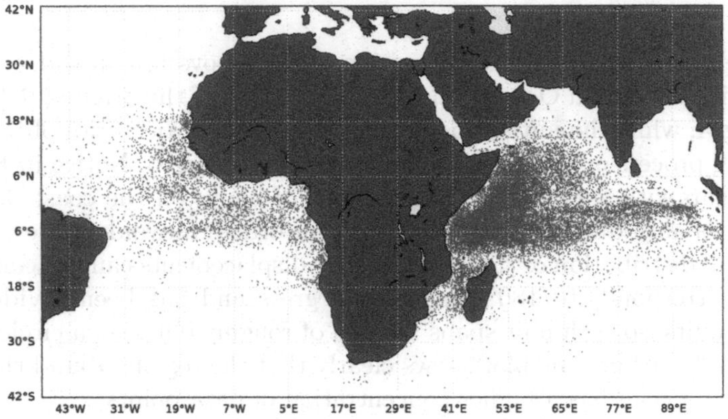

Fig. 5. Initial positions of the 29,241 new trajectories, after filtering our anomalous daily displacements

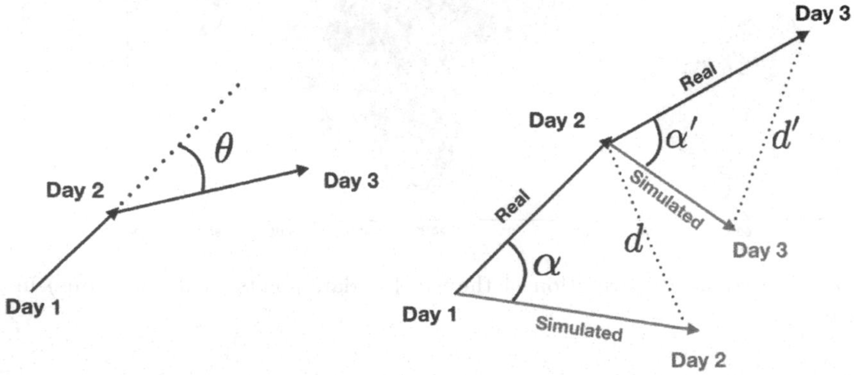

Fig. 6. Left: Angle θ between two consecutive daily displacements. Right: Angle α and distance d between the real and simulated daily displacements.

to a mean current speed of 0.32 m/s) with a standard deviation of 20.88 km, and the complete distribution can be seen in Fig. 7. For convenience, we have plotted alongside the distribution of daily real displacements, also the daily displacements predicted by Lagrangian particle simulation with velocity fields taken from the HYCOM and Copernicus models. Their geographical distribution is shown in Fig. 8, where we show at each cell the average of the daily displacements with origin in that cell (see Fig. 3). It is clear that larger daily displacements concentrate on the areas corresponding to the most intense current systems, namely the Western Boundary Currents and the Equatorial Current systems, where mean kinetic energy values are larger. Local maxima appear corresponding to the North Brazil Current in the tropical Atlantic, at the Agulhas Current around the southern tip of Africa and at the Somalia Current, Madagascar Current system and Equatorial Current system in the Indian Ocean.

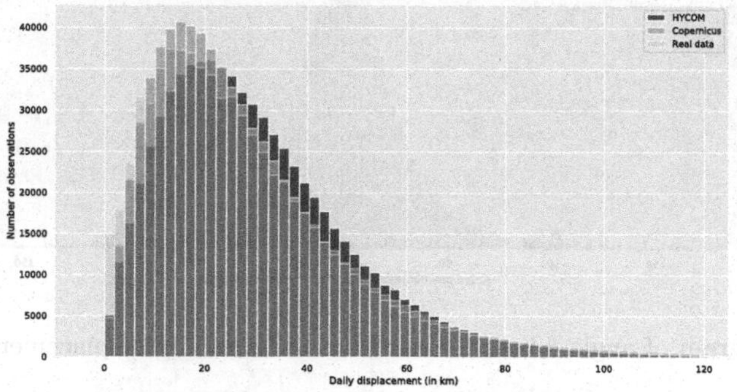

Fig. 7. Histogram of real and simulated daily displacements

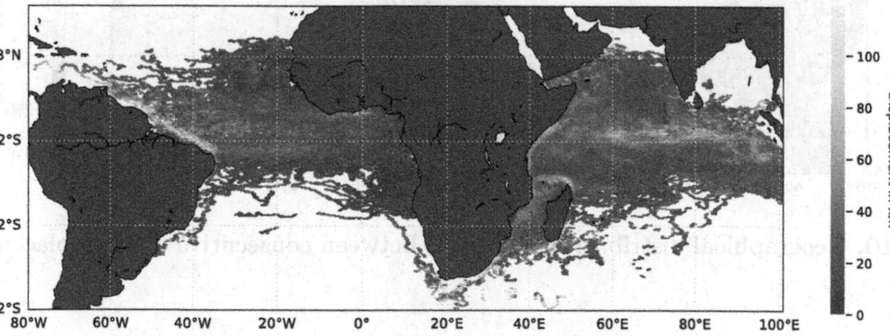

Fig. 8. Geographical distribution of daily displacements

Similarly, we have studied the daily change in direction, i.e. the angle θ between two consecutive daily displacement vectors, as shown in the left panel of Fig. 6. The distribution of such angles is clearly centered around zero (has mean value $-0.8°$) with standard deviation 42°. Here, the geographical distribution of such angles is also quite informative (see Fig. 10). We see a higher concentration of negative angle shifts (clockwise rotation) above the equator, and positive angle shifts (counter-clockwise rotation) below the equator, due to the Coriolis effect (Fig. 9).

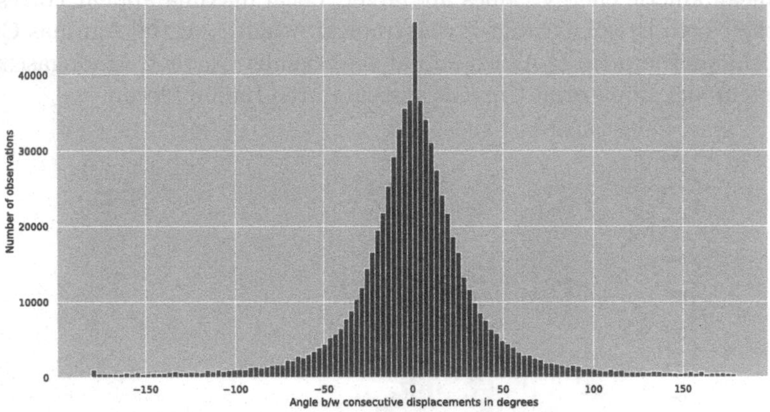

Fig. 9. Histogram of angle θ between two consecutive daily displacements in degrees

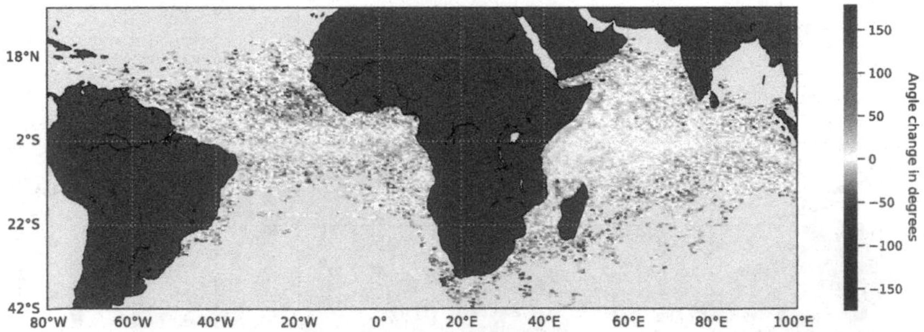

Fig. 10. Geographical distribution of angle θ between consecutive daily displacements

2.2 Simulation

In order to evaluate the OGCMs performance, the Lagrangian model simulations were performed considering the dFADs as passive particles, i.e., their displacements are dictated purely by the ocean's velocity field and their presence does not affect the field itself. The Lagrangian model runs were conducted accounting only for the advection using as input the horizontal velocity fields at the ocean's surface provided by GOFS3.1 and CMEMS analyses and using a fourth-order Runge-Kutta integration. Other processes affecting the floating object motion in the ocean, such as flow-object interaction, wind drift, turbulence and wave motion are neglected in the model runs. Particle trajectories were calculated using a time step of 300 seconds in order to avoid particle displacements larger than OGCMs spatial resolution. The dFADs trajectories were simulated using composite of successive 24 h runs. After each 24 h run, the simulated position was compared to the dFADs data real position, model skill metrics were calculated, the particle position was set to the real position and the next 24 h period was simulated again. In this manner, we have 810,165 predictions for simulated daily displacements, to be tested against the real measured displacements coming from the GPS positions of the dFADs. As mentioned in the Introduction, a more thorough modelling approach is needed to study the physics of the transport process: treating the dFADs as virtual particles and propagating them with the surface current being just one possible choice to serve as baseline for this model. A more elaborate understanding of the physical properties, together with an optimal choice of external parameters will probably improve the predictive results.

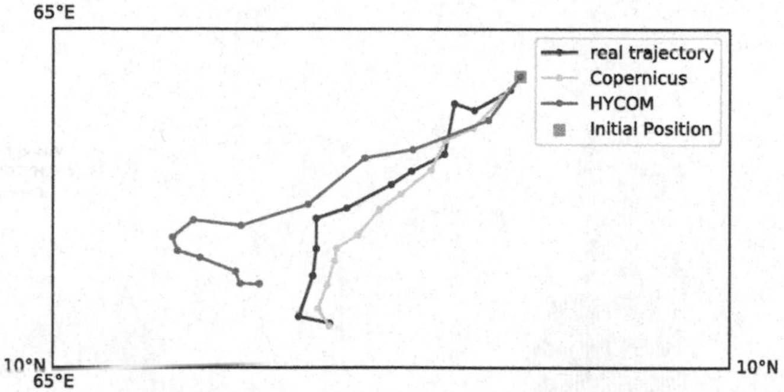

Fig. 11. Real and simulated trajectories of a Buoy for 14 days

A cherry-picked example of real and simulated trajectories can be seen in Fig. 11. It is important to stress that we are hindcasting the simulation, i.e. the velocity fields for each model have been taken from the corresponding database as the prediction for each time snapshot. Thus we compute the trajectories that the virtual particles would have taken according to HYCOM and Copernicus predictions at each time, and compare them with the real trajectories that actually took place. There are no forecasting predictions of trajectories or velocity fields in this analysis.

3 Results

We compare the predicted and real displacements with two metrics (see left panel of Fig. 6):

1. the distance d between the predicted and real final positions at the end of each day,
2. the angle α between the real and simulated daily prediction vectors.

After performing the Lagrangian simulation, we compute these magnitudes for each of the 810,165 points in our dataset. The results can be seen in Fig. 12 for the distance d and Fig. 13 for the angle α, while the main results that show the mean and standard deviations of both distributions are summarized in Table 1.

Table 1. Mean and standard deviations of the distributions depicted in Figs. 12 and 13

	α		d	
	mean	std	mean	std
HYCOM	0.4°	66°	22	14
Copernicus	−0.5°	**58°**	**17**	**12**

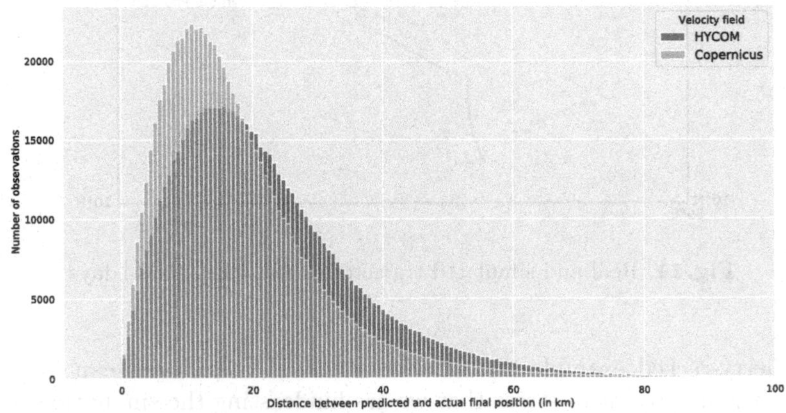

Fig. 12. Histogram of distance between predicted and actual final positions

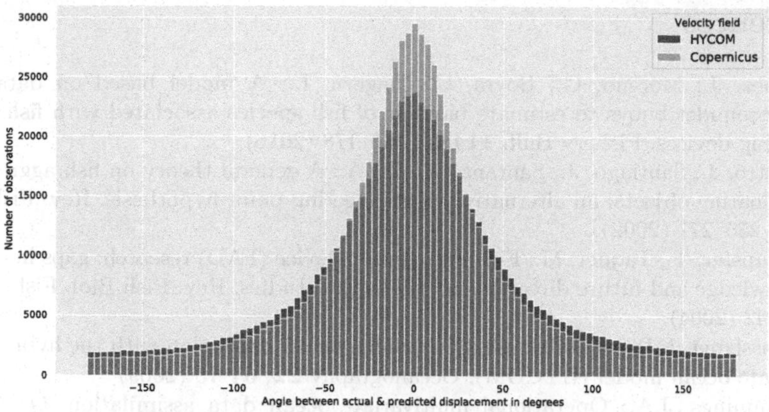

Fig. 13. Histogram of angle change in degrees

4 Conclusion

We see from the previous analysis that, all other things being equal, the daily displacement predictions based on the NEMO model by Copernicus CMEMS are slightly better (roughly 20% better) than the predictions obtained by simulating trajectories with the velocity fields predicted by the HYCOM model used by GOFS.

Large scale properties show a reasonable agreement between real and simulated data, such as the overall displacement distribution (Fig. 7). The surface circulation patterns of the tropical Indian and Atlantic ocean are also properly reflected in the dFAD data.

This being said, it is perhaps noteworthy that the predicted direction after a day of drifting motion on the ocean misses the real direction on average by 60°, which seems a rather high figure considering that we are only hindcasting rather than forecasting. Likewise, and related with this, the expected distance between the predicted and real positions after one day of drifting in the ocean is (in the best case) 17 km, almost half the expected daily displacement.

A very important remark, as a disclaimer, is that all results reported here are not a direct measure of both OGCM models, but rather an indirect measure via a specific choice of Lagrangian integration model. As it has been stressed, further study is needed to establish the best simulation model, most probably taking into account drag effects, interaction with the fluid, other numerical integration schemes, etc. However, perhaps the lesson to be learnt is that one should be very cautious about interpreting the uncertainty involved in the predictions made by these models, even if they are nowcasting. We would like to highlight also the importance of running this type of independent tests, and the relevance of dFAD data as a valuable source of information for scientific research. Given the enormous amount of dFADs, the possibilities of incorporating additional sensors, and the near-real time communication protocols there is a huge potential for their use in monitoring the tropical oceans.

References

1. Lopez, J., Moreno, G., Boyra, G., Dagorn, L.: A model based on data from echosounder buoys to estimate biomass of fish species associated with fish aggregating devices. Fishery Bull. **114**(2), 166–178 (2016)
2. Castro, J., Santiago, J., Santana-Ortega, A.: A general theory on fish aggregation to floating objects: an alternative to the meeting point hypothesis. Rev. Fish Biol. **11**, 255–277 (2002)
3. Dempster, T., Taquet, M.: Fish aggregation device (FAD) research: gaps in current knowledge and future directions for ecological studies. Rev. Fish Biol. Fish. **14**(1), 21–42 (2004)
4. Chassignet, E.P., et al.: US GODAE global ocean prediction with the hybrid coordinate ocean model (HYCOM). Oceanography **22**, 64–75 (2009)
5. Cummings, J.A.: Operational multivariate ocean data assimilation. Q. J. Roy. Meteorol. Soc. **131**, 3583–3604 (2005)
6. Madec, G., the NEMO team: NEMO ocean engine, Note du Pôle de modélisation 27, Institut Pierre-Simon Laplace (IPSL), France (2008). ISSN 1288-1619
7. Lange, M., Van Sebille, E.:Parcels v0.9: prototyping a Lagrangian ocean analysis framework for the petascale age. Geosci. Model Dev. (2017). https://doi.org/10.5194/gmd-10-4175-2017
8. Delandmeter, P., Van Sebille, E.: The parcels v2.0 Lagrangian framework: new field interpolation schemes. Geosci. Model Dev. Discuss. (2019). https://doi.org/10.5194/gmd-2018-339
9. Van Sebille, E., et al.: Lagrangian ocean analysis: fundamentals and practices. Ocean Model. **121**, 49–75 (2018)
10. Fonteneau, A., Chassot, E., Bodin, N.: Global spatio-temporal patterns in tropical tuna purse seine fisheries on drifting fish aggregating devices (DFADs): taking a historical perspective to inform current challenges. Aquat. Living Resour. **26**(1), 37–48 (2013)
11. Trygonis, V., Georgakarakos, S., Dagorn, L., Brehmer, P.: Spatiotemporal distribution of fish schools around drifting fish aggregating devices. Fish. Res. **177**, 39–49 (2016)
12. Taquet, M., et al.: Characterizing fish communities associated with drifting fish aggregating devices (FADs) in the Western Indian ocean using underwater visual surveys. Aquat. Living Resour. **20**(4), 331–341 (2007)
13. Moreno, G., et al.: Fish aggregating devices (FADs) as scientific platforms. Fish. Res. **178**, 122–129 (2016)

Implementation of a 3-Dimensional Hydrodynamic Model to a Fish Aquaculture Area in Sines, Portugal - A Down-Scaling Approach

Alexandre Correia(✉), Lígia Pinto, and Marcos Mateus

MARETEC, Instituto Superior Técnico, Universidade de Lisboa, Av. Rovisco Pais, 1049-001 Lisbon, Portugal
alexandre.c.correia@tecnico.ulisboa.pt

Abstract. Coastal zones have always been preferential areas for human settlement, mostly due to their natural resources. However, human occupation poses complex problems and requires proper management tools. Numerical models rank among those tools and offer a way to evaluate and anticipate the impact of human pressures on the environment. This work describes the implementation of a hydrodynamic 3-dimensional computational model for the coastal zone in Sines, Portugal. This implementation is done with the MOHID model which uses a finite volume approach and an Arakawa-C staggered grid for spatial equation discretization and a semi-implicit ADI algorithm for time discretization. Sines coastal area is under significant pressure from human activities, and the model implementation targets the location of a fish aquaculture. Validation of the model was done comparing model results with *in situ* data observations. The comparison shows relatively small differences between model and observations, indicating a good simulation of the hydrodynamics of this system.

Keywords: Ocean modelling · Coastal zone management · Fish aquaculture

1 Introduction

Coastal zones have always been preferential areas for human settlement mainly because they offer essential resources, but also spaces for recreational and cultural activities. They offer points of access to world-wide marine trade and transport. Consequently, population density is significantly higher in coastal areas [1] and the migration to coastal zones is expected to continue into the future [2]. Population rise in these zones also increases the number of people exposed to the risks that characterize these areas, such as extreme natural events [3] and floods due to sea-level rise [4].

Human occupation and development in the coastal zones generates high pressure on the natural resources and ecosystems [5]. A high number of activates,

J. M. F. Rodrigues et al. (Eds.): ICCS 2019, LNCS 11539, pp. 265–278, 2019.
https://doi.org/10.1007/978-3-030-22747-0_21

ranging from recreation to commercial, compete for the limited resources of the ocean and often lead to negative impacts [6]. These growing pressures require management approaches that promote a conscious use of the ocean resources, fulfilling current needs without compromising future generations and reducing impacts on the ecosystems. Due to higher spatial and temporal variability, the management of ocean and coastal areas is more demanding than terrestrial ecosystems management [7], and dynamic management approaches are necessary to match the dynamism of the system [8].

An effective dynamic management of marine systems requires real-time sources of information [9]. Numerical models are important tools and sources of information to assist in marine activities by continuously providing estimates and forecasts on the state of the ocean. In addition, models results can also be combined with other data and increase our knowledge and managing capacity of the marine space [10,11].

This work describes the implementation of a three-dimensional (3D) hydrodynamic computational model for the coastal zone in Sines, Portugal. Results from the model will be used as a management tool for the port activities, including the aquaculture. This model implementation is part of the ongoing Project PiscisMod (16-02-01-FMP-0049), financed by Mar2020.

2 Case Study

Sines is a city located on the west coast of Portugal, with approximately 15 000 inhabitants [12] and a floating population of about 5 000 people, due to economic and touristic reasons. Sines plays a major role in terms of energy production and storage. There are two large production centers of oil and gas, GALP refinery and the Repsol petro-chemical industrial complex, both connected via pipelines to oil-bearing and petro-chemical terminal of port of Sines [13].

The port of Sines, located south from the city, is one of the most important deep-water ports of the country, having geo-physical conditions to accept a wide variety of ships. It is the country's leading energetic supplier (crude and its derivatives, coal and natural gas) and an important gateway for containerized cargo. The port consists of five terminals – liquid bulk, liquid natural gas, petrochemical, container, and multipurpose – as well as fishing and leisure ports (Fig. 1) [14]. In recent studies the port of Sines was found to be the most efficient port in Portugal in terms of emissions per cargo ship, although it is also the most pollutant due only to the fact that it receives more cargo than other ports [15]. Near the container terminal, protected by the breakwater, there is a delimited area allocated for the development of aquaculture (Fig. 1) consisting of 16 cages for a yearly production of 500 metric tons of European sea bass (*Dicentrarchus labrax*). The Sines thermoelectric power plant is located around 3.5 km southeast of the aquaculture zone and uses seawater to cool the generators. The seawater intake and restitution points are protected by jetties and located close to São Torpes beach. The water is discharged by two open channels with 4.5 m depth distant about 400 m from the intake. On a yearly average the power plant system uses $40\,m^3/s$ of water [16].

This coastal area, as well as the rest of the west coast of Portugal, is frequently under the influence of upwelling, caused by the predominant north wind regime [17].

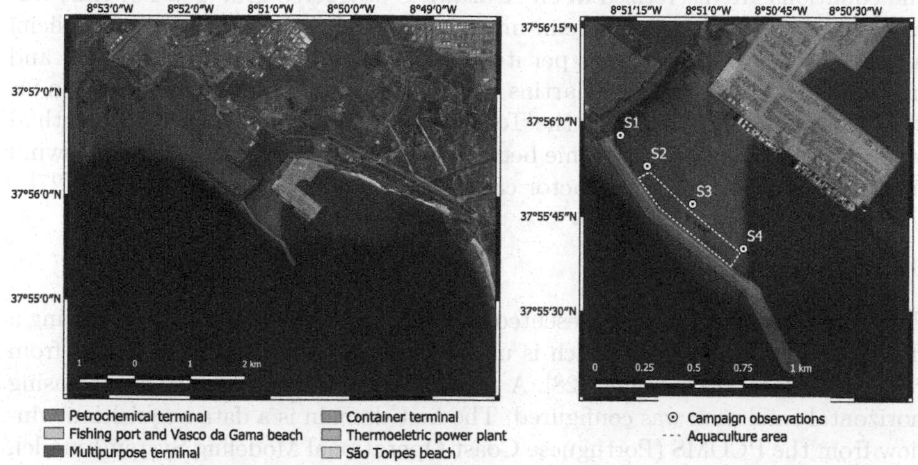

Fig. 1. Port of Sines terminals (left panel) and a close-up on the aquaculture facilities area (right panel) with the monitoring stations (S1 to S4).

3 Methodology

3.1 Model Description

The numerical model used in this work is the MOHID water modelling system (www.mohid.com), a model developed at the Marine and Environmental Technology Research Center (MARETEC). MOHID simulates physical and biogeochemical processes in the water column and the interactions between the water column with the sediment and the atmosphere. This model has already shown its ability to simulate complex systems and processes including those present in coastal and estuarine system [18,19]. And it was already proven useful as a management tool in aquaculture activities [20,21]. The model solves the three-dimensional primitive equations for the surface elevation and 3D velocity field for incompressible flows. The hydrostatic, Boussinesq and Reynolds approximation are assumed. It solves the advection-diffusion of temperature and salinity. The density is calculated using the UNESCO state equation as a function of salinity, temperature and pressure [22]. The viscosity is separated into its horizontal and vertical components, being the horizontal usually set to a constant. The vertical turbulent viscosity is handled using the General Ocean Turbulence Model (GOTM), which is integrated into the MOHID code, having its k-ε model parameterized according to Canuto et al. [23].

In the MOHID model the discretization of the equations is done using a finite volume approach. The discrete form of the governing equations is applied to a cell control volume, making the way of solving the equations independent of cell geometry and allowing the use of generic vertical coordinates [24]. Horizontally the equations are discretized in an Arakawa-C staggered grid [25]. For time discretization the model solves a semi-implicit ADI (Alternating Direction Implicit) algorithm with two time levels per iteration as proposed by Leendertse [26] and implemented as explained in Martins et al. [24]. The numerical scheme chosen for the advection and diffusion is the Total Variation Diminishing (TVD) method which consists in a hybrid scheme between a first order and a third order upwind method using a ponderation factor calculated with the SuperBee method [27].

3.2 Model Setup

The Sines modelling system presented in this study was implemented following a downscaling methodology, which is useful to provide boundary conditions from regional to local scale models [28]. A system of five nested domains of increasing horizontal resolution was configured. The first domain is a data acquisition window from the PCOMS (Portuguese Coast Operational Modeling System) model, a regional model for the Portuguese coast with a 0.06° horizontal resolution and a vertical discretization of 50 layers [29]. From domain 2 to domain 5 the horizontal resolution increases by a factor of 4 (0.015°, 0.00375°, 0.000938° and 0.000234°), the most refined nested model (domain 5) has a horizontal resolution of approximately 25 m. The bathymetric information required for the model was obtained from the European Marine Observation and Data Network (EMODnet) [30] for domains 2 and 3 while for domains 4 and 5 the information was provided by the Instituto Hidrográfico (www.hidrografico.pt) of the Portuguese navy. The result of the interpolations of the bathymetry into the different domains grids are presented in Fig. 2.

All domains are set in 3D and follow a generic vertical discretization approach [31] allowing the implementation of 7 sigma-type layers close to the surface and below a variable number of cartesian-type layers depending on the maximum depth of the corresponding domain.

The first domain model is a data acquisition window from the PCOMS model with a high temporal resolution, 900 s, enough to represent the main processes coming from the open ocean including the tide signal. PCOMS results for water level, velocity fields, temperature and salinity will be used as boundary conditions for domain 2 using a Flow Relaxation Scheme (FRS) for the velocities, temperature and salinity [32] while the water level is radiated using the Flather method [33]. On the open boundary with the atmosphere the model computes momentum and heat fluxes that interact with the free surface using results from the Mesoscale Meteorological Model 5 (MM5) atmospheric forecast model for the west Iberian Coast, running at IST (http://meteo.tecnico.ulisboa.pt). This model provides hourly results for air temperature, wind intensity and direction, atmospheric pressure, relative humidity, solar radiation and downward long wave

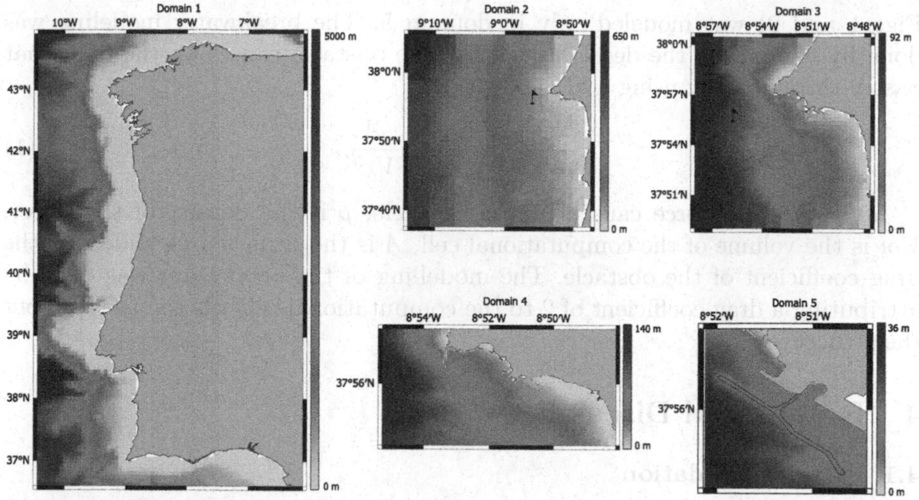

Fig. 2. Model domains: location and bathymetry. Black marker represents the buoy location.

radiation with a 9 km spatial resolution. For domain 1 to domain 4 the simulations are performed online (all at the same time) and domain 4 provides a window of results with a 900 s interval that will be used as boundary condition for domain 5 simulations. This configuration allows running domain 5 in a different computer to improve the computational time of the system. Time-step values used for the model calculations were 60.0, 30.0 and 7.5 s for domains 2, 3 and 4 respectively and 4.0 s for domain 5. Average duration for running this setup to obtain simulation results for 1 day, on a Windows computer with an Intel© Core™ i7-980 processor, was 1 h and 20 min from domain 1 to domain 4 and 50 min for domain 5. Summing to 2 h 10 min for the whole Sines modeling system.

The thermoelectric power plant intake and discharge are modeled in domains 3 and 4. In both domains the water intake was modeled by a sink term in the horizontal grid cell closest to the location of the real intake, a flow of 40 m^3/s was considered according to data from the power plant environmental declaration [16]. The discharge is modeled by a source term: in domain 4, two source terms where imposed in different horizontal grid cells in order to represent the two discharge open channels and a flow of 20 m^3/s was considered for each source term; in domain 3 only one source term with 40 m^3/s flow was considered due to the model spatial resolution. The water intake and discharges are considered along the whole water column to simulate the real conditions. A bypass function was implemented to model the 10 °C rise in temperature of the water in the discharge when compared to the intake caused by the cooling of the power plant, all other properties have the same value on both sides. The temperature increase was defined according to previous studies by Salgueiro et al. [35]. The breakwater that protects the container and multipurpose port terminals (see

Figs. 1 and 2) was modeled only in domain 5. The breakwater modeling was done by computing the deceleration that the obstacle causes on the fluid that passes through it by using a drag equation:

$$\frac{F}{\rho Vol} = \frac{1}{2} v^2 C_d \frac{A}{Vol} \tag{1}$$

F is the drag force caused by the obstacle, ρ is the density of the water, Vol is the volume of the computational cell, A is the surface area and C_d is the drag coefficient of the obstacle. The modelling of the breakwater was done by attributing a drag coefficient of 2 to the computational cells chosen to represent the structure.

4 Results and Discussion

4.1 Model Validation

Two sets of data were used to validate the Sines modelling system. Domain 2 and 3 model results were validated with temperature and water level. Temperature observations were recorded by a moored buoy located near the Sines coastal area (37°55.3' N, 008°55.7' W, black symbol in Fig. 2) and water level observation from a tide Gauge located inside the port of Sines [36]. Model results obtained in domain 5 simulations were validated with *in situ* data recorded on 29/06/2018, between 12:15 h and 14:15 h, on four stations (S1, S2, S3, S4 in Fig. 1) inside the container terminal near the aquaculture area. S1 is the station most protected by the breakwater and corresponds to a boat access platform, S2, S3 and S4 are progressively closer to the end of the breakwater. S2 is a support platform for the aquaculture activities and S3 and S4 are buoys that mark the perimeter of the aquaculture area. Depth profiles for velocity, temperature, salinity and density were recorded on all stations. Velocity was recorded using an Aanderaa acoustical current meter (model DCS4100) and the water properties were recorded with a CTD from Falmouth Scientific, Inc (model FSI NXIC CTD).

Domain 2 and 3 Validation. Model validation was made comparing time series of temperature and water level. For temperature a sample size of 1998 data points obtained between 26/12/2015 and 10/07/2016 were used. The analysis of temperature results for domain 2 and 3 show a significantly small difference from the observations (Fig. 3), having a Root Mean Square Error (RMSE) of 0.62 and 0.66 °C, respectively. Pearson correlation assumes values of 0.90 for domain 2 and 0.89 for domain 3 which indicates a good fit between models and observations. The Bias Error (BE) of −0.18 and −0.21 °C for domains 2 and 3 respectively indicate that the model tends to underestimate the temperature.

For water level a sample size of 2647 data points between 01/01/2016 and 12/05/2016 were used. Model results show a significantly small difference from observed values (Fig. 3) which is shown by the RMSE values for domain 2 and 3 of 0.07 m. Pearson correlation assumes a value of 0.997 for both domains also showing a good fit between model and observations.

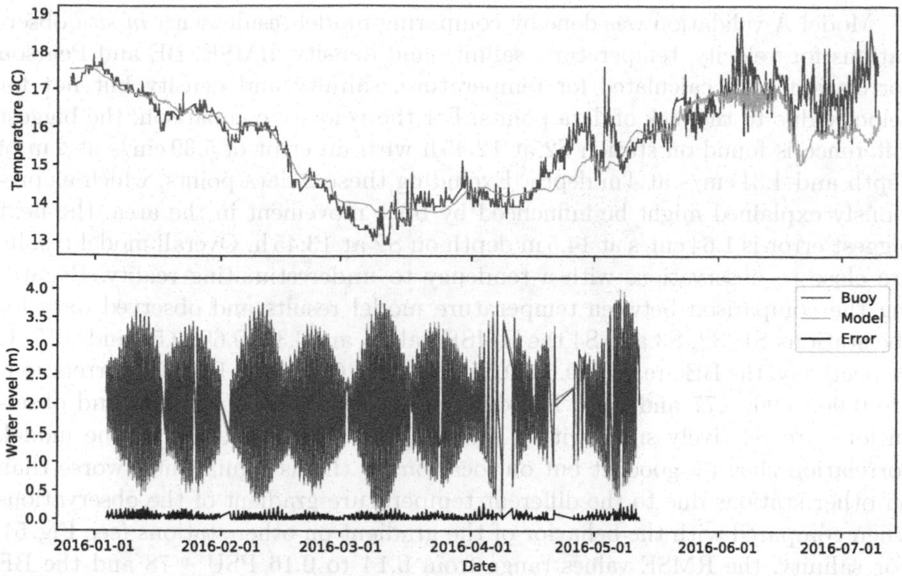

Fig. 3. Model results for domain 3 and buoy records for temperature and water level.

Domain 5 Validation. In domain 5, first the hypothesis of the breakwater being permeable to water was tested. A scenario with the breakwater modeled with a drag coefficient of 2 (model A in Figs. 4, 5, 6 and 7) was compared with a scenario in which the breakwater is impermeable by setting the horizontal grid cells that represent it to land cells (model B in Fig. 4, 5, 6 and 7). Model results from each model were compared with *in situ* data from the stations inside the container port. Differences in velocity profiles are easily noted, model A shows results closer to observations in 3 of the 5 situations compared, particularly in station S1 at 11:45 h and S2 at 13:45 h and 14:00 h (see Fig. 4). On S2 at 12:45 h although observation values are closer to model B results than model A they are significantly more distanced from either model than on other occasions, this may be explained by these two data points being closer to the surface then most. Inside the port there is movement from container ships and container ship towboats, these movements might cause waves that increase the water velocity in a way that the model can't reproduce. On S2 at 13:15 h model B is closer to observations but model A shows the same tendency of decreasing velocity with water column depth in the locations of the data points while model B shows the opposite tendency. The difference between the models for salinity were negligible (see Fig. 6). For density, model B shows, in general, lower values than model A but, on average, the deviation is not significantly higher than in model A. Notable differences between model A and B are seen for temperature with model A having results significantly closer to observation values. Model A was chosen to represent this system due to being closer to measured values.

Model A validation was done by comparing model results with *in situ* observations for velocity, temperature, salinity and density. RMSE, BE and Pearson correlation were calculated for temperature, salinity and density but not for velocity due to the lack of data points. For the velocity comparison, the biggest difference is found on station S2 at 12:45 h with an error of 5.39 cm/s at 2 m of depth and 4.31 cm/s at 4 m depth. Excluding these 2 data points, which as previously explained might be influenced by boat movement in the area, the next biggest error is 1.64 cm/s at 14.5 m depth on S2 at 13:45 h. Overall model results are close to observations with a tendency to underestimating reality. Regarding the comparison between temperature model results and observed data for the stations S1, S2, S3 and S4 the RMSE values are 0.33, 0.65, 0.58 and 0.75 °C respectively, the BE are 0.22, 0.65, 0.55 and 0.69 °C and the Pearson correlations are 0.96, 0.96, 0.77 and 0.98. The difference between model results and observations are relatively small with a tendency for overestimation by the model, correlation shows a good fit but on location S3 this is significantly worse than on other stations due to the different temperature gradient of the observations when compared with the behavior of the gradient on other stations (see Fig. 5). For salinity, the RMSE values range from 0.14 to 0.16 PSU – 78 and the BE assumes the same value as the RMSE for each location, this indicates that the model overestimates salinity on all compared data points, which is confirmed by observing the depth profiles in Fig. 6. Correlation values for stations S1 to S4 are $-0.57, 0.66, 0.73$ and 0.72, respectively, indicating a moderate fit for the data except on station S1, this value can be explained by a low number data points in S1, 12 data points, which might not be enough to accurately represent the trend that salinity follows with increasing depth, all other observations depth profile have more than 20 data points and show an increase of salinity with depth while S1 shows a decrease. Density RMSE values are 0.12, 0.04, 0.09 and 0.06 kg/m^3, BE values are 0.11, 0.03, 0.07 and 0.01 kg/m^3 and correlation values are 0.97, 0.96, 0.82 and 0.98, all from S1 to S4 respectively. Statistical indicators show a good simulation of the density by the model with the biggest difference being shown in S1. This was expected due to density results being directly dependent on temperature and salinity results. As these water properties were relatively accurately modeled, density should be too. Density depth profiles are presented in Fig. 7.

Looking at model results for surface velocity for domain 4 (Fig. 8) it is possible to confirm that the breakwater is having the desired effect of protecting the inside of the port from strong currents, while still allowing some passage of water. The strongest currents occurred from 14:00 h to 17:00 h, coincident with strong winds in the same direction.

The deviations between model results and observations may come from a variety of sources. The model relies on results from previous models, other hydrodynamics models and atmospheric models; errors from these models will propagate to the model in this study. While the variation of properties in the water is continuous in space and time in the model, values are represented discretely causing a loss of information specially when high variability exists. The breakwater is,

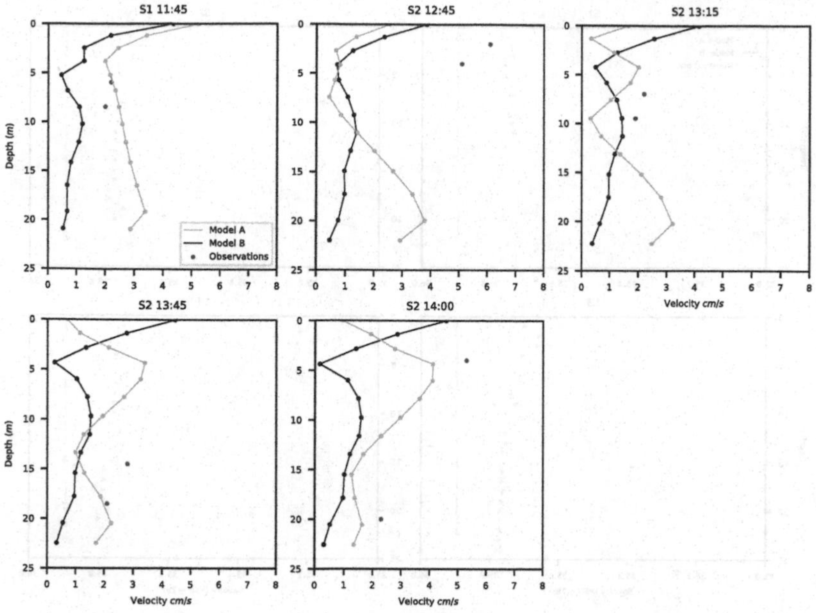

Fig. 4. Depth profiles for velocity: model results and observations.

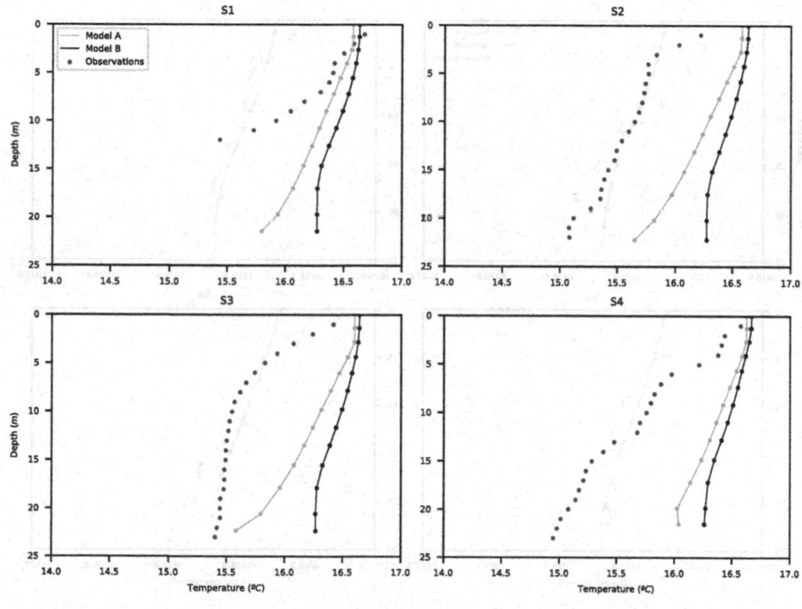

Fig. 5. Depth profiles for temperature: model results and observations.

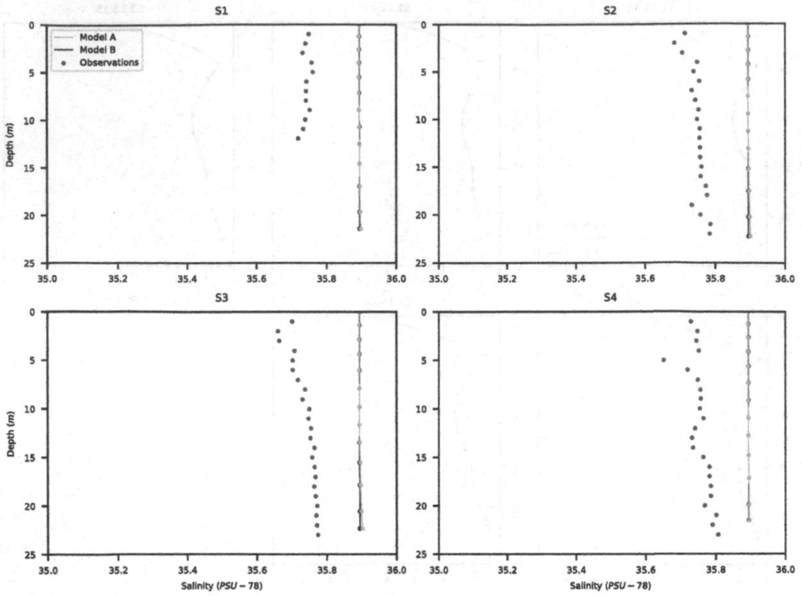

Fig. 6. Depth profiles for salinity: model results and observations.

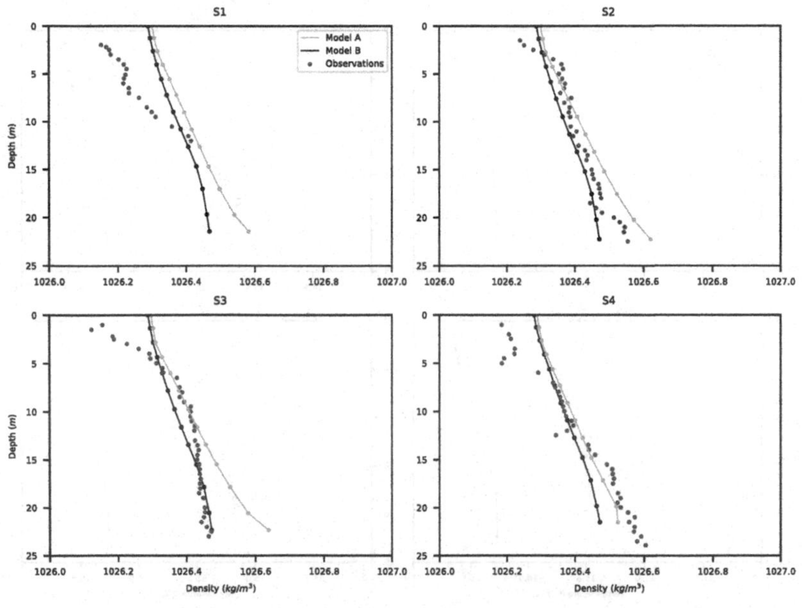

Fig. 7. Depth profiles for density: model results and observations.

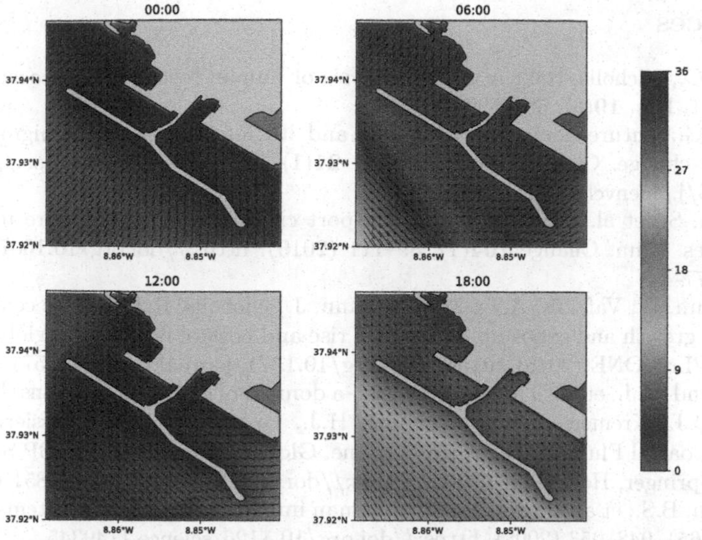

Fig. 8. Model results for velocity (intensity and direction) on 29/06/2018.

in fact, larger in width at the bottom of the water than at the surface but due to lack of information it was modeled as having the same width and the chosen drag coefficient might not be adequate. As such, more validation data is required to assess it.

5 Concluding Remarks

Model results show a good representation of the hydrodynamics and water properties in the study area. Knowing this, model results can prove useful for management of activities inside the container port, specially the aquaculture. Future work from this project will focus on this. In the next stage of this project the model will be compared with more data for further validation for different conditions. To improve accuracy, the weather forecasting model currently in use (MM5) will be replaced by a WRF (Weather Research and Forecasting) model for the Portuguese area. This model offers a better spatial resolution of 3 km, when compared to the 9 km from the MM5 model.

In the future, more processes will be added to model the water quality of the area and a bio-energetic module using DEB (Dynamic Energy Budget) theory [37] will also be included. This will allow the modelling of the fish aquaculture in order to optimize the feeding process and to asses the impact of this activity on the local ecosystem.

Acknowledgments. This project was made possible within the framework of the Mar 2020 Operational Program. Collaboration from MARE (Marine and Environmental Sciences Centre) (www.mare-centre.pt) and Jerónimo Martins Agro-Alimentar S.A. was essential for the conclusion of this work.

References

1. Small, C., Nicholls, R.J.: A global analysis of human settlement in coastal zones. J. Coast. Res. **19**(3), 584–599 (2003)
2. Hugo, G.: Future demographic change and its interactions with migration and climate change. Glob. Environ. Change **21**(1), S21–S33 (2011). https://doi.org/10.1016/j.gloenvcha.2011.09.008
3. Hanson, S., et al.: A global ranking of port cities with high exposure to climate extremes. Clim. Change **104**(1), 89–111 (2010). https://doi.org/10.1007/s10584-010-9977-4
4. Neumann, B., Vafeidis, A.T., Zimmermann, J., Nicholls, R.J.: Future coastal population growth and exposure to sea-level rise and coastal flooding - a global assessment. PLoS ONE (2015). https://doi.org/10.1371/journal.pone.0118571
5. Crossland, C.J., et al.: The coastal zone—a domain of global interactions. In: Crossland, C.J., Kremer, H.H., Lindeboom, H.J., Crossland, J.I.M., Tissier, M.D.A. (eds.) Coastal Fluxes in the Anthropocene. Global Change—The IGBP Series, pp. 1–37. Springer, Heidelberg (2005). https://doi.org/10.1007/3-540-27851-6_1
6. Halpern, B.S., et al.: A global map of human impact on marine ecosystems. Science **319**(5865), 948–953 (2008). https://doi.org/10.1126/science.1149345
7. Carr, M.H., Neigel, J.E., Estes, J.A., Andelman, S., Warner, R.R., Largier, J.L.: Comparing marine and terrestrial ecosystems: implications for the design of coastal marine reserves. Ecol. Appl. **13**(sp1), 90–107 (2003)
8. Lewison, R., et al.: Dynamic ocean management: identifying the critical ingredients of dynamic approaches to ocean resource management. BioScience **65**(5), 489–498 (2015). https://doi.org/10.1093/biosci/biv018
9. Maxwell, S.M., et al.: Dynamic ocean management: defining and conceptualizing real-time management of the ocean. Mar. Policy **58**, 42–50 (2015). https://doi.org/10.1016/j.marpol.2015.03.014
10. Neves, R., et al.: Coastal management supported by modelling: optimising the level of treatment of urban discharges into coastal waters. In: Rodriguez, G.R., Brebbia, C.A., Martell, E.P. (eds.) Environmental Coastal Regions III, vol. 43, pp. 41–49 (2000)
11. Fernandes, R., Braunschweig, F., Neves, R.: Combining operational models and data into a dynamic vessel risk assessment tool for coastal regions. Ocean Sci. **12**(1), 285–317 (2016). https://doi.org/10.5194/os-12-285-2016
12. Instituto Nacional de Estatistica: Census 2011. http://censos.ine.pt/xportal/xmain?xpid=CENSOS&xpgid=censos2011_apresentacao. Accessed 26 Feb 2019
13. Municipality of Sines. http://www.sines.pt/pages/755. Accessed 27 Feb 2019
14. Administration of the ports of Sines and Algarve. http://www.portodesines.pt. Accessed 26 Feb 2019
15. Nunes, R.A.O., Alvim-Ferraz, M.C.M., Martins, F.G., Sousa, S.I.V.: Environmental and social valuation of shipping emissions on four ports of Portugal. J. Environ. Manag. **235**(2019), 62–69 (2019). https://doi.org/10.1016/j.jenvman.2019.01.039
16. EDP - Gestão da Produção de Energia, S.A.: Update on the environmental declaration from 2016, Sines thermoeletric power plant (2017). https://a-nossa-energia.edp.pt/pdf/desempenho_ambiental/da_76_2017_cen_term.pdf
17. Kämpf, J., Chapman, P.: Upwelling Systems of the World. Springer, Cham (2016). https://doi.org/10.1007/978-3-319-42524-5
18. Abbaspour, M., Javid, A.H., Moghimi, P., Kayhan, K.: Modelling of thermal pollution in coastal area and its economical and environmental assessment. Int. J. Environ. Sci. Technol. **2**(1), 13–26 (2005). https://doi.org/10.1007/BF03325853

19. Ospino, S., Restrepo, J.C., Otero, L., Pierini, J., Alvarez-Silva, O.: Saltwater intrusion into a river with high fluvial discharge: a microtidal estuary of the Magdalena River, Colombia. J. Coast. Res. **34**(6), 1273–1288 (2018). https://doi.org/10.2112/JCOASTRES-D-17-00144.1
20. Tironi, A., Martin, V.H., Campuzano, F.: A management tool for salmon aquaculture: integrating MOHID and GIS applications for local waste management. In: Neves, R., Baretta, J.W., Mateus, M. (eds.) Perspectives on Integrated Coastal Zone Management in South America. IST Press, Lisbon (2009)
21. Neto, R.M., Nocko, H.R., Ostrensky, A.: Carrying capacity and potential environmental impact of fish farming in the cascade reservoirs of the Paranapanema river. Braz. Aquacult. Res. **48**(7), 3433–3449 (2016). https://doi.org/10.1111/are.13169
22. Fofonoff, N.P., Millard Jr., R.C.: Algorithms for the computation of fundamental properties of seawater. UNESCO Technical Papers in Marine Sciences, vol. 44 (1983). http://hdl.handle.net/11329/109
23. Canuto, V.M., Howard, A., Cheng, Y., Dubovikov, M.S.: Ocean turbulence. Part I: one-point closure model—momentum and heat vertical diffusivities. J. Phys. Oceanogr. 31, 1413–1426 (2001). https://doi.org/10.1175/1520-0485(2001)031⟨1413:OTPIOP⟩2.0.CO;2
24. Martins, F.A., Neves, R., Leitao, P.C.: A three-dimensional hydrodynamic model with generic vertical coordinate. In: 3rd International Conference on Hydroinformatics, pp. 1403–1410 (1998). http://hdl.handle.net/10400.1/124
25. Arakawa, A.: Computational design for long-term numerical integration of the equations of fluid motion: two-dimensional incompressible flow. Part I. J. Comput. Phys. **1**(1), 119–143 (1966). https://doi.org/10.1016/0021-9991(66)90015-5
26. Leendertse, J.J.: Aspects of a computational model for long-period water-wave propagation. No. RM-5294-PR. RAND CORP SANTA MONICA CALIF (1967)
27. Roe, P.L.: Characteristic-based schemes for the Euler equations. Annu. Rev. Fluid Mech. **18**(1), 337–365 (1986). https://doi.org/10.1146/annurev.fl.18.010186.002005
28. Ascione Kenov, I., et al.: Advances in modeling of water quality in estuaries. In: Finkl, C.W., Makowski, C. (eds.) Remote Sensing and Modeling. CRL, vol. 9, pp. 237–276. Springer, Cham (2014). https://doi.org/10.1007/978-3-319-06326-3_10
29. Mateus, M., et al.: An operational model for the west Iberian coast: products and services. Ocean Sci. **8**(4), 713–732 (2012). https://doi.org/10.5194/os-8-713-2012
30. EMODnet Bathymetry Portal. http://www.emodnet-bathymetry.eu. Accessed 20 Feb 2018
31. Martins, F., Leita, P., Silva, A., Neves, R.: 3D modelling in the Sado estuary using a new generic vertical discretization approach. Oceanologica Acta **24**, 51–62 (2001). https://doi.org/10.1016/S0399-1784(01)00092-5
32. Martinsen, E.A., Engedahl, H.: Implementation and testing of a lateral boundary scheme as an open boundary condition in a barotropic ocean model. Coast. Eng. **11**(5–6), 603–627 (1987). https://doi.org/10.1016/0378-3839(87)90028-7
33. Flather, R.A.: A tidal model of the north-west European continental shelf. Mem. Soc. R. Sci. Liege. **10**, 141–164 (1976)
34. Riflet, G.F.: Downscaling large-scale ocean-basin solutions in coastal tri-dimensional hydrodynamical models (2010)
35. Salgueiro, D.V., de Pablo, H., Nevesa, R., Mateus, M.: Modelling the thermal effluent of a near coast power plant (Sines, Portugal). Revista de Gestão Costeira Integrada-J. Integr. Coast. Zone Manag. **15**(4) (2015). https://doi.org/10.5894/rgci577

36. Insitituto Hidrografico Buoys. http://www.hidrografico.pt/boias. Accessed 26 Feb 2019
37. Bourlès, Y., et al.: Modelling growth and reproduction of the Pacific oyster Crassostrea gigas: advances in the oyster-DEB model through application to a coastal pond. J. Sea Res. **62**(2–3), 62–71 (2009). https://doi.org/10.1016/j.seares.2009.03.002

Numerical Characterization of the Douro River Plume

Renato Mendes[1,2](✉), Nuno Vaz[2], Magda C. Sousa[2], João G. Rodrigues[3], Maite deCastro[4], and João M. Dias[2]

[1] CIIMAR, University of Porto, 4099-002 Porto, Portugal
rpsm@ua.pt
[2] CESAM Physics Department, University of Aveiro, 3810-193 Aveiro, Portugal
[3] HIDROMOD, Rua Teles Palhinha, 4, 2740-278 Porto Salvo, Portugal
[4] EPhysLab, Facultade de Ciencias, Universidade de Vigo, Ourense, Spain

Abstract. The Douro is one of the largest rivers of the Iberian Peninsula, representing the most important buoyancy source into the Atlantic Ocean on the northwestern Portuguese coast. The main goal of this study is to contribute to the knowledge of physical processes associated with the propagation of the Douro River plume. The general patterns of dispersion in the ocean and how the plume change hydrography and coastal circulation were evaluated, considering the main drivers involved: river discharge and wind. Coastal models were implemented to characterize the propagation of the plume, its dynamics, and its impact on coastal circulation. Different numerical scenarios of wind and river discharge were analyzed. The estuarine outflow is sufficient to generate a northward coastal current without wind under moderate-to-high river discharge conditions. Under easterly winds, the propagation pattern is similar to the no wind forcing, with a northward current speed increasing. A southward coastal current is generated only by strong westerly winds. Under upwelling-favorable (northerly) winds, the plume extends offshore with tilting towards the southwest. Southerly winds increase the velocity of the northward current, being the merging of the Douro and Minho estuarine plumes a likely consequence.

Keywords: Estuary · Coastal circulation · NW Iberian Peninsula · MoHid

1 Introduction

River plumes are flow structures, which can occur in a range of sizes and shapes and that impact the coastal ocean, changing local flow and water properties concentration. They represent a significant link between terrestrial and marine systems, affecting densely populated areas. Therefore, understanding the processes that determine the transport of estuarine waters is essential to know the fate and transport of solute and particulate materials across the land-sea interface. The propagation of estuarine waters along the coast can be influenced by several external forcing, which includes Earth rotation represented by the Coriolis force, tides, winds, freshwater discharge, and ambient flows [1].

The propagation of the Douro River plume has a high impact on the coastal hydrography of NW of Portugal [2]. This plume promotes buoyant water transport in

J. M. F. Rodrigues et al. (Eds.): ICCS 2019, LNCS 11539, pp. 279–286, 2019.
https://doi.org/10.1007/978-3-030-22747-0_22

the downstream direction. Recently, the interaction between the Douro and Minho estuarine plumes under southerly winds was studied through numerical simulations [2], revealing an indirect role of the Douro plume on both plumes' fate over the shelf.

The aim of this study is to analyze the horizontal plume propagation under different wind direction and river discharge scenarios by means of validated simulations using a MOHID (www.mohid.com) model implementation developed for this coastal area.

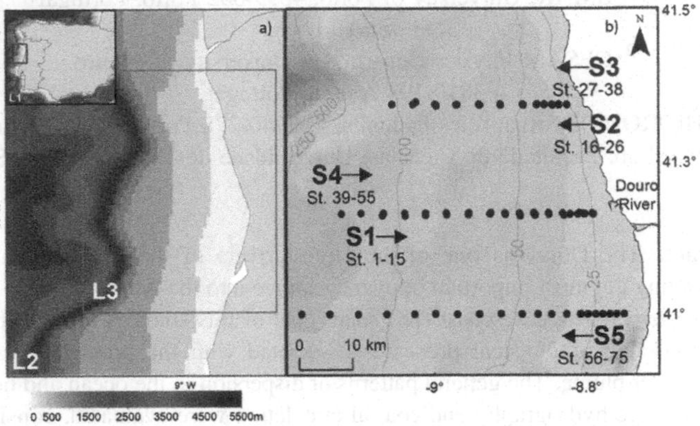

Fig. 1. (a) Study area and the MOHID nesting domains (b) Location of the CTD sampling stations (black dots) during the 2007 survey at five zonal sections (S). Note: CTD sampling stations at S1 and S4, and S2 and S3 were performed at the same locations but at different times.

2 Data and Methods

2.1 Model Configuration

The implementation of the numerical model MOHID to the region influenced by the Douro plume follows a nested downscaling methodology similar to that previously validated for the NW coast of the Iberian Peninsula [3].

This nested model comprises a large domain to compute the barotropic tide (L1) and two smaller baroclinic domains (L2 and L3). The third level comprises a smaller region with higher horizontal resolution (Fig. 1a).

The first level (L1) ranged from 13.5°W to 1°E and 33.5°N to 50°N, with a variable horizontal resolution of 0.06°(~6 km) and was constructed based on the ETOPO1 global database. This level is a 2D model using the thoroughly revised tidal solution from a global hydrodynamic model (FES2012) as a boundary condition [4], which constitutes an update to the previous coastal application [3], which used an older dataset. The tidal reference solution generated by this model is then propagated to the subsequent 3D baroclinic levels applying the radiation scheme [5]. The time step was 180 s, and the horizontal eddy viscosity was 100 $m^2\ s^{-1}$.

L2 and L3 levels are focused on the Douro Estuary and were built based on GEBCO bathymetries [6]. L2 domain ranges from 9.94°W to 8.40°W and 40.16°N to 42.10°N, with a horizontal resolution of ~2 km and L3 ranges from 9.52°W to

8.60°W and 40.50°N to 41.75°N, with a horizontal resolution of ~500 m. The initial ocean stratification in L2 and L3 was set through 3D monthly mean climatological fields of water temperature and salinity from the World of Ocean Atlas 2013 [7]. Following [3], the baroclinic forcing was slowly activated over ten inertial periods, and the biharmonic filter coefficients were set to 1 × 107 m^4 s^{-1} and 1 × 104 m^4 s^{-1} for L2 and L3, respectively. A time step of 60 s was used in L2 domain with a turbulent horizontal eddy viscosity set to 20 m^2 s^{-1}. The time step and turbulent horizontal eddy viscosity were set to 60 s and 5 m^2 s^{-1}, respectively in the L3. A z-level vertical discretization was adopted using Cartesian coordinates. Considering the surface behavior of estuarine plumes and the importance of the baroclinic processes on their dispersion, sigma coordinates were applied on the first 10 m.

The 3D momentum, heat and salt balance equations were computed implicitly in the vertical direction and explicitly in the horizontal direction. All ocean boundary conditions at L3 domain were supplied by hydrodynamic and water properties solutions from L2 domain. A Flow Relaxation Scheme [8] was applied to water level, velocity components, water temperature and salinity in baroclinic models.

Surface boundary conditions were imposed using high-resolution results from Weather Research and Forecasting Model (WRF) with a spatial resolution of 4 km [9]. WRF output spatial fields were hourly interpolated for L2 and L3 domains, using triangulation and bilinear methods in space and time, respectively. At the surface, the sensible and latent heat fluxes were calculated using the Bowen and Dalton laws, respectively. A shear friction stress was imposed for bottom boundary condition, assuming a logarithmic velocity profile.

Douro outflows were previously computed using an estuarine model developed for the inner part of the estuary [2] and directly imposed offline as momentum, water, and mass discharge to the coastal model (L3) as a discharge point. The effects of other small rivers were not considered in this study due to their neglectable freshwater inflow [2].

2.2 Experimental Design

Built on a statistical analysis of the Douro River discharge data, three scenarios were chosen accounting the 25th, 50th and 75th percentiles of annual maxima of the month with maximum daily mean inflows - January. These percentiles correspond to low (608 m^3 s^{-1}), moderate (1486 m^3 s^{-1}) and high (3299 m^3 s^{-1}) discharge winter scenarios, respectively.

Following the method proposed by [3], the idealized scenario with high river discharge starts from the average value for January (1055 m^3 s^{-1}) and then increases exponentially under 4 days until reach 3299 m^3 s^{-1}. The maximum is about three times the average. This ratio was also adopted for moderate and low discharge scenarios. The water temperature and salinity of the Douro River in the estuarine model implementation [2] were set to 8 °C and 0, respectively.

The definition of wind scenarios was also based on statistical results presented by [3]. The probability of occurrence of winds with intensity lower than 3 m s^{-1} is 32%, while the probability of moderate winds (between 3 and 6 m s^{-1}) is higher than 46% [3]. These wind intensities were used in the simulations as representative of the prevailing wind regime in this region.

After the model validation, 27 experiments to test the plume propagation were done, including four wind directions (north, south, west, and east), two wind intensities (3 and 6 m s^{-1}) and three river discharges (high, moderate and low). Moreover, three simulations were carried out without wind. In all simulations, results were analyzed for January 2010 after a spin-up of six months, from July 2009 to December 2009. This period was considered representative of local winter condition [2]. Wind forcing for each scenario starts when the river discharge reached its maximum.

3 Results and Discussion

3.1 Validation

A 1-month simulation was carried out for all domains from 20 January to 28 February 2007 to evaluate the coastal model accuracy. Salinity was set to 0 in the estuarine model, and temperature of freshwater inflow was set based on the daily smooth air temperature provided by WRF predictions. In situ dataset includes CTD hydrographic surveys acquired on January 24 (S1) and 26 (S2) and February 6 (S3, S4, and S5), 2007 (Black points in Fig. 1b). The adjustment between model predictions and observed data was evaluated using the Root Mean Square Error (RMSE) and Bias computation.

RMSE values for salinity range from 0.01 (St. 16) to 1.05 (St. 55) (Fig. 2). Bias range from −0.54 (St 55) to 0.22 (St. 27) being negative for the majority of stations, which indicates a model underestimation of the river plume. Significant exceptions are stations 27 and 28, where the bias are 0.22 and 0.08, respectively. In situ and modeled salinities reveal low salinity features in all sections, being Sect. 2 an exception, which explain a lower bias. Although the plume spreading is well simulated in front of the mouth, the model tends to overestimate the stratification, observed as a large RMSE and negative bias in Sect. 4.

A lower agreement is expected for water temperature because of its higher variability. The RMSE ranges from 0.16 °C (St. 34) to 1.10 °C (St. 42), with an average of 0.58 °C considering all stations. Bias is positive for all profiles, ranging from 0.08 °C (St. 37) to 0.95 °C (St. 49). The model tends to underestimate observed water temperature along the water column (Fig. 2). Nonetheless, the deviation range (~1 °C) is low and close to that obtained by [3].

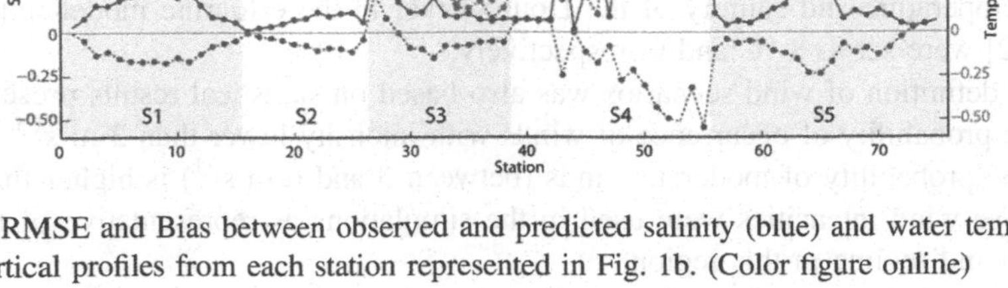

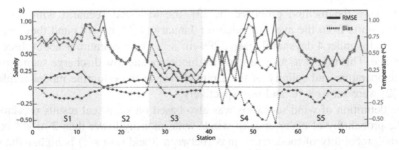

Fig. 2. RMSE and Bias between observed and predicted salinity (blue) and water temperature (red) vertical profiles from each station represented in Fig. 1b. (Color figure online)

3.2 River Discharge Influence

The spread of the Douro plume is described in Fig. 3 under the different river discharge scenarios and without wind after nine days of simulation and five from the river inflow peak.

All figures show the advection of low salinity waters to the right due to the Coriolis effect and the extension of the river plume after establishing a geostrophic balance. Additionally, all river discharge scenarios share common features such as a re-circulating bulge in the front of the mouth, a southward filament from the central bulge, and a northward coastal current following the coastline, which generates small-scale eddies promoted by the bathymetry and morphology constraints, (Fig. 3a–c).

Under low river discharge (Fig. 3a), the offshore extension of the plume is about 18 km, the northward current is weak (0.2–0.3 m s^{-1}), and the plume front reaches the Cávado River mouth. In addition, a re-circulating bulge in the near-field region is detectable, slightly tilted northward with an approximate diameter of about 17 km. Under moderate river discharge (Fig. 3b), the Douro estuarine plume presents an off-shore extension of about 22 km with a re-circulating bulge (~25 km of diameter) northward tilted and partially detached from the coast. A nearshore surface current towards the north is also observed (>0.2 m s^{-1}). Salinity surface patterns present similar features under high river discharge (Fig. 3c), but with an increase of freshwater, mainly in the bulge region. The longitudinal extension of the plume is also similar (~23 km), but the bulge is wider (~35 km in the major axis of the ellipse).

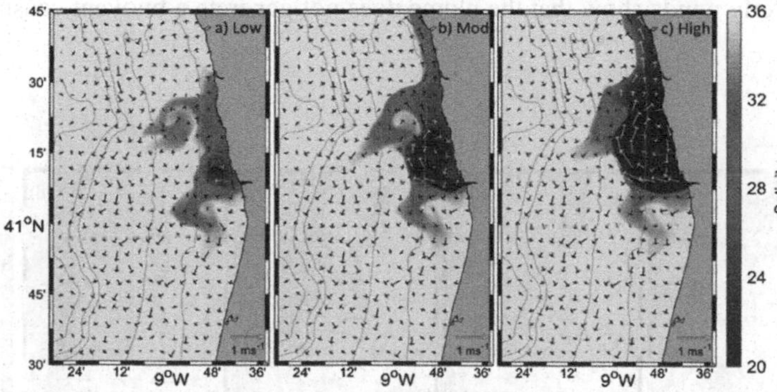

Fig. 3. Surface currents and salinity simulations 5 days after the peak river discharge (day 9) under low (a), moderate (b), and high (c) river discharges without wind.

3.3 Wind-Driven Plume Dispersion

Surface salinity fields under the wind scenarios previously defined at moderate river discharge are depicted in Fig. 4. The time percentage from day 5 to 10 during which each grid cell has a salinity value <30 is represented in the top-left corner of each panel. For the sake of simplicity, only the results under moderate river discharge and higher wind intensity are presented.

The maximum distance of offshore plume spreading is observed under upwelling-favorable (northerly) winds (Fig. 4a). In this case, the inclination observed between the main direction of the plume propagation and the river's mouth is influenced by an equilibrium between the river discharge and wind intensity. No re-circulating bulge is detectable, and the plume propagation exceeds the western boundary of the domain under moderate and high river discharges.

Surface salinity fields show plume confinement in the coastal region under southerly winds (Fig. 4b). The downwelling-favorable winds shrink the bulge near the river mouth and the current speed towards north increases in comparison with simulations without wind (Fig. 3c). The offshore extension of the plume in front of the river mouth does not exceed ~10 km and ~6 km under moderate and high southerly winds, respectively. Combined with moderate-to-high river discharge, downwelling-favorable winds increase the possibility of the Douro plume to merge with estuarine sources located north of the estuary mouth, such as Minho estuarine plume [2].

The numerical scenario under easterly winds (Fig. 4c) shows a plume core detached from the coast and a weaker northward current (~0.3 m s^{-1}). The plume detachment is noticeable under strong winds, which push the plume offshore in opposition to the pressure gradient force, stopping the water re-circulation near the coast. The plume width near the mouth is larger under moderate (~15 km) than strong winds (~11 km), which can be explained by the reminiscent freshwater from the re-circulating bulge. However, the coastal plume is wider (~15 km) under strong than moderate wind intensities (~10 km).

Simulations under westerly winds (Fig. 4d) show the confinement of river plume landward, accumulating freshwater along the coast. Simulations under moderate and strong westerly winds show that the plume does not generate a buoyant coastal current northward.

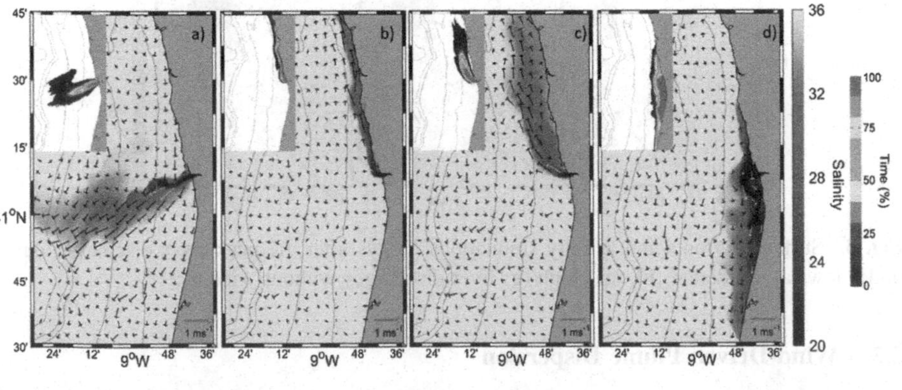

Fig. 4. Surface currents (ms^{-1}) and salinity 5 days after the peak discharge under moderate river discharge and strong northerly (a), southerly (b), easterly (c), and westerly wind (d). The top-left corner of each panel represents the time percentage during which each grid cell has a salinity value <30.

4 Conclusions

The objective of this study was to characterize numerically the Douro estuarine plume dispersion regarding its dynamics, scale, and fate under different river discharges and wind conditions. Numerical simulations were carried out by means of MOHID previously validated in the area under study. The results obtained from this analysis suggest the following:

- Without wind forcing, the plume expands offshore, creating a re-circulating bulge. Low salinity waters are advected to the right due to the Coriolis force, and then the plume water flows northward.
- Easterly winds form a similar feature to the no-wind case. However, the low salinity band is detached from the coast, with an increase of the northward current.
- Westerly winds tend to accumulate freshwater to the coast and can decrease the momentum near the mouth. A southward coastal current is identified under strong winds and moderate and high river discharges.
- Northerly winds generate a large offshore extension of the plume. There is an increase of stratification, which turns more efficient the control of the plume fate by the wind.
- Southerly winds confine the plume to the coast, which propagates only as a coastal buoyant current northward.

Acknowledgments. RM benefits from doctoral and post-doctoral grants (SFRH/BD/79555/2011 and SFRH/BPD/115093/2016) given by the Portuguese Science Foundation (FCT). MCS and NV are funded by national funds (OE), through FCT, I.P., in the scope of the framework contract foreseen in the numbers 4, 5 and 6 of the article 23, of the Decree-Law 57/2016, of August 29, changed by Law 57/2017, of July 19. Thanks are due for the financial support to CESAM (UID/AMB/50017 - POCI-01-0145-FEDER-007638) to FCT/MCTES through national funds (PIDDAC), and the co-funding by the FEDER, within the PT2020 Partnership Agreement and Compete 2020.

References

1. Hetland, R.D.: Relating river plume structure to vertical mixing. J. Phys. Oceanogr. **35**(9), 1667–1688 (2005)
2. Mendes, R., Sousa, M.C., DeCastro, M., Gómez-Gesteira, M., Dias, J.M.: New insights into the Western Iberian Buoyant Plume: Interaction between the Douro and Minho river plumes under winter conditions. Prog. Oceanogr. **141**, 30–43 (2016)
3. Sousa, M.C.: Modelling the Minho River Plume Intrusion into the Rias Baixas. Universidade de Aveiro (2013)
4. Carrère, L., Lyard, F., Cancet, M., Guillot, A., Roblou, L.: FES2012: a new global tidal model taking advantage of nearly twenty years of altimetry. In: Proceedings of the 20 Years of Progress in Radar Altimetry Symposium, pp. 1–20 (2012)
5. Flather, R.: A tidal model of the northwest European continental shelf. Mem. Soc. R. Sci. Liege **10**(6), 141–164 (1976)

6. Becker, J.J., et al.: Global bathymetry and elevation data at 30 arc seconds resolution: SRTM30_PLUS. Mar. Geod. **32**(4), 355–371 (2009)
7. Zweng, M.M., et al.: World ocean atlas 2013: salinity. In: Levitus, S., Mishonov, A. (eds.) NOAA Atlas NESDIS 74, vol. 1, 40 p. (2013)
8. Martinsen, E.A., Engedahl, H.: Implementation and testing of a lateral boundary scheme as an open boundary condition in a barotropic ocean model. Coast. Eng. **11**(5–6), 603–627 (1987)
9. Skamarock, W.C., et al.: A Description of the Advanced Research WRF Version 3 (2008)

The Impact of Sea Level Rise in the Guadiana Estuary

Lara Mills[1(✉)], João Janeiro[1], and Flávio Martins[1,2]

[1] Centro de Investigação Marinha e Ambiental (CIMA), University of Algarve, Faro, Portugal
{a60162,fmartins}@ualg.pt, janeiro.jm@gmail.com

[2] Instituto Superior de Engenharia (ISE), University of Algarve, Faro, Portugal

Abstract. Understanding the impact of sea level rise on coastal areas is crucial as a large percentage of the population live on the coast. This study uses computational tools to examine how two major consequences of sea level rise: salt intrusion and an increase in water volume affect the hydrodynamics and flooding areas of a major estuary in the Iberian Peninsula. A 2D numerical model created with the software MOHID was used to simulate the Guadiana Estuary in different scenarios of sea level rise combined with different freshwater flow rates. An increase in salinity was found in response to an increase in mean sea level in low and intermediate freshwater flow rates. An increase in flooding areas around the estuary were also positively correlated with an increase in mean sea level.

Keywords: Guadiana Estuary · Sea level rise · MOHID · Salinity intrusion

1 Introduction

A rise in mean sea level is a global concern as 10% of the world's population live within 10 m elevation of the current sea level (Carrasco et al. 2016). Based on projections by the Intergovernmental Panel on Climate Change (IPCC), the rate of sea level rise is accelerating. An acceleration in global sea level rise has been reported throughout the 20th century based on satellite altimeter data (Church and White 2006). Global mean sea level rise could be as high as 1 m by the year 2100 if greenhouse gas emissions continue to be very high (Church et al. 2013). As the number of people accommodating coastal areas increases, an acceleration in the rate of sea level rise will severely impact society and the economy. Impacts of sea level rise on coastal areas include submergence of land, increased flooding, increased erosion, changes in ecosystems and an increase in salinity (Nicholls et al. 2011). Not yet fully explored in the literature is how the dynamics of salinity in estuaries will evolve due to sea level rise. It is thus the aim of this work to quantify a relationship between sea level rise and salinity content in estuaries, as an increase in estuarine salinity has the potential to damage estuarine environments.

From a physical standpoint, an estuary is defined as a semi-enclosed body of water where river meets sea, extending into the river as far as the upper limit of the tides. Estuaries are productive environments, hosting highly diverse and complex ecosystems

J. M. F. Rodrigues et al. (Eds.): ICCS 2019, LNCS 11539, pp. 287–300, 2019.
https://doi.org/10.1007/978-3-030-22747-0_23

(Sampath et al. 2015). Fisheries, aquaculture, ecotourism and port facilities rely on estuaries to contribute revenues to national, regional and local economies.

These transitional areas where fresh water derived from land mixes with salt water from the sea are sensitive to changes in climate such as sea level rise. Estuaries adapt to sea level rise by deepening their channels. This increases the accommodation space of sediments, further enhancing ebb and flood asymmetry (Sampath et al. 2015).

A water column is considered stratified when less dense water lies over denser water. Thus, the stratification of an estuary is determined by the balance between buoyancy and mixing of denser salt water from the sea and less dense fresh water from land. An increase in sea level allows more salt water to enter the estuary (Ross et al. 2015). Thus, a major consequence of sea level rise in estuaries is the intrusion of salt from sea water into fresh ground water, which can further impact stratification (McLean 2001).

The intrusion of salt will have a profound effect on the physical properties of estuaries as the dynamic mixing of salt water and fresh water is the driving factor regulating stratification (Ross et al. 2015). Changes in oceanic salinity combined with varying flow rates of freshwater discharge into the estuary will alter the entire estuarine circulation of salinity (Ross et al. 2015). Hong and Shen (2012) found an increase in stratification in the Chesapeake Bay as a result of an increase in salinity based on the results of a three-dimensional numerical model. This change in horizontal salinity gradients will further alter estuarine circulation and cause oxygen depletion (Hong and Shen 2012). Alterations in estuarine circulation will be detrimental to marine ecosystems that cannot tolerate high salinity content (Chua and Xu 2014). Furthermore, sea level rise increases salinity upstream and impacts tidal currents (Hong and Shen 2012). A study on the impact of an increase in salinity on Louisiana estuaries by Wiseman et al. (1990) found both positive and negative trends attributed to the variability of freshwater flow rates. These authors found that an increase in salinity results in the death or decline in the productivity of marshes, which eventually leads to land loss (Wiseman et al. 1990). Further impacts of salt intrusion include contamination of water supplies for both consumption and for the industry (Hilton et al. 2008).

The aim of this article is to present the results of a numerical model simulating salt intrusion in the Guadiana Estuary, a major estuary on the Iberian Peninsula. The model simulates different sea level rise scenarios as projected by the IPCC along with different river discharge flow rates across two different bathymetries of the estuary.

2 Methods

2.1 Study Area

The Guadiana Estuary connects the Guadiana River to the Gulf of Cadiz. The Guadiana River is the fourth longest river on the Iberian Peninsula with a length of 810 km, 200 of which form a border between Portugal and Spain (Delgado et al. 2012). Beginning at its mouth in front of Vila Real de Santo António in the Algarve region of Portugal and Ayamonte, Spain, the estuary extends about 80 km north to its tidal limit at Mértola. The estuary drains a total area of 66,960 km^2 (Garel et al. 2009) in front of these highly populated regions. The Guadiana Estuary is at its widest in front of Vila Real de Santo António with a width of 800 m and is most narrow near Mértola with a width of 70 m. The average depth of the estuary is 5 m, but in some cases, it can reach 10 m.

Downstream, the estuary is in the form of a submerged delta where there is moderate wave energy (Garel et al. 2009). The estuary is classified as a rock-bound estuary due to its narrow and relatively deep channel, located on a passive margin exceeding freshwater discharge (Garel et al. 2009). The Guadiana Estuary is characterized by a semi-diurnal meso-tidal regime with tidal influence 50 km upstream from where it meets the Gulf of Cadiz (Delgado et al. 2012). The average neap tidal range is 1.28 m and the average spring tidal range is 2.56 m with a maximum spring tidal range up to 3.44 m (Garel et al. 2009).

Tides and freshwater input control the vertical mixing and stratification of the estuary, which ranges from very well-mixed to very well-stratified (Garel et al. 2009). The mixing of the estuary varies by season, with a well-mixed water column when there is low discharge from the Guadiana River due to a lack of rainfall in the summer and a partially stratified water column in the winter when there is more discharge from the river. It is only under extreme conditions that the water column can become stratified (Basos 2013). The ebb currents of the estuary are generally faster than flood currents, even in low river flow conditions, an effect most likely due to the large hydraulic depth of the estuarine channel. The ebb-dominance becomes more pronounced as the river discharge increases (Garel et al. 2009).

The freshwater discharge of the Guadiana River can range between 10 m^3/s and 4660 m^3/s. The construction of over 100 dams since the 1950s has controlled 70% of the drainage basin by the end of 2000 (Garel et al. 2009). Of relevance is the Alqueva Dam, located 60 km from the head of the estuary. It is the largest reservoir in southern and western Europe storing 4150 hm^3 of water (Garel et al. 2009). Its closure in 2002 has significantly reduced the mean river flow from an average of 143 m^3/s over the last 26 years to 16 m^3/s after 2002 (Garel et al. 2009). Downstream the dam, the river has ephemeral flows and is dry for 40% of the year (Fortunato et al. 2002). The Guadiana Estuary currently faces a reduction in river flow due to scarce rainfall and an increase freshwater storage.

2.2 The Model

The MOHID water modeling system was used to simulate the hydrodynamics of the Guadiana Estuary and the effects of different sea level rise scenarios on salinity distribution and transport. MOHID is a three-dimensional hydrodynamical modeling system that integrates diverse numerical models in an object-oriented programming approach (MARETEC 2017). MOHID simulates physical and biogeochemical processes in the water column and sediments (MARETEC 2017). MOHID water has been used to simulate the hydrodynamics of many coastal and estuarine systems, mainly the major estuaries of Portugal, but also in the Netherlands, France and Ireland (Calero Quesada et al. 2019). A recent study by Calero Quesada et al. (2019) used MOHID in 2D barotropic mode to simulate the effects of freshwater discharge and tidal forcing on tidal propagation in the Guadiana Estuary. This model accurately produced phase values of tidal elevations and currents as confirmed by a harmonic analysis (Calero

Quesada et al. 2019). The model was able to demonstrate how freshwater discharge has a stronger effect on tidal wave amplitude upstream the estuary, whereas tidal forcing controls tidal wave amplitude downstream the estuary (Calero Quesada et al. 2019).

The finite volume method in a generic computational mesh was used to simulate hydrodynamic and salt transport models. Navier-Stokes and transport equations were solved with an Alternating Direction Implicit (ADI) method. The calculation time to stabilize the salinity field is very high when the river flow is low due to the high residency time of the estuary in low river discharge conditions. For example, with a river flow of 10 m^3/s it takes two months of simulated time for the salinity field to stabilize. The computational mesh was chosen to provide a spatial resolution appropriate for the study without incurring excessive calculation time. Thus, each simulation used a Cartesian computational mesh of 1400 × 350 cells with a constant space step of 30 m to evaluate salinity distribution and transport as well as coastal flooding at the mouth of the Guadiana Estuary.

The model incorporates different sea level rise scenarios combined with different river discharge scenarios typical of the Guadiana River over two separate bathymetries. The first bathymetry was computed by triangular interpolation of measured bathymetric data on the mesh of 1400 × 350 cells with a space step of 30 m. This bathymetry is indicative of the present bathymetry of the estuary and corresponds to a case in which maximum human intervention is taken to keep the coastline unchanged. The second bathymetry represents a situation in which there is no human intervention, thus allowing maximum flooding of the saltmarshes and low-lying areas. Process-based models such as MOHID cannot evaluate long-term geomorphological dynamics so the results from a behavior-oriented model by Sampath et al. (2011) were used for this specific bathymetry. The two different bathymetries are shown in Figs. 1 and 2.

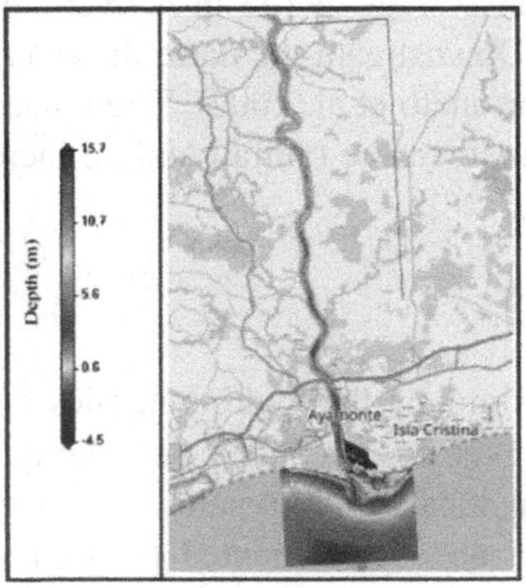

Fig. 1. Computational bathymetry of the Guadiana Estuary, if the coastline is maintained (1400 × 350 cells with a space step of 30 m).

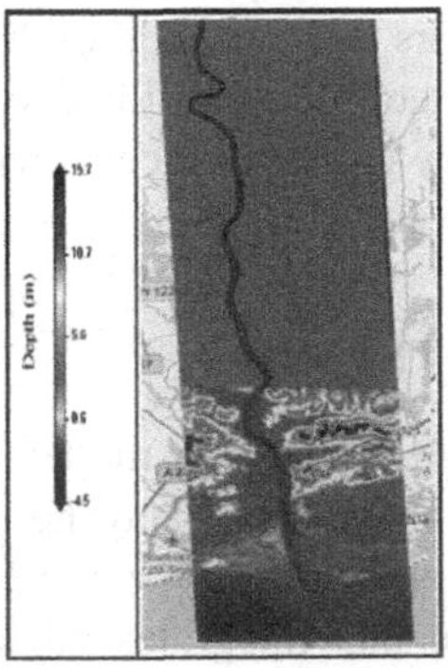

Fig. 2. Computational bathymetry of the Guadiana Estuary, allowing flooding (1400 × 350 cells with a space step of 30 m). The grey area represents land.

As previously mentioned, the flow of the Guadiana River is highly variable, so four different river discharge scenarios were simulated in the model. The first river discharge scenario of Q = 10 m^3/s represents the flow of the river in summer, when there is little rain. The second river discharge scenario of Q = 50 m^3/s and the third river discharge scenario of Q = 100 m^3/s represent intermediate conditions of the river. The fourth river discharge scenario of Q = 500 m^3/s represents a very high flow situation, which has been known to occur prior to the closure of the Alqueva Dam and will now only happen in extreme cases. A table outlining the different river discharge scenarios can be seen in Table 1.

Table 1. Guadiana River freshwater discharge scenarios

Scenario	Flow rate (m^3/s)
1	10
2	50
3	100
4	500

The model represents typical system conditions and is not associated with any specific tidal event. It was decided to force the model with an external M2 tide equivalent to the average neap and spring tide conditions, resulting in an amplitude of 1 m.

To simulate the salinity transport in the Guadiana system, the salinity of water from the Guadiana River was considered null and the salinity of the coastal region adjacent to the estuary was set to a constant value of 36. These values are typical for this region. There are no units for salinity, in agreement with recommendations by UNESCO.

The four different river discharge scenarios mentioned above were combined with different sea level rise scenarios derived from sea level rise projections provided by the 5th IPCC report for Representative Concentration Pathway (RCP) 4.5 and RCP 8.5. Representative Concentration Pathway refers to greenhouse gas concentration (Church et al. 2013). The 5th report of the IPCC projected global mean sea level rise for these two RCP values over the years 2046–2065, 2081–2100 and the sea level rise in 2100. The mean sea level rise forecasts can be seen in Table 2.

Table 2. Predictions for the rise in global mean sea level in meters based on sea level rise from 1986–2005 for RCP 4.5 and RCP 8.5 (Adapted from the 5th IPCC Report, Church et al. 2013) and scenarios used for the present study.

		RCP 4.5	Present Study	RCP 8.5	Present Study
2046–2065	P95	0.30		0.38	
	Median	0.26	0.24	0.30	0.24
	P05	0.19		0.22	0.79
2081–2100	P95	0.63		0.82	
	Median	0.47	0.48	0.63	0.48
	P05	0.32		0.45	
2100	P95	0.71		0.98	
	Median	0.53	0.48	0.74	0.79
	P05	0.36		0.52	

Due to the high computational cost of the simulations and the fact that sea level rise scenarios would be combined with the four different river discharge scenarios as shown in Table 1, it became time limiting to simulate all sea level rise forecasts from the IPCC. It can be noted in Table 2 that sea level rise projections overlap for the different RCP scenarios. It was decided to consider three different cases of sea level rise that covered all possibilities within all scenarios from the IPCC. The sea level rise projections used for the present study are outlined in the arrows in Table 2. They correspond to a sea level rise of 0.24 m for the year 2040, 0.48 m for the year 2070 and 0.79 m for the year 2100.

The three different sea level rise projections along with the current sea level were combined with the four different river discharge scenarios along the present bathymetry of the estuary for a total of 16 simulations. Each simulation evaluated the velocity of salt transport in the Guadiana Estuary as well as areas of flooding. The flooding areas were computed along ten different classes of submersion time along a tidal cycle over the bathymetry allowing maximum flooding. Table 3 outlines the different flooding classes.

Table 3. Classes of flooding time under one tidal cycle

Class	Flood interval in hours
1	0–1.2
2	1.2–2.4
3	2.4–3.7
4	3.7–4.9
5	4.9–6.1
6	6.1–7.3
7	7.3–8.6
8	8.6–9.8
9	9.8–11.0
10	11.0–12.2

For each class in Table 3 the area of flooding was computed, and a histogram was constructed for the various scenarios. Flood distribution maps of four classes of inundation time: 0 to 3 h, 3 to 6 h, 6 to 9 h and 9 to 12.2 h were also constructed.

3 Results and Discussion

3.1 Temporal Evolution of Salinity

The different sea level rise and river discharge scenarios described in the methods were imposed in the model to produce several time series and horizontal distribution maps. As the number of maps and figures produced is quite high, only the results for typical system conditions are shown. The time series demonstrate the temporal evolution of salinity in several locations of the estuary over two tidal cycles for the various scenarios of sea level rise and different freshwater flow rates. The locations used for each time series can be seen in Fig. 3.

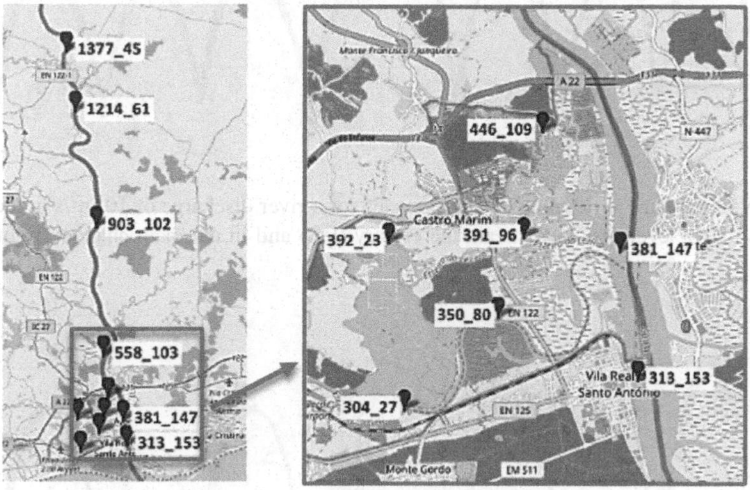

Fig. 3. Location and nomenclature of points used in the time series

The time series showed an increase in salinity in response to mean sea level rise in the main channel and lower area of the estuary for low river discharge flow rates (10 m^3/s) in the present bathymetry. The results for locations further upstream depicted an increase in salinity at both low and high tide, with salinity differences of approximately 6.

The velocity time series showed a reduction in advective transport at the mouth of the estuary due to mean sea level rise. Salt is transported mainly by diffusion, in the innermost regions where the signal of the tide is not detected. This result is consistent with other values of velocity obtained for this region. Further upstream the estuary, the salinity increased to about 4 in response to an increase in mean sea level. The increase in salinity in response to an increase in mean sea level is more pronounced closer to the coastline than it is upstream the main channel.

The time series results are similar for intermediate discharge flows (50 m^3/s) with an increase in salinity followed by an increase in mean sea level. However, the salinity differences are much less pronounced with a salinity increase of 2 in the main channel. At the streams that let out into the estuary, the salinity decreased to a value of 4, most likely due to the increase in water volume during ebb periods.

The salinity changes were much less pronounced for higher flow rates (100 m^3/s and 500 m^3/s) as the entire system is dominated by fresh water from the main channel. Velocity time series for a river discharge of 10 m^3/s and 100 m^3/s downstream the estuary and in the main channel are shown in Figs. 4 and 5. Salinity time series for a river discharge of 10 m^3/s and 100 m^3/s are shown in Figs. 6 and 7.

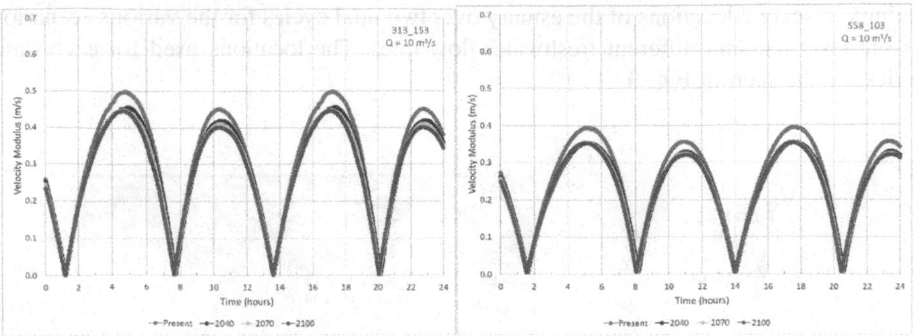

Fig. 4. Velocity time series over two tidal cycles for a river discharge of 10 m^3/s in the various scenarios of sea level rise downstream the estuary (left) and in the main channel (right).

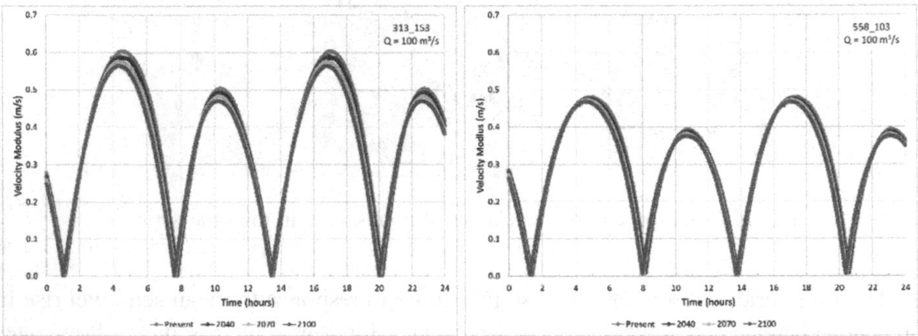

Fig. 5. Velocity time series over two tidal cycles for a river discharge of 100 m^3/s in the various scenarios of sea level rise downstream the estuary (left) and in the main channel (right).

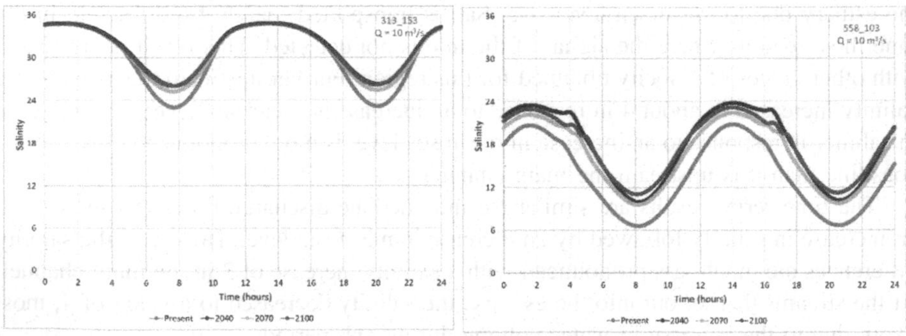

Fig. 6. Salinity time series over two tidal cycles for a river discharge of 10 m^3/s in the various scenarios of sea level rise downstream the estuary (left) and in the main channel (right).

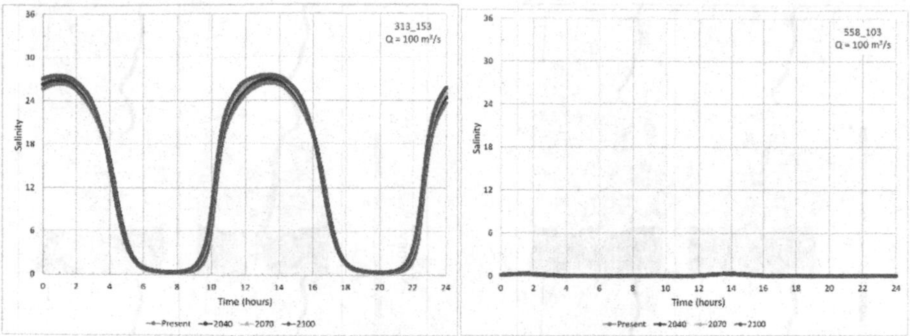

Fig. 7. Salinity time series over two tidal cycles for a river discharge of 100 m^3/s in the various scenarios of sea level rise downstream the estuary (left) and in the main channel (right).

3.2 Horizontal Distribution of Salinity Transport

The salinity distribution maps for the present bathymetry in high water conditions for the different scenarios of sea level rise can be seen in Figs. 8 and 9. Figures 10 and 11 show the salinity distribution maps for the present bathymetry in low water conditions. Only salinity maps for a river discharge of 10 m^3/s and 100 m^3/s are shown as they are most representative of typical system conditions.

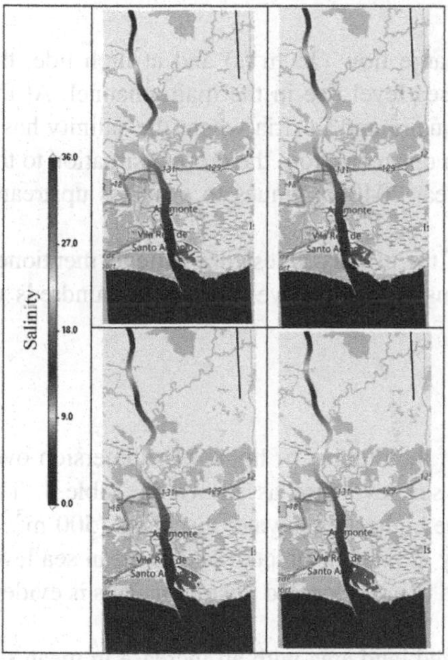

Fig. 8. Salinity distribution maps in high water conditions for a discharge flow of 10 m^3/s for the present year, 2040, 2070 and 2100.

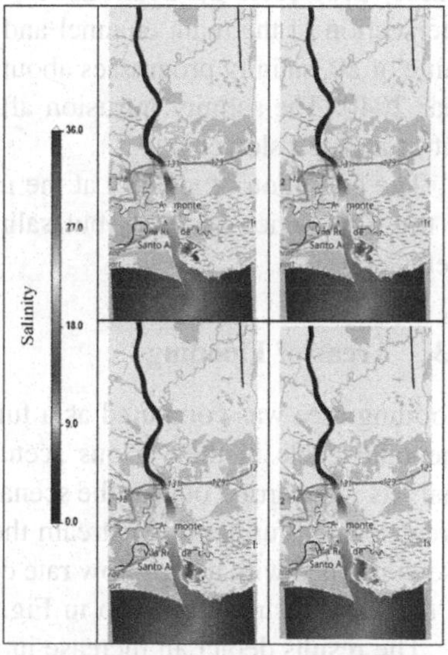

Fig. 9. Salinity distribution maps in high water conditions for a discharge flow of 100 m^3/s for the present year, 2040, 2070 and 2100.

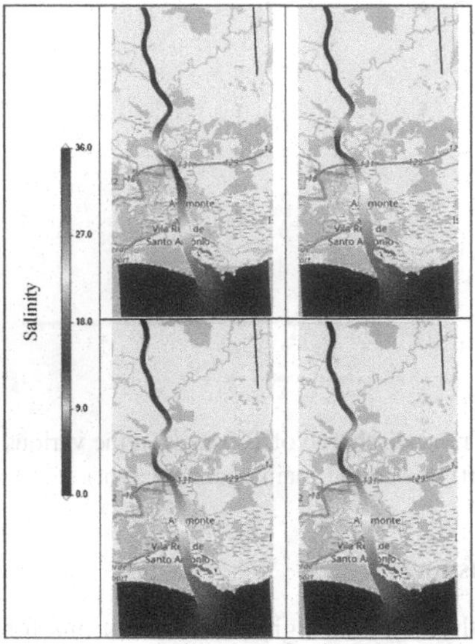

Fig. 10. Salinity distribution maps in low water conditions for a discharge flow of 10 m^3/s for the present year, 2040, 2070 and 2100.

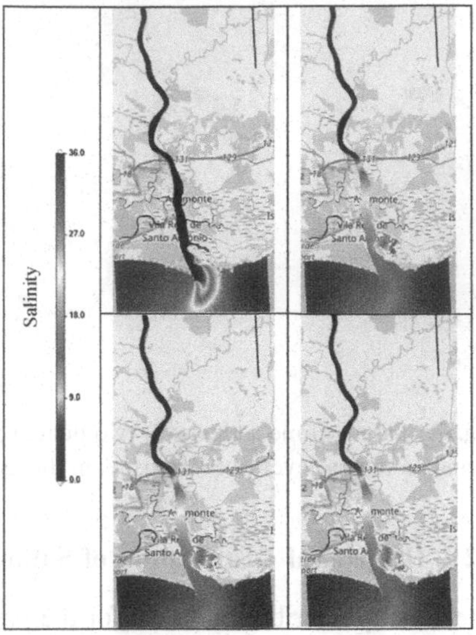

Fig. 11. Salinity distribution maps in low water conditions for a discharge flow of 100 m^3/s for the present year, 2040, 2070 and 2100.

For scenarios with low freshwater discharge flow (10 m^3/s) and at high tide, the salinity progresses upstream in response to sea level rise in the main channel. At the intersection of the main channel and Carresqueira and Lezíria where the salinity has a value of 20, salinity progresses about 3100 m upstream from the present situation to the year 2040. The salinity intrusion after the year 2040 continues to progress upstream, but at a much slower rate.

The evolution of salinity at the mouth of the estuary is less dramatic as mentioned in the time series analysis, but salinity intrusion can evolve by tens or hundreds of meters.

3.3 Areas of Flooding

Flooding area was computed as a function of the number of hours of submersion over one tidal cycle in the various scenarios of sea level rise as shown in Table 4. The analysis was carried out for the scenario of the highest freshwater discharge (500 m^3/s). Areas located further downstream the estuary are not as affected by the mean sea level rise combined with a high flow rate compared to areas located upstream. This is evident in the flood distribution map in Fig. 13.

The results depict an increase in submerged land area with an increase in mean sea level. The total intertidal submersion area increases from a value of 10 km^2 for the present year to a value of 14 km^2 in the "2100" scenario. The relationship between mean sea level and flooding area is nearly linear. Furthermore, an increase in mean sea

Table 4. Flood distribution areas based on hours of submersion over one tidal cycle in the various scenarios of sea level rise.

Class	Hours of submersion	Area (km^2)			
		Present	2040	2070	2100
1	0–1.2	0.57	0.56	0.60	0.59
2	1.2–2.4	1.39	1.45	1.65	1.45
3	2.4–3.7	1.46	1.59	1.54	1.86
4	3.7–4.9	1.82	1.93	2.04	2.11
5	4.9–6.1	1.19	1.63	1.69	1.86
6	6.1–7.3	1.29	1.46	1.97	2.17
7	7.3–8.6	0.81	1.03	1.18	1.38
8	8.6–9.8	0.82	0.99	0.97	1.22
9	9.8–11	0.63	0.54	0.72	0.85
10	11–12.2	0.24	0.24	0.31	0.44
Total		**10.23**	**11.41**	**12.67**	**13.93**

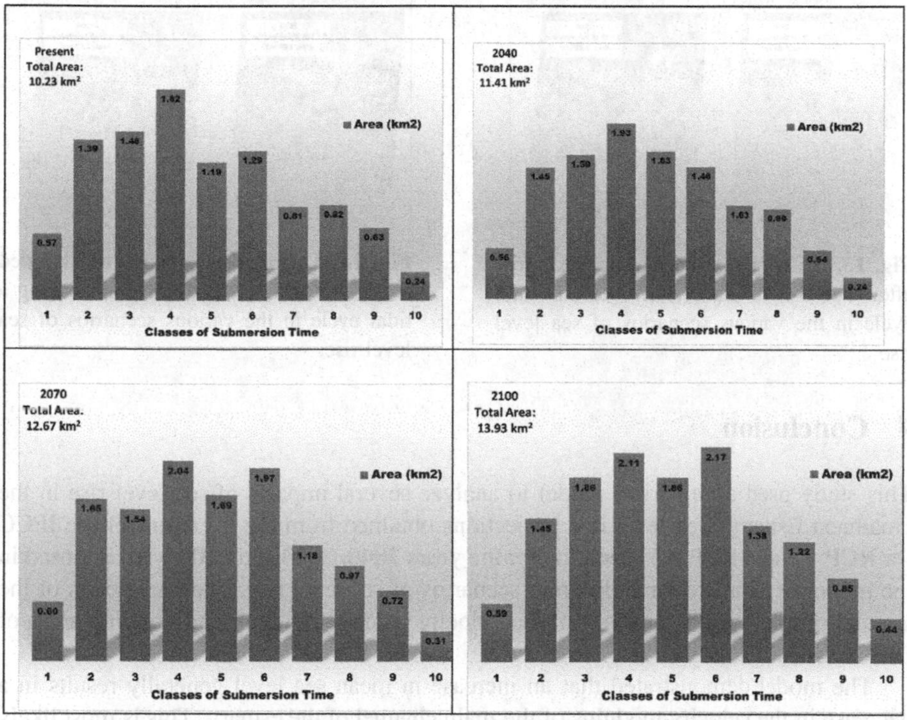

Fig. 12. Histogram of flood distribution areas as a function of the number of hours of immersion over one tidal cycle in the various scenarios of sea level rise.

level corresponds to a higher submersion time. For example, the present scenario is class 4 (3.7 to 4.9 h of submersion), but the "2100" scenario is class 6 (6.1 to 7.3 h of submersion). Figures 13 and 14 show the flooded regions over the various sea level rise scenarios by class of submersion time. The evolution of the inundated area is more pronounced in the north margin of the inlet not confined by Vila Real de Santo António. The marshes near the terminal section of Ribeira do Beliche also suffer high inundation areas as a result of sea level rise. The present study does not have bathymetry data for locations further upstream this point, so no conclusions can be drawn for the flooding areas upstream. Since the river has a narrower upstream profile the areas of flooding are expected to be lower (Fig. 12).

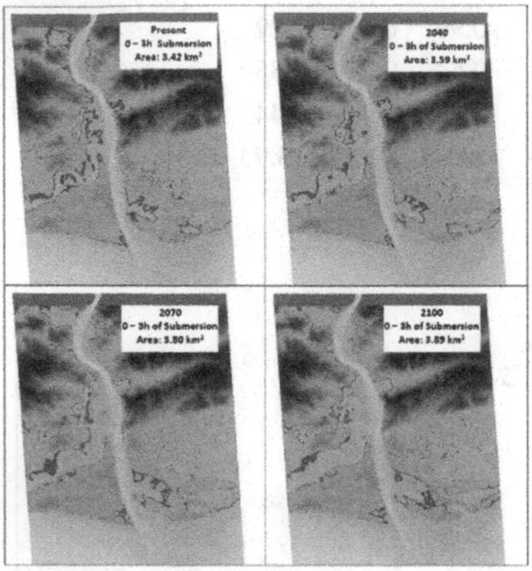

Fig. 13. Distribution maps of flooded areas after 0 to 3 h of immersion over one tidal cycle in the various scenarios of sea level rise

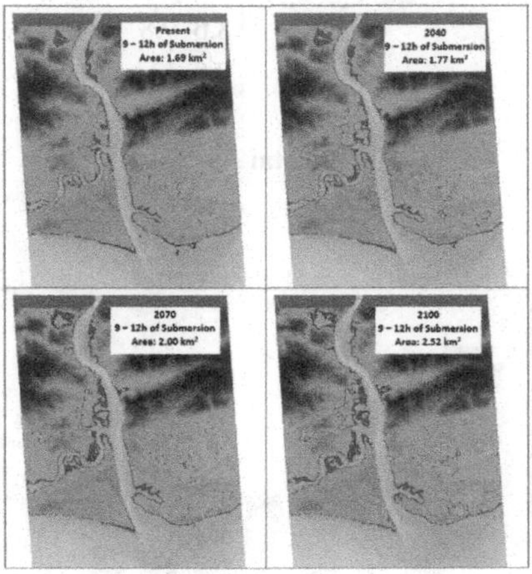

Fig. 14. Distribution maps of flooded areas after 9 to 12 h of immersion along a tidal cycle in the various scenarios of sea level rise.

4 Conclusion

This study used a numerical model to analyze several impacts of sea level rise in the Guadiana Estuary. Sea level rise projections obtained from the 5th report of the IPCC for RCP 4.5 and RCP 8.5 scenarios for the years 2040, 2070 and 2100 were imposed in the model to simulate three different scenarios of sea level rise. The parameters of the simulations allowed an analysis of the velocity modulus, salinity and flooding areas of the Guadiana Estuary.

The model demonstrated that an increase in mean sea level generally results in a decrease in the velocity modulus of the main channel of the estuary. This is most likely due to the increase in depth of the main channel as well as an increase in water volume associated with the rise in mean sea level.

The model simulations demonstrated that the mixing and transport of salinity is the result of a complex dynamic in which the lower portion of the Guadiana Estuary is dependent on both freshwater flow from the Guadiana River and the mean sea level of the adjacent coastal waters. The Guadiana Estuary alternates between purely fresh water when there is a high discharge from the Guadiana River and water of varying salinities for intermediate or low flows from the river. In the latter situation, the tide dominates the salinity with a semi-diurnal pattern.

For lower freshwater discharge rates, salinity increased in response to an increase in mean sea level with a more pronounced change in salinity in the main channel. For intermediate discharge flows, the increase in water volume due to sea level rise induced a higher penetration of fresh water from the outlet and therefore resulted in lower salinity values at the mouth of the estuary. For high discharge flows only fresh water was observed in all locations of the estuary. The results of the flood distribution areas demonstrated an increase in submerged land area due to mean sea level rise with an estimated increase of 4 km^2 for the entire intertidal area of the present situation. There is most significant flooding at Esteiro da Carrasqueira and the marshland surrounding Ribeira de Beliche.

The results portray an overall increase in both salinity and flooding area in the Guadiana Estuary with respect to an increase in mean sea level. As these two consequences are detrimental to estuarine ecosystems, society and the economy, further research must be done to further analyze the impacts of sea level rise in the Guadiana Estuary. One limitation of this study is that a two-dimensional model was used as opposed to a three-dimensional model. It is representative of most cases of the Guadiana Estuary because it is generally well-mixed, but as previously mentioned the estuary can become partially stratified in high freshwater flow conditions. A three-dimensional model could provide a more complete analysis of the hydrodynamics of the estuary. Additional numerical models should be simulated to portray a more quantifiable relationship between salinity and sea level rise and to further examine how the salinity changes with respect to other parameters, such as different tidal ranges.

Acknowledgements. This work was carried out under the “Plano Intermunicipal de Adaptação às Alterações Climáticas (PIAAC-AMAL)” project, national procedure 1472/2017.

References

Basos, N.: GIS as a tool to aid pre- and post-processing of hydrodynamic models. Application to the Guadiana Estuary Faculdade de Ciências e Tecnologia e Instituto Superior de Engenharia GIS as a tool to aid pre- and post-processing of hydrodynamic models. Applica. Faro, Portugal (2013)

Calero Quesada, M.C., García-Lafuente, J., Garel, E., Delgado Cabello, J., Martins, F., Moreno-Navas, J.: Effects of tidal and river discharge forcings on tidal propagation along the Guadiana Estuary. J. Sea Res. **146**(January), 1–13 (2019). https://doi.org/10.1016/j.seares.2019.01.006

Carrasco, A.R., Ferreira, O., Roelvink, D.: Coastal lagoons and rising sea level: a review. Earth Sci. Rev. **154**, 356–368 (2016). https://doi.org/10.1016/j.earscirev.2015.11.007

Chua, V.P., Xu, M.: Impacts of sea-level rise on estuarine circulation: an idealized estuary and San Francisco Bay. J. Mar. Syst. **139**, 58–67 (2014). https://doi.org/10.1016/j.jmarsys.2014.05.012

Church, J.A., et al.: Sea level change. In: Climate Change 2013: The Physical Science Basis. Contribution of Working Group I to the Fifth Assessment Report of the Intergovernmental Panel on Climate Change, pp. 1137–1216 (2013). https://doi.org/10.1017/CBO9781107415315.026

Church, J.A., White, N.J.: A 20th century acceleration in global sea-level rise. Geophys. Res. Lett. **33**(1), 94–97 (2006). https://doi.org/10.1029/2005GL024826

Delgado, J., et al.: Sea-level rise and anthropogenic activities recorded in the late Pleistocene/Holocene sedimentary infill of the Guadiana Estuary (SW Iberia). Quat. Sci. Rev. **33**, 121–141 (2012)

Fortunato, A.B., Oliveira, A., Alves, E.T.: Circulation and salinity intrusion in the Guadiana estuary (Portugal/Spain). Thalassas **18**(2), 43–65 (2002)

Garel, E., Pinto, L., Santos, A., Ferreira, Ó.: Tidal and river discharge forcing upon water and sediment circulation at a rock-bound estuary (Guadiana estuary, Portugal). Estuar. Coast. Shelf Sci. **84**(2), 269–281 (2009). https://doi.org/10.1016/j.ecss.2009.07.002

Hilton, T.W., Najjar, R.G., Zhong, L., Li, M.: Is there a signal of sea-level rise in Chesapeake Bay salinity? J. Geophys. Res. Oceans **113**(9), 1–12 (2008). https://doi.org/10.1029/2007JC004247

Hong, B., Shen, J.: Responses of estuarine salinity and transport processes to potential future sea-level rise in the Chesapeake Bay. Estuar. Coast. Shelf Sci. **104–105**, 33–45 (2012). https://doi.org/10.1016/j.ecss.2012.03.014

MARETEC: MOHID Water (2017). http://wiki.mohid.com/index.php?title=Mohid_Water. Accessed 4 Apr 2019

McLean, R.F., et al.: Coastal zones and marine ecosystems. In: Climate Change 2001: Impacts, Adaptation and Vulnerability, Cambridge, UK, pp. 343–379 (2001)

Nicholls, R.J., et al.: Sea-level rise and its possible impacts given a "beyond 4°C world" in the twenty-first century. Philos. Trans. Roy. Soc. A: Math. Phys. Eng. Sci. **369**(1934), 161–181 (2011)

Ross, A.C., Najjar, R.G., Li, M., Mann, M.E., Ford, S.E., Katz, B.: Sea-level rise and other influences on decadal-scale salinity variability in a coastal plain estuary. Estuar. Coast. Shelf Sci. **157**, 79–92 (2015). https://doi.org/10.1016/j.ecss.2015.01.022

Sampath, D.M.R., Boski, T., Loureiro, C., Sousa, C.: Modelling of estuarine response to sea-level rise during the holocene: application to the Guadiana Estuary-SW Iberia. Geomorphology **232**, 47–64 (2015). https://doi.org/10.1016/j.geomorph.2014.12.037

Sampath, D.M.R., Boski, T., Silva, P.L., Martins, F.A.: Morphological evolution of the Guadiana estuary and intertidal zone in response to projected sea-level rise and sediment supply scenarios. J. Quat. Sci. **26**(2), 156–170 (2011). https://doi.org/10.1002/jqs.1434

Wiseman, W.J., Swenson, E.M., Power, J.: Trends in Louisiana estuaries salinity. Estuaries **13**(3), 265–271 (1990)

Estuarine Light Attenuation Modelling Towards Improved Management of Coastal Fisheries

Marko Tosic[1](✉), Flávio Martins[2], Serguei Lonin[3], Alfredo Izquierdo[4], and Juan Darío Restrepo[1]

[1] School of Sciences, Department of Earth Sciences, Universidad EAFIT, Carrera 49 #7S-50, A.A.3300, Medellín, Colombia
marko.tosic7@gmail.com

[2] Instituto Superior de Engenharia, Universidade do Algarve, Campus da Penha, 8000 Faro, Portugal

[3] Escuela Naval de Cadetes "Almirante Padilla", Isla Naval Manzanillo, Cartagena de Indias, Colombia

[4] Faculty of Marine and Environmental Sciences, Applied Physics Department, University of Cádiz, Puerto Real, 11510 Cádiz, Spain

Abstract. The ecosystem function of local fisheries holds great societal importance in the coastal zone of Cartagena, Colombia, where coastal communities depend on artisanal fishing for their livelihood and health. These fishing resources have declined sharply in recent decades partly due to issues of coastal water pollution. Mitigation strategies to reduce pollution can be better evaluated with the support of numerical hydrodynamic models. To model the hydrodynamics and water quality in Cartagena Bay, significant consideration must be dedicated to the process of light attenuation, given its importance to the bay's characteristics of strong vertical stratification, turbid surface water plumes, algal blooms and hypoxia. This study uses measurements of total suspended solids (TSS), turbidity, chlorophyll-*a* (Chl*a*) and Secchi depth monitored in the bay monthly over a 2-year period to calculate and compare the short-wave light extinction coefficient (Kd) according to nine different equations. The MOHID-Water model was used to simulate the bay's hydrodynamics and to compare the effect of three different Kd values on the model's ability to reproduce temperature profiles observed in the field. Simulations using Kd values calculated by equations that included TSS as a variable produced better results than those of an equation that included Chl*a* as a variable. Further research will focus on evaluating other Kd calculation methods and comparing these results with simulations of different seasons. This study contributes valuable knowledge for eutrophication modelling which would be beneficial to coastal zone management in Cartagena Bay.

Keywords: Light attenuation · 3D hydrodynamic modelling · Tropical estuaries · Coastal management

J. M. F. Rodrigues et al. (Eds.): ICCS 2019, LNCS 11539, pp. 301–314, 2019.
https://doi.org/10.1007/978-3-030-22747-0_24

1 Introduction

The coastal zone of Cartagena, Colombia is inhabited by small coastal communities which have traditionally depended on artisanal fishing for their livelihood. However, these communities are vulnerable to environmental changes that have occurred in Cartagena Bay in terms of water and sediment pollution [1]. These fishing communities have witnessed drastic declines in their fishing resources over the past decades which have led to reduced catch and fish-size [2], while studies have also demonstrated impacts of metal contamination in the fish [3, 4]. As a result, these populations have gradually reduced their diet of fish, though this can have repercussions in terms of health indicators, and they have also shifted their economy towards beach tourism [4]. While tourism may be economically beneficial to the communities now, the sustainability of beach tourism in coastal Cartagena is fragile due to a rapidly increasing number of beach users and limited management infrastructure, while the decline of fishing activities threatens the loss of the traditional culture of artisanal fishing.

There are various issues of water quality in Cartagena Bay that could be affecting the fisheries, including turbid freshwater plumes, eutrophication, hypoxia, bacteria and metal pollution [1], which come from multiple sources of continental runoff, domestic and industrial wastewater [5], making mitigation a complex and challenging task. Such impacts affect fisheries due to the effect of degraded water quality on coral reefs [6], which provide fish with important habitats for feeding, refuge and reproduction. Fish are also impacted directly by insufficient oxygen levels in the water column which limit their respiration [7, 8]. Recent research of oxygen in the water column of Cartagena Bay [1] shows hypoxic conditions as levels of dissolved oxygen and oxygen saturation were found below recommended threshold levels of 4 mg/l [9] and 80% [10], respectively, at depths of just 5–10 m from the surface (Fig. 1). A large input of freshwater in the bay leads to strong vertical stratification which inhibits vertical mixing and thus limits the oxygenation of the water column [11]. Though this problem is alleviated during the dry/windy season (Jan-April), hypoxic conditions are found for most of the year and worst during the transitional season (June-August) when heightened temperatures reduce oxygen levels even further (Fig. 1).

Monthly monitoring in the bay has found average concentrations of biological oxygen demand (BOD) of 1.15 ± 0.90 mg/l [1], exceeding the maximum threshold value of 1 mg/l established for fishing resources in Cuba [12]. While these conditions are due in part to large amounts of organic matter flowing into the bay from runoff and wastewater discharges [5], a substantial input of nutrients also leads to algal blooms [1, 13, 14]. However, the persistence of turbid plumes during most of the year causes primary production in the bay to be light-limited, rather than nutrient-limited, resulting in a lack of productivity below the pycnocline until the dry season when the plumes subside and algal blooms occur (Fig. 2) [1, 13, 15–17]. This seasonal dynamic of primary productivity is also reflected by higher levels of BOD and turbidity found in the bay's bottom waters during the dry/windy season and the transitional season that follows [1]. These seasonal blooms, in combination with the bay's stratification (weak vertical mixing) and morphology (small seaward straits, deep bay), result in oxygen depletion in the bottom waters [13, 16, 17].

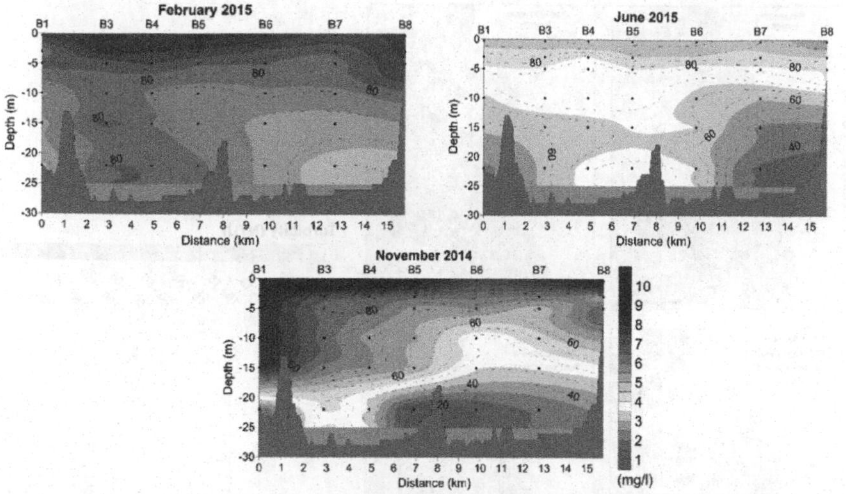

Fig. 1. Measurements of oxygen profiles in Cartagena Bay during the dry/windy (Feb.), transitional (June) and rainy (Nov.) seasons. Dissolved oxygen concentration (mg/l) is shown as a colour gradient. Oxygen saturation (%) is shown as dashed contour lines. Points (•) represent measurement locations at stations (B1, B3–B8) shown in Fig. 3 (see Tosic et al. [1] for details).

To better understand these complex processes leading to hypoxic conditions and evaluate strategies to mitigation the loss of fisheries, numerical models can be used to simulate the hydrodynamic and water quality conditions of the bay. Various studies have been done in Cartagena Bay to model its hydrodynamics and sediment transport [15, 18, 19], eutrophication and oxygen regimes [13, 16–18]. Recent modelling work has since applied the MOHID-Water model in Cartagena Bay [11].

Light attenuation in a system such as Cartagena Bay is particularly important for the successful application of a hydrodynamic-water quality model. Given the bay's characteristics of high surface turbidity and strong vertical stratification, light attenuation plays an important role in the processes of short-wave radiation penetration and the resulting heat flux, which are significant factors for the effective functioning of a hydrodynamic model. Furthermore, the penetration of light is also essential to primary production in such a light-limited system, and thus understanding light attenuation is an important step towards applying a eutrophication model.

In this study, we investigate the processes of light attenuation through monthly field measurements of temperature, salinity, total suspended solids (TSS), turbidity, chlorophyll-a and Secchi depth in Cartagena Bay and apply these data to numerical simulations with the MOHID-Water model. Multiple methods of calculating the short-wave light extinction coefficient (Kd) are carried out utilizing the field measurements and tested in model simulations. By comparing modelled thermoclines to field observations, the best fitting Kd calculation method is identified for further modelling. This first step towards eutrophication modelling in the bay provides an improved understanding of the water quality processes impacting coastal fisheries in Cartagena Bay.

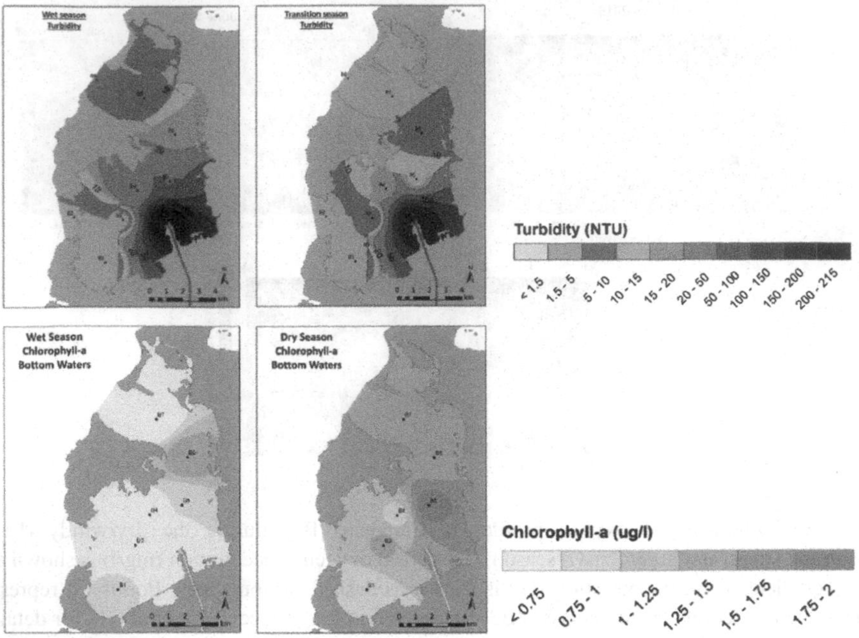

Fig. 2. Interpolated maps of surface water turbidity (NTU; above) and bottom water chlorophyll-*a* (μg/l; below) measured in Cartagena Bay and averaged over the 2014 rainy (left) and 2015 dry/windy (right) seasons (see Tosic et al. [1] for details).

2 Methods

2.1 Study Area

The tropical semi-closed estuary of Cartagena Bay is situated in the southern Caribbean Sea on the north coast of Colombia (10°20’ N, 75°32’ W, Fig. 3). The bay has an average depth of 16 m, a maximum depth of 32 m and a surface area of 84 km^2, including a small internal embayment situated to the north. Water exchange with the Caribbean Sea is governed by wind-driven circulation and tidal movement through its two seaward straits: “Bocachica” to the south and “Bocagrande” to the north [11]. Movement through Bocagrande strait is limited by a defensive colonial seawall 2 m below the surface. Bocachica strait consists of a shallow section with depths of 1–3 m, including the Varadero channel, and the Bocachica navigation channel which is 100 m wide and 24 m deep [13, 19]. The tides in the bay have a mixed, mainly diurnal signal with a micro-tidal range of 20–50 cm [20].

Estuarine conditions in the bay are generated by the Dique Canal to the south which discharges approximately 50–250 m^3/s of freshwater into the bay, the variability of which is strongly related to seasonal runoff from the Magdalena watershed [5, 13]. The flow of freshwater and sediments into the bay generates a highly stratified upper water column with a pronounced pycnocline in the upper 4 m of depth, above which turbid freshwater is restricted from vertical mixing and fine suspended particles tend to remain in the surface layer [11, 13, 16, 17].

2.2 Data Collection

Water quality was monitored monthly in the field from Sept. 2014 to Nov. 2016 between the hours of 9:00–12:00. Measurements were taken from 11 stations (Fig. 3), including one station in the Dique Canal (C0), eight stations in Cartagena Bay (B1–B8) and two stations at the seaward end of Barú peninsula (ZP1–ZP2). At all stations, Secchi depth was measured and CTD casts were deployed using a YSI Castaway measuring salinity and temperature every 30 cm of depth. Grab samples were taken from surface waters while bottom waters (22 m depth) were sampled with a Niskin bottle. Surface samples were collected in triplicate at station C0. A triplicate sample was also taken at a different single station in the bay each month to estimate sample uncertainty. Samples were analyzed at the nearby Cardique Laboratory for total suspended solids (TSS), turbidity and chlorophyll-*a* by standard methods [21].

At station C0 in the Dique Canal, discharge was measured with a Sontek mini-ADP (1.5 MHz) along a cross-stream transect three times per sampling date from Sept. 2015 to Nov. 2016. Bathymetric data with 0.1 m vertical resolution were digitized from georeferenced nautical maps (#261, 263, 264) published by the Centre for Oceanographic and Hydrographic Research (CIOH-DIMAR). In the 3 × 2 km area of Bocachica strait, the digitized bathymetry was updated with high-precision (1 cm) bathymetric data collected in the field on 17 Nov. 2016.

Hourly METAR data of wind speed, wind direction, air temperature and relative humidity were obtained from station SKCG at Rafael Núñez International Airport (approximately 10 km north of the bay; Fig. 3). Tidal components were obtained for numerous locations offshore of the bay from the finite element tide model FES2004 [22] using the MOHID Studio software. Hourly measurements of water level at a location within the bay were also obtained from the Centre for Oceanographic and Hydrographic Research (CIOH-DIMAR).

2.3 Model Configuration

The hydrodynamics of Cartagena Bay were simulated using the MOHID Water Modelling System [23, 24]. The MOHID Water model is a 3D free surface model with complete thermodynamics and is based on the finite volume approach, assuming hydrostatic balance and the Boussinesq approximation. It also implements a semi-implicit time-step integration scheme and permits combinations of Cartesian and terrain-following sigma coordinates for its vertical discretization [25]. Vertical turbulence is computed by coupling MOHID with the General Ocean Turbulence Model [26].

Model configuration for Cartagena Bay was based on an equally-spaced Cartesian horizontal grid with a resolution of 75 m and a domain area of 196 km^2 (Fig. 3). This included an offshore area extending 2.3 km off the bay, though only the results inside the bay are considered within the limits of the monitoring stations used for calibration. A mixed vertical discretization of 22 layers was chosen to reproduce the mixing processes of the highly stratified bay by incorporating a 7-layer sigma domain for the top five meters of depth and a variably-spaced (depth-incrementing) Cartesian domain below that depth. An optimal time step of 20 s was chosen.

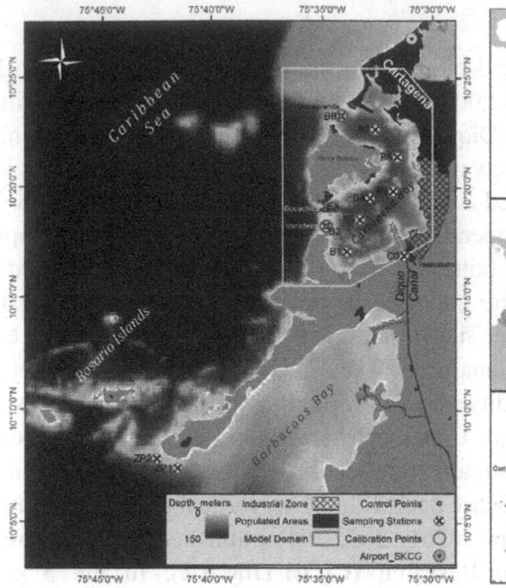

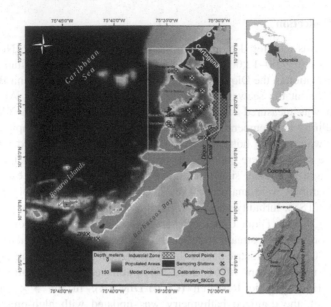

Fig. 3. Principal panel: study area showing sampling stations, model calibration points, control points, weather station (SKCG), bathymetry and model domain. Secondary panels: location of Colombia (upper panel); location of the Magdalena River (middle panel); flow of the Magdalena into the Caribbean Sea and along the Dique Canal into Cartagena Bay (lower panel).

The collected data of temperature, salinity, canal discharge, bathymetry, tides, winds and other meteorological data were used to force the hydrodynamic model. This process included a sensitivity analysis to identify the system's most effective calibration parameters (horizontal viscosity and bottom roughness), and subsequent calibration and validation of the model. For more details on the hydrodynamic model's configuration, calibration and results, see Tosic et al. [11].

2.4 Evaluation of Kd Calculation Methods

Multiple calculation methods of the short-wave light extinction coefficient (Kd) were compared utilizing the field measurements of total suspended solids (TSS), chlorophyll-a and Secchi depth. These included previously established relationships between Kd-TSS (Eq. 1) and Kd-Secchi (Eq. 7) in Cartagena Bay [15], Kd-TSS and Kd-Secchi relationships in coastal (Eqs. 2, 8) and offshore (Eqs. 3, 9) waters around the United Kingdom [27], a Kd-TSS relationship developed for the Tagus estuary, Portugal (Eq. 4) [28], a relationship for Kd-chlorophyll-*a* (Eq. 5) in oceanic waters [29], and a combined Kd-TSS-chlorophyll-*a* relationship (Eq. 6) based on Portela et al. [28] and Parsons et al. [29]. Kd values were computed for all surface stations and months of the monitoring program in order to compare the variability of values obtained using the different calculation methods.

Equation 1. $Kd = 1.31 * (TSS)^{0.542}$
Equation 2. $Kd = 0.325 + 0.066 * (TSS)$

Equation 3. $Kd = 0.039 + 0.067 * (TSS)$
Equation 4. $Kd = 1.24 + 0.036 * (TSS)$
Equation 5. $Kd = 0.04 + 0.0088 * (Chla) + 0.054 * (Chla)^{2/3}$
Equation 6. $Kd = (0.04 + 0.0088 * (Chla) + 0.054 * (Chla)^{2/3}) * (0.7 + 0.018 * TSS)$
Equation 7. $Kd = 2.3/(Secchi)$
Equation 8. $Kd = e^{(0.253-1.029*Ln(Secchi))}$
Equation 9. $Kd = e^{(-0.01-0.861*Ln(Secchi))}$

Hydrodynamic simulations were run using the different Kd values to compare the resulting thermoclines with measurements. Due to limitations of computing time, model simulations completed at the time of writing were limited the use of Eqs. 2, 4 and 5 during the dry/windy season (27 Jan. – 24 Feb. 2016), while continued research will evaluate the remaining Eqs. (1, 3, 6–9) and seasons (rainy, transitional) in the near future. Windy season simulations using Eqs. 2, 4 and 5 were chosen as the first simulations for multiple reasons: The windy season was chosen due to the presence of both suspended sediments and increased chlorophyll-a concentrations during this season, whereas the other seasons tend to have high suspended sediments but low Chla; Eq. 4 was chosen as it has already been included as code within the MOHID model, while Eq. 2 was chosen due to its similarity with Eq. 4; Eq. 5 was chosen to evaluate the importance of chlorophyll-a on Kd, as Chla in the only variable in Eq. 5.

Start and end times for each the dry/windy season simulation coincided with the dates of monthly field sessions. Initial conditions of salinity and temperature were defined by CTD measurements made on the corresponding start date. The Kd values used for simulations were computed using the mean value of the given parameter (TSS, Secchi, Chla) over all surface stations measured at both the start-month and end-month. Kd values were thus held spatio-temporally constant during the simulation period. To avert numerical instabilities, a spin-up period of one day was applied to gradually impose wind stress and open boundary forcings [30]. Outputs from the simulations were compared with field measurements of temperature profiles on the respective end date at six stations points in the bay (Fig. 3). Model performance was quantified by calculating the root mean squared error (RMSE).

3 Results

3.1 Field Observations

Concentrations of total suspended solids (TSS) at the bay's surface had a range of 2.1–64.8 mg/l with an average of 19.9 ± 13.4 mg/l. Greater concentrations were observed during the months of April-May, Aug. and Oct.–Nov. (Figure 4). Stations B3, B4 and B5 yielded higher concentrations in general due to their proximity to the Dique Canal's outlet.

Average concentrations of chlorophyll-a (Chla) of 3.5 ± 4.0 μg/l, and a median value of 2.0 μg/l, were found in the surface waters of Cartagena Bay. A wide range of concentrations of 0.5–22.0 μg/l was observed with peak values of 10–22 μg/l measured between the months of Feb.–May. The occurrence of these peaks during the dry season demonstrates the process of algal blooms occurring following the rainy season when sediment plumes recede and transparency improves. Greater Chla concentrations were found to the north of the bay at stations B4–B8, likely due to improved transparency further from the Dique Canal.

The range of turbidity values found in the bay's surface waters was 0.7–65.3 NTU with an average of 7.4 ± 9.8 NTU. Greater turbidity levels were found at stations B3, B4 and B5, similar to the spatial pattern of TSS. Turbidity was particularly high in the months of May, June, Aug., and Nov. Increased turbidity during the months of May, Aug. and Nov. reflects the findings of TSS. However, higher turbidity in the month of June occurs after the increases of TSS and Chla found in May.

Measurements with the Secchi disk yielded depths of transparency ranging from 0.1–6.0 m, with an average of 1.8 ± 1.3 m. Greater depths were observed at the stations B7–B8 to the north of the bay, where the lowest turbidity levels were also found. The smallest transparency was observed during the months of May, June, Aug., and Nov., in close agreement with turbidity results. Meanwhile, the peak transparency was observed by Secchi depths in the month of Jan., when the lowest levels of turbidity, TSS and Chla were all found as well.

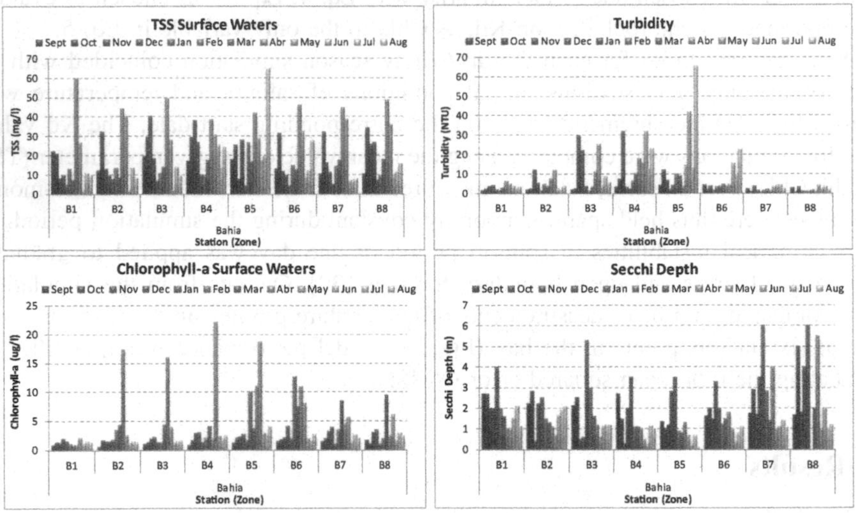

Fig. 4. Monthly surface measurements of total suspended solids (TSS), turbidity, Secchi depth and chlorophyll-*a* at 8 stations in Cartagena Bay (see Fig. 3 for locations).

3.2 Short-Wave Light Extinction Coefficients

The average short-wave light extinction coefficients calculated from the various equations ranged from almost zero to 6.3. The highest coefficient values resulted from Eq. 1 based on TSS with an average Kd value of 6.3 ± 2.1. Also based on TSS, Eqs. 2, 3 and 4, all produced similar average Kd values between 1.4–1.9 with standard deviations between 0.5–0.9. The inclusion of Chl*a* as a variable in Eqs. 5 and 6 resulted in much lower average Kd values of 0.2 ± 0.1.

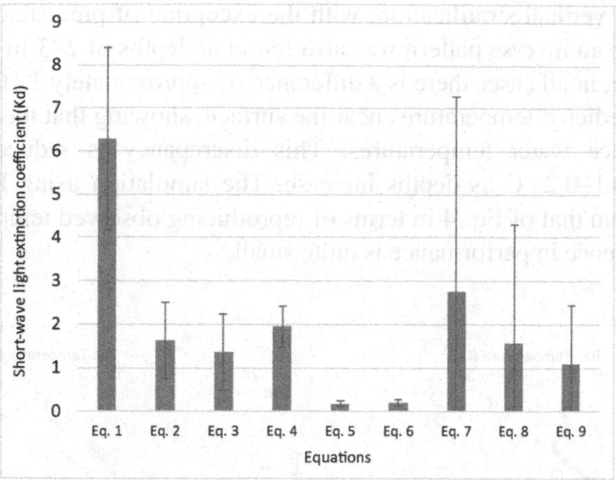

Fig. 5. Average values of the short-wave light extinction coefficient (Kd) calculated from Eqs. 1–9 using observed data of TSS, Chl*a* and Secchi depth. Error bars represent standard deviations.

Calculations of the short-wave light extinction coefficient with Eqs. 7–9 using measurements of Secchi depth resulted in high variability as standard deviations were greater than the average values themselves. An average coefficient of 2.7 ± 4.5 was calculated with Eq. 7, while lower Kd values of 1.6 ± 2.7 and 1.1 ± 1.3 were found with Eqs. 8 and 9, respectively.

3.3 Modelling Results

Results of the model simulations yielded relatively good predictions of temperatures when compared to observations, with RMSE values of 0.33 °C, 0.35 °C and 0.37 °C for simulations using Eqs. 2, 4 and 5, respectively. Results of the model simulation using Eq. 5 for calculation of the Kd coefficient based on concentrations of chlorophyll-*a* were found to be least effective in reproducing the temperature profiles of the dry/windy season (RMSE = 0.37 °C; Fig. 6). Temperature profiles generated by this simulation at different stations in Cartagena Bay were less vertically stratified than measurements. Within the top meter of surface water, the predicted temperature profiles were vertically uniform at stations B4–B7 suggesting a lack of light attenuation at the surface, in sharp contrast to the observed temperature profiles. At stations B1 and B3, the temperature profiles predicted by the simulation using Eq. 5 were inverse in the top 4 and 2 m of water, respectively, showing a drastic dissimilarity to the observations. However, model results for the subsurface seemed to improve significantly using Eq. 5, which could reflect an agreement between this Chla-based equation and the observations of increased Chla in subsurface waters during the dry/windy season (Fig. 2).

Predicted results of temperature from the simulations using Eqs. 2 (RMSE = 0.33 °C) and 4 (RMSE = 0.35 °C) were more similar to observations than those of Eq. 5. With Eqs. 2 and 4, the predicted temperature profiles were more similar to observations in terms of the pattern of vertical stratification, with the exception of predicted temperatures at station B1 where an inverse pattern was also found at depths of 2–3 m in the simulated results. However, in all cases there is a difference of approximately 1 °C between in the observed and predicted temperatures near the surface, showing that the model is underestimating surface water temperatures. This discrepancy is reduced gradually to approximately 0.1–0.2 °C as depths increase. The simulation using Eq. 2 performed slightly better than that of Eq. 4 in terms of reproducing observed temperature profiles, though the difference in performance is quite small.

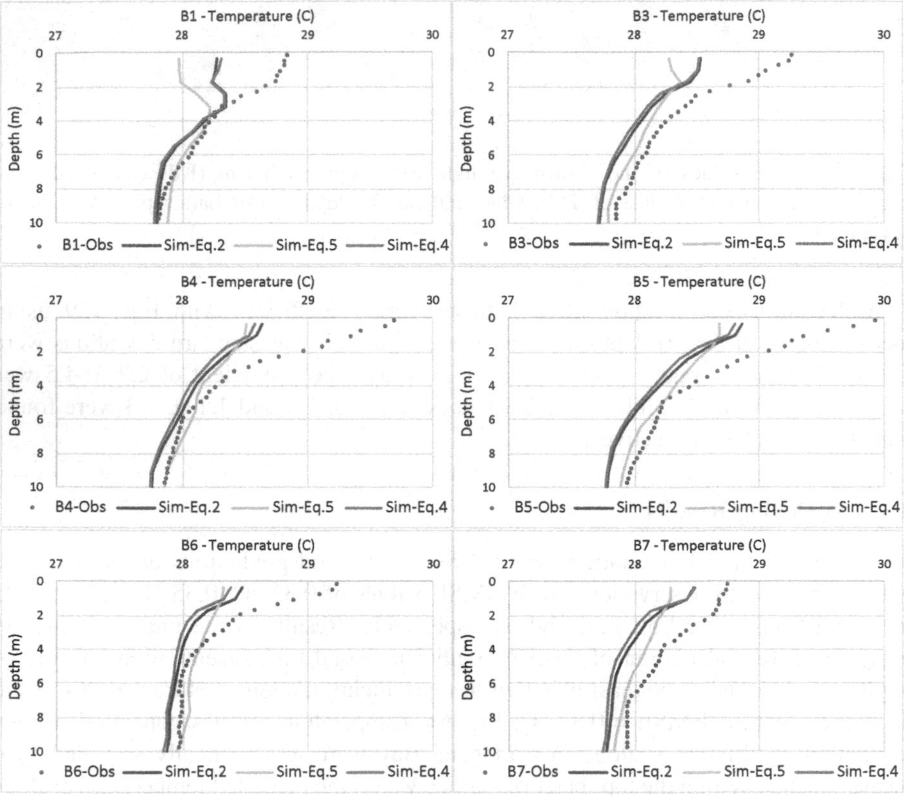

Fig. 6. Model results (Sim) using different Kd equations (Eqs. 2, 4, 5) compared to measurements (Obs) of temperature profiles at the end of the dry/windy season simulation at six stations in Cartagena Bay (B1, B3–B7).

4 Discussion

4.1 Comparison of Kd Calculation Methods

The improved performance found using Eqs. 2 and 4 when compared to Eq. 5 may be expected given the high concentrations of TSS found in the bay. As Eqs. 2 and 4 include TSS as a variable, and Eq. 5 does not, it would be reasonable to expect that the model would be more dependent on concentrations of TSS than those of Chl*a*, which is the variable in Eq. 5. Perhaps Chl*a* might be a more important variable in simulations run between the months of Feb.–May when higher Chl*a* concentrations were observed. In this regard, additional simulations should be run during the months of Feb.–May using Eq. 5 based on the Kd-Chla relationship or Eq. 6 based on a combination of Kd-Chla and Kd-TSS relationships.

Additional simulations should also be run using Eq. 1, as it was previously developed for Cartagena Bay. Initial simulations were focused on using Eqs. 2 and 4 because of the similarity in average Kd values that they produced (Fig. 5) and given that Eq. 4 is integrated in the MOHID Studio software. However, considering the discrepancy found between observed and predicted temperatures, perhaps the higher Kd values produced by Eq. 1 will produce results that are more consistent with the steep vertical temperature gradients observed in the bay's surface layers.

The use of Eqs. 7–9 may be expected to produce results similar to Eq. 2 and 4 due to the similarity in the average Kd values calculated by these equations (Fig. 5). The large standard deviations calculated using Eqs. 7–9 show the high variability in Kd values generated with these equations. Therefore, it may be illogical to use an average Kd value calculated with such equations that result in such high variability. In this regard, a more realistic approach with all of the equations would be to use a model configuration that permits a continuous calculation of Kd in space and time (for each cell and at each time-step) according to the variability of the parameters TSS, Chl*a* or Secchi depth.

Limitations in the model's prediction of the thermocline could also be due to inaccuracies in predicting the river plume with the model's present configuration. As the freshwater discharge is based on interpolation of monthly measurements and the winds are taken hourly from a single location, these parameters forcing the model could generate inconsistencies with the field observations. These issues could possibly even be exacerbated using a model configuration that permits a continuous calculation of Kd in space and time, as the model's accuracy in predicting surface water variables such as TSS or Chla (which are strongly dependent on plume dispersion patterns) would then play an even greater role in determining thermocline predictions.

One variable that could possibly contribute to the underestimation of surface temperature of the model could be additional freshwater input not considered in the model's present configuration. Additional sources of freshwater not considered include a series of smaller canals flowing from the industrial sector along the bay's east coast, surface runoff from the small catchment area of the bay itself, and occasional discharges from the city's sewerage system through an outdated submarine outfall and backup outlets along the coast when the system overflows. As this freshwater is warmer than seawater but not accounted for in the model, it may result in a bias for the model to underestimate surface temperatures in the bay,

Additional simulations should also be run for the rainy season and traditional seasons when conditions in turbidity differ from the dry/windy season. Ideally a Kd-relationship that satisfies the varying conditions of the different seasons could be found. However, it may also be possible that the strong seasonal variability in water quality parameters like turbidity, TSS and Chl*a* could result in a seasonal variability in light attenuation processes that require different Kd calculation methods depending on the season.

4.2 Social Implications

Though a better understanding of light attenuation properties in Cartagena Bay may not result in an immediate improvement in the livelihood of artisanal fishermen, an estuarine water model would provide environmental authorities with a powerful tool for managing the coastal zone. Mitigation strategies are needed to improve the coastal zone's water quality which is essential for fishing resources. Such strategies should be evaluated with model simulations prior to implementation in order to assess the consequences on coastal water quality, as they may be contrary to the intended result [11]. Given the importance of light attenuation on the processes of primary productivity and oxygen in the bay, this research provides valuable knowledge for eutrophication modeling which would be beneficial to mitigation planning.

However, it should also be noted that in addition to mitigating water pollution, there is also a need for improved fishery management. Research done by Escobar et al. [2] on fish populations in the study area showed that genetic diversity in fish populations was high, which supports the sustainability of the resource. However, the authors also found that the mean body length for all species was significantly smaller than body length at maturity. While this latter finding implies that fish stocks in this coastal zone are low, it also indicates a need for improved fishery management, as fishermen should not be consciously capturing immature fish. Given the recent findings of Tosic et al. [1, 5] and those of Escobar et al. [2], the cause of this low abundance in the fish stock is likely due to both pollution and overfishing.

5 Conclusions

Estuarine modelling in a semi-enclosed water body such as Cartagena Bay requires an improved understanding of processes of light attenuation. The bay is characterized by turbid surface waters and hypoxic conditions throughout most of the year. However, during the dry/windy season (Jan.–April) when greater vertical mixing improves oxygen conditions and receding sediment plumes result in increased transparency and primary productivity in the water column.

Calculation of the short-wave light extinction coefficient can vary greatly depending on the equation used. Average Kd values calculated for Cartagena Bay from nine different equations ranged in values from almost zero to 6.3. Numerical modelling simulations showed that the selection of the appropriate method for Kd calculation is important for the model's functionality. Among the three equations evaluated, the two equations that included TSS as a variable generated temperature profiles that were more similar to field observations than the profiles simulated using an equation including Chl*a* as a variable. However, further research is needed to evaluate the remaining Eqs. (1, 3, 6–9) and seasons (rainy, transitional).

An estuarine water model presents environmental authorities with a powerful tool for managing the coastal zone as it allows for the evaluation of mitigation strategies needed to improve coastal zone water quality. Such strategies are particularly important to the artisanal fisheries of Cartagena Bay and the vulnerable coastal communities that depend on them. As the economies, traditional culture and public health of these societies may be impacted by water contamination, the use of water quality models by coastal management in Cartagena Bay would be beneficial to the effective planning of pollution mitigation.

Acknowledgements. This work was carried out with the aid of a grant from the International Development Research Centre, Ottawa, Canada (grant number 108747-001). Financial support was also provided by EAFIT University, Corporación Autónoma Regional del Canal del Dique (CARDIQUE; agreement number 15601), as well as a scholarship granted to the lead author by the Erasmus Mundus Doctoral Programme in Marine and Coastal Management (MACOMA).

References

1. Tosic, M., Restrepo, J.D., Lonin, S., Izquierdo, A., Martins, F.: Water and sediment quality in Cartagena Bay, Colombia: seasonal variability and potential impacts of pollution. Estuar. Coast. Shelf Sci. **216**, 187–203 (2019). ISSN 0272-7714
2. Escobar, R., Luna-Acosta, A., Caballero, S.: DNA barcoding, fisheries and communities: What do we have? Science and local knowledge to improve resource management in partnership with communities in the Colombian Caribbean. Mar. Policy **99**, 407–413 (2019)
3. Olivero-Verbel, J., Caballero-Gallardo, K., Torres-Fuentes, N.: Assessment of mercury in muscle of fish from Cartagena Bay, a tropical estuary at the north of Colombia. Int. J. Environ. Health Res. **19**(5), 343–355 (2009)
4. Restrepo, J.D., Tosic, M. (eds.): Executive summary of the Project Basin Sea Interactions with Communities. EAFIT University, Medellín, June 2017, 31 p. (2017). http://www.basic-cartagena.org/boletines/BASIC%20Cartagena%20-%20Resumen%20Ejecutivo.pdf
5. Tosic, M., Restrepo, J.D., Izquierdo, A., Lonin, S., Martins, F., Escobar, R.: An integrated approach for the assessment of land-based pollution loads in the coastal zone demonstrated in Cartagena Bay, Colombia. Estuar. Coast. Shelf Sci. **211**, 217–226 (2018)
6. Fabricius, K.E.: Effects of terrestrial runoff on the ecology of corals and coral reefs: Review and synthesis. Mar. Pollut. Bull. **50**(2), 125–146 (2005)
7. Díaz, R.J., Rosenberg, R.: Marine benthic hypoxia: a review of its ecological effects and the behavioral responses of benthic macrofauna. Oceanogr. Mar. Biol. Annu. Rev. **33**, 245–303 (1995)
8. Correll, D.L.: The role of phosphorus in the eutrophication of receiving waters: a review. J. Environ. Qual. **27**(2), 261–266 (1998)
9. MinSalud - Ministerio de Salud: Decreto No. 1594 del 26 de junio. Por el cual se reglamenta parcialmente el Título I de la Ley 9 de 1979, así como el Capítulo II del Título VI - Parte III - Libro II y el Título III de la Parte III - Libro I - del Decreto - Ley 2811 de 1974 en cuanto a usos del agua y residuos líquidos, 61 p. (1984)
10. Newton, A., Mudge, S.M.: Lagoon-sea exchanges, nutrient dynamics and water quality management of the Ria Formosa (Portugal). Estuar. Coast. Shelf Sci. **62**, 405–414 (2005)
11. Tosic, M., Martins, F., Lonin, S., Izquierdo, A., Restrepo, J.D.: Hydrodynamic modelling of a polluted tropical bay: assessment of anthropogenic impacts on freshwater runoff and estuarine water renewal. J. Environ. Manag. **236**, 695–714 (2019)
12. NC (Oficina Nacional de Normalización de Cuba): Evaluación de los objetos hidricos de uso pesquero. Norma Cubana 25/1999. Ciudad de La Habana, Cuba (1999)

13. Tuchkovenko, Y., Lonin, S.: Mathematical model of the oxygen regime of Cartagena Bay. Ecol. Model. **165**(1), 91–106 (2003). ISSN 0304-3800
14. Cañón-Páez, M.L., Tous, G., Lopez, K., Lopez, R., Orozco, F.: Variación espaciotemporal de los componentes fisicoquímico, zooplanctónico y microbiológico en la Bahía de Cartagena. Boletín Científico CIOH **25**, 120–134 (2007)
15. Lonin, S.: Cálculo de la transparencia del agua en la bahía de Cartagena. Boletín Científico CIOH **18**, 85–92 (1997)
16. Tuchkovenko, Y., Lonin, S., Calero, L.A.: Modelación ecológica de las bahías de Cartagena y Barbacoas bajo la influencia del Canal del Dique. Avances en Recursos Hidráulicos **7**, 76–94 (2000)
17. Tuchkovenko, Y., Lonin, S., Calero, L.A.: Modelo de eutroficación de la bahía de Cartagena y su aplicación práctica. Boletín Científico CIOH **20**, 28–44 (2002)
18. Lonin, S.A., Tuchkovenko, Y.S.: Modelación matemática del régimen de oxígeno en la Bahía de Cartagena. Avances en Recursos Hidráulicos **5**, 1–16 (1998)
19. Lonin, S., Parra, C., Andrade, C., Thomas, I.: Patrones de la pluma turbia del canal del Dique en la bahía Cartagena. Boletín Científico CIOH **22**, 77–89 (2004)
20. Molares, R.: Clasificación e identificación de las componentes de marea del Caribe colombiano. Boletín Científico CIOH **22**, 105–114 (2004)
21. APHA – American Public Health Association: Standard Methods for the Examination of Water and Wastewater. American Public Health Association, Washington DC (1985)
22. Lyard, F., Lefèvre, F., Letellier, T., Francis, O.: Modelling the global ocean tides: modern insights from FES2004. Ocean Dyn. **56**(5), 394–415 (2006)
23. Leitão, P.C., Mateus, M., Braunschweig, F., Fernandes, L., Neves, R.: Modelling coastal systems: the MOHID water numerical lab. In: Neves, R., Baretta, J., Mateus, M. (eds.) Perspectives on Integrated Coastal Zone Management in South America, pp. 77–88. IST Press, Lisbon (2008)
24. Mateus, M., Neves, R. (eds.): Ocean Modelling for Coastal Management: Case Studies with MOHID, 165 p. IST Press, Lisbon (2013)
25. Martins, F., Leitão, P.C., Silva, A., Neves, R.: 3D modelling in the Sado estuary using a new generic vertical discretization approach. Oceanol. Acta **24**, 551–562 (2001)
26. Burchard, H.: Applied Turbulence Modelling in Marine Waters, vol. 100. Springer, Heidelberg (2002)
27. Devlin, M.J., et al.: Relationships between suspended particulate material, light attenuation and Secchi depth in UK marine waters. Estuar. Coast. Shelf Sci. **79**(3), 429–439 (2008)
28. Portela, L.I.: Mathematical modelling of hydrodynamic processes and water quality in Tagus estuary. Ph.D. thesis, Universidade Técnica de Lisboa, Instituto Superior Técnico, Lisboa, Portugal (1996)
29. Parsons, T., Takahashi, M., Hargrave, G.: Biological Oceanographic Processes, 330 p. Pergamon Press, New York (1984)
30. Franz, G.A.S., Leitão, P., Santos, A.D., Juliano, M., Neves, R.: From regional to local scale modelling on the south-eastern Brazilian shelf: case study of Paranaguá estuarine system. Braz. J. Oceanogr. **64**(3), 277–294 (2016)

The NARVAL Software Toolbox in Support of Ocean Models Skill Assessment at Regional and Coastal Scales

Pablo Lorente[1(✉)], Marcos G. Sotillo[1], Arancha Amo-Baladrón[1], Roland Aznar[1], Bruno Levier[2], Lotfi Aouf[3], Tomasz Dabrowski[4], Álvaro De Pascual[1], Guillaume Reffray[2], Alice Dalphinet[3], Cristina Toledano[1], Romain Rainaud[3], and Enrique Álvarez-Fanjul[1]

[1] Puertos del Estado, Madrid, Spain
plorente@puertos.es
[2] Mercator Ocean, Toulouse, France
[3] MeteoFrance, Toulouse, France
[4] Marine Institute, Galway, Ireland

Abstract. The significant advances in high-performance computational resources have boosted the seamless evolution in ocean modelling techniques and numerical efficiency, giving rise to an inventory of operational ocean forecasting systems with ever-increasing complexity. The skill of the Iberia-Biscay-Ireland (IBI) regional ocean forecasting system, implemented within the frame of the Copernicus Marine Environment Monitoring Service (CMEMS), is routinely evaluated by means of the NARVAL (North Atlantic Regional VALidation) web-based toolbox. Multi-parameter comparisons against observational sources (encompassing both in situ end remote-sensing platforms) are regularly conducted along with model intercomparisons in the overlapping areas. Product quality indicators and skill metrics are automatically computed not only averaged over the entire IBI domain but also over specific sub-regions of particular interest in order to identify strengths and weaknesses of each model. The primary goal of this work is three-fold. Firstly, to provide a flavor of the basic functionalities of NARVAL software package in order to elucidate the accuracy of IBI near real time forecast components (physical, biogeochemical and waves); secondly, to showcase a number of the practical applications of NARVAL; finally, to present the future roadmap to build a new upgraded version of this software package, which will include the quality assessment of multi-year and interim products, the computation of long-term skill metrics or the evaluation of event-oriented multi-model intercomparison exercises. This synergistic approach, based on the integration of numerical models and multi-platform observational networks, should be useful to comprehensively characterize the highly dynamic sea states and the dominant modes of spatio-temporal variability.

Keywords: Model skill assessment · Ocean forecasting · Model intercomparison

J. M. F. Rodrigues et al. (Eds.): ICCS 2019, LNCS 11539, pp. 315–328, 2019.
https://doi.org/10.1007/978-3-030-22747-0_25

1 Introduction

Over recent decades, substantial advances have been achieved in the discipline of operational oceanography thanks to the significant increase in high-performance computational resources, among others: multicore processor-based cluster architectures, massive storage capabilities, optimized parallelization and efficient scalability strategies. Such progresses have boosted the seamless evolution in ocean modelling techniques and numerical efficiency, thereby giving rise to an inventory of operational ocean forecasting systems (OOFSs) with ever-increasing complexity. Nowadays, more sophisticated and memory-demanding simulations can be effectively run at shorter time intervals and finer spatio-temporal resolutions for coupled forecast systems that might include entangled air-sea, wave-current and/or biophysical interactions. In this context, the implementation of operational data assimilation schemes has constituted a quantum leap in terms of realistic forecast predictions since they maximize the interconnection of multi-platform ocean observing systems and OOFSs. This integrated approach provides critical oceanographic information to support wise decision-making in the marine environment with subsequent societal benefits.

In the framework of the Copernicus Marine Environment Monitoring Service (CMEMS), a global ocean model together with a wealth of nested regional OOFSs are currently running in different areas of the European seas and providing paramount oceanographic forecast products. Since the comparison of OOFSs against independent observations constitutes a core activity in CMEMS, the development of skill assessment software packages and dedicated web applications is an active theme. In particular, the accuracy of Iberia-Biscay-Ireland (IBI) regional OOFS is routinely evaluated by means of the NARVAL (North Atlantic Regional VALidation) system [1, 2], a web-based toolbox that provides a number of skill metrics automatically computed. NARVAL tool has been implemented to routinely monitor IBI performance and to objectively evaluate model's veracity and prognostic capabilities. Both real-time comparisons ('online mode') and regular-scheduled 'delayed-mode' comparisons (for longer time periods) are performed using a wealth of independent observational sources as benchmark, among others: in situ observations (from moorings, tide-gauges, drifters, gliders and ARGO floats networks) and remote-sensed estimations (provided by satellites and High-Frequency radars -HFR hereinafter-).

NARVAL is modular and flexible enough to assess the quality of a variety of near real-time (NRT) forecast products, encompassing the physical (IBI-NRT-PHY), biogeochemical (IBI-NRT-BIO) and waves (IBI-NRT-WAV) components. Furthermore, model intercomparisons are regularly conducted in the overlapping areas to elucidate pros and cons of each model performance. Product quality indicators and skill metrics are automatically computed not only averaged over the entire IBI domain but also over specific sub-regions of particular interest from a user perspective (i.e. coastal or shelf areas) in order to infer IBI accuracy and the spatiotemporal uncertainty levels (Fig. 1a).

The main goal of this contribution is to showcase: (i) the current practical applications of NARVAL software toolbox to evaluate the performance of IBI-NRT forecasting system and (ii) the future roadmap to build a new upgraded version here named NARVAL-PRO, which will include a number of novelties such as the accuracy assessment of multi-year (MY) and interim (INT) products or the computation of long-term skill metrics (Fig. 1b).

The paper is organized as follows: Sect. 2 provides further details about the model quality assessment framework and the workflow of NARVAL software toolbox. Section 3 describes the diverse NARVAL components and outlines a wealth of illustrative examples. Finally, main conclusions, ongoing work and future prospects are summarized in Sect. 4.

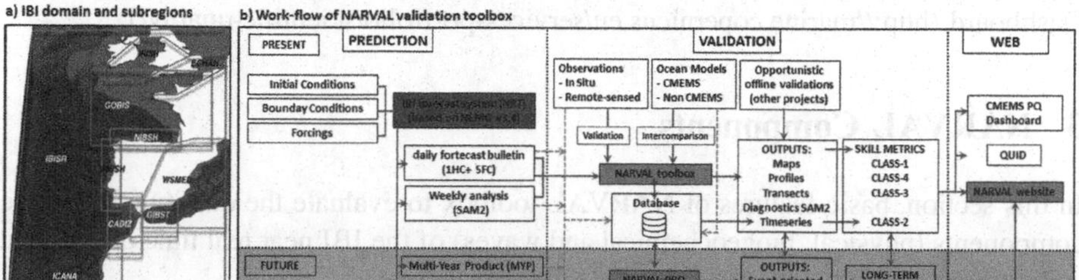

Fig. 1. (a) IBI regional service domain (IBISR) and defined sub-regions: Irish Sea (IRISH), English Channel (ECHAN), Gulf of Biscay (GOBIS), North Iberian Shelf (NIBSH), Western Iberian Shelf (WIBSH), Gulf of Cadiz (CADIZ), Strait of Gibraltar (GIBST), Western Mediterranean (WSMED) and Canary Islands (ICANA); (b) Workflow of NARVAL toolbox and future prospects.

2 Quality Assessment Framework and Workflow of NARVAL

The comparison of OOFSs against independent quality-controlled measurements constitutes a core activity in oceanographic operational centers (Fig. 1b) since it aids: (i) to infer the relative strengths and weaknesses in the modelling of several key physical processes; (ii) to compare different versions of the same OOFS and evaluate potential improvements and degradations before a new version is transitioned into operational status; (iii) to compare coarse resolution 'father' and nested high-resolution 'son' systems to quantify the added value of the downscaling approach adopted. With regards to the third aspect, IBI forecast products are regularly intercompared not only against other CMEMS regional model solutions (e.g. its parent system, the GLOBAL) in the overlapping areas but also against other non CMEMS models by means of NARVAL tool. Complementarily, opportunistic intercomparisons are conducted in the frame of diverse EU-funded projects such as MEDESS-4MS [3].

The agreement between both in situ and remote-sensing instruments and the ocean forecasting system is evaluated by means of computation of a set of statistical metrics traditionally employed in this framework: histograms, bias, root mean squared differences (RMSE), scalar and complex correlation coefficients, current roses, histograms, quantile-quantile (QQ) plots and the best linear fit of scatterplots. Skill metrics have been defined in four different types, including gridded model output (CLASS-1), time-series at specified locations and sections (CLASS-2), transports through sections and other integrated quantities (CLASS-3), and metrics of forecast capability (CLASS-4).

The statistical metrics regularly generated by NARVAL are online delivered to inform end-users and stakeholders about the quality and reliability of the marine forecast products routinely delivered, fostering downstream services and user uptake. This is achieved thanks to the QUality Information Document - QUID- [4], which is periodically updated and freely available in CMEMS website (http://marine.copernicus.eu/). Equally, the skill metrics can be also displayed through the CMEMS Product Quality Dashboard (http://marine.copernicus.eu/services-portfolio/scientific-quality/).

3 NARVAL Components

In this section, basic features of NARVAL toolbox to evaluate the quality of the three components (physical, biogeochemical and waves) of the IBI near real time operational suite are described.

3.1 IBI-NRT-WAV

The operational IBI near real time wave forecast system (IBI-NRT-WAV), based on MF-WAM, provides a 5-day regional wave forecast, which is updated twice a day (cycles at 00 z and 12 z). The model performs a partitioning technique on wave spectra that allows the separation between sea wind and primary and secondary wave swell systems [5]. The model was implemented on the IBI domain with a grid size of 10 km and with a spectral resolution of 24 directions and 30 frequencies, starting from 0.035 Hz. The IBI-WAV runs are driven by 3-hourly analyzed winds provided by the European Center for Medium-Range Weather Forecasts (ECMWF). The boundary conditions (wave spectra) are provided by the global CMEMS wave system, which uses the assimilation of altimeter wave data.

A specific module has been implemented in NARVAL toolbox to compare this forecast system against all the available observations on both online and delayed mode (Fig. 2a). The coverage area of IBI regional domain includes an array of deep and coastal buoys which provide hourly-averaged quality-controlled in situ measurements of significant wave height (SWH), mean wave period (MWP), the wave period at spectral peak (PKP), and the mean wave direction (WD). Such buoys are used in concert as a robust benchmark to conduct a multi-parameter skill assessment of the IBI-NRT-WAV. An example of annual (2017) comparison focused on the Irish Sea is here provided: scatter and Quantile-Quantile (QQ) plots for SWH and MWP, along with the associated skill metrics, clearly proved that the model properly captured basic features of the wave regime in a region with one of the most energetic wave climate in Europe, with a number of wave height events clearly above 11 m (Fig. 2b).

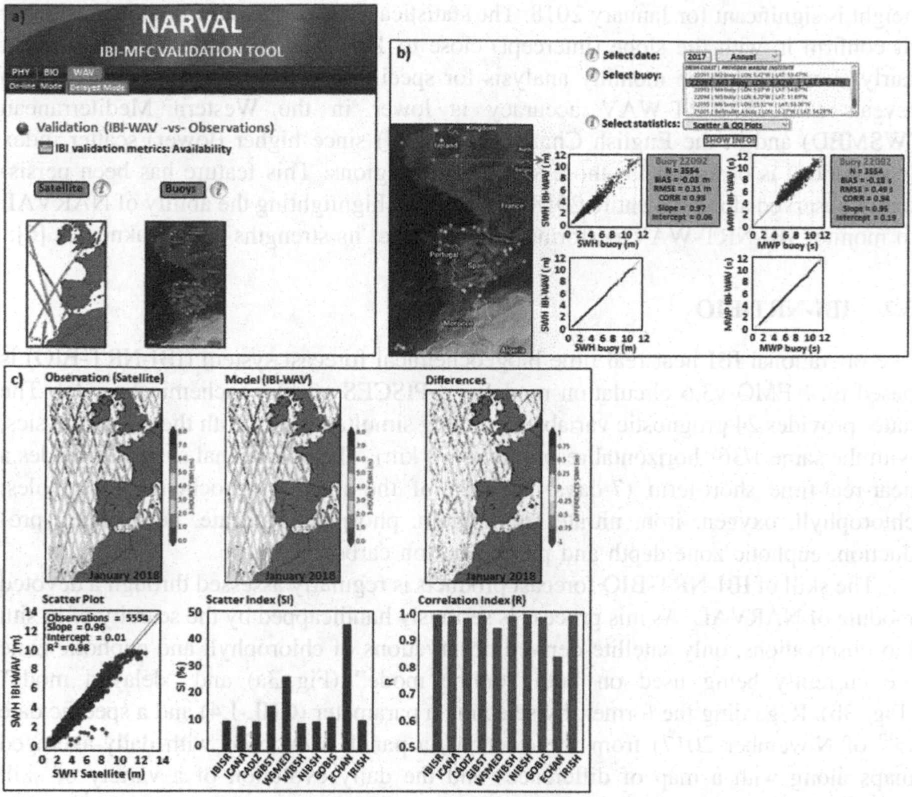

Fig. 2. (a) Snapshot of the NARVAL website, focused on the skill assessment of IBI-NRT-WAV on "delayed-mode"; (b) NARVAL section devoted to compare the model against coastal (orange dots) and deep-water (green dots) buoys: Annual comparison (2017) of SWH (left) and MWP (right) at the model grid point closest to buoy 22092 (red circle): scatter and QQ plots. N represents the number of hourly observations; (c) Monthly comparison against satellite-derived observations (January 2018): map of SWH differences, scatter plot and analysis for different sub-regions are presented. (Color figure online)

Complementarily, the wave altimetry product used to quantify the skill of IBI-NRT-WAV model comprises a pool of three different satellite missions (Jason-2, Saral/Altika, and Cryosat-2) that is subsequently merged and prepared by Meteo-France. The satellite-sensed SWH estimations have been spatially averaged on a 0.1° grid and chosen for specific three-hourly time steps (00–21 h) in order to associate the measured and simulated significant wave heights and objectively assess the quality of IBI-NRT-WAV on a daily and monthly basis. Maps of differences and the best linear fit of scatterplots are routinely computed through NARVAL. According to the maps shown in Fig. 2c, the resemblance between observed and simulated significant wave height is significant for January 2018. The statistical results derived from the best linear fit confirm it, with the slope (intercept) close to 1 (0) and the correlation coefficient fairly above 0.95. The monthly analysis for specific sub-regions (defined in Fig. 1a) reveals that IBI-NRT-WAV accuracy is lower in the Western Mediterranean (WSMED) and in the English Channel (ECHAN) since higher (lower) scatter index (correlation) is observed for those specific sub-regions. This feature has been persistently observed along the entire 2018 (not shown), highlighting the ability of NARVAL to monitor IBI-NRT-WAV performance and detect its strengths and weaknesses [6].

3.2 IBI-NRT-BIO

The operational IBI near real time biogeochemical forecast system (IBI-NRT-BIO) is based on NEMO v3.6 circulation model and PISCES v2 biogeochemical model. The latter provides 24 prognostic variables, running simultaneously with the ocean physics, with the same 1/36° horizontal resolution (~2 km). The operational system provides a near-real-time short-term (7-days) forecast of the main biogeochemical variables: chlorophyll, oxygen, iron, nitrate, ammonium, phosphate, silicate, net primary production, euphotic zone depth and phytoplankton carbon.

The skill of IBI-NRT-BIO forecast products is regularly assessed through a devoted module of NARVAL. As this process is seriously handicapped by the scarcity of in situ bio observations, only satellite-derived observations of chlorophyll and euphotic zone are currently being used on both "online mode" (Fig. 3a) and "delayed mode" (Fig. 3b). Regarding the former, by selecting a parameter (CHL-L4) and a specific date (5th of November 2017) from the calendar, a panel is exposed with daily-averaged maps along with a map of differences and the daily evolution of a variety of skill metrics for the last 15 days (Fig. 3a). In order to provide a deeper insight into IBI model performance during 2017, a qualitative model-observation comparison was performed: Hovmöller diagrams were computed at two selected transects of constant latitude with the main aim of analyzing the temporal evolution of the daily chlorophyll concentration in key regions like the Strait of Gibraltar or the Galician upwelling system. In this example, a relevant model-observation resemblance is observed in the Strait of Gibraltar and the Alboran Sea (6°W–1°W), where quasi-permanent peaks of chlorophyll were satisfactorily reproduced by IBI along the entire year 2017 (Fig. 3c–e). The mean absolute difference (MAD) remained moderate along the transect, with a relative peak detected over the Alboran Sea, indicating a model overestimation (Fig. 3f). On the other hand, IBI-NRT-BIO appeared to capture basic characteristics of the NW Iberian upwelling system such as the intensification of the chlorophyll concentration during specific summer coastal upwelling events when northerly winds are predominant (Fig. 3g–i), although the MAD was also higher in this region (Fig. 3j).

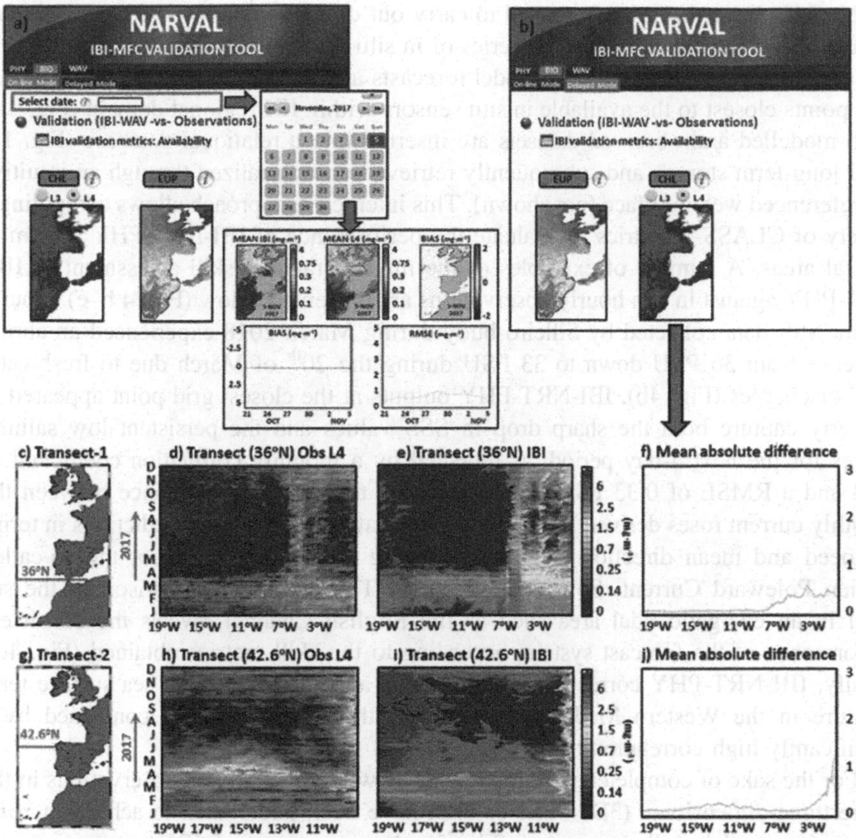

Fig. 3. (a–b) Snapshot of NARVAL web, focused on the "online mode" and "delayed mode" quality assessment of IBI-NRT-BIO products, respectively; (c–j) Transects of constant latitude and the associated Hovmöller diagrams and mean absolute differences.

3.3 IBI-NRT-PHY

The operational IBI near real time physical forecast system (IBI-NRT-PHY) provides a short-term 5-day hydrodynamic 3D forecast of a range of physical parameters (currents, temperature, salinity and sea level) since 2011 [1]. The system is based on an eddy-resolving NEMO v3.6 model application, run at 1/36° horizontal resolution on an Arakawa-C grid and 50 geopotential vertical levels, assuming hydrostatic equilibrium and Boussinesq approximation. Final products are routinely delivered in a service domain extending between 19°W–5°E and 26°N–56°N. The IBI run is forced every 3 h with up-to-date high-frequency meteorological forecasts provided by ECMWF. Lateral open boundary data are interpolated from the daily outputs of the CMEMS GLOBAL system. These are complemented by 11 tidal harmonics built from FES2004 and TPXO7.1 tidal models solutions. A SAM2-based data assimilation scheme was recently introduced (April 2018) in order to enhance IBI predictive skills but will not be further described here.

NARVAL has been implemented to carry out direct comparison of model outputs with quality-controlled hourly time series of in situ observations. To this aim, the skill assessment software ingests daily model forecasts and extracts the time series on those grid points closest to the available in situ sensors within IBI regional domain (Fig. 4a). Both modelled and observed datasets are inserted into a relational database (Fig. 1b) for a long-term storage and subsequently retrieved and visualized through an intuitive georeferenced web interface (not shown). This interactive approach allows computing a variety of CLASS-2 metrics to evaluate the performance of IBI-NRT-PHY system in coastal areas. A number of examples of the multi-parameter skill assessment of IBI-NRT-PHY against in situ hourly observations are presented below (Fig. 4 b–e). Hourly in situ SSS data collected by Silleiro buoy during March 2018 experienced an abrupt decrease from 36 PSU down to 33 PSU during the 20th of March due to freshwater river discharges (Fig. 4b). IBI-NRT-PHY outputs at the closest grid point appeared to properly capture both the sharp drop in SSS values and the persistent low salinity values for the next 4-day period, as reflected by a monthly correlation coefficient of 0.92 and a RMSE of 0.33 PSU. There is also a noticeable resemblance between the monthly current roses derived from in situ observations and model predictions in terms of speed and mean direction (Fig. 4c), showing the predominance of the so-called Iberian Poleward Current, flowing northwards. The monthly comparison of the sea level in an energetic tidal area such as the English Channel reveals the consistent performance of the forecast system, according to the skill metrics obtained (Fig. 4d). Finally, IBI-NRT-PHY correctly reproduced the annual cycle of the sea surface temperature in the Western Mediterranean during the entire 2018, as confirmed by a significantly high correlation of 0.99 (Fig. 4e).

For the sake of completeness, supplementary works with in situ observations in the entire three-dimensional (3D) water column have been undertaken to achieve a comprehensive model skill assessment (Fig. 5). A comparison of IBI temperature and salinity profiles against ARGO floats within IBI regional domain are routinely conducted on a monthly basis thanks to NARVAL web tool. CLASS-4 metrics are computed for the whole water column and for different layers, being the specific levels considered: 0–5 m, 5–200 m, 200–600 m, 600–1500 m, 1500–2000 m (not shown). According to the skill metrics derived from the monthly comparison against ARGO floats, IBI-NRT-PHY seems to properly capture the vertical distribution of temperature and salinity (Fig. 5a). Furthermore, the resemblance between both datasets is significantly high and the Temperature-Salinity (TS) diagrams look rather alike (Fig. 5b). As a result of strong alliances with local partners, IBI-NRT-PHY model solution is verified against specific glider missions. In particular, the exercises performed in the Ibiza Channel by SOCIB are used to assess the model 3D consistency. As reflected in Fig. 5c, the model performance seems to be rather consistent (especially in lower depth levels), whereas moderate discrepancies are mainly found in the first 200 m.

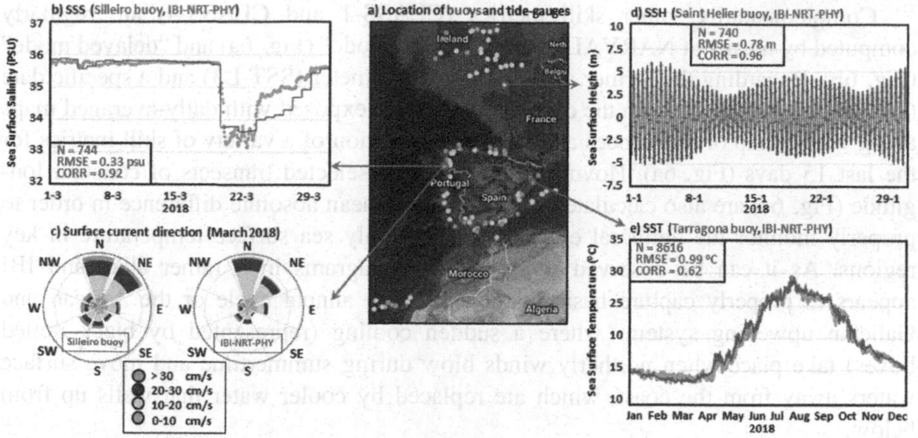

Fig. 4. (a) Location of buoys (purple dots) and tide-gauges (green dots) used to evaluate the quality of IBI-NRT-PHY forecast system; (b–e) CLASS-2 multi-parameter comparison at selected model grid points closest to moorings location: timeseries comparison, surface current roses and skill metrics are automatically computed on a different time basis, ranging from monthly to seasonal and annual frequencies. (Color figure online)

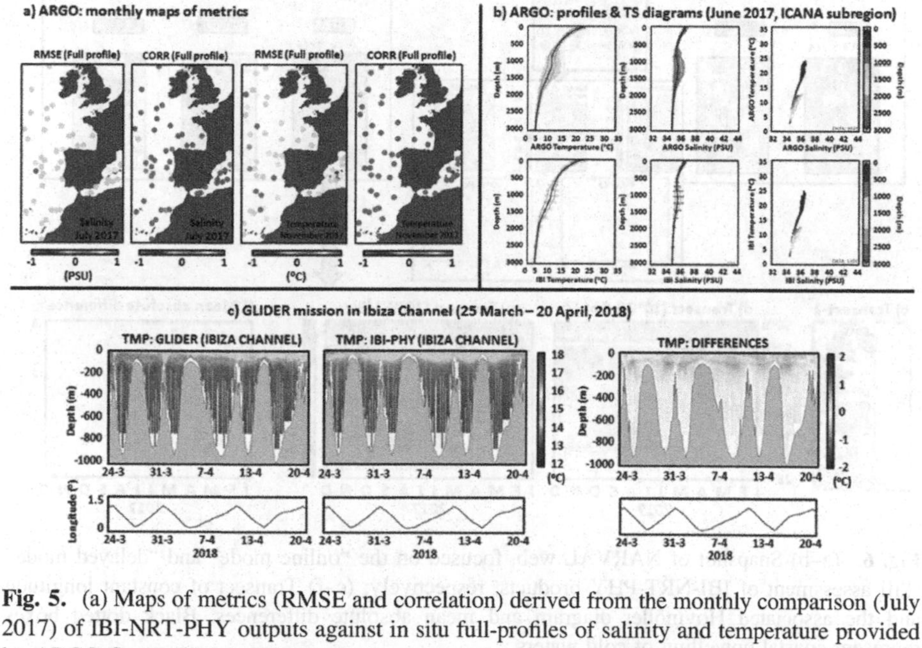

Fig. 5. (a) Maps of metrics (RMSE and correlation) derived from the monthly comparison (July 2017) of IBI-NRT-PHY outputs against in situ full-profiles of salinity and temperature provided by ARGO floats; (b) Monthly qualitative comparison between daily model outputs and in situ observations in ICANA sub-region (June 2017): profiles of temperature and salinity along with TS diagrams; (c) Glider mission in Ibiza Channel: temperature profiles observed, modelled and differences (courtesy of SOCIB) during March–April 2018.

Complementarily, other skill metrics (CLASS-1 and CLASS-4) are regularly computed by means of NARVAL on both "online mode" (Fig. 6a) and "delayed mode" (Fig. 6b). Regarding the former, by selecting a parameter (SST-L3) and a specific date (20th of October 2017) from the calendar, a panel is exposed with daily-averaged maps along with a map of differences and the daily evolution of a variety of skill metrics for the last 15 days (Fig. 6a). Hovmöller diagrams at selected transects of constant longitude (Fig. 6c) are also calculated along with the mean absolute difference in order to properly monitor the temporal evolution of the daily sea surface temperature in key regions. As it can be observed in Fig. 6(d–f), diagrams look rather alike and IBI appears to properly capture basic features like the annual cycle or the African and Galician upwelling systems where a sudden cooling (represented by black dotted boxes) take place when northerly winds blow during summertime and move surface waters away from the coast, which are replaced by cooler water that wells up from below.

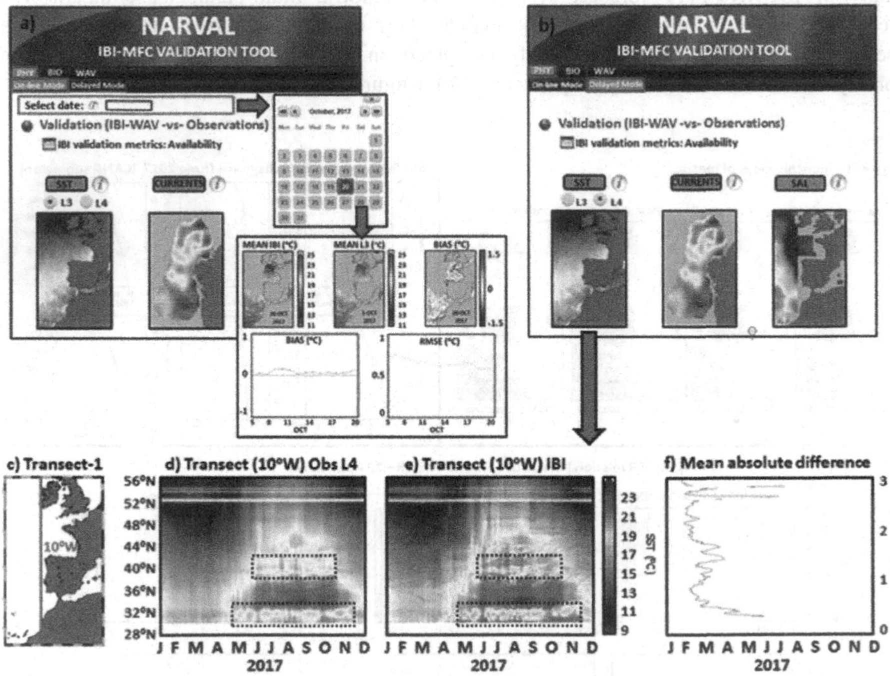

Fig. 6. (a–b) Snapshot of NARVAL web, focused on the "online mode" and "delayed mode" skill assessment of IBI-NRT-PHY products, respectively; (c–f) Transect of constant longitude and the associated Hovmöller diagram and mean absolute differences. Black dotted boxes represent coastal upwelling of cold waters.

An additional aspect addressed in NARVAL toolbox is the multi-parameter intercomparison of diverse ocean forecast models in the overlapping regions, conducted at the sea surface, ranging from global to regional and local scales [7]. Here we present a multi-model intercomparison exercise for August 2018 in the Strait of Gibraltar among three CMEMS forecast systems (GLOBAL, IBI and MED – being the last two systems nested to GLOBAL) and the SAMPA high-resolution coastal forecast system (embedded in IBI) in order to elucidate the accuracy of each system to characterize the Atlantic Jet (AJ) inflow dynamic (Fig. 7). To this end, a HFR system has been used as benchmark since it regularly provides quality-controlled hourly maps of the surface currents of the Strait [8]. The qualitative inspection of monthly-averaged circulation maps reveals that each forecast system reproduces reasonably well the eastward AJ inflow into the Mediterranean as previously observed in HFR estimations (Fig. 7a), but differ in the intensity and direction of the mean surface inflow. Whilst GLOBAL (Fig. 7c) and MED (Fig. 7e) appear to underestimate the speed in the Strait of Gibraltar, IBI seem to overestimate it, exhibiting besides a more zonal surface flow (Fig. 7g).

However, SAMPA outperforms the parent systems (IBI and GLOBAL) and MED by better replicating the orientation and strength of the inflow (Fig. 7i). A quantitative CLASS-2 comparison at the selected grid point (indicated by a black square in Fig. 7a) was assessed (Fig. 7, right panels). The scatter plot of HFR-derived hourly current speed versus direction (taking as reference the North and positive angles clockwise) revealed that the AJ flowed predominantly eastwards, forming an angle of 79° (Fig. 7b). The current velocity, on average, was 82 cm s^{-1} and reached peaks above 200 cm s^{-1}. Speeds below 50 cm s^{-1} were registered along the entire range of directions. Westwards currents, albeit minority, were also observed. In the case of the scatter plot derived from GLOBAL estimations, substantial discrepancies were detected as the variability of both the AJ direction and speed were clearly limited (Fig. 7d). No flow reversals were detected and peak velocities of the eastward flow were underestimated. In the monthly scatter plot of regional MED (Fig. 7f) and IBI estimations (Fig. 7h), surface current velocities below 20 cm s^{-1} were barely replicated and the AJ inversion was not observed. Despite IBI appeared to properly portray the mean characteristics of the eastwards flow, the model tended to privilege zonal flow directions. By contrast, the scatter plot of SAMPA estimations presented a significant resemblance in terms of prevailing current velocity and direction (Fig. 7j). The main features of the AJ were qualitatively reproduced and surface flow reversals to the west were properly captured. Accordingly, the skill metrics obtained for SAMPA coastal system are better than those derived for regional and global model solutions.

In summary, this exercise reflects the added value of the dynamical downscaling performed through the SAMPA coastal system with respect to IBI regional solution (in which SAMPA is nested). Overall, a steady improvement in the AJ characterization is evidenced in model performance when zooming from global to coastal configurations, highlighting the benefits of the downscaling approach adopted and also the potential relevance of a variety of factors at local scale, among others: a more refined horizontal resolution, a tailored bathymetry and/or a higher spatio-temporal resolution of the atmospheric forcing.

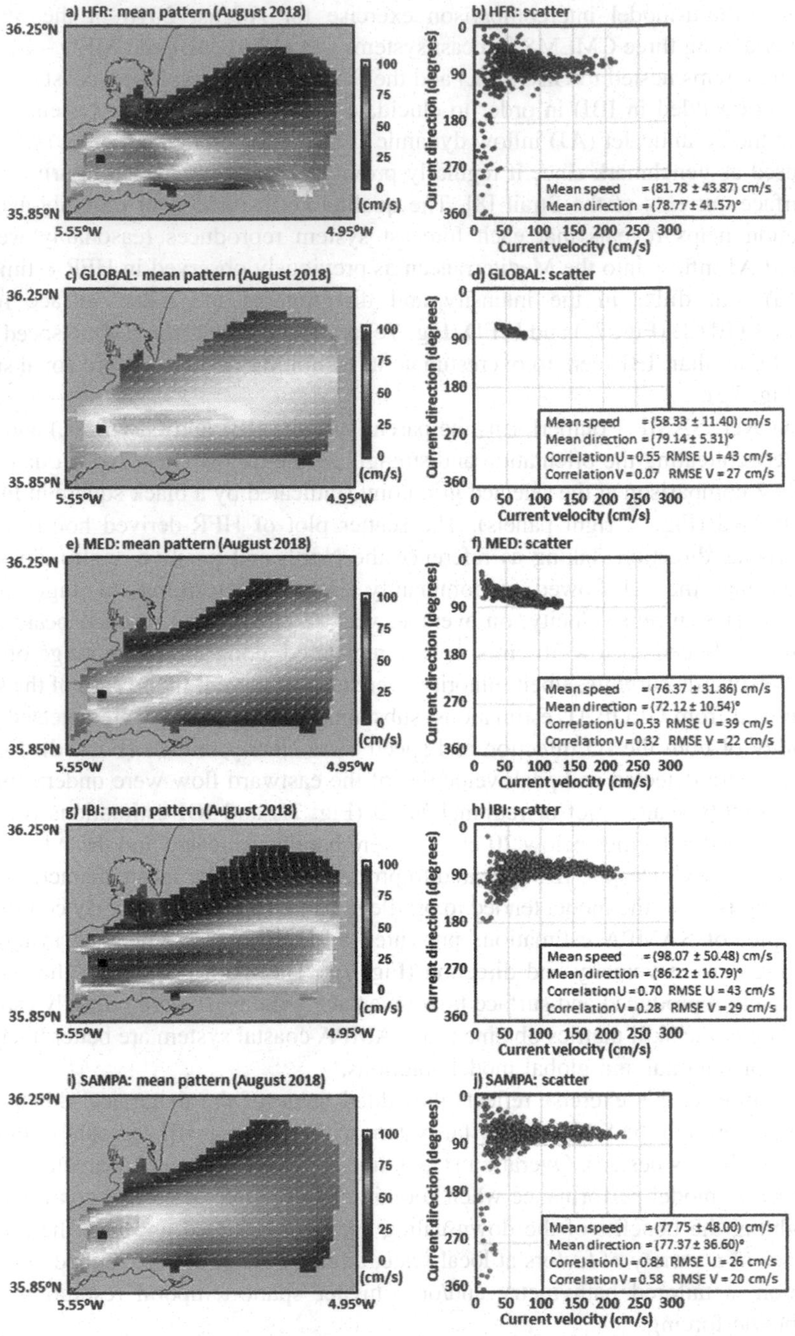

Fig. 7. (left) Monthly-averaged surface circulation patterns for August 2018. Black squares indicate the grid point selected to conduct the comparison; (right) Scatter of hourly current direction (taking as reference the North and positive angles clockwise) versus current speed for August 2018 at the selected grid point.

4 Conclusions and Future Work

In this work, a general overview of the main features of NARVAL toolbox has been presented, with special emphasis on the rigorous skill assessment of a wealth of CMEMS IBI near real time forecasting products (WAV, BIO and PHY). Some of the current practical applications of NARVAL have been showcased, highlighting the benefits of a synergistic approach based on the integration of numerical models (CMEMS IBI) and observational networks, used in tandem to comprehensively characterize the highly dynamic sea states and the dominant modes of spatio-temporal variability.

With the advent of new technologies (coastal altimetry, autonomous underwater vehicles, BIO ARGO floats, etc.), a combined use of multi-platform, multi-scale observing systems encompassing both in situ (buoys, tide gauges, etc.) and remote (HFR, satellite, etc.) sensors will provide further insight into the comprehensive characterization of the shelf's surface circulation and will also contribute positively to a more exhaustive model accuracy assessment.

The future roadmap to build a new upgraded version denominated NARVAL-PRO includes the extension of its capabilities to evaluate the quality of multi-year and interim products or the computation of long-term skill metrics in order to gain insight into the evolution of IBI model performance (Fig. 1b).

NARVAL is a live software package in terms of seamless evolution and continuous upgrades. Successive versions will be developed with a focus on the inclusion of: (i) novel skill metrics; (ii) new observational platforms; and (iii) new ocean forecasting systems, implemented at both regional and coastal scales. In the framework of MyCoast and OCASO projects, several regional systems are currently being intercompared by means of NARVAL. Likewise, this toolbox will also play a relevant role in order to evaluate the benefits of downscaling approaches in coastal areas since a variety of operational port-scale forecasts products have been recently developed under the umbrella of SAMOA project, aimed at implementing a fully integrated monitoring service to increase safety and efficiency of marine operations in the Spanish harbors.

Ancillary verification exercises should focus on the evaluation of ocean models ability to accurately reproduce singular oceanographic processes. Since the NARVAL tool is devoted to intercompare model solutions on a monthly, seasonal or annual basis, part of the picture is missing due to traditional time averaging. An event-oriented multi-model intercomparison methodology would allow better quantifying the skill of each system to capture small-scale coastal processes. Those oceanographic events subject of further insight might encompass, among others: (i) coastal upwelling, downwelling and relaxation episodes; (ii) fronts and submesoscale eddies; (iii) extreme events.

References

1. Sotillo, M.G., et al.: The MyOcean IBI ocean forecast and reanalysis systems: operational products and roadmap to the future copernicus service. J. Oper. Oceanogr. **8**(1), 1–18 (2015)
2. Lorente, P., et al.: Ocean model skill assessment in the NW Mediterranean using multi-sensor data. J. Oper. Oceanogr. (2016). https://doi.org/10.1080/1755876x.2016.1215224

3. Sotillo, M.G., et al.: How is the surface Atlantic water inflow through the strait of gibraltar forecasted? A lagrangian validation of operational oceanographic services in the Alboran sea and Western Mediterranean. Deep Sea Res. **133**, 100–117 (2016)
4. Sotillo, M.G., Levier, B., Lorente, P.: Quality information document for the Atlantic -Iberian Biscay Irish- ocean physics analysis and forecasting product: IBI_ANALYSIS_FORECAST_PHYS_005_001. CMEMS Technical report, pp. 1–94 (2018)
5. Rainaud, R., et al.: The impact of wave physics in the CMEMS-IBI ocean system Part A: wave forcing validation. Ocean Sci. Discuss. https://doi.org/10.5194/os-2018-165 (2019)
6. Lorente, P., et al.: Extreme wave height events in NW Spain: a combined multi-sensor and model approach. Remote Sens. **10**(1) (2018). https://doi.org/10.3390/rs10010001
7. Lorente, P., et al.: Skill assessment of global, regional and coastal circulation forecast models: evaluating the benefits of dynamical downscaling in IBI surface waters. Ocean Sci. Discuss. (2019). https://doi.org/10.5194/os-2018-168
8. Lorente, P., Piedracoba, S., Sotillo, M.G., Álvarez-Fanjul, E.: Long-term monitoring of the Atlantic jet through the strait of gibraltar with HF radar observations. J. Mar. Sci. Eng. **7**, 3 (2019). https://doi.org/10.3390/jmse7010003

Salinity Control on Saigon River Downstream of Dautieng Reservoir Within Multi-objective Simulation-Optimisation Framework for Reservoir Operation

Okan Aygun[1], Andreja Jonoski[2], and Ioana Popescu[2(✉)]

[1] Cold Region Environmental Research Laboratory, Department of Environmental Sciences, University of Quebec, Trois-Rivières, QC, Canada
okan.aygun@uqtr.ca

[2] IHE Delft Institute for Water Education, PO Box 3015, 2601 DA Delft, The Netherlands
{a.jonoski,i.popescu}@un-ihe.org

Abstract. This research proposes a modelling framework in which simulation and optimisation tools are used together in order to obtain optimal reservoir operation rules for the multi-objective Dautieng reservoir on the Saigon River (Vietnam), where downstream salinity control is the main objective. In this framework, hydrodynamic and salinity transport modelling of the Saigon River is performed using the MIKE 11 modelling system. In the first optimisation step this simulation model is coupled with the population simplex evolution (PSE) algorithm from the AUTOCAL optimisation utility (available as a part of MIKE 11) to estimate the discharge required to meet salinity standards at the downstream location of Hoa Phu pumping station for public water supply. In the second optimisation step, with the use of MATLAB optimisation toolbox, an elitist multi-objective genetic algorithm is coupled with a simple water balance model of the Dautieng reservoir to investigate how the optimised discharges obtained from the first optimisation step can be balanced with the other objectives of the reservoir.

The results indicate that optimised releases improve the performance of the reservoir especially on controlling salinity at Hoa Phu pumping station. In addition, the study demonstrates that use of smaller time steps in optimisation gives a closer match between varying demands and releases.

1 Introduction

Most reservoirs are designed and constructed for multiple tasks including water supply for agriculture and industry, hydropower production, flood control, navigation etc. These objectives sometimes conflict with each other so the operation and management of the reservoir system becomes more complicated for decision makers and reservoir operators. As it has been pointed out by many experts, there is a problem of ineffective operation of existing reservoirs mainly due to highly subjective management practices (Chen 2003; Hejazi et al. 2008). Hence, there is a continuous need for methodologies to

J. M. F. Rodrigues et al. (Eds.): ICCS 2019, LNCS 11539, pp. 329–345, 2019.
https://doi.org/10.1007/978-3-030-22747-0_26

optimise the reservoir operation in order to obtain satisfactory results in terms of meeting each demand from the reservoir (Labadie 2004; Le Ngo et al. 2007).

This particular study investigates the optimal operation of a reservoir located in Vietnam where it is necessary to have many multi-objective reservoirs due to highly variable rainfall caused by tropical monsoon climate. In Vietnam, the reservoirs are used to manage different water requirements that occur both during the rainy and the dry season. While there are many reservoirs mainly used as flood control pools during the wet season, meeting irrigation and other water demands are major tasks of the reservoirs during the dry season. However, 'many difficulties are encountered in reservoir operation due to too much water in the wet season and too little water during the dry season' (Le Ngo et al. 2006). Whilst, there is a risk of flooding (and sometimes even dam break) in the rainy season, water levels in the reservoirs drop considerably in the dry season with the higher demands of the water users. Sea level rise, triggers the salinity intrusion in coastal areas. Based on the report prepared by UNEP (2009), sea level along the coastline of Vietnam has risen by about by 20 cm and it is expected to rise by 65–100 cm by the year 2100 (VCAPS-Consortium 2013). Especially during the dry season, due to decreased natural discharges of rivers and low releases from reservoirs, salinity penetrates more in upstream direction. This migration decreases the quality of freshwater, which brings serious problems for the ecosystem and the other users of freshwater, e.g. public water supply intakes. In this context, although there are structural measures such as building control structures at river mouths to control the salinity intrusion, operational changes of existing upstream reservoirs are considered as alternative solutions such that more discharge is released in order to push back the salinity towards the sea.

In analysing a reservoir system for setting operating policies, simulation and optimisation are the most commonly used tools. In simulation models, the behaviour of the catchment together with the operation of the reservoir system are analysed given many variable conditions. On the other hand, optimisation algorithms are applied to obtain optimal operating rules for the reservoir system by using different objective functions and constraints.

One of the earlier application of this simulation-optimisation approach to control salinity intrusion was of Waite (1980), who developed a mathematical model of salt water intrusion in conjunction with a model of reservoir operation for water control scheme on the River Abary, Guyana. It was proved that the levels of salinity in the estuary can be decreased during the dry season by controlling the releases from the reservoir. Le Ngo et al. (2007) developed a framework for adaptive management which combines simulation model with optimisation model to obtain trade-off solutions between flood control and hydropower generation objectives of Hoa Binh Reservoir, Vietnam. MIKE 11 hydrodynamic model was applied to simulate the reservoir releases. It was demonstrated that coupling of simulation with optimisation gave optimum solutions in this multi-objective task.

Dhar and Datta (2008) proposed a GA-based linked simulation and optimisation methodology to determine the optimal releases for controlling water quality at downstream locations. They used CE-QUAL-W2 model developed by US Army Corps of Engineers (USACE) to simulate the hydrodynamics and water quality downstream. This simulation model was linked externally to an elitist GA.

Abdullah et al. (2018) developed a methodology based on coupled simulation-optimisation approach for setting water allocation strategies, given the salinity from estuary into Shat el Arab river.

Given the recent successful applications of use of simulation and optimisation models together, this approach is adopted in the present research in order to obtain optimum operational policies for the Dautieng Reservoir which is located on the upstream part of Saigon River, Vietnam. The main objective is to provide optimum amount of discharge from the reservoir to keep the balance between sea water intrusion and fresh water allocations, especially during the dry season when the salinity in the river exceeds the fresh water salinity standards.

2 Case Study Description

The Dongnai River basin is the largest national river basin in southern Vietnam (see Fig. 1), with a total surface area of 48471 km^2 within Vietnam, which constitutes 15% of country's total land surface area (Ringler et al. 2002). The basin includes Dongnai, Saigon and Vamco rivers. The lower part of the basin covers one of the most important economic development zones including Ho Chi Minh City, which contributes more than half of total GDP of the country (ADB 2009). At the same time, this basin plays a large role in the agricultural sector of the country, producing coffee, fruits, and vegetables. The economy of this productive basin highly depends on proper management of water resources for different uses (Ringler et al. 2002).

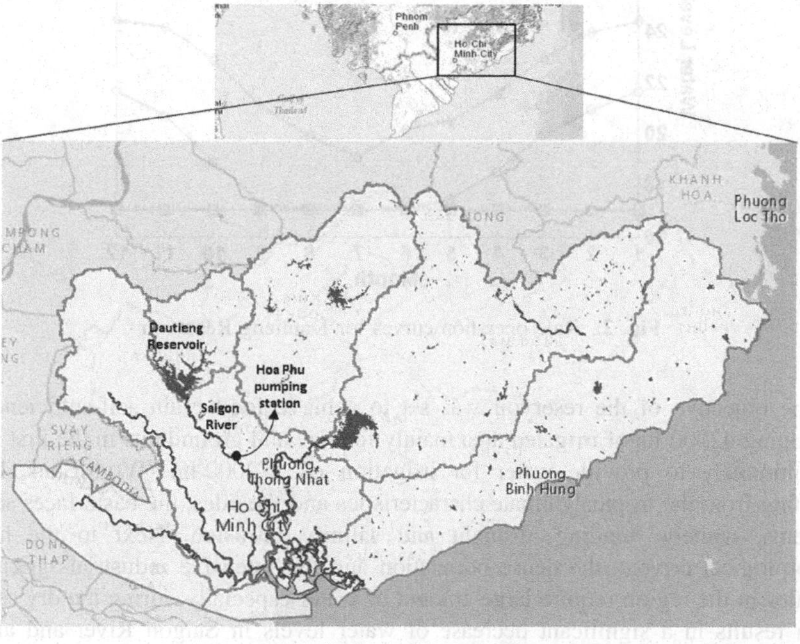

Fig. 1. Dongnai River Basin

There are two main seasons in the basin, namely wet and dry season. The wet season lasts from May to December, causing serious flooding problems, while the dry season takes place between December and May, bringing water scarcity.

Saigon River, is one of the main water suppliers for towns and industries and need a careful salinity control management. It is the second largest river in the Dongnai River basin. It originates in south-eastern Cambodia and after flowing 280 km in southeast direction it joins in Dongnai River 40 km from the East Sea.

The largest irrigation reservoir of Vietnam, Dautieng Reservoir, was completed in 1983 with the support of World Bank. This reservoir is located on the Saigon River, 120 km from the confluence point with the Dongnai River, and has a reservoir capacity of 1050×10^6 m^3.

The existing rule curves for the reservoir operation are given in Fig. 2. Among these curves, crest level represents the highest level of the reservoir (28 m), whereas dead storage level (17 m) is the level below which the reservoir does not operate. Flood control curve is formed by the maximum water levels above which the water has to be spilled in order to control flooding. On the other hand, critical operation curve is the minimum operation level of the reservoir below which the operation of the reservoir is restricted.

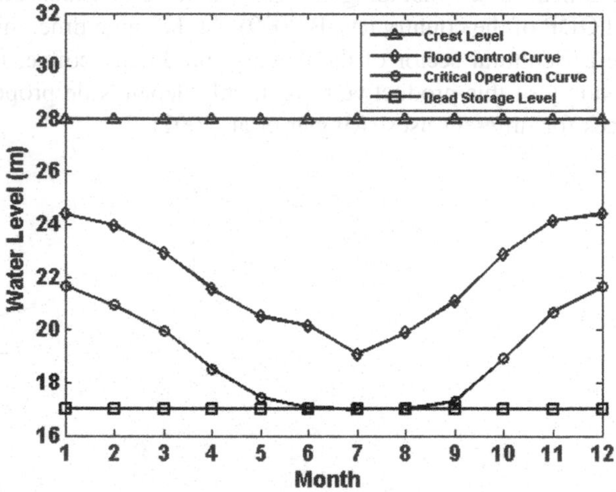

Fig. 2. Rule operation curves for Dautieng Reservoir

The objective of the reservoir was set to achieve food grain self-sufficiency by developing 42000 ha of irrigated area mainly for rice and groundnuts in the first stage, and ultimately to provide water for irrigation of 172000 ha (WorldBank 1989). Resulting from the tropical climate characteristics and the tides, the basin faces serious problems, namely flooding, drought and salinity intrusion. Next to the hydro-meteorological drivers, the dense population and the intensive industrial-agricultural activities in the region require large amount of water especially during the dry season, which results in a significant decrease of water levels in Saigon River and also in

Dautieng Reservoir. In addition, the water quality in the river is highly affected by the discharges from the industrial and agricultural zones.

Apart from heavy pollution due to industrial and agricultural wastes in Saigon River, higher salinity levels threaten the downstream communities located in lower lying reaches of Saigon River. In addition to different demands from the river, Saigon River is one of the major sources of public water supply for Ho Chi Minh City.

A pumping station located in Tan Hiep, Hoa Phu, used for domestic water supply is taken as a control point to check salinity levels in this research. This station is located along Saigon River about 70 km from the Dautieng Reservoir and about 90 km from the sea (see Fig. 1).

The competition between agriculture, industry and other water demands is causing salinity intrusion further upstream, especially during the dry season. Exceeding the salinity limits would cause suspension of operation of the pumping station and water treatment will not be performed, which could result in a shortage of clean water for Ho Chi Minh City. This is why there is an urgent need for operational strategies that will keep salinity levels under a certain limit.

In this present study, operational changes for Dautieng reservoir are explored in order to control salinity at the Hoa Phu pumping station, which will prevent the shortage of drinking water for the region. All data used in this research were obtained from Ho Chi Minh University of Technology, Vietnam.

3 Methodology

This study searches for an optimal operation of an existing reservoir by using simulation and optimisation techniques together. In the simulation part, hydrodynamic and salinity transport modelling of Saigon River is performed. For the optimisation process, a two step approach is adopted. The first step optimisation is carried out to determine the optimal/minimal discharges in Saigon River to control the salinity at Hoa Phu pumping station. The hydrodynamic simulation and salinity intrusion model of the river is linked with a generic tool, normally used for parameter optimisation. Here, the optimisation problem is formulated with river discharge as a decision variable and only one objective function, namely minimisation of the difference between the actual and target salinity at the location of Hoa Phu pumping station. In the second step optimisation, a multi-objective optimisation is performed by considering the other demands from the reservoir next to salinity control, together with additional constraints regarding the reservoir operation. At the end of this step, optimal reservoir releases with respect to each objective are obtained. The final step is to evaluate the level of satisfaction of the individual demands with these releases.

MIKE 11 modelling system developed by Danish Hydraulics Institute (DHI) is chosen to simulate hydrodynamics and salinity intrusion in Saigon River. The schematics of the model is represented in Fig. 3. In order to select a critical year for salinity control, out of the years with available data, HD and AD modules are run simultaneously and the salinity levels at the control point are simulated. The criterion for selecting the critical year is the extent at which salinity exceeds the permissible limit, which is set as 0.3 PSU. This analysis shows that the only year in which salinity at the control point above the limit is 2001, with a maximum value of 0.510 PSU. Simulation of flow for the year 2001, indicates that salinity has significant values and variations only during the dry period between January and May, whereas it is low during the rest of the year. Salinity exceeds the critical value only for the months March and April, which is reasonable since these are the driest months of the dry season when the natural discharge in the river reaches its minimum values, triggering the salinity migration upstream. Since salinity exceeds the limit only in March and April in 2001, the optimisation is performed for these two months.

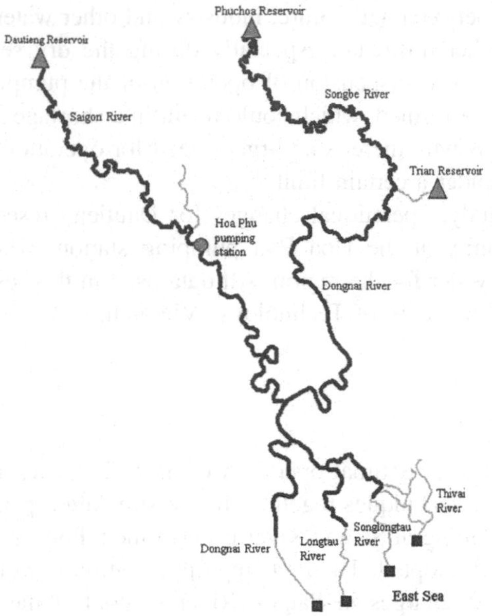

Fig. 3. Mike 11 schematisation of the river, with location of reservoirs and the salinity control point at Hoa Phu

3.1 First Step Optimisation

The aim of the first step of optimisation is to obtain optimal/minimum discharge to keep the salinity under the permissible limit of 0.3 PSU at Hoa Phu pumping station. To achieve this objective, the MIKE 11 model is coupled with a parameter optimisation tool available in Mike 11, the AUTOCAL. In this step the hydrodynamic and advection dispersion models are executed together with optimisation algorithms for determining optimal parameter sets.

AUTOCAL is mainly used for the purposes of automatic calibration, parameter optimisation and sensitivity analysis of MIKE 11 models. Parameter optimisation option of this tool is applied by Le Ngo et al. (2007) in order to obtain optimised rules curves for Hoa Binh Reservoir, Vietnam. In this particular study, a kind of parameter optimisation option is performed as well.

AUTOCAL includes two different parameter optimisation methods, namely Shuffled Complex Evolution (SCE) and Population Simplex Evolution (PSE). Out of these two methods, PSE is chosen since it is especially suited for simultaneous simulations that take place in AUTOCAL (DHI 2014).

The objective function in this first step optimisation aims to minimise the maximum squared difference between simulated salinity at the control point and permissible salinity limit, which is formulated as:

$$F = MIN[Max\{S, S\varepsilon[0, T]\} - Target]^2 \tag{1}$$

With

S = Simulated salinity at Hoa Phu pumping station
$Target$ = 0.3 PSU (permissible salinity limit)
$[0, T]$ = Simulation period, T = 30 days for monthly optimisation;
T = 10 days for ten-daily optimisation

The decision variables are the discharges of Saigon River, which are modified during the optimisation process in order to obtain the best value of the objective function. Once one of the stopping criteria is met, the optimal discharge, corresponding to the best (minimum) objective function is obtained. With this optimal discharge, MIKE 11 model is executed one more time using AUTOCAL in order to simulate the final salinity at the control point. The chosen population size is 100 and optimisation is run for a simulation period of 30 days for purposes of monthly optimisation in the second step (see next section) and 10 days for ten-daily optimisation (also in the second step). This leads to 2 monthly and 6 ten-daily results in terms of critical salinity discharge.

3.2 Second Step Optimisation

In the second step of the optimisation next to salinity control purpose, the other objectives of the reservoir are taken into account to perform a multi-objective optimisation. The aim is to meet all the demands as much as possible during the optimisation period. This second step optimisation is performed for monthly and ten-daily time steps during the two months; March and April.

The general form of the objective function is defined as follow:

$$F = MIN\left[\sum_{t=1}^{N} (Rt - Dt)^2\right] \tag{2}$$

where

Dt = Demand for a specific purpose of the reservoir during time t
Rt=Reservoir release to meet a specific demand during time t
N = Number of time steps for the optimisation process

The main objective of Dautieng reservoir is to supply enough water for the agricultural areas. Next to this, the reservoir release should control the salinity at Hoa Phu pumping station, which is the key interest of this research and this is taken as the second objective for the reservoir operation. In addition to these two objectives, domestic and industrial water supply are also objectives for the reservoir operation. On the other hand, since the optimisation is carried out only during the dry season of the year, flood control objective is not taken into account.

The objective functions for this second step optimisation are defined as follows:

$$F(I) = MIN\left[\sum_{t=1}^{N} (R_{I_t} - D_{I_t})^2\right] \tag{3}$$

$$F(S) = MIN\left[\sum_{t=1}^{N} (R_{S_t} - D_{S_t})^2\right] \tag{4}$$

$$F(Ind) = MIN\left[\sum_{t=1}^{N} (R_{Ind_t} - D_{Ind_t})^2\right] \tag{5}$$

$$F(Dom) = MIN\left[\sum_{t=1}^{N} (R_{Dom_t} - D_{Dom_t})^2\right] \tag{6}$$

The terms given in Eqs. 3 to 6 are explained in Table 1.

Table 1. The terms in objective function equations

$\boldsymbol{F(I)}$: Irrigation objective function	$\boldsymbol{F(S)}$: Salinity control objective function	$\boldsymbol{F(Ind)}$: Industry objective function	$\boldsymbol{F(Dom)}$: Domestic water objective function
$\boldsymbol{R_{I_t}}$: Irrigation release during time t	R_{S_t}: Salinity control release during time t	R_{Ind_t}: Industry release during time t	R_{Dom_t}: Domestic water release during time t
$\boldsymbol{D_{I_t}}$: Irrigation demand during time t	D_{S_t}: Salinity control demand during time t	D_{Ind_t}: Industry demand during time t	D_{Dom_t}: Domestic water demand during time t
N: Number of time steps for the optimisation process (2 for monthly, 6 for ten-daily optimisation)			

The demands for salinity control, in monthly and ten-daily forms, are taken from the results of the first step optimisation. Since the irrigation, domestic and industrial water demands are defined as monthly average values, for both monthly and ten-daily optimisation the same average values are used.

The reservoir continuity/mass balance equation which is used in the optimisation process is given by Eq. 7.

$$S_{t+1} = S_t + I_t - R_t - E_t \tag{7}$$

where

S_{t+1} = Final storage at time t + 1
S_t = Storage at time t
I_t = Inflow to reservoir during time t

R_t = Outflow from reservoir during the time t
E_t = Evaporation loss from reservoir during time t

The constraints include storage and release constraints as follows:

– Storage constraints:

$$S_{min} \leq S_t \leq S_{max} \tag{8}$$

where

S_{min} = Allowable minimum storage in the reservoir
S_t = Storage in the reservoir at time t
S_{max} = Allowable maximum storage in the reservoir

– Release constraints:

$$0 \leq R_t \leq D_t \tag{9}$$

Where

R_t = Reservoir release for a specific objective during time t
D_t = Demand for a specific objective during time t

In order to calculate the storage at time t by using water balance/continuity equation given in Eq. 7, there is a need for assessing the inflow values for the reservoir. However, there is a lack of observed inflow data for the last 25 years. Therefore, it is decided to use the measured inflows from the available data on monthly average inflows in the period of 1960–1984, for three characteristic years, dry, average and wet, which are given in Table 2.

Table 2. Monthly average inflow data for dry, average and wet year

Year	Month											
	I	II	III	IV	V	VI	VII	VIII	IX	X	XI	XII
1982 (dry)	16.2	12.8	11.2	13.7	12.3	32.5	41.3	80.5	171	107	69.9	25.6
1979 (average)	25.9	23.4	20.2	19	23.8	26.6	68.5	117	170	185	73.1	38
1966 (wet)	34	32.8	26.1	21.9	25.9	26	36.1	134	173	161	92	43.5

The average monthly average inflows from the year of 1982, which is the driest year in terms of observed inflows, are chosen to be used as representative for the most critical scenario analysed during the optimisation process. In addition to the problem of not having the real inflow discharges in the reservoir for the period of interest, the operation rules of the reservoir releases could not be obtained. Therefore, instead of building a reservoir operation model, the already defined critical reservoir operation curve below which water supply restricted, is taken as the lowest water level/storage target for the reservoir operation (Fig. 2). Since optimisation is performed for the dry season only, the flood control curve which specifies the maximum water levels that can be reached is not taken into account.

Next to this most critical scenario in which the inflow data of the driest year is used as input for the reservoir, reservoir operation is tested with the monthly inflow data of the average year and the wet year, namely 1979 and 1966 (Table 2).

Reservoir optimisation was carried out to fulfil the two objectives, irrigation and industry, where the following constraints are defined for meeting the salinity control and domestic water demands:

$$R_{S_t} \geq D_{S_t} \tag{10}$$

$$R_{Dom_t} \geq D_{Dom_t} \tag{11}$$

With these hard constraints, it is ensured that salinity control and domestic water demands are always fully met.

The optimisation is carried out by applying an elitist multi-objective genetic algorithm using the genetic algorithm solver available in the Matlab optimisation toolbox. This elitist multi-objective genetic algorithm, a variant of NSGAII, searches for a Pareto front which is a subset of feasible solutions for multi-objective minimisation. It favours the individuals even if they have a lower objective function value in order to increase the diversity of the population. In this research, a population of 1000 is specified and the number of generations is taken as 350 for each run.

Among all the points on the Pareto set, best option for each objective and the trade-off solution between the objectives are analysed in terms of their effects on meeting demands. The best option for each objective is the point which makes the value of objective function minimum. However, there are different techniques in choosing a particular trade-off solution between the objectives. In this research, the Pseudo-Weight Approach proposed by Deb (2001) is adopted to select a trade-off solution, as in the Eq. 12 below:

$$w_i = \frac{(f_i^{\max} - f_i(x))/(f_i^{\max} - f_i^{\min})}{\sum_{m=1}^{M}(f_m^{\max} - f_m(x))/(f_m^{\max} - f_m^{\min})} \tag{12}$$

f_i^{max} = Maximum value for the ith objective function
f_i^{min} = Minimum value for the ith objective function

The summation of all the weight vectors for a particular solution is equal to one. A 100% weight for objective function 1 (f_1) and 0% for objective 2 (f_2) gives the best option for f_1 and vice versa. In this research three points from the Pareto set are selected for further analysis, of which two are such extreme points and one is the trade-off solution that gives 50% weight to both objectives.

4 Results and Discussions

The initial average releases from the reservoir for the months March and April in 2001 are given in Table 3.

Table 3. Initial releases from the reservoir in 2001

Month	Irrigation release (m^3/s)	Salinity control release (m^3/s)	Industry release (m^3/s)	Domestic water release (m^3/s)
March	62.75	2	Not known	Not known
April	51.55	2	Not known	Not known

As a result of initial releases from the reservoir to control the salt water intrusion, the salinity at the control point exceeds the limit of 0.3 PSU for *800* h in the two months period considered. On the other hand, by looking at the irrigation demands and irrigation releases for the months March and April, the irrigation deficit is calculated as *32.40%* for March and *28.60%* for April 2001, where the deficiency describes the percentage of demand (in m^3 of water) that cannot be met by the current release.

4.1 First Step Optimisation

The optimisation process aimed to determine river discharges to keep the salinity under 0.3 PSU at the control point. First, the monthly average releases for March and April where observed to have salinity exceeding the limit. These two month duration is divided in six ten-daily operation periods to perform the optimisation with ten-daily time steps. In order to keep the salinity under 0.3 PSU during March, monthly average river discharge of the Saigon River is considered between 01/03/2001 and 01/04/2001. As a result, the optimal/minimum river discharge obtained is 9.21 m^3/s, which has the minimum objective function value of zero. With this optimal discharge, MIKE 11 model is run and the salinity at the Hoa Phu pumping station is simulated (Fig. 4).

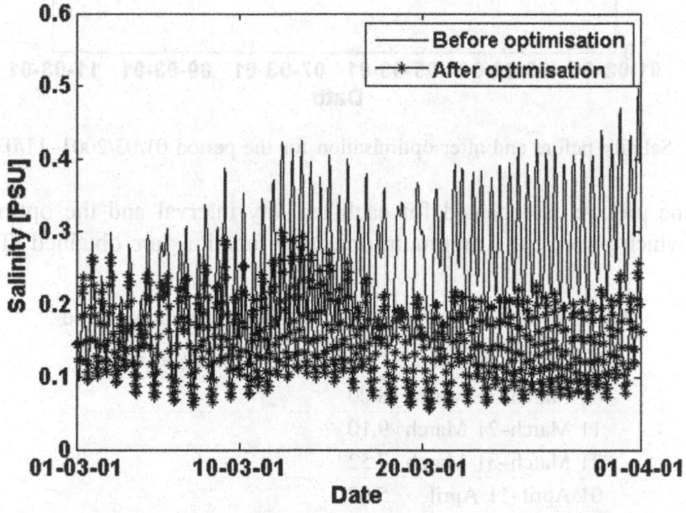

Fig. 4. Simulated salinity before and after optimisation in March

Monthly average discharge of the Saigon River for April is optimised in a same way as the previous case. The simulation date is defined between 01/04/2001 and 01/05/2001. The results in terms of hydrodynamics and advection-dispersion from the previous optimisation process in March are provided as initial conditions for the optimisation in April. In this optimisation, the optimal river discharge is obtained as 5.79 m^3/s, which gives minimum objective function value of zero.

As a next step, optimisation of the river discharges to control salinity is performed for ten-daily intervals. In order to keep the salinity under 0.3 PSU during the first ten days of March, ten-daily average river discharge for this period is considered over dates between 01/03/2001 and 11/03/2001. This optimisation gives an optimum discharge as 8 m^3/s. With this optimal discharge, MIKE 11 model is run and then the salinity at the control point is simulated. Figure 5 compares the salinity before and after the optimisation in March.

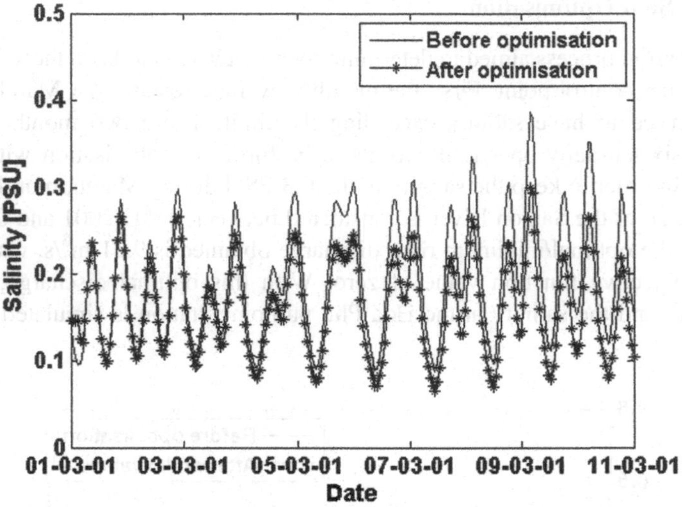

Fig. 5. Salinity before and after optimisation for the period 01/03/2001–11/03/2001

The same process is repeated for each ten day interval and the optimised river discharges which provide the minimum objective functions are obtained (Table 4).

Table 4. Optimal river discharges in each ten-daily period

Simulation period	Optimal river discharge (m^3/s)
01 March–11 March	8.00
11 March–21 March	9.10
21 March–31 March	7.52
01 April–11 April	5.62
11 April–21 April	7.24
21 April–30 April	6.43

In comparison to the initial case of 2001 in which the average discharges for the salinity control is 2 m^3/s for both March and April, both monthly and ten daily time-step simulations give optimal discharges with higher values, which ensures that salinity is below than 0.3 PSU at the Hoa Phu pumping station. It is important to note that for both simulation time-step, salinity before and after optimisation follows the same pattern, there is only a a difference in the discharges of the river.

4.2 Second Step Optimisation

The results of the second step optimisation are presented here only for the ten-daily time step. The obtained Pareto front is presented in Fig. 6 and the obtained deficits for the three characteristic points are represented in Fig. 7.

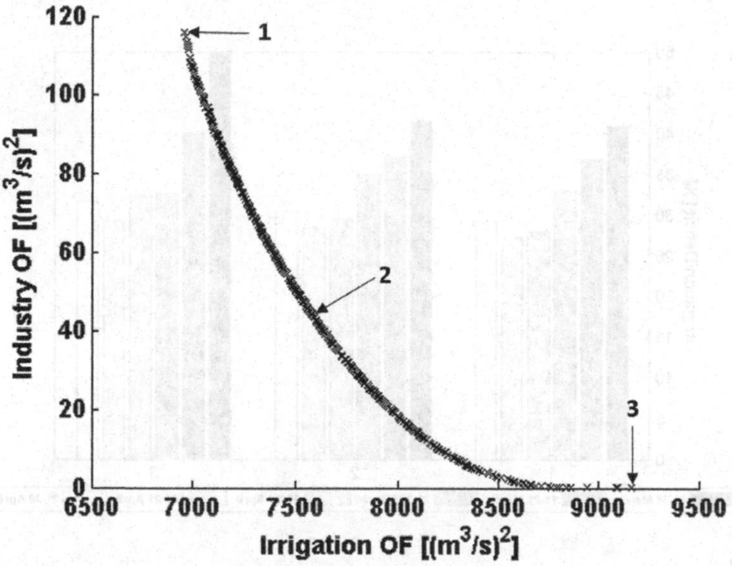

Fig. 6. Pareto set for the irrigation and industry objectives

The demands for salinity control and domestic water are always satisfied in this case since they are introduced as hard constraints.

4.3 Sensitivity Analysis with Different Inflow Scenarios

The optimisation results were further tested for two other inflow scenarios namely the inflows of average year (1979) and the inflows of wet year (1966).

Average Year (1979)

The year 1979 is a year with average inflows, hence an average volume between the curves of critical reservoir operation and flood control is assigned as an initial volume. An initial check is performed to see if all the demands can be met without compromise between the objectives of the reservoir. Once all the inflows and outflows, which refer to irrigation, salinity control, industrial and domestic water demands, are inserted in reservoir continuity equation, the water levels at the end of each ten-daily period are computed. The obtained calculated water levels in the reservoir are greater than the critical operation curve, which means that there is no need for an optimisation in case of an average year if the initial water level is assumed as the average of flood control curve level and critical operation curve level. However, it is important to check the reservoir operation if the initial water level is at the critical operation level. Such a check shows that the calculated water level drops under the critical operation curve, which means that not all the demands can be fully supplied, hence the need for reservoir releases optimisation.

The best option for Irrigation (Point 1), trade-off solution (Point 2) and the best option for Industry (Point 3) are chosen from the Pareto set. Corresponding to these points, irrigation and industry deficits are calculated as presented in Fig. 7.

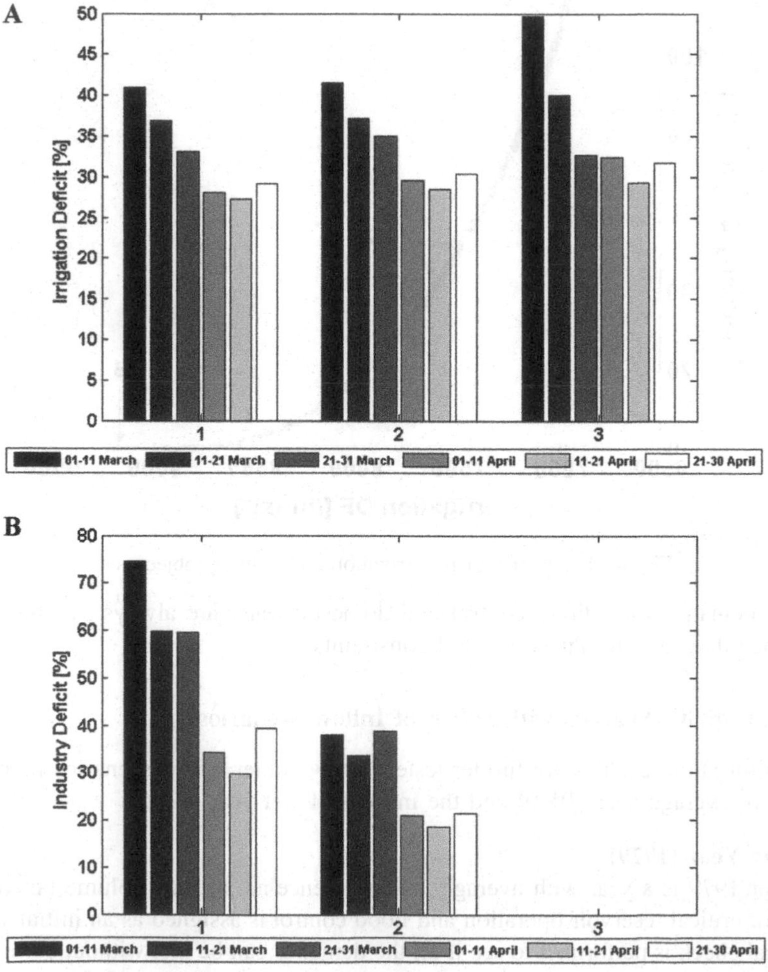

Fig. 7. Year 1979: Irrigation deficits (A) and industry deficits (B) for Point 1, Point 2 and Point 3

Wet Year (1966)

For a wet year there is a need to check the situation if the initial water level is at the critical level. Same as in the average year case, the inflows to the reservoir and outflows, which refer to irrigation, salinity control, industrial and domestic water demands, are inserted in reservoir continuity equation. The calculated water level drops under the critical operation curve, which indicates that not all demands from the reservoir can be satisfied completely and reservoir optimisation is required also for wet year case if the initial water level is at the critical level. Irrigation and industry deficits regarding the best option for irrigation (Point 1), trade-off (Point 2) and the best option for industry (Point 3) are assessed as shown in Fig. 8.

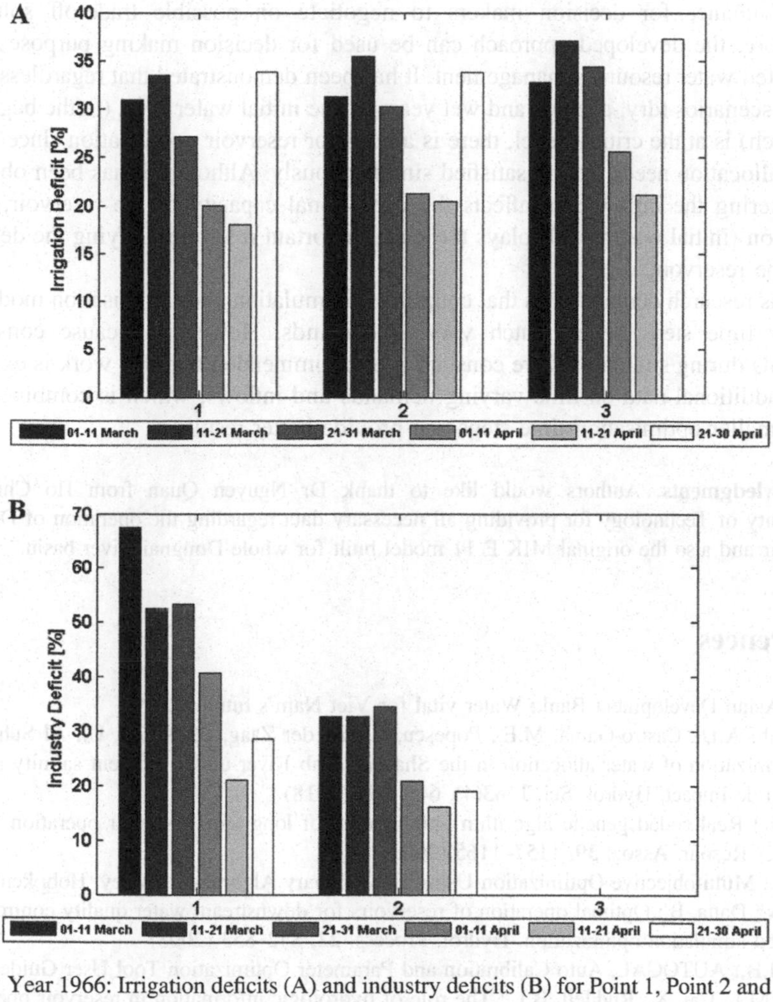

Fig. 8. Year 1966: Irrigation deficits (A) and industry deficits (B) for Point 1, Point 2 and Point 3

The results obtained for average year and wet year demonstrate that with the increase in inflows to the reservoir, more satisfactory results are obtained on satisfying the demands from the reservoir.

5 Conclusions

The main objective of this research was to find an optimal reservoir operation strategy for downstream salinity control in Saigon River Basin, Vietnam. Therefore, it is proposed to develop a simulation-optimisation framework for finding the optimal solutions to control salinity by satisfying the other objectives of the reservoir as much as possible. The final outputs of the multi-objective optimisations are the Pareto sets which give a chance for decision makers to negotiate on possible trade-off solutions. Therefore, the developed approach can be used for decision making purposes in an integrated water resources management. It has been demonstrated that regardless of the inflow scenarios (dry, average and wet years) if the initial water level (in the beginning of March) is at the critical level, there is a need for reservoir optimisation since not all water allocation needs can be satisfied simultaneously. Although it has been observed that altering the inflow data affects the operational capacity of the reservoir, initial condition (initial water level) plays the most important role on satisfying the demands from the reservoir.

This research demonstrates that coupling of simulation and optimisation models for smaller time steps better match varying demands. However, because considered demands during simulations are constant it is recommended that this work is extended using additional data on time-varying demands and inflows, which is combined with even smaller optimisation time steps can provide better results.

Acknowledgments. Authors would like to thank Dr Nguyen Quan from Ho Chi Minh University of Technology for providing all necessary data regarding the operation of Dautieng reservoir and also the original MIK E 11 model built for whole Dongnai River basin.

References

ADB, Asian Development Bank: Water vital for Viet Nam's future (2009)

Abdullah, A.D., Castro-Gama, M.E., Popescu, I., van der Zaag, P., Karim, U., Al Suhail, Q.: Optimization of water allocation in the Shatt al-Arab River under different salinity regimes and tide impact. Hydrol. Sci. J. **63**(4), 646–656 (2018)

Chen, L.: Real coded genetic algorithm optimization of long term reservoir operation. J. Am. Water Resour. Assoc. **39**, 1157–1165 (2003)

Deb, K.: Multi-objective Optimization Using Evolutionary Algorithms. Wiley, Hoboken (2001)

Dhar, A., Datta, B.: Optimal operation of reservoirs for downstream water quality control using linked simulation optimization. Hydrol. Process. **22**, 842–853 (2008)

DHI, M.B.: AUTOCAL, Auto Calibration and Parameter Optimization Tool User Guide (2014)

Hejazi, M.I., Cai, X., Ruddell, B.L.: The role of hydrologic information in reservoir operation–learning from historical releases. Adv. Water Resour. **31**, 1636–1650 (2008)

Labadie, J.W.: Optimal operation of multireservoir systems: state-of-the-art review. J. Water Resour. Plann. Manag. **130**, 93–111 (2004)

Le Ngo, L., Madsen, H., Rosbjerg, D.: Simulation and optimisation modelling approach for operation of the Hoa Binh reservoir, Vietnam. J. Hydrol. **336**, 269–281 (2007)

Le Ngo, L., Rosbjerg, D., Madsen, H.: Optimising reservoir operation: a case study of the Hoa Binh reservoir, Vietnam. Technical University of DenmarkDanmarks Tekniske Universitet, Department of Hydrodynamics and Water ResocurcesStrømningsmekanik og Vandressourcer (2006)

Ringler, C., Cong, N.C., Huy, N.V.: Water allocation and use in the Dong Nai River Basin in the context of water institution strengthening. In: Integrated Water-resources Management in a River-basin Context: Institutional Strategies for Improving the Productivity of Agricultural Water Management, p. 215 (2002)

UNEP, I.A.: Viet Nam assessment report on climate change (2009)

VCAPS-Consortium, V.C.A.P.: Climate adaptation strategy Ho Chi Minh City moving towards the sea with climate change adaptation (2013)

Waite, P.J.: Control of salt water intrusion in estuaries by means of a dual purpose reservoir. In: Hydrological Forecasting -Proceedings of the Oxford Symposium, April 1980. IAHS-AISH, Publ. no. 129 (1980)

WorldBank: Dau Tieng irrigation project, 8239 (1989)

Clustering Hydrographic Conditions in Galician Estuaries

David E. Losada[1(✉)], Pedro Montero[2], Diego Brea[1], Silvia Allen-Perkins[2], and Begoña Vila[2]

[1] Centro Singular de Investigación en Tecnoloxías da Información (CiTIUS), Universidade de Santiago de Compostela, Santiago de Compostela, Spain
david.losada@usc.es, diebrea@gmail.com
[2] Instituto Tecnolóxico para o Control do Medio Mariño de Galicia (INTECMAR), Vilagarcía de Arousa, Spain
{pmontero,scaceres,bvila}@intecmar.gal

Abstract. In this paper we describe our endeavours to explore the role of unsupervised learning technology in profiling marine conditions. The characterization of the marine environment with hydrographic variables allows, for example, to make technical and health control of sea products. However, the continuous monitoring of the environment produces large amounts of data and, thus, new information technology tools are needed to support decision-making. We present here a first contribution to this area by building a tool able to represent and normalize hydrographic conditions, cluster them using unsupervised learning methods, and present the results to domain experts. The tool, which implements visualization methods adapted to the problem at hand, was developed under the supervision of specialists on monitoring marine environment in Galicia (Spain). This software solution is promising to early identify risk factors and to gain a better understanding of sea conditions.

1 Introduction

The Ría de Vigo is the southernmost of Galician Rías (NW Spain), several inlets placed at the northern boundary of the NW Africa upwelling system [21]. The Ría de Vigo is a 32 km long v-shaped, 40 m average depth estuary connected to the shelf by a 52 m deep southern channel and a 23 m deep northern mouth, separated by the Cies islands. In the inner part of the ria, the main river, Oitaven-Verdugo is placed. The runoff of this river is mainly seasonal with high flow in winter and low in summer [14].

The wind-driven barotropic flow is the main driving force of the residual circulation [17], with time responses of local winds and remote winds within 6 h and 12 h [7]. From March-April to September-October, due to southward winds, cold and nutrient-rich Eastern North Atlantic Central Water (ENACW) upwells onto the shelf and is introduced to the ria [1,6]. This entrance of ocean water mainly comes from the bottom. During the rest of the year, SW winds provoke the entrance of warm and nutrients-depleted water by the surface, blocking the

J. M. F. Rodrigues et al. (Eds.): ICCS 2019, LNCS 11539, pp. 346–360, 2019.
https://doi.org/10.1007/978-3-030-22747-0_27

circulation of the ria. This condition is related to a high runoff of the river and an exit of surface fresh water, producing a convergent front in the ria. However, this stational scheme gives only a general picture since the events with frequencies <30 days explains $>70\%$ of the variability [13]. As a general picture, the winter condition consists in a stratified column maintained by the fresh water discharge, in spite of the thermal inversion by heat losing through the surface. In summer, the upwelling colds down the bottom layers. This fact along the warm surface water (because of radiation) causes a strong thermal stratification. The rest of the year, the column mix is dominant because of downwelling events.

Galician Rias are one of the most productive oceanic regions of the world [3]. Subtidal dynamics is important since it is the main responsible for the net export and import of water, nutrients, contaminants, plankton, to and from the ria of Vigo [8]. Hydrography is a fundamental tool to understand the dynamics of the ria.

The key contribution of this research is to design a tool able to cluster and visualize hydrographic conditions in Galician rias. An automatic categorization of hydrographic conditions is fundamental to determine the target *typical* state of the ria and its anomalies, which could be used as descriptor 7 of European Marine Strategy Framework Directive (MSFD). Any permanent alteration of these conditions could be used as an indicator of loss of good environmental status. Moreover, the determination of these typical conditions is the first stage to obtain an environmental impact assessment. As a matter of fact, it is important to understand how anthropogenic activity results in deviations from the usual marine conditions. On the other hand, the behavior of most phytoplankton species is influenced by the physical conditions, and knowing how much conditions deviate from the average could be a proxy of the bloom or decay of these species. There is also a direct relationship between the subtidal circulation and the hydrography and, thus, the knowledge derived from our tool can be also indicative of the capability of natural cleaning. The clustering of marine conditions can also help to understand where the marine litter will go to.

This is a preliminary research project that aims at exploring the possibilities of unsupervised learning technology. We therefore selected an initial sample of Galician stations (see Fig. 1) and we represented the data extracted from these stations at different points in time.

2 Materials and Methods

2.1 Collection of Observations

In order to control the quality of the water in the Galician shellfish harvesting areas, INTECMAR[1] weekly monitors the hydrography of Galician coast. These weekly campaigns have been running since 1992. The current oceanographic network is formed by 43 oceanographic stations distributed along Rias Baixas and the Ría de Ares. Eight of these stations are located in the Ría de Vigo.

[1] www.intecmar.gal.

Among other measurements, salinity and temperature profiles are recorded using a SBE25 CTD (conductivity-temperature-depth) profiler. Conductivity measurements are converted into salinity values using the UNESCO equation [20]. Every week, the obtained raw CTD data are processed, filtered and bin averaged using the standard prescriptions of the CTD manufacturer [9,10]. All data are downloaded, processed and saved on the INTECMAR data center (and distributed through www.intecmar.gal).

In order to make an initial prototype, only the samples of two years (2015–2016) obtained from the Ría de Vigo were considered. The profiles of temperature and salinitiy were used to represent the hydrographic conditions.

2.2 Data Representation

The main aim is to automatically discover associations among campaigns and, thus, for each campaign, the information collected from all the stations is represented into a single *campaign representational unit*. This is a vector of numerical values (temperature and salinity) obtained from all the stations at different levels of depth.

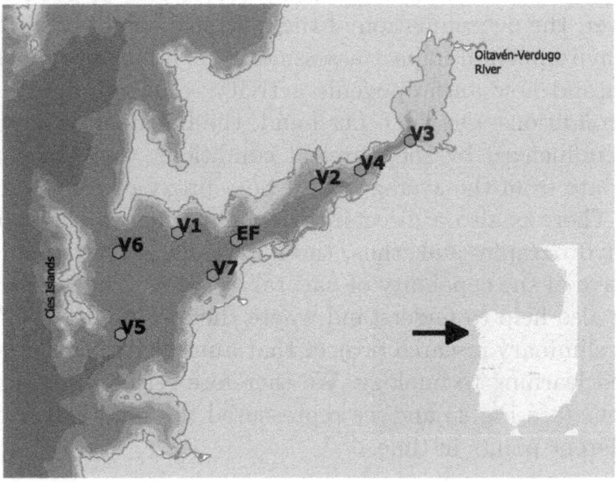

Fig. 1. Stations in one Galician estuary ("Ría de Vigo")

The information obtained from each station comes originally in the form of triples: $(depth, temperature, salinity)$ from INTECMAR data center. For example, (2.35 m, 12.34 °C, 35.28), (2.47 m, 12.38 °C, 35.48), and so forth[2]. The depth values are not uniform over campaigns. For example, campaign 1 might have measures at depth levels of 2.35 m, 2.47 m, ... while campaign 2 might have

[2] Temperature is measured in degrees centigrades and practical salinity is dimensionless.

measures at depth levels of 2.15 m, 2.87 m, and so forth. Such inconsistencies come from the characteristics of the measuring devices and the type of bin average routinely used in the procedure. From a data representational perspective, this demands some normalization strategy. Following domain expert knowledge, superficial measures (all measures taken at depths lower than 2 m) were discarded and the remaining measures were bin averaged at 1 m intervals. This leads to the following intervals: [2 m,3 m), [3 m–4 m),... For each interval, all measures whose depth falls in the interval are aggregated by weighted average. This results in a representation that has two values for each interval: weighted average of temperatures and weighted average of salinities in the interval. Given l_i, the number of intervals for station i, the overall vectorial representation of a given campaign is:

$$station_i = (T_{1,i}, S_{1,i}, T_{2,i}, S_{2,i}, \cdots, T_{l_i,i}, S_{l_i,i}) \quad (1)$$

$$campaign_k = (station_1, station_2, ...) \quad (2)$$

where $T_{n,i}$ ($S_{n,i}$) is the temperature (salinity) of the n-th interval in station i. The geographical locations of the stations (located at different points of the estuaries) make that the maximum depths are different and, thus, different stations contribute with different number of values to this representation. Additionally, since the maximum depth of each cast can vary among campaigns (e.g., due to the sea-weather, tide and surveyor skill), the deepest measures of each cast were discarded.

In this way, $campaign_k$ represents the hydrographic condition of the estuary at the k-th campaign (on a given date). This condition is modelled by the sequences of temperatures and salinities obtained at different depths from stations located at strategic points in the estuaries. By automatically associating $campaign_k$ with other campaigns, we can relate current campaigns with conditions seen in the past, we can profile marine conditions and we can try to anticipate risk factors.

2.3 Clustering Campaign Data

Each campaign is represented by a vector of values (Eq. 2) and the main purpose of this new marine application is to cluster campaigns into groups. To meet this aim, we employ k-means [11,12]. K-means is a well-known clustering algorithm that finds clusters and cluster centroids in a set of unlabelled data. The number of desired clusters (k) has to be chosen in advance and k-means proceeds iteratively by moving the cluster centroids in order to minimize the total within cluster variance. Given an initial set of centroids, k-means alternates between (i) identifying the data points that are closer to each cluster centroid and (ii) updating the centroids by computing the average of the points in each cluster (each cluster centroid is the vector of the feature means for the points in the cluster). The algorithm iterates until convergence. K-means aims to find a good set of non-overlapping clusters. And the main intuition is that a good cluster is one for which its points do not differ much from each other.

Cluster Quality. There are different methods for choosing the optimal number of clusters. Next, we describe some of them. The Elbow method [19] runs k-means for a range of values of k and for each value of k computes the total within-cluster sum of squares (WSS). Such an approach estimates the compactness of the clustering from the pairwise squared Euclidean distances between the points in the cluster. The Elbow method plots WSS against the number of clusters and suggests to choose the number of clusters so that adding another cluster does not reduce much the total WSS. To meet this aim, the presence of an elbow or knee in the plot is considered as an indicator of the ideal number of clusters.

Silhouette plots [18] are alternative displays for interpreting and validating clusterings. They graphically represent clusters by a *silhouette*, which depicts the tightness and separation of the clusters. For each point, its silhouette score measures how similar the point is to its own cluster (cohesion) compared to the other clusters (separation). This score ranges into $[-1, +1]$ and a score close to 1 means that the point fits well with its cluster and it is dissimilar to the other clusters.

Caliński and Harabasz [4] proposed another criterion to evaluate the quality of clusterings. It evaluates cluster validity based on the mean between- and within-cluster sum of squares. Davies and Bouldin [5] presented a measure that can be also used to infer the appropriateness of cluster partitions. Their measure incorporates well-accepted features employed in cluster analysis and its design was driven by certain heuristic criteria.

Our tool implements the four clustering quality measures described above. These four estimates can be used to automatically filter out bad partitions. However, the output of a given clustering configuration requires human interpretation and, thus, our tool is flexible and allows the user to specify the number of clusters, analyze the results, visualize the campaigns associated to each cluster, etc. As a matter of fact, this subjective analysis, done by the domain expert, should shed light on what is to be considered a good cluster of marine conditions.

Dimensionality Reduction. The high number of dimensions or features in the campaign vectors makes it difficult to visualize clustering results. We therefore adopted some standard dimensionality reduction methods that are used for presenting the output of the clustering in three-dimensional graphs.

Principal Component Analysis (PCA) [15] is a traditional way to do dimensionality reduction. It is a statistical method based on orthogonal transformation that converts a set of points represented with possibly correlated features into a set of points represented with a set of linearly uncorrelated features (known as principal components). The transformation is performed in such a way that the first component accounts for as much of the variability in the data as possible, the next component has the highest variance under the constraint that it is orthogonal to the first component, and so forth.

In exploratory data analysis, PCA is often employed for visualization purposes. High-dimensional datasets cannot be easily explored and analyzed by humans. PCA supplies the user with low-dimensional representations. These representations, which can retain as much of the variance of the original representation as possible, can be plotted on informative graphs. Our tool uses PCA to generate visually amenable graphs that better communicate the clusterings to the domain expert.

3 Experiments

The dataset was built from the measures obtained from eight stations in one Galician estuary ("Ría de Vigo"). More specifically, we got data from the stations labelled as EF, V1, V2, V3, V4, V5, V6, and V7 in Fig. 1. We analyzed the database provided by INTECMAR and selected an initial sample of dates (years 2015 and 2016). The overall number of campaigns in this sample (e.g. the number of points to be clustered) is 80. This is a small sample but it helps us to make initial tests with the tool. In the future, we plan to extend this cluster analysis to many more data points (larger range of dates, more campaigns and more stations from other Galician locations).

Given the characteristics of the eight stations (maximum depths of the measuring exercises), each data point was represented by a vector with 254 features (127 temperatures + 127 salinities). On average, each station contributed with about 16 depth levels.

The tool we developed is written in Python. This facilitates the incorporation of multiple data analysis libraries and toolkits. Furthermore, it is a language that is currently employed in several INTECMAR projects and, thus, the tool can be later adapted and maintained by INTECMAR analysts. Unsupervised learning is driven by a number of libraries and classes from scikit-learn [16].

3.1 Preliminary Tests

We first performed a number of experiments designed to set the main configuration options of the clustering algorithm. We worked with two versions of k-means, k-means and MiniBatchKMeans[3]. The first is a standard k-means implementation, while the latter is a variant that uses mini-batches to reduce the computation time. MiniBatchKMeans optimizes the same objective function as k-means but drastically reduces the computational effort required to converge to a solution. Mini-batches are sub-samples of the input data that are randomly selected at each training iteration. In contrast to other solutions that reduce k-means' computational time, mini-batch k-means outputs results that are generally only slightly worse than k-means' results. With the current dataset, computational time is not a major concern and, therefore, we did not observe substantial differences between both algorithms. We decided to adopt k-means for the subsequent experiments. However, our tool can be easily configured to work with MiniBatchKMeans (if needed for performing large-scale experiments with massive datasets of marine conditions).

[3] The corresponding scikit-learn classes are sklearn.cluster.KMeans and sklearn.cluster.MiniBatchKMeans, respectively.

Next, we varied a number of parameters and observed the results obtained. More specifically, we tested some initialization parameters and a parameter related to the maximum number of iterations. K-means finds a local optimum rather than a global optimum. As a result, the final output depends on the initial set of centroids. For this reason, it is customary to run k-means multiple times from different initial configurations. This is governed by the parameter `n_init`, which we set to 1000. This means that the final results reported will be the best output (minimum within-cluster sum of squares) of 1000 consecutive runs of k-means. For each execution of the algorithm, the maximum number of iterations was set to 1000 (`max_iter` = 1000). We also experimented with different initialization choices: (i) a random selection of centroids, which chooses k data points at random for the initial centroids, and (ii) `k-means++` [2], a more sophisticated selection of seed centroids, which selects initial centroids in a smart way to speed up convergence. Although there was no much overall difference, we finally selected the following configuration for the subsequent experiments:

```
KMeans(init='k-means++', n_init=1000, max_iter=1000)
```

3.2 Ideal Number of Clusters

First, the experimentation focused on selecting the number of clusters. To meet this aim, we experimented with clusterings with up to 12 clusters and we computed the metrics described in Sect. 2.3. Figure 2 depicts the results of the Elbow method. Although there is not a clear knee, it appears that the most consistent solutions are clusterings with a number of clusters in the range from 3 to 7. Next, we proceeded to compute the Silhouette, Caliński-Harabasz and Davies-Bouldin metrics. Table 1 reports the suggested number of clusters of the three best configurations according to these metrics. The results are a mixed bag. They seem to suggest a high number of clusters (greater than 5) but there is not a clear choice.

Table 1. The three best configurations according to Silhouette, Caliński-Harabasz and Davies-Bouldin metrics. The figures in the table are numbers of clusters.

	Best configuration	2nd best configuration	3rd best configuration
Silhouette	8	5	10
Caliński-Harabasz	6	7	8
Davies-Bouldin	5	10	11

Figures 3, 4 and 5 show the Silhouette graphs for clustering configurations from 2 to 10 clusters. This graphical presentation helps to shed light on the ideal number of clusters. For each plot, the X axis represents the Silhouette scores of the data points (the higher the better). Data points with Silhouette scores close to 0 are on the border between two clusters. The dashed vertical line represents the average silhouette score of all the values in the plot. Each plot contains a certain number of clusters, where each cluster is represented by a bar graph with

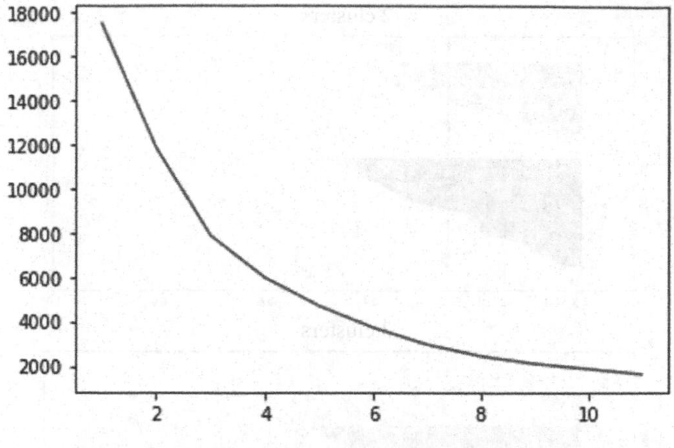

Fig. 2. Elbow method. The X axis represents the number of clusters and the Y axis represents the corresponding sum of squared errors.

the Silhouette scores of the data points in the cluster. The bar graph of each cluster is labeled with a number (from 0 to the number of clusters minus 1). The thicker the bar graph, the larger the cluster (more data points were assigned to the cluster). A number of observations can be derived from this visual analysis. First, cluster configurations with 5 or more clusters tend to produce clusterings where some clusters have very few data points. For example, with 5 clusters, cluster #3 is tiny. The same happens for cluster #5 (6 clusters plot) or cluster #0 (8 clusters plot). These partitions do not seem reliable as these tiny clusters hardly reflect a real group of similar marine conditions. Furthermore, many of these cluster configurations (for example, all cases with 7 or more clusters) show data points with negative Silhouette scores, reflecting far from ideal cluster partitions (some points do not fit well with their cluster). Clusterings with 2, 3 or 4 clusters look better. However, the 2 and 4-cluster plots show also some negative scores. This suggests that the 3-cluster partition is the most natural choice. This outcome was discussed with a domain expert, who analyzed the groups and confirmed that a 3-cluster solution is rational and the three groups could be associated with three typical environmental conditions. We therefore adopted 3 clusters as our configuration for the rest of the analysis.

3.3 3D Visualization

To further analyze the quality of this clustering, we proceeded to apply PCA on the data. This transformation was done for visualization purposes and, thus, it is performed *after clustering* the original dataset (the dataset is clustered using the original set of features and, next, the data points are transformed into a reduced PCA space of features). With three principal components, the PCA transformation of the dataset maintains 86% of the variance. This suggests that visualizing the clusters using the three PCA principal components does not lose

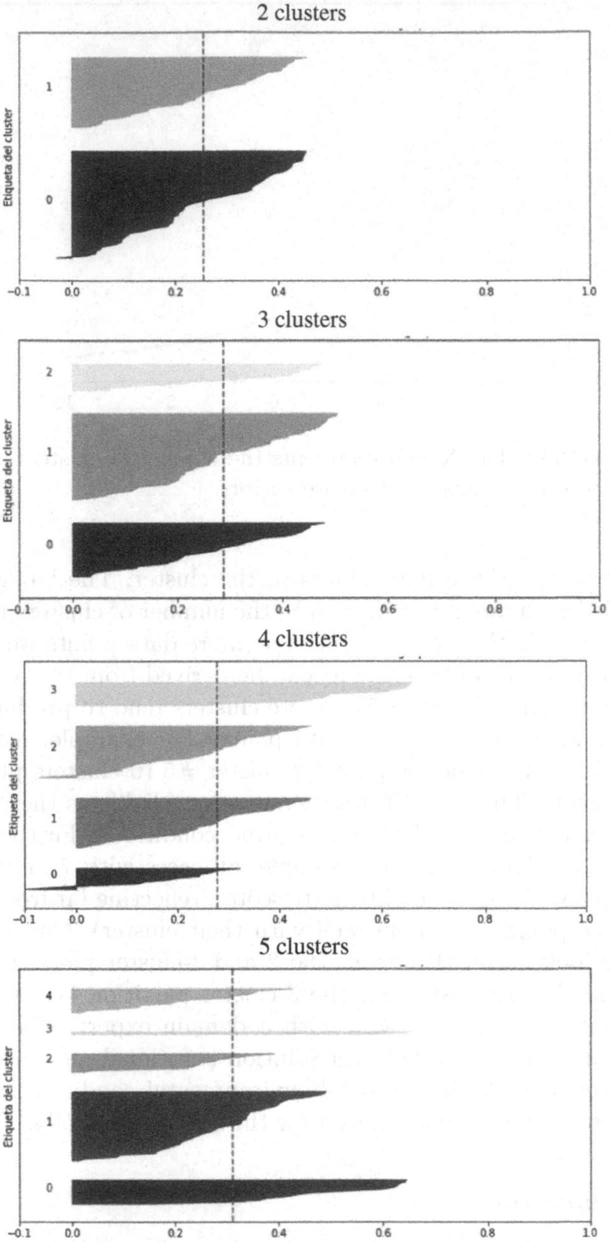

Fig. 3. Silhouette graphs (2–5 clusters)

much information. Figure 6 shows the results. This graph confirms that a 3-cluster configuration partitions the data in a rational way, with no much overlapp among the three cluster *regions*.

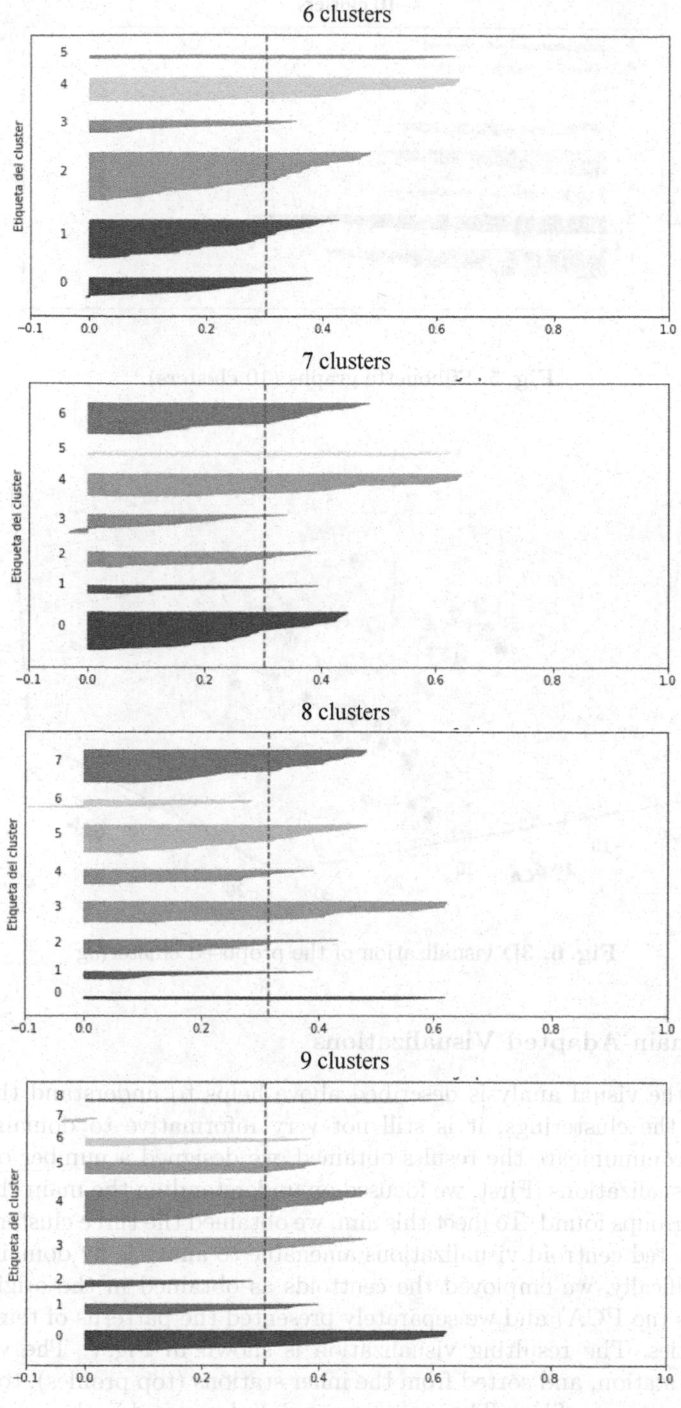

Fig. 4. Silhouette graphs (6–9 clusters).

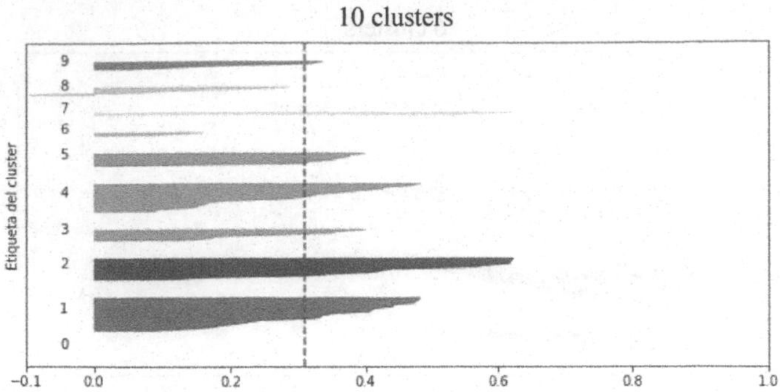

Fig. 5. Silhouette graphs (10 clusters).

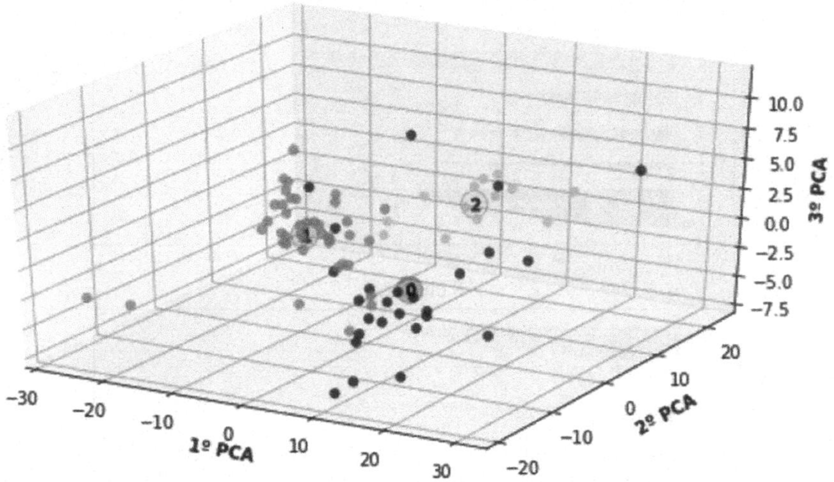

Fig. 6. 3D visualization of the proposed clustering

3.4 Domain-Adapted Visualizations

Although the visual analysis described above helps to understand the relative quality of the clusterings, it is still not very informative to domain experts. To better communicate the results obtained, we designed a number of domain-adapted visualizations. First, we focused on understanding the main characteristics of the groups found. To meet this aim, we obtained the three cluster centroids and we created centroid visualizations amenable to analysis by domain experts. More specifically, we employed the centroids as obtained in the original set of dimensions (no PCA) and we separately presented the patterns of temperatures and salinities. The resulting visualization is shown in Fig. 7. The vector was divided by station, and sorted from the inner stations (top profiles), to the outer stations (bottom profiles). The casts are presented as usual in the oceanographic style, where depth is shown in the y-axis. Left profiles correspond to temperature

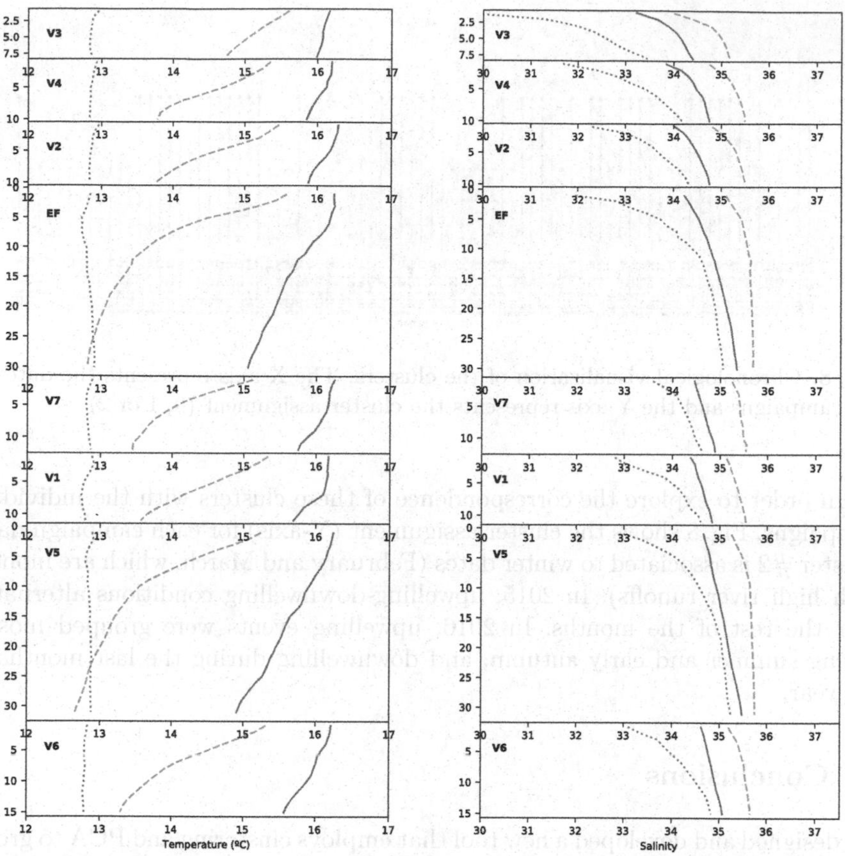

Fig. 7. Temperature (left graph) and Salinity (right graph) patterns associated to the three cluster centroids. (Color figure online)

and right ones to salinity. Each cluster is represented by a line. This graph was analyzed by the domain expert and he found it highly informative. For example, we found that cluster #2 (green dotted line) represents a typical winter condition, where temperatures are lower and constant for each profile. The salinity is lower for the surface layers, mainly in the inner stations, due to the high river discharges during this season. The orange dashed line (cluster #1) can be representing upwelling events, with temperatures in the bottom layers very low (even lower than in the winter cluster #2 condition in the deepest stations: EF and V5) and high temperatures at the surface (because of sun radiation). The salinity profiles are mostly constant and high as corresponding with a lower river runoff. Only in the inner stations, the surface fresh water signal is significant. The blue solid line (cluster #0) is related to a downwelling/mixed condition, warmer than the condition described above, mainly in the surface layers and with significant presence of freshwater in the inner stations.

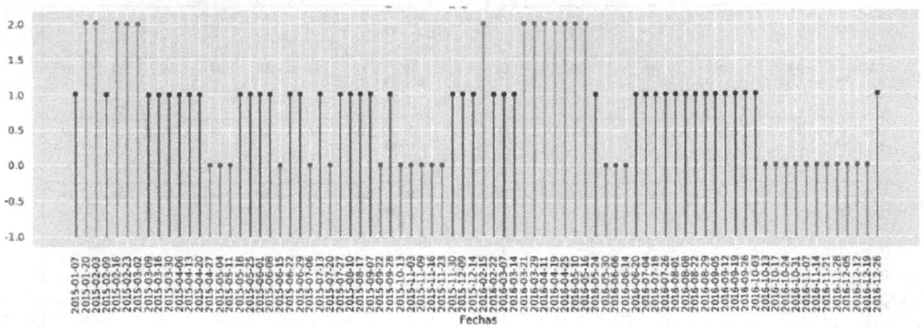

Fig. 8. Chronological visualization of the clusters. The X axis represents the dates of the campaigns and the Y axis represents the cluster assignment (0, 1 or 2)

In order to explore the correspondence of these clusters with the individual campaigns, Fig. 8 shows the cluster assignment (Y-axis) for each campaign date. Cluster #2 is associated to winter dates (February and March, which are months with high river runoffs). In 2015, upwelling-downwelling conditions alternated over the rest of the months. In 2016, upwelling events were grouped mostly during summer and early autumn, and downwelling during the last months of the year.

4 Conclusions

We designed and developed a new tool that employs clustering and PCA to group hydrographic events. Using this new tool, K-means clustering effectively organized data measures obtained from marine campaigns. With a 3-clustering setting and using data from two years of CTD weekly campaigns in the Ría de Vigo, the tool recognized typical conditions described in the literature (upwelling, downwelling events and transition between them [13]). Furthermore, the influence of the river runoff during winter and the surface radiation during summer was also detected and visualized.

This work represents our first attempt to use machine learning to understand hydrographic conditions. This research needs to be extended in a number of ways: (a) to enlarge the study to datasets associated to larger periods of time; (b) to apply this methodology to other rias and (c) to test a higher number of clusters. By considering and analyzing a larger number of clusters, we might be able to discover specific conditions that are not fully described in the literature. Furthermore, the centroids of these classes could be indicative of unknown circumstances, such as deviations or defects in CTD campaigns or new trends in the ria. The potential identification of unknown conditions is a promising feature that can help to early identify risk factors and to further understand the conditions of the sea.

Acknowledgements. This research has received financial support from the Galician Ministry of Education (grants ED431C 2018/29 and ED431G/08, co-funded by the ERDF), and the European Union MarRISK project: "Adaptación costera ante el Cambio Climático: conocer los riesgos y aumentar la resiliencia" (0262_MarRISK_1_E), through EP-INTERREG V A España-Portugal (POCTEP) program. www.poctep.eu/es/2014-2020/marrisk. The third author thanks the support obtained from Xunta de Galicia through the "BECAS EXCELENCIA XUVENTUDE EXTERIOR" program.

References

1. Alvarez-Salgado, X.A., Gago, J., Miguez, B.M., Gilcoto, M., Perez, F.F.: Surface waters of the NW Iberian margin: upwelling on the shelf versus outwelling of upwelled waters from the Rías Baixas. Estuar. Coast. Shelf Sci. **51**(6), 821–837 (2000)
2. Arthur, D., Vassilvitskii, S.: k-means++: the advantages of careful seeding. In: Proceedings of the 18th Annual ACM-SIAM Symposium on Discrete Algorithms (2007)
3. Blanton, J.O., Atkinson, L.P., Castillejo, F.F., Lavin-Montero, A.: Coastal upwelling off the Rias Bajas, Galicia, northwest Spain I: Hydrographic studies. Rapp. P.- v. Rèun. Cons. Int. Explor. Mer. **183**, 79–90 (1984)
4. Caliński, T., Harabasz, J.: A dendrite method for cluster analysis. Commun. Stat. **3**(1), 1–27 (1974)
5. Davies, D.L., Bouldin, D.W.: A cluster separation measure. IEEE Trans. Pattern Anal. Mach. Intell. **1**(2), 224–227 (1979)
6. Fraga, F.: Upwelling off the Galician Coast, Northwest Spain, pp. 176–182. American Geophysical Union (AGU) (2013)
7. Gilcoto, M., et al.: Rapid response to coastal upwelling in a semienclosed bay. Geophys. Res. Lett. **44**(5), 2388–2397 (2017)
8. Gilcoto, M., Pardo, P.C., Álvarez Salgado, X.A., Pèrez, F.F.: Exchange fluxes between the Ría de Vigo and the shelf: a bidirectional flow forced by remote wind. J. Geophys. Res.: Oceans **112**(C6) (2007)
9. Seabird Inc.: SBE 25 Sealogger CTD, User's Manual, 13th edn. Seabird Inc., Bellevue (2005)
10. Seabird Inc.: Seasoft V2: SBE Data Processing, Software Manual. Seabird Inc., Bellevue (2017). (7.26.8 edition)
11. Lloyd, S.P.: Least squares quantization in PCM. Technical report, Bell Laboratories (1957)
12. MacQueen, J.: Some methods for classification and analysis of multivariate observations. In: Proceedings Fifth Berkeley Symposium on Mathematical Statistics and Probability (1967)
13. Nogueira, E., Pàrez, F.F., Ríos, A.F.: Seasonal patterns and long-term trends in an estuarine upwelling ecosystem (Ría de Vigo, NW Spain). Estuar. Coast. Shelf Sci. **44**(3), 285–300 (1997)
14. Otero, P., Ruiz-Villarreal, M., Peliz, Á., Cabanas, J.M.: Climatology and reconstruction of runoff time series in northwest Iberia: influence in the shelf buoyancy budget off Ría de Vigo. Scientia Marina **74**(2), 247–266 (2010)
15. Pearson, K.: On lines and planes of closest fit to systems of points in space. Lond. Edinb. Dublin Philos. Mag. J. Sci. **2**(11), 559–572 (1901)
16. Pedregosa, F., et al.: Scikit-learn: machine learning in Python. J. Mach. Learn. Res. **12**, 2825–2830 (2011)

17. Piedracoba, S., Álvarez Salgado, X.A., Rosón, G., Herrera, J.L.: Short-timescale thermohaline variability and residual circulation in the central segment of the coastal upwelling system of the Ría de Vigo (northwest Spain) during four contrasting periods. J. Geophys. Res.: Oceans **110**(C3) (2005)
18. Rousseeuw, P.J.: Silhouettes: a graphical aid to the interpretation and validation of cluster analysis. J. Comput. Appl. Math. **20**, 53–65 (1987)
19. Thorndike, R.L.: Who belongs in the family? Psychometrika **18**(4), 267–276 (1953)
20. UNESCO: The international system of units (SI) in oceanography. UNESCO Technical Papers in Marine Science, vol. 45, p. 131 (1985)
21. Wooster, W.S., Bakun, A., McLain, D.R.: Seasonal upwelling cycle along the eastern boundary of the north Atlantic. J. Mar. Res. **34**(2), 131–141 (1976)

Early Warning Systems for Shellfish Safety: The Pivotal Role of Computational Science

Marcos Mateus[1(✉)], Jose Fernandes[2], Marta Revilla[2], Luis Ferrer[2], Manuel Ruiz Villarreal[3], Peter Miller[4], Wiebke Schmidt[5], Julie Maguire[6], Alexandra Silva[7], and Lígia Pinto[1]

[1] MARETEC, Instituto Superior Técnico, Universidade de Lisboa, Lisbon, Portugal
marcos.mateus@tecnico.ulisboa.pt
[2] AZTI, Herrera Kaia, Portualdea, z/g, 20110 Pasaia, Gipuzkoa, Spain
[3] Instituto Español de Oceanografía, 15011 A Coruña, Galicia, Spain
[4] Plymouth Marine Laboratory, Prospect Place, Plymouth PL1 3DH, UK
[5] Marine Environment & Food Safety Services, Marine Institute Rinville, Oranmore H91 R673, Co. Galway, Ireland
[6] Bantry Marine Research Station, Gearhies, Bantry P75AX07, Co. Cork, Ireland
[7] IPMA, I.P - Instituto Português do Mar e da Atmosfera, R. Alfredo Magalhães Ramalho 6, 1495-006 Lisbon, Portugal

Abstract. Toxins from harmful algae and certain food pathogens (*Escherichia coli* and Norovirus) found in shellfish can cause significant health problems to the public and have a negative impact on the economy. For the most part, these outbreaks cannot be prevented but, with the right technology and know-how, they can be predicted. These Early Warning Systems (EWS) require reliable data from multiple sources: satellite imagery, *in situ* data and numerical tools. The data is processed and analyzed and a short-term forecast is produced. Computational science is at the heart of any EWS. Current models and forecast systems are becoming increasingly sophisticated as more is known about the dynamics of an outbreak. This paper discusses the need, main components and future challenges of EWS.

Keywords: Shellfish safety · Early warning systems · Aquaculture

1 Introduction

Shellfish harvesting and production in aquaculture has been steadily growing over the past few years, both in quantity and value, a pattern mostly driven by the constant increase in human needs of fish protein. In 2013, for instance, shellfish amounted up to almost 25% (≈4.9 kg per capita) of fish consumption

Supported by the Interreg Atlantic Area Operational programme, Grant Agreement No.: EAPA_182/2016. A. Silva supported by Grant IPMA-BCC-2016-35.

J. M. F. Rodrigues et al. (Eds.): ICCS 2019, LNCS 11539, pp. 361–375, 2019.
https://doi.org/10.1007/978-3-030-22747-0_28

worldwide [1]. The response of the aquaculture sector to this gradual increase in demand has been the intensification in production, which has been rewarded with a growing willingness to pay by the consumer and the consequent escalating financial revenue for the sector (see Fig. 1). However, several natural and human-related factors hinder these efforts, with impacts to both the economy and human health. Besides the many polluting sources (e.g. heavy-metals, hydrocarbons, etc.), a number of biological agents, which may or may not be associated to human activities, pose increasing risks to human health and to the seafood industry worldwide. Harmful algal blooms (HABs), enteric bacteria such as *Escherichia coli* (*E. coli*) and marine viruses (e.g. norovius, NoV) are currently seen as the major threats to shellfish production and safety [2].

The presence of these agents in the water and in shellfish has been the reason for persistent closures in production areas, sometimes for long periods, resulting in heavy monetary losses. In the EU, the annual cost of HAB events is estimated to be more than 850 million USD [3]. Also, the consumption of contaminated shellfish has caused health problems, occasionally leading to human death. For this reason, governments, management agencies and producers are seriously committed to address this problem and find a set of adequate tools to prevent exposure to these agents. So far, the best options rely on monitoring and early warning systems (EWS) providing timely information for the industry to prevent or minimize exposure and, failing at this, to mitigate the impacts.

Limited knowledge about natural mechanisms triggering toxic HABs or the processes involved in outbreaks of microbiological events poses serious problems to scientists and engineers involved in developing such warning systems. Besides, these efforts rely on a significant variety of computational systems and methodologies, ranging from traditional data loggers used in monitoring programs, to more sophisticated computational approaches such as the processing of satellite remote sensing imagery and complex computational models.

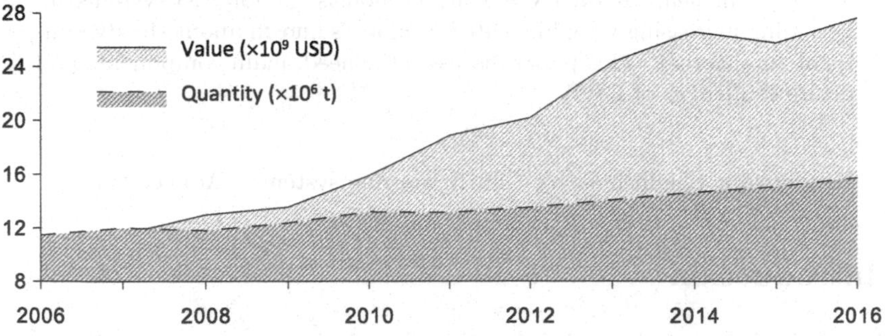

Fig. 1. Shellfish aquaculture in quantity and value for brackish and marine waters worldwide (2006–2016). Values for abalones, winkles, conchs, clams, cockles, arkshells, mussels, oysters, scallops and pectens. Data source: FAO - Fisheries and Aquaculture Information and Statistics Branch - 20/02/2019.

Over the past few years some forecast systems have been proposed, such as the experimental HAB forecast system in Lake Erie by the National Oceanic and Atmospheric Administration (NOAA) in the US, or the forecasting of the onset of HABs in the coastal waters surrounding Charlotte County, Florida [4]. EWS prototypes were recently developed and other systems already in place were matured during project ASIMUTH (supported by the EU FP7 Programme, Space Theme, Grant Agreement No. 261860) for some European countries in the Atlantic Arc (France, Ireland, Portugal, Scotland and Spain) [5,6]. Many other countries are pursuing similar systems [7]. However, only recently a few have focused on shellfish safety and try to tackle the negative impacts of microbiological contaminants or viruses. Some of these systems are currently providing warnings including such additional elements in their forecast.

This paper addresses some of the benefits of EWS for the shellfish aquaculture sector, and how computational science is central in their development. As such, this work engages EWS from a computational perspective mostly, briefly addressing some of the essential steps from harvesting field data, up to the delivery of information (early warnings) to end-users, tackling issues such as data gathering, algorithms, modeling approaches, forecast systems and platforms to integrate data. The work described here is being developed by an international consortium, as part of the work plan of the project PRIMROSE – Predicting Risk and Impact of Harmful Events on the Aquaculture Sector (information available at www.shellfish-safety.eu).

2 Aims of Early Warning Systems

Anticipating risks for public health is essential to communicate, promote and regulate public health measures. The huge amount of data collected by monitoring programs, remote sensing and ocean models is essential in this identification, but the very nature of the amount and type of data requires integrative computational tools to merge, process, interpret and transform data into something useful, minimizing human intervention in the process. EWS for impending threats to shellfish safety aim at such demanding task; they provide a short window of opportunity for producers and regulating entities to take preventive actions against impending threats, safekeeping the financial revenue of the sector and human health. For that, they must necessarily address the threats that may lead to major impacts on health and the economy: harmful algae, bacteria and viruses. Their major causes and impacts are briefly described.

2.1 Algal Toxicity

Algal blooms, the excessive growth of phytoplankton species triggered by optimal environmental conditions in the water, are natural phenomena and a frequent occurrence in the ocean, but mostly in coastal areas, upwelling systems, estuaries, bays and coastal lagoons. The main mechanisms triggering algal blooms are well known (light, nutrients and water column stability) and the conditions favoring different taxa understood [8]. However, predicting which species will bloom, and where, is still elusive. This might be due to the lack of accurate information about the environmental conditions resulting in the bloom (both in terms of conditions and temporal evolution of the conditions), but also to a fundamental unpredictability due to competition between species and chaotic effects [9]. In addition, under certain circumstances which are not yet well understood, some species flourish and produce toxins, giving rise to harmful algal blooms with ecotoxicological consequences [10].

The main syndromes usually associated with the ingestion of shellfish contaminated with the toxins produced by some HAB species include ciguatera poisoning, paralytic shellfish poisoning (PSP), neurotoxic shellfish poisoning (NSP), amnesic shellfish poisoning (ASP), and diarrhetic shellfish poisoning (DSP) [11]. Death is also a possible outcome of the exposure to such toxins. Other symptoms associated with contact with toxic algae include gastroenteritis, respiratory problems, skin irritation and liver failure [12]. Besides the impacts on human health, HABs also negatively affect the marine ecosystems through hypoxia/anoxia events, decreased water clarity, and altered feeding behavior and toxicosis of marine fauna [13], leading to mortality of sea birds, marine mammals, fish and sea turtles [14]. Consequently, HABs are associated with detrimental effects on marine biota and to significant economic damage to the aquaculture industry by making shellfish unsafe to eat and, ultimately, challenging to commercialize.

2.2 Microbiological Contamination

Most shellfish harvesting grounds for human consumption are located on inshore coastal areas. Due to their proximity to land and frequently by being within range of heavily occupied coastal strips, these shellfish producing areas are subjected to human fecal pollution from a number of point and diffusive sources. Significant quantities of fecal pathogens are introduced into the marine environment by the discharges of wastewater treatment plants, by septic tanks and pits, and by the overflows from such systems. Also, fecal pollution may reach shellfish waters by land runoff or by watercourses contaminated higher in the catchment.

The control of shellfish-borne disease related to microbiological agents has traditionally been based on the classification of production areas by the monitoring of fecal indicator bacteria, mostly *E. coli* [15]. *E. coli* may cause diseases in gastrointestinal, urinary, or central nervous systems, with symptoms such as nausea, abdominal pain, vomiting, diarrhea and cramps. *E. coli* outbreaks are mainly produced by poor management of water quality. Regulatory and market requirements for supply of safe shellfish products to consumers imposes serious restrictions on contaminated shellfish growing waters (e.g., Regulation (EC) No 854/2004 of the European Parliament and of the Council of 29 April 2004). In addition, the risk of illness associated with the ingestion of shellfish exposed to fecal pollution raises concern to the aquaculture industry and food authorities.

2.3 Viruses Infections

Shellfish accumulates NoV in a similar way to fecal pathogens, and may cause outbreaks with substantial impacts on human health. However, shellfish require a longer period to purge NoV than fecal indicator bacteria, when transferred to uncontaminated waters. NoV outbreaks are one of the leading causes of acute gastroenteritis and responsible for substantial morbidity and mortality. NoV poses the major viral risk to human health associated with shellfish consumption, though Hepatitis A virus (HAV) is also a threat. NoV outbreak symptoms are usually expressed in the form of diarrhea, nausea, vomiting, and abdominal cramps [16]. Secondary transmission from person to person is also likely to occur.

NoV are frequently present in oysters growing in contaminated waters, especially after heavy rainfall, which often results in contaminated overland run-off, combined sewer overflow, or hydraulic overload in sewage treatment plants [17], the same input routes as for *E. coli.* These infections lead to obvious healthcare costs. In the U.S. alone, a total of $184 million has been estimated as the annual cost of illness attributed to seafood contamination with NoV [18]. In Australia, for instance, 525 NoV cases were identified in March 2013, originating from consumption of contaminated oysters [19]. Given the threat to human health, the presence of NoV in oyster production areas may lead to the closure of the harvesting waters and costly oyster recalls, resulting in serious financial losses.

3 Main Components of the System

An EWS is typically set in a threefold structure or steps (see Fig. 2) combining multiple computational resources and methodologies (e.g. computational models, algorithms to process remote sensing data, machine learning approaches, etc.). The first step usually involves the gathering of information, or input data that can be potentially used by the system. The second steps deals with both the interpretation of data and its selection to run predictive models (when applicable). The last step is for the final assembly of the forecast (e.g. web services, apps, bulletins, etc.) and its dissemination to end users. Some steps can be partly or fully automatized, and are repeated according to the desired frequency of the bulletin and/or warnings.

3.1 Data Acquisition

EWS are primarily information systems: acquiring available information on the state of the ocean, meteorological conditions, health reports, etc., and processing it to generate specific information to assist end users in the decision-making process. As such, they rely on data and on many retrieval approaches including satellite imagery (remote sensing), field observations (monitoring programs) and numerical tools (computer models) (see Fig. 3). Data format and size varies significantly depending on their source and acquisition method (see Table 1).

The EWS starts with the collection of data from remote sensing and ongoing monitoring programs, some from long term coastal surveillance stations that monitor phytoplankton communities and other physical, chemical and biological parameters. Data usually relate to phytoplankton biomass and composition at study sites, and the presence of toxins in the water and shellfish. Automatic download and processing of satellite images or use of processed data services frequently take place at this stage (e.g. Copernicus, https://www.copernicus.eu/en). If the EWS include models that are not part of an operational modeling forecast system, the necessary data to force those models (e.g. meteo data) can also be gathered at this stage. Data sources for specific parameters used in the preparation of EWS are summarized in Table 2.

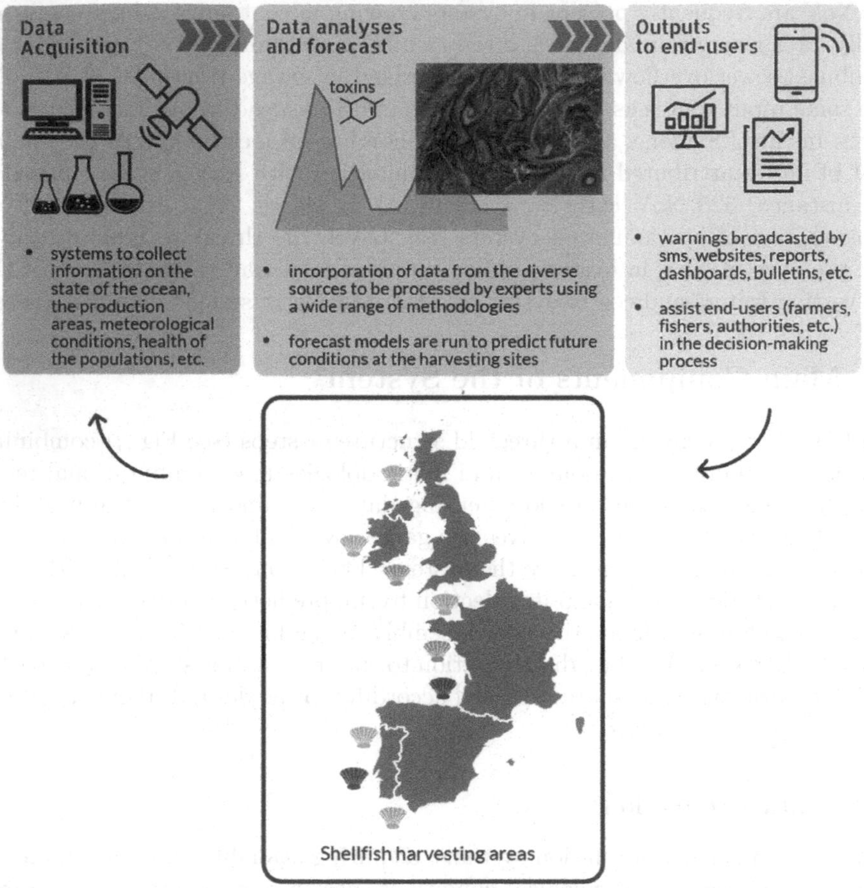

Fig. 2. Production pipeline of an EWS for shellfish harvesting areas. The outcome of the EWS can range from a simple traffic light system to classify particular areas according to shellfish safety (illustrated in the picture), to more elaborate bulletins with detailed information.

Monitoring Programs. Monitoring is an essential part of the implementation of the body of laws and directives that drive/force management strategies of natural resources. Monitoring programs can be defined as the establishment and operation of appropriate devices, methods, systems and procedures necessary to monitor, compile, and analyze data on the condition of a target systems. As such, monitoring can range from simple systematic observation and recording of current and changing conditions of a few parameters at a local scale, to a wide range of parameters over a significant wide area. These programs usually include the assessment of data leading to an evaluation on the state and evolution of the target systems, thus supporting the decision-making and planning processes.

Well designed and executed monitoring and assessment programs are a critical components in water resources management and protection. They allow to establish a baseline in the monitored systems condition and function, detect change, assess value, and characterize trends over time. Monitoring programs must ensure appropriate field sampling techniques, to obtain accurate data for HAB species such as *Dinophysis* spp. that can cause shellfish poisoning even when they comprise a small percentage of the microplankton community [20].

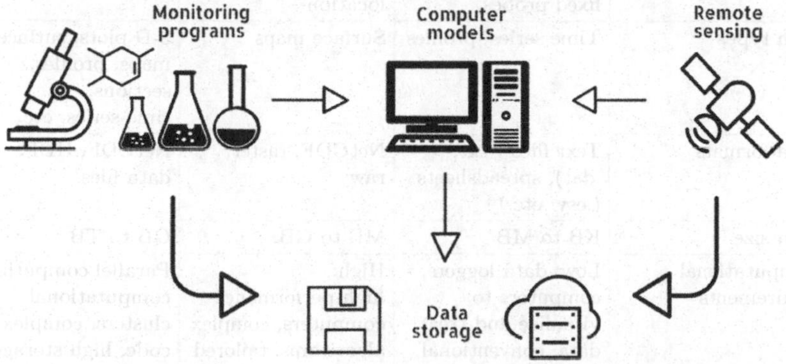

Fig. 3. Data sources for early warning system. Data from field observations and satellite imagery are also used by computer models to generate new layers of information.

Remote Sensing. Remote sensing, or data acquisition by satellite Earth observation (EO) technology, provides the capacity to monitor several parameters with reasonable accuracy, for significant expanses of territory. The strength of remote sensing techniques lies in their ability to provide both spatial and temporal views of surface water quality and atmospheric parameters that is typically not possible from *in situ* measurements. EO can observe microalgae blooms due to their coloring of the sea when in high concentrations [21,22]. Although not all plankton species that produce water discoloration are toxic (e.g. *Mesodinium rubrum*), some can be harmful to the ecosystems due to high biomass related effects and can be harmful also for the mariculture industry [23]. Consequently,

Table 1. Summary description of the characteristics of data according to its origin.

	Monitoring programs	Remote sensing	Computer models
Parameters/variables	Meteorological data, oceanographic physical and biogeochemical parameters, pathogens, pollutants, etc	Salinity, sea surface temperature, chlorophyll, ocean color	Several physical and biogeochemical param-eters, and some pollu-tants
Spatial coverage	One sampling point to several sampling stations over a vast area	Global	Local (areas of a few km^2) to regional (ocean basin scale)
Spatial representation	1-D (point samples) and 2-D (profiles)	2-D (ocean top layer)	3-D
Temporal resolution	Instant (e.g. water sam-ples) to continuous monitoring (e.g. fixed probes)	Depending on the revisiting frequency of a satellite sensor for a specific location	Continuous outputs with temporal resolutions from a few minutes to days
Data types	Time series, profiles	Surface maps	3-D plots, surface maps, profiles, sections, time-series, etc.
Data formats	Text files (.txt, .dat), spreadsheets (.csv, etc.)	NetCDF, raster, raw	NetCDF, HDF, data files
Data size	KB to MB	MB to GB	GB to TB
Computational requirements	Low: data loggers, computers to visualize and treat data, conventional software	High: high-performance computers, complex algorithms, tailored programs	Parallel computing, computational clusters, complex code, high storage requirements
Required skills	Elementary knowledge on generic software (e.g. excel); basic programing skills	Proficient users: good programing skills (e.g. python)	Advanced programing skills (several computer languages)

satellite-related technologies may be integrated in monitoring programs defined by national or local authorities. As such, remote sensing is central in the development, implementation and control of management strategies.

Currently there are several sources of remote sensing data, covering a wide range of products available to be accessed or downloaded. The Copernicus Services (https://www.copernicus.eu/en) is the most relevant EO data provider for European seas, incorporating a range of sensors and temporal and spatial resolutions. A significant number of these products are freely available for registered users, and additional products can be available for national governments or local authorities. Novel Copernicus Climate Change Service (C3S) products will also provide indicators for future scenarios (https://climate.copernicus.eu/), including indicators relevant to aquaculture.

Computer Models. Computational models offer unparalleled capabilities for studying the ocean by simulating its physical, biogeochemical and ecological processes. Their unique contribution to EWS is the ability to forecast ocean conditions from a known state, usually characterized by data retrieved from monitoring programs and remote sensing. In this sense, numerical models integrate data from field observations and satellite imagery and initial and forcing conditions for the simulations, but also for calibration and validation, contributing to produce additional layers of information that would be impossible to achieve by any other means [5].

Numerical modeling approaches comprise a significant range of methodologies, algorithms and computational tools [24]. They can be purely hydrodynamic and simulate the physical structures of the ocean to describe them in the form of tridimensional velocity fields and thermal distribution. Such models can be coupled to Lagrangian models to simulate the passive transport of particles, without necessarily incorporating any biological processes. In the modeling of HABs, for instance, this approach is commonly used when physical processes dominate over biological ones [25]. During ASIMUTH, HAB forecasts based on hydrodynamic and particle tracking models showed skill in predicting and characterizing transports of HABs alongshore and in and out of harvesting areas in Ireland, Portugal, Galicia and Scotland [26,27]. In a more complex approach, ocean models can also couple physical and ecological algorithms, typically Eulerian in nature, calculating biogeochemical properties, with these variables being subject to physical (advection, diffusion) and biological processes (growth, mortality, mineralization, etc.).

Statistically based models are also used to identify potential relationships between variables and processes, such as phytoplankton abundance and potentially causative environmental conditions for NoV outbreaks and patterns of human health. Finally, there are also risk assessment models that are empirical in nature and, from a computational perspective, significantly simpler than the modeling approaches previously mentioned. These models infer cause and effect relationships between HABs or NoV outbreaks and environment parameters, identifying possible trigger thresholds.

3.2 Data Analyses and Forecasts

This stage involves the incorporation of data from the diverse sources and may include current synoptic, recent trends (past month) and historical patterns (10 years) from phytoplankton and shellfish toxin monitoring programmes, satellite temperature and chlorophyll, results from 3D primitive equation coastal hydrodynamic models and particle tracking models. These data are processed by experts using a wide range of methodologies:

Table 2. Data sources for specific parameters used in the preparation of EWS.

	Monitoring programs	Remote sensing	Computer models
Algal biomass, composition and toxins	Yes	Ocean color (algorithms to estimate algal biomass from chlorophyll *a* pigments)	Yes, models can simulate several phytoplank-ton groups but only a few include toxins
Microbiological agents	Yes	—	Yes
Viruses	No	—	—
Temperature	Yes	Sea surface temperature	Yes (thermal structure)
Currents	Yes	—	Yes (2-D, 3-D)
Nutrients	Yes	—	Yes
Diss. oxygen	Yes	—	Yes

(1) Processing software to extract temporal and spatial variations of remote sensing-based variables over the study area. Remotely sensed data provide insight into the possible locations and extent of the blooms. For instance, remote sensing data in near real-time from Medium Resolution Imaging Spectrometer (MERIS) and Moderate Resolution Imaging Spectroradiometer (MODIS) like sensors has been used to detect and trace HABs [28,29]. Satellite images are also used to assess thermal patterns at the surface, enabling the identification of phenomena that may be related with the occurrence of HABs (e.g. upwelling episodes and fronts);

(2) Numerical models to identify oceanographic structures and conditions associated with the onset of HABs [30]. Numerical models can also simulate biogeochemical processes and include several functional groups of phytoplankton and may predict the timing and place of the formation of HABs. Another modeling approach that is frequently used consists in the use of particle tracers to simulate the transport patterns of HAB, once they have been identified by monitoring or remote sensing. While the development of operational biological models of HAB dynamics remains a major challenge, physical models to simulate the Lagrangian transport of detected HABs have proven to be a useful tool for HAB early warning [25–27];

(3) Statistical approaches that are characterized by having an explicit underlying probability model, which provides a probability of the outcome, rather than simply forecast without uncertainty [31]. The Probabilistic Graphical Models (PGMs) paradigm, based on probability theory and graph theory, can be used. PGMs include Bayesian networks, which are suitable to deal with uncertainty. Their intuitive properties and the explicit consideration of uncertainty enhance experts' confidence on forecasts [32,33].

3.3 Outputs to End-Users

At the final stage of the process, experts select the graphics, sometimes relying on automated routines, to produce the short term forecast. Frequently a short text is also prepared, highlighting some aspects found relevant by the experts. If a bulletin is produced, it is usually sent to producers and regulators via email or uploaded to a designated site (usually from local or national authorities), becoming accessible to the general public. The actual content of the EWS depends on the site characteristics, type of shellfish production, used methodologies and data, specific threats, etc. As such, EWS do not have a specific template, and can either be quite simplistic (using a simple traffic light system to classify production areas, for instance) or highly complex, with elements such as current and sea surface forecasts, historical trend analyses for the presence of toxins or site closures, plots for hydrodynamic patterns, etc. Either way, EWS aim to be relevant and effective and provide simple answers regarding shellfish safety, however sophisticated their methodologies.

4 Current and Future Challenges

4.1 Computational Tools and the New Possibilities

Computational science is at the core of EWS. In that respect, significant developments are expected to occur in the hardware and software, ultimately leading to improved performance of both machines and algorithms. The implications to EWS of these continuous developments are significant: new approaches to gather and process data may arise, and more computational power may become available to face the ever increasing complexity of forecast ocean models [30].

While HABs and microbiological agents such as *E. coli* impacts in shellfish areas have been the focus of intense research (even though their dynamics are not entirely understood), the study of NoV is comparatively recent. Consequently, one of the main challenges at the moment lies in addressing the dynamics of NoV and its inclusion in EWS for shellfish safety. Computational tools are also playing a major role here, with considerable effort applied to better understand the epidemiology and control of NoV, especially in the development of mathematical models to describe their transmission dynamics [34,35]. These models range from purely statistical [36], to more sophisticated approaches, such as probability based Artificial Neural Network models to predict NoV outbreaks in oyster cultures [37]. Such models can be combined with other approaches to provide the necessary time frame for a timely intervention and/or appropriate management decisions to reduce or even prevent norovirus related risks.

4.2 Knowledge Gaps

The need to continue optimizing early warning systems for shellfish safety is an ongoing challenge, not just to scientists, but also to the shellfish industry and regulators. Although knowledge has increased considerably in recent years, and algorithms, models and computational systems keep developing, further research and technical developments are required to address the following:

(1) Predicting the onset of algal blooms from identified conditions has become possible by using new algorithms to analyze remote sensing data and model forecasts. However, predicting the presence of toxins in such blooms is still elusive, as many of the underlying mechanisms inducing toxicity are still unknown;
(2) While modeling the dynamics of generic types of phytoplankton (diatoms, flagellates, picoalgae, etc.) has become a rather straightforward task, as seen by the myriad of models available today, achieving the same goal for harmful groups (e.g., toxin-producing species) has proved to be significantly more challenging;
(3) Development of effective NoV monitoring programs for commercial production areas to better understand contamination patterns and improve understanding on the hydrographical relationships between NoV inputs and consequential impacts on shellfisheries to better model risk [38];
(4) Monitoring data for phytoplankton composition and biomass, or even toxicity levels, and fecal contamination have an abundant spatial-temporal coverage, whereas NoV data are relatively limited [39,40];
(5) Exchange of data among different national monitoring programs is required for accurate HAB forecast when transnational alongshore transport takes place [6]. For example, early warning for the risk of autumn toxic dinoflagellate blooms in the Galician Rías is only feasible if the system is combined with a similar system for the Portuguese waters [27].

4.3 Climate Change

Expected changes in climate conditions poses additional challenges to the shellfish aquaculture industry, as key environmental parameters are expected to shift from their mean values. Water temperature, that strongly regulates the metabolism of organisms, is one of these parameters. The increased frequency of warm water events in recent decades has been reported [41]. Besides affecting the shellfish metabolic rates, these shifts may also change phytoplankton productivity and composition in coastal waters, potentially promoting the formation, and even the dominance of HAB species [42]. Under this scenario, an increased impact of HABs on shellfish can be expected. Furthermore, ocean acidification can exacerbate the impact on shellfish species [43].

Climate change may also bring new challenges associated with fecal pollution. More frequent flood events and rainwater discharge will increase the exposure of shellfish to *E. coli* and NoV (and other potentially harmful agents), imposing adaptive strategies in design and capacity of wastewater treatment plants in shellfish production areas. Considering that there is a recognized link between winter seasonality and NoV outbreaks, climate change has the potential to influence the transmissibility, host susceptibility and virus resistance to environmental conditions [44]. Climate change will drive developments in EWS but, at the same time, will also provide the opportunity to prove their value. If these systems maintain their links to the latest knowledge and state-of-the-art computation, they will surely become critical or even mandatory forecast tools, in the management of shellfish harvesting areas.

References

1. FAO: The State of World Fisheries and Aquaculture 2016, Rome (2016)
2. Fleming, L.E., et al.: Oceans and human health: a rising tide of challenges and opportunities for Europe. Mar. Environ. Res. **99**, 16–19 (2014)
3. Bernard, S., et al.: Developing global capabilities for the observation and prediction of harmful algal blooms. Cambridge Scholars Publishing (2014)
4. Karki, S., et al.: Mapping and forecasting onsets of harmful algal blooms using MODIS data over coastal waters surrounding Charlotte county. Florida. Remote Sens. **10**(10), 1656 (2018)
5. Davidson, K., et al.: Forecasting the risk of harmful algal blooms. Harmful Algae **53**, 1–7 (2016)
6. Maguire, J., et al.: Applied simulations and integrated modelling for the understanding of toxic and harmful algal blooms (ASIMUTH): integrated HAB forecast systems for Europe's Atlantic Arc. Harmful Algae **53**, 160–166 (2016)
7. Cuellar-Martinez, T., et al.: Addressing the problem of harmful algal blooms in Latin America and the Caribbean - a regional network for early warning and response. Front. Mar. Sci. **5**, 409 (2018)
8. Glibert, P.M.: Margalef revisited: a new phytoplankton mandala incorporating twelve dimensions, including nutritional physiology. Harmful Algae **55**, 25–30 (2016)
9. Huisman, J., Weissing, F.J.: Fundamental unpredictability in multispecies competition. Am. Nat. **157**(5), 488–494 (2001)
10. McCabe, R.M., et al.: An unprecedented coastwide toxic algal bloom linked to anomalous ocean conditions. Geophys. Res. Lett. **43**(19), 10366–10376 (2016)
11. Grattan, L.M., et al.: Harmful algal blooms and public health. Harmful Algae **57**, 2–8 (2016)
12. Rose, J.B., et al.: Climate variability and change in the United States: potential impacts on water- and foodborne diseases caused by microbiologic agents. Environ. Health Perspect. **109**(Suppl 2), 211–221 (2001)
13. Zingone, A., Oksfeldt Enevoldsen, H.: The diversity of harmful algal blooms: a challenge for science and management. Ocean Coast. Manag. **43**(8), 725–748 (2000)
14. Amaya, O., et al.: Large-scale sea turtle mortality events in El Salvador attributed to paralytic shellfish toxin-producing algae blooms. Front. Mar. Sci. **5**, 411 (2018)
15. Butt, A.A., et al.: Infections related to the ingestion of seafood Part I: viral and bacterial infections. Lancet Infect. Dis. **4**(4), 201–212 (2004)
16. Le Guyader, F.S., et al.: Comprehensive analysis of a norovirus-associated gastroenteritis outbreak, from the environment to the consumer. J. Clin. Microbiol. **48**(3), 915–920 (2010)
17. Flannery, J., et al.: Concentration of norovirus during wastewater treatment and its impact on oyster contamination. Appl. Environ. Microbiol. **78**(9), 3400–3406 (2012)
18. Batz, B., et al.: Ranking the risks: the 10 pathogen-food combinations with the greatest burden on public health. University of Florida (2011)

19. Lodo, K.L., et al.: An outbreak of norovirus linked to oysters in Tasmania. Commun. Dis. Intell. **38**, 1 (2014)
20. Reguera, B., et al.: Harmful Dinophysis species: a review. Harmful Algae **14**, 87–106 (2012)
21. Davidson, K., et al.: A large and prolonged bloom of Karenia mikimotoi in Scottish waters in 2006. Harmful Algae **8**(2), 349–361 (2009)
22. Kurekin, A.A., et al.: Satellite discrimination of Karenia mikimotoi and Phaeocystis harmful algal blooms in European coastal waters: merged classification of ocean colour data. Harmful Algae **31**, 163–176 (2014)
23. Yih, W., et al.: The red-tide ciliate Mesodinium rubrum in Korean coastal waters. Harmful Algae **30**, S53–S61 (2013)
24. McGillicuddy Jr., D.J.: Models of harmful algal blooms: conceptual, empirical, and numerical approaches. J. Mar. Syst. **83**(3–4), 105–107 (2010)
25. Pinto, L., et al.: Modeling the transport pathways of harmful algal blooms in the Iberian coast. Harmful Algae **53**, 8–16 (2016)
26. Cusack, C., et al.: Harmful algal bloom forecast system for SW Ireland. Part II: are operational oceanographic models useful in a HAB warning system. Harmful Algae **53**, 86–101 (2016)
27. Ruiz-Villarreal, M., et al.: Modelling the hydrodynamic conditions associated with Dinophysis blooms in Galicia (NW Spain). Harmful Algae **53**, 40–52 (2016)
28. Hu, C., et al.: Red tide detection and tracing using MODIS fluorescence data: a regional example in SW Florida coastal waters. Remote Sens. Environ. **97**(3), 311–321 (2005)
29. Wang, G., et al.: Multi-spectral remote sensing of phytoplankton pigment absorption properties in cyanobacteria bloom waters: a regional example in the Western Basin of Lake Erie. Remote Sens. **9**(12), 1309 (2017)
30. Aleynik, D., et al.: A high resolution hydrodynamic model system suitable for novel harmful algal bloom modelling in areas of complex coastline and topography. Harmful Algae **53**, 102–117 (2016)
31. Blondeau-Patissier, D., et al.: A review of ocean color remote sensing methods and statistical techniques for the detection, mapping and analysis of phytoplankton blooms in coastal and open oceans. Progr. Oceanogr. **123**, 123–144 (2014)
32. Fernandes, J.A., et al.: Supervised pre-processing approaches in multiple class variables classification for fish recruitment forecasting. Environ. Model. Softw. **40**, 245–254 (2013)
33. Fernandes, J.A., et al.: Evaluating machine-learning techniques for recruitment forecasting of seven North East Atlantic fish species. Ecol. Inform. **25**, 35–42 (2015)
34. Gaythorpe, K.A.M., et al.: Norovirus transmission dynamics: a modelling review. Epidemiol. Infect. **146**(2), 147–158 (2018)
35. Towers, S., et al.: Quantifying the relative effects of environmental and direct transmission of norovirus. Roy. Soc. Open Sci. **5**(3), 170602–170602 (2018)
36. Matsuyama, R., et al.: The transmissibility of noroviruses: statistical modeling of outbreak events with known route of transmission in Japan. PLOS ONE **12**(3), e0173996 (2017)
37. Chenar, S.S., Deng, Z.: Development of genetic programming-based model for predicting oyster norovirus outbreak risks. Water Res. **128**, 20–37 (2018)
38. Campos, C.J.A., Lees, D.N.: Environmental transmission of human noroviruses in shellfish waters. Appl. Environ. Microbiol. **80**(12), 3552–3561 (2014)
39. Wyn-Jones, A.P., et al.: Surveillance of adenoviruses and noroviruses in European recreational waters. Water Res. **45**(3), 1025–1038 (2011)

40. Kim, M.S., et al.: Distribution of human norovirus in the coastal waters of South Korea. PLOS ONE **11**(9), e0163800 (2016)
41. González-Pola, C., et al.: ICES Report on Ocean Climate 2017. ICES Cooperative Research Report No. 345 (2018)
42. Henson, S.A., et al.: Observing climate change trends in ocean biogeochemistry: when and where. Glob. Change Biol. **22**(4), 1561–1571 (2016)
43. Fernandes, J.A., et al.: Estimating the ecological, economic and social impacts of ocean acidification and warming on UK fisheries. Fish Fish. **18**(3), 389–411 (2017)
44. Ahmed, S.M., et al.: A systematic review and meta-analysis of the global seasonality of norovirus. PLOS ONE **8**(10), e75922–e75922 (2013)

Track of Multiscale Modelling and Simulation

Creating a Reusable Cross-Disciplinary Multi-scale and Multi-physics Framework: From AMUSE to OMUSE and Beyond

Inti Pelupessy[1](✉), Simon Portegies Zwart[2], Arjen van Elteren[2], Henk Dijkstra[3], Fredrik Jansson[3], Daan Crommelin[4,5], Pier Siebesma[6,7], Ben van Werkhoven[1], and Gijs van den Oord[1]

[1] Netherlands eScience Center, Amsterdam, The Netherlands
i.pelupessy@esciencecenter.nl
[2] Leiden Observatory, Leiden, The Netherlands
[3] Institute for Marine and Atmospheric Research Utrecht, Utrecht, The Netherlands
[4] Centrum Wiskunde & Informatica, Amsterdam, The Netherlands
[5] University of Amsterdam, Amsterdam, The Netherlands
[6] Royal Netherlands Meteorological Institute, De Bilt, The Netherlands
[7] Delft University of Technology, Delft, The Netherlands

Abstract. Here, we describe our efforts to create a multi-scale and multi-physics framework that can be retargeted across different disciplines. Currently we have implemented our approach in the astrophysical domain, for which we developed AMUSE (github.com/amusecode/amuse), and generalized this to the oceanographic and climate sciences, which led to the development of OMUSE (bitbucket.org/omuse). The objective of this paper is to document the design choices that led to the successful implementation of these frameworks as well as the future challenges in applying this approach to other domains.

Keywords: Multi-scale simulations · Coupling framework · Multi-physics

1 Introduction

The current frontier in computational modelling is the simulation of complex phenomena involving different physical processes interacting on vastly different scales. The advent of massively parallel machines and GPU accelerated solvers, has meant that memory and CPU time bounds are less of a limitation as before, the difficulty shifting instead to the intrinsic complexity of the calculations.

A recurring challenge involves the interaction of processes acting on widely different scales. For example, when modelling the formation of planetary systems in a stellar cluster one needs to follow the collapse of interstellar gas, down to the formation of proto-stellar systems. Another example occurs when modelling

J. M. F. Rodrigues et al. (Eds.): ICCS 2019, LNCS 11539, pp. 379–392, 2019.
https://doi.org/10.1007/978-3-030-22747-0_29

the dynamical effects of clouds and convection on the atmospheric circulation. Atmospheric convection and cloud formation are physical processes with small spatial scales, however they affect the properties of the large-scale atmospheric flow, for example through their impact on the distribution of moisture and heat as well as on radiative transfer in the atmosphere.

Many more examples could be given and the conventional approach to these multi-scale problems, i.e. building a single, monolithic program with as much physics as possible, is expensive and difficult to scale. Building multi-scale and multi-physics simulation codes becomes increasingly complex with each new physical ingredient that is added. Furthermore, one is presented with the prospect of duplicating much of this work when a different solver or method is needed, a situation that often arises when a slightly different regime is accessed than that originally envisaged or when results need to be verified with a different method.

Different strategies that attempt to simplify this problem by compartmentalizing processes and combining the resulting building blocks exist. For example [8,18] are coupling frameworks geared towards earth system modelling. Other examples are the more general approaches taken in the toolkits [1,2]. See [6] for a more thorough review. Most of these can be roughly divided into *integrated* and *coupling library* approaches [19]. In the integrated approach, the functionality provided by the components (e.g. by subroutines of the code) is separated out and joined in a new single executable. In the library approach the original codes themselves are adapted to communicate with each other using an Application Programming Interface (API), linking against the coupling library.

We recently developed a promising alternative, especially when the target solvers use completely independent computational methods or discretizations. The fundamental idea of AMUSE [13,15] and OMUSE [14] is the abstraction of the functionality of existing simulation codes - which are often highly specialized and optimized for their domain of application - into physically motivated interfaces and bind these into a modern and flexible scripting language. Our approach has the benefit of the parallelism and flexibility provided by a coupling library approach, and the benefit of abstracting much of the bookkeeping inherent to code couplings using modern high-level constructs. In this way, complex simulations can be described in compact scripts, that can be easily understood and communicated between peers.

2 Genesis

AMUSE was conceived within MODEST, a tight knit astrophysical community of modellers interested in dense stellar system. AMUSE was envisioned as the need of going beyond purely gravitational N-body calculations became apparent, and the MODEST community sought ways to incorporate the effects of stellar evolution and the dynamics of the interstellar gas into their models. It quickly became evident that many simulation codes already existed, specifically developed to study these processes separately. More practically speaking, the

MODEST community realized that they did not have the manpower nor the expertise to develop these from scratch.

Even so, such a body of existing scientific codes, which we refer to as the community code base, is not trivial to interface with. The codes are written in different languages, have different requirements and are not necessarily written in a way that allows easy interfacing.

The way this problem was solved in AMUSE was by defining a thin interface layer in Python for these codes, which integrates with a framework layer tying the various components together, minimizing the necessary changes to the community codes. The use of these codes is simplified by standardizing the interface for the different relevant domains (e.g. gravitational dynamics, gas dynamics or radiative transfer) and a high degree of automation. The framework is designed with parallel simulation codes in mind and allows for running in a distributed environment. In practice the computational effort is in the highly tuned codes, allowing for high performance. In this way, AMUSE allowed codes to be retargeted for novel interactions and couplings with other component codes.

AMUSE was developed by a small team of astrophysicist and software engineers over a couple of years. In our experience, it is crucial to seek active involvement from the community early on, by organizing workshops and tutorials. This allows for early feedback, fosters involvement and helps creating a forgiving user base.

2.1 Development of OMUSE

While the original goal of AMUSE was to allow for realistic simulations of star cluster formation and evolution, no limits were imposed on the design and many published results using AMUSE had no relation to star cluster physics. In discussion with researchers in other scientific disciplines it became apparent that they struggled with fundamentally the same multi-scale coupling problems. At this point (around 2013), the Netherlands eScience Center, the national center for academic research software, funded a project to generalize the AMUSE framework, which resulted in OMUSE.

OMUSE was developed as an extension of the AMUSE framework, exposing a `omuse` name space with similar structure as for AMUSE, using the underlying infrastructure of AMUSE (Practically speaking this means that to use OMUSE, the user first has to install AMUSE, which is not ideal, since by default many component of AMUSE are installed that are not needed by OMUSE). The development of OMUSE involved transplanting the experiences gained in astrophysics to another scientific fields, which is as much a cultural challenge as it is a technical one.

To support earth science applications a number of features were added to the AMUSE framework: the data model was extended with a hierarchy of grid types, such that codes with various grid types, ranging from regular Cartesian grids to unstructured grids, can be supported (within AMUSE only Cartesian grids were available). For the data transfer between these grids data channels can be defined which perform grid remapping and functional transforms (in addition to simple copy channels for use between equal formed grids). A number of domain specific units and utility functions were also added. As the initial focus was on oceanographic applications, the codes that are included range from simple conceptual ocean models to global circulation models.

3 About the Design

Here we will illustrate the design of AMUSE stepping through an example script to evolve a model star cluster using pure N-body dynamics.

As mentioned above, AMUSE and OMUSE are implemented in Python. The requirement of the high level interactions defined in the framework layer is not so much performance but one of algorithmic flexibility and ease of programming. This suggests the use of a modern interpreted scripting language with object oriented features. Python also provides for excellent integration with existing scientific computation tools and libraries.

The following example could be typed in an interactive Python session, but usually saved in a script or Jupyter notebook. As usual for a Python script, an AMUSE script starts with the necessary imports,

```
from amuse.units import nbody_system
from amuse.ic.plummer import new_plummer_model
from amuse.community.huayno.interface import Huayno
```

In this case three modules are imported from the AMUSE, a module for scaleless N-body units[1], an initial condition generator and a basic N-body gravitational integrator.

The following two lines instantiate the simulation code and prepare the code for the problem at hand by setting (one of) its parameters:

```
code=Huayno()
code.parameters.epsilon_squared= (0.01 | nbody_system.length)**2
```

At this point a separate worker process for the integrator code is started and running in the background. The worker consists of the original simulation code with a layer of native code to capture calls from the framework (see Fig. 1 for a general schematic of the interface design as discussed below).

The standardized set of methods on the `Huayno` object defines the interface to the code. It is designed to communicate the *physical* quantities relevant to that domain, as opposed to numerical concepts. Codes from a given physical domain conform to the same interface, and the interface to different domains are developed along similar concepts.

The communication with the code uses a remote function protocol with different transport channels available. The default is a channel based on MPI for computations on a local compute cluster, but a channel based on the eStep platform for distributed computing is also available.

[1] The N-body unit system is useful to interact with codes that internally use a unit system where the gravitational constant $G = 1$. Other codes, for example stellar evolution codes or radiative transfer codes often work with a definite set of units, and then the normal SI unit system is used.

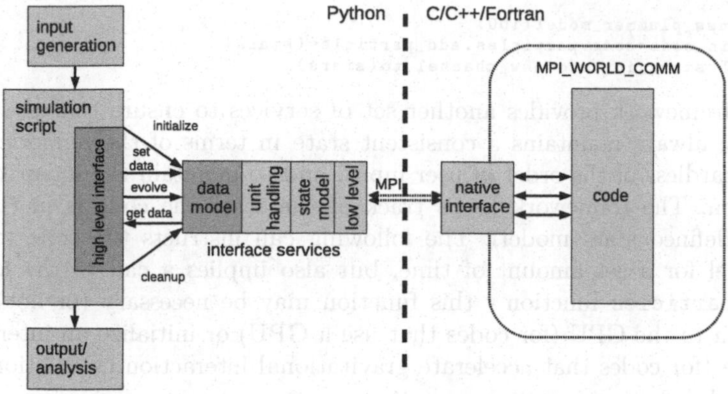

Fig. 1. Design of the AMUSE/OMUSE framework. This schematic representation shows the way a simulation code (the "code") is accessed from the framework. The code has a thin layer of interface functions in its native language (e.g. Fortran) which communicates through a message channel with the Python host process. The framework layer provides a number of services such as unit conversion, maintaining the simulation in a consistent state and converting the internal state data of the code to a common object oriented data representation. The user script ("simulation script") makes only generic calls to the high-level interface. Adapted from [14].

At the lowest level, the interface functions provide means to set and query model state variables and code parameters, as well as functions that change the state of the code (like initialization, triggering construction of data structures internal to the code) and functions that affect the state of the physical model (evolve forward in time).

On the basis of this low level interface a high level interface is build that provides a number of services that minimize the burden placed on the user in interacting with the code. The goal of these services is to eliminate common sources of error when running numerical simulations, by hiding the particulars of a given code and automating the interactions with the codes as much as possible.

First, the framework provides for the conversion of units and data structures. Every physical quantity that is used needs to have an attached unit. In this example the `epsilon_squared` parameter (describing the amount of smoothing applied to the gravitational force) is specified in the nbody unit of length.

The internal data representation of codes is translated to modern object oriented data structures. For this, two basic data stores are available, particle sets and grids. In the current example we proceed by initializing our model with a star cluster of 100 stars, distributed according to Plummer model (a theoretical equilibrium model used in stellar astrophysics). We send these particles to the code and define a channel to efficiently copy back the results to the particle set in the memory space of the Python script:

```
stars=new_plummer_model(100)
stars_in_code=code.particles.add_particles(stars)
channel=stars_in_code.new_channel_to(stars)
```

The framework provides another set of services to ensure that the simulation code always maintains a consistent state in terms of the numerics of the code, regardless of the order of user input, and to maintain the integrity of the simulation. The framework keeps track of the state the code is in (in terms of a predefined state model). The following call instructs the code to evolve the model for a set amount of time, but also implies a call to the low level `commit_particles` function - this function may be necessary (or not) to e.g. copy data to the GPU (for codes that use a GPU) or initialize an internal tree structure (for codes that accelerate gravitational interaction calculations using tree based data structures):

```
code.evolve_model( 0.5 | nbody_system.time)
channel.copy_attributes(["x","y","z","vx","vy","vz"])
```

At the end of the script, data is copied back to the master script's memory where it remains available for further analysis (or to be written to disk).

The high level interface in this example is the preferred way to interact with codes within AMUSE. At this level the interactions with the code are standardized as much as possible, with much of the tedious bookkeeping automated. This minimizes the possibility of programming errors and makes it easier to switch to a different model.

Currently AMUSE is distributed with more than 50 codes drawn from various astrophysical application domains (gravitational dynamics, stellar evolution, hydrodynamics and radiative transfer). In addition to this, it contains domain specific support and utility functions. These allow a researcher or student to get started quickly. Examples are: generators for initial conditions (e.g. generators for various particle models for solar systems, stellar clusters, galaxies), common analysis tools (e.g. identifying binaries, determining orbital elements) and file input/output functions. A number of common coupling schemes are also included.

4 Experiences

Using the AMUSE/OMUSE interfaces allows researchers (students and experienced researchers alike) to quickly develop, run and analyze computational experiments. The fact that the interfaces are homogeneous allows trivially switching between codes. Experiments are written as concise Python scripts which can easily be communicated between peers. Other use cases of the interfaces are the scripting of simulations for parameter searches and optimization, or the detection of special events. The access to the internal state of the simulation allows the integration of data analysis with a running simulation. In addition, the interfaces expose enough of the internal state such that new solvers can be developed. These can combine different physics and/or to bridge different scales. Below, we present two case studies detailing our experiences with the framework.

4.1 Online Data Analysis

Large-scale simulations are capable of generating enormous amounts of data. Usually, it is only possible to store a limited subset for offline analysis. The alternative is online data analysis, where the analysis code runs at the same time as the model. This offers several opportunities, including inspecting the model internal data at spatial and temporal resolutions that are not available offline.

We have created an example application for OMUSE that uses the Parallel Ocean Program (POP) and an ocean eddy tracking analysis code. The POP model is a parallel global circulation model for ocean flows [17]. POP is often used to calculate strongly eddying ocean circulation models.

The interest in ocean eddies comes from the fact that eddies transport considerable amounts of energy and mass and thus influence the dynamics of large-scale ocean circulation and the climate e.g. [5,20]. To understand eddy properties and variability, several mesoscale eddy tracking algorithms have been proposed in recent years. We have adapted a sea surface height-based eddy tracking code by Mason et al. [11]. The interface to this code allows the user to interact with the code using the high-level data structures, such as grids and units that are used in OMUSE.

Figure 2 shows the Python code of our example application for online data analysis. The application first instantiates the POP interface for high-resolution to run on a large computing cluster. After that we set the analysis interval for the eddy tracker to 7 days of simulation.

The EddyTracker is initialized using the same grid as is used in POP for the sea surface height values. From this grid object, the EddyTracker automatically extracts the coordinates of the grid points and the sea surface height values, and performs unit conversions if needed. Note that while POP is running a global simulation, the eddy tracker is set to only track the eddies in a particular region. Our application alternatingly runs the POP model for 7 simulation days and calls EddyTracker to track the eddies at the current time in the model for a full simulation year.

Figure 3 shows the output of our online eddy tracking program. In this image, we can clearly see the large anticyclonic eddies that result from the retroflection of the Agulhas Current, as well as many smaller eddies being tracked over time by the eddy tracker algorithm. The data generated by the online eddy tracker can, for example, be used to compare the statistics of the simulated eddies to the statistics of eddies found in altimetry data.

4.2 Multi-scale Coupling of Atmospheric LES Models to OpenIFS

As an example of the use of OMUSE for multi-scale coupling, we present here the use of the framework in a project on cloud-resolving atmospheric modelling. The project aims to couple the global atmospheric model OpenIFS [3] with a local, high-resolution (cloud-resolving) Large Eddy Simulation (LES) model, DALES [7]. The reason for this coupling is that global atmospheric models, such

```
from omuse.units import units
from omuse.ext.eddy_tracker.interface import EddyTracker
from omuse.community.pop.interface import POP

# start and initialize POP
p=POP(channel_type="distributed", mode="3600x2400x42",
      number_of_workers=592)
p.parameters. ...  # set all input files needed by POP

# set the analysis interval & initialize the EddyTracker
dt_analysis = 7 | units.day
tracker = EddyTracker(grid=p.nodes, domain="Regional",
            lonmin=0. | units.deg, lonmax=50. | units.deg,
            latmin=-45. | units.deg, latmax=-20. | units.deg,
            dt_analysis)

# evolve the simulation until the set end time
tend = p.model_time + (1 | units.yr)
while (p.model_time < tend):
    p.evolve_model(p.model_time + dt_analysis)
    tracker.find_eddies(ssh=p.nodes.ssh, rtime=p.model_time)

tracker.stop()
p.stop()
```

Fig. 2. This example demonstrates how to build an application that analyses data from a running simulation using OMUSE. This code executes an eddy tracking program that tracks the eddies based on sea surface height every seven days of simulation time of a running POP model.

as OpenIFS, typically cannot resolve individual clouds, as the clouds are smaller than the model grid size. The global models instead rely on parameterizations to account for processes on sub-grid scales.

Replacing a parameterization scheme (e.g. for convection) by a full microscopic model that resolves, rather than parameterizes, the small-scale process is in this context known as a superparameterization [4]. More specifically, superparameterization concerns the two-way nesting of a high-resolution model with limited spatial domain (in our case, DALES) in model columns of a global model of lower resolution (e.g. OpenIFS). Separate instances (or copies) of the high-resolution model are nested in the different model columns of the global model. This approach is seen as one possible route to understanding the feedbacks between cloud processes and climate [16] - one of the largest remaining uncertainties in climate modelling. Figure 4 shows a snapshot of a superparameterized simulation, with 72 DALES models over the Netherlands coupled to the global OpenIFS.

Both OpenIFS and DALES are implemented mainly in Fortran. For coupling the two models we considered a number of different strategies. The most straightforward might have been to directly embed DALES in the physics routines of OpenIFS. However, it was desirable to create the option of using superparameterization only for a selected number of OpenIFS model columns, to allow using a high resolution for the local models at a reasonable total computational cost. The selective superparameterization would have made load balancing in the existing parallelization of OpenIFS complicated. For this reason, and also for increased

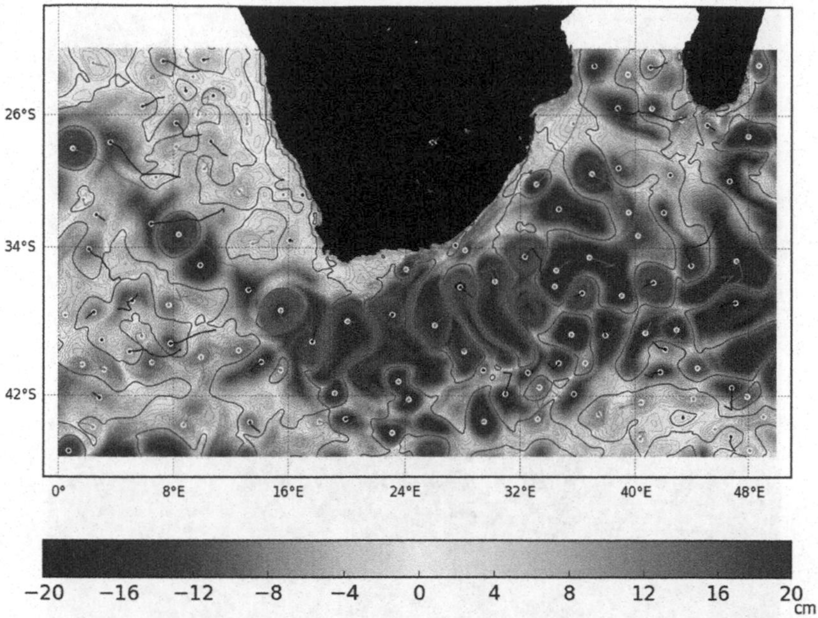

Fig. 3. Output of the online eddy tracking application using data from a running POP simulation, showing a region around the southern tip of Africa. The green lines show the contours between areas of different sea level anomaly values. Red indicates areas of elevated sea level, and is used to indicate anticyclonic eddies. Similarly, blue indicates a lower sea level, and is used to identify cyclonic eddies. The red or blue lines indicate the track that an eddy has traveled since it was first detected. (Color figure online)

flexibility, we chose to keep the two models separate, give each a library interface, and implement the coupling as a separate program.

A direct benefit is the convenience of writing the coupling code in Python. Performance-wise, typically more than 90% of the time of the whole simulation is spent in the DALES models. Neither the Python code nor the communication has been a significant bottle-neck so far, helped by the fact that the coupling is formulated in terms of vertical profiles and thus does not require the exchange of 3D fields.

The DALES models in our setup are time-stepped in parallel. The asynchronous function call mechanism in AMUSE works very well for this - the DALES interface contains a function to perform a time step, the coupler makes an asynchronous call to this function for every DALES model, then waits for them all to complete.

We have mainly used the Cray system at ECMWF for the simulations. One practical difficulty we encountered there is that the Cray MPI so far does not support spawning new MPI processes, which is how AMUSE normally launches its worker codes. Support for MPI spawning is scheduled to be available this year, until then we are using a work-around where all workers are launched at

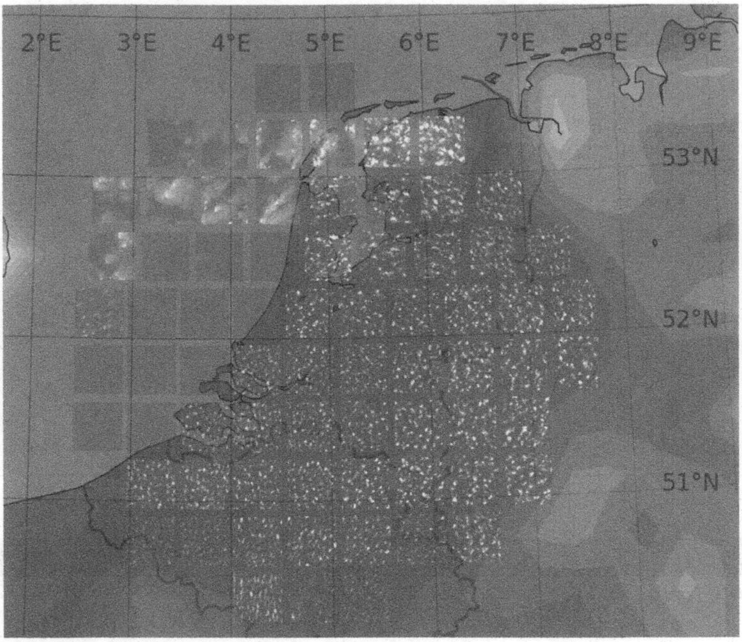

Fig. 4. Superparameterized weather simulation over the Netherlands, with the global OpenIFS model in purple and local, cloud-resolving DALES models in blue. Both models show the cloud fields in shades of white in the form of liquid water path. (Color figure online)

the start of the simulation as a regular MPI job, after which the appropriate MPI communicators are created. This solution is possible for us since we know how many workers are needed for a particular simulation prior to starting the simulation.

4.3 Testing and Validation

AMUSE and OMUSE try to foster good scientific computation practices: framework code and interfaces are tested and basic verification/ validation is done. A natural question remains is to what extent we can guarantee the correctness of the resulting simulations (see also the related discussion in [9]). Our philosophy is that this remains the responsibility of the researcher. A new AMUSE/OMUSE application should be thoroughly tested, especially where it involves new interactions between components. This is unavoidable since, while the framework does simplify the design and implementation of couplings (the "engineering" aspect), developing e.g. a new coupling involves an element of scientific inquiry (which couplings are physically sound). For example, in the OpenIFS-DALES coupling there are various ways to downscale the input variations from the global circulation model to the local LES simulation. This has a significant impact on the results, and uncovers subtle issues about the numerical representation of both models.

4.4 Performance

An important concern of any coupling framework is the computational performance. The architecture of the AMUSE framework is designed with a high degree of parallelism, and individual simulation codes are often highly optimized. In the applications we have developed, the performance of the AMUSE/OMUSE framework is rarely a concern: the overhead imposed by the framework is measured to be rather small (less than a few percent [15]), However, this is strongly problem dependent. In particular massively parallel codes that use thousands of processes can generate big data transfers, and it is not difficult to formulate problems where the strength of the coupling is intrinsically so strong that very frequent communication between the component solvers is necessary.

The current design of AMUSE and OMUSE uses parallel data structures for storing the data inside parallel models. A limitation of the current design is the fact that the communication between solvers is handled by a single process user script. One of the main performance challenges that will be addressed in the near future is the development of distributed data transport. Note that such distributed communication channels would not change the semantics of the use of a channel between data structures in the user script.

5 Challenges

While the current implementations of AMUSE and OMUSE provide a core functionality, challenges remain for more general applications. These range from technical ones to challenges related to organizational and human factors.

Generalization and Extension. Currently we are working on making sure the core framework is domain agnostic, such that an implementation for another field can be more easily developed. The concrete motivation stems from the development of an hydrological version ("HyMUSE") as computational core of the eWaterCycle project [10].

OMUSE was built as an extension of AMUSE, with some of the development related to OMUSE done inside the code base of AMUSE. This brought a clearer picture of the difference between general framework components within AMUSE, and astrophysics specific code. It also brought to light that this model of development would not scale to multiple domains, as the AMUSE framework framework code would have to support and intermingle code of multiple domains. With the application to a third domain (Hydrology) it becomes more urgent to refactor the code in clearly separated general framework code and domain specific packages. This generalization of the framework involves new challenges in the coordination and prioritization of the development goals.

Interoperability. Over the years a number of frameworks have been developed that have some overlap and are conceptually close to the AMUSE/OMUSE system (e.g. CSDMS, [12,18]). One of the future goals is to attain a level of interoperability. For example, within the CSDMS the Basic Model Interface (BMI) has been defined as a minimal code interface to interact with a model code. BMI shares some characteristics with the AMUSE low-level interface. We have developed an automatic interface generator for codes that support BMI. This makes it possible to automatically generate low-level AMUSE interfaces and serves as a starting point for building a high-level AMUSE interface.

Maintainability. Another challenge involves the maintainability of the framework. This involves adapting the framework to keep it up to date with the evolving software ecosystem in on which it depends, i.e. making sure breakages from compiler updates etc. are fixed and keeping up with changes in software development practices and trends in software usage and distribution.

For sustainability of the interfaces, it is important to consider how the code is archived, especially when the AMUSE interface depends on a modified version of a third-party code (such as OpenIFS). In other cases the code of the model itself is open source and the modifications needed in the code to support the AMUSE interface are rather small.

6 Conclusions

The AMUSE/OMUSE system is a unique tool for scientific discovery. It encapsulates existing legacy codes into a modern simulation environment. This allows researchers to quickly translate conceptual ideas into numerical experiments. These are documented in portable scripts that can easily be communicated among peers. This lowers the barrier for verification and validation, and the framework thus aids in making simulations more reproducible. We have also shown here that the system is quite general and is easily retargeted for different scientific disciplines.

During the development of AMUSE and OMUSE we have found that the engagement of the community and domain scientist is crucial. Since the resources available for development in an academic setting are quite modest the involvement of domain scientist helps developers to focus on features that have most immediate use.

The generalization of the framework provides an opportunity - different disciplines sharing the same code base should lower the burden in the development and maintaining the framework, as well as a challenge: it becomes more difficult to coordinate development efforts and different fields may have conflicting needs - and thus attention must be paid to prioritizing the development of new features. Within our project, we found that the involvement of the Netherlands eScience Center, as an entity dedicated to the generalization of research software, provides a good focal point for these tasks. This way, scientists can spend more of their limited time on developing and testing scientific ideas.

Acknowledgments. FJ, DC, PS and GvdO acknowledge support from the Netherlands eScience Center (grant 027-015-G03) for their project *Towards Large-Scale Cloud-Resolving Climate Simulations.* Furthermore, acknowledgment is made for the use of ECMWF's computing and archive facilities in this project.

References

1. Borgdorff, J., et al.: Distributed multiscale computingwith muscle 2, the multiscale coupling library and environment. J. Comput. Sci. **5**(5), 719 – 731 (2014). https://doi.org/10.1016/j.jocs.2014.04.004. http://www.sciencedirect.com/science/article/pii/S1877750314000465
2. Buis, S., Piacentini, A., Déclat, D.: The PALM Group: PALM: a computational framework for assembling high-performance computing applications. Concurrency Comput. Pract. Exp. **18**(2), 231–245 (2006)
3. Carver, G., et al.: The ECMWF OpenIFS numerical weather prediction model release cycle 40r1: description and use cases. To be submitted to GMDD (2018, in preparation)
4. Grabowski, W.W.: An improved framework for superparameterization. J. Atmos. Sci. **61**(15), 1940–1952 (2004) https://doi.org/10.1175/1520-0469(2004)061⟨1940:AIFFS⟩2.0.CO;2
5. Griffies, S.M., et al.: Impacts on ocean heat from transient mesoscale eddies in a hierarchy of climate models. J. Clim. **28**(3), 952–977 (2015)
6. Groen, D., Zasada, S.J., Coveney, P.V.: Survey of multiscale and multiphysics applications and communities. Comput. Sci. Eng. **16**(2), 34–43 (2014). https://doi.org/10.1109/MCSE.2013.47
7. Heus, T., et al.: Formulation of the Dutch atmospheric large-eddy simulation (DALES) and overview of its applications. Geosci. Model Dev. **3**(2), 415–444 (2010). https://doi.org/10.5194/gmd-3-415-2010
8. Hill, C., DeLuca, C., Suarez, M., Da Silva, A., et al.: The architecture of the earth system modeling framework. Comput. Sci. Eng. **6**(1), 18–28 (2004)
9. Hoekstra, A., Chopard, B., Coveney, P.: Multiscale modelling and simulation: a position paper. Philos. Trans. Roy. Soc. A: Math. Phys. Eng. Sci. **372**(2021), 20130377 (2014). https://doi.org/10.1098/rsta.2013.0377
10. Hut, R., van de Giesen, N., Drost, N.: The future of global is local. eWaterCycle II: bridging the gap between catchment hydrologists and global hydrologists (2018). https://meetingorganizer.copernicus.org/EGU2018/EGU2018-10614.pdf
11. Mason, E., Pascual, A., McWilliams, J.C.: A new sea surface height-based code for oceanic mesoscale eddy tracking. J. Atmos. Ocean. Technol. **31**(5), 1181–1188 (2014)
12. Peckham, S.D., Hutton, E.W., Norris, B.: A component-based approach to integrated modeling in the geosciences: the design of CSDMS. Comput. Geosci. **53**, 3–12 (2013). https://doi.org/10.1016/j.cageo.2012.04.002
13. Pelupessy, F.I., van Elteren, A., de Vries, N., McMillan, S.L.W., Drost, N., Portegies Zwart, S.: The astrophysical multipurpose software environment. Astron. Astrophys. **557**, 84 (2013)
14. Pelupessy, I., et al.: The oceanographic multipurpose software environment (omuse v1.0). Geosci. Model Dev. **10**(8), 3167–3187 (2017). https://doi.org/10.5194/gmd-10-3167-2017. https://www.geosci-model-dev.net/10/3167/2017/

15. Portegies Zwart, S., McMillan, S.L.W., van Elteren, E., Pelupessy, I., de Vries, N.: Multi-physics simulations using a hierarchical interchangeable software interface. Comput. Phys. Commun. **183**, 456–468 (2013)
16. Schneider, T., et al.: Climate goals and computing the future of clouds. Nat. Clim. Change **7**, 3 (2017). https://doi.org/10.1038/nclimate3190
17. Smith, R.D., et al.: The Parallel Ocean Program (POP) Reference Manual. Los Alamos National Laboratory, LAUR-10-01853 (2010)
18. Valcke, S.: The OASIS3 coupler: a European climate modelling community software. Geosci. Model Dev. **6**(2), 373–388 (2013). https://doi.org/10.5194/gmd-6-373-2013
19. Valcke, S., et al.: Coupling technologies for earth system modelling. Geosci. Model Dev. **5**(6), 1589–1596 (2012). https://doi.org/10.5194/gmd-5-1589-2012
20. Viebahn, J., Eden, C.: Towards the impact of eddies on the response of the southern ocean to climate change. Ocean Model. **34**(3–4), 150–165 (2010)

A Semi-Lagrangian Multiscale Framework for Advection-Dominant Problems

Konrad Simon(✉) and Jörn Behrens

Center for Earth System Research and Sustainability (CEN),
Department of Mathematics, University of Hamburg,
Grindelberg 5, 20144 Hamburg, Germany
{konrad.simon,joern.behrens}@uni-hamburg.de

Abstract. We introduce a new parallelizable numerical multiscale method for advection-dominated problems as they often occur in engineering and geosciences. State of the art multiscale simulation methods work well in situations in which stationary and elliptic scenarios prevail but are prone to fail when the model involves dominant lower order terms which is common in applications. We suggest to overcome the associated difficulties through a reconstruction of subgrid variations into a modified basis by solving many independent (local) inverse problems that are constructed in a semi-Lagrangian step. Globally the method looks like a Eulerian method with multiscale stabilized basis. The method is extensible to other types of Galerkin methods, higher dimensions, nonlinear problems and can potentially work with real data. We provide examples inspired by tracer transport in climate systems in one and two dimensions and numerically compare our method to standard methods.

Keywords: Multiscale simulation · Semi-Langrangian method · Advection-dominance · Multiscale finite elements · Advection-diffusion equation

1 Introduction

1.1 Motivation and Overview

Simulating complex physical processes at macroscopic coarse scales poses many problems to engineers and scientists. Such simulations strive to reflect the effective behavior of observables involved at large scales even if the processes are partly driven by highly heterogeneous micro scale behavior. On the one hand resolving the microscopic processes would be the safest choice but such a strategy is often prohibitive since it would be computationally expensive. On the other hand microscopic processes significantly influence the macroscopic behavior and can not be neglected.

Incorporating micro scale effects into macro simulations in a mathematically consistent way is a challenging task. There exist many scenarios in different disciplines of science that are faced with such challenges. In fully coupled paleo

J. M. F. Rodrigues et al. (Eds.): ICCS 2019, LNCS 11539, pp. 393–409, 2019.
https://doi.org/10.1007/978-3-030-22747-0_30

climate simulations, i.e., climate simulations over more than hundred thousand years a typical grid cell can have edge lengths around 200 km and more. Consequently, subgrid processes such as heterogeneously distributed and moving ice shields are not or just insufficiently resolved [32]. These subgrid processes are usually taken care of by so-called parametrizations. One can imagine this as small micro scale simulations that are then coupled to the prognostic variables such as wind speed, temperature and pressure on the coarse grid (scale of the dynamical core). This coupling from fine to coarse scales is being referred to as upscaling and is unfortunately often done in rather heuristic ways. This leads to wrong macroscopic quantities like wrong pressures and eventually even to wrong wind directions and many more undesired effects such as phase errors. This demands for mathematically rigorously justified multiscale methods.

Many multiscale methods have been proposed in the past two decades. Among them are the multiscale Eulerian Lagrangian localized adjoint methods (MsELLAM) which constitute a space-time finite element framework, see [9,10,20,35]. Homogenization is an originally analytical tool to find effective models of otherwise heterogeneous models in the limit of a large scale separation [5,8,27]. The heterogeneous multiscale method (HMM) was pioneered by Weinan and Enquist [13,36] and refers to a rich zoo of methods [1–3,12,18,19]. Variational multiscale methods (VMM) have been developed in the 1990s by Hughes and collaborators, see [4,24,25,33]. The spirit of the method lies in a decomposition of the solution space and in the design of variational forms that reflect the relevant scale interactions between the spaces.

The present work is inspired by multiscale finite element methods (MsFEM) which can be seen as a special variant of the VMM. The idea of this method is to introduce subgrid variations into basis functions and can be dated back to works by Babuška, Caloz and Osborn [6]. It shares ideas with the partition of unity method [31]. The MsFEM in its current form was introduced in [15,22,23]. The essential idea of the method is to capture the local asymptotic structure of the solution through adding problem dependent bubble correctors to a standard basis and use these as ansatz and trial functions. Many variations of this method exist and we refer the reader to [14,17] for a review.

Many of the aforementioned methods have the advantage that they work well for elliptic or parabolic problems and that they are accessible to an analysis. The difficulty in many applications on the other hand is their advection or reaction dominated character, i.e., the dynamics is often driven by first or zero order terms. This poses major difficulties to numerical multiscale methods. Multiscale finite element methods naively applied will not converge to any reasonable solution since basis functions will exhibit artificial boundary layers that are not present in the actual physical flow. Ideas to tackle this problem are based on combining transient multiscale methods with Lagrangian frameworks [34] or with stabilization methods for stationary problems, see [28] for an overview. A HMM based idea for incompressible turbulent flows can be found in [29]. For a method based on the VMM see [30]. None of the mentioned methods fully remedies the difficulty of the undesired loss of the localization principle (inherent to advection-dominated problems) that is necessary to obtain weakly coupled problems on the subscale. This is the starting point of our work.

1.2 Contribution

Our main contribution is a framework of numerical methods for advection-dominated flows which by reconstructing subgrid variations on local basis functions aims at reflecting the local asymptotic structure of solutions correctly. The idea combines multiscale Galerkin methods [14,17] with semi-Lagrangian methods [7,11] by locally solving an inverse problem for the basis representation of solutions that is adapted to the actual flow scenario. We demonstrate the idea on one and two-dimensional advection-diffusion equations with heterogeneous background velocities and diffusivities in both non-conservative and conservative form.

Conformal MsFEM techniques for advection-dominated tracer transport were already explored in our previous work [34] in one spatial dimension on a transient advection-diffusion equation. The finding is that one has to follow a Lagrangian point of view on coarse scales such that flow is "invisible". On fine scales one can then simplify Lagrangian transforms in order to make advective effects locally milder without going to a fully Lagrangian setting. This amounts to prescribing Dirichlet boundary conditions on coarse flow characteristics and has the effect that basis functions do not develop boundary layers that are not there in the actual flow. While this work gave some useful insights it is unfortunately not feasible for practical applications since it suffers from several weaknesses. First, it is not directly generalizable to higher dimensions. Secondly, it needs assumptions on the background velocity that are not necessarily fulfilled in practical applications to ensure that coarse scale characteristics do not come to close.

In order to circumvent these problems we suggest a new idea based on a semi-Lagrangian framework that locally in time constructs a multiscale basis on a fixed Eulerian grid via many local and independent inverse problems. The construction is done in a semi-Lagrangian fashion on the subgrid scale whereas the macroscopic scale is conveniently treated as completely Eulerian. This is in complete contrast to our previous work but still respects that information in advection-dominant flows is "mostly" propagated along flow characteristics.

All reconstructions of the basis cells are independent from each other and can be performed in parallel while the global time step with the modified basis is numerically cheap. Note that although we formulate the algorithm globally in an implicit form it can well be formulated explicitly which can make the global step more efficient.

Our new approach is generic, conceptionally new, and has several advantages. First, we can effectively incorporate subgrid behavior of the solution into the multiscale basis via the solution of local inverse problems (and possibly even real data). Secondly, the idea works in any dimension. Numerical tests show that it is accurate in both L^2 and H^1 since it represents subgrid features correctly and can handle problems that involve an additional reaction term. This consequently includes conservation problems.

2 Semi-Lagrangian Multiscale Reconstruction

We will briefly outline and demonstrate our ideas on an advection-diffusion equation (ADE) with periodic boundaries as a model problem. For the sake of brevity we will focus on $d = 1$ dimension for the presentation of the method and mention strategies for its generalization to $d = 2$ dimensions on

$$\begin{aligned} \partial_t u + c_\delta \partial_x u &= \partial_x \left(A_\varepsilon \partial_x u \right) + f \quad \text{in } [0,1] \times [0,T] \\ u(x,0) &= u_0(x) \end{aligned} \tag{1}$$

and

$$\begin{aligned} \partial_t u + \partial_x \left(c_\delta u \right) &= \partial_x \left(A_\varepsilon \partial_x u \right) + f \quad \text{in } [0,1] \times [0,T] \\ u(x,0) &= u_0(x) \end{aligned} \tag{2}$$

where $c_\delta(x,t)$ is the background velocity, $A_\varepsilon(x,t)$ is a diffusivity coefficient, and f and u_0 are some smooth external forcing and initial condition.

The indices $\delta > 0$ and $\varepsilon > 0$ indicate that both quantities may have large variations on small scales that are not resolved on coarse scales $H > 0$ of our multiscale method. We will also work locally on a scale $h \ll H$ that can resolve the variations in the coefficients. Furthermore, we assume that $c_\delta \gg A_\varepsilon$ (see remark below) and that c_δ is well-behaved, for example $C^{1,1}$ in space and continuous in time. This assumption is still practical since, for example, in long term climate simulations c_δ represents as a relatively smooth coarse grid function. Depending on the application variations in c_δ may be resolved on scale H or not. The latter is often the case in subsurface flow problems whereas the former is the usual case in tracer transport in climate simulations. The diffusivity which we assume to be $A_\varepsilon \in L_t^\infty L_x^\infty$ often comes from parametrized processes and is furthermore assumed to be positive definite (uniformly in ε and point-wise in $\mathbf{x}$). Note, that (1) does not conserve the tracer u in contrast to Eq. (2).

Remark. Advection-dominance of a flow (what we sloppily expressed by $c_\delta \gg A_\varepsilon$) is usually expressed by a dimensionless number – the Péclet number Pe – which is essentially the ratio between advective and diffusive time scales. There exist several versions of this number [26]. Since for large variations of the coefficients on the subgrid scale advection-dominance is a very local property we need to be more precise with what me mean by that. Here we assume that Pe is high on average, i.e., $\mathrm{Pe} = \frac{\|c_\delta\|_{L^2} L}{\|A_\varepsilon\|_{L^2}}$ where L is a characteristic length. We take L to be the length of the computational domain and construct our test examples such that the Pe is high on average but also such that relevant phenomena will occur far below the reference.

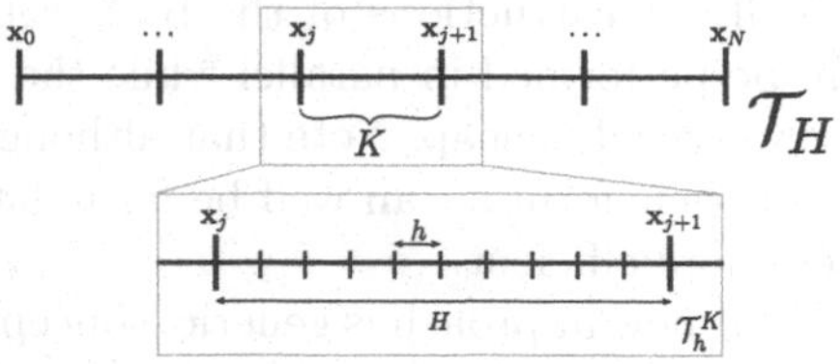

Fig. 1. Global coarse mesh and local fine mesh in 1D.

We focus on outlining our method in one dimension since the idea is simple and avoids complications that arise in higher dimensions. The idea is to represent non-resolved fine scale variations of the solution locally on a set of non-polynomial basis functions in each coarse cell. That is we fix two (Eulerian) meshes: A coarse mesh $\mathcal{T}_H$ of width H and on each cell $K \in \mathcal{T}_H$ of the coarse mesh we have a fine mesh $\mathcal{T}_h^K$ of width $h \ll H$. On the coarse mesh we have a multiscale basis $\varphi_i^{H,\mathrm{ms}} : [0,1] \times [0,T] \to \mathbb{R}$, $i = 1, \ldots, N_H$, where N_H is the number of nodes of $\mathcal{T}_H$. This basis depends on space and time and will be constructed so that we will obtain a spatially H^1-conformal (multiscale) finite element space. Note that this is a suitable space for problems (1) and (2) since they have unique solutions $u \in L^2([0,T], H^1([0,1]))$ with $\partial_t u \in L^2([0,T], H^{-1}([0,1]))$ and hence in $u \in C([0,T], L^2([0,1]))$ [16]. The initial condition u^0 can therefore be assumed to be in $L^2([0,1])$. The fine mesh on each cell $K \in \mathcal{T}_H$ is used to represent the basis locally, see Fig. 1.

Standard MsFEM methods are designed so that the basis functions solve the PDE model locally with appropriate boundary conditions (often Dirichlet). Their choice is crucial. If we now replace the standard basis by these functions the true local asymptotic structure of the global solution is represented since the global solution is a linear combination of modified local basis functions that do reflect the asymptotics. This works for stationary elliptic problems and for parabolic problems as long as there is no advective term involved. The reason is that advective terms like in our model problems prevent a basis constructed by a standard MsFEM to contain the correct asymptotics since flow of information is artificially blocked at coarse cell boundaries. This forms artificial steep boundary layers in basis functions and displays local behavior that is not there in the actual global flow. For transient problems another difficulty is that the local asymptotics around a point depends on the entire domain of dependence of this point and hence there must be some "memory" on the basis.

The results of [34] suggest that a coarse numerical splitting of the domain must correspond to a reasonable physical splitting of the problem. Instead of a fully Lagrangian method on coarse scales a semi-Lagrangian method is used to construct a basis to circumvent the difficulties of pure Lagrangian techniques. But these are only local in time and therefore they do not take into account the entire domain of dependence of a point. We show how to deal with this in the following. First, we start with the global problem.

2.1 The Global Time Step

Suppose we already know a set of multiscale basis functions in a conformal finite element setting, i.e., we approximate the global solution at each time step in a spatially coarse subspace $V^H(t) \subset H^1([0,1])$ in which the solution u is sought. We denote this finite-dimensional subspace as

$$V^H(t) = \operatorname{span}\left\{ \varphi_j^{H,\mathrm{ms}}(\cdot,t) \,\middle|\, i = 1, \ldots, N_H \right\} . \qquad (3)$$

First, we expand the solution $u^H(x,t)$ in terms of the basis at time $t \in [0,T]$, i.e.,

$$u^H(x,t) = \sum_{j=0}^{N_H} u_j^H(t)\varphi_j^{H,\mathrm{ms}}(x,t) . \tag{4}$$

Then we test with the modified basis and integrate by parts. Therefore, the spatially discrete version of both problem (1) and (2) becomes the ODE

$$\begin{aligned} \mathbf{M}(t)\frac{\mathrm{d}}{\mathrm{d}t}\mathbf{u}^H(t) + \mathbf{N}(t)\mathbf{u}^H(t) &= \mathbf{A}(t)\mathbf{u}^H(t) + \mathbf{f}^H(t) \\ \mathbf{u}^H(0) &= \mathbf{u}^{H,0} \end{aligned} \tag{5}$$

where

$$\mathbf{A}_{ij}(t) = \int_{[0,1]} \varphi_i^{H,\mathrm{ms}}(x,t)\cdot A_\varepsilon(x,t)\partial_x\varphi_j^{H,\mathrm{ms}}(x,t) - \varphi_i^{H,\mathrm{ms}}(x,t)c_\delta(x,t)\partial_x\varphi_j^{H,\mathrm{ms}}(x,t)\mathrm{d}x \tag{6}$$

for (1) and

$$\mathbf{A}_{ij}(t) = \int_{[0,1]} \varphi_i^{H,\mathrm{ms}}(x,t)\cdot A_\varepsilon(x,t)\partial_x\varphi_j^{H,\mathrm{ms}}(x,t) + \partial_x\varphi_i^{H,\mathrm{ms}}(x,t)c_\delta(x,t)\varphi_j^{H,\mathrm{ms}}(x,t)\mathrm{d}x \tag{7}$$

for (2). The mass matrix is given by

$$\mathbf{M}_{ij}(t) = \int_{[0,1]} \varphi_i^{H,\mathrm{ms}}(x,t)\varphi_j^{H,\mathrm{ms}}(x,t)\,\mathrm{d}x , \tag{8}$$

$\mathbf{f}^H(t)$ contains forcing and boundary conditions and the initial condition $\mathbf{u}^{H,0}$ is the projection of $u^0 \in L^2([0,1])$ onto $V^H(0)$. Note that (5) contains a derivative of the mass matrix:

$$\mathbf{N}_{ij}(t) = \int_{[0,1]} \varphi_i^{H,\mathrm{ms}}(x,t)\partial_t\varphi_j^{H,\mathrm{ms}}(x,t)\,\mathrm{d}x . \tag{9}$$

This is necessary since the basis functions depend on time and since we discretized in space first.

For the time discretization we simply use the implicit Euler method. The discrete ODE then reads

$$\mathbf{M}(t^n)\mathbf{u}^{n+1} = \mathbf{M}(t^n)\mathbf{u}^n + \delta t\left[\mathbf{A}(t^{n+1})\mathbf{u}^{n+1} - \mathbf{N}(t^{n+1})\mathbf{u}^{n+1} + \mathbf{f}^H(t^n)\right] . \tag{10}$$

Other time discretization schemes, in particular, explicit schemes are of course possible but may involve conditions on the time step size that originate from the (space-time local) transformation to Lagrangian coordinates. We pass on elaborating this here for the sake of brevity. For didactic reasons we therefore choose to present the algorithm in an implicit version. The next step is to show how to construct the multiscale basis.

Convention. Quantities marked with a tilde like $\tilde{x}$ signalize (semi-)Lagrangian quantities.

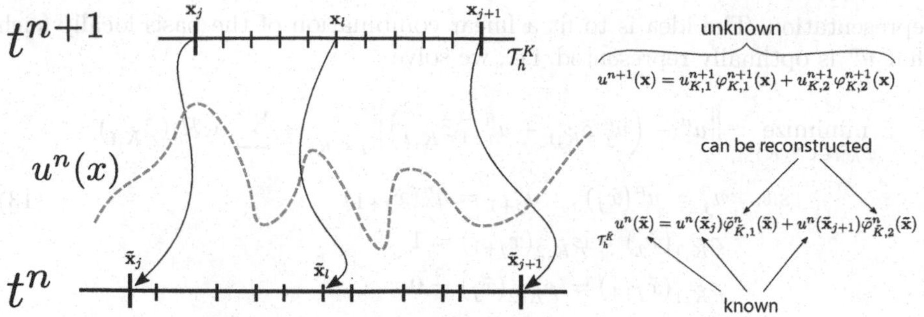

Fig. 2. The fine mesh in each cell $K \in \mathcal{T}_H$ is traced back one time step where the known solution can be used to reconstruct a basis representation of the solution.

2.2 The Reconstruction Mesh

Our idea combines the advantage of both semi-Lagrangian and multiscale methods to account for dominant advection. The reconstruction method is based on the simple observation that local information of the entire domain of dependence is still contained in the global solution at the previous time step. This can be used to construct an Eulerian multiscale basis: we trace back an Eulerian cell $K \in \mathcal{T}_H$ at time t^{n+1} on which the solution and the basis are unknown to the previous time step t^n. This gives a distorted cell $\tilde{K}$ over which the solution u^n is known but not the multiscale basis $\tilde{\varphi}_i$, $i = 1, 2$.

In order to find the points where transported information originates we trace back all nodes in $\mathcal{T}_H^K$ from time t^{n+1} to t^n. For this one simply needs to solve an ODE with the time-reversed velocity field that reads

$$\begin{aligned} \frac{\mathrm{d}}{\mathrm{d}t}\tilde{x}_l(t) &= -c_\delta(\tilde{x}_l(t), -t)\,, \quad t \in [-t^{n+1}, -t^n] \\ \tilde{x}_l(-t^{n+1}) &= x_l \end{aligned} \tag{11}$$

for each x_l and then take $\tilde{x}_l = \tilde{x}_l(-t^n)$, see Fig. 2 for an illustration. This procedure is standard in semi-Lagrangian schemes and can be parallelized.

2.3 Basis Reconstruction

After tracing back each point x_l of $K \in \mathcal{T}_H$ to its origin $\tilde{x}_l$ in a distorted coarse cell $\tilde{K} \in \mathcal{T}_H$ we need to reconstruct a local representation of the (known) solution u^n on $\tilde{K}$:

$$u^n(x)|_{\tilde{K}} = u^n(\tilde{x}_j)\tilde{\varphi}_{K,1}(x) + u^n(\tilde{x}_{j+1})\tilde{\varphi}_{K,2}(x) \tag{12}$$

where $\tilde{x}_j$ and $\tilde{x}_{j+1}$ are the boundary points of $\tilde{K}$. In one dimension one can of course choose a representation using the standard basis of hat functions but this would not incorporate subgrid information at step t_n at all. We solve this problem by solving an inverse problem for the basis to modify the local basis representation. The idea is to fit a linear combination of the basis locally such that u^n is optimally represented, i.e., we solve

$$\begin{aligned} \underset{\tilde{\varphi}_{K,i}\in C^0(\tilde{K})}{\text{minimize}} \quad & \left\| u^n - \left(u_j^n \tilde{\varphi}_{\tilde{K},1} + u_{j+1}^n \tilde{\varphi}_{\tilde{K},2} \right) \right\|_{L^2(\tilde{K})}^2 + \sum_i \alpha_i \mathcal{R}_i(\tilde{\varphi}_{\tilde{K},i}) \\ \text{s.t.} \quad & u_j^n = u^n(\tilde{x}_j)\,, \quad u_{j+1}^n = u^n(\tilde{x}_{j+1}) \\ & \varphi_{\tilde{K},1}(\tilde{x}_j) = \varphi_{\tilde{K},2}(\tilde{x}_{j+1}) = 1 \\ & \varphi_{\tilde{K},1}(\tilde{x}_{j+1}) = \varphi_{\tilde{K},2}(\tilde{x}_j) = 0\,. \end{aligned} \tag{13}$$

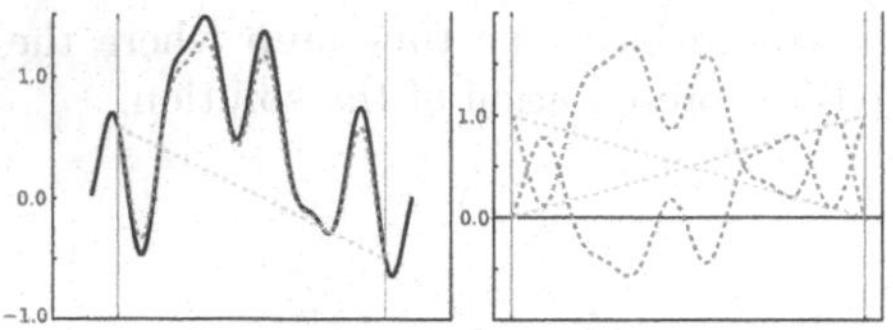

Fig. 3. Left: An oscillatory function (black) is being approximated by a standard linear basis (red) on an interval $[a, b]$ compared to a modified basis (blue) that solves (13). The regularization parameters were taken as $\alpha_i = 0.1$. Right: Comparison of the standard basis to the modified basis. The modified basis neither constitutes a partition of unity nor is it positive. (Color figure online)

The operators $\mathcal{R}_i : C^0(\tilde{K}) \to \mathbb{R}$ denote regularizers weighted by positive numbers $\alpha_i \in \mathbb{R}$. A simple regularizer that we found useful in one spatial dimension is a penalization of the quadratic mean deviation of the modified basis function from the standard linear basis function, i.e., we use

$$\mathcal{R}_i(\tilde{\varphi}_{\tilde{K},i}) = \left\| \tilde{\varphi}_{\tilde{K},i} - \tilde{\varphi}_{\tilde{K},i}^0 \right\|_{L^2}^2 \tag{14}$$

where $\tilde{\varphi}_{\tilde{K},i}$ denotes the t-th standard (linear) basis on $\tilde{K}$. In a spatially discrete version this system is linear and small and will be cheap to solve. A suitable choice of a regularizer depends on the problem at hand. Figure 3 illustrates the effect of a local reconstruction of a basis compared to a representation with a standard basis.

2.4 Basis Propagation

After having reconstructed a suitable basis on each coarse cell $\tilde{K}$ we have an H^1-conformal basis. This basis, however is a basis at time step t^n and does not live on the coarse Eulerian grid $\mathcal{T}_H$ that we initially fixed. The step to take now is to construct a basis at t^{n+1} on $\mathcal{T}_H$. This is done by evolving the basis according to the model at hand with a vanishing external forcing. Note, however, that we compute the basis at t^{n+1} along Lagrangian trajectories starting from t^n, i.e., we need to transform the original model. Eq. (1) becomes

$$\begin{aligned} \partial_t \varphi_{K,i} &= \tilde{\partial}_x \left(\tilde{A}_\varepsilon \tilde{\partial}_x \varphi_{K,i} \right) \quad \text{in } \tilde{K} \times [t^n, t^{n+1}] \\ \varphi_{K,i}(\tilde{x}_j, t) &= \tilde{\varphi}_{\tilde{K},i}(\tilde{x}_j)\,, \quad t \in [t^n, t^{n+1}] \\ \varphi_{K,i}(\tilde{x}_{j+1}, t) &= \tilde{\varphi}_{\tilde{K},i}(\tilde{x}_{j+1})\,, \quad t \in [t^n, t^{n+1}] \\ \varphi_{K,i}(\tilde{x}, t^n) &= \tilde{\varphi}_{\tilde{K},i}(\tilde{x}) \end{aligned} \tag{15}$$

and Eq. (2) transforms into

$$
\begin{aligned}
\partial_t \varphi_{K,i} + \left(\tilde{\partial}_x \tilde{c}_\delta\right) \varphi_{K,i} &= \tilde{\partial}_x \left(\tilde{A}_\varepsilon \tilde{\partial}_x \varphi_{K,i}\right) \quad \text{in } \tilde{K} \times [t^n, t^{n+1}] \\
\varphi_{K,i}(\tilde{x}_j, t) &= \tilde{\varphi}_{\tilde{K},i}(\tilde{x}_j)\,, \quad t \in [t^n, t^{n+1}] \\
\varphi_{K,i}(\tilde{x}_{j+1}, t) &= \tilde{\varphi}_{\tilde{K},i}(\tilde{x}_{j+1})\,, \quad t \in [t^n, t^{n+1}] \\
\varphi_{K,i}(\tilde{x}, t^n) &= \tilde{\varphi}_{\tilde{K},i}(\tilde{x})\,.
\end{aligned}
\tag{16}
$$

These evolution equations are solved on $\tilde{K}$, i.e., on the element $K \in \mathcal{T}_H$ traced back in time. Advection is "invisible" in these coordinates. The end state $\varphi_{K,i}(\tilde{x}, t^{n+1})$ on $\tilde{K}$ can then be transformed onto the Eulerian element $K \in \mathcal{T}_H$ to obtain the desired basis function $\varphi_{K,i}(x, t^{n+1}) \sim \varphi_{K,i}^{n+1}(x)$ at the next time step. Corresponding basis functions in neighboring cells can then be glued together to obtain a modified global basis $\varphi_i^{H,\text{ms}}$, $i = 1, \ldots, N_H$. This way we get a basis of a subspace of H^1 that is neither a partition of unity nor is it necessarily positive. Nonetheless, it is adjusted to the problem and the data at hand. The propagation step is illustrated in Fig. 4.

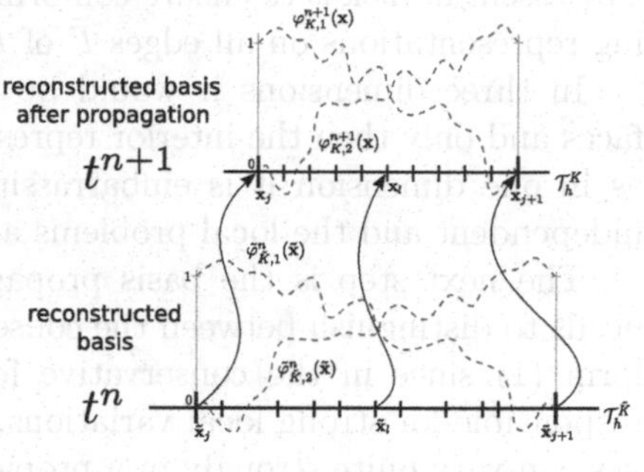

Fig. 4. The basis reconstructed according to (13) at time t^n is propagated forward to time t^{n+1} according to (15) or (16).

Using our method we reconstruct and advect the representation of the global solution first and then the solution itself using the modified representation. The global step is completely Eulerian while the local reconstruction step is semi-Lagrangian in contrast to [34] where the global step is Lagrangian and the local step is "almost"-Lagrangian. Note that the steps to reconstruct the multiscale basis are embarrassingly parallel and all constitute of small problems.

2.5 Basis Reconstruction and Propagation in 2D

The ideas of the above method can be transferred to two dimensions. The reconstruction in two dimensions though is different. We intend to briefly give the reader an idea of the differences without describing the details in order to point out difficulties in the generalization to higher dimensions.

Suppose that we want to reconstruct a basis on cell $K \in \mathcal{T}_H$ at time t^{n+1}. We trace back the cell as described in (11) and denote the distorted cell at time t^n by $\tilde{K}$ and its edges by $\tilde{\Gamma}$. As in one spatial dimension, to construct a H^1-conformal basis we need to ensure continuity of the basis across coarse cell boundaries. This can be achieved by first reconstructing the solution at the previous time step t^n with a basis representation on each edge, i.e., we solve first an inverse problem on each deformed edge $\tilde{\Gamma}$ similar to (13).

Note that the regularizer (14) needs to be replaced since the edge $\tilde{\Gamma}$ is usually curved. We use a harmonic prior

$$\mathcal{R}_i(\tilde{\varphi}_{\tilde{K},i}) = \left\| \Delta_{g(\tilde{\Gamma})} \tilde{\varphi}_{\tilde{\Gamma},i} \right\|^2_{L^2(\tilde{\Gamma})} \tag{17}$$

with a low weight α_i as in (13). The operator $\Delta_{g(\tilde{\Gamma})}$ denotes the Laplace-Beltrami operator induced by the standard Laplace operator with the trace topology of the respective edge $\tilde{\Gamma}$, i.e., $g(\tilde{\Gamma})$ is the metric tensor. We pass on providing details here.

The edge reconstruction provides boundary values for the cell basis reconstruction. The optimization problem to solve on $\tilde{K}$ can then again be designed similar to (13) just constrained by the previously reconstructed boundary values. The essential task is to ensure conformity of the global basis by first reconstructing representations on all edges $\tilde{\Gamma}$ of $\tilde{K}$ and then inside the cells $\tilde{K}$.

In three dimensions it would be necessary to first reconstruct edges then faces and only then the interior representations. This might seem expensive but as in one dimension it is embarrassingly parallel since all reconstructions are independent and the local problems are small.

The next step is the basis propagation. The observation here is that one needs to distinguish between the conservative form (2) and the non-conservative form (1) since in the conservative form an additional local reaction term is responsible for strong local variations. Hence, reconstructed edge boundary values can vary quite strongly in a propagation step similarly to the one described in Sect. 2.4. Consequently, edge boundary values need to be adjusted in the propagation. This is done by first propagating the reconstructed edge boundary values and then using these as (time-dependent) boundary values for propagation problems similar to (15) or (16). We pass on providing technical details.

3 Numerical Examples

For all 1D tests we use a Gaussian

$$u_0(x) = \frac{1}{\sigma\sqrt{2\pi}} \exp -\frac{(x-\mu)^2}{2\sigma^2} . \tag{18}$$

with variance $\sigma = 0.1$ centered in the middle of the domain $[0, 1]$, i.e., $\mu = 0.5$. The end time is set to $T = 1$ with a time step $\delta t = 1/300$. We show our semi-Lagrangian multiscale reconstruction method (SLMsR) with a coarse resolution $H = 10^{-1}$ in comparison to a standard FEM with the same resolution and high order quadrature. As a reference we choose a high-resolution standard FEM with $h_{\text{ref}} = 10^{-3}$. For the multiscale method we choose a fine mesh $\mathcal{T}_h^K$ with $h = 10^{-2}$ in each coarse cell $K \in \mathcal{T}_H$. We would like to point out that there are no standardized test cases for our type of model. Therefore, we designed our tests in such a way that small scale effects occur in the solution below the coarse resolution and compare our multiscale methods to the performance of standard methods with the same (coarse) resolution.

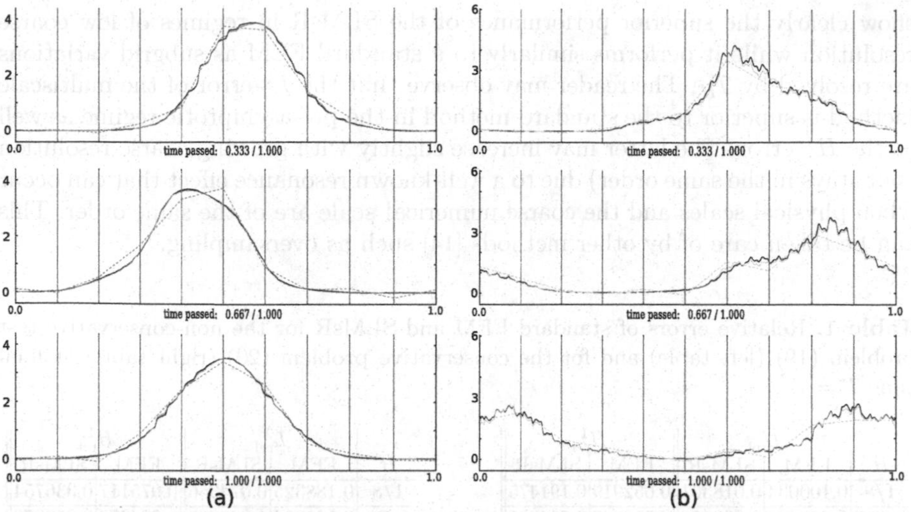

Fig. 5. Snapshots of the solution at $t = 1/3$, $t = 2/3$ and $T = 1$. The colored dashed lines show the solution of the standard FEM (10 elements), the colored line shows the SLMsR (10 coarse elements). The reference solution is shown in black (1K elements). **(a)** non-conservative equation (1) coefficients (19). **(b)** conservative equation (2) coefficients (20). (Color figure online)

Test 1. We will show two examples in a non-conservative and conservative setting according to (1) and (2), respectively, to show the capability of the SLMsR to capture subgrid variations correctly. Note that the coarse standard FEM has as many cells as the SLMsR has coarse cells and that the standard FEM does not capture subgrid variations in the following tests which can result in aliasing and phase errors. The resolution for the reference solution resolves all subgrid variations but the reader should keep in mind that practical applications do not allow the application of high-resolution methods. The coefficients

$$\begin{aligned} c_\delta(x,t) &= \frac{1}{2}\cos(2\pi t) + \frac{1}{4}\cos(6\pi t)\cos(8\pi x) + \frac{1}{8}\cos(4\pi t)\cos(62\pi x) + \frac{1}{8}\cos(150\pi x) \\ A_\varepsilon(x,t) &= 10^{-3} + 9\cdot 10^{-4}\cos(10\pi t)\cos(86\pi x) \end{aligned} \tag{19}$$

are chosen for the non-conservative equation (1) and the coefficients

$$\begin{aligned} c_\delta(x,t) &= \frac{1}{2} + \frac{1}{8}\cos(8\pi x) + \frac{1}{8}\cos(62\pi x) + \frac{1}{8}\cos(150\pi x) \\ A_\varepsilon(x,t) &= 10^{-2} + 9\cdot 10^{-3}\cos(10\pi t)\cos(86\pi x) \end{aligned} \tag{20}$$

for the conservative equation (2). The latter one is numerically more difficult when it comes to capturing fine-scale variations, i.e., if $c_\delta(x) \sim f(x/\delta)$ then $\frac{\mathrm{d}}{\mathrm{d}x}c_\delta(x) \sim \delta^{-1}f'(x/\delta)$ and one can expect very steep slopes in the solution. The results of the tests are shown in Fig. 5. The corresponding errors in Table 1 show clearly the superior performance of the SLMsR in regimes of low coarse resolution while it performs similarly to a standard FEM as subgrid variations are resolved by $\mathcal{T}_H$. The reader may observe that the L^2-error of the multiscale method is superior to the standard method in the pre-asymptotic regime as well as the H^1-error. The latter may increase slightly with growing coarse resolution (but stays in the same order) due to a well-known resonance effect that can occur when physical scales and the coarse numerical scale are of the same order. This can be taken care of by other methods [14] such as oversampling.

Table 1. Relative errors of standard FEM and SLMsR for the non-conservative test problem (19) (left table) and for the conservative problem (20) (right table) at final time $T = 1$.

	L^2_{rel}		H^1_{rel}	
H	FEM	SLMsR	FEM	SLMsR
1/8	0.100034	0.018361	0.652919	0.191475
1/32	0.026464	0.009116	0.559342	0.144821
1/128	0.011342	0.002582	0.426786	0.160761
1/512	0.000578	0.000491	0.159977	0.173261

	L^2_{rel}		H^1_{rel}	
H	FEM	SLMsR	FEM	SLMsR
1/8	0.188325	0.074356	0.975117	0.336754
1/32	0.140284	0.039869	1.024470	0.252878
1/128	0.044685	0.023380	0.782080	0.183882
1/512	0.004562	0.006220	0.290947	0.140458

Test 2. This test shows an example where both diffusion and background velocity are generated randomly. We intend to show an example of how the SLMsR behaves when data is involved that does not exhibit a clear scale separation which is a common situation in practice. For this we initially generate (fixed) mesh based functions with random nodal coefficients. In each mesh cell the functions are interpolated linearly. Note that this is not to simulate a sampled stochastic process. We simply intend not to create any scale or symmetry bias when constructing coefficient functions. The results look appealing and show a clear advantage of the SLMsR, see Fig. 6.

Test 3. Here we show one preliminary example of our SLMsR equation (1) with a dominant advection term in two dimensions. The test was carried out on the torus $\mathbb{T}^2$ (periodic unit square) in the time interval $t \in [0, 1]$. As initial value we chose a normalized super-position of two isotropic Gaussians

$$u_0(\mathbf{x}) = \frac{1}{2\sqrt{(2\pi)^2 \det(\mathbf{M})}} \sum_{i=1}^{2} \exp\left\{-\frac{1}{2}(\mathbf{x} - \boldsymbol{\mu}_i)^T \mathbf{M}^{-1} (\mathbf{x} - \boldsymbol{\mu}_i)\right\} \tag{21}$$

where

$$\mathbf{M} = \begin{bmatrix} \frac{3}{100} & 0 \\ 0 & \frac{3}{100} \end{bmatrix} \quad \text{and} \quad \boldsymbol{\mu}_i = \left[\frac{i}{3}\,,\, \frac{1}{2}\right]^T . \tag{22}$$

The test of the SLMsR was performed on a coarse unstructured uniform triangular Delaunay mesh with $n_c = 62$ coarse cells, i.e., for our triangulation $H \sim 0.3$ (maximum mean diameter of circumcircle of a cell). We compare the SLMsR to a standard low resolution FEM with the same resolution and to a standard

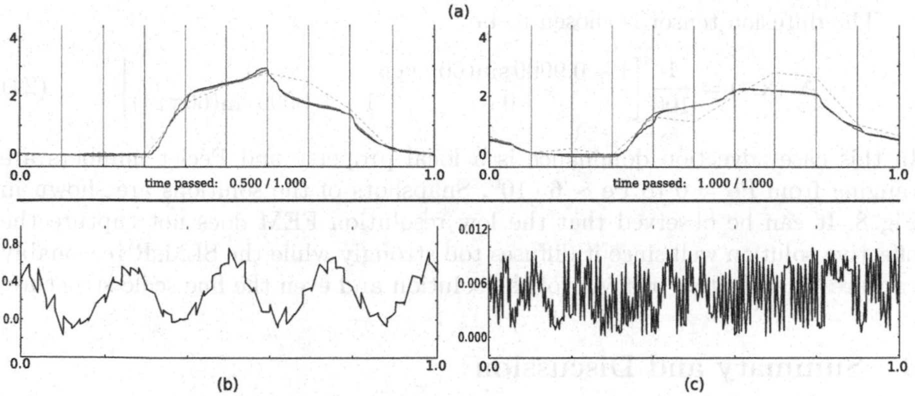

Fig. 6. Comparison of SLMsR and standard FEM for randomly generated (but fixed) coefficient functions. **(a)** Snapshots of the solution at $t = 1/2$ and $t = 1$. Solid black lines show the reference solution, dashed red lines show the standard solution and solid blue solid lines show the SLMsR. **(b)** The velocity coeffcient we chose to be a smooth function disturbed by Gaussian noise with mean zero and variance 0.1. **(c)** The diffusion coefficient was generated from a uniform distribution and scaled to have minumum 10^{-5} and maximum 10^{-2}. (Color figure online)

high resolution FEM with approximately $n_f = 63K$ cells. To get a fine mesh on each coarse cell of the SLMsR we created a triangulation such that the sum of all fine cells over all coarse cells is approximately n_f to get a fair comparison of the SLMsR to the low resolution standard FEM with respect to the reference solution that resolves all coefficients involved.

Fig. 7. Background velocity for *Test 1* and *Test 3*. Four vortices moving through the domain from left to right and come back to their starting points at $T = 1$.

We test our multiscale reconstruction method with a solenoidal field $\mathbf{c}_\delta$ described by the stream function

$$\psi(\mathbf{x}, t) = \sin(2\pi(x_1 - t))\sin(2\pi x_2) \qquad (23)$$

so that $\mathbf{c}_\delta(\mathbf{x}, t) = \nabla^T \psi$.

This background velocity describes four vortices moving in time through the (periodic) domain from left to right and get back to their starting point at $T = 1$. Note that this velocity field involves both scales that are resolved by the coarse mesh and scales that are not resolved, see Fig. 7. Also note that since $\nabla \cdot \mathbf{c}_\delta = 0$ Eqs. (1) and (2) are (analytically) identical and hence we only solve (1).

The diffusion tensor is chosen to be

$$\mathbf{A}_\varepsilon(\mathbf{x},t) = \frac{1}{100}\begin{bmatrix} 1-0.9999\sin(60\pi x_1) & 0 \\ 0 & 1-0.9999\sin(60\pi x_2)\end{bmatrix}. \quad (24)$$

In this case advection dominance is a local property and Péclet numbers are ranging from Pe $= 0$ to Pe $\sim 6\cdot 10^6$. Snapshots of the solutions are shown in Fig. 8. It can be observed that the low resolution FEM does not capture the effective solution well since it diffuses too strongly while the SLMsR reasonably captures the effective behavior of the solution and even the fine scale structure.

4 Summary and Discussion

In this work we introduced a new idea for a framework of multiscale methods demonstrated on advection-diffusion equations that are dominated by the advective term. Such a methods are of importance, for example, in reservoir modeling and tracer transport in earth system models. The main obstacles in these applications are, first, the advection-dominance and, secondly, the multiscale character of the background velocity and the diffusion tensor. The latter makes it impossible to simulate with standard methods due to computational constraints while simulating using standard methods with lower resolution that does not resolve variations in the coefficients leads to incorrect solutions on coarse meshes.

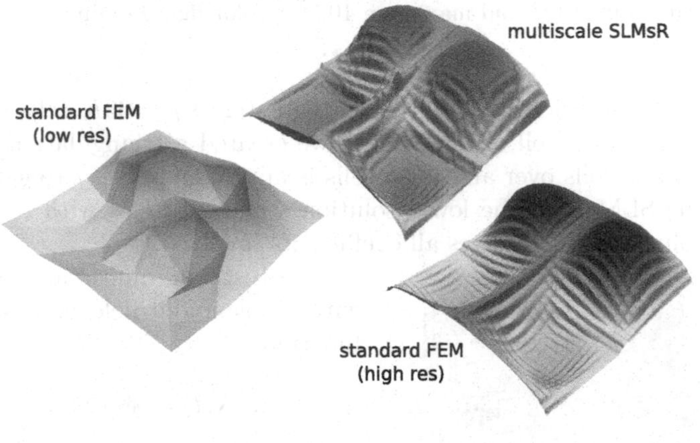

Fig. 8. Snapshots of the solution (surface plot) for *Test 3* at time $T = 1$ for the low resolution standard FEM (62 elements), the SLMsR (62 coarse elements) and the reference solution (63 K elements).

Our idea to cope with these difficulties is inspired by ideas for semi-Lagrangian methods, ideas based on "convected fluid microstucture" as described in [21], inverse problems and multiscale finite elements [14]. At each time step we reconstruct fine scale information from the solution at the previous time step. This fine scale information enters the local representation of the solution in each coarse cell, i.e., it is added as a corrector to the local basis such that the basis representation is optimal in some sense. The reconstruction is done by solving an inverse problem with a suitable regularizer and constructs a basis that does not constitute a partition of unity (PoU) and that is made for the concrete problem at hand. The idea of adding prior knowledge about the solution to a local representation in PoU methods, however, is similar, see for example [18,31]. After reconstructing the basis at the previous time step the basis is evolved with suitable boundary conditions to the time step the basis is sought for, i.e., we evolve the local representation of the solution rather than the solution itself. Note that the global framework of the SLMsR is completely Eulerian while only the local reconstruction step in each coarse cell is semi-Lagrangian.

One of the main features of the SLMsR is its scalability: Although it sounds expensive to trace back each coarse cell, then solve an inverse problem and then solve a PDE at each time step (the so-called offline phase) we would like to point out that these local problems are independent and usually small and therefore the offline phase is embarrassingly parallel, although we did not take advantage of that in our implementation. The global time step (online phase) also consists of a small problem and matrix assembly procedures can be made very efficient by using algebraic tricks, see [14,23].

We would like to further emphasize the flexibility of the SLMsR. Here we presented an implicit version but explicit time stepping is possible. The method can be transferred to higher dimensions as well as it can be extended to deal with advection-diffusion-reaction problems. Furthermore, the use of inverse problems in the local steps to adjust the basis makes it generally possible to incorporate knowledge coming from measurement data. For this a thorough understanding of the data is necessary (as for any other assimilation method). The SLMsR is promising for practical applications but lots of work needs to be done on the path towards applicability. This includes a numerical analysis which we do not aim at in this work. We would like to explore that opportunity in the future.

References

1. Abdulle, A.: The finite element heterogeneous multiscale method: a computational strategy for multiscale PDES. GAKUTO Int. Ser. Math. Sci. Appl. **31**, 135–184 (2009). (EPFL-ARTICLE-182121)
2. Abdulle, A., Weinan, E., Engquist, B., Vanden-Eijnden, E.: The heterogeneous multiscale method. Acta Numerica **21**, 1–87 (2012)
3. Abdulle, A., Engquist, B.: Finite element heterogeneous multiscale methods with near optimal computational complexity. Multiscale Model. Simul. **6**(4), 1059–1084 (2007)
4. Ahmed, N., Rebollo, T.C., John, V., Rubino, S.: A review of variational multiscale methods for the simulation of turbulent incompressible flows. Arch. Comput. Methods Eng. **24**(1), 115–164 (2017)
5. Allaire, G.: Homogenization and two-scale convergence. SIAM J. Math. Anal. **23**(6), 1482–1518 (1992)
6. Babuška, I., Caloz, G., Osborn, J.E.: Special finite element methods for a class of second order elliptic problems with rough coefficients. SIAM J. Numer. Anal. **31**(4), 945–981 (1994)
7. Behrens, J.: Adaptive Atmospheric Modeling - Key Techniques in Grid Generation, Data Structures, and Numerical Operations with Application. LNCSE, vol. 54. Springer, Berlin (2006). https://doi.org/10.1007/3-540-33383-5

8. Bensoussan, A., Lions, J.L., Papanicolaou, G.: Asymptotic Analysis for Periodic Structures, vol. 374. American Mathematical Society, Providence (2011)
9. Celia, M.A., Russell, T.F., Herrera, I., Ewing, R.E.: An Eulerian-Lagrangian localized adjoint method for the advection-diffusion equation. Adv. Water Resour. **13**(4), 187–206 (1990)
10. Cheng, A., Wang, K., Wang, H.: A preliminary study on multiscale ELLAM schemes for transient advection-diffusion equations. Numer. Methods Partial Diff. Equ. **26**(6), 1405–1419 (2010)
11. Durran, D.R.: Numerical Methods for Fluid Dynamics: With Applications to Geophysics, vol. 32. Springer, New York (2010). https://doi.org/10.1007/978-1-4419-6412-0
12. Weinan, E.: Principles of Multiscale Modeling. Cambridge University Press, Cambridge (2011)
13. Weinan, E., Engquist, B.: The heterognous multiscale methods. Commun. Math. Sci. **1**(1), 87–132 (2003)
14. Efendiev, Y., Hou, T.Y.: Multiscale Finite Element Methods: Theory and Applications, vol. 4. Springer, New York (2009). https://doi.org/10.1007/978-0-387-09496-0
15. Efendiev, Y.R., Hou, T.Y., Wu, X.H.: Convergence of a nonconforming multiscale finite element method. SIAM J. Numer. Anal. **37**(3), 888–910 (2000)
16. Evans, L.C.: Partial Differential Equations. American Mathematical Society, Providence (2010)
17. Graham, I.G., Hou, T.Y., Lakkis, O., Scheichl, R.: Numerical Analysis of Multiscale Problems, vol. 83. Springer, Berlin (2012). https://doi.org/10.1007/978-3-642-22061-6
18. Henning, P., Morgenstern, P., Peterseim, D.: Multiscale partition of unity. In: Griebel, M., Schweitzer, M.A. (eds.) Meshfree Methods for Partial Differential Equations VII. LNCSE, vol. 100, pp. 185–204. Springer, Cham (2015). https://doi.org/10.1007/978-3-319-06898-5_10
19. Henning, P., Ohlberger, M.: The heterogeneous multiscale finite element method for advection-diffusion problems with rapidly oscillating coefficients and large expected drift. NHM **5**(4), 711–744 (2010)
20. Herrera, I., Ewing, R.E., Celia, M.A., Russell, T.F.: Eulerian-Lagrangian localized adjoint method: the theoretical framework. Numer. Methods Partial Diff. Equ. **9**(4), 431–457 (1993)
21. Holm, D.D., Tronci, C.: Multiscale turbulence models based on convected fluid microstructure. J. Math. Phys. **53**(11), 115614 (2012)
22. Hou, T., Wu, X.H., Cai, Z.: Convergence of a multiscale finite element method for elliptic problems with rapidly oscillating coefficients. Math. Comput. Am. Math. Soc. **68**(227), 913–943 (1999)
23. Hou, T.Y., Wu, X.H.: A multiscale finite element method for elliptic problems in composite materials and porous media. J. Comput. Phys. **134**(1), 169–189 (1997)
24. Hughes, T.J.: Multiscale phenomena: Green's functions, the Dirichlet-to-Neumann formulation, subgrid scale models, bubbles and the origins of stabilized methods. Comput. Methods Appl. Mech. Eng. **127**(1–4), 387–401 (1995)
25. Hughes, T.J., Feijóo, G.R., Mazzei, L., Quincy, J.B.: The variational multiscale method - a paradigm for computational mechanics. Comput. Methods Appl. Mech. Eng. **166**(1–2), 3–24 (1998)
26. Huysmans, M., Dassargues, A.: Review of the use of Péclet numbers to determine the relative importance of advection and diffusion in low permeability environments. Hydrol. J. **13**(5–6), 895–904 (2005)

27. Jikov, V.V., Kozlov, S.M., Oleinik, O.A.: Homogenization of Differential Operators and Integral Functionals. Springer, Berlin (2012)
28. Le Bris, C., Legoll, F., Madiot, F.: A numerical comparison of some multiscale finite element approaches for advection-dominated problems in heterogeneous media. ESAIM: Math. Model. Numer. Anal. **51**(3), 851–888 (2017)
29. Lee, Y., Engquist, B.: Multiscale numerical methods for passive advection-diffusion in incompressible turbulent flow fields. J. Comput. Phys. **317**(317), 33–46 (2016)
30. Li, G., Peterseim, D., Schedensack, M.: Error analysis of a variational multiscale stabilization for convection-dominated diffusion equations in 2D. arXiv preprint arXiv:1606.04660 (2016)
31. Melenk, J.M., Babuška, I.: The partition of unity finite element method: basic theory and applications. Comput. Methods Appl. Mech. Eng. **139**(1–4), 289–314 (1996)
32. Notz, D., Bitz, C.M.: Sea ice in earth system models. In: Thomas, D.N. (ed.) Sea Ice, pp. 304–325. Wiley, Hoboken (2017). (Chap. 12)
33. Rasthofer, U., Gravemeier, V.: Recent developments in variational multiscale methods for large-eddy simulation of turbulent flow. Arch. Comput. Methods Eng. **25**(3), 647–690 (2018)
34. Simon, K., Behrens, J.: Multiscale finite elements through advection-induced coordinates for transient advection-diffusion equations. arXiv preprint arXiv:1802.07684 (2018)
35. Wang, H., Ding, Y., Wang, K., Ewing, R.E., Efendiev, Y.R.: A multiscale Eulerian-Lagrangian localized adjoint method for transient advection-diffusion equations with oscillatory coefficients. Comput. Vis. Sci. **12**(2), 63–70 (2009)
36. Weinan, E., Engquist, B., Huang, Z.: Heterogeneous multiscale method: a general methodology for multiscale modeling. Phys. Rev. B **67**(9), 092101 (2003)

A Multiscale Model of Atherosclerotic Plaque Development: Toward a Coupling Between an Agent-Based Model and CFD Simulations

Anna Corti[1], Stefano Casarin[2,3](✉), Claudio Chiastra[1,4], Monika Colombo[1], Francesco Migliavacca[1], and Marc Garbey[2,3,5]

[1] LABS, Department of Chemistry, Materials and Chemical Engineering "Giulio Natta", Politecnico di Milano, Milan, Italy

[2] Center for Computational Surgery, Houston Methodist Research Institute, Houston, TX, USA

scasarin@houstonmethodist.org

[3] Department of Surgery, Houston Methodist Hospital, Houston, TX, USA

[4] PoliTo[BIO]Med Lab, Department of Mechanical and Aerospace Engineering, Politecnico di Torino, Turin, Italy

[5] LASIE UMR 7356, CNRS, University of La Rochelle, La Rochelle, France

Abstract. Computational models have been widely used to predict the efficacy of surgical interventions in response to Peripheral Occlusive Diseases. However, most of them lack a multiscale description of the development of the disease, which, in our hypothesis, is the key to develop an effective predictive model. Accordingly, in this work we present a multiscale computational framework that simulates the generation of atherosclerotic arterial occlusions. Starting from a healthy artery in homeostatic conditions, the perturbation of specific cellular and extracellular dynamics led to the development of the pathology, with the final output being a diseased artery. The presented model was developed on an idealized portion of a Superficial Femoral Artery (SFA), where an Agent-Based Model (ABM), locally replicating the plaque development, was coupled to Computational Fluid Dynamics (CFD) simulations that define the Wall Shear Stress (WSS) profile at the lumen interface. The ABM was qualitatively validated on histological images and a preliminary analysis on the coupling method was conducted. Once optimized the coupling method, the presented model can serve as a predictive platform to improve the outcome of surgical interventions such as angioplasty and stent deployment.

Keywords: Agent-Based Model · Computational Fluid Dynamics · Peripheral Occlusive Diseases · Atherosclerosis · Multiscale model

1 Introduction

Peripheral Arterial Occlusive Diseases (PAODs) hold a high incidence worldwide with more than 200 million people affected annually [1]. Their etiology is mainly attributable to atherosclerosis, a chronic inflammation of the arterial wall that causes the

J. M. F. Rodrigues et al. (Eds.): ICCS 2019, LNCS 11539, pp. 410–423, 2019.
https://doi.org/10.1007/978-3-030-22747-0_31

narrowing of the lumen through the build-up of a fatty plaque and that preferentially develops in sites affected by a low or oscillatory Wall Shear Stress (WSS) profile [2].

The consequences of severe PAODs range from limb gangrene insurgence, which ultimately requires the amputation of the foot or leg, to the spreading of atherosclerosis to other parts of the body up to exposing the patient to high risk of heart attack and stroke [3].

Percutaneous Transluminal Angioplasty (PTA), with or without stent, and Vein Graft Bypass (VGB) are the preferred interventions aimed to restore the physiological circulation. These procedures suffer of a high rate of long-term failure, with a 3-years patency of 60% and a 5-years patency of 70% for PTA and VGB respectively [4].

Mathematical models, combined with computational simulations, are powerful tools that can offer a virtual environment to test clinical hypotheses and to drive the improvement of surgical interventions.

In this optical, many works simulated the post-surgical follow-up of both PTA and VGB procedures by using a heterogeneous thread of computational techniques [5–13]. Specifically, Agent-Based Models (ABMs) effectively describe pathophysiological processes in which spatial interactions play a major role [14], even though most of them perform their simulations by starting from healthy arteries [5, 7, 8, 10–12]. In some, the atherosclerotic plaque was introduced to provide a more realistic geometrical configuration for the pre-intervention condition, but without being generated as consequence of the pathological agents' dynamics [13]. To the best of our knowledge, only few works addressed the modeling of atherosclerotic plaque formation and progression with an ABM approach, most of them by focusing on the inflammatory and immune events occurring in early atherosclerosis [15–17].

Our hypothesis is that an effective predictive model of surgical outcome must replicate the multiscale dynamics that lead to the formation of the atherosclerotic plaque itself, so that the simulation of an intervention is performed on a system whose current dynamics are already altered by the pathological status, which thus influences the outcome. The multiscale character of our approach is of crucial importance as triggering mechanisms and pathology development operate on different time and space scales, respectively seconds vs. weeks and tissue vs. cellular.

Accordingly, this work presents a computational model in which, starting from an artery in homeostatic conditions, the perturbation of specific cellular and extracellular dynamics led to the development of the pathology, ending up with a diseased artery.

The presented model was developed on an idealized portion of a Superficial Femoral Artery (SFA), where an ABM, locally replicating the plaque development, was coupled to Computational Fluid Dynamics (CFD) simulations defining the WSS profile at the wall.

2 Methods

Figure 1 shows the structure of the computational framework. Starting from an idealized 3D model of SFA, CFD simulations compute the WSS distribution along the entire lumen interface. Then, a discrete number ($M = 10$) of 2D cross-sections is selected and an ABM is implemented for each of them to locally replicate the cellular and extra-cellular dynamics driving the wall remodeling. The resulting geometrical alterations, described by the ensemble of the *M* ABMs outputs, imply a variation of the flow/pressure pattern, and consequently of the WSS profile, which needs to be re-evaluated by coupling back the ABMs to the CFD simulation. For this purpose, a new 3D geometry is reconstructed starting from the *M* newly obtained 2D cross-sections. The CFD simulation is finally re-run with the same modalities and the loop re-starts.

The CFD simulation and the ABM, embedded in the dashed orange box in Fig. 1, constitute the core of the multiscale model. The first simulates at tissue level the average hemodynamics in a heartbeat, which is on the scale of seconds, while the second, replicating plaque growth, operates at cellular level on the scale of weeks.

The continuous interaction between wall remodeling and hemodynamics is captured by an iterative four-step cycle consisting in: (i) geometry preparation and meshing, (ii) CFD simulation, (iii) ABM simulation, and (iv) coupling and retrieval of the new 3D geometry. Cited steps are separately described below.

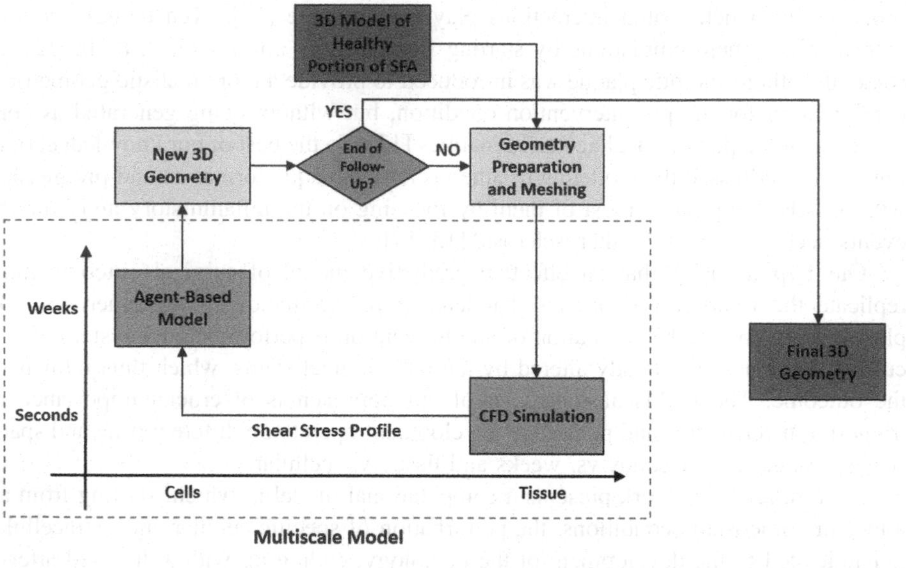

Fig. 1. Computational Workflow. Starting from an idealized 3D model of SFA, the geometry is prepared for the CFD simulation that gives in input to the ABM the WSS profile at the lumen interface. The ABM performs the cellular and extracellular dynamics that modify the geometrical profile. A new 3D geometry is retrieved in output from the ABM module and the loop re-starts.

2.1 Geometry Preparation and Meshing

An idealized 3D geometry resembling a tortuous portion of SFA was initially built by using Rhinoceros® software (v. 6.0, Robert McNeel & Associates, Seattle, WA, USA).

The geometry presents a centerline with length $L_{cent} = 84.023$ mm, and circular inlet and outlet sections with diameters $D_{inlet} = 4.1$ mm and $D_{outlet} = 3.708$ mm respectively.

The model was imported in ICEM CFD® (v. 18.0, Ansys, Inc., Canonsburg, PA, USA) and discretized with tetrahedral elements and 5 boundary layers of prism elements close to the walls. The hybrid tetrahedral and prism mesh was created with the Octree method and globally smoothed by imposing 5 smoothing iterations and a quality criterion up to 0.4.

The 3D model was finally exported to Fluent® (v. 18.0, Ansys, Inc. Canonsburg, PA, USA) to perform the CFD simulations (Fig. 2).

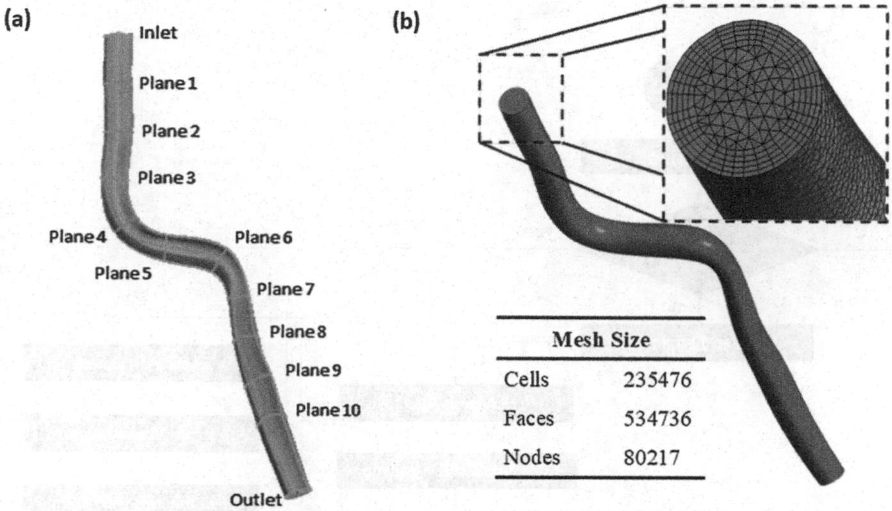

Fig. 2. 3D geometry preparation and meshing. A 3D portion of SFA is built in Rhinoceros® 6.0 (a) and $M = 10$ 2D cross-sections selected for the future ABM analysis. The geometry is then discretized and meshed in preparation for the CFD simulation (b).

2.2 CFD Simulation

Steady-state CFD simulations were performed on the 3D geometry presented in Sect. 2.1. The choice of a steady-state simulation is since at this developmental stage, an inlet pulsatile waveform-based analysis would have been too much time consuming. In addition, the replicated cellular events occur in the time scale of weeks, while the cardiac output waveform is in the order of the seconds. We can then accept that the current ABM implementation is insensitive to the transient of WSS within a single heartbeat.

At the inlet cross-section, a parabolic velocity profile, whose mean velocity was derived from the analysis of patient's Doppler ultrasound image at the femoral artery level [18], was imposed. A reference zero pressure was applied at the outlet cross-section, while no-slip condition was chosen for the arterial walls, here supposed as a rigid structure. Blood was modeled as a non-Newtonian fluid with a density of 1060 kg/m^3 and a viscosity defined by Caputo et al., as in [19].

A pressure-based couple method was adopted, with a least square cell-based scheme for the spatial discretization of the gradient, a second-order scheme for the pressure and a second-order upwind scheme for the momentum spatial discretization.

2.3 ABM Simulation

The following description refers to a single implementation of the ABM, whose basic principles are the same for each loop iteration and for each cross-section selected.

The simulation flowchart is showed in Fig. 3.

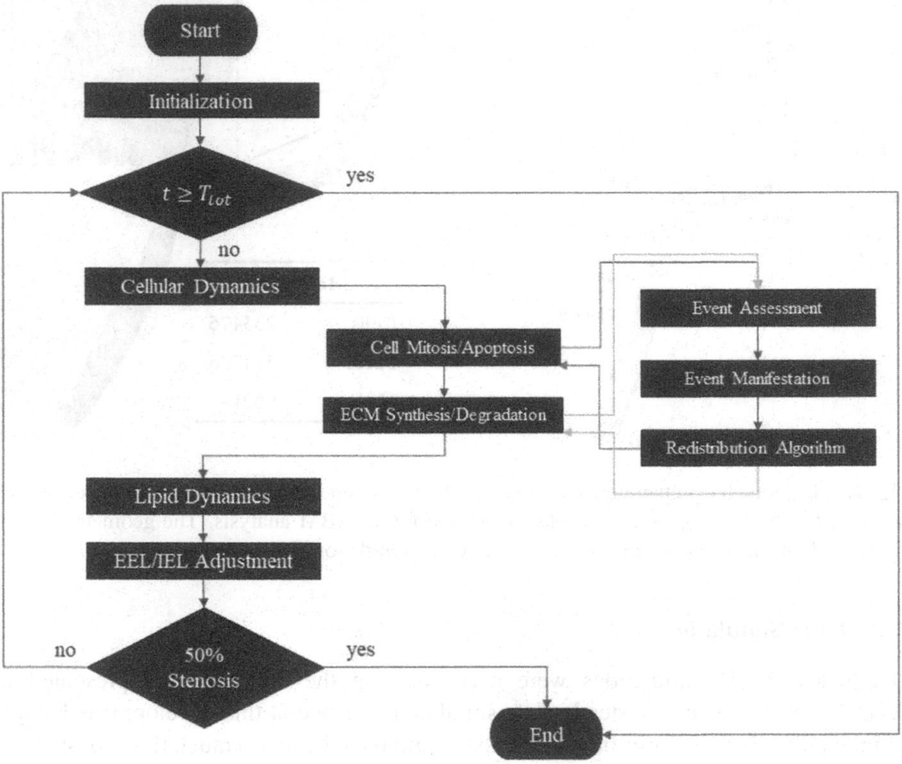

Fig. 3. ABM simulation flowchart.

The ABM, implemented in Matlab® (v. 2016b, MathWorks, Natick, MA, USA), locally simulates the alterations due to atherosclerotic insurgence at cell-tissue level starting from a homeostatic condition. The model was developed assuming that the risk factors promoting the disease were already present. Accordingly, plaque growth in a specific region was exclusively associated with the alteration of WSS profile.

The main structure of the ABM follows the one developed by Garbey et al. [20] in the study of VGB post-surgical adaptation, differing however in agent types and simulated events.

The model is initialized to reflect the initial 2D geometry. Then, the cellular, extra-cellular, and lipid dynamics are performed. The latter are strongly influenced by the hemodynamic profile, which can promote either the maintaining of a homeostatic condition or the progression to a pathological one. Cited dynamics imply a certain level of perturbation within the arterial wall, which adjusts to the new geometrical/compositional conformation.

The simulation arrests either if a level of stenosis greater than 50% is recorded, condition after which a revascularization treatment is commonly performed, or at the end of the chosen follow-up time.

The dynamics of each agent is described with a probability law and the simulation is performed with a Monte Carlo method to introduce a level of randomness that simulates the noise characterizing a true experimental setup. The resulting stochastic nature of the ABM makes necessary to run N independent simulations starting from the same initial condition and taking as output the average of them. A reasonable tradeoff between computational time and minimization of the standard deviation was reached by setting $N = 10$.

Initialization. The ABM lays on a <130 × 130> hexagonal grid such as each site is surrounded by 6 neighbors that, determining its local environment, affect its state.

According to the vessel's anatomy, a 2D circular cross-section of the artery is generated with the wall composed by 3 concentric layers, i.e. tunica intima, media and adventitia. Each site within the wall is occupied by an agent randomly assigned to be a cell or Extra-Cellular Matrix (ECM), according to the relative tissue density. Sites belonging to the lumen or to the portion external to the wall are instead empty.

The process of atherosclerotic formation is triggered in regions interested by a low WSS profile where the altered flow damages the endothelial layer, which consequently allows for lipid infiltration and subsequent accumulation within the arterial wall [2].

Accordingly, the ABM is informed with an initial WSS distribution obtained from CFD simulations and used to associate to each site i at the lumen wall a level of endothelial dysfunction, defined as it follows:

$$D(WSS)^i = D^i = \begin{cases} 1 - \frac{WSS^i}{WSS_0}, & \text{if } WSS^i < WSS_0 \\ 0, & \text{otherwise} \end{cases}. \quad (1)$$

WSS^i is the shear stress recorded at site i and $WSS_0 = 1\,Pa$ is the threshold of WSS below which the atherosclerotic process is initiated. WSS_0 was set in accordance with findings of plaque localization in areas exposed to WSS < 1 [21] and the identification of the physiological range of WSS in the SFA between 1.5–2 Pa [22].

The level of endothelial dysfunction drives the state of alteration, labeled as $A^{i,k}(D^i, d)$, responsible for the atherosclerosis initiation. $A^{i,k}(D^i, d)$ represents the level of alteration originated by the site i and perceived by site k. It is computed for each site belonging to the wall and it propagates instantaneously across it from the i-th site of origin with isotropic diffusion and with a peak of intensity identified by D^i, in formula:

$$A^{i,k}\left(D^i, d\right) = A^{i,k} = D^i * e^{-\frac{1}{2}\left(\frac{d}{4\phi t}\right)^2}, \tag{2}$$

where d is the distance between sites i and k, and ϕ the diffusion constant.

The ensemble of the alteration states defines the level of inflammation I that interests each site k of the wall. The resulting state of inflammation for a given k writes

$$I^k = \sum\nolimits_{i=1}^{N_L} A^{i,k}, \tag{3}$$

where N_L is the number of sites initially belonging to the lumen wall.

Although the endothelial layer was not explicitly modeled, its key role in the pathogenesis of atherosclerosis was considered through the implementation of Eqs. (1)–(3). In presence of an atherogenic WSS profile, the mechanisms of plaque formation are activated, i.e. lipid dynamics and increased cellular activity in the intima, while, with an atheroprotective profile, the homeostatic condition of a healthy artery is maintained.

Agents Dynamics. For designing and tuning purposes, the ABM was first run on a single cross-section for a 2 months follow-up and with a 1-h time step, in correspondence of which the grid is randomly scanned to compute the agents' dynamics. Once the ABM will be coupled to the CFD module, the running time will be the chosen coupling time and the ABM analysis obviously extended to all the M cross-sections.

Every site containing a cell is accessed with a frequency of $T_{cell} = 12\,\text{h}$, while the ones containing ECM every $T_{matrix} = 2\,\text{h}$. To determine whether the potential event is happening, a Monte Carlo method is applied, for which a random number $test \in [0; 1]$ is generated by the CPU and compared with the probability of the event itself, here labeled as p_{event}. If $p_{event} > test$, the event happens, otherwise the algorithm explores the next active site.

The investigated events were cellular mitosis/apoptosis, ECM deposition/degradation and lipids infiltration. Their probability densities are reported below.

First, a baseline density of probability was defined for both mitosis/apoptosis and ECM deposition/degradation with Eqs. (4) and (5) respectively:

$$p_{mit} = p_{apop} = \alpha_1 \tag{4}$$

$$p_{prod} = \beta * p_{deg} = \alpha_4 \tag{5}$$

α_1, α_4 and β were chosen to maintain a homeostatic balance between mitosis/apoptosis and ECM production/degradation. Accordingly, α_1 and α_4 were set to compensate the different cellular and extracellular timeframe, i.e. $\alpha_1 = \frac{1}{T_{matrix}} = 0.05$ and $\alpha_4 = \frac{1}{T_{cell}} = 0.008$. Finally, β was calibrated for each layer to compensate the tendency of the model to preferentially degrade ECM due to the prevalence of ECM on cells. To this aim, the model was run for a 2 months follow-up with several tentative β values and the ratio between final and initial ECM concentration, $\frac{ECM_f}{ECM_i}$, was computed. As the choice must ensure the maintaining of a homeostatic condition, β was chosen by interpolating the $\frac{ECM_f}{ECM_i}$ vs. β plot in correspondence of $\frac{ECM_f}{ECM_i} = 1$, obtaining $\beta = \{1.57, 1.57, 2.5\}$ for intima, media and adventitia, respectively.

Equations (4) and (5) represent the minimum set of equations to reproduce a homeostatic condition. Within the process of plaque formation, i.e. in presence of at least one site at the lumen wall exposed to $WSS < WSS_0$, while media and adventitia maintain baseline activities, cellular mitosis is perturbed within the intima as follows:

$$p_{mit} = \begin{cases} \alpha_1 \cdot (1 + \alpha_2 I^k) \; if \; n_{lip} = 0 \\ \alpha_1 \cdot (1 + \alpha_2 I^k)(1 + \alpha_3 n_{lip})\{1 + \exp(-d^k_{lumen})\} \; if \; n_{lip} \neq 0 \end{cases} \tag{6}$$

Similarly, the production of ECM in intima writes

$$p_{prod} = \begin{cases} \alpha_4 \cdot (1 + \alpha_2 I^k) \; if \; n_{lip} = 0 \\ \alpha_4 \cdot (1 + \alpha_2 I^k)(1 + \alpha_3 n_{lip})\{1 + \exp(-dist^k_{lumen})\} \; if \; n_{lip} \neq 0 \end{cases} \tag{7}$$

$\alpha_2 = 1.5$ drives the perturbation due to the inflammation state I^k of the specific site and $\alpha_3 = 0.1$ accounts for the influence of the surrounding lipids, n_{lip}. Finally, d^k_{lumen} is the distance of the site k from the lumen. With this formulation, the probability of mitosis and ECM production increases with the number of surrounding lipids and the closeness to the lumen.

The process of lipids infiltration only allows one lipid per time step to extravasate with a probability defined as it follows:

$$p_{lipid} = \alpha_5 (1 + I^k)\left\{1 + \alpha_6 \cdot \exp\left(-dist^k_{lip}\right)\right\}\left(1 + \frac{n_{lip}}{\alpha_7}\right) \tag{8}$$

where $\alpha_5 = 0.05$ drives the probability of infiltration, while $\alpha_6 = 10$ accounts for the increased probability of a lipid to occupy a site k near to another lipid agent in the wall, whose distance is $dist^k_{lip}$. Finally, the term $\left(1 + \frac{n_{lip}}{\alpha_7}\right)$ allows for the formation of a lipid cluster and subsequent buildup of the fatty plaque, with $\alpha_7 = 6$ being a normalization constant to maintain the ratio in the interval [0; 1].

Tissue Plasticity and Geometrical Regularization. The perturbation carried by the agents at the level of the wall is recovered by following a principle of free energy minimization. The sites belonging to the wall are re-arranged accordingly.

For example, to free a site for a newly introduced element, the mitotic/synthetic cell moves the surrounding elements either towards the lumen or the external support, respectively if the cell belongs to the intima or to the media/adventitia. The same principle applies in case of vacation of a site, i.e. cell apoptosis or ECM degradation, but with a reverse movement. Finally, the structure of the model is regularized and adjusted at each time step to obtain a smooth profile at each interface.

2.4 Retrieval of the New 3D Geometry

In an optical of an ABM-CFD coupling, $N = 10$ independent simulations were conducted for each of the $M = 10$ pre-selected cross-sections of the artery. The new 3D geometry was generated from the ensemble of the ABM solutions of each cross-section that mostly resembled the corresponding average solution about (i) lumen radius, (ii) external radius, and (iii) plaque size.

For each ABM output labeled as i = {1,..., N} lumen and external radii and plaque thickness as function of ϑ were computed once assumed that they can be represented in polar coordinates. Being ϑ the angle in degrees, the quantities were indicated as $R_j^i(\vartheta)$, with j = 1, 2, 3, respectively. The corresponding deviation, Δ^i, from the average configuration, $\overline{R_j(\vartheta)}$, was defined in Eq. (9) as:

$$\Delta^i = \sum_{j=1}^{3} \int_0^{2\pi} w_j \sqrt{(R_j(\vartheta) - \overline{R_j(\vartheta)})^2} d\vartheta, \quad (9)$$

where w_j is the weight of the j-th measurement.

The cross-section in output of the ABM associated with the lowest Δ was selected and the criterion repeated for all the $M = 10$ cross-sections to obtain the skeleton of the new geometry. The 3D geometry was finally reconstructed in Rhinoceros® by lofting the lumen profiles of said sections.

Finally, a sensitivity analysis was performed to define the best coupling time between CFD simulations and ABM, i.e. at which time step the ABM simulation needs to be paused to update the hemodynamic conditions according to the geometrical changes. A reasonable coupling time should constitute a good compromise between accuracy of the results and computational effort required.

Three different cases were tested on a 14 days follow-up period. In the first two cases, coupling times of 7 and 3.5 days were adopted for the whole follow-up. In the third case, an adaptive method was used, in which the first coupling was performed after 7 days, and then every 3.5 days.

3 Results and Discussion

3.1 Atherosclerotic Plaque Generation

The ability of the ABM to accurately replicate the plaque formation was evaluated on a single cross-section and on a 2 months follow-up. To reach a fast evaluation of the present model, the dynamics of plaque development was on purpose accelerated.

By using a 16.00 GB RAM CPU, Intel® Core™ i7-4790, with 4 Cores and 8 Logical Processors, the mean time of computation was $t\bar{c}$ = 25.87 h for each simulation.

Basing on the ABM results, (i) output morphology of the model, and (ii) its sensitivity to the WSS profile were analyzed.

Output Morphology and Robustness. Under atherogenic conditions, the ABM is able to generate an asymmetric lipid-rich atherosclerotic plaque with features resembling histological evidences, as shown in Fig. 4.

In the histology shown in Fig. 4(b), a thick layer of fibrous intimal tissue covers the lipid core, while it is not present in the cross-section generated by the ABM in Fig. 4 (a). This is since said layer is thought to form once the lipid infiltration process stops and the only active dynamics gets back to be cells and ECM related only. However, in the presented simulations, at the end of the follow-up, lipids were still actively migrating into the intima and building the core up.

Figure 4(a) also highlights that, differently from the ABM developed by Bhui et al. [17], the plaque volumetric growth was due to lipid accumulation, SMCs proliferation, as also reported in [24], and ECM production, being ECM the major extracellular component of fibroatheromas after lipids [25].

In Fig. 5, the temporal dynamics of intimal, Fig. 5(a), lumen, Fig. 5(b), medial, Fig. 5(c), and adventitial area, Fig. 5(d), normalized on their initial value are provided. As expected, a decrease in lumen area is appreciable. Its monotonic trend is the consequence of having replicated only the inward thickening of the intima, neglecting the initial enlargement of the vessel in response to early stages of plaque formation [26], which, instead, was considered in [17]. Also, coherently with the implemented baseline cellular and ECM dynamics, medial and adventitial areas remained stable.

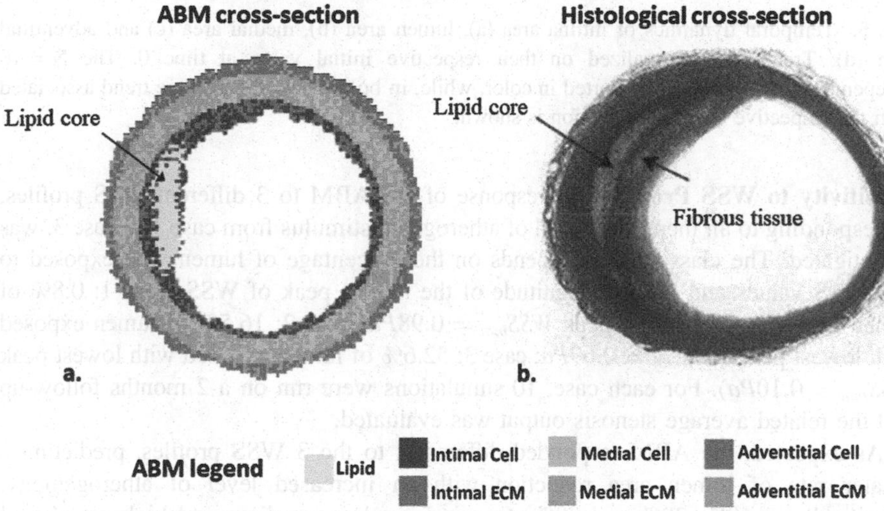

Fig. 4. ABM-Histology Comparison. (a) An ABM cross-section after 2 months of follow-up is compared with (b) a histology of coronary fibrous cap atheroma in a 24-years-old man [23].

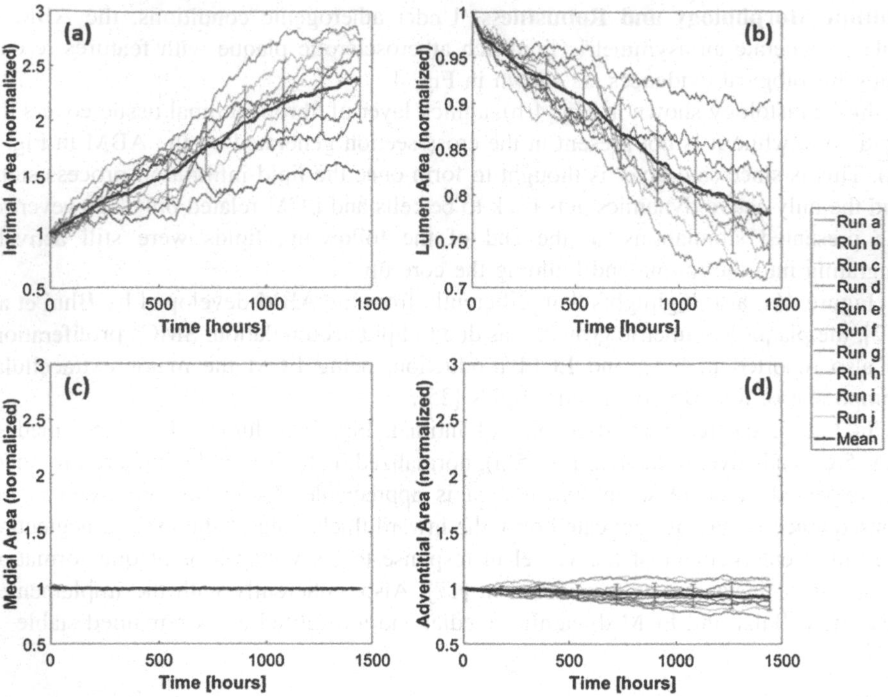

Fig. 5. Temporal dynamics of intima area (a), lumen area (b), medial area (c) and adventitial area (d). Trends are normalized on their respective initial value at time 0. The N = 10 independent simulations are reported in color, while, in bold black, the average trend associated with the respective standard deviation is shown.

Sensitivity to WSS Profile. The response of the ABM to 3 different WSS profiles, corresponding to an increasing level of atherogenic stimulus from case 1 to case 3, was investigated. The classification depends on the percentage of lumen wall exposed to low WSS values and on the magnitude of the lowest peak of WSS (case 1: 0.8% of lumen exposed with lowest peak $WSS_{min} = 0.98Pa$; case 2: 16.5% of lumen exposed with lowest peak $WSS_{min} = 0.69Pa$; case 3: 52.6% of lumen exposed with lowest peak $WSS_{min} = 0.10Pa$). For each case, 10 simulations were run on a 2 months follow-up and the related average stenosis output was evaluated.

As expected, the ABM responded differently to the 3 WSS profiles, predicting a greater rate of lumen area reduction with an increased level of atherogenicity. Specifically, a 10%, 20% and 80% stenosis for a low, medium and high severity of WSS distribution were obtained as reported in Fig. 6, where case 1 is the least prone to generate the plaque, while case 3 the most prone.

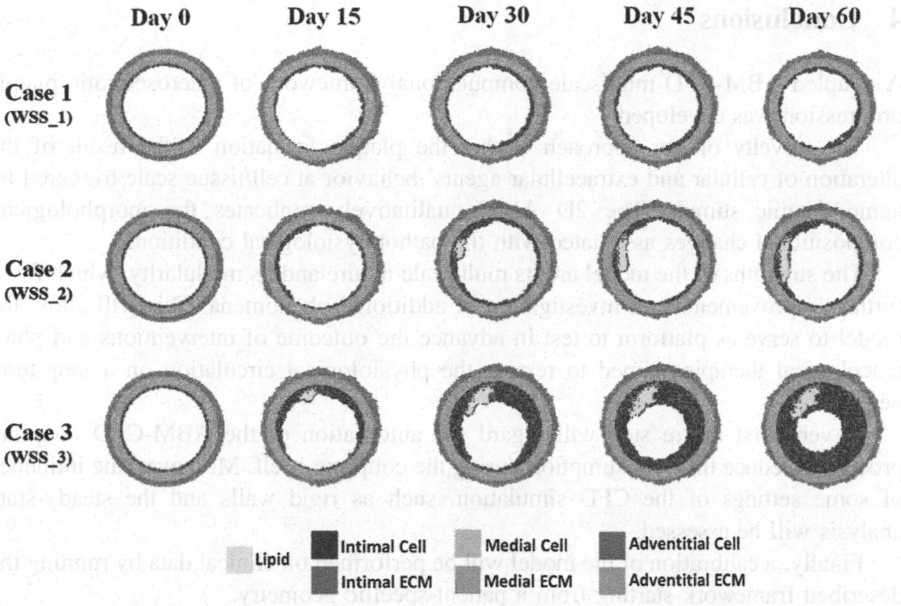

Fig. 6. Sensitivity to WSS Profile. The geometrical evolution of an ABM cross-section on a 2 months follow-up with 15 days-time step is provided for an increasing level of atherogenicity. For each case, one representative cross-section is chosen among 10 independent simulations.

3.2 ABM-CFD Coupling

The most suitable coupling profile between ABM and CFD was studied by running the complete framework described in Fig. 1 with a 14 days follow-up and with the modalities outlined in Sect. 2.4.

The solution of the complete framework at day 14 was obtained in about 20 h, 21.5 h and 20.5 h respectively when coupling every 7 days, 3.5 days and when applying the adaptive coupling time. Considering the average computation time of the ABM to be about 30 min per 1 simulated day, the different time consumption among the tested cases lies in the time required at each coupling step to generate the updated 3D geometry, mesh the model and perform the CFD simulation, estimated to be about 50 min.

From a preliminary analysis based on the temporal dynamics of lumen area, the choice of the coupling period did not imply any difference in the output. However, some differences were locally recorded in planes originally exposed to a WSS profile close to 1 Pa. In these planes, being the boundary between non-pathological and pathological condition very thin, a longer coupling time is not suitable to catch potential switching behaviors between the two conditions. Consequently, the choice of the coupling should follow the trends of these planes, identified as uncertain situations.

The latter suggests the use of a hybrid coupling time where the early phase of plaque development is simulated with a 3.5 days-coupling time, which is then extended to 7 days when the system has stabilized. The latter coupling scheme will be tested in future developments.

4 Conclusions

A coupled ABM-CFD multiscale computational framework of atherosclerotic plaque progression was developed.

The novelty of our approach is that the plaque formation is the result of the alteration of cellular and extracellular agents' behavior at cell/tissue scale triggered by hemodynamic stimuli. The 2D ABM qualitatively replicates the morphological/compositional changes associated with the pathophysiological condition.

The strengths of the model are its multiscale nature and its modularity, which favor further improvements and investigation of additional phenomena. This will allow the model to serve as platform to test in advance the outcome of interventions and pharmacological therapies aimed to restore the physiological circulation on a long-term perspective.

A very first future step will regard the automation of the ABM-CFD coupling process to reduce time consumption during the coupling itself. Moreover, the influence of some settings of the CFD simulation, such as rigid walls and the steady-state analysis will be assessed.

Finally, a calibration of the model will be performed on clinical data by running the described framework starting from a patient-specific geometry.

Acknowledgments. National Institutes of Health (NIH) grant U01HL119178-01 has supported this work. Claudio Chiastra and Monika Colombo have been supported by Fondazione Cariplo, Italy (Grant number 2017-0792, TIME).

References

1. Kakkar, A., et al.: Percutaneous versus surgical management of lower extremity peripheral artery disease. Curr. Atheroscler. Rep. **17**(2), 479–487 (2015)
2. Bentzon, J., et al.: Mechanisms of plaque formation and rupture. Circ. Res. **114**(12), 1852–1866 (2014)
3. Mohler II, E.R.: Peripheral arterial disease – identification an implications. Arch. Intern. Med. **163**, 2306–2314 (2003)
4. Malas, M.B., et al.: Comparison of surgical bypass with angioplasty and stenting of superficial femoral artery disease. J. Vasc. Surg. **59**(1), 129–135 (2014)
5. Tahir, H., et al.: Multi-scale simulations of the dynamics of in-stent restenosis: impact of stent deployment and design. Interface Focus **1**(3), 365–373 (2011)
6. Garbey, M., et al.: A versatile hybrid agent-based, particle and partial differential equations method to analyze vascular adaptation. Biomech. Model. Mechanobiol. (2018). https://doi.org/10.1007/s10237-018-1065-0
7. Caiazzo, A., et al.: A complex automata approach for in-stent restenosis: two-dimensional multiscale modeling and simulations. J. Comput. Sci. **2**(1), 9–17 (2011)
8. Boyle, C., et al.: Computational simulation methodologies for mechanobiological modelling: a cell-centered approach to neointima development in stents. Philos. Trans. R. Soc. A: Math. Phys. Eng. Sci. **368**(1921), 2919–2935 (2010)
9. Sankaranarayanan, M., et al.: Computational model of blood flow in the aorto-coronary bypass graft. Biomed. Eng. Online **4**, 4–14 (2005)

10. Nolan, D.R., et al.: An investigation of damage mechanisms in mechanobiological models of in-stent restenosis. J. Comput. Sci. **24**, 132–142 (2018)
11. Tahir, H., et al.: Modelling the effect of a functional endothelium on the development of in-stent restenosis. PLoS One **8**(6), e66138 (2013). https://doi.org/10.1371/journal.pone.0066138
12. Zun, P., et al.: A comparison of fully-coupled 3D in-stent restenosis simulations to in-vivo data. Front. Physiol. **8**, 284 (2017)
13. Curtin, A.E., et al.: An agent-based model of the response to angioplasty and bare-metal stent deployment in an atherosclerotic blood vessel. PLoS One **9**(4), e94411 (2014). https://doi.org/10.1371/journal.pone.0094411
14. An, G., et al.: Agent-based models in translational systems biology. Wiley Interdiscip. Rev. Syst. Biol. Med. **1**(2), 159–171 (2013)
15. Poston, R.N., et al.: Typical atherosclerotic plaque morphology produced in silico by an atherogenesis model based on self-perpetuating propagating macrophage recruitment. Math. Model. Nat. Phenom. **2**(2), 142–149 (2007)
16. Pappalardo, F., et al.: Modeling immune system control of atherogenesis. Bioinformatics **24** (15), 1715–1721 (2008)
17. Bhui, R., et al.: An agent-based model of leukocyte transendothelial migration during atherogenesis. PLoS Comput. Biol. **13**(5), e1005523 (2017). https://doi.org/10.1371/journal.pcbi.1005523
18. Colombo, M., et al.: Hemodynamic analysis of patient-specific superficial femoral arteries: from computed tomography images to computer simulations. Master's thesis (2017). https://www.politesi.polimi.it/handle/10589/138002
19. Caputo, M., et al.: Simulation of oxygen transfer in stented arteries and correlation with in-stent restenosis. Int. J. Numer. Methods Biomed. Eng. **29**, 1373–1387 (2013)
20. Garbey, M., et al.: Vascular adaptation: pattern formation and cross validation between an agent based model and a dynamical system. J. Theor. Biol. **429**, 149–163 (2017)
21. Samady, H., et al.: Coronary artery wall shear stress is associated with progression and transformation of atherosclerotic plaque and arterial remodeling in patients with coronary artery disease. Circulation **124**, 779–788 (2011)
22. Schlager, O., et al.: Wall shear stress in the superficial femoral artery of healthy adults and its response to postural changes and exercise. Eur. J. Vasc. Endovasc. Surg. **41**(67), 821–827 (2011)
23. Rhodin, JAG.: Architecture of the Vessel Wall. Comprehensive Physiology, Supplement 7: Handbook of Physiology, The Cardiovascular System, Vascular Smooth Muscle, pp. 1–31 (2014)
24. Amanda, C., et al.: Role of smooth muscle cells in the initiation and early progression of atherosclerosis. Arterioscler. Thromb. Vasc. Biol. **28**(5), 812–819 (2008)
25. Stary, H., et al.: A definition of advanced types of atherosclerotic lesions and a histological classification of atherosclerosis. A report from the Committee on Vascular Lesions of the Council on Arteriosclerosis, American Heart Association. Circulation **92**(5), 1355–1374 (1995)
26. Glagov, S., et al.: Compensatory enlargement of human atherosclerotic coronary arteries. N. Engl. J. Med. **316**(22), 1371–1375 (1987)

Special Aspects of Hybrid Kinetic-Hydrodynamic Model When Describing the Shape of Shockwaves

Yurii Nikitchenko, Sergei Popov, and Alena Tikhonovets(✉)

Moscow Aviation Institute, Volokolamskoe shosse, 4, 125993 Moscow, Russia
nikitchenko7@ya.ru, flowmech@mail.ru,
tikhonovets.a.v@gmail.com

Abstract. A mathematical model of the flow of a polyatomic gas containing a combination of the Navier-Stokes-Fourier model (NSF) and the model kinetic equation of polyatomic gases is presented. At the heart of the hybrid components is a unified physical model, as a result of which the NSF model is a strict first approximation of the model kinetic equation. The model allows calculations of flow fields in a wide range of Knudsen numbers (*Kn*), as well as fields containing regions of high dynamic nonequilibrium. The boundary conditions on a solid surface are set at the kinetic level, which allows, in particular, to formulate the boundary conditions on the surfaces absorbing or emitting gas. The hybrid model was tested. The example of the problem of the shock wave profile shows that up to Mach numbers $M \approx 2$ the combined model gives smooth solutions even in those cases where the sewing point is in a high gradient region. For the Couette flow, smooth solutions are obtained at $M = 5$, $Kn = 0.2$. A model effect was discovered: in the region of high nonequilibrium, there is an almost complete coincidence of the solutions of the kinetic region of the combined model and the "pure" kinetic solution.

Keywords: Polyatomic gases · Navier-Stokes-Fourier model · Model kinetic equation · Hybrid model · Dynamic nonequilibrium · Sorption surfaces

1 Introduction

Modern aerospace and nanotechnologies are in need of the improved computational methods and mathematical models of gas flow in a wide range of Mach and Knudsen numbers. One of the areas to which the present paper belongs is related to the development of hybrid or composed flow models. These models involve the combined use of the methods of molecular kinetic theory and continuum mechanics.

A number of models suggest the separation of the computational domain of geometric space into hydrodynamic and kinetic subdomains, for example [1–3]. In the hydrodynamic subdomain, the Navier – Stokes equations are used; in the kinetic one, the model BGK kinetic equation with a certain numerical implementation, or statistical models, are used.

J. M. F. Rodrigues et al. (Eds.): ICCS 2019, LNCS 11539, pp. 424–435, 2019.
https://doi.org/10.1007/978-3-030-22747-0_32

The models of [4, 5] distinguish hydrodynamic and kinetic subdomains in the space of velocities: hydrodynamic for "slow" molecules and kinetic for "fast" molecules. In the subdomain of "slow" molecules, the Euler or Navier-Stokes models are used, and for fast molecules the BGK equation.

The BGK model was obtained for the weight function (the velocity distribution function of molecules) of a monatomic gas. Its collision integral corresponds to a gas with a Prandtl number $Pr = 1$ [6]. Thus, this model implies some hypothetical gas. The continuous transition from this model to hydrodynamics is very difficult without the use of artificial smoothing procedures [1].

The model [7] in the kinetic subdomain of the geometric space uses the S-model [6], which distinguishes it favorably from the models cited above. Prandtl number of the S-model is $Pr = 2/3$, which corresponds to monatomic gas. When calculating the flow near a rough surface, satisfactory results were obtained even in the transition region of the flow. The limitation of this model lies in the fact that the consistency of the kinetic and hydrodynamic description exists only for monatomic gases.

The present work has as its goal the development of a hybrid model of the flow of polyatomic gases. The Navier-Stokes-Fourier model (NSF) [8] is composed with the model kinetic equation of polyatomic gases (MKE) [9]. The NSF model is a rigorous first approximation of the system of moment equations for polyatomic gases [10]. When obtaining this approximation, nonequilibrium quantities (stress deviator, heat fluxes, difference of translational and rotational temperatures) in their moment equations were set so small that their second and higher degrees can be neglected. Flows that satisfy the conditions of the first approximation will be called weakly nonequilibrium.

The relaxation terms of MKE are obtained using the polyatomic gas system used in the development of the NSF model. The coefficient of bulk viscosity of the NSF model is presented in such a way that this model is the first approximation in the above sense of the MKE model. Thus, both composable models are based on a single physical model.

Section 2 provides basic assumptions and notations used in the paper. Section 3 provides a general description of the hybrid model. Section 4 provides the method of composing the kinetic and hydrodynamic models. Section 5 discusses a particular case of using the model to describe the profile of a plane shock wave.

2 Basic Assumptions and Notations

We consider the flow of monocomponent perfect gases. All expressions are written for polyatomic gases. In the case of monatomic gases, the expressions remain valid after obvious transformations.

The index writing of tensor expressions is used. A repeated Greek index indicates a summation from 1 to 3. The velocity space integral is denoted by $\int \ldots d\mathbf{c} \equiv \int_{-\infty}^{+\infty} dc_1 \int_{-\infty}^{+\infty} dc_2 \int_{-\infty}^{+\infty} \ldots dc_3$ (Table 1).

Table 1. The notations used

δ_{ij}	Kronecker delta
$t,\ x_i,\ \xi_i$	Time, geometric coordinate and molecular velocity
$m_0,\ n,\ \rho = m_0 n$	Molecular mass, molecular concentration and gas density
$u_i,\ c_i = \xi_i - u_i$	Group (macroscopic) and thermal molecular velocities
f	Weight function (molecular velocity distribution function)
$c_p,\ c_v,\ \gamma,\ \mathrm{k},\ R = \mathrm{k}/m_0$	Heat capacities at constant pressure and volume, specific heat ratio, Boltzmann constant, specific gas constant
$T_t,\ T_r,$ $T = 1.5(\gamma-1)T_t + 0.5(5-3\gamma)T_r$	Temperatures of translational and rotational degrees of freedom of molecules, thermodynamic temperature
$P_{ij},\ p_{ij} = P_{ij} - \delta_{ij} R\rho T_t$	Total and nonequilibrium stresses
$p = P_{\alpha\alpha}/3 = R\rho T_t$	Mechanical pressure
$\phi_i,\ \omega_i$	Heat fluxes caused by the transport of energy of thermal motion on the translational and rotational degrees of freedom of molecules
$q_i = \phi_i + \omega_i$	Full heat flux
M, Kn, Pr	Mach, Knudsen and Prandtl numbers

In considered models the following is accepted: $Pr = 4\gamma/(9\gamma-5)$.

3 Hybrid Model

3.1 Hydrodynamic Model

The NSF model, which differs from the traditional system of conservation equations in the Navier-Stokes approximation by the presence of the coefficient of bulk viscosity in the equations of nonequilibrium stress, is considered as a hydrodynamic model. In terms of [8], the system of equations of this model has the following form:

$$\begin{cases} \frac{\partial \rho}{\partial t} + \frac{\partial \rho u_\alpha}{\partial x_\alpha} = 0 \\ \frac{\partial u_i}{\partial t} + u_\alpha \frac{\partial u_i}{\partial x_\alpha} + \frac{1}{\rho}\frac{\partial P_{i\alpha}}{\partial x_\alpha} = 0 \qquad i = 1,\ 2,\ 3 \\ \frac{\partial T}{\partial t} + u_\alpha \frac{\partial T}{\partial x_\alpha} + (\gamma-1)\frac{P_{\alpha\beta}}{\rho}\frac{\partial u_\alpha}{\partial x_\beta} + \frac{1}{c_v \rho}\frac{\partial q_\alpha}{\partial x_\alpha} = 0 \end{cases} \tag{1}$$

where

$P_{ij} = \delta_{ij} R\rho T - \mu\left(\frac{\partial u_i}{\partial x_j} + \frac{\partial u_j}{\partial x_i}\right) + \delta_{ij}\frac{2}{3}\left(1 - \frac{5-3\gamma}{2}Z\right)\mu\frac{\partial u_\alpha}{\partial x_\alpha},\ q_i = -\frac{c_p}{\mathrm{Pr}}\mu\frac{\partial T}{\partial x_i}.$

The viscosity coefficient μ and the parameter Z are determined by the dependencies which are used in the MKE model [9], but with preservation of the order of approximation of the NSF model, i.e.$\mu = \mu(T_t = T),\ Z = Z(T_t/T_r = 1)$.

3.2 Kinetic Model

As a kinetic model of the flow, the MKE model [9] is used, built for a single-particle weight function, the phase space of which is supplemented by the energy of rotational degrees of freedom ε: $f(t, \mathbf{x}, \xi, \varepsilon)$. After reducing the dimension of the weight function, the kinetic equations of the model take the form:

$$\frac{\partial}{\partial t}\begin{vmatrix} f_t \\ f_r \end{vmatrix} + \xi_\alpha \frac{\partial}{\partial x_\alpha}\begin{vmatrix} f_t \\ f_r \end{vmatrix} = \frac{p}{\mu}\begin{vmatrix} f_t^+ - f_t \\ f_r^+ - f_r \end{vmatrix} \tag{2}$$

where

$$f_t = \int f d\varepsilon, f_r = \int \varepsilon f d\varepsilon,$$

$$f_t^+ = \frac{n}{(2\pi R T_t^+)^{3/2}} \exp\left(-\frac{c^2}{2RT_t^+}\right)\left(1 + \frac{\phi_\alpha c_\alpha}{3\rho (RT_t^+)^2}\left(\frac{c^2}{5RT_t^+} - 1\right)\right)$$

$$f_r^+ = \frac{5-3\gamma}{2(\gamma-1)} kT_r^+ f_t^+, T_t^+ = T + 0.5(5-3\gamma)(1-1/Z)(T_t - T_r)$$

$$T_r^+ = T - 1.5(\gamma-1)(1-1/Z)(T_t - T_r), T_t = (3Rn)^{-1}\int c^2 f_t d\mathbf{c}, \phi_i = 0.5 m_0 \int c_i c^2 f_t d\mathbf{c}$$

$$T_r = 2(\gamma-1)((5-3\gamma)R\rho)^{-1}\int f_r d\mathbf{c}, \phi_i = 0.5 m_0 \int c_i c^2 f_t d\mathbf{c},$$

The viscosity coefficient $\mu = \mu(T_t)$ and the parameter $Z = Z(T_t, T_r)$ showing the amount of intermolecular collisions per one inelastic collision are considered as free parameters of the model.

The remaining moments of the weight function required for the hybrid model are defined as:

$$n = \int f_t d\mathbf{c}, \tag{3}$$

$$u_i = n^{-1}\int \xi_i f_t d\mathbf{c}, \tag{4}$$

$$P_{ij} = m_0 \int c_i c_j f_t d\mathbf{c}, \tag{5}$$

$$\omega_i = \int c_i f_r d\mathbf{c}. \tag{6}$$

In [9], the results of calculations using MKE were compared with the well-studied R-model [11, 12] and experimental data [13, 14]. We obtained a satisfactory agreement between the calculated profile of a plane hypersonic shock wave and experimental data even for such a "thin" parameter as the temperature of the rotational degrees of freedom. In the test calculations of this work, the MKE model that is not composed with the hydrodynamic model is considered as a reference.

It should be noted that the models are well consistent, since the NSF model is a strong first approximation of the kinetic model.

4 The Method of Composing Kinetic and Hydrodynamic Models

One of the applications of the hybrid model involves the application of the kinetic model in strongly nonequilibrium domains of the flow field and the hydrodynamic model in other domains.

Another application relates to weakly nonequilibrium flows near active (gas-absorbing or gas-emitting) surfaces. In this case, the kinetic model is necessary only for the formation of boundary conditions on the surface. In Fig. 1, the schemes of the computational domain for both cases are presented: variants A and B, respectively. The vertical line on variant B denotes a streamlined surface.

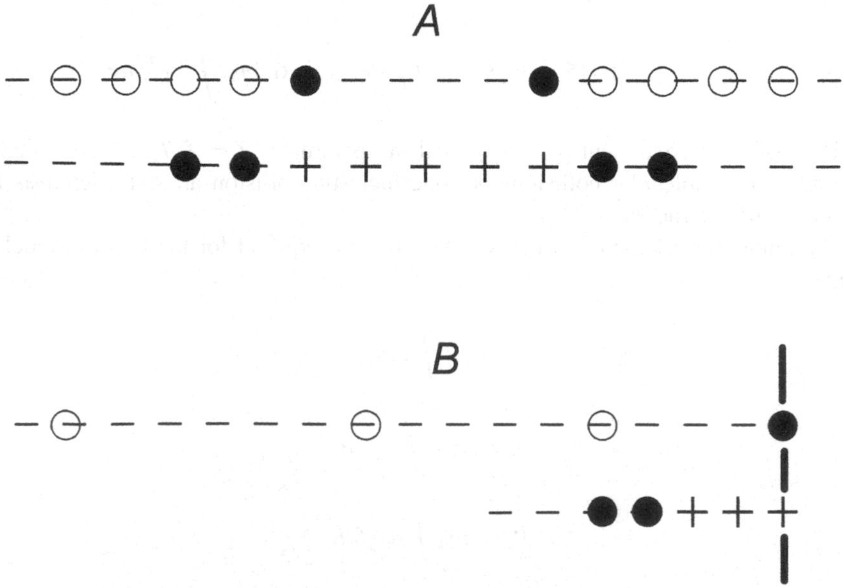

Fig. 1. Schemes of computational domains. A – flow region that does not interact with a solid surface, B – near-wall flow region, O – nodes of the hydrodynamic model, + – nodes of the kinetic model, ● – models joining nodes

Without loss of generality, we consider a one-dimensional stationary flow field with a geometrical coordinate $x_i \equiv x$, and a velocity coordinate corresponding to it. It is supposed that a finite-difference method is used for a numerical solution. Derivatives in system (1) are approximated by central differences on three nodes, in system (2) by one-sided differences upstream, also on three nodes. Note that the direction of flow in the kinetic equations is determined by the direction of the molecular velocity. In this case, it is a speed ξ_x that has two opposite directions: $\xi_x > 0$ and $\xi_x < 0$. Consequently, there are two multidirectional difference schemes. Such a discrete analogue of the computational domain will later be used for numerical tests.

In both variants on Fig. 1, the computational domain is shown twice, separately for the hydrodynamic (open circles) and kinetic (crosses) models. In option A, the solution area of the hydrodynamic solution is divided into two subdomains. The boundary conditions of the left subdomain are formed in the node (the joining node), indicated by a black dot and belonging to the area of the kinetic solution. For the selected differential template, one node is sufficient. Values ρ, u_x, T in this node are defined as the moments of the weight function calculated in the kinetic domain. Similar is the solution in the right subdomain of the hydrodynamic solution. When describing a flow in the near-wall region, one hydrodynamic subdomain is sufficient. In variant A, a kinetic subdomain is located between the hydrodynamic subdomain and the wall (not shown in Fig. 2).

The boundary conditions of the kinetic solution are formed in the nodes of the hydrodynamic domain (black circles): two nodes in each hydrodynamic subdomain for the corresponding ($\xi_x > 0$ or $\xi_x < 0$) difference patterns. Since the hydrodynamic model is less informative than the kinetic model, an approximating weight function is used to reconstruct the weight function in the nodes. In the case of a near-wall flow, the weight function is determined in the node boundary with the wall, which is determined by the law of interaction of molecules with a solid surface.

Taking into account the order of approximation of the hydrodynamic model, as an approximating weight function, it is advisable to take the expansion of the equilibrium, Maxwell function. Such an expansion is used in a number of works, for example [7], for monatomic gases. In the case of polyatomic gas for functions f_{At} и f_{Ar} similar expansions lead to the expressions [9]:

$$f_{At} = f_M \left(1 + \frac{1}{\rho (RT_t)^2} \left(\frac{1}{2} p_{\alpha\beta} c_\alpha c_\beta + \phi_\alpha \left(\frac{c^2}{5RT_t} - 1 \right) c_\alpha \right) \right), \tag{7}$$

$$f_{Ar} = kT_r \left(\frac{5 - 3\gamma}{2(\gamma - 1)} f_{At} + f_M \frac{\omega_\alpha c_\alpha}{\rho R^2 T_t T_r} \right), \tag{8}$$

$$f_M = \frac{n}{(2\pi RT_t)^{3/2}} \exp\left(-\frac{c^2}{2RT_t} \right) \tag{9}$$

The macroparameters of these expressions are determined by the hydrodynamic model and are considered in the appropriate approximation:

$$T_r = T, p_{ij} = -\mu\left(\frac{\partial u_i}{\partial x_j} + \frac{\partial u_j}{\partial x_i}\right) + \delta_{ij}\frac{2}{3}\left(1 - \frac{5-3\gamma}{2}Z\right)\mu\frac{\partial u_\alpha}{\partial x_\alpha}\varphi = -\frac{15}{4}R\mu\frac{\partial T}{\partial x_i},$$
$$\omega_i = -\frac{5-3\gamma}{2(\gamma-1)}R\mu\frac{\partial T}{\partial x_i} \tag{10}$$

Variant B in Fig. 1 assumes a solution of the NSF model in the entire near-wall computational region. This allows to build computational grids with hydrodynamic steps, which is fundamentally important for small Kn. The grid of the kinetic solution with a step of the order of the mean free path of the molecules is constructed within the last, near-wall step of the hydrodynamic grid. Macroparameters, ρ, u_x, T in the kinetic model joining nodes are obtained by interpolating the hydrodynamic solution. In this case, the role of the kinetic model is reduced only to the formation of boundary conditions on a solid surface for the hydrodynamic model.

After calculating the weight function of molecules moving to the surface, the weight function of the reflected molecules is restored at the boundary node. For this, any law of interaction of molecules with the surface can be used. For example, for chemo or cryosorbing surfaces, the diffuse reflection law can be used, taking into account the mass absorption coefficient [15].

At the next stage, the weight function of the reflected molecules is calculated and macroparameters are calculated in the kinetic domain. For the general, hydrodynamic solution, only the boundary node macroparameters are used. The kinetic model is used only to form the boundary conditions of the hydrodynamic model.

Variant B was considered in [16] for the Couette flow. It is shown that the solution is smooth in a wide range of Knudsen and Mach numbers and agrees well with the "pure" kinetic solution. In this work, in connection with its orientation, the near-wall flows were not considered.

5 Numerical Tests. The Problem of the Structure of the Shockwave

The problem is solved in a stationary formulation and is formulated as follows. On the boundaries of the computational domain, the Rankine-Hugoniot conditions are set. The size of the computational domain is about one hundred of free paths of the molecule in the undisturbed flow:

$$\lambda_\infty = 3.2\mu_\infty(2\pi RT_\infty)^{-1/2}/\rho_\infty \tag{11}$$

The system of equations of the *MKE* model is transformed as follows:

$$\xi_x\frac{\partial}{\partial x}\begin{vmatrix} f_t \\ f_r \end{vmatrix} = \frac{p}{\mu}\begin{vmatrix} f_t^+ - f_t \\ f_r^+ - f_r \end{vmatrix}. \tag{12}$$

The transformation of the functions and parameters entering into (12) is obvious. The system of equations of the *NSF* model for this problem:

$$\begin{cases} \rho u_x = \rho_\infty u_{x\infty} \\ \rho u_x \partial u_x/\partial x + \partial P_{xx}/\partial x_x = 0 \\ \rho u_x \partial T/\partial x + (\gamma - 1)P_{xx}\partial u_x/\partial x_x + c_v^{-1}\partial q_x/\partial x = 0 \end{cases} \quad (13)$$

In all calculations, approximations of the viscosity coefficient were taken as $\mu = \mu(T_t^s)$ for the kinetic model and $\mu = \mu(T^s)$ for the hydrodynamic one. The power s was chosen from considerations of the best coincidence of the density profile with the experimental profiles of [13, 14]. The parameter Z approximations for various flow regimes are taken from [9, 12, 17]. Difference schemes are as described in Sect. 4, variant A. The grid pitch was $\approx 0.1\lambda_\infty$.

To solve the system of hydrodynamic equations, we used the stationary Thomas method (sweep method) on a three-diagonal matrix. For the kinetic equations, a stationary solution method was also used with differences up the molecular flow.

The calculated shock wave profiles of the hybrid model were compared with the shock wave profiles of the *MKE* and *NSF* models. Calculations showed that the greatest disarrangement between the shock wave profiles of the hybrid model and the *MKE*

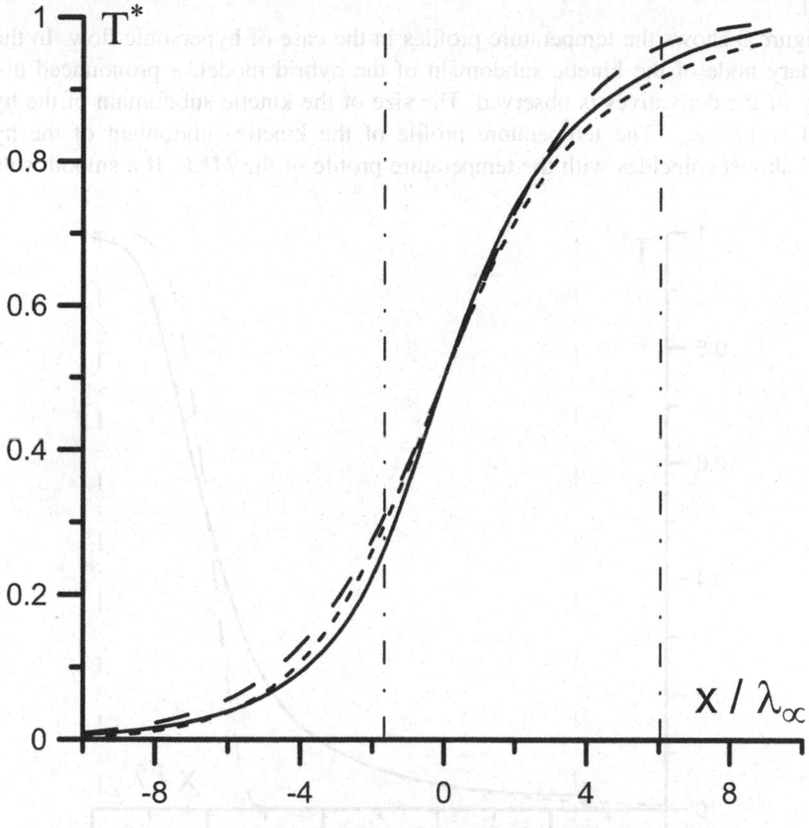

Fig. 2. The referenced temperature profiles in a plane shock wave of a diatomic gas. $M_\infty = 1.55$. The solid line is the hybrid model; fine dotted line is model *MKE*; large dotted line is *NSF* model; vertical dash-dotted lines are the boundaries of the kinetic subdomain of the hybrid model

model takes place on the temperature profiles. Density and group velocity profiles agreed much better. In the following, only temperature profiles referenced to a single segment T^* will be considered.

In the region of moderate Mach numbers, the hybrid model produced smooth temperature profiles, although in the kinetic solution subdomain, some difference was observed from the temperature profiles of the *MKE* model. Figure 2 shows temperature profiles for $M_\infty = 1.55$.

The size of the kinetic subdomain of the hybrid model was $7.8\lambda_\infty$. The joining nodes of the models were in the high gradient subdomain. With an increase in the size of the kinetic subdomain, the corresponding temperature profile became closer to the temperature profile of the *MKE* model. Analysis of the second temperature derivatives at the joining nodes did not reveal a discontinuity of the first derivatives, that is, a break in the graph.

This type of solution was observed until $M_\infty \approx 2$. For larger Mach numbers, even for sufficiently large sizes of the kinetic subdomain ($20 \div 30\ \lambda_\infty$), a discontinuity of derivatives appeared at its boundary nodes. At the same time, the temperature profile of the kinetic region of the hybrid model came close to the temperature profile of the *MKE* model.

Figure 3 shows the temperature profiles in the case of hypersonic flow. In the left boundary node of the kinetic subdomain of the hybrid model, a pronounced discontinuity of the derivatives is observed. The size of the kinetic subdomain of the hybrid model is $17.2\ \lambda_\infty$. The temperature profile of the kinetic subdomain of the hybrid model almost coincides with the temperature profile of the *MKE*. If a smooth solution

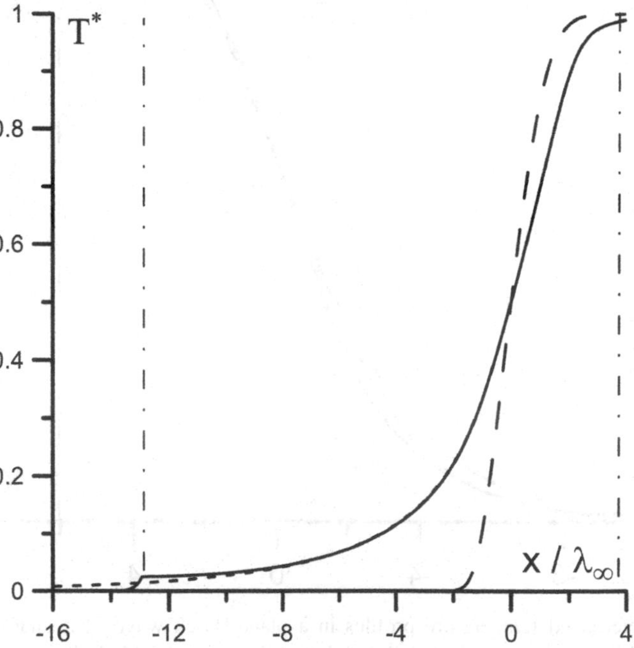

Fig. 3. Same as Fig. 2, $M_\infty = 7$

in the field of moderate Mach numbers was quite expected, then the indicated coincidence of temperature profiles seems to be quite paradoxical.

Another characteristic feature of the hybrid model is that throughout the computational domain, including the derivative discontinuity node, the conservation laws are fulfilled up to the error of numerical integration of the weight function over the velocity space.

This feature of the hybrid model is related to the fact that the components of its *NSF* and *MKE* models require the exact fulfillment of the conservation laws, and the non-equilibrium parameters of the *NSF* model are determined only in the first approximation.

From Fig. 3 it also follows that despite the discontinuity of the derivative in the boundary node of the kinetic subdomain, the hybrid model can significantly improve the hydrodynamic solution.

A quantitative estimate of the maximum relative error in calculating the temperature depending on the size of the kinetic region of the combined model is shown in Fig. 4. This data allows one to select a compromise solution in terms of accuracy and efficiency. As mentioned above, the largest relative error in the calculation of a plane shock wave was observed on the temperature profile.

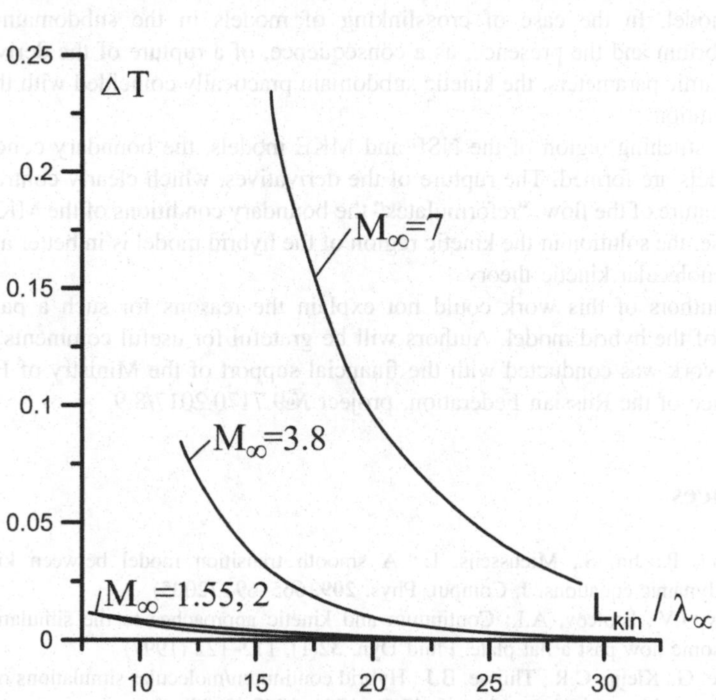

Fig. 4. Maximum relative error of temperature calculation depending on the length of the kinetic subdomain of the hybrid model

6 Conclusion

With regard to the presented hybrid model, we note the following. The results of this work, together with the results of [16], show that the combination of hydrodynamic and kinetic models of flow can be quite successful if both models are based on the same physical model.

By analogy with the proposed model, a hybrid model can be constructed in which the moment equations of any arbitrarily high order are used to hydrodynamically describe the flow. This removes the question of setting boundary conditions on a solid surface. For all moments included in the system of differential equations, Dirichlet boundary conditions hold.

Additional research requires the development of an algorithm to isolate highly non-equilibrium domains in a flow field, for example, the selection of a shock wave front. The upper estimate of the size of the kinetic subdomain of the model in the case of shock waves can serve as data for a plane shock wave presented in Fig. 4.

According to the authors, the important result of the research is a model effect obtained for highly non-equilibrium flows. It has been established that with a small flow nonequilibrium, including the joining region, the kinetic subdomain of the combined model differs, albeit insignificantly, from the subdomain of the "pure" kinetic model. In the case of crosslinking of models in the subdomain of high nonequilibrium and the presence, as a consequence, of a rupture of the derivatives of hydrodynamic parameters, the kinetic subdomain practically coincided with the "pure" kinetic solution.

In the stitching region of the NSF and MKE models, the boundary conditions of these models are formed. The rupture of the derivatives, which clearly contradicts the physical nature of the flow, "reformulates" the boundary conditions of the MKE model. In this case, the solution in the kinetic region of the hybrid model is in better agreement with the molecular kinetic theory.

The authors of this work could not explain the reasons for such a paradoxical behavior of the hybrid model. Authors will be grateful for useful comments.

This work was conducted with the financial support of the Ministry of Education and Science of the Russian Federation, project №9.7170.2017/8.9.

References

1. Degond, P., Jin, S., Mieussens, L.: A smooth transition model between kinetic and hydrodynamic equations. J. Comput. Phys. **209**, 665–694 (2005)
2. Egorov, I.V., Erofeev, A.I.: Continuum and kinetic approaches to the simulation of the hypersonic flow past a flat plate. Fluid Dyn. **32**(1), 112–122 (1997)
3. Abbate, G., Kleijn, C.R., Thijsse, B.J.: Hybrid continuum/molecular simulations of transient gas flows with rarefaction. AIAA J. **47**(7), 1741–1749 (2009)
4. Crouseilles, N., Degond, P., Lemou, M.: A hybrid kinetic-fluid model for solving the gas dynamics Boltzmann-BGK equations. J. Comput. Phys. **199**, 776–808 (2004)
5. Crouseilles, N., Degond, P., Lemou, M.: A hybrid kinetic-fluid model for solving the Vlasov-BGK equations. J. Comput. Phys. **203**, 572–601 (2005)

6. Shakhov, E.M.: Metod issledovaniia dvizhenii razrezhennogo gaza, 207 s. Nauka, Moscow (1975)
7. Rovenskaya, O.I., Croce, G.: Numerical simulation of gas flow in rough micro channels: hybrid kinetic–continuum approach versus Navier-Stokes. Microfluid. Nanofluid. **20**, 81 (2016)
8. Nikitchenko, Y.: On the reasonability of taking the volume viscosity coefficient into account in gas dynamic problems. Fluid Dyn. **2**, 128–138 (2018)
9. Nikitchenko, Y.: Model kinetic equation for polyatomic gases. Comput. Math. Math. Phys. **57**(11), 1843–1855 (2017)
10. Yu.A. Nikitchenko, Modeli neravnovesnykh techenii, 160 s. Izd-vo MAI, Moscow (2013)
11. Rykov, V.A.: A model kinetic equation for a gas with rotational degrees of freedom. Fluid Dyn. **10**(6), 959–966 (1975)
12. Larina, I.N., Rykov, V.A.: Kinetic model of the Boltzmann equation for a diatomic gas with rotational degrees of freedom. Comput. Math. Math. Phys. **50**(12), 2118–2130 (2010)
13. Alsmeyer, H.: Density profiles in argon and nitrogen shock waves measured by the absorption of an electron beam. J. Fluid Mech. **74**(Pt. 3), 497–513 (1976)
14. Robben, F., Talbot, L.: Experimental study of the rotational distribution function of nitrogen in a shock wave. Phys. Fluids **9**(4), 653–662 (1966)
15. Glinkina, V.S., Nikitchenko, Y., Popov, S.A., Ryzhov, Y.: Drag coefficient of an absorbing plate set transverse to a flow. Fluid Dyn. **51**(6), 791–798 (2016)
16. Berezko, M.E., Nikitchenko, Yu.A., Tikhonovets, A.V.: Sshivanie kineticheskoi i gidrodinamicheskoi modelei na primere techeniya Kuetta, Trudy MAI, Vyp. №94, http://trudymai.ru/published.php?ID=80922

Computational Analysis of Pulsed Radiofrequency Ablation in Treating Chronic Pain

Sundeep Singh[1] and Roderick Melnik[1,2](✉)

[1] MS2Discovery Interdisciplinary Research Institute, Wilfrid Laurier University, 75 University Avenue West, Waterloo, ON N2L 3C5, Canada
rmelnik@wlu.ca

[2] BCAM - Basque Center for Applied Mathematics, Alameda de Mazarredo 14, 48009 Bilbao, Spain

Abstract. In this paper, a parametric study has been conducted to evaluate the effects of frequency and duration of the short burst pulses during pulsed radiofrequency ablation (RFA) in treating chronic pain. Affecting the brain and nervous system, this disease remains one of the major challenges in neuroscience and clinical practice. A two-dimensional axisymmetric RFA model has been developed in which a single needle radiofrequency electrode has been inserted. A finite-element-based coupled thermo-electric analysis has been carried out utilizing the simplified Maxwell's equations and the Pennes bioheat transfer equation to compute the electric field and temperature distributions within the computational domain. Comparative studies have been carried out between the continuous and pulsed RFA to highlight the significance of pulsed RFA in chronic pain treatment. The frequencies and durations of short burst RF pulses have been varied from 1 Hz to 10 Hz and from 10 ms to 50 ms, respectively. Such values are most commonly applied in clinical practices for mitigation of chronic pain. By reporting such critical input characteristics as temperature distributions for different frequencies and durations of the RF pulses, this computational study aims at providing the first-hand accurate quantitative information to the clinicians on possible consequences in those cases where these characteristics are varied during the pulsed RFA procedure. The results demonstrate that the efficacy of pulsed RFA is significantly dependent on the duration and frequency of the RF pulses.

Keywords: Chronic pain · Pulsed radiofrequency ablation · Brain and nervous system · Finite element analysis · Pennes bioheat transfer

1 Introduction

Radiofrequency ablation (RFA) of painful nerves has been gaining increasing popularity among pain management therapists for reducing certain kinds of chronic pain, which represent complex diseases ultimately involving the brain and nervous system. Among them are low back pain, hip pain, knee pain and migraine, to name just a few [1–4]. The complexity of linkages that produces pathophysiology in these diseases

J. M. F. Rodrigues et al. (Eds.): ICCS 2019, LNCS 11539, pp. 436–450, 2019.
https://doi.org/10.1007/978-3-030-22747-0_33

would ideally require multiscale modelling [5]. RFA is a minimally invasive treatment modality whereby electromagnetic energy in the frequency range of 300–500 kHz is applied to make discrete therapeutic lesions. The lesion formed interrupts the conduction of nociceptive signals and blocks the pain transmission from reaching the central nervous system. Apart from RFA, several other non-surgical ablative modalities have been used for the treatment of painful nerves, viz., cryoablation and chemical neurolysis [6]. However, RFA stands out to be the most effective and widely applied modality for alleviating chronic pain, as compared to its counterparts. Importantly, RFA is not a curative procedure and does not treat the root cause of pain, but it anesthetizes the source of patient's pain that usually lasts for about 12 months [7].

The RFA system for nerve ablation usually comprises of three main components: (a) radiofrequency (RF) generator for generation of high frequency (~500,000 Hz) alternating currents, (b) a fully insulated 20–22 gauge electrode needle with 5–10 mm exposed tip, and (c) dispersive ground plate having large surface area. During the RFA procedure, a high-frequency alternating current is delivered from the generator to the RF electrode and flows from the uninsulated electrode tip through the biological tissue to the dispersive ground electrode. The interaction of high-frequency alternating current with the biological tissue initiates the rapid oscillations of charged molecules (mainly proteins) that generates heat. The rate of this ionic heating will be higher around the uninsulated active tip of the electrode, where the current density is largest [7, 8]. The lethal temperature during the application of RFA for treating target nerve in mitigating chronic pain is considered to be at or above 45–50 °C. Moreover, it is highly undesirable, if the temperature within the biological tissue goes beyond 100 °C during the RFA procedure, as it would result in tissue boiling, vaporization and charring. Indeed, it is known that the charring results in an abrupt rise in the electrical impedance of the tissue surrounding the active tip of the electrode, limiting any further conduction of the thermal energy and acting as a barrier, restricting the energy deposition and reducing the size of ablation volume [9, 10].

The power delivery during the RFA application can be done using either continuous or pulsed modes. In the conventional continuous power delivery mode, a high-frequency alternating current is delivered from the RF generator to the electrode placed close to the target nerve to heat the neural tissue (80–90 °C). This causes protein denaturation and destruction of the axons and stops the transmission of nociceptive signals from the periphery [8]. However, in the pulsed RFA, brief 'pulses' of RF signals are applied from the RF generator to the neural tissue followed by silent phases that allow time for heat dissemination. Also, the pulsed RFA can produce far stronger electrical fields as compared to the continuous RFA. Initially, the pulsed RFA was thought to be a completely non-destructive procedure for mitigating chronic pain, but recent research in the field suggests that there are both thermal and non-thermal effects of the pulsed RFA procedure [7]. What currently known is that, during the pulsed RFA, clinical effects are not only caused by the high-frequency alternating electrical current leading to heat-generated lesion, but also by the intense electric fields causing a change in the cellular behavior [11]. Importantly, the pulsed RFA procedure is less destructive as compared to the conventional continuous RFA, as there have been no reports of neurological side effects [7].

Several feasibility studies have been reported utilizing the application of the pulsed RFA in treating some of chronic pain conditions [12–18]. Although several available clinical studies on pulsed RFA techniques portray their effectiveness, especially among patients who suffer from radicular pain and peripheral neuropathies, the exact explanation of the mechanisms of action during the pulsed RFA still remains elusive [7]. Thus, further research is needed to determine the clinical effectiveness of pulsed RFA among different chronic conditions for long term pain relief. In view of the above, the present computational study aims to evaluate the efficacy of the pulsed RFA, focusing on the key input characteristics such as frequencies and durations of the RF pulses. While at the final stage, the efficacy of any therapeutic modality needs to be evaluated and justified by clinical studies, the systematic computational analysis could play a crucial role by serving as a quick, convenient and inexpensive tool for studying the thermo-electric performance of the treatment modality under the influence of varying input parameters.

The challenges associated with modelling the heterogeneous surrounding associated with the target nerve that includes, bones, muscle and other critical structures have precluded the development of comprehensive mathematical models in this field, where the brain and nervous system becomes an important part for consideration in multiscale approaches [5]. Moreover, accurately addressing the significant variations among the thermo-electric properties of these tissues in the numerical model of RFA could become problematic, especially, if the non-linear variations of these properties are to be considered in the coupled analysis. Furthermore, the modelling of the pulsed RFA procedure could become even more challenging in those cases where we need to capture accurately the impact of small pulses of millisecond durations, rapidly altering in between lower and maximum applied energy levels, on the temperature distribution during the numerical analysis. In addressing some of the above challenges, we have developed a model to analyze the pulsed RFA procedure for treating chronic pain within the biological tissues and evaluated the impact of durations and frequencies of RF pulses delivered by the commercially available RF generators on the efficacy of the pulsed RFA procedure.

2 Materials and Methods

This section provides details of the computational geometry, mathematical and computational models, along with main materials parameters and thermo-electric characteristics.

2.1 Description of the Computational Model

Figure 1 shows the two-dimensional axisymmetric computational domain considered in the present study that comprises of the muscle tissue with a single needle RF electrode. The dimensions of this domain have been chosen consistently with those available in the literature (e.g., [19]). The pulsed RFA has been performed utilizing a 22-gauge needle electrode with the active tip length of 10 mm. The considered thermo-electric properties in the present study at the frequency range of 500 kHz are given in Table 1 [11, 19, 20].

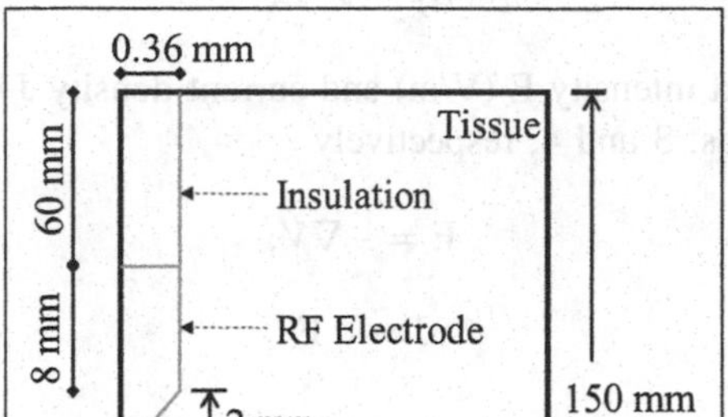

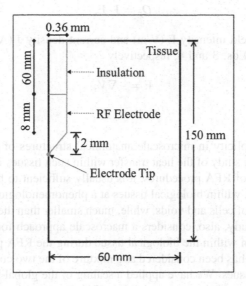

Fig. 1. Two-dimensional axisymmetric pulsed RFA model considered in the present study.

Table 1. Thermo-electric properties of different materials applied in this study [11, 19, 20].

Material (Tissue/Electrode)	Electrical conductivity σ (S/m)	Specific heat capacity c (J/kg/K)	Thermal conductivity k (W/m/K)	Density ρ (kg/m^3)
Tissue	0.446	3421	0.49	1090
Insulation	10^{-5}	1045	0.026	70
Electrode	7.4×10^6	480	15	8000

2.2 Governing Equations

The pulsed RFA procedure represents a coupled thermo-electric problem whereby electromagnetic energy is applied to heat the biological tissue. Importantly, the electric field distribution within the computational domain has been computed by using the simplified version of the Maxwell's equations, known as quasi-static approximation. The governing equation for the electrical problem is represented by

$$\nabla \cdot (\sigma \nabla V) = 0, \tag{1}$$

where σ is the electrical conductivity (S/m) and V is the electric potential (V). The volumetric heat generated Q_p (W/m^3) due to the ionic agitation [10] within the biological tissue caused by the high-frequency alternating current during the pulsed RFA procedure [11] is evaluated by

$$Q_p = \mathrm{J} \cdot \mathrm{E}, \tag{2}$$

where the electric field intensity E (V/m) and current density J (A/m^2) can be derived from the following Eqs. 3 and 4, respectively

$$\mathrm{E} = -\nabla V, \tag{3}$$

$$\mathrm{J} = \sigma \, \mathrm{E} \tag{4}$$

Due to the complicity in microscale anatomical structures of biological tissues, a detailed microscopic study of the heat transfer within such tissues is a very challenging task. In the context of RFA procedures, it is usually sufficient to have the information on the heat transport within biological tissues at a phenomenological scale much larger than the microscale of cells and voids, while, much smaller than the system length scale [21]. The present study, also, considers a macroscale approach for modelling the heat transfer phenomenon within the biological tissue during the RFA procedures, whereby the biological tissue has been considered as a mixture of the two continuum deformable media, blood and tissues. We have applied a scaling of the global balance equations to derive our equations at the macroscopic scale [21]. Furthermore, the temperature distribution during the pulsed RFA procedure has been obtained using the Fourier-law-based Pennes bioheat transfer equation

$$\rho c \frac{\partial T}{\partial t} = \nabla \cdot (k \nabla T) - \rho_b c_b \omega_b (T - T_b) + Q_m + Q_p \tag{5}$$

Where ρ (kg/m^3) is the density, c (J/kg/K) is the specific heat capacity, k (W/m/K) is the thermal conductivity, ρ_b is the density of blood (1050 kg/m^3), c_b is the specific heat capacity of blood (3617 J/kg/K), ω_b is the blood perfusion rate within the tissue (6.35×10^{-4} s^{-1}), Q_p (W/m^3) is the ionic heat generated during RFA and is computed using Eq. 2 above, Q_m (W/m^3) is the metabolic heat generation that has been neglected due to its insignificant contribution during RFA [19], T_b is the blood temperature (37 °C), T is the unknown tissue temperature to be computed from Eq. 5 and t (s) is the duration of the pulsed RFA procedure.

The novelty of the present consideration is in the evaluation of the impact of the commercially available RF generators pulse durations (10 to 50 ms) and pulse frequencies (1 to 10 Hz) on the efficacy of the pulsed RFA procedure. The present study highlights the significance of the pulsing algorithm in comparison to the continuous RF power supply. The temperature distribution within the biological tissue along with the charring temperatures at the tip of the RF electrode for different cases considered in the present study have been reported to provide a critical information about such input characteristics to the clinicians during the pulsed RFA procedure.

2.3 Initial and Boundary Conditions

The initial temperature in the model has been set to 37 °C, which resembles the internal temperature of the human body. The initial voltage, before the onset of the pulsed RFA procedure, has been considered to be zero. A pulsed voltage (with the maximum applied voltage of 45 V) has been applied at the active uninsulated portion of the RF electrode, while 0 V electric potential has been applied on the outer boundaries of the computational domain, simulating the ground pads. Further, electrical and thermal continuity boundary conditions have been imposed at each interface of our computational domain. The ablation volume has been quantified by the isotherm temperature of 50 °C (i.e. the volume within the computational domain having temperature ≥ 50 °C at the end of the pulsed RFA procedure). Figure 2 represents the schematic of the pulse train applied at the active tip of the electrode during the pulsed RFA procedure. Importantly, the train of RF burst pulses applied during the pulsed RFA procedure, as reported in previous clinician studies, varies from 10 to 20 ms in duration and from 1 to 10 Hz in repetition frequency [11, 22]. In view of the above, the present numerical study evaluates the effects of variation in the duration and frequencies of RF pulses during the pulsed RFA procedure. In what follows, we analyze two main situations: (a) the different values of the duration of RF pulses are set to 10 ms, 20 ms, 30 ms, 40 ms and 50 ms, respectively, at the pulse frequency of 2 Hz; (b) the different values of the frequency of RF pulses are set to 1 Hz, 2 Hz, 6 Hz and 10 Hz, respectively, with the pulse duration of 20 ms. Such a consideration has been motivated by [23].

2.4 Numerical Setup and Computational Details

The computational model of the pulsed RFA procedure (shown in Fig. 1) has been solved numerically after imposing the necessary initial and boundary conditions mentioned earlier. The spatial and temporal temperature distributions during the RFA procedures have been obtained by solving the Pennes bioheat transfer equation (Eq. 5). A finite-element method (FEM) based commercial COMSOL Multiphysics [24] software has been utilized to solve the coupled thermo-electric problem of the pulsed RFA procedure. The computational domain has been discretized using a heterogeneous triangular mesh elements using COMSOL's built-in mesh generator with a finer mesh size at the electrode-tissue interface, where the highest thermal and electrical gradients are expected. A convergence analysis has been carried out in order to determine the optimum mesh element size that will lead to mesh-independent solution and reduce the computational cost. The electric field of the coupled thermo-electric problem has been found by using "multifrontal massively parallel sparse direct solver" (MUMPS) with default pre-ordering algorithm. The iterative conjugate gradient method with geometric multi-grid pre-smoothers has been used to determine the temperature field. The relative tolerance has been set to 10^{-5} and the numerical convergence has been attained below this pre-specified value for all simulations. The implicit time-stepping method backward differentiation formulas (BDF) has been used to solve the time-dependent problem and care has been taken to store the data at each steps by assigning strict steps to the solvers. The convergence of the coupled thermo-electric FEM problem is further improved by using segregated algorithms that splits the solution process into sub-steps. Importantly, the coupling is achieved by iterating the entire process after the application of segregated algorithms to resolving physical fields in a sequential manner (e.g., thermal and electrical). The great advantage of chosen approach is that an optimal iterative solver can be used in each sub-step, thus solving a smaller problem in a computationally efficient way. Although this approach generally does require more iterations until convergence, each iteration takes significantly less time compared to one iteration of the fully coupled approach, thus reducing the total solution time and memory usage. All simulations have been conducted on a Dell T7400 workstation with Quad-core 2.0 GHz Intel® Xeon® processors. The mean computation time for each simulation was 3 h.

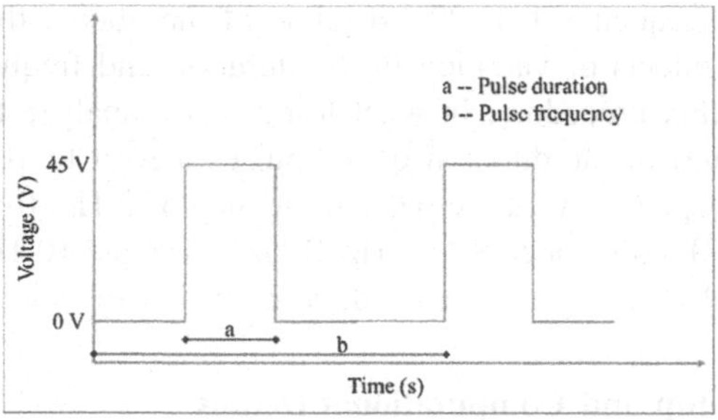

Fig. 2. Schematic view of the pulse train applied to the pulsed RFA model in this study.

3 Results and Discussion

Our main focus is on the effects of the duration and the frequency of the RF pulses and the influence of these effects on the efficacy of the pulsed RFA procedure. As mentioned earlier, a parametric study has been conducted with the applied voltage of 45 V for two different scenarios, (a) with a fixed pulse frequency of 2 Hz and different durations, viz., 10 ms, 20 ms, 30 ms, 40 ms and 50 ms, respectively, and (b) with a fixed pulse duration of 20 ms and different frequencies, viz., 1 Hz, 2 Hz, 6 Hz and 10 Hz, respectively. Importantly, the combination of 2 Hz frequency and 20 ms RF pulse duration is widely used in clinical practices with the applied voltage of 45 V during the pulsed RFA procedure [14, 19]. In what follows, we will explore different aspects of altering the critical input parameters, viz., duration and frequency of the RF pulses, during the pulsed RFA procedure and compare the results with the standard protocol (45 V–20 ms–2 Hz).

Firstly, a comparison study has been conducted between the continuous RFA and pulsed RFA (20 ms–2 Hz) to evaluate the temperature distribution with the applied voltage of 45 V. Figure 3 depicts the temperature distribution as a function of time at the tip of the RF electrode within few seconds after the initiation of RFA application for these two cases. To our surprise, with the applied voltage of 45 V the charring temperature of 100 °C has been reached at the active tip of the RF electrode within less than 1 s during continuous RFA, as shown in Fig. 3(a). The temperature distribution within the first 2.5 s of pulsed RFA (45 V–20 ms–2 Hz) procedure has been depicted in Fig. 3(b). As evident from Fig. 3(b), the pulsed RFA protocol results in the formation of alternating temperature spikes that are located between the lower and higher values pertaining to the duration and frequency of the applied pulse train, restricting or delaying the occurrence of charring. As mentioned earlier, the charring is highly undesirable phenomena during the RFA procedure that results in dramatic increase of the impedance and prevents the RF power generator from delivering further energy within the biological tissue [9, 25, 26]. Different strategies are developed and adopted in clinical practices to either mitigate or delay the occurrence of charring during the RFA procedure by either modifying the RF delivery protocol or the RF electrode [10]. However, during the pulsed RFA procedure (45 V–20 ms–2 Hz), some dissipation of even such high energy is allowed within the biological tissue without the occurrence of charring. This is due to the fact that during the pulsed RFA procedure, periods of high-energy deposition (45 V) are rapidly alternating with the periods of low-energy deposition (0 V). It allows tissue cooling adjacent to the RF electrode during the periods of minimal energy deposition and subsequently leads to a greater energy deposition which, in its turn, results in deeper heat penetration and greater tissue coagulation. The differences between the temperature distributions within the biological tissue obtained after 1 s of continuous and pulsed RFA procedures have been presented in Fig. 4. Further, Fig. 5 depicts the propagation of damage volume corresponding to the 50 °C isotherm at different time steps using the standard pulsed RFA procedure.

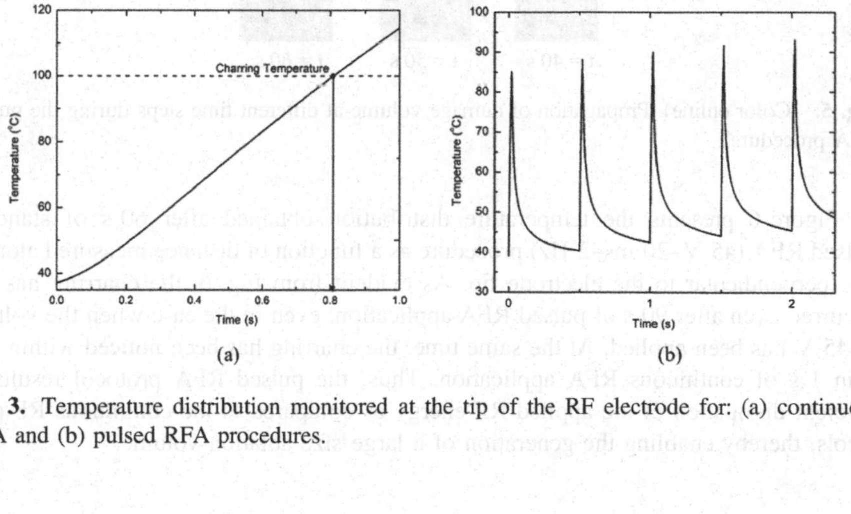

Fig. 3. Temperature distribution monitored at the tip of the RF electrode for: (a) continuous RFA and (b) pulsed RFA procedures.

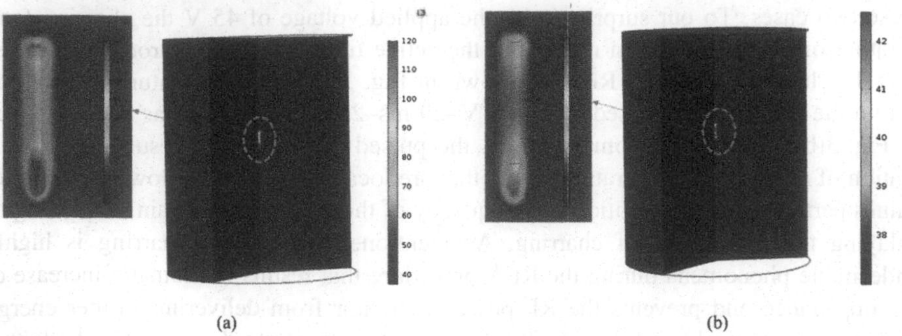

Fig. 4. (Color online) Temperature distribution within the biological tissue after 1 s of the RFA procedure: (a) for the continuous mode and (b) for the pulsed mode.

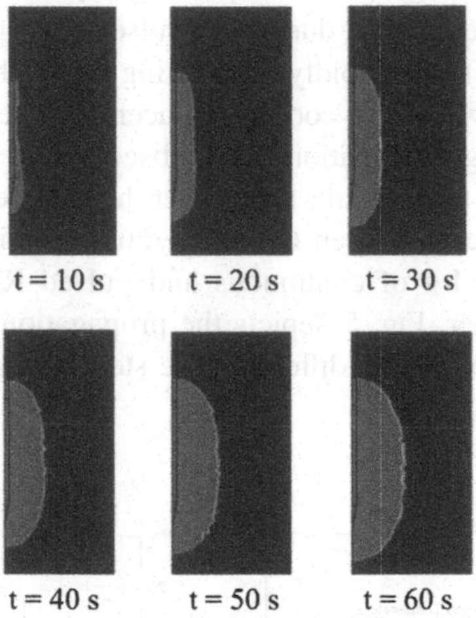

Fig. 5. (Color online). Propagation of damage volume at different time steps during the pulsed RFA procedure.

Figure 6 presents the temperature distribution obtained after 60 s of standard pulsed RFA (45 V–20 ms–2 Hz) procedure as a function of distance measured along a line perpendicular to the electrode tip. As evident from Fig. 6, the charring has not occurred even after 90 s of pulsed RFA application, even in the case when the voltage of 45 V has been applied. At the same time, the charring has been noticed within less than 1 s of continuous RFA application. Thus, the pulsed RFA protocol results in efficient dissipation of the applied RF energy as compared to the continuous RF protocols, thereby enabling the generation of a large size ablation volume.

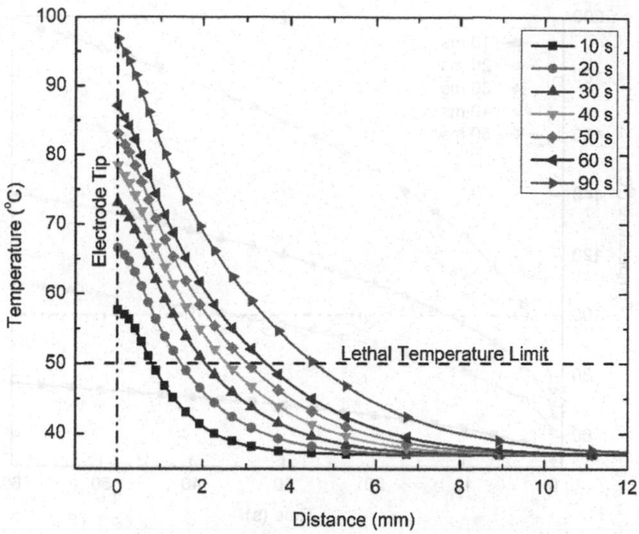

Fig. 6. (Color online). Temperature distribution as a function of distance monitored along the line perpendicular to the tip of the RF electrode during the standard pulsed RFA procedure.

The effects of different RF pulse durations (viz., 10 ms, 20 ms, 30 ms, 40 ms and 50 ms) on the average value of temperature monitored at the active tip of RF electrode as a function of time during the pulsed RFA procedure for treating chronic pain has been presented in Fig. 7. As evident from Fig. 7, there prevails a significant variation in the temperature profile among different cases considered in the present study. The target tip temperature increases with increasing the RF pulse duration and vice-versa. Furthermore, the increase in RF pulse duration beyond 20 ms makes it more susceptive to the occurence of charring within the biological tissue. Figure 8 presents the temperature distribution obtained after 60 s of the pulsed RFA procedure, along the line perpendicular to the tip of the RF electrode for fixed pulse frequency of 2 Hz, but different values of the pulse durations, viz., 10 ms, 20 ms, 30 ms, 40 ms and 50 ms, respectively. It can be seen from Fig. 8 that up to 20 ms durations of RF pulses no charring has occured after 60 s of the pulsed RFA procedure. Figure 9 depicts the temperature distribution after 60 s of the pulsed RFA procedure with different durations of the RF pulses considered in this study. As evident from Fig. 9, a small region of tissue has been charred with the RF pulse duration of 30 ms, and the charring region significantly increases for the duration of 40 ms and 50 ms, respectively.

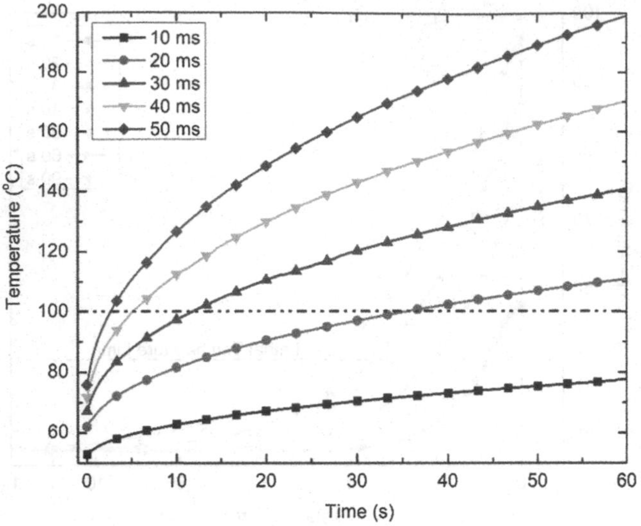

Fig. 7. (Color online). Average value of temperature profile as a function of time monitored at the active tip of the RF electrode for pulse frequency of 2 Hz and different values of pulse durations considered in the present study.

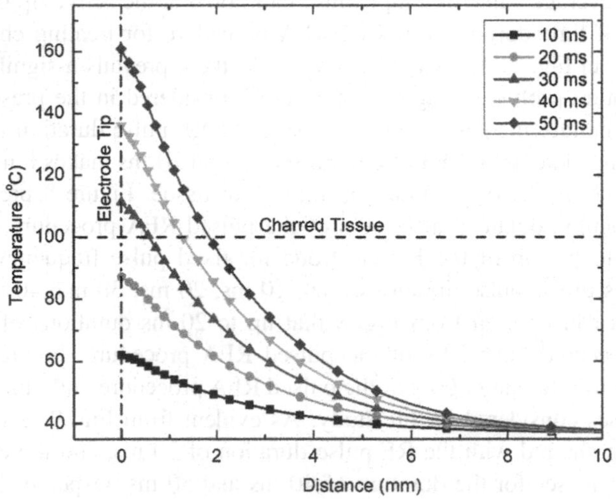

Fig. 8. (Color online). Temperature distribution as a function of distance monitored along the line perpendicular to the tip of the RF electrode for different values of RF pulse durations.

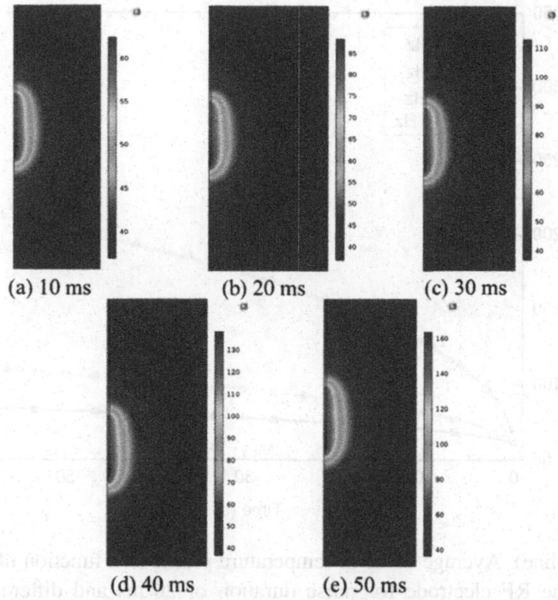

Fig. 9. (Color online). Temperature distribution obtained after 60 s of pulsed RFA application for different values of RF pulse durations considered in the present study.

Figure 10 presents the effects of different RF pulse frequencies (viz., 1 Hz, 2 Hz, 6 Hz and 10 Hz) with a fixed pulse duration of 20 ms on the average value of temperature monitored at the active tip of RF electrode as a function of time. It can be seen from Fig. 10 that significant variations prevail in the temperature profile among different cases considered in this study. Particularly, the average value of target tip temperature reaches the charring temperature within less than 5 s for the RF pulse frequencies of 6 Hz and 10 Hz, respectively. These results demonstrate the importance of variations in the RF pulses, their durations and frequencies, on the temperature distribution obtained at active tip of the electrode and within the biological tissue. Indeed, such variations can lead to significant quantitative and qualitative changes in the scenarios of the attainment of charring within the biological tissue. Thus, *a priori* reasonable choice of these parameters need to be made during the treatment planning stage of the therapeutic procedure, so as to minimise the risk and perform safe and reliable procedure. The reported results would provide the clinicians with deeper insight in the selection of pulses durations and frequencies during the pulsed RFA procedures for chronic pain relief. Importantly, the synergy between the durations and frequencies RF pulses would result in greater absorption of the RF energy and thereby increasing the ablation volume during the pulsed RFA procedure for treating chronic pain.

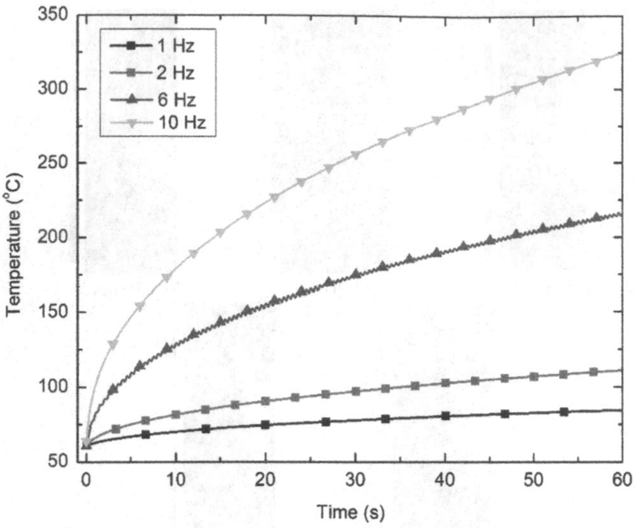

Fig. 10. (Color online). Average value of temperature profile as a function of time monitored at the active tip of the RF electrode for pulse duration of 20 ms and different values of pulse frequencies considered in the present study.

The next step in this study would be to consider a fully heterogeneous surrounding, including the target nerve and the bone, in the three-dimensional model along with the temperature-dependent thermo-electric properties and the temperature-controlled algorithms. We also plan to include the actual models of nerve damage (i.e. mitigation of pain signals). This would enhance the capability of our model even further, since the current model uses a simplified approach of defining the lesion/ablation zone with temperature isotherms during the modelling of the pulsed RFA procedure for treating chronic pain. While our current results provide the critical information on the variation of RF pulses, their durations and frequencies, the development of the model mentioned above will likely provide even better correlation with the clinical scenarios.

4 Conclusion

In this work, a comparative computational study has been conducted to evaluate the temperature distribution between the continuous and pulsed RFA procedures. The effect of durations and frequencies of short burst RF pulses on the efficacy of the pulsed RFA procedure for mitigating chronic pain has been systematically evaluated. Major challenges in developing comprehensive multiscale mathematical models in this field have been highlighted. The reported results revealed that the pulsed RFA could result in greater dissipation of the RF energy as compared to the continuous RFA. This likely occurs due to certain delay in the manifestation of charring owing to the alternating cooling cycles of the RF pulses. Further, it has been found that the increase in both duration and frequency of the RF pulses during pulsed RFA procedures could result in an increase of temperature distribution within the biological tissue and vice-versa. However, the consideration of optimized combination of duration and frequency of RF pulses would considerably reduce the chances of charring and/or would lead to a better RF energy distribution within the tissue along with generation of a large ablation volume during such procedures.

Acknowledgements. Authors are grateful to the NSERC and the CRC Program for their support. RM is also acknowledging support of the BERC 2018-2021 program and Spanish Ministry of Science, Innovation and Universities through the Agencia Estatal de Investigacion (AEI) BCAM Severo Ochoa excellence accreditation SEV-2017-0718.

References

1. Leggett, L.E., Soril, L.J., Lorenzetti, D.L., et al.: Radiofrequency ablation for chronic low back pain: a systematic review of randomized controlled trials. Pain Res. Manag. **19**(5), 146–154 (2014)
2. Bhatia, A., Yasmine, H., Philip, P., Steven, P.C.: Radiofrequency procedures to relieve chronic hip pain: an evidence-based narrative review. Reg. Anesth. Pain Med. **43**(1), 72–83 (2018)
3. Bhatia, A., Philip, P., Steven, P.C.: Radiofrequency procedures to relieve chronic knee pain: an evidence-based narrative review. Reg. Anesth. Pain Med. **41**(4), 501–510 (2016)
4. Abd-Elsayed, A., Kreuger, L., Wheeler, S., et al.: Radiofrequency ablation of pericranial nerves for treating headache conditions: a promising option for patients. Ochsner J. **18**(1), 59–62 (2018)
5. Lytton, W.W., Arle, J., Bobashev, G., et al.: Multiscale modeling in the clinic: diseases of the brain and nervous system. Brain Inform. **4**(4), 219–230 (2017)
6. Soloman, M., Mekhail, M.N., Mekhail, N.: Radiofrequency treatment in chronic pain. Expert Rev. Neurother. **10**(3), 469–474 (2010)
7. Calodney, A., Rosenthal, R., Gordon, A., Wright, R.E.: Targeted radiofrequency techniques. In: Racz, G., Noe, C. (eds.) Techniques of Neurolysis, pp. 33–73. Springer, Cham (2016). https://doi.org/10.1007/978-3-319-27607-6_3
8. Lundeland, B., Kvarstein, G.: Is there a place for pulsed radiofrequency in the treatment of chronic. Scand. J. Pain **12**, 55–56 (2016)
9. Makimoto, H., Metzner, A., Tilz, R.R., et al.: Higher contact force, energy setting, and impedance rise during radiofrequency ablation predicts charring: New insights from contact force-guided in vivo ablation. J. Cardiovasc. Electrophysiol. **29**(2), 227–235 (2018)
10. Zhang, B., Moser, M.A.J., Zhang, E.M., et al.: A review of radiofrequency ablation: large target tissue necrosis and mathematical modelling. Physica Med. **32**(8), 961–971 (2016)
11. Cosman Jr., E.R., Cosman Sr., E.R.: Electric and thermal field effects in tissue around radiofrequency electrodes. Pain Med. **6**(6), 405–424 (2005)
12. Sluijter, M.E., Cosman, E., Rittman, W., van Kleef, M.: The effect of pulsed radiofrequency fields applied to the dorsal root ganglion – a preliminary report. Pain Clin. **11**, 109–117 (1998)
13. Cahana, A., Zundert, J., Macrea, L., et al.: Pulsed radiofrequency: current clinical and biological literature available. Pain Med. **7**, 411–423 (2006)
14. Kroll, H.R., Kim, D., Danic, M.J., et al.: A randomized, double-blind, prospective study comparing the efficacy of continuous versus pulsed radiofrequency in the treatment of lumbar facet syndrome. J. Clin. Anesth. **20**, 534–537 (2008)

15. Snidvongs, S., Mehta, V.: Pulsed radiofrequency: a non-neurodestructive therapy in pain management. Curr. Opin. Support Palliat. Care **4**, 107–110 (2010)
16. Pangarkar, S., Miedema, M.L.: Pulsed versus conventional radio frequency ablation for lumbar facet joint dysfunction. Curr. Phys. Med. Rehabil. Rep. **4**(1), 61–65 (2014)
17. Gupta, A., Huettner, D.P., Dukewich, M.: Comparative effectiveness review of cooled versus pulsed radiofrequency ablation for the treatment of knee osteoarthritis: a systematic review. Pain Physician **20**(3), 155–171 (2017)
18. Chang, M.C.: Efficacy of pulsed radiofrequency stimulation in patients with peripheral neuropathic pain: a narrative review. Pain Physician **21**(3), E225–E234 (2018)
19. Ewertowska, E., Mercadal, B., Muñoz, V., et al.: Effect of applied voltage, duration and repetition frequency of RF pulses for pain relief on temperature spikes and electrical field: a computer modelling study. Int. J. Hyperth. **34**(1), 112–121 (2018)
20. Hasgall, P.A., Gennaro, F.D., Baumgartner, C., et al.: IT'IS database for thermal and electromagnetic parameters of biological tissues. Version 4.0, 15 May 2018. https://doi.org/10.13099/vip21000-04-0.itis.swiss/database
21. Fan, J., Wang, L.: A general bioheat model at macroscale. Int. J. Heat Mass Transf. **54**(1–3), 722–726 (2011)
22. Rohof, O.J.: Radiofrequency treatment of peripheral nerves. Pain Pract. **2**, 257–260 (2002)
23. Chua, N.H.L., Vissers, K.C., Sluijter, M.E.: Pulsed radiofrequency treatment in interventional pain management: mechanisms and potential indications - a review. Acta Neurochir. **153**(4), 763–771 (2011)
24. COMSOL Multiphysics® v. 5.2. COMSOL AB, Stockholm, Sweden. www.comsol.com
25. Topp, S.A., McClurken, M., Lipson, D., et al.: Saline-linked surface radiofrequency ablation: factors affecting steam popping and depth of injury in the pig liver. Ann. Surg. **239** (4), 518 (2004)
26. Fonseca, R.D., Monteiro, M.S., Marques, M.P., et al.: Roll off displacement in ex vivo experiments of RF ablation with refrigerated saline solution and refrigerated deionized water. IEEE Trans. Biomed. Eng. (2018). https://doi.org/10.1109/tbme.2018.2873141

MaMiCo: Parallel Noise Reduction for Multi-instance Molecular-Continuum Flow Simulation

Piet Jarmatz[1,3](✉) and Philipp Neumann[2]

[1] Department of Informatics, Technical University of Munich, Garching, Germany
[2] Department of Informatics, University of Hamburg, Hamburg, Germany
[3] Faculty of Mechanical Engineering, Helmut Schmidt University, Hamburg, Germany
jarmatz@hsu-hh.de

Abstract. Transient molecular-continuum coupled flow simulations often suffer from high thermal noise, created by fluctuating hydrodynamics within the molecular dynamics (MD) simulation. Multi-instance MD computations are an approach to extract smooth flow field quantities on rather short time scales, but they require a huge amount of computational resources. Filtering particle data using signal processing methods to reduce numerical noise can significantly reduce the number of instances necessary. This leads to improved stability and reduced computational cost in the molecular-continuum setting.

We extend the Macro-Micro-Coupling tool (MaMiCo) – a software to couple arbitrary continuum and MD solvers – by a new parallel interface for universal MD data analytics and post-processing, especially for noise reduction. It is designed modularly and compatible with multi-instance sampling. We present a Proper Orthogonal Decomposition (POD) implementation of the interface, capable of massively parallel noise filtering. The resulting coupled simulation is validated using a three-dimensional Couette flow scenario. We quantify the denoising, conduct performance benchmarks and scaling tests on a supercomputing platform. We thus demonstrate that the new interface enables massively parallel data analytics and post-processing in conjunction with any MD solver coupled to MaMiCo.

Keywords: Multiscale · Fluid dynamics · HPC · Noise filter · Parallel · Transient · Coupling · Software design · Molecular dynamics · Simulation · Data analytics · Multi-instance · POD · Molecular-continuum

1 Introduction

Multiscale methods in computational fluid dynamics, in particular coupled molecular-continuum simulations [6,7,12,22], allow to go beyond the limitations imposed by modelling accuracy or computational feasibility of a particular

J. M. F. Rodrigues et al. (Eds.): ICCS 2019, LNCS 11539, pp. 451–464, 2019.
https://doi.org/10.1007/978-3-030-22747-0_34

single-scale method. They are frequently applied for instance for nanostructure investigation in chemistry, especially for research in lithium-ion batteries [15].

In the molecular-continuum context, MD regions and continuum flow regions are coupled to extend the simulation capability over multiple temporal and spatial scales. This is useful for many challenging applications involving e.g. nanomembranes [20] or polymer physics [1,18].

Due to applications that often require large domains, long time spans, or fine resolution, parallelism and scalability constitute important aspects of molecular-continuum methods and software. Several codes for massively parallel execution of molecular-continuum simulations on high performance computing (HPC) systems exist, for instance the CPL library [22], MaMiCo [14] or HACPar [19].

Many coupling schemes for molecular-continuum simulations have been investigated [2,8,12,16,24], based amongst others on the internal-flow multiscale method [2] or the heterogeneous multiscale method [8]. In the steady state case, time-averaging of hydrodynamic quantities sampled from MD is sufficient for the coupling to a continuum solver [7]. But a transient simulation with short coupling time intervals can easily become unstable due to fluctuating MD flow field quantities.

One approach to tackle this is multi-instance sampling [13], where averaged information comes from an ensemble of MD systems. This approach, however, is computationally very expensive.

It has been shown that noise removal techniques, i.e. filters, are another effective approach to reduce thermal fluctuations in the molecular-continuum setting. Grinberg used proper orthogonal decomposition (POD) of MD flow field data, demonstrating also HPC applicability of their method by running a coupled atomistic-continuum simulation on up to 294,912 compute cores [11]. Zimoń et al. [26] have investigated and compared different kinds of noise filters for several particle based flow simulations, such as harmonically pulsating flow or water flow through a carbon nanotube. They pointed out that the combination of POD with one of various other noise filtering algorithms yields significant improvements in denoising quality.

One of the main challenges in the field of coupled multiscale flow simulation is the software design and implementation, since goals such as flexibility, interchangeability of solvers and modularity on an algorithmic level often contradict to hardware and high performance requirements. Some very generic solutions exist, such as MUI [23] or the MUSCLE 2 [3] software, which can be used for various multiscale settings. We recently presented the MaMiCo coupling framework [13,14]. It is system-specific for molecular-continuum coupling, but independent of actual MD and CFD solvers. MaMiCo provides interface definitions for arbitrary flow simulation software, hides coupling algorithmics from the solvers and supports 2D as well as 3D simulations.

In this paper, we present extensions of the MaMiCo tool for massively parallel particle simulation data analytics and noise reduction. We introduce a new interface that is intended primarily for MD flow quantity noise filtering, but in the same way usable for any kind of MD data post-processing for the molecular-continuum coupling. We also present an implementation of a HPC compatible and scalable POD noise filter for MaMiCo. Our interface is compatible with multi-instance MD computations in a natural way, thus it enables a novel combination of noise filtering and multi-instance sampling: A small ensemble of separate MD simulations delivers information about the behaviour of the simulated system on a molecular level, while the filtering efficiently extracts a smooth signal for the continuum solver. The goal of this paper is to investigate this new combination and to discuss the related software design issues.

In Sect. 2 we introduce the relevant theoretical background on MD (Sect. 2.1), the considered molecular-continuum coupling (Sect. 2.2) and POD (Sect. 2.3). Section 3 focuses on the implementation of the MaMiCo data analytics and filtering extension. In Sect. 4, we quantify the denoising quality of the filtering. We further analyse signal-to-noise ratios (SNR) for a three-dimensional oscillating Couette flow with synthetic noisy data (Sect. 4.1). To validate the resulting transient coupled molecular-continuum simulation, we use a Couette flow setup and examine its startup (Sect. 4.2). We also conduct performance measurements and scaling tests (Sect. 4.3) of our new coupled flow simulation on the *LRZ Linux Cluster*[1] platforms *CoolMUC-2* and *CoolMUC-3*. Finally, we give a summary and provide an outlook to future work in Sect. 5.

2 Theoretical Background

2.1 Molecular Dynamics (MD)

For the sake of simplicity we restrict our considerations to a set of Lennard-Jones molecules here, without loosing generality since the coupling and noise reduction methodology and software is compatible with other particle systems such as dissipative particle dynamics [9] systems in the same way.

For a number N of molecules with positions $\boldsymbol{x}_i$, velocities $\boldsymbol{v}_i$, mass m and interacting forces $\boldsymbol{F}_i$, the behaviour of the system is determined by a Verlet-type numerical integration, using a particle system time step width dt_{P}, of Newton's equation of motion

$$\frac{d}{dt}\boldsymbol{v}_i = \frac{1}{m}\boldsymbol{F}_i, \qquad \frac{d}{dt}\boldsymbol{x}_i = \boldsymbol{v}_i. \tag{1}$$

$\boldsymbol{F}_i$ is defined by Lennard-Jones parameters ϵ, σ and a cut-off-radius r_c, see [13] for details.

The method is implemented using linked cells [10]. The implementation is MPI-parallelized in the simulation software SimpleMD [14] which is part of MaMiCo; note that this solver choice is for simplicity only as it has been shown that MaMiCo interfaces various MD packages, such as LAMMPS, ls1 mardyn, and ESPResSo/ESPResSo++.

[1] https://www.lrz.de/services/compute/linux-cluster/overview/.

2.2 Coupling and Quantity Transfer

The method used here to couple continuum and MD solver is based on [13,16].

The simulation setup and overlapping domain decomposition is shown in Fig. 1. Nested time stepping is used, i.e. the particle system is advanced over n time steps during every time step of the continuum solver, $dt_C := n \cdot dt_P$.

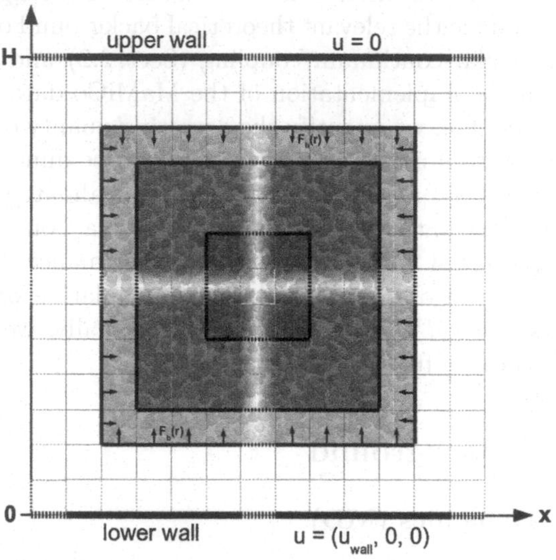

Fig. 1. Coupled Couette simulation transfer region setup. 2D slice through 3D domain, H is the wall distance. Here, only the four outer cell layers close to the MD boundary are shown. At the outer MD domain boundary, an additional boundary force $F_b(r)$ is applied to the molecules. In the green outer MD cells, mass fluxes are transferred from continuum to MD. On the two blue cell layers, velocity values are imposed to MD. Finally, inside the red cells (arbitrarily large region), velocity and density values are sampled, post-processed, and sent back to the continuum solver. (Color figure online)

Particle → continuum data transfer: Continuum quantities for the coupling, such as average velocity $\boldsymbol{u}$, are sampled cell-wise over a time interval dt_C:

$$\boldsymbol{u} = \frac{dt_P}{N dt_C} \sum_{k=0}^{\frac{dt_C}{dt_P}-1} \sum_{i=1}^{N} \boldsymbol{v}_i(t_0 + k dt_P) \tag{2}$$

Afterwards, these quantities are optionally post-processed (separately for every MD instance), accumulated from all instances together, and represent the molecular flow field at continuum scale.

Continuum → particle data transfer: Velocities coming from overlapping cells of the continuum flow solver are used to accelerate the molecules in the corresponding grid cells, they are applied via an additional forcing term with shifted time intervals, see [13] for details. Mass flux from the continuum into the particle simulation is realised by particle deletion and insertion with the USHER algorithm [5]. As boundary conditions for the MD-continuum boundary reflecting boundaries are used. We add the boundary force $F_b(r)$ proposed by Zhou et al. [25]. It is given as an analytic fitting formula and estimates missing intermolecular force contributions over a wide range of densities and temperatures.

2.3 Noise Reduction: POD

The proper orthogonal decomposition (POD), also known as principal component analysis, is a general statistical method for high-dimensional data analysis of dynamic processes and a main technique for data dimensionality reduction. It was already used and investigated for analysis of velocity fields from atomistic simulations [11,26]. Here we employ POD as a fast standard method and basic example to demonstrate the capabilities of our data analytics and noise reduction interface. However, note that various algorithms are known to yield significantly stronger denoising, such as the POD+ methods proposed by Zimoń et al. [26], e.g. POD together with wavelet transform or Wiener filtering. Also many multi-dimensional image processing filters like non-local means, anisotropic filtering or total variation denoising [4,17] are promising for CFD data. They can be applied with the noise reduction interface in the same way as POD.

Based on the method of snapshots [21] to analyse a space-time window of fluctuating data, consider a window of N discrete time snapshots $t \in \{t_1, ..., t_N\} \subset \mathbb{R}$. A finite set Ω of discrete sampling points $\boldsymbol{x} \in \Omega \subset \mathbb{R}^3$ defines a set of signal sources $u(\boldsymbol{x}, t)$. POD describes the function $u(\boldsymbol{x}, t)$ as a finite sum:

$$u(\boldsymbol{x}, t) \approx \sum_{k=1}^{k_{max}} \phi_k(\boldsymbol{x}) a_k(t) \quad (3)$$

where the orthonormal basis functions (modes) $\phi_k(\boldsymbol{x})$ and $a_k(t)$ represent the temporal and spatial components of the data, such that the first k_{max} modes are the best approximation of $u(\boldsymbol{x}, t)$. Choosing a sufficiently small k_{max} will approximate only the dominating components of the signal u and exclude noise. The temporal modes $a_k(t)$ can be obtained by an eigenvalue decomposition of the temporal auto-correlation covariance matrix C,

$$C_{ij} = \sum_{\boldsymbol{x} \in \Omega} u(\boldsymbol{x}, t_i) u(\boldsymbol{x}, t_j) \quad i, j = 1, 2, ..., N. \quad (4)$$

They in turn can be used to compute the spatial modes $\phi_k(\boldsymbol{x})$ with orthogonality relations.

In case of parallel execution, i.e. when subsets of Ω are distributed over several processors, Eq. (4) can be evaluated by communicating local versions of C in a global reduction operation. Although this is computationally expensive compared to purely local, independent POD executions, it is helpful to prevent inconsistencies between subdomains and enforce smoothness, especially in a coupled molecular-continuum setting. Every other step of the algorithm can be performed locally.

3 Implementation: MaMiCo Design and Extension

The MaMiCo tool [13,14] is a C++ framework designed to couple arbitrary massively parallel (i.e., MPI-parallel) continuum and particle solvers in a modular and flexible way, employing a Cartesian grid of so-called macroscopic cells to enable the data exchange. It also provides interfaces to coupling algorithms and data exchange routines. In this paper we employ only the built-in MD simulation, called *SimpleMD* on the microscopic side. On the macroscopic side, a simple Lattice Boltzmann (LB) implementation, the *LBCouetteSolver* and an analytical *CouetteSolver* are used. The necessary communication between *SimpleMD* ranks and the ranks of the *LBCouetteSolver*, where the respective cells are located, is performed by MaMiCo.

The extended system design of MaMiCo is shown in Fig. 2, where the latest developments have been marked in the central red box: A newly introduced interface *NoiseReduction* (Listing 1) is part of the quantity transfer and coupling algorithmics bundle. It is primarily intended for noise filtering, but able to manage arbitrary particle data analytics or post-processing tasks during the coupling, independently of the coupling scheme, actual particle and continuum solver in use. It is designed to be compatible with multi-instance MD computations in a natural way, as separate noise filters are instantiated for each MD system automatically. This yields a strong subsystem separation rather than algorithmic interdependency. However, explicit cross-instance communication is still possible if necessary.

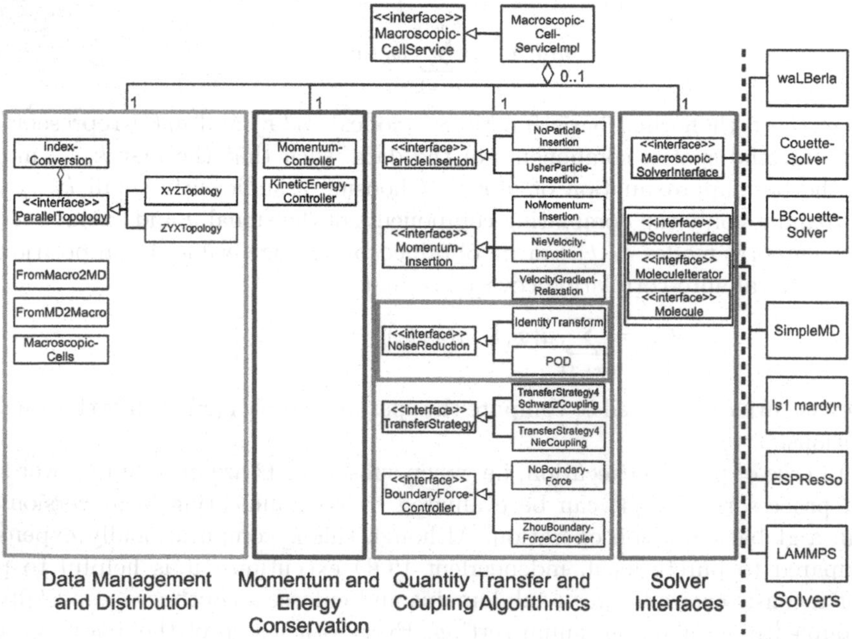

Fig. 2. Extended MaMiCo system design. (Color figure online)

We provide a dummy implementation, *IdentityTransform*, that performs no filtering at all, of the *NoiseReduction* interface, as well as our own massively parallel implementation of a particle data noise filter using POD. The POD implementation employs Eigen[2] to perform linear algebra tasks such as the eigenvalue decomposition. It involves a single invocation of a global (per MD instance) MPI reduction operation to enable detection of supra-process flow data correlation and its separation from thermal noise. It is fully configurable using XML configuration files, where you can specify k_{max} and N.

```
template<unsigned int dim> class
↪ coupling::noisereduction::NoiseReduction {
  public:  [...]
    /** Is called for every macroscopic cell right before sending the
    ↪  macroscopicMass and -Momentum data to the macroscopic solver,
    ↪  but after the call to TransferStrategy::[...] */
    virtual void processInnerMacroscopicCell(
      coupling::datastructures::MacroscopicCell<dim> &cell, const
      ↪  unsigned int &index ){}
    virtual void beginProcessInnerMacroscopicCells(){}
    virtual void endProcessInnerMacroscopicCells(){}
    [...]
}; // Excerpts from main coupling loop:
for (int cycles = 0; cycles < couplingCycles; cycles++){
  couetteSolver->advance(dt_C); // Run one continuum step.
  // Extract data from couette solver and send them to MD.
  fillSendBuffer(*couetteSolver,sendBuffer,...);
  multiMDCellService.sendFromMacro2MD(sendBuffer,...);
  for (int i = 0; i < mdInstances; i++) // Run MD instances.
    simpleMD[i]->simulateTimesteps(dt_P,...);
  // Send back data from MD instances and merge it into recvBuffer of
  ↪  this rank. This automatically calls the noise filter.
  multiMDCellService.sendFromMD2Macro(recvBuffer,...);  }
```

Listing 1: Code snippets from C++ noise reduction interface and main coupling loop, outlining its callback methodology, structure and basic usage.

4 Analysis of Simulation Results

Our test scenario, the Couette flow, consists of flow between two infinite parallel plates. A cubic simulation domain with periodic boundaries in x- and y-direction is used. The upper wall at a distance of $z = H$ is at rest, while the lower wall at $z = 0$ moves in x-direction with constant speed $u_{\text{wall}} = 0.5$. The analytical flow solution for the start-up from unit density and zero velocity everywhere can be derived from the Navier–Stokes equations and is given by:

[2] http://eigen.tuxfamily.org.

$$u_x(z,t) = u_{\text{wall}}\left(1 - \frac{z}{H}\right) - \frac{2u_{\text{wall}}}{\pi}\sum_{k=1}^{\infty}\frac{1}{k}\sin\left(k\pi\frac{z}{H}\right)e^{-k^2\pi^2\nu t/H^2} \tag{5}$$

where ν is the kinematic viscosity of the fluid and $u_y = 0, u_z = 0$.

We refer to the smallest scenario that we frequently use as *MD-30*, because it has a MD domain size of $30 \times 30 \times 30$, embedded in a continuum simulation domain with $H = 50$. We always use a continuum (and MD) cell size of 2.5 and time steps of $dt_{\text{C}} = 0.5$, $dt_{\text{P}} = 0.005$. The Lennard-Jones parameters and the molecule mass are set to $m = \sigma = \epsilon = 1.0$. The MD domain is filled with $28 \times 28 \times 28$ particles, this yields a density value $\rho \approx 0.813$ and a kinematic viscosity $\nu \approx 2.63$. *MD-60* and *MD-120* scenarios are defined by doubling or quadrupling domain sizes and particle numbers in all spatial directions, keeping everything else constant.

For a fluctuating x-velocity signal $u(\boldsymbol{x},t)$ we define the signal-to-noise ratio

$$\text{SNR} = 10\ log_{10}\left(\frac{\sum_{\forall \boldsymbol{x},t}\hat{u}(\boldsymbol{x},t)^2}{\sum_{\forall \boldsymbol{x},t}(\hat{u}(\boldsymbol{x},t) - u(\boldsymbol{x},t))^2}\right)\text{dB} \tag{6}$$

with the analytical noiseless x-velocity $\hat{u}$. As SNR is expressed using a logarithmic scale (and can take values equal to or less than zero - if noise level is equal to or greater than signal), absolute differences in SNR correspond to relative changes of squared signal amplitude ratios, so we define the gain of a noise reduction method as

$$\text{gain} = \text{SNR}_{\text{OUT}} - \text{SNR}_{\text{IN}}, \tag{7}$$

with SNR_{OUT} and SNR_{IN} denoting the signal-to-noise ratios of denoised data and original fluctuating signal, respectively.

4.1 Denoising Harmonically Oscillating Flow

To point out and quantify the filtering quality of MaMiCo's new POD noise filter component, we investigate a harmonically oscillating 3D Couette flow. We use the MD-30 scenario and set u_{wall} to a time dependent sine signal. Since we want to investigate only the noise filter here but not the coupled simulation, we use synthetic (analytical) MD data with additive Gaussian noise, without running a real MD simulation. This is not a physically valid flow profile as it disregards viscous shear forces caused by the time-dependent oscillating acceleration, but it is eligible to examine and demonstrate noise filtering performance. We show the influence of varying the POD parameters k_{max} and N in Fig. 3, where the x-component of velocity, for clarity in only one of the cells in the MD domain, is plotted over time; the SNR value is computed over all cells.

The number k_{max} of POD modes used for filtering is considered to be a fixed simulation parameter in this paper, however note that the optimal value of k_{max} depends on the flow features and several methods to choose k_{max} adaptively at runtime by analysing the eigenspectra have been proposed [11,26].

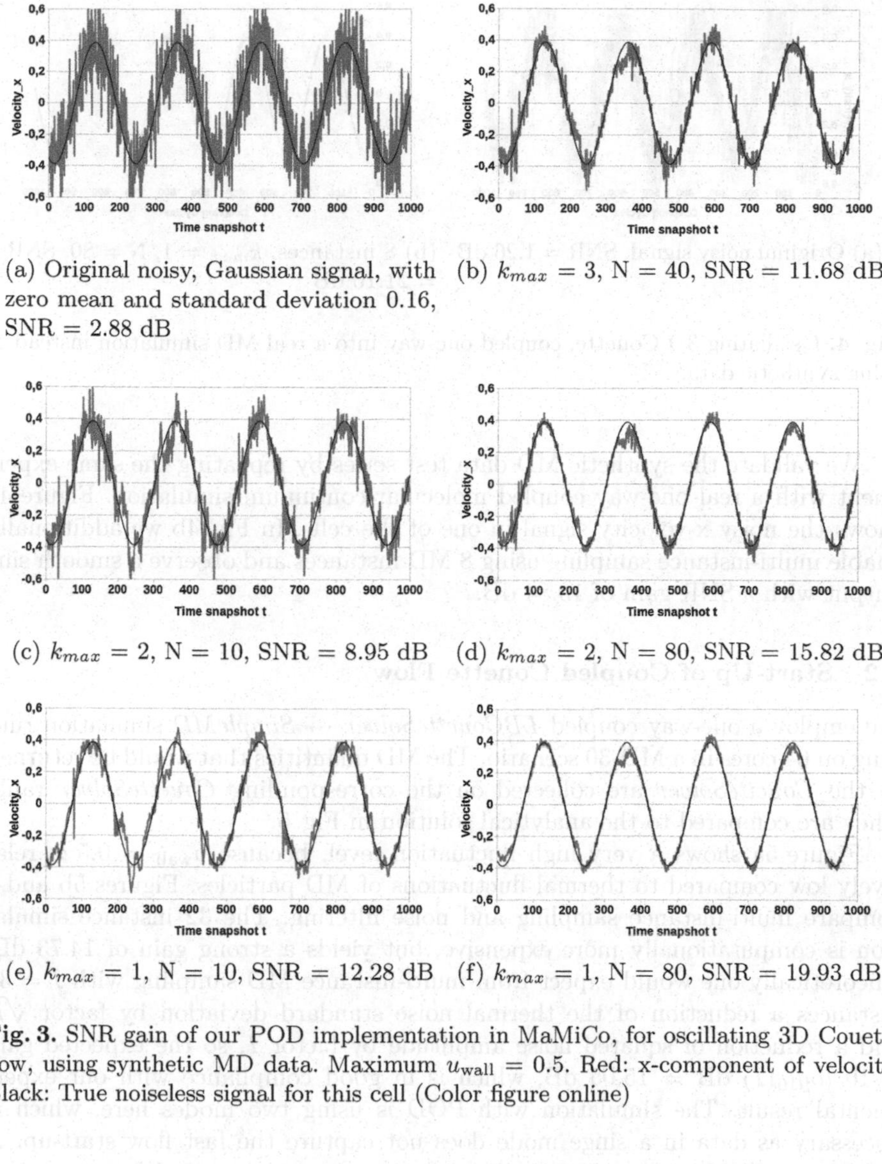

(a) Original noisy, Gaussian signal, with zero mean and standard deviation 0.16, SNR = 2.88 dB

(b) k_{max} = 3, N = 40, SNR = 11.68 dB

(c) k_{max} = 2, N = 10, SNR = 8.95 dB

(d) k_{max} = 2, N = 80, SNR = 15.82 dB

(e) k_{max} = 1, N = 10, SNR = 12.28 dB

(f) k_{max} = 1, N = 80, SNR = 19.93 dB

Fig. 3. SNR gain of our POD implementation in MaMiCo, for oscillating 3D Couette flow, using synthetic MD data. Maximum u_{wall} = 0.5. Red: x-component of velocity; Black: True noiseless signal for this cell (Color figure online)

Figure 3f shows the best reduction of fluctuations with a SNR gain of 17.05 dB compared to the input signal Fig. 3a. Only one POD mode is sufficient here, since the sine frequency is low and the flow close to a steady-state, so that higher eigenvalues already describe noise components of the input.

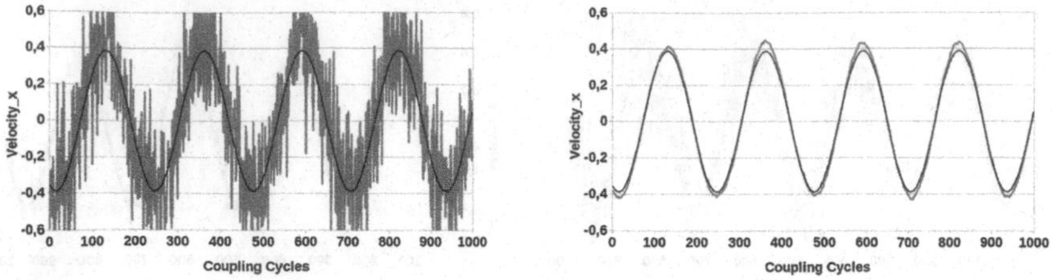

(a) Original noisy signal, SNR = 1.26 dB (b) 8 instances, $k_{max} \doteq 1$, N = 80, SNR = 21.10 dB

Fig. 4. Oscillating 3D Couette, coupled one-way into a real MD simulation instead of using synthetic data.

We validate the synthetic MD data test series by repeating the same experiment with a real one-way coupled molecular-continuum simulation. Figure 4a shows the noisy x-velocity signal in one of the cells. In Fig. 4b we additionally enable multi-instance sampling using 8 MD instances and observe a smooth sine output with a SNR gain of 19.84 dB.

4.2 Start-Up of Coupled Couette Flow

We employ a one-way coupled *LBCouetteSolver* → *SimpleMD* simulation running on 64 cores in a MD-30 scenario. The MD quantities that would be returned to the *CouetteSolver* are collected on the corresponding *CouetteSolver* rank. They are compared to the analytical solution in Fig. 5.

Figure 5a shows a very high fluctuation level, because $u_{\text{wall}} = 0.5$ is relatively low compared to thermal fluctuations of MD particles. Figures 5b and c compare multi-instance sampling and noise filtering. The 32 instance simulation is computationally more expensive, but yields a strong gain of 14.73 dB. Theoretically one would expect from multi-instance MD sampling with $I = 32$ instances a reduction of the thermal noise standard deviation by factor $\sqrt{I}$, and a reduction of squared noise amplitude by factor I, so the expected gain is $10\ log_{10}(I)$ dB = 15.05 dB, which is in good compliance with our experimental result. The simulation with POD is using two modes here, which is necessary as data in a singe mode does not capture the fast flow start-up. A smaller comparable gain of 11.15 dB is obtained here, using much less computational resources (see Sect. 4.3). The best result is achieved using a combination of multi-instance sampling and noise filtering shown in Fig. 5d. This novel approach yields a signal-to-noise gain of 22.63 dB for this test scenario, so that the experimentally produced velocity values closely match the analytical solution.

Thus the new combined coupling approach features benefits for both performance and precision. Its ability to extract considerably smoother flow field quantities permits coupling on shorter time scales.

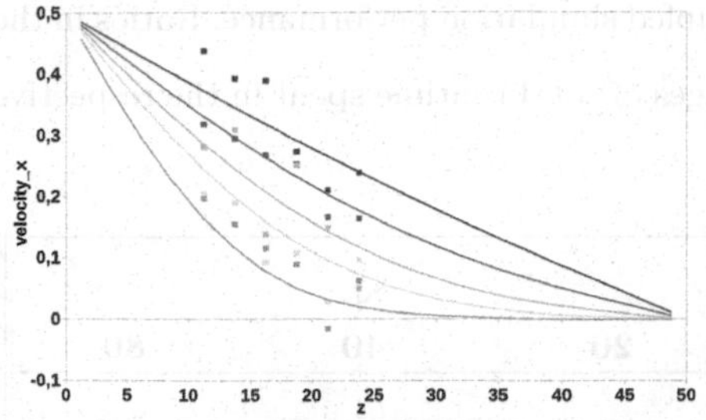

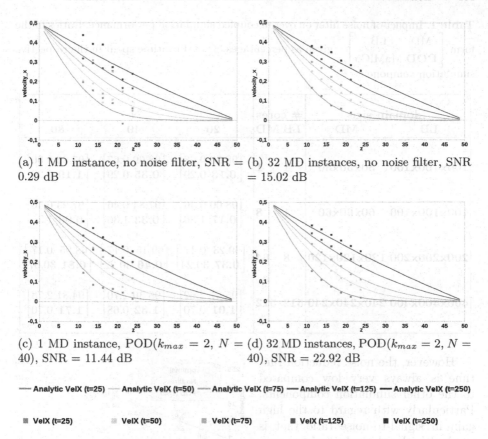

(a) 1 MD instance, no noise filter, SNR = 0.29 dB

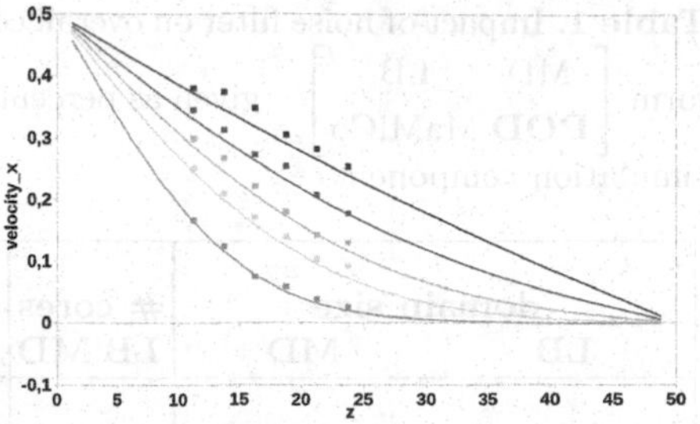

(b) 32 MD instances, no noise filter, SNR = 15.02 dB

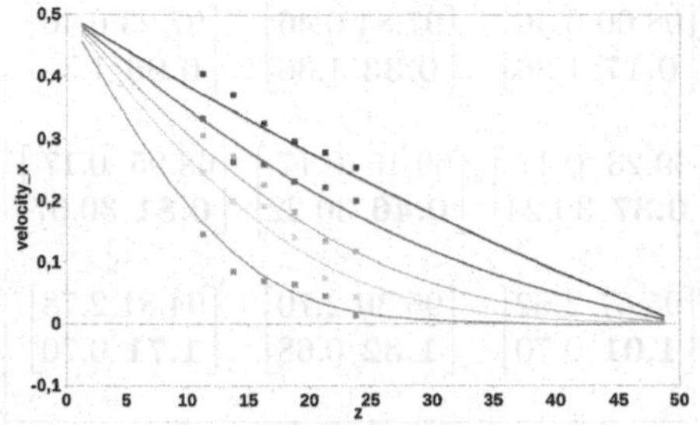

(c) 1 MD instance, POD(k_{max} = 2, N = 40), SNR = 11.44 dB

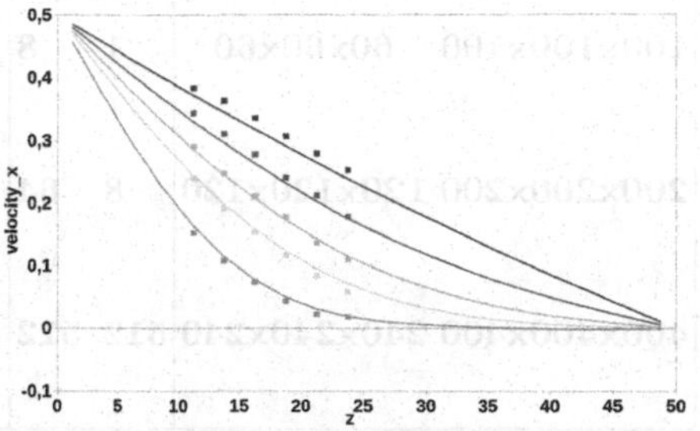

(d) 32 MD instances, POD(k_{max} = 2, N = 40), SNR = 22.92 dB

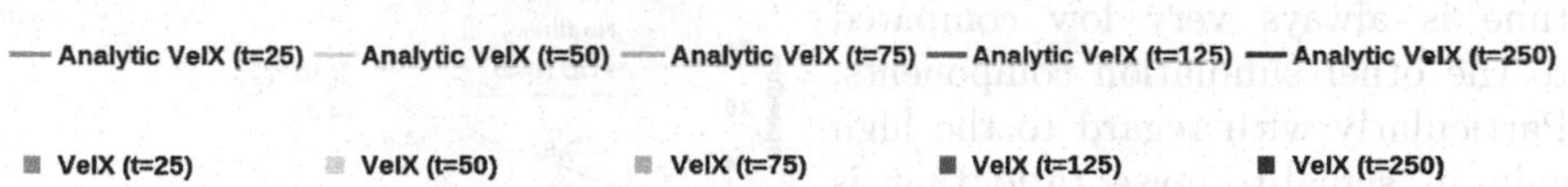

Fig. 5. Couette startup flow profiles, one-way LB → SimpleMD coupling, multi-instance MD versus noise filtering (Color figure online)

4.3 Performance and Scaling Tests

In Table 1 we investigate the performance of the noise filtering subsystem compared to the other components of the coupled simulation. POD always runs with same number of cores as MD. The MaMiCo time includes only efforts for coupling communication – particle insertion and velocity imposition is counted as MD time. The table entries are sampled over 100 coupling cycles each, excluding initializations, using a Couette flow simulation on the LRZ Linux Cluster.

The filter parameter N (time window size) strongly influences the POD runtime, as $C \in \mathbb{R}^{N \times N}$ – and communication and eigenvalue decomposition is performed on C. This leads in practice to a complexity in the order of $\mathcal{O}(N^3)$. Thus, choosing a sufficiently small N is important to limit the computational demand of the method.

Table 1. Impact of noise filter on overall coupled simulation performance. Ratios in the form $\begin{bmatrix} \text{MD} & \text{LB} \\ \textbf{POD} & \text{MaMiCo} \end{bmatrix}$ – given as percentages of total runtime spent in the respective simulation component.

domain size LB	domain size MD	# cores LB	# cores MD	N = 20	N = 40	N = 80
100x100x100	**30x30x30**	**1**	**1**	$\begin{bmatrix} 98.90 & 0.68 \\ \mathbf{0.13} & 0.29 \end{bmatrix}$	$\begin{bmatrix} 98.69 & 0.67 \\ \mathbf{0.35} & 0.29 \end{bmatrix}$	$\begin{bmatrix} 97.86 & 0.66 \\ \mathbf{1.19} & 0.29 \end{bmatrix}$
100x100x100	**60x60x60**	**1**	**8**	$\begin{bmatrix} 98.00 & 0.46 \\ \mathbf{0.17} & 1.36 \end{bmatrix}$	$\begin{bmatrix} 97.84 & 0.46 \\ \mathbf{0.33} & 1.36 \end{bmatrix}$	$\begin{bmatrix} 97.23 & 0.46 \\ \mathbf{0.96} & 1.35 \end{bmatrix}$
200x200x200	**120x120x120**	**8**	**64**	$\begin{bmatrix} 69.23 & 0.17 \\ \mathbf{0.37} & 30.24 \end{bmatrix}$	$\begin{bmatrix} 69.15 & 0.17 \\ \mathbf{0.46} & 30.22 \end{bmatrix}$	$\begin{bmatrix} 68.95 & 0.17 \\ \mathbf{0.81} & 30.07 \end{bmatrix}$
400x400x400	**240x240x240**	**512**	**512**	$\begin{bmatrix} 95.67 & 2.62 \\ \mathbf{1.01} & 0.70 \end{bmatrix}$	$\begin{bmatrix} 95.30 & 2.70 \\ \mathbf{1.32} & 0.68 \end{bmatrix}$	$\begin{bmatrix} 94.81 & 2.78 \\ \mathbf{1.71} & 0.70 \end{bmatrix}$

However, the noise reduction runtime is always very low compared to the other simulation components. Particularly with regard to the high gain in signal-to-noise ratio that is reached with relatively little computational effort here, this is a very good result.

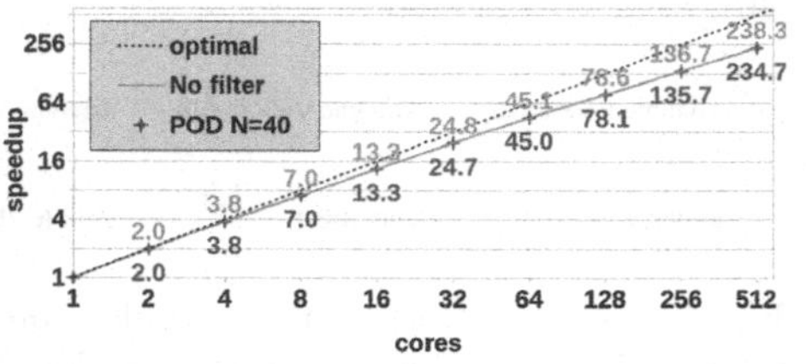

Fig. 6. Strong scaling of coupled simulation, MD-120, including noise filtering. All cores are used for LB, MD and POD, respectively.

The scalability of the coupled simulation, including the noise filter, is evaluated on LRZ CoolMUC-3. The strong scaling tests in Fig. 6 are performed in a fixed MD-120 domain with up to 512 cores used. It is found that our new POD implementation does not significantly impede the scaling, even though it employs a MPI reduction operation. This operation is executed only once per coupling cycle, i.e. every 100 MD time steps, and is restricted to a single MD instance. These results demonstrate the eligibility of our MD data post-processing approach and POD implementation for high-performance computing applications.

5 Conclusions

We have introduced a new noise filtering subsystem into MaMiCo that enables massively parallel particle data post-processing in the context of transient molecular-continuum coupling. Thanks to MaMiCo's modular design, it is compatible with any particle and continuum solver and can be utilized in conjunction with MD multi-instance sampling. Experiments with flow start-up profiles validate our coupled simulation. SNR considerations for oscillating flow demonstrate the filtering quality of our POD implementation. The noise filtering interface and implementation are very scalable and have only a minimal impact on the overall coupled simulation performance.

We observed SNR gains from the POD filter ranging roughly from 11 dB to 17 dB. This corresponds to a potential simulation performance increase by MD instance reduction by a factor of 10 to 50. However, we point out that a combination of both, POD and multi-instance sampling, yields even more smooth quantities and thus enables coupling on shorter time scales, while yielding a higher level of parallelism than a POD-only approach.

As our interface is very generic and flexible, a possible future work suggesting itself would be to conduct experiments with more noise filtering algorithms. Especially the area of image processing offers many promising methods which could be applied to CFD instead of image data. Besides, further data analytics and particle analysis tasks may be tackled in the future, such as flow profile extraction, pressure wave and density gradient detection, or specific molecular data collection modules, e.g. for determination of radial distribution functions. Resilient MD computations by a fault-tolerant multi-instance system with the ability to recover from errors are also conceivable, as well as a machine learning based extension extracting more smooth quantities by learning and detection of prevalent flow features.

Acknowledgements. The provision of compute resources through the Leibniz Supercomputing Centre is acknowledged. The authors further acknowledge financial support by the Federal Ministry of Education and Research, Germany, project *"Task-based load balancing and auto-tuning in particle simulations"* (TaLPas), grant number 01IH16008B.

References

1. Barsky, S., Delgado-Buscalioni, R., Coveney, P.V.: Comparison of molecular dynamics with hybrid continuum-molecular dynamics for a single tethered polymer in a solvent. J. Chem. Phys. **121**(5), 2403–2411 (2004)
2. Borg, M.K., Lockerby, D.A., Reese, J.M.: A hybrid molecular-continuum method for unsteady compressible multiscale flows. J. Fluid Mech. **768**, 388–414 (2015). https://doi.org/10.1017/jfm.2015.83
3. Borgdorff, J., et al.: Distributed multiscale computing with MUSCLE 2, the multiscale coupling library and environment. J. Comput. Sci. **5**(5), 719–731 (2014)
4. Buades, A., Coll, B., Morel, J.M.: A non-local algorithm for image denoising. In: IEEE Computer Society Conference on Computer Vision and Pattern Recognition 2005, CVPR 2005, vol. 2, pp. 60–65 (2005)
5. Delgado-Buscalioni, R., Coveney, P.: USHER: an algorithm for particle insertion in dense fluids. J. Chem. Phys. **119**(2), 978–987 (2003)
6. Delgado-Buscalioni, R., Coveney, P.V., Riley, G.D., Ford, R.W.: Hybrid molecular-continuum fluid models: implementation within a general coupling framework. Philos. Trans. Roy. Soc. Lond. A Math. Phys. Eng. Sci. **363**(1833), 1975–1985 (2005)

7. Dupuis, A., Kotsalis, E., Koumoutsakos, P.: Coupling lattice Boltzmann and molecular dynamics models for dense fluids. Phys. Rev. E **75**(4), 046704 (2007)
8. Weinan, E., Engquist, B., Li, X., Ren, W., Vanden-Eijnden, E.: Heterogeneous multiscale methods: a review. Commun. Comput. Phys. **2**(3), 367–450 (2007)
9. Español, P.: Hydrodynamics from dissipative particle dynamics. Phys. Rev. E **52**, 1734–1742 (1995)
10. Gonnet, P.: A simple algorithm to accelerate the computation of non-bonded interactions in cell-based molecular dynamics simulations. J. Comput. Chem. **28**(2), 570–573 (2007)
11. Grinberg, L.: Proper orthogonal decomposition of atomistic flow simulations. J. Comput. Phys. **231**(16), 5542–5556 (2012)
12. Koumoutsakos, P.: Multiscale flow simulations using particles. Annu. Rev. Fluid Mech. **37**, 457–487 (2005)
13. Neumann, P., Bian, X.: MaMiCo: transient multi-instance molecular-continuum flow simulation on supercomputers. Comput. Phys. Commun. **220**, 390–402 (2017)
14. Neumann, P., Tchipev, N.: A coupling tool for parallel molecular dynamics-continuum simulations. In: 2012 11th International Symposium on Parallel and Distributed Computing (ISPDC), pp. 111–118 (2012)
15. Ngandjong, A.C., Rucci, A., Maiza, M., Shukla, G., Vazquez-Arenas, J., Franco, A.A.: Multiscale simulation platform linking lithium ion battery electrode fabrication process with performance at the cell level. J. Phys. Chem. Lett. **8**(23), 5966–5972 (2017)
16. Nie, X., Chen, S., Robbins, M.O., et al.: A continuum and molecular dynamics hybrid method for micro-and nano-fluid flow. J. Fluid Mech. **500**, 55–64 (2004)
17. Perona, P., Malik, J.: Scale-space and edge detection using anisotropic diffusion. IEEE Trans. Pattern Anal. Mach. Intell. **12**(7), 629–639 (1990)
18. Praprotnik, M., Site, L.D., Kremer, K.: Multiscale simulation of soft matter: from scale bridging to adaptive resolution. Annu. Rev. Phys. Chem. **59**(1), 545–571 (2008)
19. Ren, X.G., Wang, Q., Xu, L.Y., Yang, W.J., Xu, X.H.: HACPar: an efficient parallel multiscale framework for hybrid atomistic-continuum simulation at the micro-and nanoscale. Adv. Mech. Eng. **9**(8), 1–13 (2017)
20. Ritos, K., Borg, M.K., Lockerby, D.A., Emerson, D.R., Reese, J.M.: Hybrid molecular-continuum simulations of water flow through carbon nanotube membranes of realistic thickness. Microfluid. Nanofluid. **19**, 997–1010 (2015)
21. Sirovich, L.: Turbulence and the dynamics of coherent structures. I. Coherent structures. Q. Appl. Math. **45**(3), 561–571 (1987)
22. Smith, E.: On the coupling of molecular dynamics to continuum computational fluid dynamics (2013)
23. Tang, Y.H., Kudo, S., Bian, X., Li, Z., Karniadakis, G.E.: Multiscale universal interface: a concurrent framework for coupling heterogeneous solvers. J. Comput. Phys. **297**, 13–31 (2015)
24. Werder, T., Walther, J.H., Koumoutsakos, P.: Hybrid atomistic-continuum method for the simulation of dense fluid flows. J. Comput. Phys. **205**(1), 373–390 (2005)
25. Zhou, W., Luan, H., He, Y., Sun, J., Tao, W.: A study on boundary force model used in multiscale simulations with non-periodic boundary condition. Microfluid. Nanofluid. **16**(3), 1 (2014)
26. Zimoń, M., et al.: An evaluation of noise reduction algorithms for particle-based fluid simulations in multi-scale applications. J. Comput. Phys. **325**, 380–394 (2016)

Projection-Based Model Reduction Using Asymptotic Basis Functions

Kevin W. Cassel(✉)

Illinois Institute of Technology, Chicago, IL 60616, USA
cassel@iit.edu

Abstract. Galerkin projection provides a formal means to project a differential equation onto a set of preselected basis functions. This may be done for the purpose of formulating a numerical method, as in the case of spectral methods, or formulation of a reduced-order model (ROM) for a complex system. Here, a new method is proposed in which the basis functions used in the projection process are determined from an asymptotic (perturbation) analysis. These asymptotic basis functions (ABF) are obtained from the governing equation itself; therefore, they contain physical information about the system and its dependence on parameters contained within the mathematical formulation. This is referred to as reduced-physics modeling (RPM) as the basis functions are obtained from a physical model-driven, rather than data-driven, technique. This new approach is tailor-made for modeling multiscale problems as the various scales, whether overlapping or distinct in time or space, are formally accounted for in the ABF. Regular- and singular-perturbation problems are used to illustrate that projection of the governing equations onto the ABF allows for determination of accurate approximate solutions for values of the "small" parameter that are much larger than possible with the asymptotic expansion alone and naturally accommodate multiscale problems in which large gradients occur in adjacent regions of the domain.

Keywords: Galerkin projection · Asymptotic methods · Reduced-order modeling

1 Introduction

Projection-based methods are frequently used as the basis for numerical methods, such as spectral methods, and formulation of reduced-order models (ROM), such as proper-orthogonal decomposition (POD). A ROM can then be used as a simplified mathematical model with a reduced number of degrees-of-freedom for systems involving complex physics in optimization, control, and system identification settings, for example. Projection methods approximate a solution in terms of a linear combination of preselected basis functions. In spectral methods, the basis functions are chosen for their ease of integration and other desirable mathematical properties. In ROM, the basis functions are computed from numerical

J. M. F. Rodrigues et al. (Eds.): ICCS 2019, LNCS 11539, pp. 465–478, 2019.
https://doi.org/10.1007/978-3-030-22747-0_35

or experimental data using an optimization procedure, such as POD, which is a "data-driven" method.

The rise in popularity of spectral numerical methods and ROM techniques has highlighted the need for determining basis functions that are appropriate for the particular problem under consideration. A "model-driven" approach is proposed here in which the basis functions are obtained directly and formally from the governing equations rather than data obtained from the system. Tailor-made for such a purpose, particularly in multiscale problems, is asymptotic (or perturbation) methods, which constitute a set of techniques for obtaining an asymptotic series in terms of a physical parameter within the system that becomes very small or large. They provide a powerful set of tools that allow one to learn a great deal about a system directly from the governing equation(s) alone without the need to solve it, simulate it, or conduct an experiment. Such techniques lead to important physical insights that would be difficult to glean numerically or experimentally alone – a need that only becomes more acute as we seek to solve increasingly complex problems in multiscale physics, multi-disciplinary design optimization, and control. The primary interest here is in applications to differential equations in which multiple spatial and/or temporal scales are present within the system.

2 Projection Methods

Projection methods, and the closely associated method of weighted residuals, have their origins in variational methods (Cassel 2013). Consider a general nonhomogeneous differential equation of the form

$$\mathcal{L}u(x) = f(x), \tag{1}$$

where the differential operator $\mathcal{L}$ may be linear or nonlinear. In the *inverse problem*, the differential Euler equation (1) is converted into its *proper variational form*. To do so, the inner product of the differential equation (1) with the variation of the dependent variable is taken, i.e. the differential equation is projected onto δu according to

$$\int_{x_0}^{x_1} (\mathcal{L}u - f)\, \delta u \, dx = 0, \tag{2}$$

which is known as the *reduced variational form*. Carrying out the necessary integration by parts leads to the proper variational form if one exists.

In the Galerkin method, an approximation to the solution $u(x)$ is devised in the form of a linear combination of a set of basis functions of the form

$$\bar{u}(x) = \phi_0(x) + \sum_{n=1}^{N} c_n\phi_n(x) = \phi_0(x) + c_1\phi_1(x) + \cdots + c_n\phi_n(x) + \ldots + c_N\phi_N(x). \tag{3}$$

The linearly-independent *basis functions* $\phi_n(x)$, $n = 0, \ldots, N$ that comprise this *trial function* account for the spatial dependence in the solution and are specified functions that satisfy the boundary conditions. Any non-homogeneous boundary conditions are satisfied by $\phi_0(x)$, such that the remaining basis functions are homogeneous at the boundaries of the domain.

Because the trial function $\bar{u}(x)$ is only an approximate solution, it does not satisfy the differential equation (1) exactly. As a proxy for the unknown error of this approximate solution, the *residual* is defined from the differential equation by

$$r(x) = \mathcal{L}\bar{u}(x) - f(x). \tag{4}$$

In general, the differential operator could be linear or nonlinear, steady or unsteady, and ordinary or partial; however, a one-dimensional framework will be used here in order to introduce the method. The reduced variational form (2) can be written in terms of the residual and trial function as

$$\int_{x_0}^{x_1} r(x)\delta\bar{u}\,dx = 0. \tag{5}$$

Although the reduced-variational form (2) is only useful as an exact representation of the so-called *weak form* of the differential equation, the analogous form (5) in terms of the trial function is simply setting the inner product of the residual with the variation of the trial function to zero, i.e. it is enforcing an orthogonal projection.

Let us more closely examine what Eq. (5) is indicating. Because the basis functions $\phi_n(x)$, $n = 0, \ldots, N$ are specified and do not vary, and it is only the coefficients c_n, $n = 1, \ldots, N$ that vary, taking the variation of the trial function (3) and substituting into Eq. (5) yields

$$\int_{x_0}^{x_1} r(x)\left[\phi_1(x)\delta c_1 + \cdots + \phi_n(x)\delta c_n + \ldots + \phi_N(x)\delta c_N\right] dx = 0.$$

Because the coefficients are arbitrary, for this sum to vanish, the expression multiplying each variation must vanish. That is,

$$\int_{x_0}^{x_1} r(x)\phi_i(x)\,dx = 0, \quad i = 1, \ldots, N. \tag{6}$$

The index is changed to i so that there is no confusion with the index n that identifies the basis functions in the residual $r(x)$. Note that each of the orthogonal projections (6) includes all coefficients c_n and basis functions $\phi_n(x)$, $n = 1, \ldots, N$ in the residual but only one of the basis functions $\phi_i(x)$. This is referred to as an orthogonal projection, not because the basis functions must be mutually orthogonal, but because orthogonality of the residual and basis functions is being enforced inherently in the method. Evaluating these N definite integrals removes the dependence on the spatial coordinate x and leads to an $N \times N$ system of algebraic equations for the coefficients c_n, $n = 1, \ldots, N$. If the problem is unsteady, then this process will lead to an $N \times N$ system of ordinary differential equations for the time-dependent coefficients $c_n(t)$, $n = 1, \ldots, N$. This system of algebraic or ordinary differential equations is the ROM.

The solution to the ROM produces the coefficients that for the given basis functions lead to the trial function that is closest to the exact stationary function $u(x)$. This is typically called the *Galerkin method* when applied as a numerical method, whereas it is referred to as *Galerkin projection* when applied in ROM. It is helpful, however, to realize that all such methods trace their roots back to the inverse variational problem. The Galerkin method is particularly straightforward when the basis functions are mutually orthogonal, in which case all of the products of basis functions $\phi_n(x)\phi_i(x)$ in Eq. (6) vanish except when $n = i$.

Within the Galerkin method, we may select the basis functions in two ways depending on our objective:

1. Preselect the basis functions for their ease of integration and orthogonality properties. This gives rise to *spectral numerical methods*, in which Fourier series or Chebyshev polynomials are typically used as the basis functions.
2. Calculate the basis functions from numerical or experimental data obtained from the system for a particular set of parameters. This is done by solving an optimization problem and gives rise to *proper orthogonal decomposition* (POD) and its extensions and is the basis of the ROM for the system's behavior.

For more on spectral methods, see Fletcher (1984) and Canuto et al. (1988), and for more on ROM, see Rowley and Dawson (2017).

3 Asymptotic (Perturbation) Methods

Clearly, the effectiveness of projection-based spectral numerical methods and ROM hinge on the choice of basis functions used in the trial function. This is where asymptotic methods may prove to be transformational. Once again, the ideal basis functions would contain as much information about the physics of the system as possible, thereby minimizing the number of modes required to obtain an accurate spectral method or ROM. While POD forms the basis functions from numerical or experimental data obtained from the solution itself for a given set of parameters, the *asymptotic basis functions* (ABF) to be put forward here contain the parametric dependence within them and thus apply over a wide parameter range. More to the point, POD is a data-driven method that does not take advantage of any knowledge of the system's mathematical model, whereas such a model is the basis for obtaining the ABF that are the centerpiece of the method introduced here.

Asymptotic (perturbation) methods are a collection of techniques, including matched asymptotic expansions, multiple scales, WKB theory, and strained coordinates for treating systems containing a small or large parameter. The analysis results in the so-called distinguished limit, gauge functions, and asymptotic series. The distinguished limit exposes the dominant balances of terms in the governing equation(s) in the limiting case and indicates the size of the various regions in a domain. The gauge functions in the small parameter allow us to quantify the level of approximation of each term in the asymptotic expansion. Finally, the terms in the asymptotic series provide increasingly higher-order approximations of the system.

An asymptotic expansion is a parametric expansion in the small parameter, say ϵ, of the form[1]

$$u(x;\epsilon) = \sum_{n=1}^{\infty} g_n(\epsilon) u_n(x;\epsilon) = g_1(\epsilon)u_1(x;\epsilon) + g_2(\epsilon)u_2(x;\epsilon) + \cdots, \tag{7}$$

where $g_n(\epsilon)$ are the *gauge functions* and show the asymptotic orders of the successive terms. The accuracy of the expansion improves as ϵ is reduced and/or as additional terms are included in the asymptotic expansion. An asymptotic expansion is local in ϵ, i.e. it applies for $\epsilon \ll 1$, but it is global in x, i.e. it applies for all x in the domain. Often only a small number of terms are necessary in an asymptotic series for a good approximation of the overall solution. The $u_n(x;\epsilon)$ functions provide useful information about the dominant behavior of the system when the parameter ϵ is small, and only a small number of terms are typically required to decipher this information.

An asymptotic sequence of gauge functions $g_1(\epsilon), g_2(\epsilon), \ldots, g_n(\epsilon), \ldots$ is an asymptotic sequence as $\epsilon \to 0$ if

$$g_{n+1}(\epsilon) \ll g_n(\epsilon) \quad \text{as} \quad \epsilon \to 0^+, \tag{8}$$

or equivalently

$$\lim_{\epsilon \to 0^+} \frac{g_{n+1}(\epsilon)}{g_n(\epsilon)} = 0, \quad n = 1, 2, 3, \ldots. \tag{9}$$

The most common situation is when the gauge functions are simply integer powers of ϵ

$$g_1(\epsilon) = \epsilon^0 = 1, \quad g_2(\epsilon) = \epsilon, \quad g_3(\epsilon) = \epsilon^2, \quad \ldots,$$

which clearly satisfy the above properties of an asymptotic sequence. Therefore, an asymptotic sequence exhibits an asymptotic convergence rate in terms of the small parameter.

In general, the small parameter ϵ could be in the equation(s), boundary or initial conditions, and/or the domain geometry (e.g. thin-airfoil theory and thin-shell theory). In regular-perturbation problems, the small parameter does not multiply the highest-order derivative term(s) in the differential equation, and a single expansion is uniformly valid over the entire domain. In singular-perturbation problems, however, the small parameter multiplies the highest-order derivative term so that the order of the equation is reduced for $\epsilon = 0$. In this case, different expansions must be obtained in separate regions of the

[1] Note that traditionally in asymptotic methods, the terms are numbered in the asymptotic expansions starting with zero. That is, the *leading-order term* is $g_0(\epsilon)u_0(x;\epsilon)$. Here, however, we start the expansion from unity in order to be consistent with our RPM nomenclature. In this way, $u_0(x)$ can be used to accommodate non-homogeneous boundary conditions as in spectral methods.

domain, each with its own dominant physics. The method of matched asymptotic expansions ensures that neighboring expansions formally match with one another. It is in this way that asymptotic analysis reveals the dominant physics within each region of the domain where there is fundamentally different, i.e. multiscale, behavior arising from the governing equation(s). That is, the effect of the small parameter is not "small" qualitatively. In particular, singular-perturbation problems are such that the solution with $\epsilon = 0$ is of a fundamentally different form and does not smoothly approach the leading-order solution as $\epsilon \to 0^+$.

A common criticism of asymptotic methods is that the resulting asymptotic expansions often only agree closely with the exact or numerical solution of the governing equation(s) for small values of the parameter ϵ. However, the contention here is that their use in projection methods holds great promise in extending the relevance of the asymptotic series to a wider parameter range for a given system or even to other similar systems. For more details on asymptotic methods, see Hinch (1991) and Weinan (2011).

4 Asymptotic Basis Functions

Presenting these brief overviews of projection and asymptotic methods side-by-side leads one to ask the seemingly obvious question, "Why not use the terms in an asymptotic expansion as the basis functions for a ROM?" Would not these modes take into account the important physical parameter(s) in the system such that the same modes could be used over a wide range of such parameters? Recall that POD modes only apply for the specific value of the parameter(s) for which the data was obtained. New values of the parameters means new data, which means new POD modes. Also recall that spectral numerical methods have difficulties dealing with solutions having large gradients within the domain. What if the chosen basis functions in the trial function actually became more accurate, rather than less so, when large gradients appear in a solution and even fewer basis functions are required?

Because ABF are obtained from the physical model, we refer to them as *reduced-physics models* (RPM); not only does it allow for reduction in the *order* of the model for the system, as with ROM, it contains valuable physical information about the system. In some sense, it is the natural progression of basis functions used in projection methods:

1. Spectral numerical methods - choose Fourier, Legendre, or Chebyshev functions for computational efficiency and ease of integration in the projection process.
2. ROM, e.g. POD - use *data-driven* modes obtained directly from experimental or numerical data for the given problem.
3. RPM, e.g. ABF - use *model-driven* modes obtained directly from the governing equation(s) for the given problem (or a similar one).

The ABF could be used in a spectral method or ROM context. That is, they could be used in any setting involving the use of preselected basis functions for modeling a system having a small parameter. Although we lose the advantages of orthogonal basis functions in RPM, this is more than made up for by the extreme reduction in the number of required basis functions enabled by use of the ABF. In large part, this is facilitated by the fact that the dependence on the small parameter is accounted for in the gauge functions, which form an asymptotic sequence, and the asymptotic basis functions themselves, which is not the case in traditional ROM techniques.

The ABF must be uniformly valid across the entire domain in order to provide global basis functions in the projection process. This is naturally the case for regular-perturbation problems or when the method of multiple scales is used. For singular-perturbation problems using matched asymptotic expansions, on the other hand, the *composite solution* must be formed from the asymptotic series in each distinct region of the domain having their own distinguished limits in terms of the small parameter.

5 Regular-Perturbation Illustration

A regular-perturbation problem is considered first for two reasons. First, it will be shown that the RPM coefficients approach unity as $\epsilon \to 0$. This confirms that the Galerkin projection process is consistent with asymptotic series expansions and preserves the asymptotic solution, which is simply the RPM with $c_n = 1$ for $n = 1, \ldots, N$. Secondly, it will be shown that the ABF can be used with Galerkin projection in the RPM to obtain accurate solutions for values of ϵ that are much larger than is possible for the asymptotic solution alone.

Consider the ordinary differential equation

$$\frac{d^2u}{dx^2} + 2\epsilon\frac{du}{dx} + u = 1, \quad u(0) = 0, \quad u\left(\frac{\pi}{2}\right) = 0, \tag{10}$$

where $0 \leq \epsilon \ll 1$, i.e. ϵ is a small, but positive, parameter. Thus, we have a differential equation of the form

$$\mathcal{L}u(x;\epsilon) = f(x),$$

where

$$\mathcal{L}u(x;\epsilon) = u''(x) + 2\epsilon u'(x) + u(x), \quad f(x) = 1.$$

The ABF are obtained through a regular-perturbation analysis followed by computing the RPM coefficients by projecting the governing equation (10) onto the ABF.

5.1 Asymptotic Basis Functions

Because the small parameter does not multiply the highest-order derivative in the governing equation (10), a regular-perturbation expansion is expected to be suitable of the form

$$u(x;\epsilon) = u_1(x) + \epsilon u_2(x) + \epsilon^2 u_3(x) + \ldots, \tag{11}$$

where the gauge functions are integer powers of ϵ. Substituting the asymptotic expansion (11) into the differential equation (10) and equating like powers of ϵ results in a series of differential equations and boundary conditions for each order in the asymptotic series. Solving each of these equations in succession leads to the first three orders:

$$\begin{aligned} u_1(x) &= 1 - \sin x - \cos x, \\ u_2(x) &= x \sin x - \frac{\pi}{2} \sin x + x \cos x, \\ u_3(x) &= \frac{1}{8} \left[\left(2\pi(2x+1) - 4(x+1)x + \pi^2 \right) \sin x - 4(x-1)x \cos x \right]. \end{aligned} \tag{12}$$

Given these expressions, the first three terms in the regular-perturbation expansion (11) are

$$u(x;\epsilon) = u_1(x) + \epsilon u_2(x) + \epsilon^2 u_3(x) + O(\epsilon^3).$$

As in other ROM settings, the solution $u(x;\epsilon)$ of the differential equation is approximated in terms of a linear combination of a finite number of ABF according to:

$$u(x;\epsilon) = \sum_{n=1}^{N} c_n \phi_n(x;\epsilon) = c_1 \phi_1(x;\epsilon) + \cdots + c_n \phi_n(x;\epsilon) + \ldots + c_N \phi_N(x;\epsilon). \tag{13}$$

The ABF are the product of each order of the asymptotic solution and their corresponding gauge functions, i.e. $\phi_n(x;\epsilon) = g_n(\epsilon) u_n(x;\epsilon)$, and the c_n, $n = 1, \ldots, N$ are the RPM coefficients to be determined from the projection process. Observe that the asymptotic expansion (7) is the sum of the asymptotic basis functions, which corresponds to the RPM solution (13) with all $c_n = 1$. Therefore, the RPM coefficients provide a quantitative measure of the accuracy of the asymptotic solution without knowledge of the exact or numerical solution. In this regular-perturbation case, the ABF are given by

$$\phi_1(x;\epsilon) = u_1(x), \quad \phi_2(x;\epsilon) = \epsilon u_2(x), \ldots, \quad \phi_n(x;\epsilon) = \epsilon^{n-1} u_n(x), \ldots. \tag{14}$$

Note that rather than the basis functions being characterized by modes with increasing frequencies of oscillation or higher-order polynomials as n increases, as in traditional spectral methods, the asymptotic basis functions form an asymptotic sequence according to Eqs. (8) and (9). Moreover, adding additional terms, i.e. increasing N, does not influence the lower-order ABF already obtained; only the coefficients in the RPM need to be recomputed.

5.2 Reduced-Physics Model

Substituting the RPM expansion (13) for the solution $u(x;\epsilon)$, the spatial definite integrals in the projection Eq. (6) eliminate the explicit dependence on the spatial coordinate x and produce a system of N algebraic equations for the coefficients c_n, $n = 1, \ldots, N$ for a given value of ϵ. Observe that the continuous, infinite-dimensional ordinary differential equation $\mathcal{L}u = f$ has been converted into a

Table 1. RPM coefficients for the regular-perturbation problem (10) using Galerkin projection with $N = 4$ asymptotic basis functions; L_2-norm of error for asymptotic expansion (AE) and reduced-physics model (RPM) as compared to the exact solution.

ϵ	c_1	c_2	c_3	c_4	$\|e_{AE}\|_2$	$\|e_{RPM}\|_2$
0.01	1.000000	1.000000	0.999964	0.999963	3.14167×10^{-10}	3.0188×10^{-12}
0.1	0.999999	0.999999	0.996393	0.996393	3.13265×10^{-6}	3.00695×10^{-8}
1.0	0.992809	0.992809	0.728974	0.728974	2.45447×10^{-2}	2.12738×10^{-4}
2.0	0.939427	0.939427	0.383799	0.383799	0.249235	1.65804×10^{-3}
3.0	0.841983	0.841983	0.197772	0.197772	0.843277	4.04899×10^{-3}
4.0	0.728941	0.728941	0.107342	0.107342	1.94203	6.74403×10^{-3}
5.0	0.619753	0.619753	0.0616782	0.0616782	3.69399	9.38273×10^{-3}
10.0	0.273812	0.273812	0.00736198	0.00736198	27.8312	1.66313×10^{-2}

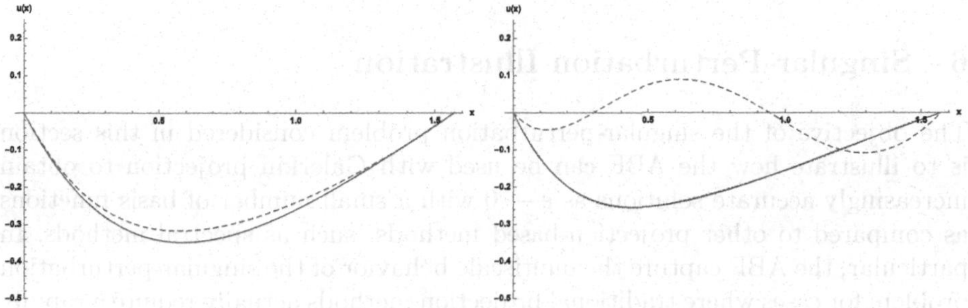

(a) $\epsilon = 1$. (b) $\epsilon = 2$.

Fig. 1. Exact (solid), asymptotic (dashed), and RPM (dotted) solutions for regular-perturbation problem with $N = 4$.

system of N algebraic equations for the RPM coefficients in the trial function. This is the RPM.

The results of the Galerkin projection for the regular-perturbation problem (10) with $N = 4$ ABF are given in Table 1 and Fig. 1. Recall that the values of the RPM coefficients provide a quantitative measure of how accurate the asymptotic expansion is. The RPM coefficients resulting from the projection for various values of ϵ show that the RPM coefficients indeed approach unity as $\epsilon \to 0$ as expected. This confirms that the projection process is consistent with the asymptotic series. Observe from the table that the coefficients begin to deviate from being close to unity for $\epsilon > 0.1$, above which the RPM and asymptotic solutions deviate substantially. The deviation between the asymptotic and RPM solutions is illustrated in Fig. 1(a) for $\epsilon = 1$. While the asymptotic solution displays a notable error compared to the exact solution, the RPM solution is indistinguishable from the exact solution.

As shown in Fig. 1(b) for $\epsilon = 2$, the asymptotic solution does not even agree qualitatively with the exact solution (as expected), whereas the RPM solution is still graphically indistinguishable from the exact solution. Also given in Table 1 is the L_2-norm of the error between the asymptotic and RPM solutions as compared to the exact solution of the ordinary differential equation (10). Observe that the L_2-norm of the error for the RPM solution is two orders of magnitude smaller than that for the underlying asymptotic solution for $\epsilon < 3$. While the error for the asymptotic solution increases dramatically thereafter, that for the RPM increases very slowly up to $\epsilon = 10$. Consequently, as the value of ϵ increases, the asymptotic series is not at all representative of the actual solution. However, the RPM solution remains quite accurate with the L_2-norm only increasing marginally up to $\epsilon = 10$. This is well beyond the range of validity of the asymptotic solution alone, which is obtained in the limit $\epsilon \to 0$. With no additional information, therefore, the accuracy of the asymptotic solution can be improved dramatically for increasing values of ϵ using Galerkin projection with ABF!

6 Singular-Perturbation Illustration

The objective of the singular-perturbation problem considered in this section is to illustrate how the ABF can be used with Galerkin projection to obtain increasingly accurate solutions as $\epsilon \to 0$ with a small number of basis functions as compared to other projection-based methods, such as spectral methods. In particular, the ABF capture the multiscale behavior of the singular-perturbation problem for cases where traditional projection methods actually require a rapidly increasing number of terms to accurately resolve. In fact, because the ABF are the exact solution in the limit as $\epsilon \to 0$, fewer ABF are actually required. In addition, just as for the regular-perturbation problem, Galerkin projection using ABF allows for accurate solutions with larger values of the small parameter than possible with the asymptotic solution alone.

Consider the ordinary differential equation

$$\epsilon \frac{d^2u}{dx^2} + 2\frac{du}{dx} + u = 1, \quad u(0) = 0, \quad u(1) = 0, \tag{15}$$

where $0 \leq \epsilon \ll 1$. Observe that this is very similar to the regular-perturbation problem (10) except that the small parameter now multiplies the highest-order derivative in the governing equation (15); therefore, a *singular-perturbation expansion* is expected to be appropriate. In fact, there is an $O(\epsilon)$ (distinguished limit) thin boundary layer near the boundary at $x = 0$. The *method of matched asymptotic expansions* then leads to separate *outer* and *inner asymptotic expansions* for $x = O(1)$ and $x = O(\epsilon)$, respectively. Matching in the overlap region between the two regions leads to a uniformly-valid *composite solution* that applies across the entire domain. In this case, an asymptotic expansion is obtained of the form

$$u(x; \epsilon) = u_1(x) + \epsilon u_2(x) + \dots. \tag{16}$$

Here, only the first two terms are considered, i.e. $N = 2$ in the ABF, in the Galerkin projection to obtain the RPM. Once again, the ABF include both the asymptotic solutions and their corresponding gauge functions.

Table 2. RPM coefficients for the singular-perturbation problem (15) using Galerkin projection with $N = 2$ asymptotic basis functions; L_2-norm of error for asymptotic expansion (AE) and reduced-physics model (RPM) as compared to the exact solution; and the number of Chebyshev polynomials N_{Cheb} required to obtain the same level of accuracy as the RPM with $N = 2$.

ϵ	c_1	c_2	$\|e_{AE}\|_2$	$\|e_{RPM}\|_2$	N_{Cheb}
0.02	0.999985	1.01089	2.21793×10^{-5}	5.17765×10^{-6}	39
0.05	0.999911	1.02711	1.28291×10^{-4}	3.06292×10^{-5}	26
0.1	0.999663	1.05360	4.43086×10^{-4}	1.09102×10^{-4}	17
0.2	0.998849	1.10231	1.25002×10^{-3}	3.49802×10^{-4}	11
0.3	0.998031	1.13953	2.97762×10^{-3}	1.85862×10^{-3}	7
0.4	0.997889	1.15391	8.71716×10^{-3}	7.99755×10^{-3}	5
0.5	0.999162	1.03925	2.01347×10^{-2}	2.02258×10^{-2}	5

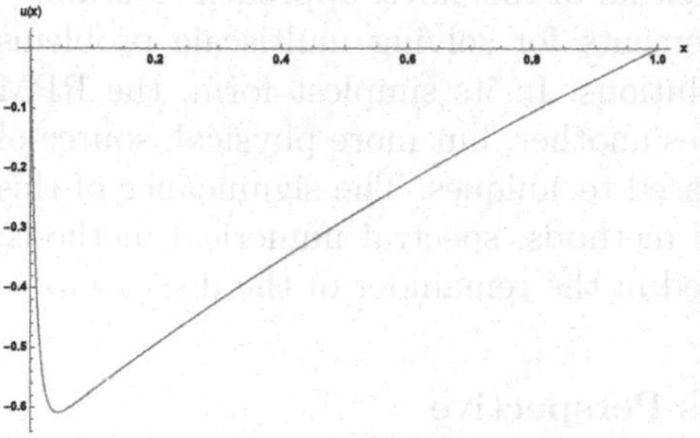

(a) $\epsilon = 0.02$.

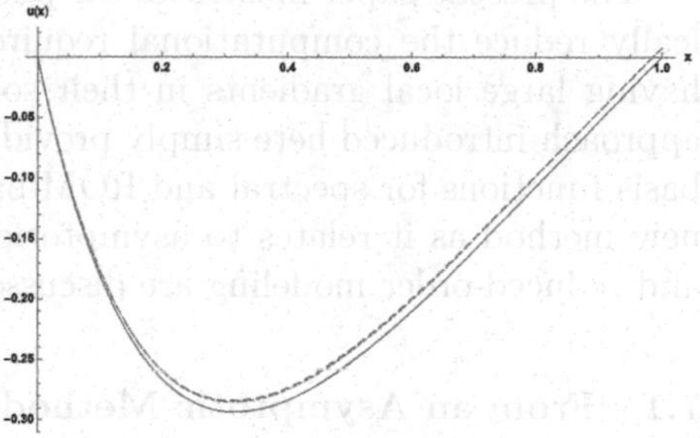

(b) $\epsilon = 0.4$.

Fig. 2. Exact (solid), asymptotic (dashed), and RPM (dotted) solutions for singular-perturbation problem with $N = 2$. Observe the thin boundary layer near $x = 0$.

The results for Galerkin projection with the $N = 2$ ABF for the singular-perturbation problem are shown in Table 2 and Fig. 2. Once again, the values of the RPM coefficients approach unity as $\epsilon \to 0$. Although not as dramatic as for the regular-perturbation problem, the RPM solution with ABF is found to be more accurate than the asymptotic solution alone for values of the small parameter that are larger than would be possible otherwise. Unlike in the regular-perturbation case, the small parameter ϵ must remain sufficiently small to maintain the integrity of the asymptotic structure. That is, ϵ must remain sufficiently small such that the boundary-layer thickness remains smaller than the overall domain size. The right column of the table shows the number of Chebyshev polynomials required in a traditional spectral method to obtain the same level of accuracy as the RPM solution with only two ABF. Observe the dramatic increase in the number of Chebyshev polynomials required as the boundary layer gets thinner and the associated gradients get larger as $\epsilon \to 0$.

7 Conclusions and Discussion

A novel approach to projection methods has been developed in which the terms in an asymptotic (perturbation) expansion are used as basis functions within a Galerkin projection framework. Using regular- and singular-perturbation problems based on ordinary differential equations, it is illustrated how (1) the projection method is consistent with the asymptotic expansion in the sense that as the small parameter goes to zero, the RPM coefficients approach unity, (2) the projection process allows a small number of ABF to form the basis for very accurate solutions for values of the "small" parameter that are well beyond those for the asymptotic series itself, and (3) Galerkin projection with only two ABF for the singular-perturbation problem result in increasingly accurate solutions for decreasing values of the small parameter, while the number of Chebyshev polynomials required to obtain an equally accurate solution increases dramatically as $\epsilon \to 0$.

The present paper illustrates the potential of this novel approach to dramatically reduce the computational requirements for solving multiscale problems having large local gradients in their solutions. In its simplest form, the RPM approach introduced here simply provides another, but more physical, source of basis functions for spectral and ROM-based techniques. The significance of this new method as it relates to asymptotic methods, spectral numerical methods, and reduced-order modeling are discussed in the remainder of the discussion.

7.1 From an Asymptotic Methods Perspective

The primary strength of asymptotic methods is their ability to illuminate the dominant physics in various regions of multiscale problems for small (or large) values of a physical parameter. Their primary weakness is that the range of validity of the resulting asymptotic expansion is generally very limited and not known unless a separate analytical or numerical solution is available. RPM with ABF addresses both of these issues. By projecting the governing equation(s) onto the ABF, the resulting RPM coefficients provide a quantitative measure of the range of validity of the asymptotic solution that is self-consistent and does not require any external means of verification – the closer the coefficients are to unity, the better the asymptotic expansion.

While determining the range of validity of an asymptotic expansion is primarily of academic interest, applying Galerkin projection with ABF to dramatically increase the range of validity of an asymptotic expansion to $O(1)$ values of the "small" parameter has a great deal of practical usefulness. Crucially, no additional information is required to do so. In addition to extending the validity of an asymptotic expansion, the physical insight inherent in the asymptotic method allows one to see how $\epsilon = O(1)$ cases relate to the limiting behavior to determine to what extent the dominant balances obtained as $\epsilon \to 0$ persist to larger ϵ. All of this comes from the original differential equation(s) governing the physics of the system without requiring any numerical solutions or experiments.

Essentially, the ABF provide another source of basis functions for numerical methods, such as spectral methods, ROM, such as proper-orthogonal decomposition, system identification, such as SINDy, and other modeling tasks as discussed below. Rather than simply being chosen for their mathematical properties, such as in spectral methods, or being obtained from numerical or experimental data, such as in POD, however, the ABF are obtained from the governing equations themselves. Therefore, they provide a *model-driven*, rather than *data-driven*, approach to obtaining basis functions for further calculation or analysis. Consequently, an inherently physically-based multiscale method is derived with the characteristic scales being accommodated in the ABF.

7.2 From a Spectral Numerical Methods Perspective

Recall that spectral numerical methods are based on Galerkin projection (or another weighted residuals method) and utilize basis functions preselected for their ease of integration and other mathematical properties, such as orthogonality, for example. Typically, Fourier modes, Legendre polynomials, or Chebyshev polynomials are used. Spectral methods lead to very accurate global solutions with very fast "spectral" convergence rates. The latter is the case, however, only when the solution is sufficiently smooth. Solutions containing large gradients require a large number of spectral modes to accurately resolve the solution.

Singular-perturbation asymptotic expansions display the opposite behavior. They provide increasingly accurate solutions for increasingly singular problems as $\epsilon \to 0$. Spectral methods with ABF, which contain the dependence on the small parameter through the gauge functions, allow one to take advantage of this behavior. Incorporating the model-driven ABF into spectral numerical methods hold the potential to dramatically reduce the number of basis functions required for an accurate solution, particularly for singular problems. Therefore, incorporating ABF directly into spectral methods could significantly extend their usefulness to problems with sharp gradients and singularities, thereby addressing their primary weakness.

It is also possible that hybrid spectral methods could be developed that combine ABF and traditional basis functions within the Galerkin projection framework. The ABF could account for any singular behavior, while the traditional basis functions resolve the remaining smooth details of the solution. Similarly, a hybrid method could be developed in which ABF are used in the coordinate direction(s) containing singular behavior, and traditional spectral modes are used in the other direction(s), where the solution is smooth.

7.3 From a Reduced-Order Modeling Perspective

Instead of seeking a numerical solution of a differential equation, as with spectral numerical methods, the objective of ROM is to obtain a simplified mathematical model of a system that contains its essential features but involves a finite, and small, number of degrees-of-freedom. The modes are calculated using an optimization procedure, such as POD and its variants, applied to experimental or numerical data from the full system. The resulting ROM can then be used in optimization, closed-loop control, system identification, multi-disciplinary design optimization, and multiscale modeling, for example.

Interestingly, the projection methods used to obtain the ROM utilize the governing equation(s) of the system, while the POD approach to determining the basis functions does not. That is, POD is a *data-driven* method. While this is advantageous for data sets obtained from systems for which a mathematical model is not available, in a Galerkin projection context, where a mathematical model is known, this model does not come into play in formation of the POD basis functions themselves. Consequently, the primary advantage of POD analysis in generating basis functions directly from data is also its primary shortcoming in ROM.

In addition, because the optimal basis functions are determined from the actual data set, they provide the best representation of the original data with the fewest POD modes for the values of the parameter(s) used to obtain the data. However, this means that they are problem – and parameter – dependent, requiring one to obtain a new set of basis functions each time the data set changes, whether from consideration of a different dynamical model or a different set of data from the same model. RPM with ABF addresses this weakness of ROM with POD by obtaining the basis functions directly from the mathematical model of the system. Recall that the ABF incorporate both the gauge functions as well as the asymptotic solutions for each order. Consequently, both the dependence on the small parameter ϵ and the spatial coordinate x are accounted for by the ABF, which is not the case in POD analysis. Thus, this model-driven approach provides a complement and enhancement to the data-driven methods that are gaining traction in many fields today.

Finally, the majority of ROM techniques are based on linear theory, which renders them straightforward to apply but not always ideal for nonlinear system. RPM, on the other hand, accounts for the inherent nonlinearity of the system when present. Although some of the mathematical advantages of using linear system theory is lost, a dramatic reduction in the number of basis functions is anticipated when using ABF.

References

Canuto, C., Hussain, M.Y., Quarteroni, A., Zang, T.A.: Spectral Methods in Fluid Dynamics. Springer, Berlin, Heidelberg (1988). https://doi.org/10.1007/978-3-642-84108-8

Cassel, K.W.: Variational Methods with Applications in Science and Engineering. Cambridge University Press, Cambridge (2013)

Weinan, E.: Principles of Multiscale Modeling. Cambridge University Press, Cambridge (2011)

Fletcher, C.A.J.: Computational Galerkin Methods. Springer, New York (1984). https://doi.org/10.1007/978-3-642-85949-6

Hinch, E.J.: Perturbation Methods. Cambridge University Press, Cambridge (1991)

Rowley, C.W., Dawson, S.T.M.: Model reduction for flow analysis and control. Annu. Rev. Fluid Mech. **49**, 387–417 (2017)

Introducing VECMAtk - Verification, Validation and Uncertainty Quantification for Multiscale and HPC Simulations

Derek Groen[1(✉)], Robin A. Richardson[2], David W. Wright[2], Vytautas Jancauskas[10], Robert Sinclair[2], Paul Karlshoefer[9], Maxime Vassaux[2], Hamid Arabnejad[1], Tomasz Piontek[5], Piotr Kopta[5], Bartosz Bosak[5], Jalal Lakhlili[4], Olivier Hoenen[4], Diana Suleimenova[1], Wouter Edeling[6], Daan Crommelin[6,7], Anna Nikishova[8], and Peter V. Coveney[2,3,8]

[1] Department of Computer Science, Brunel University London, London, UK
Derek.Groen@brunel.ac.uk
[2] Centre for Computational Science, University College London, London, UK
[3] Centre for Mathematics and Physics in the Life Sciences and Experimental Biology, University College London, London, UK
[4] Max-Planck Institute for Plasma Physics - Garching, Munich, Germany
[5] Poznań Supercomputing and Networking Center, Poznań, Poland
[6] Centrum Wiskunde & Informatica, Amsterdam, The Netherlands
[7] Korteweg-de Vries Institute for Mathematics, Amsterdam, The Netherlands
[8] University of Amsterdam, Amsterdam, The Netherlands
[9] BULL/ATOS, Paris, France
[10] LRZ, Garching, Germany

Abstract. Multiscale simulations are an essential computational method in a range of research disciplines, and provide unprecedented levels of scientific insight at a tractable cost in terms of effort and compute resources. To provide this, we need such simulations to produce results that are both robust and actionable. The VECMA toolkit (VECMAtk), which is officially released in conjunction with the present paper, establishes a platform to achieve this by exposing patterns for verification, validation and uncertainty quantification (VVUQ). These patterns can be combined to capture complex scenarios, applied to applications in disparate domains, and used to run multiscale simulations on any desktop, cluster or supercomputing platform.

Keywords: Multiscale simulations · Verification · Validation · Uncertainty quantification

1 Introduction

The overarching goal of computational modeling is to provide insight into questions that in the past could only be addressed by costly experimentation, if at all. In order for the results of computational science to impact decision making

J. M. F. Rodrigues et al. (Eds.): ICCS 2019, LNCS 11539, pp. 479–492, 2019.
https://doi.org/10.1007/978-3-030-22747-0_36

outside of basic science, for example in industrial or clinical settings, it is vital that they are accompanied by a robust understanding of their degree of validity. In practice, this can be decomposed into checks of whether the codes employed are solving equations in an accurate manner (*verification*), solving the correct equations to begin with (*validation*), and providing estimates that comprehensively capture uncertainty (*uncertainty quantification*) [1,2]. These processes, collectively known as VVUQ, provide the basis for determining our level of trust in any given model and the results obtained using it [3]. Recent advances in the scale of computational resources available, and the algorithms designed to exploit them mean that it is increasingly possible to conduct the additional sampling required by VVUQ even for highly complex calculations and workflows. The goal of the VECMA project (www.vecma.eu) is to provide an open source toolkit containing a wide range of tools to facilitate the use of VVUQ techniques in multiscale, multiphysics applications. At this initial stage, these range from fusion and advanced materials through climate and forced population displacement, to drug discovery and personalised medicine.

Multiscale computing presents particular difficulties as such applications frequently consist of complex workflows where uncertainty propagates through highly varied components, some of which may only be executed conditionally. Additionally, uncertainties may be associated with the process of transforming the output variables from one scale to another, e.g. coarse to fine scale or vice versa. Although a wide range of toolkits exist to facilitate multiscale computing (see the review by Groen et al. [4]), applying rigorous VVUQ is a major challenge that still needs to be addressed in this domain. The goal of the VECMA toolkit (VECMAtk) is to provide open source tools which implement VVUQ approaches that range from those which treat components or workflows as an immutable 'black box' to semi-intrusive methods in which components of the workflow may be replaced by statistically representative, but cheaper, surrogate models [5].

Key to VECMA's approach is an understanding of the growing size and diversity of available supercomputers as we move to the exascale [6]. For the first time, the vast number of cores available on modern systems make it conceivable for researchers to execute the ensembles necessary to sample phase space for VVUQ analyses concurrently for even very computationally intensive simulations. Moreover, the use of ensembles of simulations offers an efficient way to use large supercomputers. However, effectively running and managing multiscale workflows composed from components with divergent computational requirements (for example with some models capable of exploiting GPUs and others not) represents a key challenge our tools must address. Nonetheless, advances in hardware continue to present opportunities that can be exploited. For example, some intrusive VVUQ methods which make extensive use of machine learning techniques are able to take advantage of hardware designed to accelerate these techniques (such as fast SSD storage). The support for compute intensive applications and workflow management within VVUQ analyses are key factors motivating the development of the VECMAtk.

The philosophy of VECMAtk is that, as far as possible, we want to be able to add VVUQ to existing scientific workflows without changing researchers conception of the problem they are investigating. This goal informs the conception of the toolkit which is designed (i) to be highly modular, (ii) have minimal installation requirements and, (iii) provide control over where (locally or on remote resources) and at what time analysis takes place.

This means that VECMAtk is a collection of elements that can be reused and recombined in different workflows. Our aim is to define stable interfaces, data formats and APIs that facilitate VVUQ in the widest range of applications. The modularity of the toolkit allows us to account for the differing requirements of VVUQ sampling and analysis and provide tools tailored for the needs of both steps of the process. Several software packages or libraries are already available for performing VVUQ (such as OpenTurns [7], UQLab [8], Uncertainpy [9], Chaospy [10], etc.), but in many cases these are closed source and none of them provide the separation of concerns needed to allow the analysis of both small local computations and highly compute intensive kernels (potentially using many thousands of cores and GPUs on HPC or cloud resources). We aim to reuse existing tools where appropriate to provide robust and optimised code for sampling and analysis.

To enable researchers to use the toolkit and derive custom approaches, we will provide a range of exemplar workflows and documentation. Essential to the design of these exemplars is the need to lower the barriers to usage and, naturally, to be transferable to a large range of applications with minimal modification.

2 The VECMA Toolkit

The goal of the toolkit is to facilitate the process of researchers mapping the requirements of VVUQ to their specific scientific problem and existing workflow(s). The main factors that shape the development approach of VECMAtk are: (a) the need to fit into existing applications with minimal modification effort, (b) the wide range of target application domains, (c) the flexible and recombinable nature of the toolkit itself, (d) the geographically distributed nature of the users and particularly the developers.

2.1 Development and Prototyping Process

To support these needs and characteristics, we have chosen to adopt an evolutionary prototyping approach. In this approach, a existing application developers initially establish VVUQ techniques using their own scripts, which they share with the VECMAtk developers together with additional needs that they have not been able to easily address themselves. These scripts and requirements, together with existing libraries, form the base from which the development team develops initial prototypes. The prototypes are then tested and refined at frequent intervals, with the user feedback and integration testing guiding further developments. As a result, some prototypes are reduced in scope, simplified, or removed altogether; and some prototypes are being refined into more advanced, flexible, scalable and robust tools. Regular development meetings within the project help to monitor and disseminate progress as well as providing a venue in which we ensure that all component development teams are following best development practices (for example making use of version control and continuous integration). Although we work closely with a group of application developers, both FabSim3 and EasyVVUQ are publicly available, and anyone can independently install, use and modify these tools to suit their own purposes.

As part of our prototyping process, we identify common workflow patterns and software elements (for example those needed to encode complex parameter distributions) that can be abstracted for re-use in a wide range of application scenarios. We label patterns found in verification and validation contexts *VVPs*, and those associated with uncertainty quantification or sensitivity analysis *UQPs*. The definition of a VVP or UQP should never require the use of any specific execution management platform, as the toolkit is envisioned to provide multiple solutions that facilitate workflows. Examples include diverse sampling algorithms and job types running on heterogeneous resources. Within VECMAtk, we categorize procedures that treat underlying applications as a black box (*non-intrusive*), that account for the coupling mechanisms between submodels (*semi-intrusive*) or the algorithms of the submodels (*fully intrusive*).

2.2 Key Components

QCG-Broker/Computing: Easy and efficient access to computing power is crucial when a single run of an application is demanding or a large number of application replicas has to be executed to guarantee reliable VVUQ. To fulfil this requirement, VECMAtk uses QCG[1] which provides advanced capabilities for the unified execution of complex jobs across single or multiple computing clusters. The QCG infrastructure, which is presented in Fig. 1, uses the QCG-Broker service to manage execution of computational experiments, e.g. through multi-criteria selection of resources, while several QCG-Computing services offer unified remote access to underlying resources. The QCG services can be accessed with numerous user-level tools [11], of which a few examples are provided in the aforementioned figure.

FabSim3: The combination of different UQPs and VVPs in one application also leads to a cognitively complex workflow structure, where different sets of replicas need to be constructed, organized, executed, analyzed, and actioned upon (i.e. triggering subsequent execution and/or analysis activities). FabSim3 [12][2] is a freely available tool that supports the construction, curation and execution of these complex workflows, and allows users to invoke them on remote resources using one-liner commands. In contrast to its direct predecessor FabSim, FabSim3 inherently supports the execution of job ensembles, and provides a plug-in system

[1] http://www.qoscosgrid.org.
[2] https://github.com/djgroen/FabSim3.

which allows users to customize the toolkit in a modular and lightweight manner (e.g., evidenced by the minimalist open-source FabDummy plugin). We provide an overview of the FabSim3 architecture in Fig. 3. In the context of VECMA FabSim3 plays a key role in introducing application-specific information in the Execution Layer, and in conveniently combining different UQPs and VVPs.

QCG Pilot Job: A *Pilot Job*, is a container for many subjobs that can be started and managed without having to wait individually for resources to become available. Once the Pilot Job is submitted, it may service a number of defined VVUQ subtasks (as defined by e.g. EasyVVUQ or FabSim3). The QCG Pilot Job mechanism provides two interfaces that may be used interchangeably. The first one allows to specify a file with the description of sub-jobs and execute the scenario in a batch-like mode, conveniently supporting static scenarios. The second interface is offered with the REST API and it can be accessed remotely in a more dynamic way. It will be used to support scenarios where a number of replicas and their complexity dynamically changes at application runtime.

EasyVVUQ is a Python library, developed specifically for VECMA, designed to simplify the implementation of creation of (primarily blackbox) VVUQ workflows involving existing applications. The library is designed around a breakdown of such workflows into four distinct stages; sampling, simulation execution, result aggregation, and analysis. The execution step is deemed beyond the remit of the package (it can be handled for instance by FabSim3 or QCG Client), whilst the other three stages are handled separately. A common data structure, the *Campaign*, which contains information on the application being analyzed alongside the runs mandated by the sampling algorithm being employed, is used to transfer information between each stage. The architecture of EasyVVUQ is shown in Fig. 2.

The user provides a description of the model parameters and how they might vary in the sampling phase of the VVUQ pattern, for example specifying the distribution from which they should be drawn and physically acceptable limits on their value. This is used to define a *Sampler* which populates the Campaign with a set of run specifications based on the parameter description provided by the user. The Sampler may employ one of a range of algorithms such as the Monte Carlo or Quasi Monte Carlo approaches [13]. At this point all of the information is generic in the sense that it is not specific to any application or workflow. The role of the *Encoder* is to create input files which can be used in a specific application. Included in the base application is a simple templating system in which values from the Campaign are substituted into a text input file. For many applications it is envisioned that specific encoders will be needed and the framework of EasyVVUQ means that any class derived from a generic Encoder base class is picked up and may be used. This enables EasyVVUQ to be easily extended for new applications by experienced users.

The simulation input is then used externally to the library to execute the simulations. The role of the *Decoder* is twofold, to record simulation completion in the Campaign and to extract the output information from the simulation runs. Similarly to the *Encoder* the *Decoder* is designed to be user extendable to

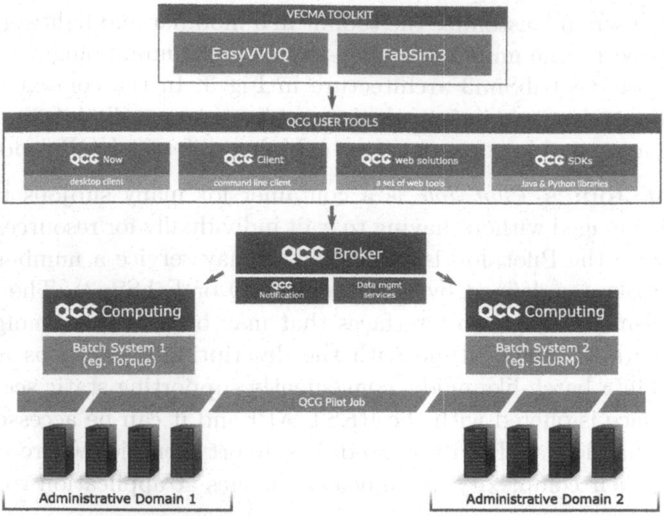

Fig. 1. Simplified overview of QCG usage in VECMA. Jobs requested by the toolkit layer may be farmed out to one or more queues on different computing resources.

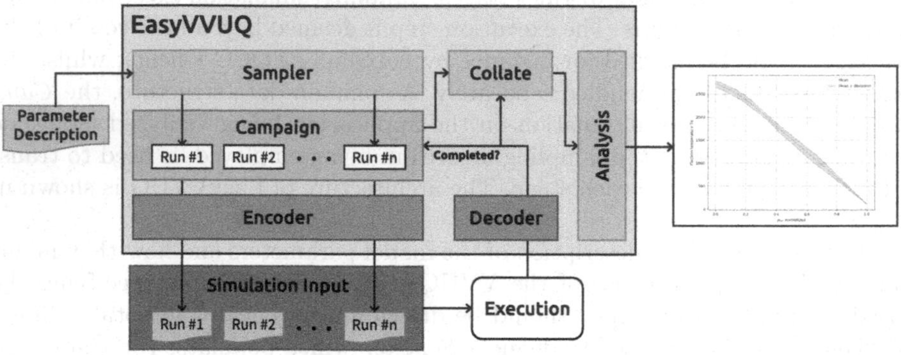

Fig. 2. The architecture of EasyVVUQ. The UQ workflow is split into a sampling, execution and analysis stage, orchestrated via a (persistent) 'Campaign' object.

facilitate analysis of a wide range of applications. The Decoder is used in the collation step to provide a combined and generic expression of the simulation output for further analysis (for example the default is to bring together output from all simulation runs in a Pandas dataframe). Following the output collation we provide a range of analysis algorithms which produce the final output data. Whilst the library was originally designed for acyclic 'blackbox' VVUQ workflows, development is ongoing to allow the library to be used in more complex patterns.

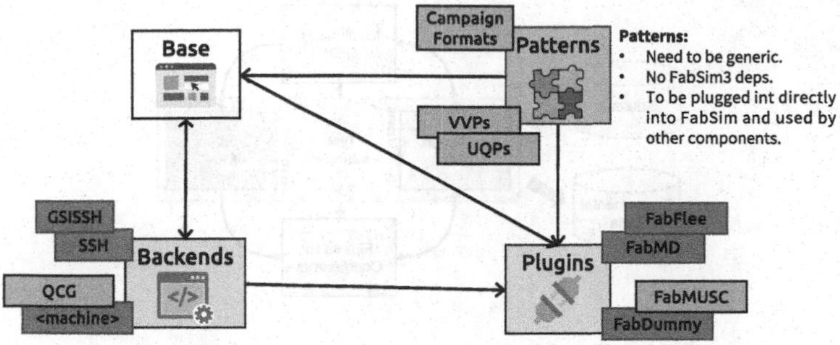

Fig. 3. Essential building blocks in the FabSim3 component of VECMAtk, and their interdependencies. Components in yellow are under development as of June 2019

2.3 How the Components Work Together

The components in VECMAtk (FabSim3, EasyVVUQ, QCG Pilot Job and other QCG components) can be combined in a variety of ways, enabling users to combine their added values while retaining a limited deployment footprint. At time of writing, we are working on the following combinations:

- FabSim3 has been integrated with QCG-Client to enable job submission to QCG Broker, allowing users to schedule their jobs across multiple remote (QCG-supporting) machines.
- EasyVVUQ can use FabSim3 to facilitate automated execution. Users can convert their EasyVVUQ campaigns to FabSim3 ensembles using a one-liner (`campaign2ensemble`), and FabSim3 output is ordered such that it can be directly moved to EasyVVUQ for further decoding and analysis.
- Integration between QCG Pilot Job and FabSim3 is under way, enabling users to create and manage pilot jobs using FabSim3 automation.
- Integration between QCG Pilot Job and EasyVVUQ is under way, enabling EasyVVUQ users to execute their tasks directly using pilot jobs.

3 Initial Applications

The following section gives an overview of applications that currently use the VECMAtk and are guiding the development of new features for the toolkit. We present a detailed look at a fusion calculation and brief description of climate, population displacement, materials, force field, and cardiovascular modeling.

3.1 Fusion Example

Heat and particle transport in a fusion device play a major role in the performance of the thermo-nuclear fusion reaction. Current understanding is that turbulence arising at the micro space and time scales is a key factor for such transport which has effects on much larger scales. Multiscale simulation are

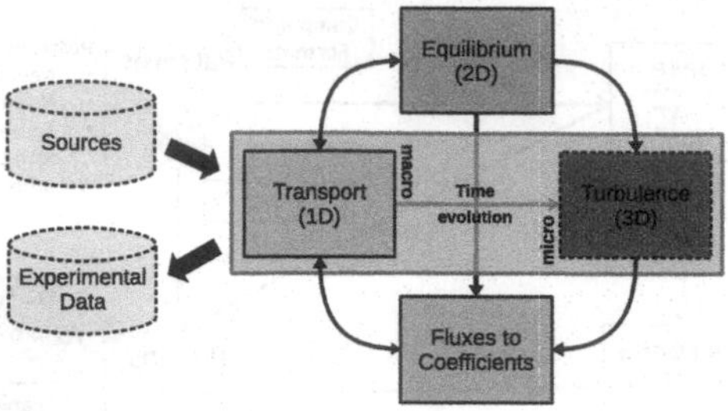

Fig. 4. Schematic view of the targeted fusion application.

developed to bridge such disparate spatiotemporal scales, for instance Luk et al. [14] couple a gyro-fluid 3D turbulence submodel to a 1D transport solver that evolves the temperature profiles over the macro scales. The turbulence submodel provides heat fluxes from which the transport coefficients are derived, but its output is inherently "noisy". Hence, the calculated profiles are exposed to this noisy signal, and uncertainties will propagate from one submodel to the next. Additional uncertainties come from external sources and from experimental measurements against which the simulation could be validated. This leads to a complex scenario as depicted in Fig. 4.

The goal here is to produce simulated quantities of interest (temperature profile, density, etc.) and their confidence intervals and propagate this additional information through a complex cyclic workflow that involves several components with different properties, computational costs and uncertain inputs. The confidence intervals allow for better validation of the interpretative simulation against experimental results (for existing tokamaks machines), and give insight into the confidence of predictive simulations.

Following previous work on quantifying the propagation of uncertainty through a 'black-box' model in fusion plasma [15], we started to apply a polynomial chaos expansion method (PCE) to the cheaper single-scale models from Fig. 4, but taken separately (each one becomes a black-box). The method doesn't require changes to the underlying model equations and provides a quantitative measure for which combinations of inputs have the most important impact on the results. The PCE coefficients are used to compute statistical metrics that are essential for the basic descriptions of the output distribution. To compute those coefficients we use a quadrature scheme which depends on the polynomial type which itself depends on the probability distribution of uncertain inputs. As a result, integration rules along each axis can be calculated using a tensor product.

Even with a limited number of uncertain parameters resulting from the turbulence code (e.g 8 for a fluid code using a flux-tube approximation), and assuming these parameters are not correlated, this method will necessitate approximately 1.7 millions of runs of *Fluxes to Coefficients* and *Transport* codes if we want to calculate their propagation through these models. As these are the cheapest codes in the application, this represents only 512 CPU-hours of an embarrassingly parallel job which remains tractable with traditional means. But when the complexity of the models and the number of parameters increase, such quantities of runs with potentially large range of run times becomes very challenging and will require using advanced capabilities of 'smart' pilot-job software.

3.2 Variety of Other Applications

Climate. Computational models for atmospheric and oceanic flows are central to climate modeling and weather forecasting. As these models have finite resolutions, they employ simplified representations, so-called parameterizations, to account for the impact of unresolved processes on larger scales. An example is the treatment of atmospheric convection and cloud processes, which are important for the atmospheric circulation and hydrological cycle, but are unresolved in global atmospheric models. These parameterizations are a source of uncertainties: they have parameters that can be difficult to determine, and even their structural form can be uncertain.

A computationally very expensive approach to improve parameterizations and reduce uncertainty is by locally replacing the parameterization with a high-resolution simulation. In [16] this is applied regionally, by nesting the Dutch Atmospheric Large-Eddy Simulation (DALES) model in a selected number of global model columns, replacing the convection parameterization in these columns. The local DALES models run independently from each other and only exchange mean profiles with the global model. While this set-up allows for the use of massively parallel computer systems, running a cloud-resolving simulation in every single model column of a global model remains computationally unfeasible.

Within VECMA we are therefore developing tools for surrogate modeling. More specifically, we aim for statistically representative, data-driven surrogate models that account for the uncertainties in subgrid-scale responses due to their internal nonlinear and possibly chaotic dynamics. The surrogates are to be constructed from a (limited) set of reference data, with the goal of accurately reproducing long-term statistics, in line with the approach from [17,18].

Forced Displacement. Accurate predictions of forced population displacement can help governments and NGO's in making decisions as to how to help refugees, and efficiently allocate humanitarian resources to overcome unintended consequences [19]. We enable these simulations by establishing an automated agent-based modeling approach, FabFlee, which is a plugin to FabSim3 that uses the Flee agent-based simulation code. Flee forecasts the distribution of incoming refugees across destination camps [20], while FabFlee provides an environment to run and analyze simulation ensembles under various policy constraints, such as forced redirections, camp and border closures [21]. At time of writing, we are combining FabFlee with EasyVVUQ to more efficiently perform sensitivity analysis for varying agent awareness levels and speed limits of refugee movements [22].

Materials Modeling. Prediction of nanocomposite material properties requires multiscale workflows that capture mechanisms at every characteristic scale of the material, from chemical specificity to engineering testing conditions. For nanocomposite systems, the characteristic time and length of its nanostructure and macrostructure are so far apart, their respective dynamics can be simulated separately. We use DealLAMMPS [23,24], a new program that simulates the nanoscale with LAMMPS molecular dynamics, and the macroscale is simulated using deal.II, a finite element solver. Boundary information is passed from the FEM model to the LAMMPS simulations, the stresses arising from these changes are used to propagate the macroscale model. This workflow creates a vast number of short nanoscale, unpredictable at runtime, simulations at each macroscale timestep. Handling the execution of these task requires automation and coordination between several resources. As just the nanoscale errors due to uncertainty in the boundary condition and even the stress measured for a boundary change are correlated, complex and expensive to calculate. Over several iterations, errors can proliferate and careful consideration and understanding of these is needed from tools that VECMAtk can provide.

Molecular Dynamics Force Fields. Molecular dynamics calculations are used not only in materials modeling but in a wide range of other fields. Choices made in the design of these calculations such as the parameterization of the force field describing chemical components within the system and cut-offs used for long range interactions can have a profound effect on the results obtained and their variability. One field in which this of particular interest in free energy calculations which are increasingly widely used in modern drug design and refinement workflows. The binding affinity calculator (BAC) [25] we have developed automates this class of calculation, from model building through simulations and analysis. In order to understand the impact of forcefield parameter decisions on calculations performed using the BAC we are creating new workflows which incorporate sensitivity analysis through EasyVVUQ. The use of protocols based on ensemble simulations (known as TIES and ESMACS [26]) will give us the ability to adjust the simulation duration and ensemble size in order to robustly determine the uncertainty of results for comparison. This effort builds on previous work which uses Pilot Job manager to handle the execution of ensembles of multiple runs to enable bootstrap error statistics to be applied to calculations for individual protein-ligand pairs.

Cardiovascular Simulation. Haemodynamic simulations provide a non-invasive means of estimating flow rates, pressures and wall shear stresses in the human vasculature. Through MRI and CT scans, patient specific models may be used, with clinical applications such as predicting aneurysm rupture or treatment. We use a 3D lattice-Boltzmann solver, HemeLB [27], to simulate the continuum dynamics of bloodflow through large and highly sparse vascular systems efficiently [28]. A recent validation study focused on HemeLB simulations of a real patient Middle Cerebral Artery (MCA), using transcranial Doppler measurements of the blood velocity profile for comparison, as well as exploring the effects of a change in rheology model or inlet flow rate on the results [29]. We are now running full 3D simulations of the entire human arterial tree, with an aim of also integrating the venous tree and a cyclic coupling to state-of-the-art heart models. This necessarily introduces an even greater number of input parameters which, when combined with the great computational expense of the submodels, presents a real challenge for validation and verification of our application, and particularly with regards to a computationally feasible sensitivity analysis and uncertainty quantification within such a system.

Biomedical Model. Coronary artery stenosis is a cardiovascular disease of narrowing of the coronary artery due to clustering of fatty plaque. A common treatment is to dent the fatty plaque into the artery wall and to deploy a stent in order to keep the artery open. However, up to 10% cases end up in the re-narrowing of the artery due to an excessive healing response, which is called in-stent restenosis (ISR) [30]. The multiscale in-stent restenosis model simulates this process at different time scales [31].

In [32], uncertainty of the response of the two-dimensional ISR model on the cross-sectional lumen area was estimated and analyzed. Uncertainty quantification showed up to 15% aleatory and about 25% total uncertainty in the model predictions. Additionally, sensitivity analysis identified the endothelium regeneration time as the most influential parameter on the model response. In [33], the semi-intrusive multiscale metamodeling uncertainty quantification method was applied to improve the performance of the uncertainty analysis. Depending on the surrogate used, the simulation time of the semi-intrusive method was up to five times faster than of a Monte Carlo method. In future work, the semi-intrusive metamodeling method will be applied to the three dimensional version of the ISR model employing the VECMAtk, since a black-box approach is not feasible for this application due to high computational demand [34].

4 Roadmap and Release Strategy

For the VECMA toolkit we have adopted a release schedule with two release types: Minor releases are tagged every 3 months, advertised within the project, made public without dedicated advertising, and with a limited amount of additional documentation and examples. Major releases will be made in June 2019, June 2020 and December 2020, and are public and fully advertised. They are accompanied with extensive documentation, examples, training events and dedicated uptake activities. We are at present able to guarantee formal support for the VECMA toolkit up to June 2021, and can future-proof VECMAtk at least until 6 months after the final planned release of the toolkit. The June 2019 VECMAtk release contains FabSim3, EasyVVUQ, and several functionalities provided by QCG. Later releases will likely feature additional components. In-between releases, users will be able to access the latest code for FabSim3 and EasyVVUQ, as we are maintaining an open development environment.

Containerization of VECMAtk. Containerization [35] is a method that allow us to create a virtual machine (VM) in more easily compared to the traditional virtualization approach. These type of VMs, called *containers*, share the operating system kernel, as well as resources such as storage and networking. Containers matter because they enhance reproducibility: every run using them is guaranteed to have the same settings and configurations. Additionally, containers provide application platform portability, as they can be migrated to other computing environments without requiring code changes. Each container includes the run-time components – such as files, environment variables and libraries – necessary to run the desired software on a single control host, accessing a single kernel. Among the implementations available, we choose to use Docker (docs.docker.com).

In this work, we set up the FabSim GitHub repository with Travis-CI. After each successful test executed by Travis, a docker image is configured and built, and pushed into Docker Hub. For this work, A Docker image is provided as well through the Docker Hub[3]. And, the docker bundle for easy deployment is available in Docker Hub via: `docker pull vecmafabsim3/fabsimdocker`.

However, although Docker is one of most popular container technologies for software reproducibility, it has low adoption in the HPC world since it requires users with root access to run Docker and execute a containerized applications. To support high performance computing use cases, where users should only have access to their own data, Singularity[4] can be used as a solution for container system in HPC environment. Singularity containers differ from Docker containers in several important ways, including the handling of namespaces, user privileges, and the images themselves. A singularity image, alongside docker image for this toolkit, is available via: https://singularity-hub.org/collections/2536.

5 Conclusions

In this paper we have outlined the design and development process used in the creation of the VECMA toolkit for validation, verification and uncertainty quantification (VVUQ) of multiscale HPC applications. A number of exemplar applications from a diverse range of scientific domains (fusion, climate, population displacement, materials, drug binding affinity, and cardiovascular modeling) are being used to test and guide the development of new features for the toolkit. Through this work we aim to make VVUQ certification of complex, multiscale workflows on high end computing facilities a standard practice.

Acknowledgements. We are grateful to the VECMA consortium, Scientific Advisory Board and the VECMAtk alpha users for their constructive discussions and input around this work. We acknowledge funding support from the European Union's Horizon 2020 research and innovation programme under grant agreement 800925 (VECMA project, www.vecma.eu), and the UK Consortium on Mesoscale Engineering Sciences (UKCOMES) under the UK EPSRC Grant No. EP/L00030X/1.

[3] https://hub.docker.com/r/vecmafabsim3/fabsimdocker.

[4] https://singularity.lbl.gov.

References

1. Oberkampf, W.L., DeLand, S.M., Rutherford, B.M., Diegert, K.V., Alvin, K.F.: Error and uncertainty in modeling and simulation. Reliab. Eng. Syst. Saf. **75**(3), 333–357 (2002)
2. Oberkampf, W.L., Roy, C.J.: Verification and Validation in Scientific Computing. Cambridge University Press, Cambridge (2010)
3. Roy, C.J., Oberkampf, W.L.: A comprehensive framework for verification, validation, and uncertainty quantification in scientific computing. Comput. Methods Appl. Mech. Eng. **200**(25), 2131–2144 (2011)
4. Groen, D., Knap, J., Neumann, P., Suleimenova, D., Veen, L., Leiter, K.: Mastering the scales: a survey on the benefits of multiscale computing software. Philos. Trans. Roy. Soc. A **377**(2142), 20180147 (2019)
5. Nikishova, A., Hoekstra, A.G.: Semi-intrusive uncertainty quantification for multiscale models. arXiv preprint arXiv:1806.09341 (2018)
6. Alowayyed, S., Groen, D., Coveney, P.V., Hoekstra, A.G.: Multiscale computing in the exascale era. J. Comput. Sci. **22**, 15–25 (2017)
7. Baudin, M., Dutfoy, A., Iooss, B., Popelin, A.-L.: OpenTURNS: an industrial software for uncertainty quantification in simulation. In: Ghanem, R., Higdon, D., Owhadi, H. (eds.) Handbook of Uncertainty Quantification, pp. 2001–2038. Springer, Cham (2017). https://doi.org/10.1007/978-3-319-12385-1_64
8. Marelli, S., Sudret, B.: UQLab: a framework for uncertainty quantification in Matlab. In: Vulnerability, Uncertainty, and Risk: Quantification, Mitigation, and Management, pp. 2554–2563 (2014)
9. Tennøe, S., Halnes, G., Einevoll, G.T.: Uncertainpy: a Python toolbox for uncertainty quantification and sensitivity analysis in computational neuroscience. bioRxiv (2018). https://doi.org/10.1101/274779
10. Feinberg, J., Langtangen, H.P.: Chaospy: an open source tool for designing methods of uncertainty quantification. J. Comput. Sci. **11**, 46–57 (2015)
11. Piontek, T., et al.: Development of science gateways using QCG – lessons learned from the deployment on large scale distributed and HPC infrastructures. J. Grid Comput. **14**(4), 559–573 (2016)
12. Groen, D., Bhati, A.P., Suter, J., Hetherington, J., Zasada, S.J., Coveney, P.V.: FabSim: facilitating computational research through automation on large-scale and distributed e-infrastructures. Comput. Phys. Commun. **207**, 375–385 (2016)
13. Sobol, I.: On Quasi-Monte Carlo integrations. Math. Comput. Simul. **47**(2), 103–112 (1998)
14. Luk, O., Hoenen, O., Bottino, A., Scott, B., Coster, D.: ComPat framework for multiscale simulations applied to fusion plasmas. Comput. Phys. Commun. **239**, 126–133 (2019)
15. Preuss, R., von Toussaint, U.: Uncertainty quantification in ion–solid interaction simulations. Nucl. Instrum. Methods Phys. Res. Sect. B: Beam Interact. Mater. Atoms **393**, 26–28 (2017)
16. Jansson, F., van den Oord, G., Pelupessy, I., Grönqvist, J., Siebesma, A., Crommelin, D.: Regional superparameterization in a global circulation model using large eddy simulations (2018, in press)
17. Verheul, N., Crommelin, D.: Data-driven stochastic representations of unresolved features in multiscale models. Commun. Math. Sci **14**(5), 1213–1236 (2016)
18. Verheul, N., Viebahn, J., Crommelin, D.: Covariate-based stochastic parameterization of baroclinic ocean eddies. Math. Clim. Weather Forecast. **3**(1), 90–117 (2017)

19. Groen, D.: Simulating refugee movements: where would you go? Procedia Comput. Sci. **80**, 2251–2255 (2016)
20. Suleimenova, D., Bell, D., Groen, D.: A generalized simulation development approach for predicting refugee destinations. Sci. Rep. **7**, 13377 (2017)
21. Suleimenova, D., Bell, D., Groen, D.: Towards an automated framework for agent-based simulation of refugee movements. In: Chan, W.K.V., DAmbrogio, A., Zacharewicz, G., Mustafee, N., Wainer, G., Page, E., (eds.): Proceedings of the 2017 Winter Simulation Conference, Las Vegas, Nevada, IEEE, pp. 1240–1251 (2017)
22. Suleimenova, D., Groen, D.: How policy decisions affect refugee journeys in South Sudan: a study using automated ensemble simulations. J. Artif. Soc. and Soc. Simul. (2019, submitted)
23. Vassaux, M., Richardson, R.A., Coveney, P.V.: The heterogeneous multiscale method applied to inelastic polymer mechanics. Philos. Trans. Roy. Soc. A **377**, 20180150 (2019)
24. Vassaux, M., Sinclair, R.C., Richardson, R.A., Suter, J.L., Coveney, P.V.: The role of graphene in enhancing the mechanical properties of epoxy resins. Adv. Theory Simul. **2**, 1800168 (2019)
25. Sadiq, S.K., Wright, D., Watson, S.J., Zasada, S.J., Stoica, I., Coveney, P.V.: Automated molecular simulation based binding affinity calculator for ligand-bound HIV-1 proteases. J. Chem. Inf. Model. **48**(9), 1909–1919 (2008). PMID: 18710212
26. Wan, S., et al.: Rapid and reliable binding affinity prediction of bromodomain inhibitors: a computational study. J. Chem. Theory Comput. **13**(2), 784–795 (2017). PMID: 28005370
27. Mazzeo, M.D., Coveney, P.V.: HemeLB: a high performance parallel lattice-Boltzmann code for large scale fluid flow in complex geometries. Comput. Phys. Commun. **178**(12), 894–914 (2008)
28. Patronis, A., Richardson, R.A., Schmieschek, S., Wylie, B.J., Nash, R.W., Coveney, P.V.: Modelling patient-specific magnetic drug targeting within the intracranial vasculature. Front. Physiol. **9**, 331 (2018)
29. Groen, D., et al.: Validation of patient-specific cerebral blood flow simulation using transcranial Doppler measurements. Front. Physiol. **9**, 721 (2018)
30. Fernández-Ruiz, I.: Interventional cardiology: drug-eluting or bare-metal stents? Nat. Rev. Cardiol. **13**(11), 631–631 (2016)
31. Caiazzo, A., et al.: A complex automata approach for in-stent restenosis: two-dimensional multiscale modelling and simulations. J. Comput. Sci. **2**(1), 9–17 (2011)
32. Nikishova, A., Veen, L., Zun, P., Hoekstra, A.G.: Uncertainty quantification of a multiscale model for in-stent restenosis. Cardiovasc. Eng. Technol. **9**(4), 761–774 (2018)
33. Nikishova, A., Veen, L., Zun, P., Hoekstra, A.G.: Semi-intrusive multiscale metamodeling uncertainty quantification with application to a model of in-stent restenosis. Philos. Trans. A **377**(2142), 20180154 (2018)
34. Zun, P.S., Anikina, T., Svitenkov, A., Hoekstra, A.G.: A comparison of fully-coupled 3D in-stent restenosis simulations to in-vivo data. Front. Physiol. **8**, 284 (2017)
35. Docker: Docker for the Virtualization Admin. eBook (2016)

Track of Simulations of Flow and Transport: Modeling, Algorithms and Computation

deal.II Implementation of a Weak Galerkin Finite Element Solver for Darcy Flow

Zhuoran Wang[1], Graham Harper[1], Patrick O'Leary[2], Jiangguo Liu[1(✉)], and Simon Tavener[1]

[1] Colorado State University, Fort Collins, CO 80523, USA
{wangz,harper,liu,tavener}@math.colostate.edu
[2] Kitware, Inc., Santa Fe, NM 87507, USA
patrick.oleary@kitware.com

Abstract. This paper presents a weak Galerkin (WG) finite element solver for Darcy flow and its implementation on the deal.II platform. The solver works for quadrilateral and hexahedral meshes in a unified way. It approximates pressure by Q-type degree $k(\geq 0)$ polynomials separately defined in element interiors and on edges/faces. Numerical velocity is obtained in the unmapped Raviart-Thomas space $RT_{[k]}$ via postprocessing based on the novel concepts of discrete weak gradients. The solver is locally mass-conservative and produces continuous normal fluxes. The implementation in deal.II allows polynomial degrees up to 5. Numerical experiments show that our new WG solver performs better than the classical mixed finite element methods.

Keywords: Darcy flow · deal.II · Finite element methods · Hexahedral meshes · Quadrilateral meshes · Weak Galerkin

1 Introduction

The Darcy equation, although simple, plays an important role for modeling flow in porous media. The equation usually takes the following form

$$\begin{cases} \nabla \cdot (-\mathbf{K}\nabla p) + c\,p = f, & \mathbf{x} \in \Omega, \\ p|_{\Gamma^D} = p_D, \quad ((-\mathbf{K}\nabla p)\cdot \mathbf{n})|_{\Gamma^N} = u_N, \end{cases} \tag{1}$$

where Ω is a 2-dim or 3-dim bounded domain, p is the unknown pressure, $\mathbf{K}$ is a conductivity matrix that is uniformly symmetric positive definite (SPD), c is a known function, f is a known source term, p_D is a Dirichlet boundary condition, u_N is a Neumann boundary condition, and $\mathbf{n}$ is the outward unit normal vector on $\partial\Omega$, which has a nonoverlapping decomposition $\Gamma^D \cup \Gamma^N$.

Harper, Liu, and Wang were partially supported by US National Science Foundation grant DMS-1819252. We thank Dr. Wolfgang Bangerth for the computing resources.

J. M. F. Rodrigues et al. (Eds.): ICCS 2019, LNCS 11539, pp. 495–509, 2019.
https://doi.org/10.1007/978-3-030-22747-0_37

The elliptic boundary value problem (1) can be solved by many types of finite element methods. But in the context of Darcy flow, *local mass conservation* and *normal flux continuity* are two most important properties to be respected by finite element solvers.

- The continuous Galerkin (CG) methods [5] use the least degrees of freedom but do not possess these two properties and hence cannot be used directly. Several post-processing procedures have been developed [7,8].
- Discontinuous Galerkin (DG) methods are locally conservative by design and gain normal flux continuity after post-processing [4].
- The enhanced Galerkin (EG) methods [19] possess both properties but need to handle some minor issues in implementation.
- The mixed finite element methods (MFEMs) [2,6] have both properties by design but result in indefinite discrete linear systems, for which hybridization needs to be employed to convert them into definite linear systems.
- The weak Galerkin (WG) methods [11,13,15–17,20] have both properties and result in SPD linear systems that are easier to solve.

In this paper, we investigate efficient implementation of WG Darcy solvers in `deal.II`, a popular finite element package [3], with the intention to make WG finite element methods practically useful for large-scale scientific computation.

2 A WG Finite Element Solver for Darcy Flow

WG solvers can be developed for Darcy flow on simplicial, quadrilateral or hexahedral, and more general polygonal or polyhedral meshes. These finite element schemes may or may not contain a stabilization term, depending on choices of the approximating polynomials for pressure in element interiors and on edges/faces. Through integration by parts, these polynomial basis functions are used for computing discrete weak gradients, which are used to approximate the classical gradient in the variational form for the Darcy equation. Discrete weak gradients can be established in a general vector polynomial space [18] or a specific one like the Raviart-Thomas space [11,17] that has desired approximation properties.

This paper focuses on quadrilateral and hexahedral meshes, in which faces are or very close to being flat. We use $Q_k (k \geq 0)$-type polynomials in element interiors and on edges/faces for approximating the primal variable pressure. Their discrete weak gradients are established in local unmapped Raviart-Thomas $RT_{[k]} (k \geq 0)$ spaces, for which we do not use the Piola transformation. We use the same form of polynomials as that for rectangles and bricks in the classical MFEMs [6].

To illustrate these new ideas, we consider a quadrilateral E centered at (x_c, y_c). We define the local unmapped Raviart-Thomas space $RT_{[0]}(E)$ as

$$RT_{[0]}(E) = \text{Span}(\mathbf{w}_1, \mathbf{w}_2, \mathbf{w}_3, \mathbf{w}_4), \tag{2}$$

where

$$\mathbf{w}_1 = \begin{bmatrix} 1 \\ 0 \end{bmatrix}, \quad \mathbf{w}_2 = \begin{bmatrix} 0 \\ 1 \end{bmatrix}, \quad \mathbf{w}_3 = \begin{bmatrix} X \\ 0 \end{bmatrix}, \quad \mathbf{w}_4 = \begin{bmatrix} 0 \\ Y \end{bmatrix}, \tag{3}$$

and $X = x - x_c$, $Y = y - y_c$ are the normalized coordinates.

Now we introduce a new concept of 5 discrete weak functions $\phi_i (0 \le i \le 4)$.
- ϕ_0 is for element interior: It takes value 1 in the interior E° but 0 on the boundary E^∂ (all 4 edges);
- $\phi_1, \phi_3, \phi_3, \phi_4$ are for the four sides respectively: $\phi_i (1 \le i \le 4)$ takes value 1 on the i-th edge but 0 on all other three edges and in the interior.

Any such function ϕ has two independent parts: ϕ° is defined in E°, whereas ϕ^∂ is defined on E^∂, together written as $\phi = \{\phi^\circ, \phi^\partial\}$. Its discrete weak gradient $\nabla_w \phi$ can be specified in $RT_{[0]}(E)$ via integration by parts [20]:

$$\int_E (\nabla_w \phi) \cdot \mathbf{w} = \int_{E^\partial} \phi^\partial (\mathbf{w} \cdot \mathbf{n}) - \int_{E^\circ} \phi^\circ (\nabla \cdot \mathbf{w}), \quad \forall \mathbf{w} \in RT_{[0]}(E). \tag{4}$$

This attributes to solving a size-4 SPD linear system. Note that

(i) For a quadrilateral, ϕ° or ϕ^∂ each can also be a degree $k \ge 1$ polynomial and the discrete weak gradient $\nabla_w \phi$ is then established in the local unmapped Raviart-Thomas space $RT_{[k]}(k \ge 1)$.
(ii) For a hexahedron with nonflat faces, we can use the averaged normal vectors in (4). The Jacobian determinant is still used in computation of the integrals.

For a rectangle $E = [x_1, x_2] \times [y_1, y_2]$ ($\Delta x = x_2 - x_1, \Delta y = y_2 - y_1$), we have

$$\begin{cases} \nabla_w \phi_0 = \quad 0\mathbf{w}_1 + \quad 0\mathbf{w}_2 + \frac{-12}{(\Delta x)^2}\mathbf{w}_3 + \frac{-12}{(\Delta y)^2}\mathbf{w}_4, \\ \nabla_w \phi_1 = \frac{-1}{\Delta x}\mathbf{w}_1 + \quad 0\mathbf{w}_2 + \frac{6}{(\Delta x)^2}\mathbf{w}_3 + \qquad 0\mathbf{w}_4, \\ \nabla_w \phi_2 = \frac{1}{\Delta x}\mathbf{w}_1 + \quad 0\mathbf{w}_2 + \frac{6}{(\Delta x)^2}\mathbf{w}_3 + \qquad 0\mathbf{w}_4, \\ \nabla_w \phi_3 = \quad 0\mathbf{w}_1 + \frac{-1}{\Delta y}\mathbf{w}_2 + \qquad 0\mathbf{w}_3 + \frac{6}{(\Delta y)^2}\mathbf{w}_4, \\ \nabla_w \phi_4 = \quad 0\mathbf{w}_1 + \frac{1}{\Delta y}\mathbf{w}_2 + \qquad 0\mathbf{w}_3 + \frac{6}{(\Delta y)^2}\mathbf{w}_4. \end{cases} \tag{5}$$

Let $\mathcal{E}_h$ be a shape-regular quadrilateral mesh. Let Γ_h^D be the set of all edges on the Dirichlet boundary Γ^D and Γ_h^N be the set of all edges on the Neumann boundary Γ^N. Let S_h be the space of discrete shape functions on $\mathcal{E}_h$ that are degree k polynomials in element interiors and also degree k polynomials on edges. Let S_h^0 be the subspace of functions in S_h that vanish on Γ_h^D. For (1), we seek $p_h = \{p_h^\circ, p_h^\partial\} \in S_h$ such that $p_h^\partial|_{\Gamma_h^D} = Q_h^\partial(p_D)$ (the L^2-projection of Dirichlet boundary data into the space of degree k polynomials on Γ_h^D) and

$$\mathcal{A}_h(p_h, q) = \mathcal{F}(q), \quad \forall q = \{q^\circ, q^\partial\} \in S_h^0, \tag{6}$$

where

$$\mathcal{A}_h(p_h, q) = \sum_{E \in \mathcal{E}_h} \int_E \mathbf{K} \nabla_w p_h \cdot \nabla_w q + \sum_{E \in \mathcal{E}_h} \int_E c p q, \tag{7}$$

$$\mathcal{F}(q) = \sum_{E \in \mathcal{E}_h} \int_E f q^\circ - \sum_{\gamma \in \Gamma_h^N} \int_\gamma u_N q^\partial. \tag{8}$$

This results in a symmetric positive-definite discrete linear system [17].

Note $\nabla_w p_h$ is in the local Raviart-Thomas space, but $-\mathbf{K}\nabla_w p_h$ may not be. A local L^2-projection $\mathbf{Q}_h$ is needed [11,13,17] to get it back into the RT space:

$$\mathbf{u}_h = \mathbf{Q}_h(-\mathbf{K}\nabla_w p_h). \tag{9}$$

This is the numerical Darcy velocity for subsequent applications, e.g., transport simulations. Clearly, this process is readily parallelizable for large-scale computation. This numerical velocity is locally mass-conservative and the corresponding normal flux is continuous across edges or faces, as proved in [11,17].

As shown in [17], this Darcy solver is easy to be implemented and results in a symmetric positive-definite system that can be easily solved by a conjugate-gradient type linear solver. The WG methodology has connections to but is indeed different than the classical mixed finite element methods, especially the hybridized MFEMs [13,14].

3 `deal.II` Implementation of WG Solver for Darcy Flow

`deal.II` is a popular C++ finite element package [3]. It uses quadrilateral and hexahedral meshes instead of simplicial meshes. The former may involve less degrees of freedom than the latter. The resulting linear systems may have smaller sizes, although the setup time for these linear systems may be longer. The setup time is spent on bilinear/trilinear mappings from the reference square/cube to general quadrilaterals/hexahedra and computation of various integrals.

3.1 Quadrilateral and Hexahedral Meshes

`deal.II` handles meshes by the **GridGenerator** class. All mesh information, such as the number of active cells, degrees of freedom, are stored in this class. For any integer $k \geq 0$, our WG($Q_k, Q_k; RT_{[k]}$) solver is locally mass-conservative and produces continuous normal fluxes regardless of the quality of quadrilateral and hexahedral meshes. In order to obtain the desired order k convergence rate in pressure, velocity, and normal fluxes, we require meshes to be asymptotically parallelogram or parallelopiped [11,17].

3.2 Finite Element Spaces

The WG($Q_k, Q_k; RT_{[k]}$) solver involves three finite element spaces. The first two spaces are for the pressure unknowns, the third one (RT space) is used for discrete weak gradients and numerical velocity. In `deal.II` implementation, the first two are combined as

```
FESystem<dim> fe;
```

The third one (with `dim` being 2 or 3) is

```
FE_RaviartThomas<dim> fe_rt;
```

Raviart-Thomas Spaces for Discrete Weak Gradients and Velocity. WG allows use of unmapped RT spaces on quadrilaterals and hexahedra [11, 17]. These spaces use the same polynomials for shape functions as those in the classical RT spaces for 2-dim or 3-dim rectangles [6]. They are respectively,

$$RT_{[k]}(E) = Q_{k+1,k} \times Q_{k,k+1}, \tag{10}$$

$$RT_{[k]}(E) = Q_{k+1,k,k} \times Q_{k,k+1,k} \times Q_{k,k,k+1}. \tag{11}$$

In deal.II, we use degree for k in Eqs. (10) or (11) and have

```
fe_rt(degree);
```

Two Separate Polynomial Spaces for Pressure. Note that for the WG($Q_k, Q_k; RT_{[k]}$) finite element solver for Darcy flow, the pressure is approximated separately in element interiors by Q_k-type polynomials and on edges/faces by Q_k-type polynomials also. Note that the 2nd group of Q_k-type polynomials are defined locally on each edge/face. For the deal.II implementation, we have

```
fe(FE_DGQ<dim>(degree), 1, FE_FaceQ<dim>(degree), 1);
```

where degree is k, that is, the degree of the polynomials, "1" means these two groups of pressure unknowns are just scalars. Note that

- **FE_DGQ** is a finite element class in deal.II that has no continuity across faces or vertices, i.e., every shape function lives exactly in one cell. So we use it to approximate the pressure in element interiors.
- **FE_FaceQ** is a finite element class that is defined only on edges/faces.

However, these two different finite element spaces are combined into one finite element system, we split these shape functions as

```
const FEValuesExtractors::Scalar interior(0);
const FEValuesExtractors::Scalar face(1);
```

Here "0" corresponds to the 1st finite element class **FE_DGQ** for the interior pressure; "1" corresponds to the 2nd finite element class **FE_FaceQ** for the face pressure. Later on, we will just use **fe_values[interior].value** and **fe_values[face].value** for assembling the element-level matrices.

3.3 Gaussian Quadratures

Finite element computation involves various types of integrals, which are discretized via quadratures, e.g., Gaussian quadratures. For example, we consider

$$\int_E f \approx \sum_{k=1}^{K} w_k f(x_k, y_k) J_k, \tag{12}$$

where K is the number of quadrature points, (x_k, y_k) is the k-th quadrature point, J_k is the corresponding Jacobian determinant, and w_k is the weight. In deal.II, this is handled by the **Quadrature** class. In particular, the Jacobian determinant value and weight for each quadrature point are bundled together as

```
fe_values.JxW(q_k);
```

where q_k is the k-th quadrature point.

3.4 Linear Solvers

`deal.II` provides a variety of linear solvers that are inherited from `PETSc`. The global discrete linear systems obtained from the weak Galerkin finite element discretization of the Darcy equation are symmetric positive-definite. Thus we can choose a conjugate-gradient type linear solver for them.

3.5 Graphics Output

In our $\mathrm{WG}(Q_k, Q_k; RT_{[k]})$ solver for Darcy flow, the scalar pressures are defined separately in element interiors and on edges/faces of a mesh. These values are output separately in `deal.II`. The interior pressures are handled by **DataOut**, whereas the face pressures are handled by **DataOutFace**. Specifically,

```
data_out.build_patches(fe.degree);
data_out_face.build_patches(fe.degree);
```

are used to subdivide each cell into smaller patches, which provide better visualization if we use higher degree polynomials. The post-processed data are saved as `vtk` files for later visualization in `VisIt`.

4 Code Excerpts with Comments

This section provides some code excerpts with comments. More details can be found in `deal.II` tutorial Step-61 (subject to minor changes) [1].

4.1 Construction of Finite Element Spaces

Note that `FE_RaviartThomas` is a Raviart-Thomas space for vector-valued functions, `FESystem` defines finite element spaces in the interiors and on edges/faces. Shown below is the code for the lowest order WG finite elements.

```
FE_RaviartThomas<dim> fe_rt;
DoFHandler<dim> dof_handler_rt;
FESystem<dim> fe;
DoFHandler<dim> dof_handler;
```

```
fe_rt (0);
dof_handler_rt (triangulation);
fe (FE_DGQ<dim>(0), 1, FE_FaceQ<dim>(0), 1);
dof_handler (triangulation);
```

4.2 System Setup

The following piece distributes degrees of freedom for finite element spaces.

```
dof_handler_rt.distribute_dofs (fe_rt);
dof_handler.distribute_dofs (fe);
```

The following piece sets up matrices and vectors in the system.

```
DynamicSparsityPattern dsp(dof_handler.n_dofs());
DoFTools::make_sparsity_pattern(dof_handler, dsp, constraints);
sparsity_pattern.copy_from(dsp);
system_matrix.reinit(sparsity_pattern);
solution.reinit(dof_handler.n_dofs());
system_rhs.reinit(dof_handler.n_dofs());
```

4.3 System Assembly

The following piece uses extractors to extract components of finite element shape functions.

```
const FEValuesExtractors::Vector velocities (0);
const FEValuesExtractors::Scalar interior (0);
const FEValuesExtractors::Scalar face (1);
```

The following pieces calculates the Gram matrix for the RT space.

```
for (unsigned int q = 0; q < n_q_points_rt; ++q) {
  for (unsigned int i = 0; i < dofs_per_cell_rt; ++i) {
    const Tensor<1,dim> phi_i_u =
          fe_values_rt[velocities].value(i,q);
    for (unsigned int j = 0; j < dofs_per_cell_rt; ++j) {
      const Tensor<1,dim>  phi_j_u =
            fe_values_rt[velocities].value (j, q);
              cell_matrix_rt(i,j) += phi_i_u * phi_j_u
                                   * fe_values_rt.JxW(q);
} } }
```

The following piece handles construction of WG local matrices.

```
for (unsigned int q = 0; q < n_q_points_rt; ++q) {
 for (unsigned int i = 0; i<dofs_per_cell; ++i) {
   for (unsigned int j = 0; j<dofs_per_cell; ++j) {
    for (unsigned int k = 0; k<dofs_per_cell_rt; ++k) {
```

```
        const Tensor<1,dim> phi_k_u =
              fe_values_rt[velocities].value(k,q);
        for (unsigned int l = 0; l < dofs_per_cell_rt; ++l) {
          const Tensor<1,dim> phi_l_u =
                fe_values_rt[velocities].value(l,q);
                local_matrix(i,j) += coefficient_values[q] *
                    cell_matrix_C[i][k] * cell_matrix_C[j][l] *
                    phi_k_u * phi_l_u * fe_values_rt.JxW(q);
} } } } }
```

The following piece calculates the local right-hand side.

```
for (unsigned int q = 0; q < n_q_points; ++q) {
  for (unsigned int i = 0; i < dofs_per_cell; ++i) {
   cell_rhs(i) += (fe_values[interior].value(i, q) *
     right_hand_side.value(fe_values.quadrature_point(q)) *
     fe_values.JxW(q));
} }
```

The following piece distributes entries of local matrices into the system matrix and also incorporates the local right-hand side into the system right-hand side.

```
cell->get_dof_indices(local_dof_indices);
constraints.distribute_local_to_global(
            local_matrix, cell_rhs, local_dof_indices,
            system_matrix, system_rhs);
```

5 Numerical Experiments

This section presents three numerical examples (Eq. (1) with $c = 0$) to demonstrate accuracy and robustness of our novel WG solver for Darcy flow.

Example 1 (A smooth example for convergence rates). Here we have domain $\Omega = (0,1)^2$, conductivity $\mathbf{K} = \mathbf{I}_2$, and a known solution for the pressure:

$$p(x,y) = \sin(\pi x)\sin(\pi y).$$

A homogeneous Dirichlet boundary condition is posed on the entire boundary.

The WG$(Q_k, Q_k; RT_{[k]})$ solver is tested on Example 1 for $k = 0, 1, 2$ on a sequence of uniform rectangular meshes. As shown in Table 1, the solver exhibits order k convergence rates for the L^2-norms of the errors in the interior pressure, velocity, and normal flux. Shown in Fig. 1 are the profiles of the numerical pressure obtained from applying the WG$(Q_1, Q_1; RT_{[1]})$ solver. In the right panel, the edge pressures are plotted as grey line segments. The graphical results in both panels demonstrate nice monotonicity in the numerical pressure produced by our WG solver.

Table 1. Ex.1: Convergence rates of WG($Q_k, Q_k; RT_{[k]}$) solver on rectangular meshes

$1/h$	$\|p - p_h^\circ\|$	Rate	$\|\mathbf{u} - \mathbf{u}_h\|$	Rate	$\|\mathbf{u} \cdot \mathbf{n} - \mathbf{u}_h \cdot \mathbf{n}\|$	Rate
WG($Q_0, Q_0; RT_{[0]}$)						
4	1.5870E−01	—	5.1289E−01	—	7.0500E−01	—
8	7.9980E−02	0.988	2.5309E−01	1.018	3.5523E−01	0.988
16	4.0058E−02	0.997	1.2608E−01	1.005	1.7796E−01	0.997
32	2.0037E−02	0.999	6.2977E−02	1.001	8.9020E−02	0.999
64	1.0020E−02	0.999	3.1481E−02	1.000	4.4516E−02	0.999
128	5.0099E−03	1.000	1.5740E−02	1.000	2.2258E−02	1.000
WG($Q_1, Q_1; RT_{[1]}$)						
4	1.6130E−02	—	5.0989E−02	—	7.1588E−02	—
8	4.0560E−03	1.991	1.2762E−02	1.998	1.8016E−02	1.990
16	1.0155E−03	1.997	3.1915E−03	1.999	4.5113E−03	1.997
32	2.5396E−04	1.999	7.9792E−04	1.999	1.1283E−03	1.999
64	6.3496E−05	1.999	1.9948E−04	1.999	2.8210E−04	1.999
128	1.5874E−05	2.000	4.9871E−05	1.999	7.0528E−05	1.999
WG($Q_2, Q_2; RT_{[2]}$)						
4	1.0719E−03	—	3.3764E−03	—	4.7589E−03	—
8	1.3465E−04	2.992	4.2331E−04	2.995	5.9814E−04	2.992
16	1.6852E−05	2.998	5.2952E−05	2.998	7.4870E−05	2.998
32	2.1072E−06	2.999	6.6203E−06	2.999	9.3620E−06	2.999
64	2.6342E−07	2.999	8.2757E−07	2.999	1.1703E−06	2.999
128	3.2928E−08	2.999	1.0344E−07	2.999	1.46298E−07	2.999

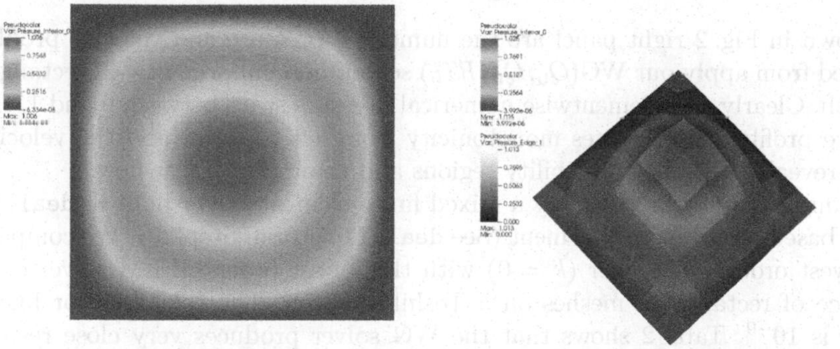

Element interior pressure for $h = \frac{1}{16}$ Interior/edge pressure 3d view ($h = \frac{1}{8}$)

Fig. 1. Ex.1: Numerical pressure by WG($Q_1, Q_1; RT_{[1]}$) solver on rectangular meshes

Example 2 (Heterogeneous permeability). The permeability profile is adopted from [9]. We consider a simple Darcy flow problem on the unit square. Dirichlet boundary conditions are posed on the left and right sides: $p = 1$ for

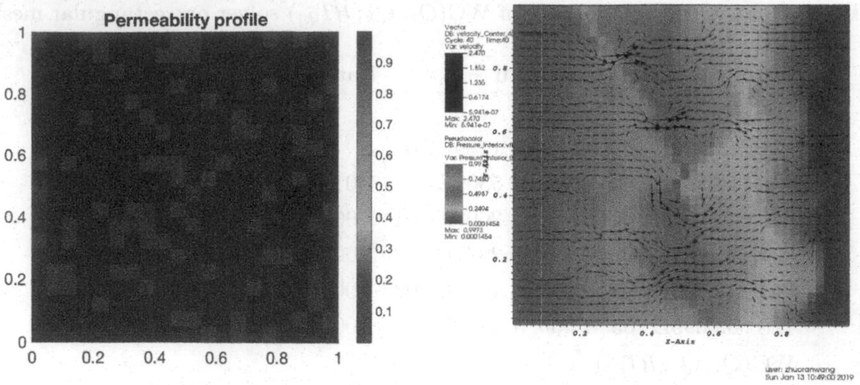

Fig. 2. Example 2 (Heterogeneous permeability): Numerical pressure and velocity

Table 2. Example 2: Comparison between WG and MFEM solvers

tol $= 10^{-9}$	WG			MFEM		
Mesh	$p_{\min}$	$p_{\max}$	Runtime	$p_{\min}$	$p_{\max}$	Runtime
20 × 20	1.21321E–4	0.995113	0.857 s	1.21320E–4	0.995113	1.410 s
40 × 40	1.45401E–4	0.997289	6.759 s	1.45401E–4	0.997289	13.833 s
80 × 80	8.73042E–5	0.998587	59.070 s	8.73043E–5	0.998587	103.141 s
160 × 160	4.59350E–5	0.999281	607.556 s	4.59345E–5	0.999281	877.648 s

$x = 0$; and $p = 0$ for $x = 1$. The other two sides have a homogeneous Neumann (no-flow) boundary condition. The problem was also tested using `Matlab` in [17].

Shown in Fig. 2 right panel are the numerical pressure and velocity profiles obtained from apply our WG($Q_0, Q_0; RT_0$) solver on a uniform 40×40 rectangular mesh. Clearly, the elementwise numerical pressure stays between 0 and 1, the pressure profile demonstrates monotonicity from left to right, and the velocity profile reveals the low-permeability regions and channels for fast flow.

Example 2 was also solved by a mixed finite element solver built in `deal.II` that is based on Schur complement (See `deal.II` tutorial Step-20). We compare the lowest order WG solver ($k = 0$) with the lowest order MFEM solver on a sequence of rectangular meshes on a Toshiba laptop. The tolerance for linear solvers is 10^{-9}. Table 2 shows that the WG solver produces very close results with significantly less runtime.

Example 3 (Permeability profile in SPE10 Model 2). SPE10 was developed as a benchmark for upscaling methods, but the 2nd dataset is becoming a popular testcase for comparing different numerical methods. The dataset is a 3-dim geo-statistical realization from the Jurassic Upper Brent formations [12]. The model has geometric dimensions 1200 (ft) × 2200 (ft) × 170 (ft). The dataset is provided on a $60 \times 220 \times 85$ Cartesian grid, in which each block has a size

20 (ft) $\times$ 10 (ft) $\times$ 2 (ft). The top 70 ft (35 layers) are for the shallow-marine Tarbert formation, the bottom 100 ft (50 layers) are for the fluvial Ness formation. The SPE10 model is structurally simple but highly heterogeneous in porosity and permeability Fig. 3. It poses significant challenges to numerical simulators.

The SPE 10 dataset is publicly available at http://www.spe.org/web/csp/. The original data assume the z-axis pointing downwards but use a right-hand coordinate system. A conversion of ordering in blocks is needed for the original data items. We use the code in `Matlab Reservoir Simulation Toolbox` (MRST) [12] to acquire the needed data.

In this paper, we focus on the Darcy flow part. We use the original permeability data and consider a flow problem. Dirichlet boundary conditions are posed on two boundary faces: $p = 1$ for $y = 0$; and $p = 0$ for $y = 1200$. All other four boundary faces have a homogeneous Neumann (no-flow) boundary condition.

We test the WG solver on three meshes (coarse, medium, fine). For better visualization, we tripled the z dimension.

(i) A **coarse mesh** with 12 $\times$ 44 $\times$ 17 partitions. For the WG($Q_0, Q_0; RT_{[0]}$) solver, there are 8,976 pressure degrees of freedom (DOFs) for element interiors; 28,408 pressure DOFs for all faces, and 37,384 total DOFs. The local $RT_{[0]}$ spaces are used to compute the discrete weak gradients of the pressure basis functions, but they do not constitute any DOFs.
(ii) A **medium mesh** with $30 \times 110 \times 85$ partitions. We use WG($Q_0, Q_0; RT_{[0]}$) again. There are 280,500 DOFs for the pressure in element interiors, 856,700 DOFs for all faces, and totally 1,137,200 (about 1M) DOFs.
(iii) A **fine mesh** with $60 \times 220 \times 85$ partitions, which is the same as the original gridblock. Again WG($Q_0, Q_0; RT_{[0]}$) is used. There are 1,122,000 interior DOFs; 3,403,000 face DOFs; and a total 4,525,000 (about 4M) DOFs.

As shown in Fig. 4, the coarse mesh is too coarse to reveal the reservoir geological features. The medium-mesh result is good enough to reflect the channel

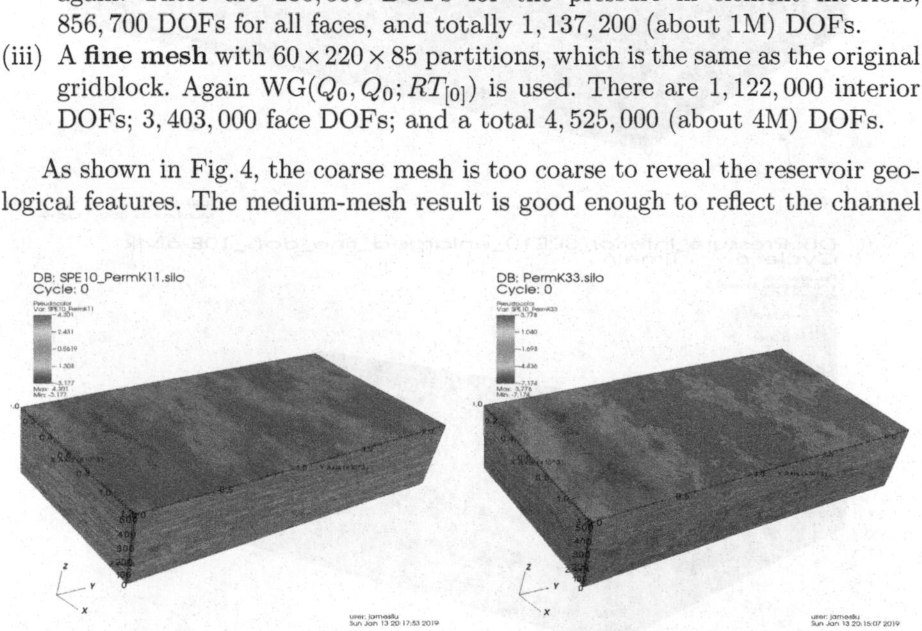

(a) Permeability $K_{11} = K_{22}$ (a) Permeability K_{33}

Fig. 3. Example 3 (SPE10 Model 2): Permeability profiles on log_{10} scales

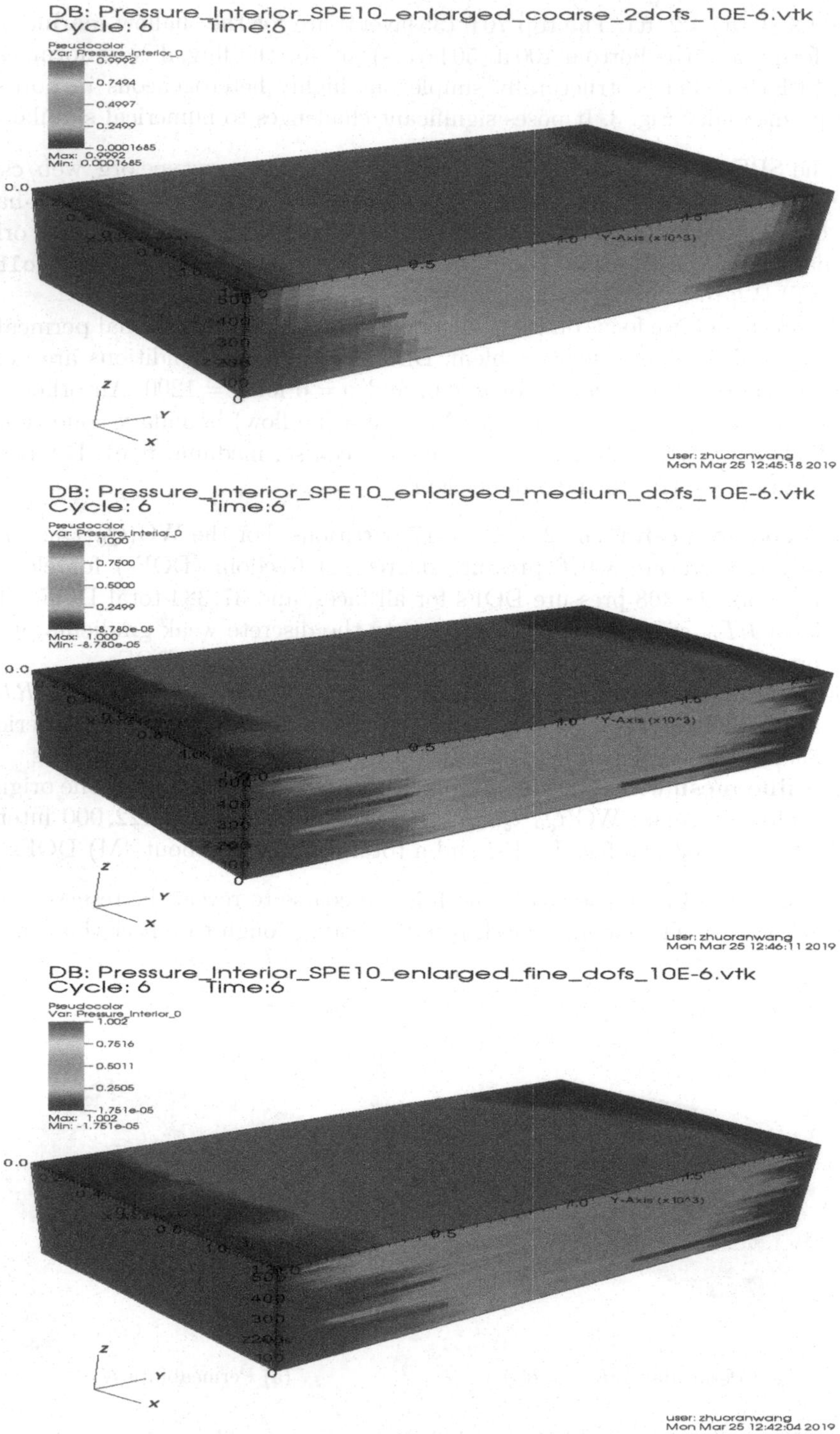

Fig. 4. Example 3 (SPE10): Numerical pressure for coarse, medium, and fine meshes

Table 3. Example 3: SPE10_Darcy by WG($Q_0, Q_0; RT_{[0]}$) on 3 meshes

tol = 10^{-6}	MaxItrs	#Itrs	$p_{\min}$	$p_{\max}$	Runtime
coarse	2*DOFs	40,383	1.684E−4	0.999,151	2 m 45 s
medium	DOFs	207,704	−8.779E−5	1.000,003	1 h 34 m
fine	DOFs	241,492	−1.750E−5	1.002,136	6 h 42 m
tol = 10^{-9}	MaxItrs	#Itrs	$p_{\min}$	$p_{\max}$	Runtime
coarse	6*DOFs	213,316	1.682E−4	0.999,151	5 m 24 s
medium	DOFs	901,327	−8.782E−5	1.000,002	3 h 20 m
fine	DOFs	1,371,887	−1.750E−5	1.000,083	17 h 38 m

Table 4. Example 3: SPE10_Darcy by MFEM($RT_{[0]}, Q_0$) on 3 meshes

tol	$p_{\min}$	$p_{\max}$	Runtime
coarse mesh			
10^{-3}	1.693E−4	0.998,574	8 m 39 s
10^{-6}	1.682E−4	0.999,150	34 m 02 s
10^{-9}	1.682E−4	0.999,151	1 h 16 m
medium mesh			
10^{-3}	−8.712E−5	1.001,850	6 h 04 m
10^{-6}	−8.782E−5	1.000,002	31 h 25 m
10^{-9}	−8.782E−5	1.000,002	82 h 58 m
fine mesh			
10^{-3}	−1.745E−5	1.028,902	71 h 16 m
10^{-6}	DidNotTry		
10^{-9}	DidNotTry		

features of the fluvial Ness formation. The fine-mesh result is smoother and exposes further details about the heterogeneity, but requires expensive computation. Tables 3 and 4 (results on a server with 40 Intel CPUs) together demonstrate that our new WG solver is more efficient than the classical MFEM.

6 Concluding Remarks

The novel weak Galerkin finite element methods represent a different type of methodology for solving many real-world problems modeled by partial differential equations. There have been efforts on implementing WG FEMs in `Matlab` and `C++`. But the work reported in this paper represents the first ever attempt for implementing WG FEMs in a popular finite element package like `deal.II`. This shall provide open access to the scientific community for examining usefulness of the WG methodology for large-scale scientific computing tasks.

Listed below are some projects for further research.

(i) Preconditioning and parallelization of the WG solver for Darcy flow;
(ii) `deal.II` implementation for coupled WG Darcy solvers and transport solvers for the full problem of SPE10 and alike;
(iii) `deal.II` implementation for both 2-dim and 3-dim for the 2-field poroelasticity solver developed in [10];
(iv) Implementation of WGFEMs for triangular/tetrahedral meshes on `FEniCS`;
(v) Comparison with the hybridizable discontinuous Galerkin (HDG) methods.

References

1. https://github.com/dealii/dealii/tree/master/examples/step-61
2. Arbogast, T., Correa, M.: Two families of mixed finite elements on quadrilaterals of minimal dimension. SIAM J. Numer. Anal. **54**, 3332–3356 (2016)
3. Bangerth, W., Hartmann, R., Kanschat, G.: deal.II- a general purpose object oriented finite element library. ACM Trans. Math. Softw. **33**, 24–27 (2007)
4. Bastian, P., Riviere, B.: Superconvergence and $H(div)$ projection for discontinuous Galerkin methods. Int. J. Numer. Meth. Fluids **42**, 1043–1057 (2003)
5. Brenner, S., Scott, L.: The Mathematical Theory of Finite Element Methods, Texts in Applied Mathematics, vol. 15, 3rd edn. Springer, New York (2008). https://doi.org/10.1007/978-0-387-75934-0
6. Brezzi, F., Fortin, M.: Mixed and Hybrid Finite Element Methods. Springer, New York (1991). https://doi.org/10.1007/978-1-4612-3172-1
7. Bush, L., Ginting, V.: On the application of the continuous Galerkin finite element method for conservation problems. SIAM J. Sci. Comput. **35**, A2953–A2975 (2013)
8. Cockburn, B., Gopalakrishnan, J., Wang, H.: Locally conservative fluxes for the continuous Galerkin method. SIAM J. Numer. Anal. **45**, 1742–1770 (2007)
9. Durlofsky, L.: Accuracy of mixed and control volume finite element approximations to Darcy velocity and related quantities. Water Resour. Res. **30**, 965–973 (1994)
10. Harper, G., Liu, J., Tavener, S., Wang, Z.: A two-field finite element solver for poroelasticity on quadrilateral meshes. In: Shi, Y., et al. (eds.) ICCS 2018. LNCS, vol. 10862, pp. 76–88. Springer, Cham (2018). https://doi.org/10.1007/978-3-319-93713-7_6
11. Harper, G., Liu, J., Zheng, B.: The THex algorithm and a simple Darcy solver on hexahedral meshes. Procedia Comput. Sci. **108C**, 1903–1912 (2017)
12. Lie, K.A.: An introduction to reservoir simulation using MATLAB/GNU Octave. Cambridge University Press (2019). ISBN 9781108492430
13. Lin, G., Liu, J., Mu, L., Ye, X.: Weak Galerkin finite element methdos for Darcy flow: anistropy and heterogeneity. J. Comput. Phys. **276**, 422–437 (2014)
14. Lin, G., Liu, J., Sadre-Marandi, F.: A comparative study on the weak Galerkin, discontinuous Galerkin, and mixed finite element methods. J. Comput. Appl. Math. **273**, 346–362 (2015)
15. Liu, J., Sadre-Marandi, F., Wang, Z.: DarcyLite: a Matlab toolbox for Darcy flow computation. Procedia Comput. Sci. **80**, 1301–1312 (2016)
16. Liu, J., Tavener, S., Wang, Z.: Lowest-order weak Galerkin finite element method for Darcy flow on convex polygonal meshes. SIAM J. Sci. Comput. **40**, B1229–B1252 (2018)

17. Liu, J., Tavener, S., Wang, Z.: The lowest-order weak Galerkin finite element method for the Darcy equation on quadrilateral and hybrid meshes. J. Comput. Phys. **359**, 312–330 (2018)
18. Mu, L., Wang, J., Ye, X.: A weak Galerkin finite element method with polynomial reduction. J. Comput. Appl. Math. **285**, 45–58 (2015)
19. Sun, S., Liu, J.: A locally conservative finite element method based on piecewise constant enrichment of the continuous Galerkin method. SIAM J. Sci. Comput. **31**, 2528–2548 (2009)
20. Wang, J., Ye, X.: A weak Galerkin finite element method for second order elliptic problems. J. Comput. Appl. Math. **241**, 103–115 (2013)

Recovery of the Interface Velocity for the Incompressible Flow in Enhanced Velocity Mixed Finite Element Method

Yerlan Amanbek[1,2(✉)], Gurpreet Singh[1], and Mary F. Wheeler[1]

[1] Center for Subsurface Modeling, Oden Institute for Computational Engineering and Sciences, University of Texas at Austin, Austin, TX, USA
{gurpreet,mfw}@ices.utexas.edu
[2] Nazarbayev University, Astana, Kazakhstan
yerlan.amanbek@nu.edu.kz

Abstract. The velocity, coupling term in the flow and transport problems, is important in the accurate numerical simulation or in the *posteriori* error analysis for adaptive mesh refinement. We consider Enhanced Velocity Mixed Finite Element Method (EVMFEM) for the incompressible Darcy flow. In this paper, our aim is to study the improvement of velocity at interface to achieve the better approximation of velocity between subdomains. We propose the reconstruction of velocity at interface by using the post-processed pressure. Numerical results at the interface show improvement on convergence rate.

Keywords: Domain decomposition · Enhanced Velocity · Velocity improvement

1 Introduction

The numerical reservoir simulations have been utilized in many subsurface applications such as groundwater remediation, reservoir well evaluation, and contaminate transport problems. For such applications, it is common to deal with the flow and transport problem. The main component or coupling term of the flow and transport systems is the velocity and its accuracy the mostly achieved by employing classical mixed finite element system. Due to the heterogeneity of porous media multiphysics problems could be categorized systematically in which one physical phenomena influences within a subdomain and another physical phenomena dominates within another subdomain. Such solutions are coupled through continuity of normal flux at interface, shared region between differently discretized subdomains. To deal with these problems there are the well-known methods such as Multiscale Mortar and Enhanced Velocity schemes that are established in various applications. Recently, a novel adaptive method was studied in subsurface applications [2–4,14] using Enhanced Velocity scheme. The main idea is here to utilize the EVMFEM as domain decomposition method to

J. M. F. Rodrigues et al. (Eds.): ICCS 2019, LNCS 11539, pp. 510–523, 2019.
https://doi.org/10.1007/978-3-030-22747-0_38

couple different discretized subdomains with more accurate upscaled subsurface parameters.

In the simulation of flow with adaptivity, the results obtained in [9] suggest that pressure values could be interpolated using neighboring elements values to approximate auxiliary pressure values within provided elements. Selection of interpolants is based on convex combinations of vertical and horizontal oriented pressure values.

In related reference [6], it was studied that the interface error of solution between subdomains for different numerical methods including Mortar Multi-scale MixedFEM which provided better approximation for second-order elliptic problems. One of reasons is the iterative procedure in the mortar scheme that is a key in coupling two subdomains physics. According to author in [6] mortar scheme is general method in coupling for practical multiphysics problems. On the other hand, the efficient Enhanced Velocity scheme has not been investigated from the point of view of the improvement solution including velocity at interface in the previous studies.

The challenge here is to construct the velocity approximation of EVMFEM and specifically at interface to have a better velocity between subdomains that leads accurate approximation in the flow and transport problems. In [17], a *priori* error analysis states that the global error is

$$\|\mathbf{u}-\mathbf{u}_h\|_{\Omega} \leq C\left(\|p\|_{1,\infty,\Omega^*} + \|\mathbf{u}\|_{1,\Omega}\, h^{1/2}\right) h^{1/2} \tag{1}$$

and away from the interface Γ the velocity error convergence rate is better, since

$$\|\mathbf{u}-\mathbf{u}_h\|_{\Omega'} \leq C_{\varepsilon}\left(\|p\|_{1,\infty,\Omega^*} + \|\mathbf{u}\|_{1,\Omega}\right) h^{r-\varepsilon} \tag{2}$$

where $\varepsilon > 0$, $r = 1$ if $d = 2$ and $r = 5/6$ if $d = 3$, and Ω_i' is compactly contained in Ω_i, $\Omega' = \bigcup_{i=1}^{N_b} \Omega_i'$. This implies that the discrete velocity should be approximated more precise near interface region Ω^*. On the question of pressure approximation, the convergence rate of pressure approximation is $\mathcal{O}(h^1)$, if $d = 2$, and $\mathcal{O}(h^{5/6})$, if $d = 3$ [15,17]. If one compares the error of velocity (1) and pressure approximation these results indicate that the velocity convergence rate is not strong as pressure in Ω. Similar a *priori* error result was shown in [4] for transient problems. Nevertheless, there are still problems including the velocity approximation at the interface to be addressed.

In this paper, we introduce the way to improve velocity accuracy at interface in the Enhanced Velocity MFEM for incompressible flow using the post-processed pressure from [5]. This improvement is important in flow coupled with transport problems and it also can be a good candidate for a recovery-based error estimate evaluation. In a recent work [2], a *posteriori* error analysis was shown for the incompressible flow problems without recovery of velocity.

The remaining part of the paper proceeds as follows. Section 2 of this paper will describe model formulation with different view of EVMFEM. In Sect. 3, the proposed numerical method will be discussed. Section 4 shows numerical results. Section 5 summarizes the results of this work and draws conclusions.

2 Model Formulation

We start by giving the model formulation for the incompressible single-phase flow. For the convenience of reader we repeat the relevant material of domain decomposition method, discrete formulation with Enhanced Velocity from [17]. We next describe the proposed different view of Enhanced Velocity Discrete Scheme with projection operator.

2.1 Governing Equations of the Incompressible Flow

We consider the incompressible single-phase flow model for pressure p and the Darcy velocity $\mathbf{u}$:

$$\mathbf{u} = -\mathbf{K}\nabla p \quad \text{in} \quad \Omega, \tag{3}$$
$$\nabla \cdot \mathbf{u} = f \quad \text{in} \quad \Omega, \tag{4}$$
$$p = g \quad \text{on} \quad \partial\Omega, \tag{5}$$

where $\Omega \in \mathbb{R}^d (d = 2 \text{ or } 3)$ is multiblock domain, $f \in L^2(\Omega)$ and $\mathbf{K}$ is a symmetric, uniformly positive definite tensor representing the permeability divided by the viscosity with $L^\infty(\Omega)$ components, for some $0 < k_{min} < k_{max} < \infty$ $k_{min}\xi^T\xi \leq \xi^T\mathbf{K}(x)\xi \leq k_{max}\xi^T\xi\ , \forall x \in \Omega\ , \forall \xi \in \mathbb{R}^d$, under the Dirichlet boundary condition.

A weak variational form of the fluid flow problem (3)–(5) is to find a pair $\mathbf{u} \in \mathbf{V}, p \in W$

$$(\mathbf{K}^{-1}\mathbf{u}, \mathbf{v}) - (p, \nabla \cdot \mathbf{v}) = -\langle g, \mathbf{v} \cdot \boldsymbol{\nu} \rangle_{\partial\Omega} \qquad \forall \mathbf{v} \in \mathbf{V}, \tag{6}$$
$$(\nabla \cdot \mathbf{u}, w) = (f, w) \qquad \forall w \in W, \tag{7}$$

where $\boldsymbol{\nu}$ is the outward unit normal to $\partial\Omega$, $\mathbf{V}$ is $H(\text{div}; \Omega) = \{\mathbf{v} \in \left(L^2(\Omega)\right)^d :$ $\nabla \cdot \mathbf{v} \in L^2(\Omega)\}$ and equipped with the norm $\|\mathbf{v}\|_V = \left(\|\mathbf{v}\|^2 + \|\nabla \cdot \mathbf{v}\|^2\right)^{\frac{1}{2}}$ and the pressure the space is $W = L^2(\Omega)$ and the corresponding norm $\|w\|_W = \|w\|$..

Discrete Formulation. Let Ω be decomposed into non-overlapping small subdomains, see Fig. 1. We consider

$$\Omega = \left(\bigcup_{i=1}^{N_b} \bar{\Omega}_i\right)^o, \ \Gamma_{i,j} = \partial\Omega_i \bigcap \partial\Omega_j, \ \Gamma = \left(\bigcup_{i,j=1}^{N_b} \bar{\Gamma}_{i,j}\right)^o, \ \Gamma_i = \Omega_i \bigcap \Gamma = \partial\Omega_j \setminus \partial\Omega.$$

This implies that the domain is divided into N_b subdomains, the interface between i^{th} and j^{th} subdomains$(i \neq j)$, the interior subdomain interface for i^{th} subdomain and union of all such interfaces, respectively. Let $\mathcal{T}_{h,i}$ be a conforming, quasi-uniform and rectangular partition of Ω_i, $1 \leq i \leq N_b$, with maximal element diameter h_i. We then set $\mathcal{T}_h = \cup_{i=1}^n \mathcal{T}_{h,i}$ and denote h the maximal element diameter in $\mathcal{T}_h$; note that $\mathcal{T}_h$ can be nonmatching as neighboring meshes

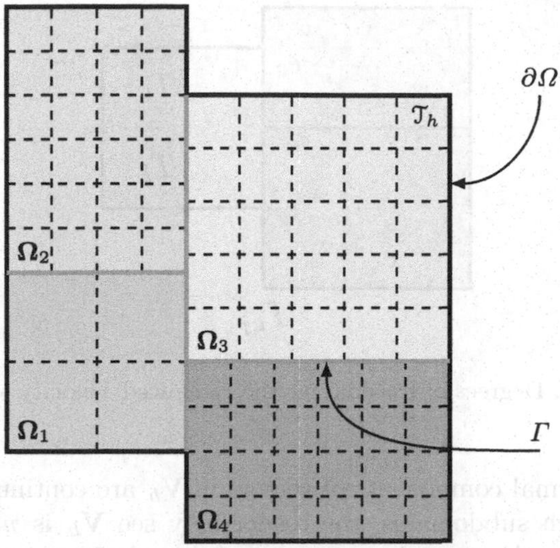

Fig. 1. Illustration of a domain Ω with subdomains Ω_i and non-matching mesh discretization $\mathcal{T}_h$.

$\mathcal{T}_{h,i}$ and $\mathcal{T}_{h,j}$ need not match on $\Gamma_{i,j}$. We assume that all mesh families are shape-regular.

In Enhanced Velocity scheme setting, the velocity basis functions are based on the traditional Raviart-Thomas spaces of lowest order on rectangles for $d = 2$ and bricks for $d = 3$. The RT_0 spaces are defined for any element $T \in \mathcal{T}_h$ by the following spaces:

$$\mathbf{V}_h(T) = \{\mathbf{v} = (v_1, v_2) \text{ or } \mathbf{v} = (v_1, v_2, v_3) : v_l = \alpha_l + \beta_l x_l : \alpha_l, \beta_l \in \mathbb{R}; l = 1, ..d\},$$
$$W_h(T) = \{w = \text{constant}\}.$$

The pressure finite element approximation space on Ω is taken to be as $W_h(\Omega) = \{w \in L^2(\Omega) : w\Big|_E \in W_h(T), \forall T \in \mathcal{T}_h\}$. In addition, a vector function in $\mathbf{V}_h$ can be determined uniquely by its normal components $\mathbf{v} \cdot \nu$ at midpoints of edges (in 2D) or face (in 3D) of T. The degrees of freedom of $\mathbf{v} \in \mathbf{V}_h(T)$ were created by these normal components. The degree of freedom for a pressure function $p \in W_h(T)$ is at center of T and piecewise constant inside of T. Let us formulate RT_0 space on each subdomain Ω_i for partition $\mathcal{T}_h$

$$\mathbf{V}_{h,i} = \{\mathbf{v} \in H(\text{div}; \Omega_i) : \mathbf{v}\Big|_T \in \mathbf{V}_h(T), \forall T \in \mathcal{T}_{h,i}\} \qquad i \in \{1, ...n\}$$

and then

$$\mathbf{V}_h = \bigoplus_{i=1}^{n} \mathbf{V}_{h,i}.$$

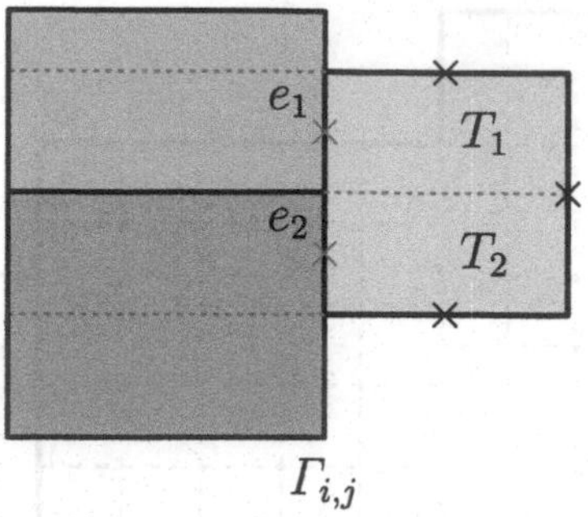

Fig. 2. Degrees of freedom for the Enhanced Velocity space.

Although the normal components of vectors in $\mathbf{V}_h$ are continuous between elements within each subdomains, the reader may see $\mathbf{V}_h$ is *not* a subspace of $H(\text{div}; \Omega)$, because the normal components of the velocity vector may not match on subdomain interface Γ. Let us define $\mathcal{T}_{h,i,j}$ as the intersection of the traces of $\mathcal{T}_{h,i}$ and $\mathcal{T}_{h,j}$, and let $\mathcal{T}_h^{\Gamma} = \bigcup_{1 \le i \le j \le N_b} \mathcal{T}_{h,i,j}$. We require that $\mathcal{T}_{h,i}$ and $\mathcal{T}_{h,j}$ need to align with the coordinate axes. Fluxes are constructed to match on each element $e \in \mathcal{T}_h^{\Gamma}$. We consider any element $T \in \mathcal{T}_{h,i}$ that shares at least one edge with the interface Γ, i.e., $T \cap \Gamma_{i,j} \neq \emptyset$, where $1 \le i,j \le N_b$ and $i \neq j$. Then newly defined interface grid introduces a partition of the edge of T. This partition may be extended into the element T as shown in Fig. 2.

Such partitioning helps to construct fine-scale velocities that is in $H(\,\mathbf{div}, \Omega)$. So we represent a basis function $\mathbf{v}_{T_k}$ in the $\mathbf{V}_h(T_k)$ space (RT_0) for given T_k with the following way:

$$\mathbf{v}_{T_k} \cdot \nu = \begin{cases} 1, & \text{on } e_k \\ 0, & \text{other edges} \end{cases}$$

i.e. a normal component $\mathbf{v}_{T_k} \cdot \nu$ equal to one on e_k and zero on all other edges(faces) of T_k. Let $\mathbf{V}_h^{\Gamma}$ be span of all such basis functions defined on all subelements induced the interface discretization $\mathcal{T}_{h,i,j}$. Thus, the enhanced velocity space $\mathbf{V}_h^*$ is taken to be as

$$\mathbf{V}_h^* = \bigoplus_{i=1}^{n} \mathbf{V}_{h,i}^0 \bigoplus \mathbf{V}_h^{\Gamma} \cap H(\text{div}; \Omega),$$

where $\mathbf{V}_{h,i}^0 = \{\mathbf{v} \in \mathbf{V}_{h,i} : \mathbf{v} \cdot \nu = 0 \text{ on } \Gamma_i\}$ is the subspace of $\mathbf{V}_{h,i}$. The finer grid velocity allows to velocity approximation on the interface and then form the $H(\text{div}, \Omega)$ conforming velocity space. Some difficulties arise, however, in analysis of method and implementation of robust linear solver for such modification of RT_0 velocity space at all elements, which are adjacent to the interface Γ. We now formulate the discrete variational form of Eqs. (3)–(5) as: Find $\mathbf{u}_h \in \mathbf{V}_h^*$ and $p_h \in W_h$ such that

$$(K^{-1}\mathbf{u}_h, \mathbf{v}) = (p_h, \nabla \cdot \mathbf{v}) - \langle g, \mathbf{v} \cdot \boldsymbol{\nu} \rangle_{\partial\Omega} \qquad \forall \mathbf{v} \in \mathbf{V}_h^*, \tag{8}$$

$$(\nabla \cdot \mathbf{u}_h, w) = (f, w) \qquad \forall w \in W_h. \tag{9}$$

2.2 A Different View of the EVMFEM in the Discrete Variational Formulation

We consider the discrete variational form that is given in (8)–(9). Find $\mathbf{u}_h \in \mathbf{V}_h^*$ and $p_h \in W_h$ such that

$$\left(K^{-1}\mathbf{u}_h, \mathbf{v}\right)_{M,T} = (p_h, \nabla \cdot \mathbf{v}) - \langle g, \mathbf{v} \cdot \boldsymbol{\nu} \rangle_{\partial\Omega} \qquad \forall \mathbf{v} \in \mathbf{V}_h^*, \tag{10}$$

$$(\nabla \cdot \mathbf{u}_h, w) = (f, w) \qquad \forall w \in W_h. \tag{11}$$

We exploit the approximation inner product and for $\mathbf{v}, \mathbf{q} \in \mathbb{R}^d$

$$(\mathbf{v}, \mathbf{q})_{M,T} = \begin{cases} (v_x, q_y)_{T_x, M_y} + (v_y, q_y)_{M_x, T_y} & \text{if d } = 2, \\ (v_x, q_y)_{T_x, M_y, M_z} + (v_y, q_y)_{M_x, T_y, M_z} + (v_z, q_z)_{M_x, M_y, T_z} & \text{if d } = 3. \end{cases}$$

where $T_{(\cdot)}$ and $M_{(\cdot)}$ denote the trapezoidal and midpoint quadrature rules in each coordinate direction respectively, see [13]. In particularly, we take $\mathbf{v} = \mathbf{K}^{-1}\mathbf{u}_h$ and $\mathbf{q} = \mathbf{v}$. It is easily proven that the finite variational form (10)–(11) is equivalent to finding $\mathbf{u}_h \in \mathbf{V}_h^*$, $p_h \in W_h$, $1 \leq i \leq N_b$, such that

$$\left(\mathbf{K}^{-1}\mathbf{u}_h, \mathbf{v}\right)_{\Omega_i, M,T} - (p_h, \nabla \cdot \mathbf{v})_{\Omega_i} = -\langle g, \mathbf{v} \cdot \boldsymbol{\nu} \rangle_{\partial\Omega_i \cap \Gamma_D} \qquad \forall \mathbf{v} \in \mathbf{V}_{h,i}^0, \tag{12}$$

$$(\nabla \cdot \mathbf{u}_h, w)_{\Omega_i} = (f, w)_{\Omega_i} \qquad \forall w \in W_{h,i}, \tag{13}$$

$$\sum_{i=1}^{N_b} \left\{ \left(\mathbf{K}^{-1}\mathbf{u}_h, \mathbf{v}^{EV}\right)_{\Omega_i, M,T} - \left(p_h, \nabla \cdot \mathbf{v}^{EV}\right)_{\Omega_i} \right\} = 0 \qquad \forall \mathbf{v}^{EV} \in \mathbf{V}^{\Gamma}. \tag{14}$$

We note that the discrete variational formulation was similarly proposed in [10] with conjugate gradient method. We want to share the idea for small number of discretization elements that can be applied for a large number of elements. Thus, we consider two subdomains, i.e., $\Omega = \bar{\Omega}_1 \cup \bar{\Omega}_2$ and Γ is the interface. Then

$$\mathbf{V}_h^* = \left(\mathbf{V}_{h,1}^0 \oplus \mathbf{V}_{h,2}^0 \oplus \mathbf{V}_h^{\Gamma}\right).$$

Consider equations

$$\left(\mathbf{K}^{-1}\mathbf{u}_h, \mathbf{v}\right)_{M,T} = (p_h, \nabla \cdot \mathbf{v}) \qquad \forall \mathbf{v} \in \mathbf{V}_h^{\Gamma}. \tag{15}$$

These allow us to express $\mathbf{u}_h^{\Gamma}$ in terms of the one-element layers along Γ, it is shown in Fig. 3:

$$\mathbf{u}_h^{\Gamma} = A_1 p_L + A_2 p_R. \tag{16}$$

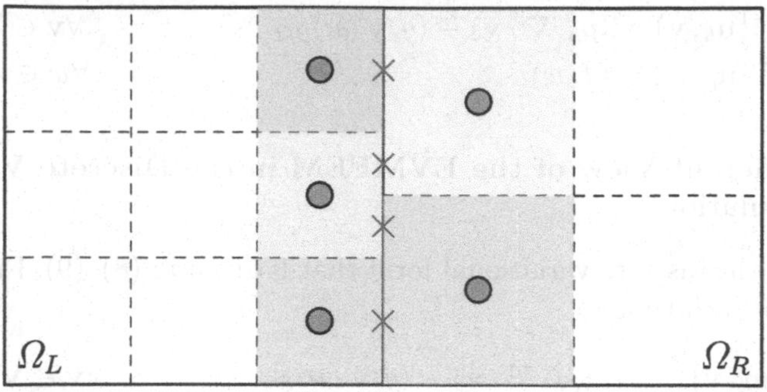

Fig. 3. The spatial domain and illustration of Enhanced Velocity values on the interface.

Now we consider each subdomain separately with ghost layers. We define L_2-projection of Enhanced Velocity space at interface $\Gamma_{i,j}$ to each subdomain space $\partial\Omega_i \cap \Gamma_{i,j}$ such that $\mathcal{P}_h^i : \mathbf{V}_h^* \to \mathbf{V}_{h,i}$.

$$\mathcal{P}_h^i : \mathbf{V}_h^*(\Gamma) \to \mathbf{V}_{h,i}(\Gamma_i) \qquad for\ \psi \in L^2(\Gamma), \qquad \langle(\psi - \mathcal{P}_h^i\psi)\cdot\boldsymbol{\nu}_i, \mathbf{v}\cdot\boldsymbol{\nu}_i\rangle_\Gamma = 0 \qquad \forall \mathbf{v} \in \mathbf{V}_{h,i}.$$

We denote

$$\mathbf{u}_{h,i}^\Gamma = \mathcal{P}_h^i \mathbf{u}_h^\Gamma, \qquad i = L\ or\ R$$

In subdomain Ω_i, we define p_i^e in the following way

$$\left(\mathbf{K}^{-1}\tilde{\mathbf{u}}_h, \mathbf{v}\right)_{M,T,\Omega_i} = (p_h, \nabla\cdot\mathbf{v})_{\Omega_i} - \langle p_i^e, \mathbf{v}\cdot\boldsymbol{\nu}\rangle_\Gamma \qquad \forall \mathbf{v} \in \mathbf{V}_{h,i}^\Gamma \text{ s.t.} \mathbf{v}\cdot\boldsymbol{\nu} = 0 \text{ on } \partial\Omega_i^*, \tag{17}$$

where Ω^* is union of all elements T that shares edge (2D) or face (3D) with Γ_i and p_i^e ghost layers pressure values, and $\tilde{\mathbf{u}}_h = \mathcal{P}_h^i(\mathbf{u}_h)$. Such ghost layers are depicted in the Fig. 4. Then, for $i = L$, we have

$$\mathbf{u}_{h,L}^\Gamma = A_1^L p_L + A_2^L p_L^e. \tag{18}$$

We compare Eq. (18) and the pressure Eq. (16) which is projected to Ω_i:

$$\mathcal{P}_h^L \mathbf{u}_h^\Gamma = \mathcal{P}_h^L A_1 p_L + \mathcal{P}_h^L A_2 p_R. \tag{19}$$

Since $\mathbf{u}_{h,L}^\Gamma = \mathcal{P}_h^L \mathbf{u}_h^\Gamma$, $A_1^L = \mathcal{P}_h^L A_1$, we have the following

$$A_2^L p_L^e = \mathcal{P}_h^L A_2 p_R. \tag{20}$$

A_2^L is non-singular and diagonal matrix, since $\mathbf{K}$ is SPD.

$$p_L^e = \left(A_2^L\right)^{-1} \mathcal{P}_h^L A_2 p_R. \tag{21}$$

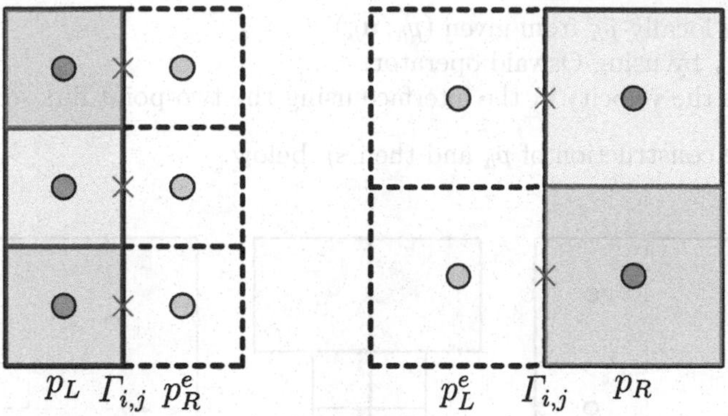

Fig. 4. Example of left (Ω_L) and right (Ω_R) domains with ghost layers.

Similarly, we can obtain

$$p_R^e = \left(A_1^R\right)^{-1} \mathcal{P}_h^R A_1 p_L. \tag{22}$$

In non-linear problems including slightly compressible flow or multiphase flow in heterogeneous porous media, this approach could be applied analogously by taking into account ghost layers values arising from e_i in each Newton iteration. So during Block Jacobi iteration variables $p_L^{e,k-1}$, $p_R^{e,k-1}$ is computed by utilizing given p_L^{k-1}, p_R^{k-1} and then solve decoupled subdomain problems with Dirichlet boundary conditions $p_i^{e,k-1}$, $i = L, R$ to find u^k, p^k.

3 Methods

We use the postprocessing procedure associated to pressure and velocity. We first apply locally postprocessing algorithm for given pressure p_h and velocity $\mathbf{u}_h$ which was previously proposed in [5] and then Oswald interpolation operator [1,11,12,16] to have better pressure values. At the interface, we use two-point flux computation method in order to have better approximation of pressure. As a result, the Enhanced Velocity scheme solution of velocity can be improved by using a post-processed pressure. The key idea is illustrated in Fig. 5 for resulting approximation of EV scheme that is shown in Fig. 3.

The velocity at the edge or face is computed by using pressure values between subdomains Ω_i and Ω_j. To be specific, $p_h \in \Omega^*$ is required in the original velocity for constructing in Enhanced Velocity MFEM. However, the post-processed pressure leads to the improved velocity and the visual representation is in Fig. 5. In case of multiscale setting, it is important to be able to approximate better pressure values nearby the interface. The recovery of velocity computation requires three steps

1. Compute locally $\tilde{p}_h$ from given $(p_h, \mathbf{u}_h)$,
2. Obtain s_h by using Oswald operator,
3. Compute the velocity at the interface using the two-point flux scheme for s_h.

We describe construction of $\tilde{p}_h$ and then s_h below.

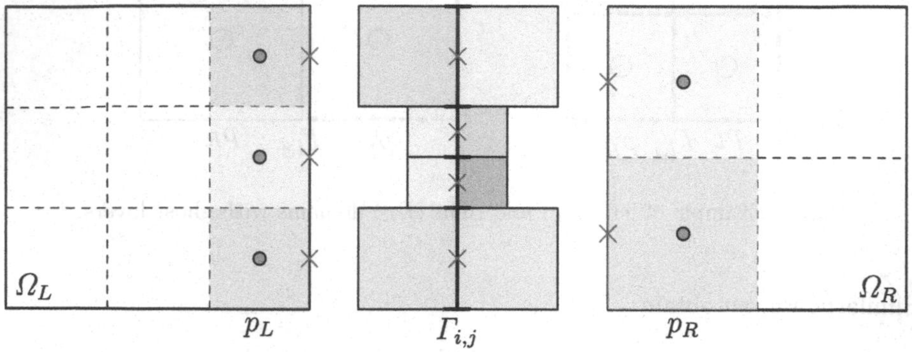

Fig. 5. The illustration of the velocity improvement at the interface using postprocessing.

Construction of $\tilde{p}_h$. In the Enhance Velocity setting, we may identify $\widehat{\mathbf{V}}_h$ be spaces omitting interface constraints $\mathbf{V}^{\Gamma}$, so $\widehat{\mathbf{V}}_{h,i} := \bigoplus_{i=1}^{n} \mathbf{V}_{h,i}(T)$ and then $\widehat{\mathbf{V}}_h := \bigoplus_{i=1}^{n} \widehat{\mathbf{V}}_{h,i}$. Let $\mathbf{u}_h$, p_h be the solution of Eqs. (8)–(9). Initially, Lagrange multipliers can be computed in each element. In other words, we define $\lambda_{h,T} \in \Lambda_h$, which is piecewise constant polynomials at edge or face,

$$\langle \lambda_{h,T}, \mathbf{v}_h \cdot \mathbf{n}_T \rangle_e := \left(\mathbf{K}^{-1}\mathbf{u}_h, \mathrm{v}_h\right)_T - (p_h, \nabla \cdot \mathbf{v}_h)_T \qquad \forall \mathbf{v}_h \in \widehat{\mathbf{V}}_h(T), \tag{23}$$

where the element $T \in \mathcal{T}_h$ and its side e. We employ the L^2 projected velocity from the interface, which has a finer enhanced velocity approximation, to the edge or face of subdomain element and the formulation is provided in Subsect. 2.2. We denote polynomial space $\widetilde{W}_h$ in the following manner

$$\widetilde{W}_h = \{\varphi_h : \langle [\![\varphi_h]\!], \psi_h \rangle_e = 0 \qquad \forall e \in \mathcal{E}_h^{int} \cup \mathcal{E}_h^{ext}, \forall \psi_h \in \mathbb{Q}_m(e)\}, \tag{24}$$

where $\mathbb{Q}_m$ is standard polynomial space that is defined in [5,8,12]. We next set the post-processed $\tilde{p}_h$ which is proposed in [5] and the construction is performed with the following properties, for each $T \in \mathcal{T}_h$

$$(\tilde{p}_h, w_h)_T = (p_h, w_h)_T \qquad \forall w_h \in \widetilde{W}_h(T) \tag{25}$$

$$\langle \tilde{p}_h, \mu_h \rangle_e = \langle \lambda_h, \mu_h \rangle_e \qquad \forall \mu_h \in \Lambda_h(e), \forall e \in \partial T. \tag{26}$$

Construction of s_h. We propose to construct s_h in each subdomain Ω_i that has the conforming mesh in order to be a computational efficient. Construction of s_h involves the averaging operator $\mathcal{I}_{\rm av} : \mathbb{Q}_k(\mathcal{T}_h) \to \mathbb{Q}_k(\mathcal{T}_h) \cap H_0^1(\Omega_i)$. For definition of $\mathbb{Q}_m$ we refer reader to [8]. The operator is called Oswald operator and appeared in [1,11,12,16] and the analysis can be found in [7,11]. It is interesting to note that the mapping of the gradient of pressure through Oswald operator also considered in [18]. For given $\varphi_h \in \mathbb{Q}_m(\mathcal{T}_h)$, we regard the values of $\mathcal{I}_{\rm av}(\varphi_h)$ as being defined at a Lagrange node $V \in \Omega$ by averaging φ_h values associated this node,

$$\mathcal{I}_{\rm av}(\varphi_h)(V) = \frac{1}{|\mathcal{T}_h|} \sum_{T \in \mathcal{T}_h} \varphi_h|_T(V), \tag{27}$$

where $|A|$ is cardinality of sets A and $\mathcal{T}_h$ is all collection of $T \in \mathcal{T}_h$ for fixed V. One can see that $\mathcal{T}_h(V) = \varphi(V)$ at those nodes that are inside of given $T \in \mathcal{T}_h$. We set the value of $\mathcal{I}_{\rm av}(\varphi_h)$ is zero at boundary nodes. Now in our setting we define recovered pressure s_h for the locally post-processed $\tilde{p}_h$ as follows:

$$s_h := \mathcal{I}_{\rm av}(\tilde{p}_h).$$

3.1 Implementation Steps

For simplicity, we provide key steps of numerical implementation of post-processed pressure in two dimensional case. However, it can be extended for general cases. Based on piecewise pressure and velocity from the lowest order Raviart-Thomas spaces over rectangles our aim to reconstruct smoother pressure s_h. For given element $T \in \mathcal{T}_h(\Omega_i)$, the main steps are

1. Evaluate $\lambda_{h,T}$ at edge e_j, $j = 1, ..4$ based on $(\mathbf{u}_h, p_h)$,
2. Compute $\tilde{p}_h$ from known $\lambda_{h,T}$, and p_h by using (23),
3. Based on $\tilde{p}_h$ compute s_h Eq. (27) at Lagrange nodes in Ω_i.

Step 1 is standard computation of Lagrange multiplier for each element. In step 2, we are relying on higher order polynomial, in our case, it is Span$\{1, x, y, x^2, y^2\}$. It is sufficient to store coefficients of polynomials. In step 3, we use Span$\{1, x, y, x^2, y^2, xy, x^2y, xy^2, x^2y^2\}$ and 9 Lagrange nodes of rectangle elements that are four rectangle nodes, four midpoints at edge and center of rectangle. This case each node requires to find neighboring elements values to compute coefficients of s_h.

Remark 1. It would be interesting to see the possibility of extension of the proposed method to high order polynomial approximation.

4 Numerical Examples

In this section, numerical results are presented to demonstrate challenging problems of velocity approximation at the interface of non-matching multiblock grids. We have conducted tests for several examples and we concentrate our attention on the interface error for heterogeneous permeability coefficients. We set same domain $\Omega = (0,1) \times (0,1)$ for all tests and for some the ratio is $H/h = 4$. Initial subdomains grids $\mathcal{T}_h$ are chosen in way that has a checkerboard pattern for subdomains. Example of such discretization is shown in Fig. 6. The discrete L^2 velocity error $e_{\mathbf{u}_h,\Gamma}$ is based on the values of the normal component at the midpoint of the edges and is normalized by the analytical solution.

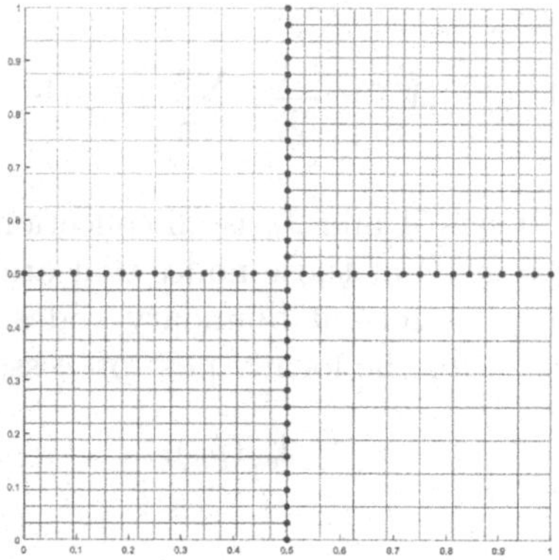

Fig. 6. Example of non-matching grids for subdomains.

Numerical Test 1

We consider the a diagonal oscillating tensor coefficient as follows.

$$\mathbf{K} = \begin{bmatrix} 15 - 10\sin(3\pi x)\sin(3\pi y) & 0 \\ 0 & 15 - 10\sin(3\pi x)\sin(3\pi y) \end{bmatrix}$$

We impose the source term f and Dirichlet boundary condition according to the analytical solution

$$p(x,y) = \sin(2\pi x)\sin(2\pi y).$$

We set the ratio $H/h = 4$ for the result that is shown in Table 1. We reported the velocity error and the improved velocity error.

From Table 1, we see a significant increase on the convergence rate for recovered velocity $O(h^{1.5})$ while the convergence rate of provided velocity stays $\mathcal{O}(h^{1.0})$. We observe that the numerical method is an effective way to improve velocity at the interface between subdomains.

Numerical Test 2

We consider the heterogeneous porous media and impose the no-flow boundary conditions for the flow problem, which is formulated in Eqs. (3)–(5). The permeability distribution profile in the log scale is shown in Fig. 7.

Table 1. Convergence test 2: velocity and recovered velocity error using the post-processed pressure at interface.

n	$e_{\mathbf{u}_h,\Gamma}$		$e_{\tilde{\mathbf{u}}_h,\Gamma}$	
	Error	Order	Error	Order
8	1.78e−01	—	3.78e−01	—
16	8.89e−02	1.00	1.00e−01	1.91
32	4.43e−02	1.00	2.87e−02	1.81
48	2.96e−02	1.00	1.56e−02	1.51

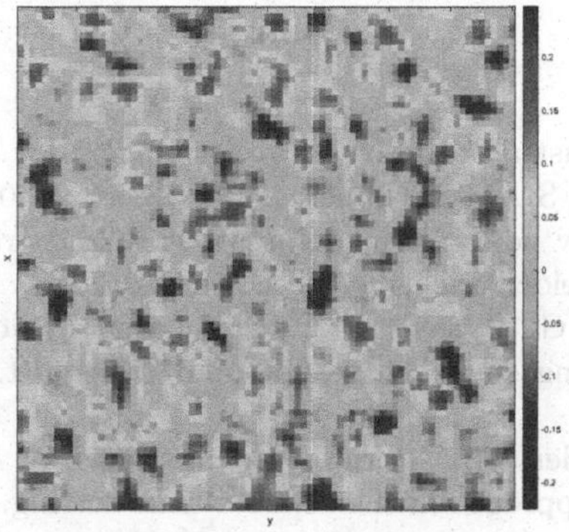

Fig. 7. The fine permeability distribution in log scale.

In our simple scenario, a rate specified injection well and a pressure specified production well are located at the bottom left and top right corners, respectively. The fine scale subdomain dimension is 40×40 grid-blocks and the coarse scale subdomain dimension is 20×20 grid-blocks. The injection rate was assumed to be $1\,\mathrm{m}^3/\mathrm{s}$. The initial reservoir pressure is taken to be 1000 Pa. As reference solution we consider fine scale solution of the flow problem with 80×80 grid blocks. In the coarse scale subdomains, the permeability distribution was evaluated or upscaled using numerical homogenization. We set the ratio $H/h = 2$, see Fig. 6.

The overall error at interface is $e_{\mathbf{u}_h,\Gamma} = 0.254656$ and the post-processed error is $e_{\tilde{\mathbf{u}}_h,\Gamma} = 0.143014$. We note that the error of post-processed velocity at interface is less than the error of the original velocity, which is computed by the EV scheme.

5 Conclusion

The present study of velocity in Enhanced Velocity Mixed Finite Element Method was designed to investigate the effect of the post-processed pressure on velocity in the interface of subdomains. In this paper, the focus of attention is on the incompressible Darcy flow in the non-matching multiblock grid setting.

Multiple numerical results demonstrate that the interface velocity approximation can be improved with using the post-processed pressure. These findings can contribute in several ways to our approximation of velocity and provide a good construction of velocity for a *posteriori* error analysis such as the recovery-based estimate.

Acknowledgments. First author would like to thank Drs. I. Yotov and T. Arbogast for discussions on formulation of the different view of EVMFEM. This research is supported by Faculty Development Competitive Research Grant (Grant No. 110119FD4502), Nazarbayev University.

References

1. Ainsworth, M.: Robust a posteriori error estimation for nonconforming finite element approximation. SIAM J. Numer. Anal. **42**(6), 2320–2341 (2005)
2. Amanbek, Y.: A new adaptive modeling of flow and transport in porous media using an enhanced velocity scheme. Ph.D. thesis (2018)
3. Amanbek, Y., Singh, G., Wheeler, M.F., van Duijn, H.: Adaptive numerical homogenization for upscaling single phase flow and transport. J. Comput. Phys. **387**, 117–133 (2019)
4. Amanbek, Y., Wheeler, M.: A priori error analysis for transient problems using Enhanced Velocity approach in the discrete-time setting. J. Comput. Appl. Math. (2019). https://doi.org/10.1016/j.cam.2019.05.009
5. Arbogast, T., Chen, Z.: On the implementation of mixed methods as nonconforming methods for second-order elliptic problems. Math. Comput. **64**(211), 943–972 (1995)
6. Arbogast, T., Estep, D., Sheehan, B., Tavener, S.: A posteriori error estimates for mixed finite element and finite volume methods for problems coupled through a boundary with nonmatching grids. IMA J. Numer. Anal. **34**(4), 1625–1653 (2014)
7. Burman, E., Ern, A.: Continuous interior penalty *hp*-finite element methods for advection and advection-diffusion equations. Math. Comput. **76**(259), 1119–1140 (2007)
8. Chen, Z., Huan, G., Ma, Y.: Computational Methods for Multiphase Flows in Porous Media, vol. 2. SIAM, New Delhi (2006)
9. Gerritsen, M., Lambers, J.: Integration of local-global upscaling and grid adaptivity for simulation of subsurface flow in heterogeneous formations. Comput. Geosci. **12**(2), 193–208 (2008)
10. Glowinski, R., Wheeler, M.F.: Domain decomposition and mixed finite element methods for elliptic problems. In: First International Symposium on Domain Decomposition Methods for Partial Differential Equations, pp. 144–172 (1988)
11. Karakashian, O.A., Pascal, F.: A posteriori error estimates for a discontinuous Galerkin approximation of second-order elliptic problems. SIAM J. Numer. Anal. **41**(6), 2374–2399 (2003)
12. Pencheva, G.V., Vohralík, M., Wheeler, M.F., Wildey, T.: Robust a posteriori error control and adaptivity for multiscale, multinumerics, and mortar coupling. SIAM J. Numer. Anal. **51**(1), 526–554 (2013)
13. Russell, T.F., Wheeler, M.F.: Finite element and finite difference methods for continuous flows in porous media, pp. 35–106. SIAM (1983)

14. Singh, G., Amanbek, Y., Wheeler, M.F.: Adaptive homogenization for upscaling heterogeneous porous medium. In: SPE Annual Technical Conference and Exhibition. Society of Petroleum Engineers (2017)
15. Thomas, S.G., Wheeler, M.F.: Enhanced velocity mixed finite element methods for modeling coupled flow and transport on non-matching multiblock grids. Comput. Geosci. **15**(4), 605–625 (2011)
16. Vohralík, M.: Unified primal formulation-based a priori and a posteriori error analysis of mixed finite element methods. Math. Comput. **79**(272), 2001–2032 (2010)
17. Wheeler, J.A., Wheeler, M.F., Yotov, I.: Enhanced velocity mixed finite element methods for flow in multiblock domains. Comput. Geosci. **6**(3–4)
18. Zienkiewicz, O.C., Zhu, J.Z.: A simple error estimator and adaptive procedure for practical engineerng analysis. Int. J. Numer. Meth. Eng. **24**(2), 337–357 (1987)

A New Approach to Solve the Stokes-Darcy-Transport System Applying Stabilized Finite Element Methods

Iury Igreja(✉)

Computer Science Department and Computational Modeling Graduate Program,
Federal University of Juiz de Fora, Juiz de Fora, MG, Brazil
iuryigreja@ice.ufjf.br

Abstract. In this work we propose a new combination of finite element methods to solve incompressible miscible displacements in heterogeneous media formed by the coupling of the free-fluid with the porous medium employing the stabilized hybrid mixed finite element method developed and analyzed by Igreja and Loula in [10] and the classical Streamline Upwind Petrov-Galerkin (SUPG) method presented and analyzed by Brooks and Hughes in [2]. The hydrodynamic problem is governed by the Stokes and Darcy systems coupled by Beavers-Joseph-Saffman interface conditions. To approximate the Stokes-Darcy coupled system we apply the stabilized hybrid mixed method, characterized by the introduction of the Lagrange multiplier associated with the velocity field in both domains. This choice naturally imposes the Beavers-Joseph-Saffman interface conditions on the interface between Stokes and Darcy domains. Thus, the global system is assembled involving only the degrees of freedom associated with the multipliers and the variables of interest can be solved at the element level. Considering the velocity fields given by the hybrid method we adopted the SUPG method combined with an implicit finite difference scheme to solve the transport equation associated with miscible displacements. Numerical studies are presented to illustrate the flexibility and robustness of the hybrid formulation. To verify the efficiency of the combination of hybrid and SUPG methods, computer simulations are also presented for the recovery hydrological flow problems in heterogeneous porous media, such as continuous injection.

Keywords: Stabilized methods · Hybrid mixed methods · Stokes-Darcy flow · Coupled problems · Heterogeneous media

1 Introduction

Numerical methods to simulate the incompressible viscous fluid flows coupling Stokes-Darcy problems has been widely developed due to various applications in physiological phenomena like the blood motion in vessels, hydrological systems in which surface water percolates through rocks and sand, petroleum engineering where are find fractured media containing vugs and caves as the naturally

J. M. F. Rodrigues et al. (Eds.): ICCS 2019, LNCS 11539, pp. 524–537, 2019.
https://doi.org/10.1007/978-3-030-22747-0_39

fractured carbonate karst reservoirs and industrial processes involving filtration [8,14,17,21]. This coupled problem is characterized by the coexistence of the free fluid governed by the Stokes equations and the porous medium modeled by the Darcy problem connected by the interface conditions that guarantee continuity of mass and momentum across the interface [1,18].

Numerically, among the several methods proposed for the coupled problem, we highlight the stable and stabilized methods introduced in [3–5,13,19]; using a Lagrange multiplier to impose the interface restrictions, we can cite [7,11,20]; and employing discontinuous Galerkin (DG) methods, we indicate [16] and [21]. Recently, hybridizations of DG methods have been successfully exploited to derive new finite element methods with improved stability and reduced computational cost but still preserving the robustness and flexibility of DG methods [6,9,10].

In this paper, in order to obtain efficiently the velocity field, we use the stabilized hybrid mixed method to solve the coupled Stokes-Darcy problem developed by [10]. This method is characterized by the introduction of a Lagrange multiplier associated with the velocity field to weakly impose continuity on each edge of the elements. Moreover, this approach naturally imposes the interface conditions between porous medium and free fluid through the Lagrange multiplier. This methodology allows the elimination of the local problems at each element level in favor of the Lagrange multiplier. Thus, the system involves only global degrees of freedom associated with the multiplier, reducing significantly the computational cost. The accuracy of this method is presented through convergence studies.

Once the hydrodynamic problem is calculated we supply the velocity field to the convection-dominated parabolic equation to obtain the concentration field in the coupled Stokes-Darcy domain. These results can, for example, characterize a reservoir through continuous or tracer injection processes, informing the preferred direction of flow [12,14] or study the spread of pollution released in the water and assess the danger [21]. In order to illustrate the performance of the hybrid method applied to the coupled Stokes-Darcy-transport problem, where the Streamline Upwind Petrov-Galerkin (SUPG) method [2] combined with a backward finite difference scheme in time is employed to approximate the concentration equation, numerical simulations are demonstrated for the miscible transport problem using a five-spot pattern for different heterogeneous scenarios through continuous injection processes.

This paper is organized as follow. The Stokes-Darcy-transport model problem is introduced in Sect. 2. In Sect. 3, notations and definitions required to present the hybrid method are described. The stabilized mixed hybrid method for the coupled Stokes-Darcy problem is presented in Sect. 4. The Sect. 5 is devoted to convergence study and continuous injection simulations in a five-spot pattern for different heterogeneous scenarios. And finally, in Sect. 6, we present the concluding remarks of this work.

2 Model Problem

Let $\Omega \subset \mathbb{R}^d$ ($d = 2$ or 3) be the domain composed by two subdomains Ω_S and Ω_D related to free fluid and porous medium, respectively. In the subdomain Ω_S, with outward unit normal $\mathbf{n}_S$, the flow is governed by the Stokes problem and in porous medium Ω_D, with outward unit normal $\mathbf{n}_D$, the Darcy's law holds. These subdomains are separated by a smooth interface $\Gamma_{SD} = \partial\Omega_S \cap \partial\Omega_D$, where $\mathbf{t}_j$ defines an orthonormal basis of tangent vectors on Γ_{SD}. Moreover, let $\Gamma = \Gamma_S \cup \Gamma_D$ with $\Gamma_i = \partial\Omega_i \setminus \Gamma_{SD}$ ($i = S, D$). The Fig. 1 represents a sketch of the described domain.

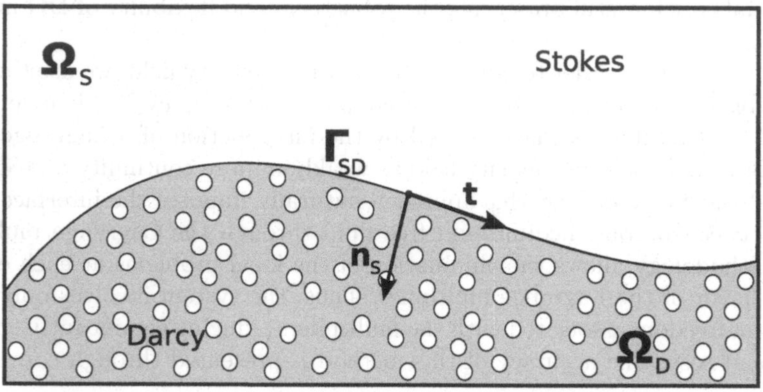

Fig. 1. A sketch of coupled Stokes-Darcy domain.

Denoting $\mathbf{u}^i = \mathbf{u}|_{\Omega_i}$ and $p^i = p|_{\Omega_i}$, with $i = S, D$, the free fluid domain Ω_S is modeled by the Stokes problem that can be written as follows

Given the viscosity ν *and the source* $\mathbf{f}$, *find the pressure* $p^S : \Omega_S \to \mathbb{R}$ *and the velocity field* $\mathbf{u}^S : \Omega_S \to \mathbb{R}^d$, *such that*

$$-\nu \operatorname{div} \nabla \mathbf{u}^S + \nabla p^S = \mathbf{f} \quad \text{in} \quad \Omega_S, \tag{1}$$

$$\operatorname{div} \mathbf{u}_S = 0 \quad \text{in} \quad \Omega_S, \tag{2}$$

$$\mathbf{u}^S = \mathbf{0} \quad \text{on} \quad \Gamma_S, \tag{3}$$

where div and ∇ denote, respectively, the divergent and gradient operators. On the other hand, in porous medium the flow is given by the Darcy problem

Given the hydraulic conductivity $\mathbf{K}$ *and the source* f, *find the hydrostatic pressure* $p_D : \Omega_D \to \mathbb{R}$ *and the Darcy velocity* $\mathbf{u}_D : \Omega_D \to \mathbb{R}^d$, *such that*

$$\mathbf{u}_D = -\mathbf{K}\nabla p_D \quad \text{in} \quad \Omega_D, \tag{4}$$

$$\operatorname{div} \mathbf{u}_D = f \quad \text{in} \quad \Omega_D, \tag{5}$$

$$\mathbf{u}_D \cdot \mathbf{n}_D = 0 \quad \text{on} \quad \Gamma_D, \tag{6}$$

we define $\mathbf{K} = \mathbf{k}/\mu$ where $\mathbf{k}$ is the permeability of the porous medium. The solvability condition, which the source function f must satisfy

$$\int_{\Omega_D} f \, d\mathbf{x} = 0.$$

On the interface free fluid/porous medium Γ_{SD} the following conditions are imposed

$$\mathbf{u}_S \cdot \mathbf{n}_S + \mathbf{u}_D \cdot \mathbf{n}_D = 0 \quad \text{on} \quad \Gamma_{SD}, \tag{7}$$

$$p_S - 2\mu\varepsilon(\mathbf{u}_S)\,\mathbf{n}_S \cdot \mathbf{n}_S = p_D \quad \text{on} \quad \Gamma_{SD}, \tag{8}$$

$$\mathbf{u}_S \cdot \mathbf{t}_j = -2\frac{\sqrt{\mathbf{k}}}{\alpha}\varepsilon(\mathbf{u}_S)\,\mathbf{n}_S \cdot \mathbf{t}_j, \quad j = 1, d-1, \quad \text{on} \quad \Gamma_{SD}. \tag{9}$$

The conditions (7) and (8) impose the continuity of flux and normal stress, respectively. The slip condition (9) is known as Beavers-Joseph-Saffman law [1,18], where $\alpha > 0$ is an experimentally determined dimensionless constant. The coupled problem, modeled by the Eqs. (1)–(9), is analyzed in detail by [11], where existence and uniqueness of the solution is demonstrated.

The Stokes-Darcy coupled problem provides the velocity field for the diffusive-convective-reactive transport equation defined on the domain $\Omega = \Omega_S \cup \Omega_D$ whose problem is given by

Given the Stokes or Darcy velocity field $\mathbf{u}$, *the porosity* ϕ, *the diffusion-dispersion tensor* $\mathbf{D}$, *the sources* $\hat{f}$ *and* g *and the function* c_0, *find the concentration* $c(\mathbf{x}, t) : \Omega \times (0, T) \to \mathbb{R}^d$, *such that*

$$\phi\frac{\partial c}{\partial t} + \mathbf{u} \cdot \nabla c - \operatorname{div}(\mathbf{D}\nabla c) + \hat{f}c = g \quad \text{in} \quad \Omega \times (0, T), \tag{10}$$

$$c(\mathbf{x}, 0) = c_0(\mathbf{x}) \quad \text{in} \quad \Omega, \tag{11}$$

$$\mathbf{D}\nabla c \cdot \mathbf{n} = 0 \quad \text{on} \quad \Gamma \times (0, T). \tag{12}$$

In the Stokes domain Ω_S

$$\phi = 1 \text{ and } \mathbf{D} = \alpha_m \mathbf{I}, \tag{13}$$

where α_m is a molecular diffusion coefficient and $\mathbf{I}$ the identity tensor. In the porous medium Ω_D, the tensor $\mathbf{D}$ can be defined as

$$\mathbf{D} = \mathbf{D}(\mathbf{u}_D) = \alpha_m \mathbf{I} + \|\mathbf{u}_D\| \left[\alpha_l \mathbf{E}(\mathbf{u}_D) + \alpha_t(\mathbf{I} - \mathbf{E}(\mathbf{u}_D))\right], \quad \mathbf{E}(\mathbf{u}) = \frac{\mathbf{u} \otimes \mathbf{u}}{\|\mathbf{u}\|^2},$$

with $\|\mathbf{u}\|^2 = \sum_{i=1}^d u_i^2$, $\otimes$ denoting the tensorial product, α_l being the longitudinal dispersion and α_t the transverse dispersion. In miscible displacement of a fluid through another in a reservoir, the dispersion is physically more important than the molecular diffusion [15]. Thus, we assume the following properties

$$0 < \alpha_m \le \alpha_l, \quad \alpha_l \ge \alpha_t > 0 \quad \text{and} \quad 0 < \phi \le 1 \quad \text{in} \quad \Omega_D.$$

3 Notations and Definitions

To introduce the stabilized hybrid formulation we first recall some notations and definitions. Let $H^m(\Omega)$ be the usual Sobolev space equipped with the usual norm $\|\cdot\|_{m,\Omega} = \|\cdot\|_m$ and seminorm $|\cdot|_{m,\Omega} = |\cdot|_m$, with $m \geq 0$. For $m = 0$, we consider $L^2(\Omega) = H^0(\Omega)$ as the space of square integrable functions and $H_0^1(\Omega)$ denotes the subspace of functions in $H^1(\Omega)$ with zero trace on $\partial\Omega$.

For a given function space $V(\Omega)$, let $[V(\Omega)]^d$ and $[V(\Omega)]^{d\times d}$ be the spaces of all vectors and tensor fields whose components belong to $V(\Omega)$, respectively. Without further specification, these spaces are furnished with the usual product norms (which, for simplicity, are denoted similarly as the norm in $V(\Omega)$). For vectors $\mathbf{v}, \mathbf{w} \in \mathbb{R}^d$ and matrices $\boldsymbol{\sigma}, \boldsymbol{\tau} \in \mathbb{R}^{d\times d}$ we use the standard notation.

Restricting to the two-dimensional case ($d = 2$), we define a regular finite element partition $\mathcal{T}_h$ of the domain Ω:

$$\mathcal{T}_h = \{K\} := \text{the union of all elements } K.$$

In cases where Ω is divided into subdomains Ω_i with smooth boundary $\partial\Omega_i$ and $\Gamma_i = \partial\Omega \cap \partial\Omega_i$, we have for each subdomain the following regular partition

$$\mathcal{T}_h^i = \{K \in \mathcal{T}_h \cap \Omega_i\},$$

and the following set of edges

$$\mathcal{E}_h^i = \{e; e \text{ is an edge of } K, \text{ for at least one } K \in \mathcal{T}_h^i\},$$

$$\mathcal{E}_h^{\partial,i} = \{e \in \mathcal{E}_h^i; e \subset \Gamma_i\}$$

$$\mathcal{E}_h^{0,i} = \{e \in \mathcal{E}_h^i; e \,\text{is an interior edge of } \Omega_i\},$$

$$\mathcal{E}_h^{ij} = \mathcal{E}_h^{0,i} \cap \mathcal{E}_h^{0,j}.$$

This last case denotes the edges that compose the interface between the subdomains, where Ω_i and Ω_j are two adjacent subdomains.

We assume that the domain Ω is polygonal. Thus, there exists $c > 0$ such that $h \leq ch_e$, where h_e is the diameter of the edge $e \in \partial K$ and h, the mesh parameter, is the maximum element diameter. For each element K we associate a unit normal vector $\mathbf{n}_K$. Let $\mathbf{V}_h^l$ and Q_h^m denote broken function spaces on $\mathcal{T}_h$ given by

$$\mathcal{V}_h^k = \{\mathbf{v} \in \mathbf{L}^2(\Omega); \mathbf{v}_h|_K \in [\mathbb{Q}_k(K)]^2, \ \forall K \in \mathcal{T}_h\}, \tag{14}$$

$$\mathcal{Q}_h^l = \{q \in L^2(\Omega); q_h|_K \in \mathbb{Q}_l(K), \ \forall K \in \mathcal{T}_h\}, \tag{15}$$

where $\mathbb{Q}_k(K)$ and $\mathbb{Q}_l(K)$ denote the space of polynomial functions of degree at most k and l, respectively, on each variable. To introduce the hybrid method, we define the following spaces associated with the Lagrange multiplier

$$\mathcal{M}_h^k = \{\boldsymbol{\mu} \in \mathbf{L}^2(\mathcal{E}_h) : \boldsymbol{\mu}|_e = [p_k(e)]^2, \ \forall e \in \mathcal{E}_h^0, \ \boldsymbol{\mu}|_e = \mathbf{0}, \ \forall e \in \mathcal{E}_h^\partial\}, \tag{16}$$

$$\mathcal{W}_h^k = \{\boldsymbol{\mu} \in \mathbf{L}^2(\mathcal{E}_h) : \boldsymbol{\mu}|_e = [p_k(e)]^2, \ \forall e \in \mathcal{E}_h^0, \ \boldsymbol{\mu}|_e \cdot \mathbf{n}_e = 0, \ \forall e \in \mathcal{E}_h^\partial\}, \tag{17}$$

Similarly, $p_k(e)$ is the space of polynomial functions of degree at most k on an edge e.

Moreover, we consider the following finite element spaces

$$V_h^k(\mathcal{T}_h) = \mathcal{V}_h^k(\mathcal{T}_h^S) \cup \mathcal{V}_h^k(\mathcal{T}_h^D) \quad (18)$$

$$Q_h^l(\mathcal{T}_h) = \mathcal{Q}_h^l(\mathcal{T}_h^S) \cup \mathcal{Q}_h^l(\mathcal{T}_h^D) \quad (19)$$

for velocity and pressure fields, respectively,

$$M_h^k(\mathcal{E}_h) = \mathcal{M}_h^k(\mathcal{E}_h^S \setminus \mathcal{E}_h^{SD}) \cup \mathcal{W}_h^k(\mathcal{E}_h^D) \quad (20)$$

for the multiplier in both Stokes and Darcy domains, the product space $\mathbf{V}_h = V_h^k(\mathcal{T}_h) \times Q_h^l(\mathcal{T}_h) \times M_h^k(\mathcal{E}_h)$ and set $\mathbf{X}_h = [\mathbf{u}_h, p_h, \boldsymbol{\lambda}_h] \in \mathbf{V}_h$.

4 Stabilized Hybrid Mixed Method for Stokes-Darcy Flow

Unlike the numerical methods employing Lagrange multipliers only in the interface free fluid/porous medium to solve the coupled problem [7,11], Igreja and Loula developed in [10] a stabilized hybrid mixed method, with Lagrange multipliers in all domain, where the interface conditions are naturally imposed, yielding a symmetric, robust and stable formulation. This formulation can be viewed below

Find $\mathbf{X}_h \in \mathbf{V}_h$ *such that,*

$$A_{SD}(\mathbf{X}_h, \mathbf{Y}_h) = F_{SD}(\mathbf{Y}_h), \quad \forall \mathbf{Y}_h \in \mathbf{V}_h, \quad (21)$$

with

$$\begin{aligned} A_{SD}(\mathbf{X}_h, \mathbf{Y}_h) &= \sum_{K \in \mathcal{T}_h} A_{SD}^K([\mathbf{u}_h, p_h]; [\mathbf{v}_h, q_h]) \\ &+ \sum_{K \in \mathcal{T}_h} \sum_{e \in \partial K} A_{SD}^e([\mathbf{u}_h, p_h, \boldsymbol{\lambda}_h]; [\mathbf{v}_h, q_h, \boldsymbol{\mu}_h]) \\ F_{SD}(\mathbf{Y}_h) &= \sum_{K \in \mathcal{T}_h} F_{SD}^K([\mathbf{v}_h, q_h]) \end{aligned}$$

where the local bilinear and linear forms are given in the Darcy domain by

$$\begin{aligned} A_{SD}^K([\mathbf{u}_h, p_h]; [\mathbf{v}_h, q_h]) &= A_D^K([\mathbf{u}_h^D, p_h^D]; [\mathbf{v}_h, q_h]), \ \forall K \in \mathcal{T}_h^D, \quad (22) \\ F_{SD}^K([\mathbf{v}_h, q_h]) &= F_D^K([\mathbf{v}_h, q_h]), \ \forall K \in \mathcal{T}_h^D, \\ A_{SD}^e([\mathbf{u}_h, p_h, \boldsymbol{\lambda}_h]; [\mathbf{v}_h, q_h, \boldsymbol{\mu}_h]) &= A_D^e([\mathbf{u}_h^D, p_h^D, \boldsymbol{\lambda}_h^D]; [\mathbf{v}_h, q_h, \boldsymbol{\mu}_h]), \forall e \notin \mathcal{E}_h^{SD} \end{aligned}$$

and in the Stokes domain by

$$\begin{aligned} A_{SD}^K([\mathbf{u}_h, p_h]; [\mathbf{v}_h, q_h]) &= A_S^K([\mathbf{u}_h^S, p_h^S]; [\mathbf{v}_h, q_h]), \ \forall K \in \mathcal{T}_h^S, \quad (23) \\ F_{SD}^K([\mathbf{v}_h, q_h]) &= F_S^K([\mathbf{v}_h, q_h]), \ \forall K \in \mathcal{T}_h^S, \\ A_{SD}^e([\mathbf{u}_h, p_h, \boldsymbol{\lambda}_h]; [\mathbf{v}_h, q_h, \boldsymbol{\mu}_h]) &= A_S^e([\mathbf{u}_h^S, p_h^S, \boldsymbol{\lambda}_h^S]; [\mathbf{v}_h, q_h, \boldsymbol{\mu}_h]), \ \forall e \notin \mathcal{E}_h^{SD}. \end{aligned}$$

plus the Beavers-Joseph-Saffman interface condition on Γ_{SD}

$$A^e_{SD}([\mathbf{u}_h, p_h, \boldsymbol{\lambda}_h]; [\mathbf{v}_h, q_h, \boldsymbol{\mu}_h]) = A^e_{BJS}([\mathbf{u}_h, p_h, \boldsymbol{\lambda}_h]; [\mathbf{v}_h, q_h, \boldsymbol{\mu}_h]), \ \forall e \in \mathcal{E}_h^{SD}$$

Electing the Lagrange multiplier $\boldsymbol{\lambda}_h^{SD} = \boldsymbol{\lambda}_h^D = \mathbf{u}_h^D|_{\partial K} = \boldsymbol{\lambda}_h^S = \mathbf{u}_h^S|_{\partial K}$ and the stabilization parameter $\beta_{SD} = \beta_S = \beta_D$ on the interface $\mathcal{E}_h^{SD}$, we have [10]

$$\begin{aligned} A^e_{BJS} = & \int_e \frac{\nu\alpha}{\sqrt{\mathbf{k}}}(\mathbf{u}_h^S \cdot \mathbf{t})(\mathbf{v}_h^S \cdot \mathbf{t})\, ds - \int_e (p_h^S - p_h^D - \nu \nabla \mathbf{u}_h^S \mathbf{n}_S \cdot \mathbf{n}_S)\boldsymbol{\mu}_h \cdot \mathbf{n}_S\, ds \\ & + \int_e \boldsymbol{\lambda}_h^{SD} \cdot q_h^D \mathbf{n}_S\, ds + \beta_{SD} \int_e (\mathbf{u}_h^D - \boldsymbol{\lambda}_h^{SD}) \cdot (\mathbf{v}_h^D - \boldsymbol{\mu}_h)\, ds \\ & + \int_e (p_h^S - \nu \nabla \mathbf{u}_h^S \mathbf{n}_S \cdot \mathbf{n}_S)\mathbf{v}_h^S \cdot \mathbf{n}_S\, ds + \int_e q_h^S\, (\mathbf{u}_h^S - \boldsymbol{\lambda}_h^{SD}) \cdot \mathbf{n}_S\, ds \\ & - \int_e \nu \nabla \mathbf{v}_h^S \mathbf{n}_S \cdot \mathbf{n}_S (\mathbf{u}_h^S - \boldsymbol{\lambda}_h^{SD}) \cdot \mathbf{n}_S\, ds \\ & + \beta_{SD} \int_e (\mathbf{u}_h^S - \boldsymbol{\lambda}_h^{SD}) \cdot \mathbf{n}_S (\mathbf{v}_h^S - \boldsymbol{\mu}_h) \cdot \mathbf{n}_S\, ds. \end{aligned}$$

Moreover, the local bilinear and linear forms for the Darcy problem

$$\begin{aligned} A_D^K([\mathbf{u}_h^D, p_h^D]; [\mathbf{v}_h, q_h]) = & \int_K \mathbf{A}\mathbf{u}_h^D \cdot \mathbf{v}_h\, d\mathbf{x} + \int_K \nabla p_h^D \cdot \mathbf{v}_h\, d\mathbf{x} + \int_K \mathbf{u}_h^D \cdot \nabla q_h\, d\mathbf{x} \\ & + \delta_1 \int_K \mathbf{K}(\mathbf{A}\mathbf{u}_h^D + \nabla p_h^D) \cdot (\mathbf{A}\mathbf{v}_h + \nabla q_h)\, d\mathbf{x} \\ & + \delta_2 \int_K A \operatorname{div} \mathbf{u}_h^D \operatorname{div} \mathbf{v}_h\, d\mathbf{x} \\ & + \delta_3 \int_K \kappa \operatorname{rot}(\mathbf{A}\mathbf{u}_h^D) \operatorname{rot}(\mathbf{A}\mathbf{v}_h)\, d\mathbf{x}, \end{aligned} \tag{24}$$

$$\begin{aligned} A_D^e([\mathbf{u}_h^D, p_h^D, \boldsymbol{\lambda}_h^D]; [\mathbf{v}_h, q_h, \boldsymbol{\mu}_h]) = & - \int_e \boldsymbol{\lambda}_h^D \cdot q_h \mathbf{n}_K\, ds - \int_e \boldsymbol{\mu}_h \cdot p_h^D \mathbf{n}_K\, ds \\ & + \beta_D \int_e (\mathbf{u}_h^D - \boldsymbol{\lambda}_h^D) \cdot (\mathbf{v}_h - \boldsymbol{\mu}_h)\, ds, \end{aligned} \tag{25}$$

$$F_D^K([\mathbf{v}_h, q_h]) = \delta_2 \int_K A\, f \operatorname{div} \mathbf{v}_h\, d\mathbf{x} - \int_K f\, q_h\, d\mathbf{x}, \tag{26}$$

For the Stokes problem, we have

$$\begin{aligned} A_S^K([\mathbf{u}_h^S, p_h^S]; [\mathbf{v}_h, q_h]) = & \int_K \nu \nabla \mathbf{u}_h^S : \nabla \mathbf{v}_h\, d\mathbf{x} \\ & - \int_K \operatorname{div} \mathbf{u}_h^S\, q_h\, d\mathbf{x} - \int_K p_h^S \operatorname{div} \mathbf{v}_h\, d\mathbf{x} \end{aligned}$$

$$F_S^K([\mathbf{v}_h, q_h]) = \int_K \mathbf{f} \cdot \mathbf{v}_h d\mathbf{x}$$

and

$$A_S^e([\mathbf{u}_h^S, p_h^S, \boldsymbol{\lambda}_h^S]; [\mathbf{v}_h, q_h, \boldsymbol{\mu}_h]) = -\int_e \nu \nabla \mathbf{u}_h^S \, \mathbf{n}_K \cdot (\mathbf{v}_h - \boldsymbol{\mu}_h) ds$$
$$- \int_e \nu \nabla \mathbf{v}_h \, \mathbf{n}_K \cdot (\mathbf{u}_h^S - \boldsymbol{\lambda}_h^S) ds + \int_e p_h^S (\mathbf{v}_h - \boldsymbol{\mu}_h) \cdot \mathbf{n}_K \, ds$$
$$+ \int_e q_h (\mathbf{u}_h^S - \boldsymbol{\lambda}_h^S) \cdot \mathbf{n}_K \, ds + \beta_S \int_e (\mathbf{u}_h^S - \boldsymbol{\lambda}_h^S) \cdot (\mathbf{v}_h - \boldsymbol{\mu}_h) ds. \qquad (27)$$

The stabilization parameters is given by

$$\beta_S = \nu \frac{\beta_0^S}{h}, \quad \text{with} \quad \beta_0^S > 0 \quad \text{and} \quad \beta_D = A \frac{\beta_0^D}{h}. \qquad (28)$$

To solve this problem, the formulation (21) is splited in a set of local problems defined at the element level and a global problem associated with the multipliers. The degrees of freedom of the variables in the local problem are condensed, through the static condensation technique, and a global system is assembled in terms of the multipliers. Then, the global problem is solved leading to the approximate solution of the multipliers, which is plugged into the local problems to recover the discontinuous approximation of the velocity and pressure fields. For more details see [10].

4.1 Concentration Approximation

Given the velocity field calculated through the hybrid method (21), we can obtain the concentration field using the SUPG method [2] to approximate the transport equation (10)–(12). For this, let the time step $\Delta t > 0$, such that $N = T/\Delta t$ and $t_n = n\Delta t$ with $n = 1, 2, ..., N$ and let $I_h = \{0 = t_0 < t_1 < ... < t_N = T\}$ be a partition of the interval $I = [0, T]$. The term involving the time derivative of the concentration is approximated by backward Euler finite difference operator

$$\frac{\partial c}{\partial t}(\mathbf{x}, t_n) = \frac{\partial c^n}{\partial t} = \frac{c^{n+1} - c^n}{\Delta t}.$$

Therefore, a semi-discrete approximation for the transport equation for each $n = 1, 2, ...N$, given $c^0(\mathbf{x}) = c_0(\mathbf{x})$, can be written as

$$\phi \frac{c^{n+1} - c^n}{\Delta t} + \mathbf{u} \cdot \nabla c^{n+1} - \text{div}(\mathbf{D}(\mathbf{u}) \nabla c^{n+1}) + \hat{f} c^{n+1} = g \quad \text{in} \quad \Omega. \qquad (29)$$

Combining the semi-discrete approximation (29) with a stabilized finite element method in space (SUPG), we introduce the following fully discrete approximation for the concentration equation: for time levels $n = 1, 2, ...N$, find $c_h^{n+1} \in \mathcal{C}_h^k$, where $\mathcal{C}_h^k$ is a C^0 Lagrangean finite element space of degree at most k, such that

$$A_{SUPG}(c_h^{n+1}; \varphi_h) = F_{SUPG}(c_h^n; \varphi_h), \quad \forall \varphi_h \in \mathcal{C}_h^k, \qquad (30)$$

with

$$A_{SUPG}(c_h^{n+1};\varphi_h) = \phi \int_\Omega c_h^{n+1}\varphi_h \, d\mathbf{x} + \Delta t \int_\Omega \mathbf{u}_h \cdot \nabla c_h^{n+1}\varphi_h \, d\mathbf{x}$$
$$+ \Delta t \int_\Omega \mathbf{D}(\mathbf{u}_h)\nabla c_h^{n+1} \cdot \nabla\varphi_h \, d\mathbf{x} + \Delta t \int_\Omega \hat{f} c_h^{n+1}\varphi_h \, d\mathbf{x}$$
$$+ \sum_{K\in\mathcal{T}_h} \int_K \left(\phi c_h^{n+1} + \Delta t \mathbf{u}_h \cdot \nabla c_h^{n+1} + \Delta t \hat{f} c_h^{n+1}\right)(\delta_K \mathbf{u}_h \cdot \nabla\varphi_h) \, ds$$
$$+ \sum_{K\in\mathcal{T}_h} \int_K (-\Delta t \, \mathrm{div}(\mathbf{D}(\mathbf{u}_h)\nabla c_h^{n+1}))(\delta_K \mathbf{u}_h \cdot \nabla\varphi_h) \, ds \tag{31}$$

and

$$F_{SUPG}(c_h^n;\varphi_h) = \phi \int_\Omega c_h^n \varphi_h \, d\mathbf{x} + \Delta t \int_\Omega g\varphi_h \, d\mathbf{x}$$
$$+ \sum_{K\in\mathcal{T}_h} \int_K (\phi c_h^n + \Delta t \, g)(\delta_K \mathbf{u}_h \cdot \nabla\varphi_h) \, ds. \tag{32}$$

In the system (30) the velocity field $\mathbf{u}_h$ is given by the solution of the hybrid formulation (21). The stabilization parameter δ_K is defined on each $K \in \mathcal{T}_h$ as described in [2,14].

5 Numerical Results

In this section we present numerical experiments to evaluate the rates of convergence of the stabilized hybrid mixed formulation (21). Moreover, we use the approximate velocity field obtained by the hybrid method, which is responsible for the flow displacement, to find the concentration field calculated by a predominantly convective Eq. (10) that is numerically solved via SUPG method applied to continuous injection process in a quarter of a repeated five-spot pattern for different heterogenous scenarios [14].

5.1 Convergence Study

In this test problem, we solve a simple problem with $\mathbf{K} = \mathbf{I}$ and $\mu = 1.0$ in a square domain $\Omega = \Omega_D \cup \Omega_S = (0.0, 1.0)^2$, with respective Stokes and Darcy sources

$$\mathbf{f} = \begin{bmatrix} (1/2 + 1/(8\pi^2))\sin(\pi x)\exp(y/2) \\ (\pi - 3/(4\pi))\cos(\pi x)\exp(y/2) \end{bmatrix}, \qquad f = \left(\frac{1}{2\pi} - 2\pi\right)\cos(\pi x)\exp(y/2),$$

with the exact solution presented in [4,10].

In the convergence study we adopt h-refinement strategy taking a sequence of $n \times n$ uniform meshes, with $n = 4, 8, 16, 32, 64$, using quadrilateral elements $\mathbb{Q}_k\mathbb{Q}_l - p_m$, where k, l and m denote, respectively, the degree of polynomial spaces for velocity, pressure and multiplier, considering equal order approximations for all fields $k = l = m = 1$ and 2 with the respectives stabilization parameters for the Stokes and Darcy multipliers $\beta_0^S = 12.0$ and 24.0 and $\beta_0^D = 1.0$ and 15.0. For the least square stabilization parameters defined in the interior of the elements we adopt in all simulations

$$\delta_1 = -0.5, \quad \delta_2 = 0.5, \quad \delta_3 = 0.5. \tag{33}$$

In Fig. 2 we can see the h-convergence study for the velocity and pressure in the $L^2(\Omega)$ norm compared to the interpolant for $\mathbb{Q}_1\mathbb{Q}_1 - p_1$ and $\mathbb{Q}_2\mathbb{Q}_2 - p_2$ elements, respectively. The results demonstrate optimal convergence rates for all fields studied, except for the pressure field approximated by biquadratic elements (Fig. 2(d)), in which case the potential loses accuracy.

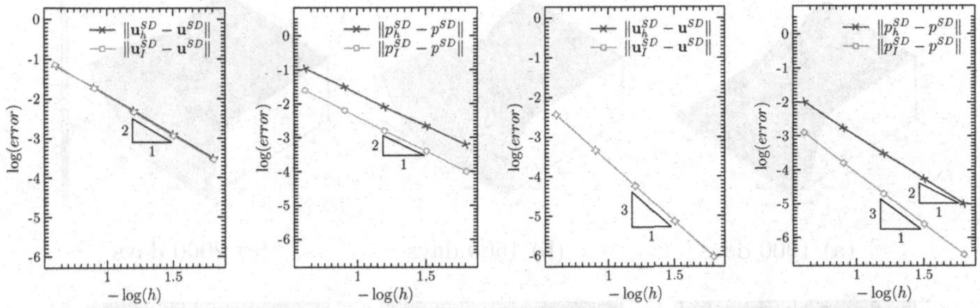

(a) $\mathbf{u}_h$ ($\mathbb{Q}_1\mathbb{Q}_1 - p_1$) (b) p_h ($\mathbb{Q}_1\mathbb{Q}_1 - p_1$) (c) $\mathbf{u}_h$ ($\mathbb{Q}_2\mathbb{Q}_2 - p_2$) (d) p_h ($\mathbb{Q}_2\mathbb{Q}_2 - p_2$)

Fig. 2. h-convergence study of the Stokes-Darcy approximations ($\mathbf{u}_h^{SD}$ and p_h^{SD}) comparing the hybrid method with the respective interpolant ($\mathbf{u}_I$ or p_I) in $L^2(\Omega)$ norm for $\mathbb{Q}_1\mathbb{Q}_1 - p_1$ and $\mathbb{Q}_2\mathbb{Q}_2 - p_2$ elements.

5.2 Continuous Injection Simulation

Here we simulate a quarter of a repeated five-spot pattern in two dimension consisting of a square domain (unit thickness) with side $L = 1000.0\,ft$. The injector well is located at the lower-left corner ($x = y = 0$) and the producer well at the upper-right corner ($x = y = L$). For this, we use the hybrid formulation (21) to approximate the hydrodynamic problem, then we supply the velocity field to the transport equation that is numerically solved by the SUPG method combined with an implicit finite difference scheme in three different scenarios described in Fig. 3.

These three cases are considered for a porous medium with homogeneous permeability $\kappa = 10.0mD$, where $\mathbf{K} = (\kappa/\mu)\mathbf{I}$, viscosity of the resident fluid is $\mu = 1.0\,cP$, porosity $\phi = 0.1$, molecular diffusion $\alpha_m = 0.0$, longitudinal dispersion $\alpha_l = 10.0\,ft^2/day$, transverse dispersion $\alpha_t = 1.0\,ft^2/day$ and the flow rate is 800 square feet per day. For the Stokes region the diffusion tensor is chosen to be $\mathbf{D} = \alpha_m\mathbf{I}$ with $\alpha_m = 1.0\,ft^2/day$. We fix the same values of the numerical stabilization parameters presented in (33) and $\alpha = 1.0$. Moreover, a time step of

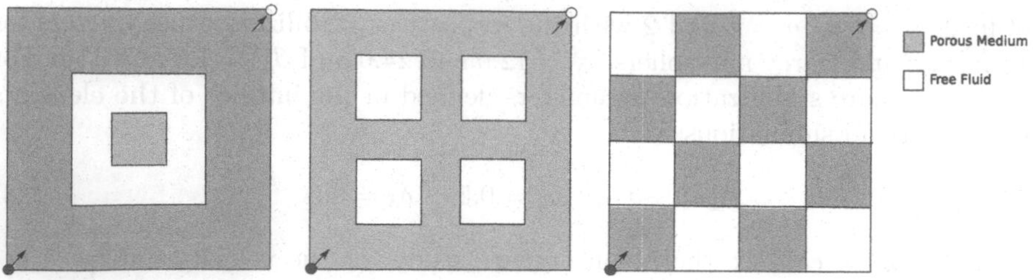

(a) Scenario 1 (b) Scenario 2 (c) Scenario 3

Fig. 3. Three coupled porous medium (shaded) free fluid (white) domains used to simulate the five-spot problem.

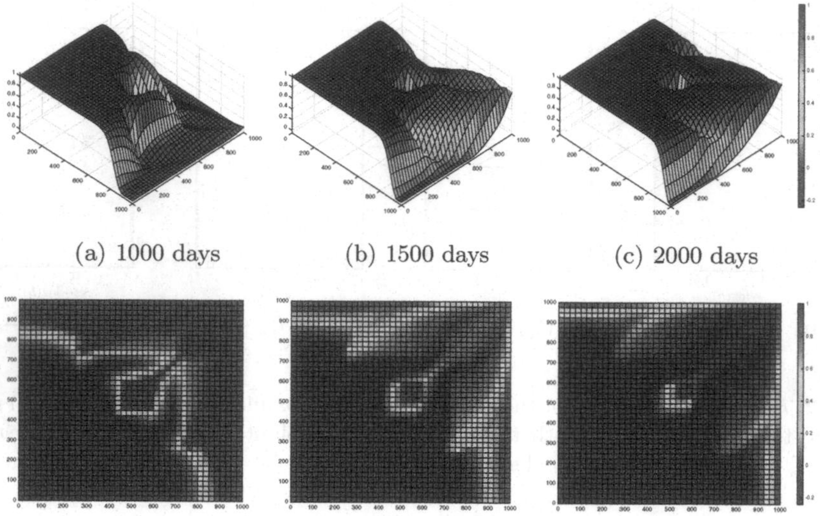

(a) 1000 days (b) 1500 days (c) 2000 days

(d) 1000 days (e) 1500 days (f) 2000 days

Fig. 4. Scenario 1: front propagation of the concentration in five-spot problem.

5 days and uniform meshes of 40×40 bilinear quadrilateral elements are adopted employing equal order approximations for all fields (velocity, pressure, multiplier and concentration).

The Figs. 4, 5 and 6 show the concentration maps and concentration contours for the proposed scenarios. In these graphs we can clearly observe the effect of the barrier on the continuous injection transport generated by the low permeability of the porous medium. The continuous injection concentration in the scenario 3 (Fig. 6) takes longer time to reach the producer well due to the higher heterogeneity of the medium, because it presents more discontinuities generated by the free fluid/porous medium interfaces, which reduces the flow velocity.

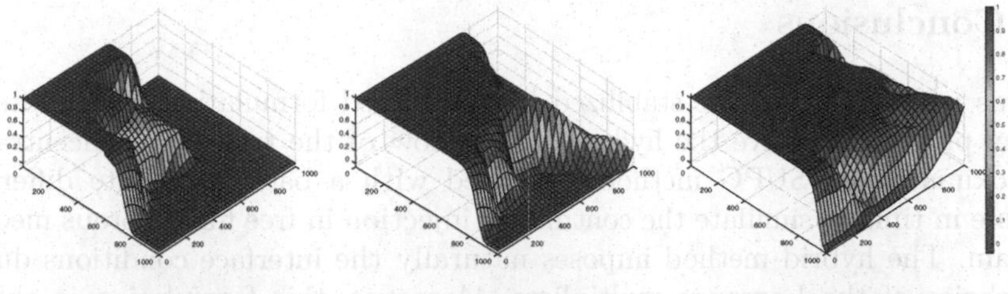

(a) 500 days (b) 1000 days (c) 2000 days

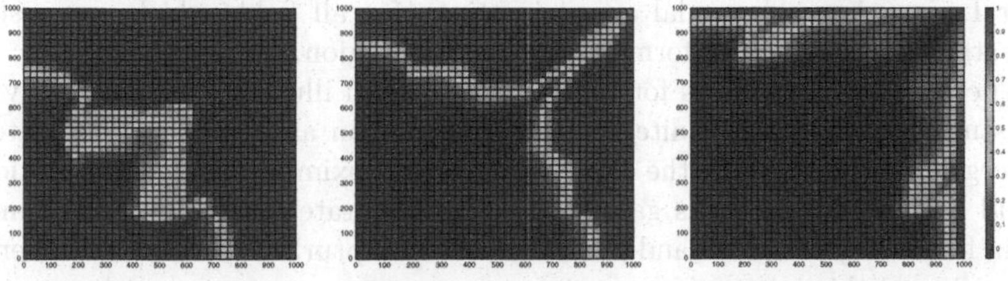

(d) 500 days (e) 1500 days (f) 2000 days

Fig. 5. Scenario 2: front propagation of the concentration in five-spot problem.

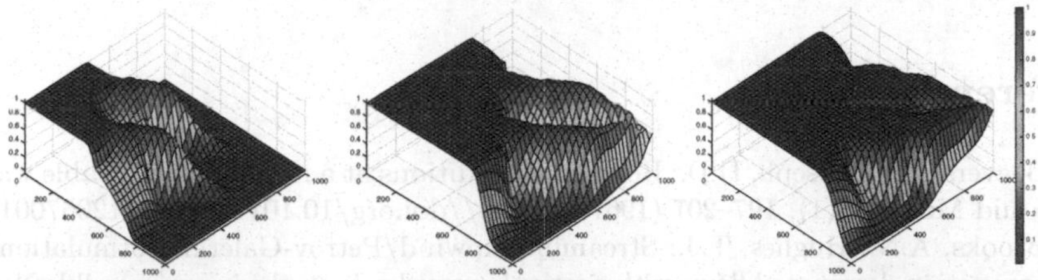

(a) 1000 days (b) 2000 days (c) 4000 days

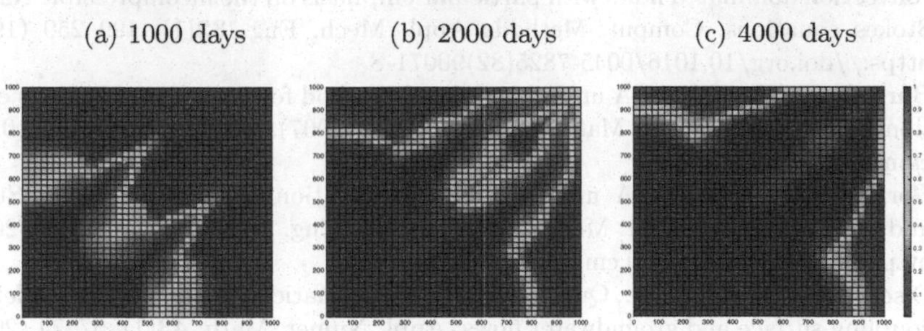

(d) 1000 days (e) 2000 days (f) 4000 days

Fig. 6. Scenario 3: front propagation of the concentration in five-spot problem.

6 Conclusions

In this work we recall the stabilized hybrid mixed formulation for the Stokes-Darcy problem to solve the hydrodynamic flow of the transport concentration approximated by SUPG method combined with a backward finite difference scheme in time to simulate the continuous injection in free fluid/porous medium domain. The hybrid method imposes naturally the interface conditions due to the choice of the Lagrange multipliers. Moreover, this formulation is able to recover stability of very convenient finite element approximations, such as equal order Lagrangian polynomial approximations for all fields which are unstable with standard dual mixed formulation for each region.

The convergence results for the hybrid method illustrate the flexibility and robustness of the hybrid finite element formulation and show optimal rates of convergence. With respect the concentration approximation, the combination of hybrid and SUPG methods gave stable and accurate results in heterogeneous media formed by free fluid and porous medium capturing precisely the phenomena arising of this interaction.

Acknowledgement. This study was financed in part by the Coordenação de Aperfeiçoamento de Pessoal de Nível Superior - Brasil (CAPES) - Finance Code 001. This work was also partially supported by CNPq and UFJF.

References

1. Beavers, G.S., Joseph, D.D.: Boundary conditions at a naturally permeable wall. J. Fluid Mech. **30**(1), 197–207 (1967). https://doi.org/10.1017/S0022112067001375
2. Brooks, A.N., Hughes, T.J.: Streamline upwind/Petrov-Galerkin formulations for convection dominated flows with particular emphasis on the incompressible Navier-Stokes equations. Comput. Methods Appl. Mech. Eng. **32**(1), 199–259 (1982). https://doi.org/10.1016/0045-7825(82)90071-8
3. Burman, E., Hansbo, P.: A unified stabilized method for Stokes' and Darcy's equations. J. Comput. Appl. Math. **198**(1), 35–51 (2007). https://doi.org/10.1016/j.cam.2005.11.022
4. Correa, M., Loula, A.: A unified mixed formulation naturally coupling Stokes and Darcy flows. Comput. Methods Appl. Mech. Eng. **198**(33), 2710–2722 (2009). https://doi.org/10.1016/j.cma.2009.03.016
5. Discacciati, M., Miglio, E., Quarteroni, A.: Mathematical and numerical models for coupling surface and groundwater flows. Appl. Numer. Math. **43**(1), 57–74 (2002). https://doi.org/10.1016/S0168-9274(02)00125-3
6. Egger, H., Waluga, C.: A hybrid discontinuous Galerkin method for Darcy-Stokes problems. In: Bank, R., Holst, M., Widlund, O., Xu, J. (eds.) Domain Decomposition Methods in Science and Engineering XX, pp. 663–670. Springer, Berlin Heidelberg (2013). https://doi.org/10.1007/978-3-642-35275-1_79
7. Gatica, G.N., Oyarzúa, R., Sayas, F.J.: A residual-based a posteriori error estimator for a fully-mixed formulation of the Stokes-Darcy coupled problem. Comput. Methods Appl. Mech. Eng. **200**(21), 1877–1891 (2011). https://doi.org/10.1016/j.cma.2011.02.009

8. Hanspal, N.S., Waghode, A.N., Nassehi, V., Wakeman, R.J.: Numerical analysis of coupled Stokes/Darcy flows in industrial filtrations. Transp. Porous Media **64**(1), 73 (2006). https://doi.org/10.1007/s11242-005-1457-3
9. Igreja, I., Loula, A.F.D.: Stabilized velocity and pressure mixed hybrid DGFEM for the Stokes problem. Int. J. Numer. Meth. Eng. **112**(7), 603–628 (2017). https://doi.org/10.1002/nme.5527
10. Igreja, I., Loula, A.F.: A stabilized hybrid mixed DGFEM naturally coupling Stokes-Darcy flows. Comput. Methods Appl. Mech. Eng. **339**, 739–768 (2018). https://doi.org/10.1016/j.cma.2018.05.026
11. Layton, W.J., Schieweck, F., Yotov, I.: Coupling fluid flow with porous media flow. SIAM J. Numer. Anal. **40**(6), 2195–2218 (2002). https://doi.org/10.1137/S0036142901392766
12. Malta, S.M., Loula, A.F., Garcia, E.L.: Numerical analysis of a stabilized finite element method for tracer injection simulations. Comput. Methods Appl. Mech. Eng. **187**(1), 119–136 (2000). https://doi.org/10.1016/S0045-7825(99)00113-9
13. Masud, A.: A stabilized mixed finite element method for Darcy-Stokes flow. Int. J. Numer. Meth. Fluids **54**(6–8), 665–681 (2007). https://doi.org/10.1002/fld.1508
14. Núñez, Y., Faria, C., Loula, A., Malta, S.: Um método híbrido de elementos finitos aplicado a deslocamentos miscíveis em meios porosos heterogêneos. Revista Internacional de Métodos Numéricos para Cálculo y Diseño en Ingeniería **33**(1), 45–51 (2017). https://doi.org/10.1016/j.rimni.2015.10.002
15. Peaceman, D.W.: Fundamentals of Numerical Reservoir Simulation. Elsevier Science Inc., New York NY, USA (1991)
16. Rivière, B., Yotov, I.: Locally conservative coupling of Stokes and Darcy flow. SIAM J. Numer. Anal. **42**(5), 1959–1977 (2005). https://doi.org/10.1137/S0036142903427640
17. Rui, H., Zhang, J.: A stabilized mixed finite element method for coupled Stokes and Darcy flows with transport. Comput. Methods Appl. Mech. Eng. **315**, 169–189 (2017). https://doi.org/10.1016/j.cma.2016.10.034
18. Saffman, P.G.: On the boundary condition at the surface of a porous medium. Stud. Appl. Math. **50**(2), 93–101 (1971). https://doi.org/10.1002/sapm197150293
19. Salinger, A., Aris, R., Derby, J.: Finite element formulations for large-scale, coupled flows in adjacent porous and open fluid domains. Int. J. Numer. Meth. Fluids **18**(12), 1185–1209 (1994). https://doi.org/10.1002/fld.1650181205
20. Urquiza, J., N'Dri, D., Garon, A., Delfour, M.: Coupling Stokes and Darcy equations. Appl. Numer. Math. **58**(5), 525–538 (2008). https://doi.org/10.1016/j.apnum.2006.12.006
21. Vassilev, D., Yotov, I.: Coupling Stokes-Darcy flow with transport. SIAM J. Sci. Comput. **31**(5), 3661–3684 (2009). https://doi.org/10.1137/080732146

Energy Stable Simulation of Two-Phase Equilibria with Capillarity

Shuyu Sun[(✉)]

Computational Transport Phenomena Laboratory,
Division of Physical Science and Engineering,
King Abdullah University of Science and Technology,
Thuwal 23955-6900, Kingdom of Saudi Arabia
shuyu.sun@kaust.edu.sa
http://web.kaust.edu.sa/faculty/shuyusun/

Abstract. We consider the affect of capillary pressure on the Van der Waals fuid and on the Peng-Robinson fluid by minimizing total Helmholtz energy in given total volume, temperature, and total moles. We propose simple but conditionally energy stable numerical schemes, and we provide interesting numerical examples. We compare our numerical results with the prediction of Kelvin's equation, indicating that Kelvin's equation works well only when the temperature is not too low.

Keywords: Interfacial tension · The Kelvin equation · Phase diagrams · Peng-Robinson equation of state · Van der Waals equation of state

1 Introduction

Two-phase and multi-phase flows are important and common phenomena in petroleum industry, where oil, gas and water are often produced and transported together. In particular, engineers and researchers in reservoir engineering study drainage problems arising during the development and production of oil and gas reservoirs so as to obtain a high economic recovery, by developing, conducting, and interpolating the simulation of subsurface flows of reservoir fluids, including water, hydrocarbon, CO_2, H_2S for example in porous geological formation. Field-scale (Darcy-scale) simulation has conventionally and routinely used for this purpose [2,4,12,16,17]. A number of parameters like relative permeability and capillary pressure are taken as given functions in Darcy-scale simulation [14]. To study these parameters as well as to obtain deep understanding of porous media flow and transport, researchers develop and utilize pore-scale simulation of two-phase [1,3,5–9,11,13], which has been shown to be a great research tool to understand the complex hydrodynamic and phase behaviors of the systems.

This work was supported by funding from King Abdullah University of Science and Technology (KAUST) through grants BAS/1/1351-01.

J. M. F. Rodrigues et al. (Eds.): ICCS 2019, LNCS 11539, pp. 538–550, 2019.
https://doi.org/10.1007/978-3-030-22747-0_40

2 Mathematical Modeling Framework

2.1 Mathematical Model of Bulk Properties

In this paper we consider the Van der Waals equation of state (EOS) and the Peng-Robinson EOS [15] to model the bulk properties of our fluid system. We note that the Van der Waals EOS is one of the simplest equations of state that allow phase splitting. On the other hand, the Peng-Robinson EOS is the most popular equation of state (EOS) to model and compute the fluid equilibrium property of hydrocarbon fluid and other petroleum fluids, and it is widely used in reservoir engineering and oil industries. Some of the material in this section can be found in many textbooks, but we list them here for completeness of information only.

The PVT-form of the Van der Waals EOS appears as

$$\left(p + \frac{a}{v^2}\right)(v - b) = RT. \tag{1}$$

At the critical condition, we have $\left(\frac{\partial p}{\partial v}\right)_T = 0$ and $\left(\frac{\partial^2 p}{\partial v^2}\right)_T = 0$. These two conditions together with the above Van der Waals EOS lead to $v_c = 3b$, $p_c = \frac{a}{27b^2}$, $T_c = \frac{8a}{27Rb}$, and $Z_c = \frac{3}{8}$. we substitute the definitions of the reduced pressure, reduced molar volume and reduced temperature with the values of critical properties (in terms of a and b) into the above Van der Waals EOS to obtain the following reduced form of the Van der Waals equation

$$\left(p_r + \frac{3}{v_r^2}\right)(3v_r - 1) = 8T_r. \tag{2}$$

From the PVT-form of the Van der Waals EOS, we can derive the bulk Helmholtz free energy for the Van der Waals fluid as $F(T, V, \mathbf{N}) = F_{\text{ideal}}(T, V, \mathbf{N}) + F_{\text{excess}}(T, V, \mathbf{N})$, where the ideal gas contribution (note that $N = \sum_i N_i$) is $F_{\text{ideal}}(T, V, \mathbf{N}) = RT \sum_{i=1}^{M} N_i \ln\left(\frac{N_i}{V}\right) + NC_{\text{intg}}(T)$, with one choice of C_{intg} as $C_{\text{intg}}(T) = -RT\left(\ln(\frac{n_Q}{N_A}) + 1\right)$ and $n_Q = \left(\frac{2\pi m k_B T}{h^2}\right)^{3/2}$. The excess part of Helmholtz free energy is: $F_{\text{excess}}(T, V, N) = -NRT \ln\left(1 - \frac{Nb}{V}\right) - \frac{N^2 a}{V}$.

We define the Helmholtz free energy density f as the Helmholtz free energy per unit volume of fluid. It is clear that

$$f(T, \mathbf{n}) = f_{\text{ideal}}(T, \mathbf{n}) + f_{\text{excess}}(T, \mathbf{n}), \tag{3}$$

$$f_{\text{ideal}}(T, \mathbf{n}) = RT \sum_{i=1}^{M} n_i \ln(n_i) + nC_{\text{intg}}(T), \tag{4}$$

$$f_{\text{excess}}(T, n) = -nRT \ln(1 - bn) - an^2. \tag{5}$$

With the Peng-Robinson EOS, the bulk Helmholtz free energy density $f(T, \mathbf{n})$ of a bulk fluid is determined by $f(T, \mathbf{n}) = f_{\text{ideal}}(T, \mathbf{n}) + f_{\text{excess}}(T, \mathbf{n})$ and

$$f^{\text{ideal}}(T, \mathbf{n}) = RT \sum_{i=1}^{M} n_i (\ln n_i - 1), \tag{6}$$

$$f^{\text{excess}}(T, n) = -nRT \ln(1 - bn) + \frac{a(T)n}{2\sqrt{2}b} \ln\left(\frac{1 + (1 - \sqrt{2})bn}{1 + (1 + \sqrt{2})bn}\right). \tag{7}$$

The two parameters a and b can be computed as follows. For a mixture, these parameters can be calculated from the ones of the pure fluids by mixing rules:

$$a(T) = \sum_{i=1}^{M} \sum_{j=1}^{M} y_i y_j a_i a_j^{1/2} (1 - k_{ij}), \quad b = \sum_{i=1}^{M} y_i b_i, \tag{8}$$

where $y_i = n_i/n$ is the mole fraction of component i, and a_i and b_i are the Peng-Robinson parameters for pure-substance component i. We often use experimental data to fit the binary interaction coefficient k_{ij} of Peng-Robinson. For convenience, k_{ij} is usually assumed to be constant for a fixed species pair.

Even though the pure-substance Peng-Robinson parameters a_i and b_i can also be fit by using experimental data, they can also be computed from the critical properties of the species:

$$a_i = a_i(T) = 0.45724 \frac{R^2 T_{c_i}^2}{P_{c_i}} \left(1 + m_i \left(1 - \sqrt{\frac{T}{T_{c_i}}}\right)\right)^2,$$

$$b_i = 0.07780 \frac{RT_{c_i}}{P_{c_i}}.$$

As intrinsic properties of the species, the critical temperature T_{c_i} and critical pressure P_{c_i} of a pure substance are available for most substances encountered in engineering practice. In the above formula for a_i, we need also to specify the parameter m_i for modeling the influence of temperature on a_i. It was suggested that one may correlates the parameter m_i experimentally to the accentric parameter ω_i of the species by the following equations:

$$m_i = 0.37464 + 1.54226\omega_i - 0.26992\omega_i^2, \quad \omega_i \le 0.49,$$
$$m_i = 0.379642 + 1.485030\omega_i - 0.164423\omega_i^2 + 0.016666\omega_i^3, \quad \omega_i > 0.49.$$

The accentric parameter can be fit by using experimental data, but if we lack data, we can also compute it by using critical temperature T_{c_i}, critical pressure P_{c_i} and the normal boiling point T_{b_i}:

$$\omega = \frac{3}{7}\left(\frac{\log_{10}\left(\frac{P_{c_i}}{14.695\ \text{PSI}}\right)}{\frac{T_{c_i}}{T_{b_i}} - 1}\right) - 1 = \frac{3}{7}\left(\frac{\log_{10}\left(\frac{P_{c_i}}{1\ \text{atm}}\right)}{\frac{T_{c_i}}{T_{b_i}} - 1}\right) - 1.$$

Based on the fundamental relation on thermodynamic variables, the pressure of homogeneous fluids p and the Helmholtz free energy $f(\mathbf{n})$ can be linked in the following way

$$\begin{aligned} p &= p(\mathbf{n},T) = -\left(\frac{\partial F(\mathbf{n},T,\Omega)}{\partial V}\right)_{T,\mathbf{N}} = -\left(\frac{\partial\left(f\left(\frac{\mathbf{N}}{V},T\right)V\right)}{\partial V}\right)_{T,\mathbf{N}} \\ &= -f - V\sum_{i=1}^{M}\left(\frac{\partial f}{\partial n_i}\right)_{T,n_1,\cdots,n_{i-1},n_{i+1},\cdots n_M}\left(\frac{\partial \frac{N_i}{V}}{\partial V}\right)_{N_i} \\ &= \sum_{i=1}^{M} n_i\left(\frac{\partial f}{\partial n_i}\right)_{T,n_1,\cdots,n_{i-1},n_{i+1},\cdots n_M} - f = \sum_{i=1}^{M} n_i\mu_i - f. \end{aligned}$$

Substitution of the Peng-Robinson expression of f leads to

$$p = \frac{nRT}{1-bn} - \frac{n^2a(T)}{1+2bn-b^2n^2} = \frac{RT}{v-b} - \frac{a(T)}{v(v+b)+b(v-b)}.$$

2.2 Modeling the Two-Phase Systems with Interfaces

The total Helmholtz energy F^{tot} has two contributions, one from the homogeneous bulk fluid, and another one from the interface between the two phases:

$$\begin{aligned} F^{\text{tot}}(\mathbf{n}) &= F_{\text{bulk}}(\mathbf{n}; T, \Omega) + F_{\text{interface}}(\mathbf{n}; T, A_I) \\ &= \int_{\Omega\backslash A_I} f(\mathbf{n}; T)d\mathbf{x} + \int_{A_I}\sigma([\mathbf{n}],\{\mathbf{n}\}; T)ds, \end{aligned}$$

where σ is the interfacial tension (or the interfacial Helmholtz energy per unit area), which is a function of the jump and average of $\mathbf{n}$ across the interface. In the paper, for convenience, we assume that σ is a given constant. With this assumption, we have

$$F^{\text{tot}} = f\left(\mathbf{n}^L\right)V^L + f\left(\mathbf{n}^G\right)V^G + \sigma A_I,$$

where $\mathbf{n}^L$ and V^L are the molar density and volume of liquid phase, respectively. Meanings of $\mathbf{n}^G$ and V^G are similar to these.

We impose the total volume V^{tot} and the total moles $\mathbf{N}^{\text{tot}}$ such as $V^{\text{tot}} = V^L + V^G$ and $N_i^{\text{tot}} = n_i^L V^L + n_i^G V^G$, $i = 1, 2, \cdots, M$. The interface A_I is also a function of V^G; for example, if there is only one single bubble, then $A_I = 4\pi r^2$ while $V^G = \frac{4\pi r^3}{3}$. Considering $\mathbf{n}^G$ and V^G as the primary variables, we can write the total Helmholtz energy F^{tot} as

$$F^{\text{tot}}(\mathbf{n}^G, V^G) = f\left(\frac{\mathbf{N}^{\text{tot}} - \mathbf{n}^G V^G}{V^{\text{tot}} - V^G}\right)(V^{\text{tot}} - V^G) + f\left(\mathbf{n}^G\right)V^G + \sigma A_I(V^G).$$

When F^{tot} achieves its minimum and the function F^{tot} is smooth, we have

$$\frac{\partial F^{\text{tot}}(\mathbf{n}^G, V^G)}{\partial n_i^G} = 0, \quad \text{and} \quad \frac{\partial F^{\text{tot}}(\mathbf{n}^G, V^G)}{\partial V^G} = 0,$$

which implies respectively $\mu(\mathbf{n}^G) = \mu(\mathbf{n}^L)$, and $p^G - p^L = \sigma\frac{dA_I}{dV^G}$. If there is only one single bubble, then $A_I = 4\pi r^2$ and $V^G = \frac{4\pi r^3}{3}$; the above condition simplifies to

$$p^G - p^L = \frac{2\sigma}{r},$$

which is known as the Young-Laplace equation.

3 Numerical Methods

To find the minimum of the $F^{\text{tot}}(\mathbf{n}^G, V^G)$, we design the following ordinary differential equation (ODE) system

$$\frac{\partial N_i^G}{\partial t} = -\frac{\partial N_i^L}{\partial t} = k_{N_i} N_i^{\text{tot}}(\mu(\mathbf{n}^L) - \mu(\mathbf{n}^G)), \quad i = 1, 2, \cdots, M,$$

$$\frac{\partial V^G}{\partial t} = -\frac{\partial V^L}{\partial t} = k_V V^{\text{tot}}(p^G(\mathbf{n}^G) - p^L(\mathbf{n}^L) - \sigma\frac{dA_I}{dV^G}).$$

We solve the above ODE using the following explicit Euler method, which is conditionally stable provided that the time step is less than a certain value.

$$\frac{N_i^{G,k+1} - N_i^{G,k}}{t^{k+1} - t^k} = k_{N_i} N_i^{\text{tot}}(\mu(\mathbf{n}^{L,k}) - \mu(\mathbf{n}^{G,k})), \quad i = 1, 2, \cdots, M,$$

$$\frac{V^{G,k+1} - V^{G,k}}{t^{k+1} - t^k} = k_V V^{\text{tot}}(p^G(\mathbf{n}^{G,k}) - p^L(\mathbf{n}^{L,k}) - \sigma\frac{dA_I}{dV^G}),$$

where $\mathbf{n}^{G,k}$ and $\mathbf{n}^{L,k}$ can be determined from $V^{G,k}$, $N_i^{G,k}$, $i = 1, 2, \cdots, M$.

Existence, uniqueness, and energy-decay property of the solution to the above ODE equation system as well as the existence, uniqueness, and the conditional energy-decay property of the numerical solution defined above can be proved using techniques similar to the ones used in our previous work [10].

4 Numerical Examples

4.1 Effect of Capillary Pressure on the Van der Waals Fluid

We first consider the single-component two-phase fluid system modeled by the Van der Waals EOS. If we write the single-component Van der Waals EOS using reduced temperature, reduced pressure and reduced molar volume, we then may obtain an universal dimensionless Eq. (2) for the EOS. That is, all single-component two-phase fluids behave in the same way after certain linear transformation. Since we will report results in reduced quantities, it does not matter the parameters a and b we choose, but in the implementation, the choice of k_N and k_V might depend on the specific values of a and b. Without loss of generality, we choose $a = 3$ and $b = \frac{1}{3}$; in this way, $p_c = 1$ and $v_c = 1$ and thus

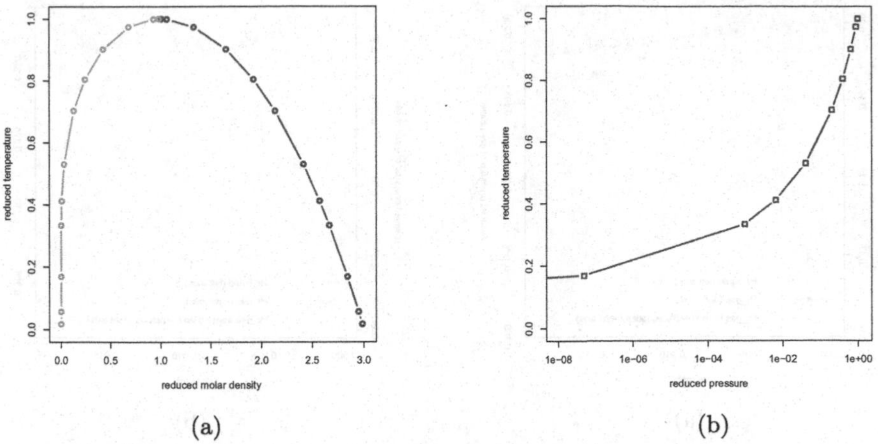

Fig. 1. Phase diagrams of the single-component Van der Waals fluid: (a) reduced molar volumes of gas and liquid phases at equilibrium as a function of reduced temperature; (b) reduced boiling temperature as a function of reduced pressure

$p_r = p$ and $v_r = v$. The units of these quantities can be any fixed units, as long as consistent units are used, such as the SI unit system.

For comparison and verification, we first let $\sigma = 0$ in our model, and we calculate the liquid-vapor phase behavior of the Van der Waals fluid without capillarity. We generate phase diagrams numerically and plot them in Figs. 1(a) and (b). Figure 1(a) is the volume-temperature phase envelope while Fig. 1(b) displays reduced boiling temperature as a function of reduced pressure.

We then consider two cases with capillary pressure. The first case is a single bubble of radius r immersed in the liquid. In this case, $V^G = \frac{4\pi r^3}{3}$ and $dA_I/dV^G = 2/r$. For this single-component two-phase fluid system, our modeling ODE reduces to

$$\frac{\partial N^G}{\partial t} = k_N N^{\text{tot}}(\mu(n^L) - \mu(n^G)),$$

$$\frac{\partial r}{\partial t} = \frac{k_V V^{\text{tot}}}{4\pi r^2}(p^G(n^G) - p^L(n^L) - \frac{2\sigma}{r}).$$

The values of k_N and k_V are manually tuned for one typical simulation, and are then fixed for all other runs. In all numerical examples in this subsection, we choose $k_N = 0.05$ and $k_V = 0.02$. We use the unit time step $\Delta t = 1$ for all numerical runs in this paper.

The effect of capillary pressure on the saturation pressure of the liquid phase and the vapor (gas) phase is provided in Figs. 2(a)–(d) under various condition of reduced temperatures and capillary pressures. The horizontal axis of these plots are $\frac{\sigma}{r p_c}$. The ratio of interfacial tension to the radius of the gas bubble $\frac{\sigma}{r}$ is the influencing factor, we divide this ratio by the critical pressure of the fluid p_c so that we get a dimensionless quantity.

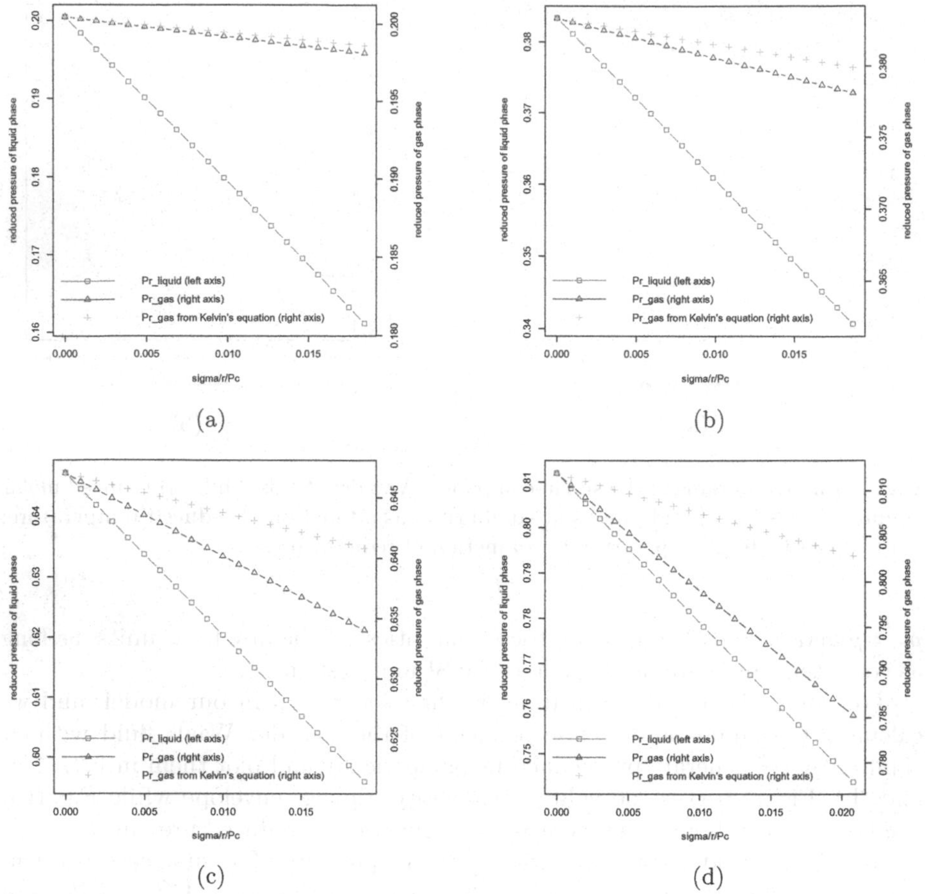

Fig. 2. The effect of capillary pressure on the saturation pressures of the liquid phase and the vapor (gas) bubble modeled by the single-component Van der Waals EOS: (a) $T_r = 0.7$ (b) $T_r = 0.8$ (c) $T_r = 0.9$ (d) $T_r = 0.95$

We also compare the our numerical prediction of vapor pressure with the one calculated by the Kelvin equation. The Kelvin equation can be derived by using the two equilibrium conditions $dp^G - dp^L = d(2\sigma/r)$ and $d\mu^G = d\mu^L$, and the two fundamental relations $d\mu^G = v^G dp^G$ and $d\mu^L = v^L dp^L$, which yield $dp^G = \frac{d(\frac{2\sigma}{r})}{(1-\frac{v^G}{v^L})}$. By assuming the ideal gas law for the v^G and assuming a constant molar density for v^L, we can integrate the above differential equation to obtain the following Kelvin equation

$$p^G = p^\infty \exp(-\frac{2\sigma v^L}{rRT}),$$

which quickly reveals that the vapor pressure decreases with increasing interface curvature and increasing interfacial tension.

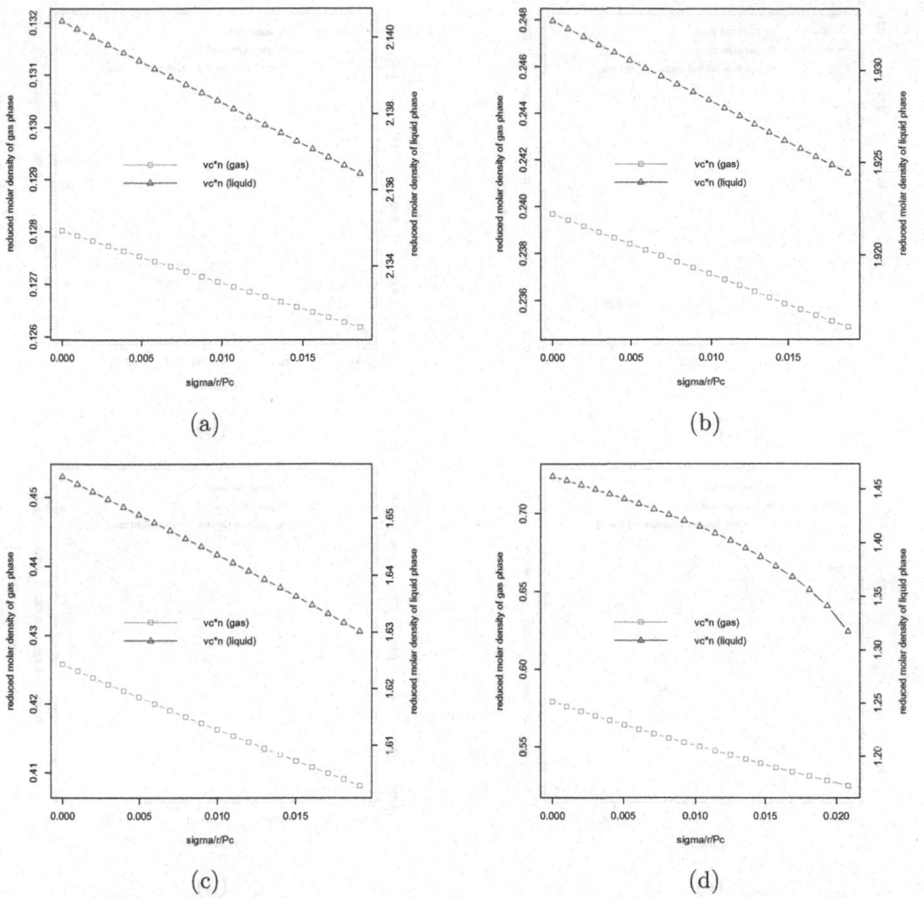

Fig. 3. The effect of capillary pressure on the molar densities of the liquid phase and the vapor (gas) bubble modeled by the single-component Van der Waals EOS: (a) $T_r = 0.7$ (b) $T_r = 0.8$ (c) $T_r = 0.9$ (d) $T_r = 0.95$

Figures 2(a)–(d) indicates that the vapor pressure predicted by the Kelvin equation agree well with our numerical simulation when the temperature is not too low. When the temperature is low, the vapor pressure predicted by the Kelvin equation has pronounced derivation, because the ideal gas law is no loner a good approximation at low temperature. From Figs. 2(a)–(d), we also see that the saturation liquid pressure departs from its zero-capillary value much more significantly than the saturation vapor pressure.

We observe the number of time steps required for convergence varies with the reduced temperature. When $T_r = 0.7$, the number of time steps required for convergence is about 10 to 30 depending on the value of $\frac{\sigma}{rp_c}$, while when $T_r = 0.95$, the number of time steps required for convergence is can be more than 2000.

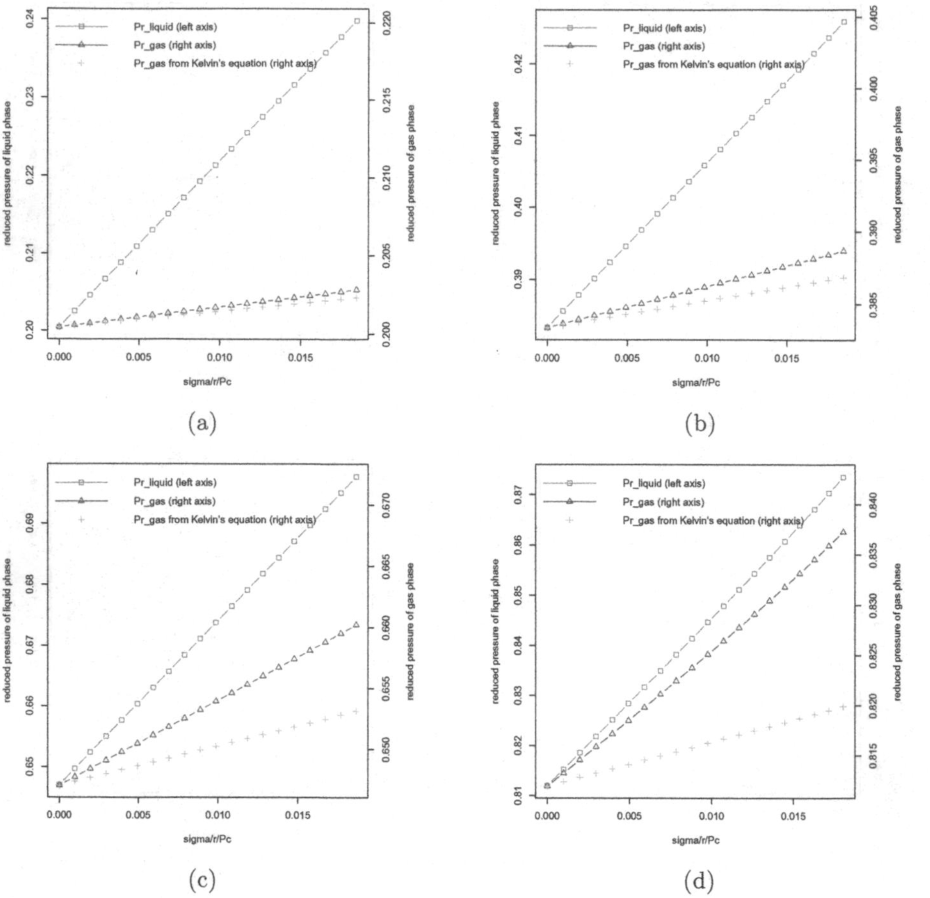

Fig. 4. The effect of capillary pressure on the saturation pressures of the liquid droplet and the vapor (gas) phase modeled by the single-component Van der Waals EOS: (a) $T_r = 0.7$ (b) $T_r = 0.8$ (c) $T_r = 0.9$ (d) $T_r = 0.95$

In Figs. 3(a)–(d), we show the variation of molar densities of the saturated liquid phase and saturated vapor phase as influenced by capillarity. It is evident that the molar densities of both phases decrease with increasing interface curvature and increasing interfacial tension. This is likely because the decrease of saturation pressure causes the decrease of molar densities.

The second case is a single liquid droplet of radius r immersed in the vapor phase. In this case, $V^G = V^{\text{tot}} - \frac{4\pi r^3}{3}$ and $dA_I/dV^G = -2/r$. For this single-component two-phase fluid system, our modeling ODE reduces to

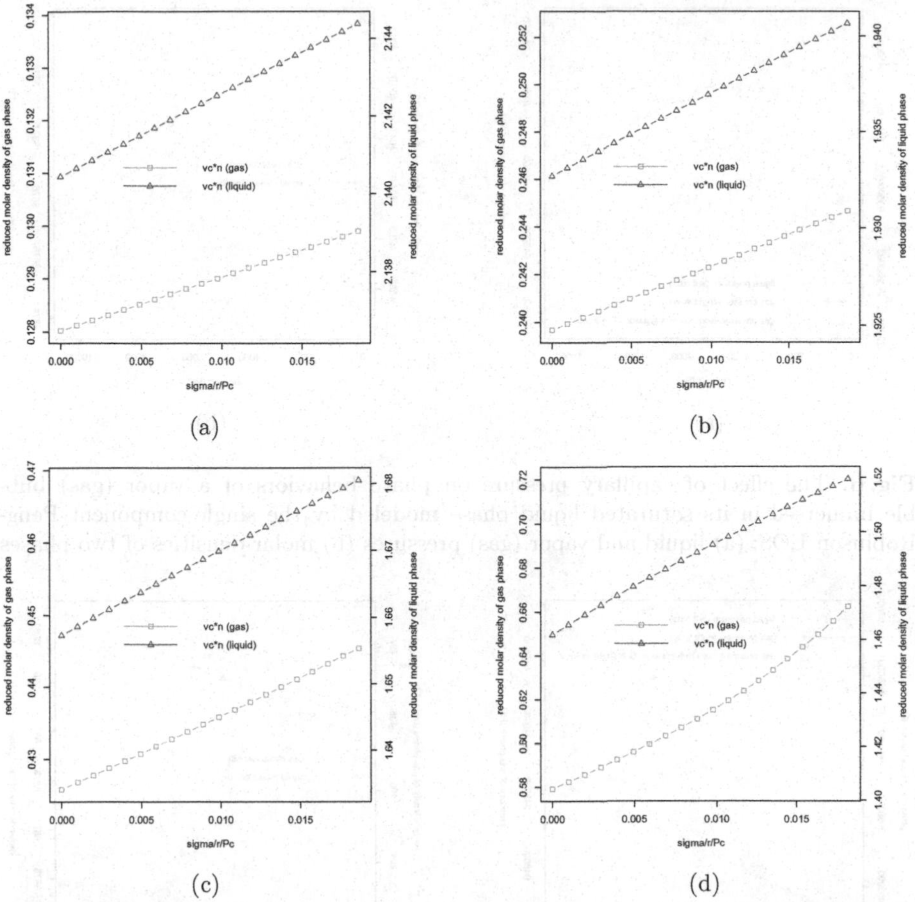

Fig. 5. The effect of capillary pressure on the molar densities of the liquid droplet and the vapor (gas) phase modeled by the single-component Van der Waals EOS: (a) $T_r = 0.7$ (b) $T_r = 0.8$ (c) $T_r = 0.9$ (d) $T_r = 0.95$

$$\frac{\partial N^G}{\partial t} = k_N N^{\text{tot}}(\mu(n^L) - \mu(n^G)),$$
$$\frac{\partial r}{\partial t} = \frac{k_V V^{\text{tot}}}{4\pi r^2}\left(p^L(n^L) - p^G(n^G) - \frac{2\sigma}{r}\right).$$

Figures 4(a)–(d) provide the trend of the saturation liquid pressure of a single liquid droplet and the surrounding saturation vapor pressure as influenced by capillarity. Unlike the vapor bubble case, both the liquid pressure and the saturation vapor pressure increase here with increasing interface curvature and increasing interfacial tension. The trend of the saturation vapor pressure is also compared with the ones predicted by the Kelvin equation, which does a good

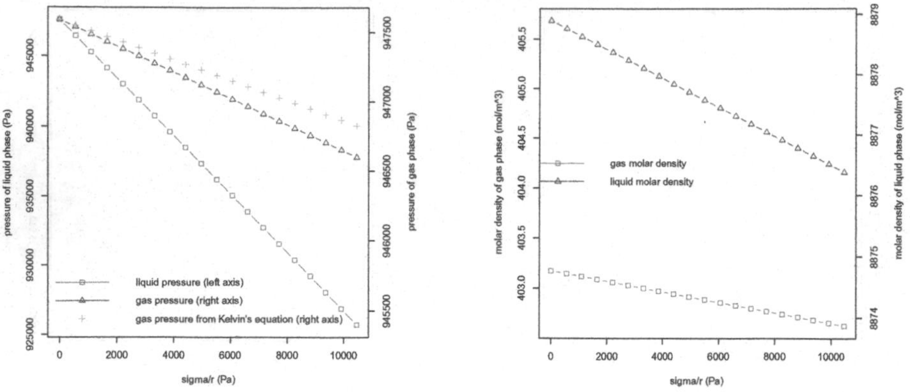

(a) (b)

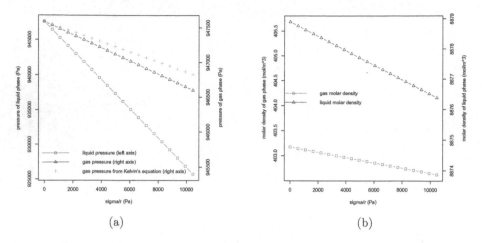

Fig. 6. The effect of capillary pressure on phase behaviors of a vapor (gas) bubble immersed in its saturated liquid phase modeled by the single-component Peng-Robinson EOS: (a) liquid and vapor (gas) pressures (b) molar densities of two phases

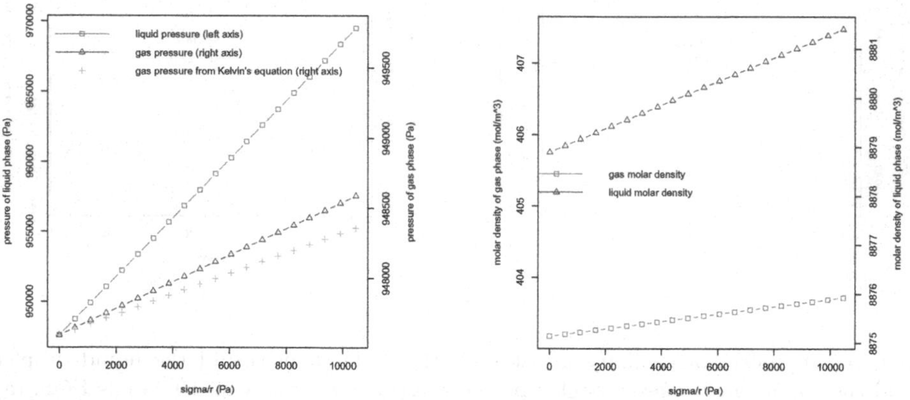

(a) (b)

Fig. 7. The effect of capillary pressure on phase behaviors of a liquid droplet immersed in its vapor (gas) phase modeled by the single-component Peng-Robinson EOS: (a) liquid and vapor (gas) pressures (b) molar densities of two phases

job only when the temperature is not too low. In Figs. 5(a)–(d), we show the variation of molar densities of the saturated liquid phase and saturated vapor phase as influenced by capillarity. Unlike the vapor bubble case, the molar densities of both phases increase with increasing interface curvature and increasing interfacial tension.

Table 1. Relevant data of isobutane (excerpted from Table 3.1 of [4], Page 141)

Component	Symbol	T_c, K	P_c, MPa	T_b, K
Isobutane	nC_4	425.18	3.797	272.64

4.2 Effect of Capillary Pressure on the Peng-Robinson Fluid

The Peng-Robinson EOS, unlike the single-component Van der Waals EOS, cannot transformed into a same reduced from by using reduced temperature, reduced pressure and reduced molar volume, because a third parameter, the accentric parameter appears in the Peng-Robinson EOS. We thus consider a specific example below with physical units. In the example below, we consider the species of isobutane (nC_4) at the temperature of 350 K. For convenience of readers, we provide in Table 1 its critical properties and its normal boiling point. The values of k_N and k_V are manually tuned for one typical simulation, and are then fixed for all other runs. In all runs in this subsection, we choose $k_N = 1 \times 10^{-3}$ mol J^{-1} s^{-1} and $k_V = 1 \times 10^{-9}$ Pa^{-1} s^{-1}. We again use the unit time step $\Delta t = 1$ s.

Figure 6(a) displays the effect of capillary pressure on the saturation pressures of both phases when a single gas bubble is in equilibrium on its liquid phase. The corresponding variation of molar densities of both phases are plotted in Fig. 6(b). The trend of the saturation vapor pressure is also compared with the ones predicted by the Kelvin equation, which does a reasonable job in the condition simulated in this example. Clearly, like the case modeled by the Van der Waals EOS, both the liquid pressure and the saturation vapor pressure decrease here with increasing interface curvature and increasing interfacial tension, which also leads a decrease of molar density of both phases.

Figure 7(a) and (b) show the variation of the saturation pressures and molar densities of both phases when a single liquid droplet is in equilibrium on its vapor phase under various conditions of interface curvature and increasing interfacial tension. The trend seems similar to the case predicted by the single-component Van der Waals EOS qualitatively.

5 Conclusion

In this paper, we formulate a framework to model a two-phase fluid with a (sharp) interface using Van der Waals EOS and Peng-Robinson EOS. Our model is able to predict the affect of capillary pressure on phase behaviors. We propose simple but conditionally energy stable numerical schemes to solve the minimization problem of total Helmholtz energy in given total volume, temperature, and total moles. We consider bubbles in liquid as well as droplets in vapor in our numerical examples. Our numerical agree well with the prediction of Kelvin's equation when the temperature is not too low. Due to the limitation of the space, we provide only single-component examples. Our model and schemes, however, are expected to work well with multi-component systems, which will be investigated and reported in a separate paper elsewhere.

References

1. Breure, B., Peters, C.: Modeling of the surface tension of pure components and mixtures using the density gradient theory combined with a theoretically derived influence parameter correlation. Fluid Phase Equilib. **334**, 189–196 (2012)
2. Dawson, C., Sun, S., Wheeler, M.F.: Compatible algorithms for coupled flow and transport. Comput. Methods Appl. Mech. Eng. **193**, 2565–2580 (2004)
3. Fan, X., Kou, J., Qiao, Z., Sun, S.: A component-wise convex splitting scheme for diffuse interface models with Van der Waals and Peng-Robinson equations of state. SIAM J. Sci. Comput. **39**(1), B1–B28 (2017)
4. Firoozabadi, A.: Thermodynamics of Hydrocarbon Reservoirs. McGraw-Hill, New York (1999)
5. Kou, J., Sun, S.: Multi-scale diffuse interface modeling of multi-component two-phase flow with partial miscibility. J. Comput. Phys. **318**, 349–372 (2016)
6. Kou, J., Sun, S.: Entropy stable modeling of non-isothermal multi-component diffuse-interface two-phase flows with realistic equations of state. Comput. Methods Appl. Mech. Eng. **341**, 221–248 (2018)
7. Kou, J., Sun, S.: A stable algorithm for calculating phase equilibria with capillarity at specified moles, volume and temperature using a dynamic model. Fluid Phase Equilib. **456**, 7–24 (2018)
8. Kou, J., Sun, S.: Thermodynamically consistent modeling and simulation of multi-component two-phase flow model with partial miscibility. Comput. Methods Appl. Mech. Eng. **331**, 623–649 (2018)
9. Kou, J., Sun, S.: Thermodynamically consistent simulation of nonisothermal diffuse-interface two-phase flow with Peng-Robinson equation of state. J. Comput. Phys. **371**, 581–605 (2018)
10. Kou, J., Sun, S., Wang, X.: An energy stable evolutional method for simulating two-phase equilibria of multi-component fluids at constant moles, volume and temperature. Comput. Geosci. **20**(1), 283–295 (2016)
11. Kou, J., Sun, S., Wang, X.: Linearly decoupled energy-stable numerical methods for multi-component two-phase compressible flow. SIAM J. Numer. Anal. **56**(6), 3219–3248 (2018)
12. Lake, L.W.: Enhanced Oil Recovery. Prentice Hall, Englewood Cliffs, New Jersey (1989)
13. Li, Y., Kou, J., Sun, S.: Numerical modeling of isothermal compositional grading by convex splitting methods. J. Nat. Gas Sci. Eng. **43**, 207–221 (2017)
14. Moortgat, J., Sun, S., Firoozabadi, A.: Compositional modeling of three-phase flow with gravity using higher-order finite element methods. Water Resour. Res. **47**, W05511 (2011)
15. Peng, D.Y., Robinson, D.: A new two-constant equation of state. Ind. Eng. Chem. Fundam. **15**(1), 59–64 (1976)
16. Sun, S., Wheeler, M.F.: Symmetric and nonsymmetric discontinuous Galerkin methods for reactive transport in porous media. SIAM J. Numer. Anal. **43**(1), 195–219 (2005)
17. Sun, S., Wheeler, M.F.: Local problem-based a *posteriori* error estimators for discontinuous Galerkin approximations of reactive transport. Comput. Geosci. **11**(2), 87–101 (2007)

Effects of Numerical Integration on DLM/FD Method for Solving Interface Problems with Body-Unfitted Meshes

Cheng Wang[1], Pengtao Sun[2](✉), Rihui Lan[2], Hao Shi[1], and Fei Xu[3]

[1] School of Mathematical Sciences, Tongji University, Shanghai, China
wangcheng@tongji.edu.cn, sh1009@163.com

[2] Department of Mathematical Sciences, University of Nevada, Las Vegas, Las Vegas, NV, USA
pengtao.sun@unlv.edu, lanr1@unlv.nevada.edu

[3] Beijing Institute for Scientific and Engineering Computing, Beijing University of Technology, Beijing, China
xufei@lsec.cc.ac.cn

Abstract. In this paper, the effects of different numerical integration schemes on the distributed Lagrange multiplier/fictitious domain (DLM/FD) method with body-unfitted meshes are studied for solving different types of interface problems: elliptic-, Stokes- and Stokes/elliptic-interface problems. Commonly-used numerical integration schemes, compound type formulas and a specific subgrid integration scheme are presented for the mixed finite element approximation and the comparison between them is illustrated in numerical experiments, showing that different numerical integration schemes have significant effects on approximation errors of the DLM/FD finite element method for different types of interface problems, especially for Stokes- and Stokes/elliptic-interface problems, and that the subgrid integration scheme always results in numerical solutions with the best accuracy.

1 Introduction

Physical phenomena in a domain consisting of multiple materials and/or multiphase fluids, which are immiscible and are divided by distinct interfaces, are often modeled by either identical or different partial differential equations with discontinuous coefficients on both sides of interfaces. These problems are generally called interface problems, sometimes called interaction problems in some specific scenarios such as fluid-structure interaction (FSI) problems, e.g., see [5,17] and others references therein. In the past several decades, two major numerical approaches – the body-fitted mesh method and the body-unfitted mesh method – have been developed for tackling interface problems, which are classified by how the computational mesh and then the interface conditions are handled along the interface. In contrast to the body-fitted mesh method such as the arbitrary

Supported by NSF DMS-1418806 (Pengtao Sun) and NSFC-11801021 (Fei Xu).

J. M. F. Rodrigues et al. (Eds.): ICCS 2019, LNCS 11539, pp. 551–567, 2019.
https://doi.org/10.1007/978-3-030-22747-0_41

Lagrangian-Eulerian (ALE) method [8,9], which is obliged to adapt the mesh to accommodate the motion of the interface, the body-unfitted mesh method, due to its simplicity in the mesh generation, becomes more promising and more advantageous for interface problems whose interfaces may bear a large deformation/displacement, such as the immersed boundary method (IBM) [12,14], the distribute Lagrange multiplier/fictitious domain (DLM/FD) method [1,7], the immersed finite element method (IFEM) [10,11], and etc.

Among the aforementioned body-unfitted mesh methods, taking all properties of reliability, accuracy, flexibility and theoretical guarantees into consideration, the DLM/FD method has shown a lot of strengths and potentials in theoretical analyses as well as practical applications for general interface problems, and has gained considerable popularity in simulating FSI problems as well. So in this paper, we focus on the DLM/FD method, where a fictitious equation that is defined in one subdomain is introduced to cover the other subdomain, and its mesh is fixed in the entire domain as a background mesh and needs not to be updated even if the interface moves or deforms. Benefited from this feature, the DLM/FD method has became more popular in the simulation of FSI problems, especially in the case of an immersed structure with large deformation/displacements. To enforce the interface conditions, the DLM/FD method introduces the Lagrange multiplier (a pseudo body force) to weakly enforce the fictitious variables equal everywhere to the primary variables of the equation defined in the immersed domain and on the interface too. A monolithic system bearing a saddle-point structure is thus formed in regard to the Lagrange multiplier and primary variables. Therefore, the classical Babuška–Brezzi's theory [3,4] can be employed to prove the well-posedness, stability as well as convergence properties of the DLM/FD finite element method [2,15].

To implement the DLM/FD method, an accurate and also efficient numerical integration scheme is needed to calculate the integration in which the distributed Lagrange multiplier is involved. For instance, (21) can be referred to in advance to preview the significance, where, in the finite element computation on each immersed element for the Lagrange multiplier terms denoted by the dual inner product $\langle \lambda, v_h|_{\Omega} \rangle_{\Omega_2}$, the integrand function is a product of two piecewise polynomials which are defined on two non-matched meshes, $T_h(\Omega)$ and $T_H(\Omega_2)$. Although the piecewise polynomial defined on $T_h(\Omega)$, $v_h|_{\Omega}$, can be transferred to $T_H(\Omega_2)$ through the interpolation approach, it is no longer sufficiently smooth in each immersed element of $T_H(\Omega_2)$ just because of the non-matching between $T_h(\Omega)$ and $T_H(\Omega_2)$. Thus we can not conclude that the commonly-used higher order numerical integration scheme will lead to a higher accuracy for those Lagrange multiplier terms.

The object of this paper is to study the effects of various numerical integration schemes on the performance of the DLM/FD methods for solving different interface problems with jump coefficients. Essentially, we do not want to let the accuracy of numerical integration influence the overall approximation accuracy, especially when the DLM/FD method is used for simulating complex problems, e.g., FSI problems. However, we find out different numerical integration methods indeed have significant effects on approximation errors of the DLM/FD method for different types of interface problems. Three types of numerical integration schemes are considered in our study: the commonly-used numerical integration schemes (see e.g. [6]), the compound type formulas, and the subgrid integration scheme proposed in [18]. Numerical results presented in Sect. 4 shows that the performance of DLM/FD method for solving elliptic interface problems is insensitive to the numerical integration but is sensitive to Stokes- and Stokes/elliptic-interface problems, and that the subgrid integration scheme always leads to the approximation solution of the highest accuracy.

The rest of this paper is organized as follows. The DLM/FD finite element method for solving different types of interface problems are recalled in Sect. 2. Several commonly-used numerical integration schemes, the compound type formulas and the subgrid integration technique are introduced in Sect. 3. Numerical performances are shown in Sect. 4.

2 DLM/FD Method for Three Types of Interface Problems

In this section, we briefly recall the DLM/FD finite element method for solving three different types of interface problems with jump coefficients.

2.1 Three Types of Interface Problems

The first type is the elliptic interface problem, defined as

$$-\nabla \cdot (\beta_1 \nabla u_1) = f_1, \qquad \text{in } \Omega_1, \tag{1}$$

$$-\nabla \cdot (\beta_2 \nabla u_2) = f_2, \qquad \text{in } \Omega_2, \tag{2}$$

$$u_1 = u_2, \qquad \text{on } \Gamma, \tag{3}$$

$$\beta_1 \nabla u_1 \cdot \boldsymbol{n}_1 + \beta_2 \nabla u_2 \cdot \boldsymbol{n}_2 = w, \qquad \text{on } \Gamma, \tag{4}$$

$$u_1 = 0,\ u_2 = 0, \qquad \text{on } \partial\Omega \backslash \Gamma, \tag{5}$$

where, $\Omega = \Omega_1 \cup \Omega_2 \subset \mathcal{R}^d$ (see Fig. 1), the interface $\Gamma = \partial\Omega_2$ is a closed curve that divides the domain Ω into an interior region Ω_2 and an exterior region Ω_1, $\boldsymbol{n}_1$ and $\boldsymbol{n}_2$ stand for the unit outward normal vectors on $\partial\Omega_1$ and $\partial\Omega_2$, respectively. u, that is defined in Ω, satisfies $u|_{\Omega_1} = u_1$, $u|_{\Omega_2} = u_2$ which are associated with $f \in L^2(\Omega)$ and $f|_{\Omega_1} = f_1 \in L^2(\Omega_1)$, $f|_{\Omega_2} = f_2 \in L^2(\Omega_2)$. $\beta \in L^\infty(\Omega)$ satisfies $\beta|_{\Omega_1} = \beta_1 \in W^{1,\infty}(\Omega_1)$, $\beta|_{\Omega_2} = \beta_2 \in W^{1,\infty}(\Omega_2)$ and $\beta_1 \neq \beta_2$.

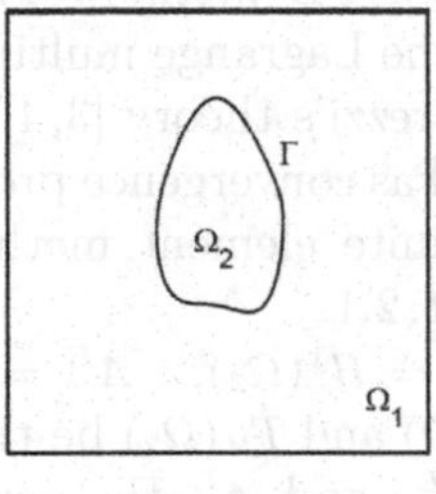

Fig. 1. Graphical depiction of the domain with an immersed interface.

The second type is the Stokes interface problem, defined as

$$-\nabla \cdot (\beta_1 \nabla \boldsymbol{u}_1) + \nabla p_1 = \boldsymbol{f}_1, \quad \text{in } \Omega_1, \tag{6}$$
$$\nabla \cdot \boldsymbol{u}_1 = 0, \quad \text{in } \Omega_1, \tag{7}$$
$$-\nabla \cdot (\beta_2 \nabla \boldsymbol{u}_2) + \nabla p_2 = \boldsymbol{f}_2, \quad \text{in } \Omega_2, \tag{8}$$
$$\nabla \cdot \boldsymbol{u}_2 = 0, \quad \text{in } \Omega_2, \tag{9}$$
$$\boldsymbol{u}_1 = \boldsymbol{u}_2, \quad \text{on } \Gamma, \tag{10}$$
$$(\beta_1 \nabla \boldsymbol{u}_1 - p_1 \boldsymbol{I})\boldsymbol{n}_1 + (\beta_2 \nabla \boldsymbol{u}_2 - p_2 \boldsymbol{I})\boldsymbol{n}_2 = \boldsymbol{w}, \quad \text{on } \Gamma, \tag{11}$$
$$\boldsymbol{u}_1 = 0,\ \boldsymbol{u}_2 = 0, \quad \text{on } \partial\Omega \backslash \Gamma, \tag{12}$$

which can be considered as the linearized model of an immiscible two-phase fluid flow problem, where the primary variable pair $(\boldsymbol{u}, p)$ satisfies $\boldsymbol{u}|_{\Omega_1} = \boldsymbol{u}_1$, $\boldsymbol{u}|_{\Omega_2} = \boldsymbol{u}_2$, $p|_{\Omega_1} = p_1$, $p|_{\Omega_2} = p_2$, which are associated with $\boldsymbol{f} \in (L^2(\Omega))^d$ satisfying $\boldsymbol{f}|_{\Omega_1} = \boldsymbol{f}_1 \in (L^2(\Omega_1))^d$, $\boldsymbol{f}|_{\Omega_2} = \boldsymbol{f}_2 \in (L^2(\Omega_2))^d$. The jump coefficient $\beta \in L^\infty(\Omega)$ satisfies $\beta|_{\Omega_1} = \beta_1 \in W^{1,\infty}(\Omega_1)$, $\beta|_{\Omega_2} = \beta_2 \in W^{1,\infty}(\Omega_2)$, $\beta_1 \neq \beta_2$.

And, the third type is the Stokes/elliptic interface problem, defined as

$$-\nabla \cdot (\beta_1 \nabla \boldsymbol{u}_1) + \nabla p_1 = \boldsymbol{f}_1, \quad \text{in } \Omega_1, \tag{13}$$
$$\nabla \cdot \boldsymbol{u}_1 = 0, \quad \text{in } \Omega_1, \tag{14}$$
$$-\nabla \cdot (\beta_2 \nabla \boldsymbol{u}_2) = \boldsymbol{f}_2, \quad \text{in } \Omega_2, \tag{15}$$
$$\boldsymbol{u}_1 = \boldsymbol{u}_2, \quad \text{on } \Gamma, \tag{16}$$
$$(\beta_1 \nabla \boldsymbol{u}_1 - p_1 \boldsymbol{I})\boldsymbol{n}_1 + \beta_2 \nabla \boldsymbol{u}_2 \boldsymbol{n}_2 = \boldsymbol{w}, \quad \text{on } \Gamma, \tag{17}$$
$$\boldsymbol{u}_1 = 0,\ \boldsymbol{u}_2 = 0, \quad \text{on } \partial\Omega \backslash \Gamma, \tag{18}$$

which can be considered as a steady state linearized model of FSI problems, where, the primary variable pair $(\boldsymbol{u}, p_1)$ satisfies $\boldsymbol{u}|_{\Omega_1} = \boldsymbol{u}_1$, $\boldsymbol{u}|_{\Omega_2} = \boldsymbol{u}_2$, $p_1 \in \Omega_1$, which are associated with $\boldsymbol{f} \in (L^2(\Omega))^d$ satisfying $\boldsymbol{f}|_{\Omega_1} = \boldsymbol{f}_1 \in (L^2(\Omega_1))^d$, $\boldsymbol{f}|_{\Omega_2} = \boldsymbol{f}_2 \in (L^2(\Omega_2))^d$. The jump coefficient $\beta \in L^\infty(\Omega)$ satisfies $\beta|_{\Omega_1} = \beta_1 \in W^{1,\infty}(\Omega_1)$, $\beta|_{\Omega_2} = \beta_2 \in W^{1,\infty}(\Omega_2)$, $\beta_1 \neq \beta_2$.

2.2 DLM/FD Method for the Elliptic Interface Problem

The main idea of the DLM/FD method is to smoothly extend one material, such as the fluid in FSI, into another material's subdomain, such as the structure in FSI, as a fictitious equation whose primary variable is constrained to equal to that of the occupied material's equation in its subdomain and on the interface too. And, such constraint is weakly imposed through the Lagrange multiplier (a pseudo body force) in the DLM/FD method. Thus, a monolithic saddle-point system is formed in regard to the Lagrange multiplier and primary variables, for which the classical Babuška–Brezzi's theory [3,4] can be employed to prove the well-posedness, stability as well as convergence properties [2,16]. In the following, we introduce the DLM/FD finite element method for each type of interface problem that is defined in Sect. 2.1.

Define $\boldsymbol{V}^E = H_0^1(\Omega)$, $\boldsymbol{V}_2^E = H^1(\Omega_2)$, $\boldsymbol{\Lambda}^E = (\boldsymbol{V}_2^E)^*$, where $(\boldsymbol{V}_2^E)^*$ denotes the dual space of $\boldsymbol{V}_2^E$. Let $T_h(\Omega)$ and $T_H(\Omega_2)$ be the meshes of Ω and Ω_2, respectively. And denote by $\boldsymbol{V}_h^E$, $\boldsymbol{V}_{2,H}^E$ and $\boldsymbol{\Lambda}_H^E$ the conforming finite element spaces of $\boldsymbol{V}^E$, $\boldsymbol{V}_2^E$ and $\boldsymbol{\Lambda}^E$, respectively. Here and hereafter, let $(\cdot,\cdot)_\omega$ be the L^2 inner product over ω, and $\langle\cdot,\cdot\rangle_\omega$ be the dual product over ω that is actually associated with the H^1 inner product [1].

Then, the DLM/FD finite element method for solving elliptic interface problem (1)–(5) can be defined as follows [1,2]: Find $(\tilde{u}_h,\ u_{2,H},\ \lambda_H) \in \boldsymbol{V}_h^E \times \boldsymbol{V}_{2,H}^E \times \boldsymbol{\Lambda}_H^E$ such that

$$(\tilde{\beta}\nabla\tilde{u}_h, \nabla v_h)_\Omega + \langle\lambda_H, v_h|_{\Omega_2}\rangle_{\Omega_2} = (\tilde{f}, v_h)_\Omega, \quad (19)$$

$$\Big((\beta_2 - \tilde{\beta})\nabla u_{2,H}, \nabla v_{2,H}\Big)_{\Omega_2} - \langle\lambda_H, v_{2,H}\rangle_{\Omega_2} = \Big(f_2 - \tilde{f}_2, v_{2,H}\Big)_{\Omega_2} + (w, v_{2,H})_\Gamma, \quad (20)$$

$$\langle\xi_H, \tilde{u}_h|_{\Omega_2} - u_{2,H}\rangle_{\Omega_2} = 0, \forall\ (v_h,\ v_{2,H},\ \xi) \in \boldsymbol{V}_h^E \times \boldsymbol{V}_{2,H}^E \times \boldsymbol{\Lambda}_H^E. \quad (21)$$

2.3 DLM/FD Method for the Stokes Interface Problem

Define $\boldsymbol{V}^S = \left(H_0^1(\Omega)\right)^d,\ \ \boldsymbol{Q}^S = L^2(\Omega),\ \ \boldsymbol{V}_2^S = \left(H^1(\Omega_2)\right)^d,\ \ \boldsymbol{\Lambda}^S = \left(\boldsymbol{V}_2^S\right)^*$, where $(\boldsymbol{V}_2^S)^*$ denotes the dual space of $\boldsymbol{V}_2^S$. Let $\boldsymbol{V}_h^S$, $\boldsymbol{V}_{2,H}^S$, $\boldsymbol{Q}_h^S$ and $\boldsymbol{\Lambda}_H^S$ be the conforming finite element spaces of $\boldsymbol{V}^S$, $\boldsymbol{V}_2^S$, $\boldsymbol{Q}^S$ and $\boldsymbol{\Lambda}^S$, respectively. Then, the DLM/FD finite element method for solving the Stokes interface problem (6)–(12) can be defined as follows [13]: Find $(\tilde{\boldsymbol{u}}_h,\ \boldsymbol{u}_{2,H},\ \tilde{p}_h,\ \boldsymbol{\lambda}_H) \in \boldsymbol{V}_h^S \times \boldsymbol{V}_{2,H}^S \times Q_h^S \times \boldsymbol{\Lambda}_H^S$ such that

$$(\tilde{\beta}\nabla\tilde{\boldsymbol{u}}_h, \nabla\boldsymbol{v}_h)_\Omega - (\tilde{p}_h, \nabla\cdot\boldsymbol{v}_h)_\Omega + \langle\boldsymbol{\lambda}_H, \boldsymbol{v}_h|_{\Omega_2}\rangle_{\Omega_2} = (\tilde{\boldsymbol{f}}, \boldsymbol{v}_h)_\Omega, \quad (22)$$

$$(\nabla\cdot\tilde{\boldsymbol{u}}_h, q_h)_\Omega = 0, \quad (23)$$

$$\Big((\beta_2 - \tilde{\beta})\nabla\boldsymbol{u}_{2,H}, \nabla\boldsymbol{v}_{2,H}\Big)_{\Omega_2} - \langle\boldsymbol{\lambda}_H, \boldsymbol{v}_{2,H}\rangle_{\Omega_2} = \Big(\boldsymbol{f}_2 - \tilde{\boldsymbol{f}}_2, \boldsymbol{v}_{2,H}\Big)_{\Omega_2} + (\boldsymbol{w}, \boldsymbol{v}_{2,H})_\Gamma, \quad (24)$$

$$\langle\boldsymbol{\xi}_H, \tilde{\boldsymbol{u}}_h|_{\Omega_2} - \boldsymbol{u}_{2,H}\rangle_{\Omega_2} = 0, \forall\ (\boldsymbol{v}_h,\ \boldsymbol{v}_{2,H},\ q_h,\ \boldsymbol{\xi}_H) \in \boldsymbol{V}_h^S \times \boldsymbol{V}_{2,H}^S \times Q_h^S \times \boldsymbol{\Lambda}_H^S. \quad (25)$$

2.4 DLM/FD Method for the Stokes/Elliptic Interface Problem

The spaces to be used for defining the weak formulation and the DLM/FD formulation of Stokes/elliptic interface problem are the same as those of Stokes interface problem. The DLM/FD finite element method is proposed and analyzed in [15] for the Stokes/elliptic interface problem (13)–(18), as described below.

Find $(\tilde{\boldsymbol{u}}_h,\ \boldsymbol{u}_{2,H},\ \tilde{p}_h,\ \boldsymbol{\lambda}_H) \in \boldsymbol{V}_h^S \times \boldsymbol{V}_{2,H}^S \times Q_h^S \times \boldsymbol{\Lambda}_H^S$ such that

$$(\tilde{\beta}\nabla\tilde{\boldsymbol{u}}_h, \nabla\boldsymbol{v}_h)_\Omega - (\tilde{p}_h, \nabla\cdot\boldsymbol{v}_h)_\Omega + \langle\boldsymbol{\lambda}_H, \boldsymbol{v}_h|_{\Omega_2}\rangle_{\Omega_2} = (\tilde{\boldsymbol{f}}, \boldsymbol{v}_h)_\Omega, \quad (26)$$

$$(\nabla\cdot\tilde{\boldsymbol{u}}_h, q_h)_\Omega - (\nabla\cdot\boldsymbol{u}_{2,H}, q_h)_{\Omega_2} = 0, \quad (27)$$

$$\Big((\beta_2 - \tilde{\beta})\nabla\boldsymbol{u}_{2,H}, \nabla\boldsymbol{v}_{2,H}\Big)_{\Omega_2} + (p_h|_{\Omega_2}, \nabla\cdot\boldsymbol{v}_{2,H})_{\Omega_2}$$
$$- \langle\boldsymbol{\lambda}_H, \boldsymbol{v}_{2,H}\rangle_{\Omega_2} = \Big(\boldsymbol{f}_2 - \tilde{\boldsymbol{f}}|_{\Omega_2}, \boldsymbol{v}_{2,H}\Big)_{\Omega_2} + (\boldsymbol{w}, \boldsymbol{v}_{2,H})_\Gamma, \quad (28)$$

$$\langle\boldsymbol{\xi}_H, \tilde{\boldsymbol{u}}_h|_{\Omega_2} - \boldsymbol{u}_{2,H}\rangle_{\Omega_2} = 0, \forall\ (\boldsymbol{v}_h,\ \boldsymbol{v}_{2,H},\ q_h,\ \boldsymbol{\xi}_H) \in \boldsymbol{V}_h^S \times \boldsymbol{V}_{2,H}^S \times Q_h^S \times \boldsymbol{\Lambda}_H^S. \quad (29)$$

3 Numerical Integration Schemes

Note that in the DLM/FD finite element methods described in Sects. 2.2, 2.3 and 2.4, some inner product terms involve an integration over the immersed domain Ω_2 with a product of two piecewise functions defined on $T_h(\Omega)$ and $T_H(\Omega_2)$, respectively. For instance,

$$\langle \lambda_H, v_h|_{\Omega_2}\rangle_{\Omega_2} \text{ in (19) and } \langle \xi_H, \tilde{u}_h|_{\Omega_2}\rangle_{\Omega_2} \text{ in (21),} \tag{30}$$

$$\langle \boldsymbol{\lambda}_H, \boldsymbol{v}_h|_{\Omega_2}\rangle_{\Omega_2} \text{ in (22) and } \langle \boldsymbol{\xi}_H, \tilde{\boldsymbol{u}}_h|_{\Omega_2}\rangle_{\Omega_2} \text{ in (25),} \tag{31}$$

$$(\nabla \cdot \tilde{\boldsymbol{u}}_{2,H}, q_h)_{\Omega_2} \text{ in (26) and } (p_h|_{\Omega_2}, \nabla \cdot \boldsymbol{v}_{2,H})_{\Omega_2} \text{ in (29).} \tag{32}$$

Since two grids $T_h(\Omega)$ and $T_H(\Omega_2)$ are constructed independently, they are generally not matched with each other, see Fig. 2. Thus, we need to employ some numerical techniques to implement the numerical integration for terms shown in (30)–(32) in order to make the DLM/FD finite element method perform well. In this paper, we restrict ourselves to the cases of triangular finite element.

It shall be pointed out that we can not simply make such a conclusion that the higher order the integration scheme, the higher accuracy the numerical integration. That is because the integrand functions shown in (30)–(32) are essentially the product of piecewise polynomials defined on two non-matched meshes, $T_h(\Omega)$ or $T_H(\Omega_2)$, inducing an insufficiently smooth integrand function in each element of $T_H(\Omega_2)$, and thus it is not able to deduce a higher order derivative in the remainder of the numerical integration scheme.

3.1 Commonly-Used Numerical Integration Schemes

The numerical integration schemes in the field of finite element method are of the form: $\int_{\hat{K}} f(\hat{x})d\hat{x} = \sum_i \omega_i f(\lambda_i)$, where $\hat{K}$ stands for the reference element that is an isosceles right triangle with two sides equal 1, that is, three vertices of $\hat{K}$ are $(0,0)$, $(1,0)$ and $(0,1)$. And, the quadrature points are denoted by the barycentric coordinates. Nine different commonly-used numerical integration schemes, from Scheme 1 to Scheme 9 as listed in the Appendix, are adopted to carry out our numerical studies in Sect. 4.

Moreover, to improve the numerical integration accuracy, we consider the compound type formulas of numerical integration, for which each triangular reference element is divided into four equilateral sub-triangles. For example, Scheme 4c1 can be obtained by applying Scheme 4 of the numerical integration to each one of four sub-triangles in the first-time refinement of the reference element, Scheme 4c2 is constructed by applying Scheme 4 to each one of 16 sub-triangles after refining the reference element twice, and Scheme 4c3 based on Scheme 4 and 64 sub-triangles after refining the reference element for three times, and so forth, which are not necessarily shown in the Appendix for the simplicity.

3.2 The Subgrid Integration Scheme

The subgrid integration scheme has been proposed and used in [18] for solving two-dimensional parabolic interface problems with the DLM/FD method. The main ingredient of this scheme is to generate a subgrid, $T_H^r(\Omega_2)$, by finding all intersections of $T_h(\Omega)$ and $T_H(\Omega_2)$ then forming a new subgrid structure in the immersed domain Ω_2. Clearly, $T_H^r(\Omega_2)$ is a matching subset of both $T_h(\Omega)$ and $T_H(\Omega_2)$, and $T_H^r(\Omega_2)$ is a finer mesh on Ω_2. Thus, all integral terms as shown in (30)–(32) can be implemented on $T_H^r(\Omega_2)$ now, and the piecewise polynomials defined on $T_h(\Omega)$ and $T_H(\Omega_2)$ must be smooth in each element of $T_H^r(\Omega_2)$. Then, the interpolations between two different meshes are avoided. An example of such a subgrid is shown in Fig. 2.

Since the integrand function on each element of the subgrid $T_H^r(\Omega_2)$ is smooth, the commonly-used integration schemes with the help of such a subgrid can yield sufficiently high accuracy for numerical integrations. In this sense, the subgrid integration scheme is of the highest accuracy among the schemes presented in this paper, as long as a subtle implementation programming can be done to eliminate every possible geometrical error when finding the intersection points of two non-matching meshes. Interested readers can refer to [18] for more details about the subgrid integration scheme and its implementation.

4 Numerical Experiments

In what follows, we will carry out two scenarios in our numerical experiments to investigate effects of different numerical integration schemes on the approximation solutions of the DLM/FD finite element method for solving three types of steady interface problems: (1) the real solution of the original interface problem is unknown; (2) the real solution of the original interface problem is known.

To that end, a fixed mesh size is chosen as $h = 1/64$ and $H = 1/64$ to attain two meshes $T_h(\Omega)$ and $T_H(\Omega_2)$, as depicted in Fig. 2. On these two fixed meshes, we will implement the DLM/FD finite element method with different numerical integration schemes for the elliptic-, Stokes-, and Stokes/elliptic-interface problems, respectively.

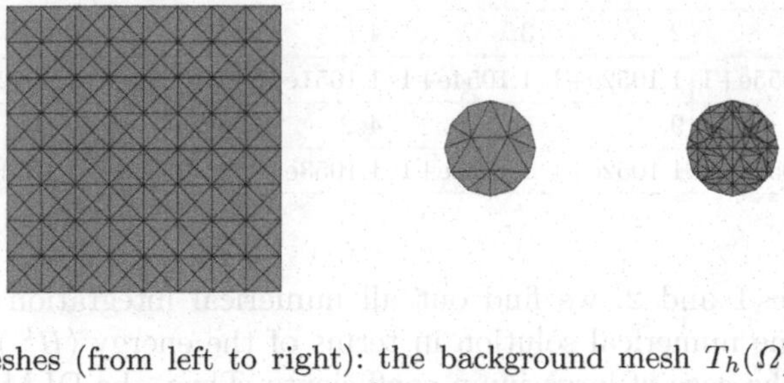

Fig. 2. The meshes (from left to right): the background mesh $T_h(\Omega)$ partitioned in a square Ω, the foreground mesh $T_H(\Omega_2)$ partitioned in a disk Ω_2, and the subgrid $T_H^r(\Omega_2)$ in Ω_2.

4.1 The Scenario of Unknown Real Solutions

In this scenario, since the real solution is unknown, we will report the energy norm of the numerical solution instead of its approximation error to the real solution, and compare all numerical solutions in the energy norm obtained from different numerical integration schemes for each type of interface problem.

The Elliptic Interface Problem with a Unknown Real Solution. Let Ω be a square with the size $(0,6) \times (0,6)$ and Ω_2 be a unit circle with the center at $(3,3)$. Consider the elliptic interface problem (1)–(5) with the following jump coefficients and right hand side functions:

- Case E1: $\beta_1 = 1$, $\beta_2 = 100$, $f_1 = f_2 = 1$,
- Case E2: $\beta_1 = 1$, $\beta_2 = 10000$, $f_1 = f_2 = 1$.

The linear (P_1) finite element space is adopted to define V_h^E, $V_{2,H}^E$ and Λ_H^E, which are used in the implementation of the DLM/FD finite element method (19)–(21). Numerical solutions in H^1-norm in regard to different numerical integration schemes are reported in Tables 1 and 2 for two cases E1–E2, where, we use the numbers from 1 to 9 to successively represent nine commonly-used non-compound type numerical integration schemes, use the notation "4*c*1", "4*c*2" and "4*c*3" to represent three compound-type formula of numerical integration, respectively, and use "*s*" to stand for the subgrid integration scheme.

Table 1. Numerical results of Case E1

schemes #	1	2	3	4	5	6	7
$\|\tilde{u}_h\|_1$	1.1057e+1	1.1057e+1	1.1057e+1	1.1057e+1	1.1057e+1	1.1057e+1	1.1057e+1
scheme #	8	9	4c1	4c2	4c3	s	
$\|\tilde{u}_h\|_1$	1.1057e+1	1.1057e+1	1.1057e+1	1.1057e+1	1.1058e+1	1.1057e+1	

Table 2. Numerical results of Case E2

schemes #	1	2	3	4	5	6	7
$\|\tilde{u}_h\|_1$	1.1055e+1	1.1052e+1	1.1054e+1	1.1051e+1	1.1052e+1	1.1052e+1	1.1052e+1
scheme #	8	9	4c1	4c2	4c3	s	
$\|\tilde{u}_h\|_1$	1.1052e+1	1.1052e+1	1.1052e+1	1.1053e+1	1.1052e+1	1.1051e+1	

From Tables 1 and 2, we find out all numerical integration schemes yield almost the same numerical solution in terms of the energy (H^1-) norm, that is even true for the case of large jump coefficients. Thus, the DLM/FD finite element method for the elliptic interface problem is insensitive with different numerical integration schemes, i.e., all commonly-used numerical integration schemes and their compound formulas can achieve the same approximation accuracy with the subgrid integration scheme which supposes to produce the highest accuracy of the numerical integration for the DLM/FD method.

The Stokes Interface Problem with a Unknown Real Solution. Let Ω be a square with the size $(0,1) \times (0,1)$ and Ω_2 be a circle with the center at $(0.3, 0.3)$ and the radius 0.1. Consider the Stokes interface problem (6)–(12) with the following jump coefficients and right hand side functions:

- Case S1: $\beta_1 = 1$, $\beta_2 = 100$, $\boldsymbol{f}_1 = (e^x, e^y)^T$, $\boldsymbol{f}_2 = (0,0)^T$.
- Case S2: $\beta_1 = 1$, $\beta_2 = 10000$, $\boldsymbol{f}_1 = (e^x, e^y)^T$, $\boldsymbol{f}_2 = (0,0)^T$.

Table 3. Numerical results of Case S1

schemes #	2	3	4	5	6	7
$\Vert\tilde{\boldsymbol{u}}_h\Vert_1$	1.7091e−2	1.8972e−2	1.8557e−2	1.8417e−2	1.8500e−2	1.8570e−2
$\Vert p_h\Vert_0$	2.0895e+0	2.0883e+0	2.0884e+0	2.0884e+0	2.0884e+0	2.0884e+0
schemes #	8	9	4c1	4c2	4c3	s
$\Vert\tilde{\boldsymbol{u}}_h\Vert_1$	1.8328e−2	1.8221e−2	1.8197e−2	1.7985e−2	1.7917e−2	1.7047e−2
$\Vert p_h\Vert_0$	2.0885e+0	2.0886e+0	2.0885e+0	2.0886e+0	2.0886e+0	2.0892e+0

Table 4. Numerical results of Case S2

schemes #	2	3	4	5	6	7
$\Vert\tilde{\boldsymbol{u}}_h\Vert_1$	1.6789e−2	1.8964e−2	1.8537e−2	1.8396e−2	1.8475e−2	1.8544e−2
$\Vert p_h\Vert_0$	2.0903e+0	2.0883e+0	2.0884e+0	2.0884e+0	2.0884e+0	2.0884e+0
schemes #	8	9	4c1	4c2	4c3	s
$\Vert\tilde{\boldsymbol{u}}_h\Vert_1$	1.8302e−2	1.8188e−2	1.8164e−2	1.7944e−2	1.7873e−2	1.7043e−2
$\Vert p_h\Vert_0$	2.0886e+0	2.0886e+0	2.0885e+0	2.0887e+0	2.0887e+0	2.0897e+0

The Taylor-Hood P_2P_1-mixed finite element is adopted to define $\boldsymbol{V}_h^S$, $\boldsymbol{V}_{2,H}^S$, Q_h^S and $\boldsymbol{\Lambda}_H^S$, which are used in the implementation of the DLM/FD finite element method (22)–(25). Numerical solutions in the energy norm (the velocity in H^1-norm and the pressure in L^2-norm) in regard to different numerical integration schemes are reported in Tables 3 and 4 for two cases S1–S2. From these tables, we observe some obvious differences on the numerical solutions of velocity between different numerical integration schemes, while the numerical pressure is insensitive with different integration schemes.

The Stokes/Elliptic Interface Problem with a Unknown Real Solution. Let Ω be a square with the size $(0,1) \times (0,1)$ and Ω_2 be a circle with the center at $(0.3, 0.3)$ and the radius 0.1. Consider the Stokes/elliptic interface problem (13)–(18) with the following jump coefficients and right hand side functions:

- Case SE1: $\beta_1 = 1$, $\beta_2 = 100$, $\boldsymbol{f}_1 = \boldsymbol{f}_2 = (1,1)^T$,
- Case SE2: $\beta_1 = 1$, $\beta_2 = 10000$, $\boldsymbol{f}_1 = \boldsymbol{f}_2 = (1,1)^T$.

The same P_2P_1-mixed finite element space is adopted to define $\boldsymbol{V}_h^S$, $\boldsymbol{V}_{2,H}^S$, Q_h^S and $\boldsymbol{\Lambda}_H^S$, which are used in the implementation of the DLM/FD finite element method (26)–(29). Numerical solutions in the energy norm in regard to different numerical integration schemes are reported in Tables 5 and 6 for two cases SE1–SE2, from which the similar conclusions can be drawn as those for the numerical results of Stokes interface problem in Cases S1–S2.

Table 5. Numerical results of Case SE1

schemes #	2	3	4	5	6	7
$\|\tilde{\boldsymbol{u}}_h\|_1$	1.8780e−2	1.8979e−2	1.8559e−2	1.8420e−2	1.8497e−2	1.8567e−2
schemes #	8	9	4c1	4c2	4c3	s
$\|\tilde{\boldsymbol{u}}_h\|_1$	1.8325e−2	1.8215e−2	1.8191e−2	1.7973e−2	1.7902e−2	1.7105e−2

Table 6. Numerical results of Case SE2

schemes #	2	3	4	5	6	7
$\|\tilde{\boldsymbol{u}}_h\|_1$	1.8763e−2	1.8964e−2	1.8537e−2	1.8396e−2	1.8475e−2	1.8544e−2
schemes #	8	9	4c1	4c2	4c3	s
$\|\tilde{\boldsymbol{u}}_h\|_1$	1.8302e−02	1.8188e−2	1.8164e−2	1.7944e−2	1.7873e−2	1.7098e−2

4.2 The Scenario of Known and Smooth Real Solutions

The Elliptic Interface Problem with a Known Real Solution. Let Ω be a square with the size $(0,6) \times (0,6)$ and Ω_2 be a unit circle with the center at $(3,3)$. Consider the elliptic interface problem (1)–(5) with the following jump coefficients and right hand side functions:

- Case E3: $\beta_1 = 1$, $\beta_2 = 100$, f_1 and f_2 are defined below;
- Case E4: $\beta_1 = 1$, $\beta_2 = 10000$, f_1 and f_2 are defined below,

where, f_1 and f_2 are chosen such that the real solution of (1)–(5) is taken as $u = (x-6)x(y-6)y((x-3)^2 + (y-3)^2 - 1)^2$.

Table 7. Numerical results of Case E3

scheme #	1	2	3	4	5	6	7
$\|u - \tilde{u}_h\|_1$	2.0333e+2	2.0330e+2	2.0330e+2	2.0329e+2	2.0329e+2	2.0329e+2	2.0329e+2
scheme #	8	9	4c1	4c2	4c3	s	
$\|u - \tilde{u}_h\|_1$	2.0329e+2	2.0329e+2	2.0329e+2	2.0329e+2	2.0338e+2	2.0329e+2	

Table 8. Numerical results of Case E4

scheme #	1	2	3	4	5	6	7
$\|u-\tilde{u}_h\|_1$	5.3984e+2	5.3830e+2	5.396e+2	5.3805e+2	5.3801e+2	5.3803e+2	5.3804e+2
scheme #	8	9	4c1	4c2	4c3	s	
$\|u-\tilde{u}_h\|_1$	5.3802e+2	5.3802e+2	5.3800e+2	5.3800e+2	5.3801e+2	5.3801e+2	

With the P_1 finite element space, the numerical results of the DLM/FD finite element method (19)–(21) are reported in Tables 7 and 8 for two cases E3-E4, from which we observe that all numerical errors in H^1-norm in regard to different numerical integration schemes are almost the same, confirming again that the DLM/FD finite element method for the elliptic interface problem is insensitive with different numerical integration schemes.

The Stokes Interface Problem with a Known Real Solution. Let Ω be a square with the size $(0,1)\times(0,1)$ and Ω_2 be a circle with the center at $(0.3, 0.3)$ and the radius 0.1. Consider the Stokes interface problem (6)–(12) with the following jump coefficients and right hand side functions:

- Case S3: $\beta_1 = 1$, $\beta_2 = 100$, $\boldsymbol{f}_1$ and $\boldsymbol{f}_2$ are defined below;
- Case S4: $\beta_1 = 1$, $\beta_2 = 10000$, $\boldsymbol{f}_1$ and $\boldsymbol{f}_2$ are defined below,

where, $\boldsymbol{f}_1$ and $\boldsymbol{f}_2$ are chosen such that the real solution of (6)–(12) can be taken as

$$\boldsymbol{u} = \begin{pmatrix} \frac{(y-0.3)((x-0.3)^2+(y-0.3)^2-0.01)}{\beta_i} \\ \frac{-(x-0.3)((x-0.3)^2+(y-0.3)^2-0.01)}{\beta_i} \end{pmatrix} \quad \text{if } (x,y)^T \in \Omega_i,$$
$$p = 0.01(x^3 - y^3).$$

Note that an inhomogeneous boundary condition is employed instead for the Stokes interface problem in the above two cases.

With the P_2P_1-mixed finite element space, the numerical results of the DLM/FD finite element method (22)–(25) are reported in Tables 9 and 10 for two cases S3-S4, from which we observe the obvious differences on numerical solutions of both the velocity and the pressure between different numerical integration schemes, and, the subgrid integration scheme produces the lowest approximation error, i.e., the best accuracy, for both the velocity and the pressure in comparison with other numerical integration schemes in all two cases.

To look into effects of other numerical integration schemes than the subgrid scheme, in terms of the following formula

$$\frac{\left|\|\boldsymbol{u}-\tilde{\boldsymbol{u}}_h^i\|_1 - \|\boldsymbol{u}-\tilde{\boldsymbol{u}}_h^s\|_1\right|}{\|\boldsymbol{u}-\tilde{\boldsymbol{u}}_h^s\|_1} \text{ or } \frac{\left|\|\boldsymbol{u}-\tilde{\boldsymbol{u}}_h^{5ci}\|_1 - \|\boldsymbol{u}-\tilde{\boldsymbol{u}}_h^s\|_1\right|}{\|\boldsymbol{u}-\tilde{\boldsymbol{u}}_h^s\|_1},$$
$$\frac{\left|\|p-p_h^i\|_0 - \|p-p_h^s\|_0\right|}{\|p-p_h^s\|_0} \text{ or } \frac{\left|\|p-p_h^{5ci}\|_0 - \|p-p_h^s\|_0\right|}{\|p-p_h^s\|_0},$$

we calculate and plot the relative differences between numerical errors obtained from other integration schemes and those from the subgrid scheme, as shown in Fig. 3, where, two cases illustrate a slow convergence tendency on numerical errors of the velocity towards those from the subgrid scheme along with the increase of numerical integration scheme number. However, such tendency does not even apply to numerical errors of the pressure, instead, Fig. 3 shows that a higher order integration scheme may lead to a worse approximation to the pressure. Furthermore, in Fig. 4, a special case is observed that the numerical solution obtained by Scheme 2 is significantly polluted inside Ω_2 while the numerical solution obtained by the subgrid integration scheme is of better accuracy in Ω_2, which implies that the Lagrange multiplier does not work well to enforce (25) hold true, reflecting from the low accuracy of Scheme 2.

Table 9. Numerical results of Case S3

scheme #	2	3	4	5	6	7
$\Vert\boldsymbol{u}-\tilde{\boldsymbol{u}}_h\Vert_1$	1.8194e−3	1.5948e−3	1.6239e−3	1.5642e−3	1.6210e−3	1.6269e−3
$\Vert p-p_h\Vert_0$	5.9637e−1	6.1253e−1	7.2035e−1	7.0451e−1	8.3067e−1	7.9560e−1
scheme #	8	9	4c1	4c2	4c3	s
$\Vert\boldsymbol{u}-\tilde{\boldsymbol{u}}_h\Vert_1$	1.6024e−3	1.5858e−3	1.6295e−3	1.5737e−3	1.5595e−3	7.8025e−4
$\Vert p-p_h\Vert_0$	7.7359e−1	7.9780e−1	8.5104e−1	8.8010e−1	8.8306e−1	4.4222e−1

Table 10. Numerical results of Case S4

scheme #	2	3	4	5	6	7
$\Vert\boldsymbol{u}-\tilde{\boldsymbol{u}}_h\Vert_1$	1.8718e−3	1.7913e−3	1.9442e−3	1.9896e−3	2.0783e−3	2.0616e−3
$\Vert p-p_h\Vert_0$	4.8900e−1	5.6678e−1	5.4602e−1	5.5996e−1	6.2313e−1	6.2266e−1
scheme #	8	9	5c1	5c2	5c3	s
$\Vert\boldsymbol{u}-\tilde{\boldsymbol{u}}_h\Vert_1$	2.0758e−3	2.0257e−3	2.0505e−3	2.0818e−3	2.0549e−3	1.2073e−3
$\Vert p-p_h\Vert_0$	5.7456e−1	5.7665e−1	6.2303e−1	6.2144e−1	6.2582e−1	4.3039e−1

The Stokes/Elliptic Interface Problem with a Known Real Solution. Let Ω be a square with the size $(0,1)\times(0,1)$ and Ω_2 be a circle with the center at $(0.3, 0.3)$ and the radius 0.1. Consider the Stokes/elliptic interface problem (13)–(18) with the following jump coefficients and right hand side functions:

- Case SE3: $\beta_1 = 1$, $\beta_2 = 100$, $\boldsymbol{f}_1$ and $\boldsymbol{f}_2$ are defined below;
- Case SE4: $\beta_1 = 1$, $\beta_2 = 10000$, $\boldsymbol{f}_1$ and $\boldsymbol{f}_2$ are defined below,

where, $\boldsymbol{f}_1$ and $\boldsymbol{f}_2$ are chosen such that the real solution of (13)–(18) is taken as

$$\boldsymbol{u} = \begin{pmatrix} \frac{(y-0.3)((x-0.3)^2+(y-0.3)^2-0.01)}{\beta_i} \\ \frac{-(x-0.3)((x-0.3)^2+(y-0.3)^2-0.01)}{\beta_i} \end{pmatrix} \quad \text{if } (x,y)^T \in \Omega_i,$$

$$p = 0.01(x^3 - y^3)((x-0.3)^2 + (y-0.3)^2 - 0.01).$$

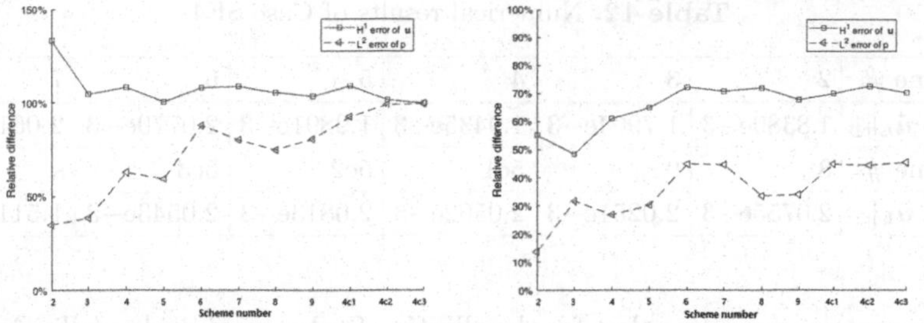

Fig. 3. Relative differences of the velocity error in H^1-norm and of the pressure error in L^2-norm for Cases S3 and S4 (from left to right).

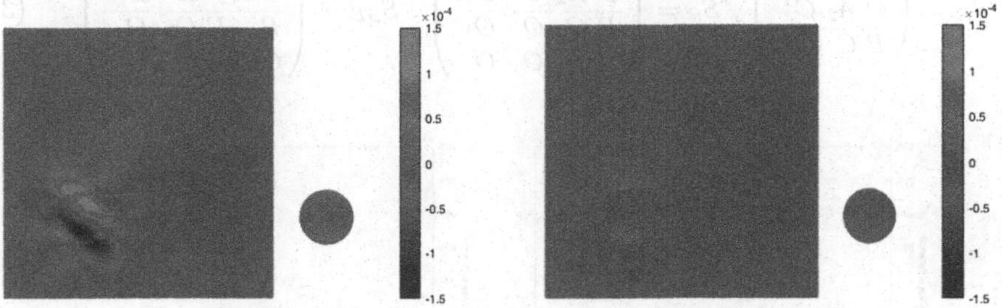

Fig. 4. The first components of $\boldsymbol{u}-\tilde{\boldsymbol{u}}_h^2$ and $\boldsymbol{u}_2-\boldsymbol{u}_{2,H}^2$ vs the first components of $\boldsymbol{u}-\tilde{\boldsymbol{u}}_h^s$ and $\boldsymbol{u}_2-\boldsymbol{u}_{2,H}^2$ for Cases S3 (from left to right).

Note that an inhomogeneous boundary condition is employed instead for the Stokes/elliptic interface problem in the above two cases.

With the P_2P_1-mixed finite element space, the numerical results of the DLM/FD finite element method (26)–(29) are reported in Tables 11 and 12 and Fig. 5 for two cases SE3–SE4, from which the same conclusions can be drawn as those for the numerical velocity errors of Stokes interface problem in Cases S3-S4, and, the slowly convergent relative differences of $\|\tilde{u}-\tilde{u}_h\|_1$, that is generally between 20% and 60%, can be observed in Fig. 5.

Table 11. Numerical results of Case SE3

scheme #	2	3	4	5	6	7
$\|\boldsymbol{u}-\tilde{\boldsymbol{u}}_h\|_1$	1.6099e−3	1.5944e−3	1.6228e−3	1.5636e−3	1.6225e−3	1.6268e−3
scheme #	8	9	5c1	5c2	5c3	s
$\|\boldsymbol{u}-\tilde{\boldsymbol{u}}_h\|_1$	1.6049e−3	1.5898e−3	1.6364e−3	1.5855e−3	1.5729e−3	1.0022e−3

Table 12. Numerical results of Case SE4

scheme #	2	3	4	5	6	7
$\|\boldsymbol{u} - \tilde{\boldsymbol{u}}_h\|_1$	1.8389e−3	1.7904e−3	1.9435e−3	1.9891e−3	2.0779e−3	2.0612e−3
scheme #	8	9	5c1	5c2	5c3	s
$\|\boldsymbol{u} - \tilde{\boldsymbol{u}}_h\|_1$	2.0755e−3	2.0251e−3	2.0502e−3	2.0813e−3	2.0543e−3	1.5111e−3

In summary, we notice that for the elliptic-, Stokes- and Stokes/elliptic- interface problem, their coefficient matrices of the linear algebra systems obtained by the DLM/FD FEM are in the following forms, respectively

$$S_E = \left(\begin{array}{cc|c} A & O & B^T \\ O & A_2 & C^T \\ \hline B & C & O \end{array}\right), \quad S_S = \left(\begin{array}{cc|cc} A & O & B^T & C^T \\ O & A_2 & O & D^T \\ \hline B & O & O & O \\ C & D & O & O \end{array}\right), \quad S_{SE} = \left(\begin{array}{cc|cc} A & O & B^T & C^T \\ O & A_2 & E & D^T \\ \hline B & -E & O & O \\ C & D & O & O \end{array}\right), \tag{33}$$

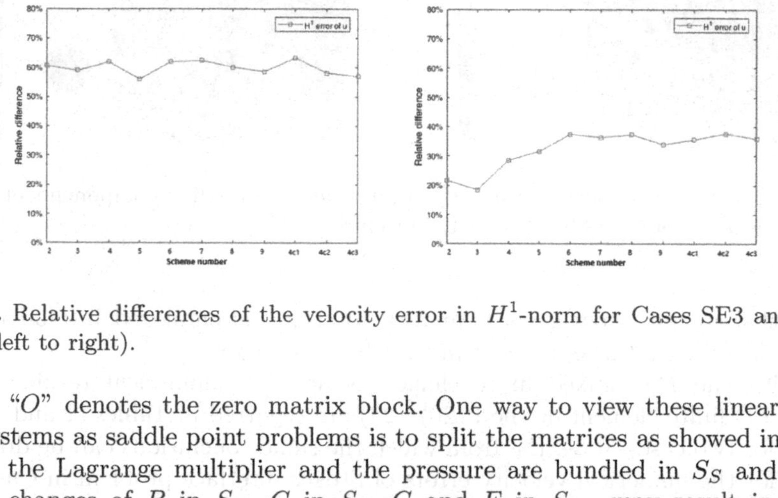

Fig. 5. Relative differences of the velocity error in H^1-norm for Cases SE3 and SE4 (from left to right).

where "O" denotes the zero matrix block. One way to view these linear algebra systems as saddle point problems is to split the matrices as showed in (33), where the Lagrange multiplier and the pressure are bundled in S_S and S_{SE}. Slight changes of B in S_E, C in S_S, C and E in S_{SE} may result in quite different effects on the corresponding numerical solutions, which is a possible reason for explaining the observations illustrated in the previous sections: the DLM/FD FEM is insensitive with various numerical integration schemes for the elliptic interface problem, however, it is sensitive with different numerical integration schemes for both the Stokes- and the Stokes/elliptic interface problems. Although only two-dimensional interface problems are considered in the numerical experiments, the DLM/FD FEM and subgrid integration technique can be used for solving the interface problems in any dimension.

5 Conclusions

In this paper, we study the effects of different numerical integration schemes, including the commonly-used integration schemes and their compound-type schemes, and the subgrid integration scheme, on the performance of the DLM/FD finite element method for solving the elliptic-, Stokes-, and Stokes/elliptic-interface problem. Numerical experiments illustrate that: (1) DLM/FD FEM is insensitive with various numerical integration schemes for the elliptic interface problem; (2) DLM/FD FEM is sensitive with different numerical integration schemes for both Stokes- and Stokes/elliptic interface problem, and sometimes, numerical solutions obtained from other integration schemes than the subgrid integration scheme may not be reliable; (3) the subgrid integration scheme always results in numerical solutions with the best accuracy.

Appendix: Numerical Integration Schemes

Scheme #	λ_i	ω_i
1	(0.333333333333333, 0.333333333333333, 0.333333333333333)	1.000000000000000
2	(0.666666666666667, 0.166666666666667, 0.166666666666667)	0.333333333333333
	(0.166666666666667, 0.666666666666667, 0.166666666666667)	0.333333333333333
	(0.166666666666667, 0.166666666666667, 0.666666666666667)	0.333333333333333
3	(0.333333333333333, 0.333333333333333, 0.333333333333333)	−0.562500000000000
	(0.600000000000000, 0.200000000000000, 0.200000000000000)	0.520833333333333
	(0.200000000000000, 0.600000000000000, 0.200000000000000)	0.520833333333333
	(0.200000000000000, 0.200000000000000, 0.600000000000000)	0.520833333333333
4	(0.108103018168070, 0.445948490915965, 0.445948490915965)	0.223381589678011
	(0.445948490915965, 0.108103018168070, 0.445948490915965)	0.223381589678011
	(0.445948490915965, 0.445948490915965, 0.108103018168070)	0.223381589678011
	(0.816847572980459, 0.091576213509771, 0.091576213509771)	0.109951743655322
	(0.091576213509771, 0.816847572980459, 0.091576213509771)	0.109951743655322
	(0.091576213509771, 0.091576213509771, 0.816847572980459)	0.109951743655322
5	(0.333333333333333, 0.333333333333333, 0.333333333333333)	0.225000000000000
	(0.059715871789770, 0.470142064105115, 0.470142064105115)	0.132394152788506
	(0.470142064105115, 0.059715871789770, 0.470142064105115)	0.132394152788506
	(0.470142064105115, 0.470142064105115, 0.059715871789770)	0.132394152788506
	(0.797426985353087, 0.101286507323456, 0.101286507323456)	0.125939180544827
	(0.101286507323456, 0.797426985353087, 0.101286507323456)	0.125939180544827
	(0.101286507323456, 0.101286507323456, 0.797426985353087)	0.125939180544827
6	(0.249286745170910, 0.249286745170910, 0.501426509658180)	0.116786275726379
	(0.249286745170910, 0.501426509658179, 0.249286745170911)	0.116786275726379
	(0.501426509658179, 0.249286745170910, 0.249286745170911)	0.116786275726379
	(0.063089014491502, 0.063089014491502, 0.873821971016996)	0.050844906370207
	(0.063089014491502, 0.873821971016996, 0.063089014491502)	0.050844906370207
	(0.873821971016996, 0.063089014491502, 0.063089014491502)	0.050844906370207
	(0.310352451033784, 0.636502499121399, 0.053145049844817)	0.082851075618374
	(0.636502499121399, 0.053145049844817, 0.310352451033784)	0.082851075618374
	(0.053145049844817, 0.310352451033784, 0.636502499121399)	0.082851075618374
	(0.636502499121399, 0.310352451033784, 0.053145049844817)	0.082851075618374
	(0.310352451033784, 0.053145049844817, 0.636502499121399)	0.082851075618374
	(0.053145049844817, 0.636502499121399, 0.310352451033784)	0.082851075618374

Scheme #	λ_i	ω_i
7	(0.333333333333333, 0.333333333333333, 0.333333333333334)	−0.149570044467682
	(0.260345966079040, 0.260345966079040, 0.479308067841920)	0.175615257433208
	(0.260345966079040, 0.479308067841920, 0.260345966079040)	0.175615257433208
	(0.479308067841920, 0.260345966079040, 0.260345966079040)	0.175615257433208
	(0.065130102902216, 0.065130102902216, 0.869739794195568)	0.053347235608838
	(0.065130102902216, 0.869739794195568, 0.065130102902216)	0.053347235608838
	(0.869739794195568, 0.065130102902216, 0.065130102902216)	0.053347235608838
	(0.312865496004874, 0.638444188569810, 0.048690315425316)	0.077113760890257
	(0.638444188569810, 0.048690315425316, 0.312865496004874)	0.077113760890257
	(0.048690315425316, 0.312865496004874, 0.638444188569810)	0.077113760890257
	(0.638444188569810, 0.312865496004874, 0.048690315425316)	0.077113760890257
	(0.312865496004874, 0.048690315425316, 0.638444188569810)	0.077113760890257
	(0.048690315425316, 0.638444188569810, 0.312865496004874)	0.077113760890257
8	(0.333333333333333, 0.333333333333333, 0.333333333333334)	0.144315607677787
	(0.081414823414554, 0.459292588292723, 0.459292588292723)	0.095091634267285
	(0.459292588292723, 0.081414823414554, 0.459292588292723)	0.095091634267285
	(0.459292588292723, 0.459292588292723, 0.081414823414554)	0.095091634267285
	(0.658861384496480, 0.170569307751760, 0.170569307751760)	0.103217370534718
	(0.170569307751760, 0.658861384496480, 0.170569307751760)	0.103217370534718
	(0.170569307751760, 0.170569307751760, 0.658861384496480)	0.103217370534718
	(0.898905543365938, 0.050547228317031, 0.050547228317031)	0.032458497623198
	(0.050547228317031, 0.898905543365938, 0.050547228317031)	0.032458497623198
	(0.050547228317031, 0.050547228317031, 0.898905543365938)	0.032458497623198
	(0.008394777409958, 0.263112829634638, 0.728492392955404)	0.027230314174435
	(0.008394777409958, 0.728492392955404, 0.263112829634638)	0.027230314174435
	(0.263112829634638, 0.008394777409958, 0.728492392955404)	0.027230314174435
	(0.728492392955404, 0.008394777409958, 0.263112829634638)	0.027230314174435
	(0.263112829634638, 0.728492392955404, 0.008394777409958)	0.027230314174435
	(0.728492392955404, 0.263112829634638, 0.008394777409958)	0.027230314174435
9	(0.333333333333333, 0.333333333333333, 0.333333333333334)	0.097135796282799
	(0.020634961602525, 0.489682519198738, 0.489682519198737)	0.031334700227139
	(0.489682519198738, 0.020634961602525, 0.489682519198737)	0.031334700227139
	(0.489682519198738, 0.489682519198738, 0.020634961602524)	0.031334700227139
	(0.125820817014127, 0.437089591492937, 0.437089591492936)	0.077827541004740
	(0.437089591492937, 0.125820817014127, 0.437089591492936)	0.077827541004740
	(0.437089591492937, 0.437089591492937, 0.125820817014126)	0.077827541004740
	(0.623592928761935, 0.188203535619033, 0.188203535619032)	0.079647738927210
	(0.188203535619033, 0.623592928761935, 0.188203535619032)	0.079647738927210
	(0.188203535619033, 0.188203535619033, 0.623592928761934)	0.079647738927210
	(0.910540973211095, 0.044729513394453, 0.044729513394452)	0.025577675658698
	(0.044729513394453, 0.910540973211095, 0.044729513394452)	0.025577675658698
	(0.044729513394453, 0.044729513394453, 0.910540973211094)	0.025577675658698
	(0.036838412054736, 0.221962989160766, 0.741198598784498)	0.043283539377289
	(0.036838412054736, 0.741198598784498, 0.221962989160766)	0.043283539377289
	(0.221962989160766, 0.036838412054736, 0.741198598784498)	0.043283539377289
	(0.741198598784498, 0.036838412054736, 0.221962989160766)	0.043283539377289
	(0.221962989160766, 0.741198598784498, 0.036838412054736)	0.043283539377289
	(0.741198598784498, 0.221962989160766, 0.036838412054736)	0.043283539377289

References

1. Auricchio, F., Boffi, D., Gastaldi, L., Lefieux, A., Reali, A.: On a fictitious domain method with distributed Lagrange multiplier for interface problems. Appl. Numer. Math. **95**, 36–50 (2015)
2. Boffi, D., Gastaldi, L., Ruggeri, M.: Mixed formulation for interface problems with distributed Lagrange multiplier. Comput. Math. Appl. **68**, 2151–2166 (2014)
3. Brezzi, F.: On the existence, uniqueness and approximation of saddle point problems arising from Lagrangian multipliers. RAIRO Analyse Numerique **8**, 129–151 (1974)
4. Brezzi, F., Fortin, M.: Mixed and Hybrid Finite Element Methods. Springer, New York (1991). https://doi.org/10.1007/978-1-4612-3172-1
5. Chakrabarti, S.K. (ed.): Numerical Models in Fluid Structure Interaction, Advances in Fluid Mechanics, vol. 42. WIT Press (2005)
6. Dunavant, D.A.: High degree efficient symmetrical Gaussian quadrature rules for the triangle. Int. J. Numer. Methods Eng. **21**(6), 1129–1148 (2010)
7. Glowinski, R., Pan, T.W., Hesla, T., Joseph, D.: A distributed Lagrange multiplier/fictitious domain method for particulate flows. Int. J. Multiph. Flow **25**, 755–794 (1999)
8. Hirt, C., Amsden, A., Cook, J.: An arbitrary Lagrangian-Eulerian computing method for all flow speeds. J. Comput. Phys. **14**, 227–253 (1974)
9. Hu, H.: Direct simulation of flows of solid-liquid mixtures. Int. J. Multiph. Flow **22**, 335–352 (1996)
10. LeVeque, R., Li, Z.: The immersed interface method for elliptic equations with discontinuous coefficients and singular sources. SIAM J. Numer. Anal. **31**, 1019–1044 (1994)
11. Li, Z., Lai, M.C.: The immersed interface method for the Navier-Stokes equations with singular forces. J. Comput. Phys. **171**, 822–842 (2001)
12. Liu, W.K., Kim, D.W., Tang, S.: Mathematical foundations of the immersed finite element method. Comput. Mech. **39**, 211–222 (2006)
13. Lundberg, A., Sun, P., Wang, C.: Distributed Lagrange multiplier-fictitious domain finite element method for Stokes interface problems. Int. J. Numer. Anal. Model. (2018). Accepted
14. Peskin, C.: The immersed boundary method. Acta Numerica **11**, 479–517 (2002)
15. Sun, P.: Fictitious domain finite element method for Stokes/elliptic interface problems with jump coefficients. J. Comput. Appl. Math. **356**, 81–97 (2019)
16. Sun, P., Wang, C.: Fictitious domain finite element method for Stokes/parabolic interface problems with jump coefficients. Appl. Numer. Math. (2018). Submitted
17. Takizawa, K., Henicke, B., Tezduyar, T.E., Hsu, M.C., Bazilevs, Y.: Stabilized space-time computation of wind-turbine rotor aerodynamics. Comput. Mech. **48**, 333–344 (2011)
18. Wang, C., Sun, P.: A fictitious domain method with distributed Lagrange multiplier for parabolic problems with moving interfaces. J. Sci. Comput. **70**, 686–716 (2017)

Application of the Double Potential Method to Simulate Incompressible Viscous Flows

Tatyana Kudryashova(✉), Sergey Polyakov, and Nikita Tarasov

Keldysh Institute of Applied Mathematics, RAS, Moscow 125047, Russia
kudryshova@imamod.ru

Abstract. In this paper we discuss an application of the double potential method for modelling flow of incompressible fluid. This method allows us to avoid the known difficulties in calculating pressure and overcome the instability of numerical solution. Also, the double potential method enables us to simplify the problem of boundary conditions setting. It arises when computing the incompressible fluid flow by the Navier-Stokes equations in the vector potential - velocity rotor variables. In the approach given, the final system of equations is approximated through applying the finite volume method. In this case, an exponential transformation of the flow terms is applied. A parallel program was developed by means of using MPI and OpenMP technologies for the purpose of the numerical method computer implementation. We used two tasks to test. One of them deals with the classical calculation of the fluid flow establishment in a long round pipe. The other one is connected with the flow calculation in the pipe that in the output region contains a separation into two symmetrical parts. To perform numerical simulation, we take into consideration the steady flow with Reynolds numbers of 50 and 100. The numerical results obtained are consistent with computational results received through using the ANSYS CFD package.

Keywords: Fluid flow · Double potential method · Navier-Stokes equations

1 Introduction

Modeling internal flow of viscous incompressible fluid is one of the most important and complex problems in continuum dynamics. This challenge implies the great potential to deal with applied tasks. For example, the problem of cleaning aquatic environment from impurities of the fine iron ions can be solved by applying the method of affecting an electromagnetic field on a collector considered in two-dimensional formulation in [1, 2]. To obtain the distribution of the impurities by the grid method we need to have a velocity field in the computational domain.

When simulating internal flow problems in 2D geometry, the Navier-Stokes model is widely used in the stream-vortex function formulation. It allows us to avoid the setting limitations in natural variables (speed-pressure). These limitations imply the discretization complexity of the equation to determine the pressure field, the high instability of the solution to this equation when using numerical simulation cellular methods and no performance guarantee of the mass conservation law [3]. This technique enables us to reduce the number of dependent variables. To receive the natural

J. M. F. Rodrigues et al. (Eds.): ICCS 2019, LNCS 11539, pp. 568–579, 2019.
https://doi.org/10.1007/978-3-030-22747-0_42

formulation, we use two coordinates of velocity and pressure. To get the current function – vortex formulation we have only two scalar variables.

Many sources suggest the lack of a generalization of such method for the three-dimensional problem [5–18], but the work [4] shows that it is possible. It is achieved through determining the vector potential and the vector vortex like the current function and the vortex for the two-dimensional variant. It should be noted that in this case, the formulation of the boundary conditions for the vector potential is nontrivial. The boundary conditions are extremely important for computational hydrodynamics and, as it is shown in [4], they have a determining influence on the resulting view of flow. To simplify the boundary conditions formulation for incompressible fluid modeling internal flow, the double potential method [19–21], was developed and successfully applied in [21]. The approach is discussed in this work.

2 Mathematical Model

To simulate the flow of viscous incompressible fluid, we use the Navier-Stokes system of equations and the incompressibility condition [3]. The dimensionless formulation of the equations has the following form:

$$\frac{\partial \mathbf{u}}{\partial t} + (\mathbf{u}, \nabla)\mathbf{u} = \frac{1}{\mathrm{Re}}\Delta\mathbf{u} - \nabla p \tag{1}$$

$$\nabla\mathbf{u} = 0 \tag{2}$$

here u is speed vector, $\frac{\partial}{\partial t}$ is a derivative of a function with respect to time, $\mathrm{Re} = \frac{u_0 \rho_0 D_0}{\eta}$ is the Reynolds number, Δ is the Laplace operator, ∇ is the Hamilton operator, p is pressure, u_0 is characteristic speed of flow, ρ_0 is medium density, D_0 is hydraulic diameter, η is coefficient of dynamic viscosity.

The system of Eqs. (1), (2) is completed with corresponding boundary and initial conditions. They will be presented further when obtaining the system of equations of the double potential method. Let's consider the main calculations of this method.

From the Eq. (2) and the rotor property $\nabla(rot[\mathbf{A}]) = 0$, we can write the formula for the vector A:

$$\mathbf{u} = rot(\mathbf{A}) \tag{3}$$

When condition (3) is fulfilled, vector A is called the vector potential of flow. A characteristic feature of the viscous fluid motion is vorticity, which is defined as follows:

$$\boldsymbol{\omega} = rot(\mathbf{u}) \tag{4}$$

Basing on the condition (3) we can write formula $rot(rot[\mathbf{A}]) = \boldsymbol{\omega}$. Then, we use the rotor property: $rot(rot[\mathbf{A}]) = \nabla(\nabla\mathbf{A}) - \Delta\mathbf{A}$, require that $\nabla\mathbf{A} = 0$ [4] and get equation to define the vector potential:

$$\Delta\mathbf{A} = -\boldsymbol{\omega} \tag{5}$$

The equations for calculating the vortex are obtained by the action of the *rot* operation on Eq. (1), then we take the final formula:

$$\frac{\partial\boldsymbol{\omega}}{\partial t} + (\mathbf{u}, \nabla)\boldsymbol{\omega} - (\boldsymbol{\omega}, \nabla)\mathbf{u} = \frac{1}{\mathrm{Re}}\Delta\boldsymbol{\omega} \tag{6}$$

So, Eqs. (3)–(6) with the boundary conditions allow us to calculate the velocity field in the computational geometry. This method is briefly described in [4] and it has applications for solving some problems of computational hydrodynamics. When using this method, there is a serious difficulty in setting the boundary condition on the vector potential. It can be seen from condition (3). In order to overcome this difficulty, a consequence of the theory of potentials was proposed in [19–21]. Write the speed in the form:

$$\mathbf{u} = rot(\mathbf{A}) - \nabla\varphi \tag{7}$$

here φ is scalar potential. To satisfy condition (2), it is necessary to require:

$$\Delta\varphi = 0 \tag{8}$$

This condition is implemented as an equation to calculate the scalar potential.

The view of the velocity in the form of (7), in accordance with the theory of potentials, allows us to represent the boundary conditions on a simply connected computational domain for the vector potential in the form [19]:

$$A_\tau = \frac{\partial A_n}{\partial\mathbf{n}} = 0, \quad \text{for } \delta\Omega \tag{9}$$

here A_τ is the tangent component of the vector potential, A_n is the normal component of the vector potential, n is normal to the border, $\delta\Omega$ is boundary of the computational area. The boundary conditions for the scalar potential are determined as follows:

$$\frac{\partial\varphi}{\partial\mathbf{n}} = \left(\mathbf{u}|_{\delta\Omega}, \mathbf{n}\right) \quad \text{for } \delta\Omega \tag{10}$$

here $\mathbf{u}|_{\delta\Omega}$ is the flow velocity of medium across the boundary of the area. It is the boundary conditions for speed. The boundary and initial conditions for the vortex are determined from expression (4) as follows:

$$\boldsymbol{\omega} = rot\left(\mathbf{u}|_{\delta\Omega}\right) \quad \text{for } \delta\Omega, \tag{11}$$

$$\boldsymbol{\omega} = rot\left(\mathbf{u}|_{t=0}\right) \quad \text{for } t = 0, \tag{12}$$

At last, the final formulation of the problem of flow modeling by the double potential method is represented by Eqs. (5)–(8) through using the boundary conditions (9)–(11) and the initial conditions (12).

3 Numerical Method for 3D Geometry

For 3D case, the grid consisting of triangular prisms is constructed in the computational domain (it is shown in Fig. 1). The finite volume method on the centers of the prismatic cells is used to approximate the equations by the double potential method (5)–(8) [22].

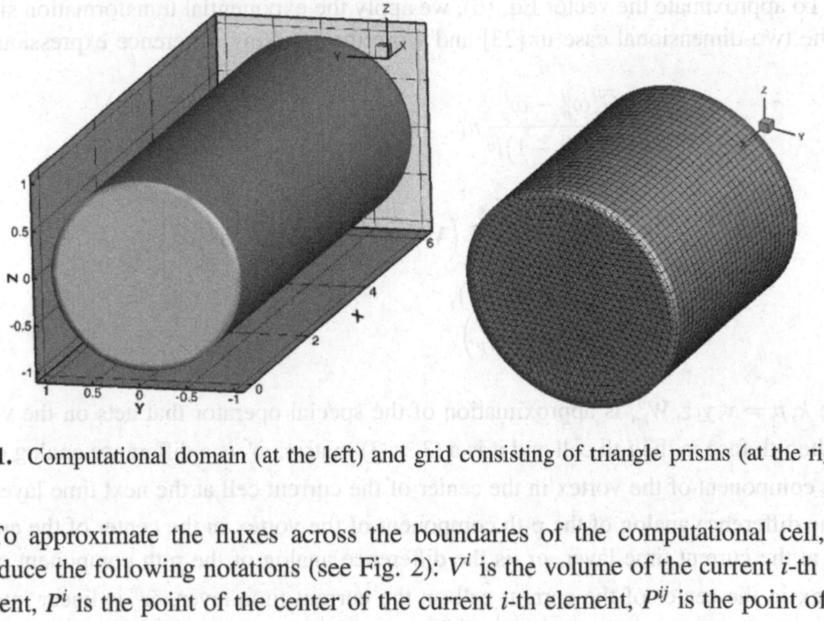

Fig. 1. Computational domain (at the left) and grid consisting of triangle prisms (at the right).

To approximate the fluxes across the boundaries of the computational cell, we introduce the following notations (see Fig. 2): V^i is the volume of the current i-th grid element, P^i is the point of the center of the current i-th element, P^{ij} is the point of the center of the j neighbor to the i-th element, l^{ij} is the distance between the centers of neighboring cells, $\mathbf{n}^{ij} = \left\{n_x^{ij}, n_y^{ij}, n_z^{ij}\right\}^T$ is the unit direction vector from the center of the current element to the center of the neighboring one, $\mathbf{S}^{ij} = \left\{S_x^{ij}, S_y^{ij}, S_z^{ij}\right\}^T$ is the directional area of the j-th edge of the i-th element. The number of adjacent elements of a prismatic cell is five.

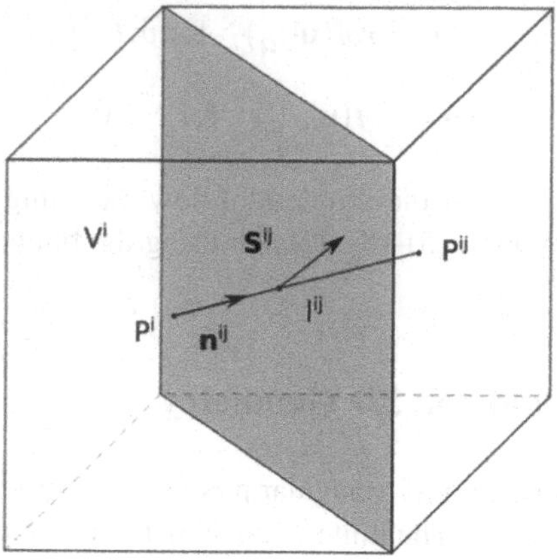

Fig. 2. Characteristic grid values for prismatic elements.

To approximate the vector Eq. (6), we apply the exponential transformation similar to the two-dimensional case in [23] and write the resulting difference expressions:

$$\bar{W}^{ij}_{kp} = \frac{\bar{G}^{ij}_k \omega^{ij}_p - \omega^i_p}{0.5\left(\bar{G}^{ij}_k - 1\right) l^{ij}} n^{ij}_k, \quad \bar{G}^{ij}_k = \exp\left(-\mathrm{Re} l^{ij} n^{ij}_k \frac{u^{ij}_k + u^i_k}{2}\right) \tag{13}$$

$$\begin{aligned} \hat{\omega}^i_p &= \omega^i_p + \frac{\tau}{V^i}\left[\sum_{j=1}^{5}\left(\mathbf{W}^{ij}_p, \mathbf{S}^{ij}\right) + \sum_{j=1}^{5}\left(\overline{\boldsymbol{\omega}}^{ij}, \mathbf{S}^{ij}\right)\bar{u}^{ij}_p\right], \\ \overline{\boldsymbol{\omega}}^{ij} &= 0.5\left(\boldsymbol{\omega}^{ij} + \boldsymbol{\omega}^i\right), \\ \bar{u}^{ij}_p &= 0.5\left(u^{ij}_p + u^i_p\right) \end{aligned} \tag{14}$$

here $k, p = x, y, z$, $\bar{W}^{ij}_{kp}$ is approximation of the special operator that acts on the vortex on the j-th face in the i-th cell and it is a (3 × 3) matrix, $\hat{\omega}^i_p$ is a difference analog of the p-th component of the vortex in the center of the current cell at the next time layer, ω^{ij}_p is the difference analog of the p-th component of the vortex in the center of the current cell at the current time layer, ω^i_p is the difference analog of the p-th component of the vortex in the center of the current cell on the current time layer, $\bar{\omega}^{ij}$ is linear interpolation of the vortex in the middle of the face, $\bar{u}^{ij}_p$ is linear interpolation of the p-th coordinate of velocity in the middle of the face. The boundary conditions for the vortex are realized by discretization of the speed rotor with the first order of accuracy.

Further, we write the difference analogues for the equation of the scalar potential (8) and the vector potential (5):

$$\hat{\varphi}^i = \sum_{j=1}^{5} \frac{\varphi^{ij}}{l^{ij}} \left(\mathbf{n}^{ij}, \mathbf{S}^{ij}\right) \left[\sum_{j=1}^{5} \frac{\left(\mathbf{n}^{ij}, \mathbf{S}^{ij}\right)}{l^{ij}}\right]^{-1} \tag{15}$$

$$\hat{A}_k^i = \left[\sum_{j=1}^{5} \frac{A_k^{ij}}{l^{ij}} \left(\mathbf{n}^{ij}, \mathbf{S}^{ij}\right) + \hat{\omega}_k^i V^i\right] \left[\sum_{j=1}^{5} \frac{\left(\mathbf{n}^{ij}, \mathbf{S}^{ij}\right)}{l^{ij}}\right]^{-1} \tag{16}$$

here $k = x, y, z$, $\hat{A}_k^i$ is difference analog of the k-th component of the vector potential in the center of the current cell on a new time layer, A_k^{ij} is difference analog of the k-th component of the vector potential in the center of a neighboring cell on the previous time layer.

So, the difference analogue of the three-dimensional velocity vector is written as follows:

$$\begin{aligned}
\hat{u}_x^i &= \frac{0.5}{V^i} \left[\sum_{j=1}^{5} \left(\hat{A}_z^i + \hat{A}_z^{ij}\right) S_y^{ij} - \sum_{j=1}^{5} \left(\hat{A}_y^i + \hat{A}_y^{ij}\right) S_z^{ij} - \sum_{j=1}^{5} \left(\hat{\varphi}^i + \hat{\varphi}^{ij}\right) S_x^{ij}\right], \\
\hat{u}_y^i &= \frac{0.5}{V^i} \left[\sum_{j=1}^{5} \left(\hat{A}_x^i + \hat{A}_x^{ij}\right) S_z^{ij} - \sum_{j=1}^{5} \left(\hat{A}_z^i + \hat{A}_z^{ij}\right) S_x^{ij} - \sum_{j=1}^{5} \left(\hat{\varphi}^i + \hat{\varphi}^{ij}\right) S_y^{ij}\right], \\
\hat{u}_z^i &= \frac{0.5}{V^i} \left[\sum_{j=1}^{5} \left(\hat{A}_y^i + \hat{A}_y^{ij}\right) S_x^{ij} - \sum_{j=1}^{5} \left(\hat{A}_x^i + \hat{A}_x^{ij}\right) S_y^{ij} - \sum_{j=1}^{5} \left(\hat{\varphi}^i + \hat{\varphi}^{ij}\right) S_z^{ij}\right]
\end{aligned} \tag{17}$$

here $\hat{\mathbf{u}}^i = \left\{\hat{u}_x^i, \hat{u}_y^i, \hat{u}_z^i\right\}$ is difference analogue of the velocity vector in the center of the current cell.

The final algorithm to calculate the velocity field on 3D prismatic grid is the sequential calculations of difference expressions (13)–(17).

4 Parallel Realization

To parallelize the numerical method, we use two-level domain decomposition technique. At the first level, we divide the computational area into domains in accordance with number of the supercomputer system nodes. At the second level, we apply splitting of domains into subdomains in accordance with threads implemented in the cores of the central processors of each node.

The resulting program code was developed in the C++ programming language using MPI [24] and OpenMP [25] parallel technologies.

MVS-10P Supercomputer of JSCC RAS (see www.jscc.ru) was chosen for testing parallel software. The system has parameters:

- 207 nodes, each includes:
- 2 x CPU Intel Xeon 2,7 GHz microprocessors with 6 cores
- 2 x VPU Intel Xeon Phi 7110X microprocessors with 61 cores
- Peak performance is 523.8 TFlops
- Transfer is FDR InfiniBand

The results of parallel computations on the prismatic grid with 2500000 cells for simple geometry (round tube) are shown in Fig. 3. They show that constructed parallel software is quite effective (the cylindrical computational domain is shown in Fig. 1).

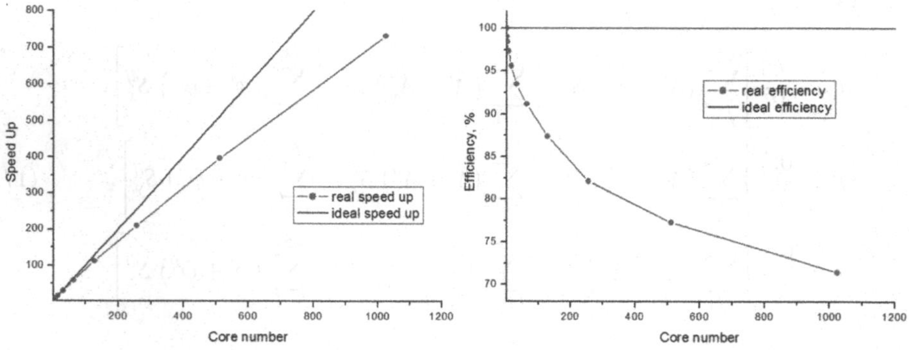

Fig. 3. Comparison of the acceleration and ideal on the prismatic grid with 2500000 cells (at the left) and the efficiency and ideal on the prismatic grid with 2500000 cells (at the right).

5 Test and Results

Two tasks were used to test our approach. One of them deals with the classical calculation of the fluid flow establishment in a long round pipe (see Fig. 1). The other one is connected with the flow calculation in the pipe that in the output region contains a separation into two symmetrical parts (see Fig. 4).

Let's consider the first of our tasks. In this case, the well-known stationary Poiseuille flow is realized in the three-dimensional cylindrical geometry. The boundary and initial conditions are written as follows:

$$\mathbf{u}|_{x=0} = \left\{1 - y^2 - z^2, 0, 0\right\}^T, \quad \mathbf{u}|_{|y^2+z^2|=1} \equiv 0, \quad \frac{\partial \mathbf{u}}{\partial x}\Big|_{x=6} \equiv 0$$
$$\mathbf{u}|_{t=0} \equiv 0.$$

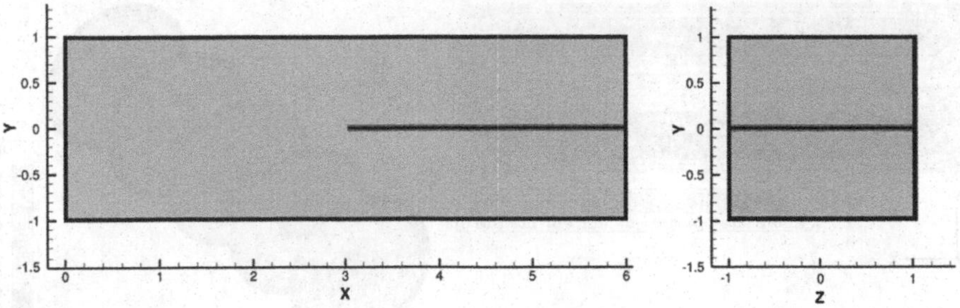

Fig. 4. Longitudinal section of the pipe for z = 0 (at the left) and the pipe cross section for $3 \leq x \leq 6$ (at the right).

The test calculation was carried out on the prismatic computational grid consisting of 133632 elements - 87 steps along the X axis, 1536 triangles at the base in the YZ plane. The Reynold number used in the calculation is Re = 100. The Figs. 5, 6, 7 and 8 show the results of calculations. As can be seen from Figs. 5 and 6, we have got steady flow of Poiseuille throughout the volume of the cylinder.

Let's consider the solution to the second problem. Here, the numerical simulation of the stationary flow was performed for the Reynolds numbers 50 and 100. The obtained numerical results are shown in Figs. 7 and 8. For comparison, we present at the same figures the calculated data obtained through using the ANSYS CFD package [26] and based on the traditional SIMPLEC method [4]. Analysis of the results shows that qualitatively and quantitatively, the calculations performed by the traditional SIMPLEC method and by the double potential method are very close. We could obtain more accurate estimates if setting parameters of the SIMPLEC method were known.

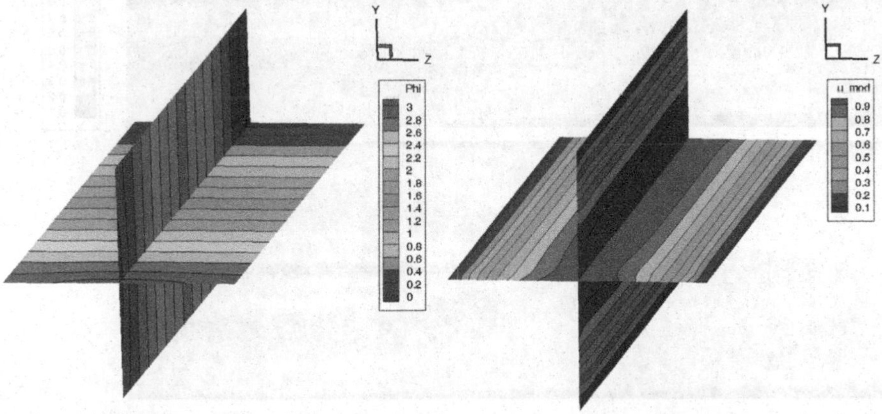

Fig. 5. The distribution of the scalar potential (at the left) and velocity modulus (at the right) in sections: XY and XZ.

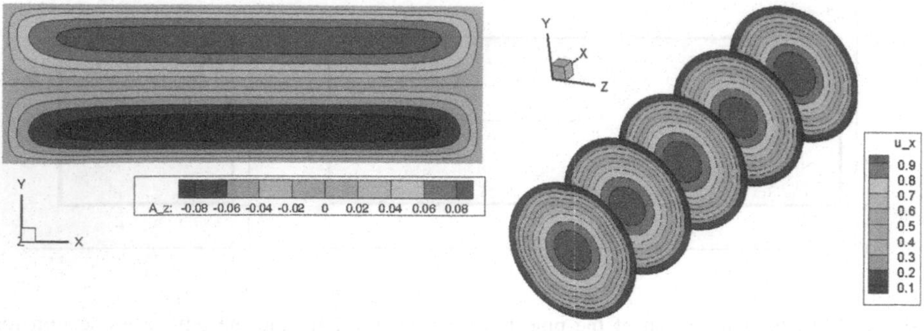

Fig. 6. The distribution of z-component of the vector potential in XY section (at the left) and the distribution of the velocity module in the array of YZ sections (at the right).

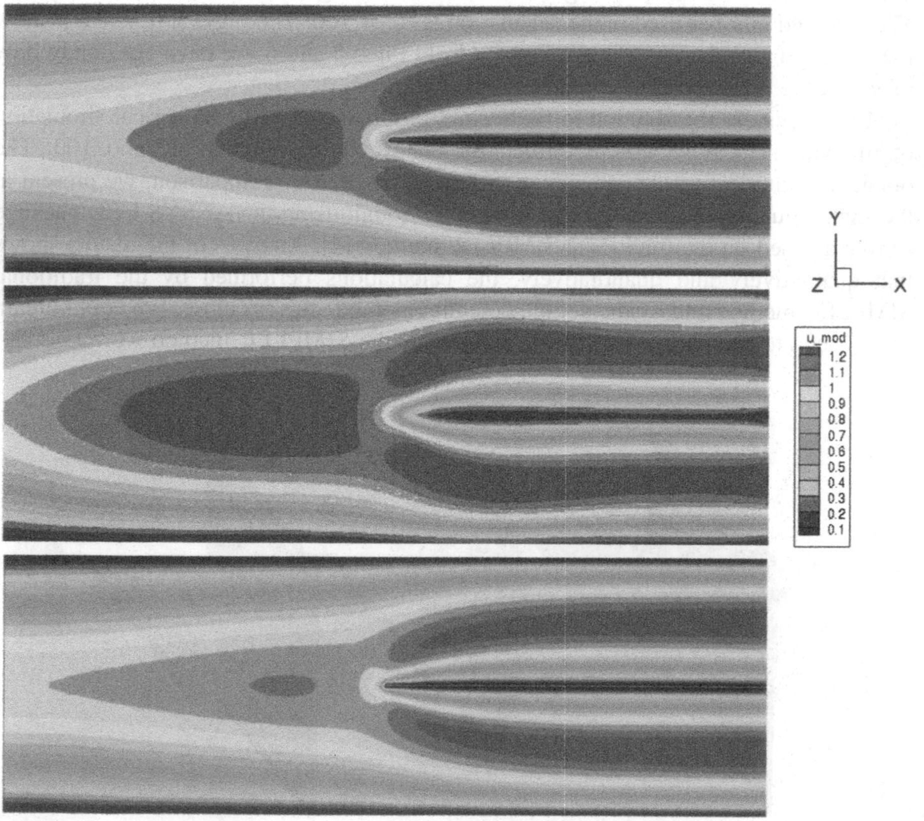

Fig. 7. The distributions of the velocity modulus in the longitudinal section $z = 0$ were calculated by SIMPLEC method of the ANSYS CFD software tools for Re = 50 (at the top) and with the help of double potential method for Re = 50 (in the middle section) and Re = 100 (below).

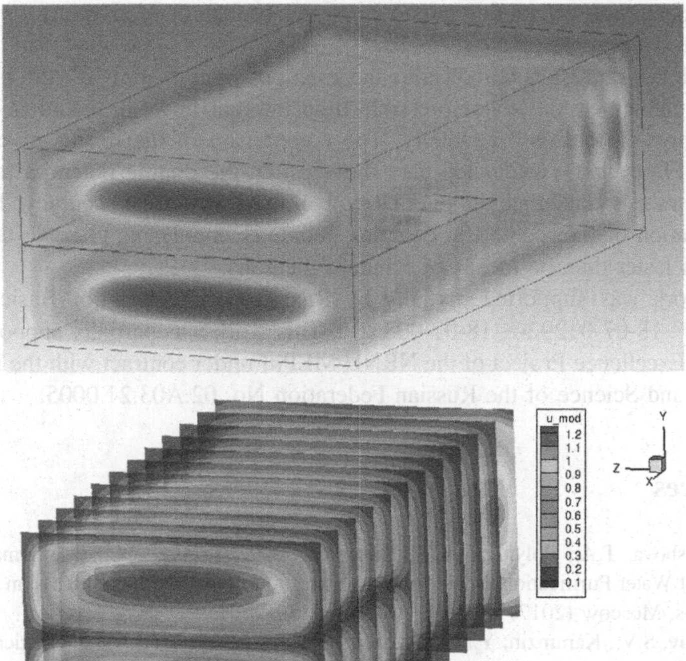

Fig. 8. The distribution of the velocity modulus in the longitudinal section $z = 0$. Distributions of the velocity modulus in cross sections YZ for Re = 50, calculated in ANSYS CFD using the SIMPLEC method (at the top) and calculated by the double potential method (at the bottom).

6 Conclusion

This paper examines the double potential method to calculate the viscous incompressible fluid internal flows. The main advantage of the method is based on the formulation preservation of the Navier-Stokes equations in the vector potential - vector vortex variables. In addition to that, it enables us to avoid setting complex boundary conditions on the vector potential.

To implement the double potential method, the original difference scheme was developed. The scheme is based on the finite volume method on cell centers through using the flux vector exponential transformation. This scheme realizes both the double potential method in the three-dimensional case and its consequence in the two-dimensional formulation.

The numerical algorithm obtained was parallelized and the software was implemented in C++ be means of MPI and OpenMP technologies. The two test problems were examined by the generated program code. One of them referring to the classical problem deals with the establishing the Poiseuille flow in the long circular pipe. The other task deals with calculating the flow in a square tube divided into two symmetrical parts in the outlet region.

The calculation results demonstrate the great potential to apply the double potential method to model viscous incompressible fluid internal flows in technical systems of complex three-dimensional geometry. The comparison of the results given with the ANSYS CFD package calculated data showed that the double potential method presented allows us to calculate the internal flow of fluid with great accuracy. The parallel implementation of the method in complex geometry enables us to solve tasks of this class much faster than be means of standard methods.

The work was supported partially by Russian Foundation for Basic Research, projects No. 18-07-01292-a, 18-07-00841-a. This work was partially supported by the Academic Excellence Project of the NRNU MEPhI under contract with the Ministry of Education and Science of the Russian Federation No. 02.A03.21.0005.

References

1. Kudryashova, T.A., Polyakov, S.V., Puzyrkov, D.V., Tarasov, N.I.: Mathematical Simulation of Water Purification Processes from Iron Impurities. Preprints of Russian Academy of Sciences, Moscow (2017)
2. Polyakov, S.V., Karamzin, Y.N., Kudryasova, T.A., Tarasov, N.I.: Mathematical modelling water treatment processes. Mathematica Montisnigri **XL**, 110–126 (2017)
3. Landau, L.D., Lifschitz, E.M.: Theoretical physics. In: Hydrodynamics, 3rd edn. Nauka, Moscow (1986)
4. Roache, P.J.: Computational Fluid Dynamics. Mir, Moscow (1980)
5. Nayfeh, A.H., Masad, J.A.: Recent advances in secondary instabilities in boundary layers. Comput. Syst. Eng. **1**(2–4), 401–414 (1990)
6. Oden, J.T., Demkowicz, L., Liszka, T., Rachowicz, W.: h–p adaptive finite element methods for compressible and incompressible flows. Comput. Syst. Eng. **1**(2–4), 523–534 (1990)
7. Rubin, S.G., Khosla, P.K.: A review of reduced Navier-Stokes computations for compressible viscous flows. Comput. Syst. Eng. **1**(2–4), 549–562 (1990)
8. Shu, C., Richard, B.E.: Parallel simulation of incompressible viscous flows by generalized differential quadrature. Comput. Syst. Eng. **3**(1–4), 271–281 (1992)
9. Chorin, A.J.: A numerical method for solving incompressible viscous flow problems. J. Comput. Phys. **135**, 118–125 (1997)
10. Garbaruk, A.V., Lapin, Y.V., Strelets, M.K.: An algebraic model of turbulence for flows with a maximum of tangential stress inside the boundary layer. High Temp. **39**(4), 548–557 (2001)
11. Garbaruk, A.V., Lapin, Y.V., Strelets, M.K.: Turbulent boundary layer under simultaneous effect of the longitudinal pressure gradient, injection (suction), and transverse surface curvature. High Temp. **40**(3), 399–404 (2002)
12. Gushchin, V.A., Matyushin, P.V.: Classification of regimes of separated fluid flows over a sphere at moderate reynolds numbers. In: Mathematical Simulation: Problems and Results, pp. 199–235. Nauka, Moscow (2003)
13. Denisikhina, D.M., Bassina, I.A., Niculin, D.A., Strelets, M.K.: Numerical simulation of self-excited oscillation of a turbulent jet flowing into a rectangular cavity. High Temp. **43**(4), 568–579 (2005)

14. Koleshko, S.B., Lapin, Y.V., Chumakov, Y.S.: Turbulent Free-convection boundary layer on a vertical heated plate: regularities of the temperature layer. High Temp. **43**(3), 429–440 (2005)
15. Gushchin, V.A., Matyushin, P.V.: Vortex formation mechanisms in the wake behind a sphere for 200 < Re < 380, R.V. Fluid Dyn. **42**, 335 (2007). https://doi.org/10.1134/S0015462807020196
16. Matyushin, P.V.: Continuous transformation of the vortex structures around a moving sphere with increasing of the fluid stratification. In: Selected Papers of the International Conference on Flows and Structures in Fluids: Physics of Geospheres, pp. 251–258 (2009)
17. Gushchin, V.A., Matyushin, P.V.: Modeling and investigation of stratified fluid flows around finite bodies. Comput. Math. Math. Phys. **56**(6), 1034–1047 (2016)
18. Matyushin, P.V.: Evolution of the diffusion-induced flow over a disk, submerged in a stratified viscous fluid. Math. Model. **30**(11), 44–58 (2018)
19. Richardson, S.M., Cornish, A.R.: Solution of three-dimensional incompressible flow problems. J. Fluid Mech. **82**(Part 2), 309–319 (1977)
20. Richardson, S.M.: Numerical solution of the three—dimensional Navier—Stokes equations. Doctoral dissertation, Department of Chemical Engineering and Chemical Technology, Imperial College of Science and Technology, London (1976)
21. Gegg, S.G.: A dual-potential formulation of the Navier-Stokes equations. Retrospective thesis and dissertations. Iowa State University (1987). https://lib.dr.iastate.edu/rtd/9040
22. Samarski, A.A., Nikolaev, E.S.: Methods for the Solution of Grid Equations. Nauka, Moscow (1978)
23. Polyakov, S.V.: Exponential difference schemes for convection-diffusion equation. Mathematica Montisnigri **XXVI**, 29–44 (2013)
24. Official documentation and manuals on MPI. http://www.mcs.anl.gov/research/projects/mpi/
25. Official documentation and manuals on OpenMP. http://www.openmp.org/resources/openmp-books
26. ANSYS. https://www.ansys.com/

A Bubble Formation in the Two-Phase System

Karel Fraňa[1(✉)], Shehab Attia[1], and Jörg Stiller[2]

[1] Technical University of Liberec, Studentská 2, 461 17 Liberec, Czech Republic
{karel.frana,shehab.attia}@tul.cz

[2] Institut für Strömungsmechanik, Technische Universität Dresden, 01062 Dresden, Germany
joerg.stiller@tu-dresden.de

Abstract. The formation of the bubbles in the liquid was examined numerically and results were successfully compared with the results provided by experiments. The study covered two different patterns defined by different Morton numbers or gas flow rates. The unsteady three dimensional calculations were carried out in code OpenFoam with the volume of fluid approach. Numerical results were in a good match to the experiments in respect to bubble shapes, diameters and Reynolds numbers. More accurate comparison was found for lower gas flow rate then for the higher one. The main reason can be that under higher gas flow rate, a complex flow behavior between gas bubbles and surrounding liquid flow is created which worsens the accuracy of calculations. The main important output of the study was a comparison of the bubble diameters in time. Especially for higher gas flow rates, bubbles are growing rapidly during its climbing. Nevertheless a satisfactory agreement was found between numerics and experiments.

Keywords: Two-phase flow · CFD · Volume of fluid · Bubbles · OpenFoam

1 Introduction

The two-phase flow problem emerges in many technical applications starting from production processes up to the energetics. The objective of this two-phase flow examination was motivated by the effort to understand and further control the bubble formation in the metal foaming processes. There are generally many technological production methods to obtain the cellular type structures e.g. the

This work was supported by the Ministry of Education, Youth and Sports of the Czech Republic and the European Union - European Structural and Investment Funds in the frames of Operational Programme Research Development and Education - project Hybrid Materials for Hierarchical Structures (HyHi, Reg. No. CZ.02.1.01/0.0/16-019/0000843).

J. M. F. Rodrigues et al. (Eds.): ICCS 2019, LNCS 11539, pp. 580–586, 2019.
https://doi.org/10.1007/978-3-030-22747-0_43

continuous gas injection in the melted materials containing mostly other stabilization components. Details about aspects of the cellular material production can be found e.g. in [1]. This paper is concerned with the study of rising gas bubble in stagnant liquid in the pursue of creating metal foam, where common parameters such as gas bubble diameters, velocity and dimensionless parameters and Morton number were examined. Other non-dimensional relevant parameters used in the problem of the bubble dynamics investigation can be found in [6]. To validate the numerical results, two different experimental examination patterns defined by the Morton numbers 1.6×10^{-11} and 5.7×10^{-9} were tested. It was found experimentally, that for the higher Morton number it was possible to trap the gas bubbles under the surface of the liquid, while for the latter it was not possible to trap any gas bubbles under the surface of the liquid fluid. The numerical approach based on the a multi-phase flow model which was built on the conservation laws of mass, momentum and energy as well as the gas-phase volume fraction advection equation was successfully adopted for the bubble formation problems [2]. In the study of bubbles in [3] involving the dynamics of a bubble, it was found that the deformation predicted using the numerical calculation was a good fit for the experimental results. It also confirmed the dependence of the bubble aspect ratio on Weber and Morton numbers for the cases of spherical and ellipse bubbles. CFD approach was successfully used for prediction of the bubble formation in the bubbly liquid metal flows. The chosen numerical method is an immersed boundary method extended to deformable bubbles. Experimental and numerical results were found to be in very good agreement both for the disperse gas phase and for the continuous liquid metal phase [5]. Other techniques used for the multicomponent two-phase fluid mixtures in a closed system at constant temperature can be found in [4].

2 Problem Formulation

Computational Domain. The computational domain is represented by the box L × B × H and it is illustrated in Fig. 1. The mesh is composed by hexahedral cells with spacing of $90 \times 90 \times 350$ and with the total number of 2.835 mil. cells. The periodic boundary conditions are applied on all sides. The inlet/outlet boundary type condition is prescribed for the top of a box having zero values for all phases except the gas phase for which the input/output velocity magnitudes are calculated. For the bottom of the gas phase, the parabolic inlet velocity profile with the U_{max} is prescribed. The mesh is generated by the SnappyHexMesh. Because the hexahedral cells are only applied for the mesh generation, the non-orthogonality stays to be zero, the max. skewness is very small 3.52×10^{-13} and max. aspect ratio is 2.34. The mesh quality is suitable in respect to the pimple solver requirements.

Mathematical Model and Numerical Approaches. The mathematical model used in the calculation is based on volume of fluid approach (VOF).

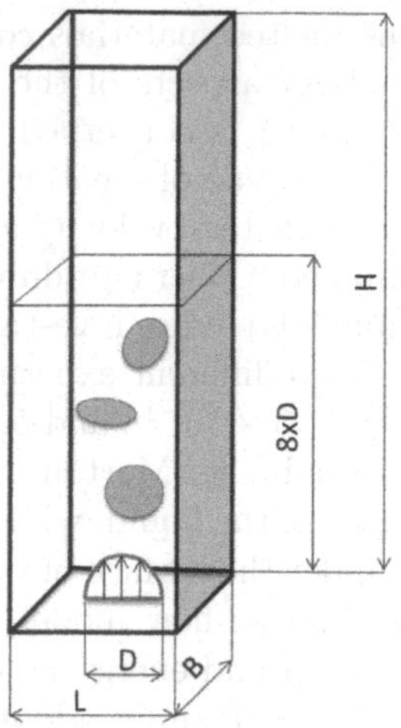

Fig. 1. Sketch of the computational domain with the size description.

Generally, the material properties of the homogeneous mixture is prescribed by the volume fraction function for the phase 1, which is defined as

$$\alpha_1 = \frac{\Omega_1}{\Omega_1 + \Omega_2} \tag{1}$$

where Ω_1 and Ω_2 are a volume of the phase 1 or 2, respectively. Analogously, the parameter for the phase 2 can be determined similarly or as simply $\alpha_2 = 1 - \alpha_1$. A conservation law of mass for each component separately must be also included. Furthermore, gravitational effects should be considered and included for liquid-phase wave problems as well. Finally, the underlying conservative part of the flow model can be expressed in the from 2.

$$\frac{\partial U}{\partial t} + \frac{\partial F_1}{\partial x} + \frac{\partial F_2}{\partial x} + \frac{\partial F_3}{\partial x} = G \tag{2}$$

where U is the vector of conservative variables, F is the flux function, and G are the source terms.

The calculation was carried out in OpenFOAM code using PIMPLE scheme. The Gauss linear scheme was applied for the most of all variables. The compressible Multiphase solver was used with varied time step determined by the stability condition based on the maximal Courant number of 0.25. The gas phase was treated as a ideal gas and the liquid phase as a perfect fluid.

Validation by Experiments. To validate numerical results, the experiments were carried out. The experiment test rig contains a glass water tank of dimensions 30 × 20 × 20 cm, that has a nozzle connected to a flow meter and an air compressor with controllable flow rate. A source of light is used to focus on the gas bubbles of which images are acquired by the means of the speed camera. A sheet of parchment paper is used as a light diffuser to decrease the contrast of the images making it possible to see the gas/liquid boundary in the bubble, using image processing software. It was possible to measure diameter of the bubble, velocity and several dimensionless parameters for several image frames, which later was imported into spreadsheet software to calculate changes through distance or time.

3 Results

Figure 2 shows numerically identified air bubble in the water formed at the nozzle located at the bottom and these bubbles were rising up due to force affects. The asymmetrical path of the bubbles is evident despite of the fact that the conditions are numerically symmetric. Similar feature of the bubble path is possible to observe experimentally. The deformation of the bubble shapes due to force acting is similar for experiment and CFD. Different calculated bubble shapes is possible to see on Fig. 2 (left) which correspond to shapes in experiment. However, the exactly match regarding bubble shapes in time and space is not possible to see it because of different time of the bubble formation.

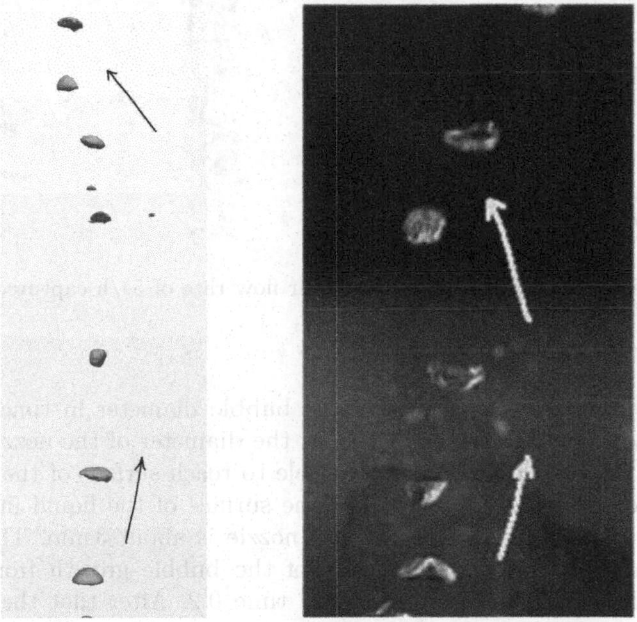

Fig. 2. Path of the bubbles for the air flow rate of 5 l/h examined numerically (left) and experimentally (right).

Figure 3 depicts the time series of bubble visualization in the liquid flow to help to identify the bubble dynamics. The air flow rate is about 5 l/h and the time step is 0.3 s. The bubble shapes change by time, it rotates and inclines. The flow of the liquid is illustrated by the vectors. At the particular position of the bubble, it is possible to observe the strong flow circulation of the liquid around the bubble. This circulating flow appears due to force imbalance acting on the bubble. The mesh used for calculations provides at least 10 points over the bubble diameter to resolve the shape of the bubbles sufficiently (this conclusion is based on the previous test calculations). The similar forms of the bubble identified by numerics were found by experiments as well.

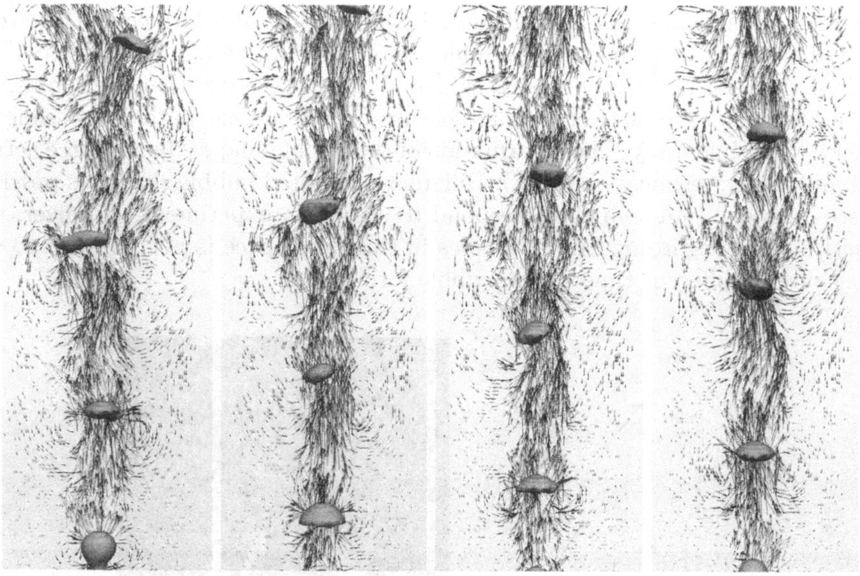

Fig. 3. Unsteady bubble dynamics for the air flow rate of 5 l/h captured by the time step of 0.3 s.

Figure 4 illustrates the change of the bubble diameter in time for the gas flow rate 5 l/h. The diameter is scaled by the diameter of the nozzle. The time is scaled by the total time needed a bubble to reach surface of the liquid level. Time required by one bubble to reach the surface of the liquid in the tank is about 0.377 s, while the diameter of the nozzle is about 3 mm. The numerical calculation overestimated the intensity of the bubble growth from the onset of the bubble up to the non-dimensional time 0.2. After that the bubble size oscillated about the non-dimensional diameter of 3. A growth of the bubble observed experimentally is significantly slower. The unique evaluation of the bubble diameters and Reynolds numbers were possible only for the cases in which the single bubbles are formed and travelled separately without interactions.

Figure 5 shows two different bubble shapes formed in the case in which the air flow rate reaches 130 l/h. In this case, no single travelling bubble is formed, but

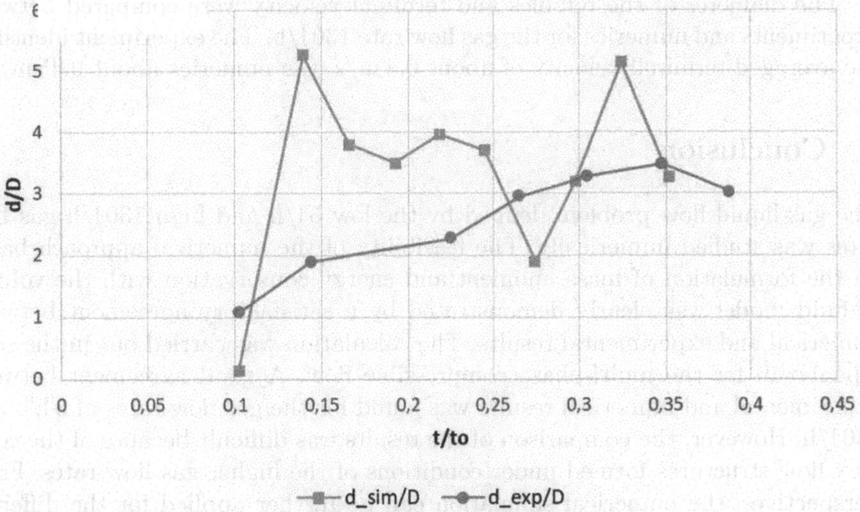

Fig. 4. Diameter of the formed bubbles in time.

the big bubble is created at the bottom which quickly moves upwards. It interacts with the previous bubbles to be formed into the big structure. To observe the dynamics of the bubble is difficult because of the complexity of the flow. At the onset of the bubble different bubble shapes are formed before detachment from the input nozzle. The numerical simulation predicted bubble shapes well, however, without typical unsymmetrical feature observed experimentally. After the onset of the bubble and its detaching, the bubble starts to grow. This feature was observed experimentally and numerically, however, numerical results embodied slow growing process. This observed feature was not still further studied in details.

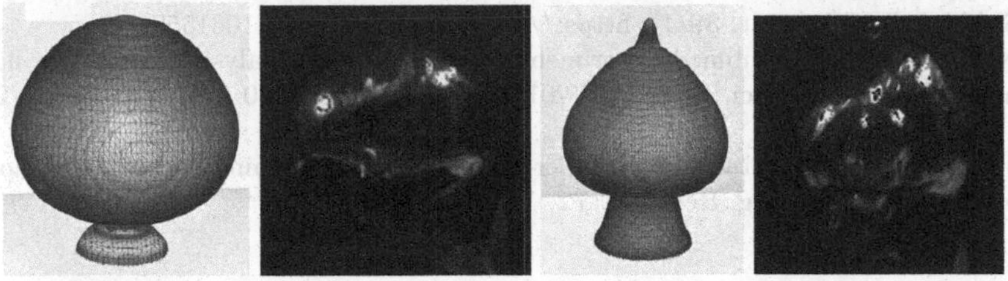

Fig. 5. Unsteady bubble dynamics for the air flow rate of 130 l/h for different time examined numerically and experimentally (figures by the black background).

The diameter of the bubbles and terminal velocity were compared between experiments and numerics for the gas flow rate 130 l/h. The experiment identified the averaged terminal velocity of about 0.4 m/s, the numerics about 0.45 m/s.

4 Conclusion

The gas-liquid flow problem defined by the low 5 l/h and high 130 l/h gas flow rates was studied numerically. The feasibility of the numerical approach based on the formulation of mass, moment and energy conservation with the volume of fluid model was clearly demonstrated by a satisfactory agreement between numerical and experimental results. The calculation was carried out in the code OpenFoam for the multi-phase compressible flow. A good agreement between experimental and numerical results was found for the gas flow rates of 5 l/h and 130 l/h. However, the comparison of the results was difficult because of the complex flow structures formed under conditions of the higher gas flow rates. From perspectives, the numerical simulation can be further applied for the different liquid phase with modified material properties (addition of the suitable chemical components) in order to find the impact of this material property change on the bubble formation and dynamics.

References

1. Rajak, D.K., Kumaraswamidhas, L.A., Das, S.: Technical overview of aluminum alloy foam. Mater. Sci. **48**, 68–86 (2007)
2. Ma, Z.H., Causon, D.M., Qian, L., Mingham, C.G., Gu, H.B., Martínez Ferrer, P.: A compressible multiphase flow model for violent aerated wave impact problems. Proc. R. Soc. A **470** (2017). https://doi.org/10.1098/20140542
3. Gumulya, M., Joshi, J.B., Utikar, R.P., Evans, G.M., Pareek, V.: Bubbles in viscous liquids: time dependent behaviour and wake characteristics. Chem. Eng. Sci. **144**, 298–309 (2016)
4. Fan, X., Kou, J., Qiao, Z., Sun, S.: A Componentwise convex splitting scheme for diffuse interface models with Van der Waals and Peng-Robinson equations of state. SIAM J. Sci. Comput. **39**(1). https://doi.org/10.1137/16M1061552
5. Krull, B., et al.: Combined experimental and numerical analysis of a bubbly liquid metal flow. Mater. Sci. Eng. **228** (2017). https://doi.org/10.1088/1757-899X/228/1/012006
6. Jamialahmadi, M., Branch, C., Muller-Steinhagen, H.: Terminal bubble rise velocity in liquids. Chem. Eng. Res. Des. **72**, 119–122 (1994)

Performance of a Two-Path Aliasing Free Calculation of a Spectral DNS Code

Mitsuo Yokokawa[1(✉)], Koji Morishita[2], Takashi Ishihara[3], Atsuya Uno[4], and Yukio Kaneda[5]

[1] Graduate School of System Informatics, Kobe University, Kobe 657-0013, Japan
yokokawa@port.kobe-u.ac.jp
[2] SOUM Corporation, Shibuya-ku, Tokyo 151-0072, Japan
[3] Graduate School of Environmental and Life Science, Okayama University, Okayama 700-8530, Japan
[4] RIKEN Center for Computational Science, Kobe 650-0047, Japan
[5] Center for General Education, Aichi Institute of Technology, Toyota 470-0392, Japan

Abstract. A direct numerical simulation (DNS) code was developed for solving incompressible homogeneous isotropic turbulence with high Reynolds numbers in a periodic box using the Fourier spectral method. The code was parallelized using the Message Passing Interface and OpenMP with two-directional domain decomposition and optimized on the K computer. High resolution DNSs with up to 12288^3 grid points were performed on the K computer using the code. Efficiencies of 3.84%, 3.14%, and 2.24% peak performance were obtained in double precision DNSs with 6144^3, 8192^3, and 12288^3 grid points, respectively. In addition, a two-path alias-free procedure is proposed and clarified its effectiveness for some number of parallel processes.

Keywords: Parallel computation · Fourier spectral method · Two-path de-aliasing method · DNS · K computer · Incompressible turbulence

1 Introduction

Recent rapid developments in the capacities and capabilities of supercomputers have led to computational approaches becoming powerful tools in turbulence studies. In particular, direct numerical simulations (DNSs) of turbulence can provide detailed turbulence data that are free from experimental uncertainties under well-controlled conditions.

In 2002, a highly efficient DNS code for incompressible turbulence based on the Fourier spectral method was developed for the Earth Simulator (ES), which was the fastest supercomputer at that time [1]. High-resolution DNSs of incompressible homogeneous isotropic turbulence with up to 4096^3 grid points and the Taylor scale Reynolds number $R_\lambda \sim 1200$ were performed on the ES

J. M. F. Rodrigues et al. (Eds.): ICCS 2019, LNCS 11539, pp. 587–595, 2019.
https://doi.org/10.1007/978-3-030-22747-0_44

using the code [2,3]. We developed a new DNS code on the basis of the original spectral code used on the ES, and optimized it on the K computer [4] in 2013 to realize larger scale DNSs of turbulence with a higher Reynolds number (*Re*) than those reported so far. DNSs with up to 12288^3 grid points were carried out on the K computer and new results on the power law were obtained [5].

Though recent high-performance computer systems are characterized by a huge number of cores and a large amount of memory capacity, they are still insufficient for performing larger DNSs. Therefore, we have to consider an efficient algorithm for DNS code using the Fourier spectral method.

In this paper, the performance of the code evaluated on the K computer is described, and a parallel de-aliasing method, two-path aliasing free calculation, is also presented for the calculation of nonlinear terms of the ordinary differential equations (ODEs) discretized from the Navier-Stokes equations and continuum equation.

2 DNS Code Implementation

We consider homogeneous isotropic turbulence that obeys the incompressible Navier-Stokes equations and the continuum equation

$$\frac{\partial \boldsymbol{u}}{\partial t} + (\boldsymbol{u} \cdot \nabla)\boldsymbol{u} = -\nabla p + \nu \nabla^2 \boldsymbol{u} + \boldsymbol{f}, \quad \text{and} \tag{1}$$

$$\nabla \cdot \boldsymbol{u} = 0, \tag{2}$$

in a periodic box of side length 2π. Here $\boldsymbol{u} = (u_1, u_2, u_3)$, p, ν, and $\boldsymbol{f} = (f_1, f_2, f_3)$ denote velocity, pressure, kinematic viscosity, and external force, respectively (see [3] for details of the external force). Since boundary condition is periodic, the Fourier spectral method can be applied for discretization.

Then, Eq. (1) are written as

$$\left(\frac{d}{dt} + \nu|\boldsymbol{k}|^2\right) \hat{u_i}(\boldsymbol{k}) = \left(\delta_{ij} - \frac{k_i k_j}{|\boldsymbol{k}|^2}\right) \hat{h_j}(\boldsymbol{k}) + \hat{f_i}(\boldsymbol{k}), \tag{3}$$

where Eq. (2) is used to eliminate the pressure term. In Eq. (3), $\boldsymbol{k} = (k_1, k_2, k_3)$ is the wavenumber vector, $\hat{h_j}(\boldsymbol{k}) = ik_l\widehat{u_j u_l}$ is the nonlinear term, and the hat $\hat{}$ denotes the Fourier coefficients. The ODEs (3) are integrated with respect to time by the fourth-order Runge-Kutta-Gill method.

In calculating the nonlinear term $\hat{h_j}(\boldsymbol{k})$ which are expressed as convolution sums in the wave vector space, the fast Fourier transforms (FFT) can be efficiently used as the left normal series in Fig. 1(a). Aliasing errors generated by the spectral transform method are removed by the so-called phase shift method (Fig. 1) and the cut-off beyond the maximum wavenumber $k_{\max} \equiv (\sqrt{2}/3)N$, where N is the number of grid points in each of the Cartesian coordinates in the physical space.

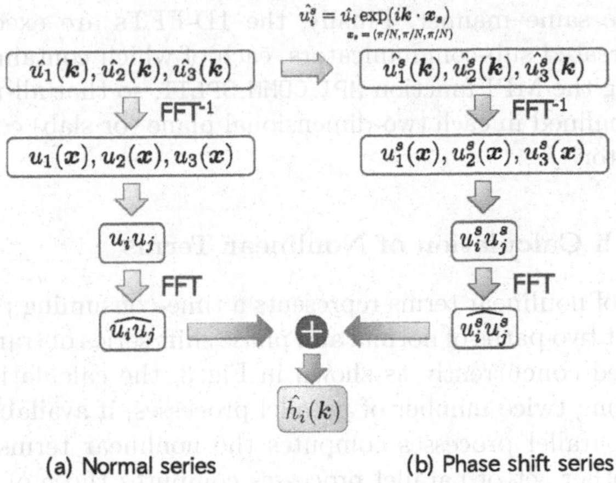

Fig. 1. Phase shift method

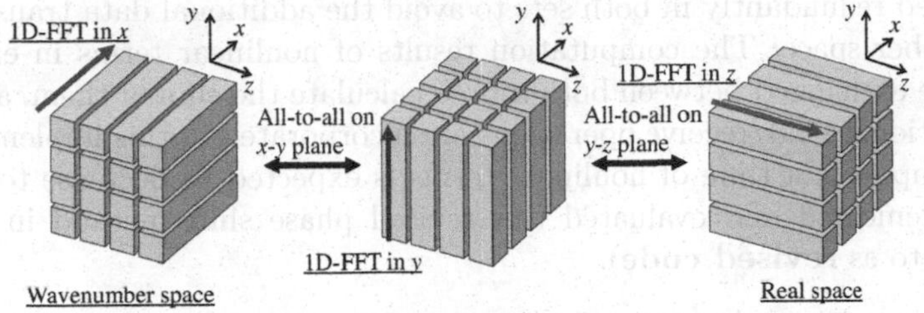

Fig. 2. Schematic diagram of the 3D-FFT procedure used in the base code.

2.1 Code Parallelization

In DNS using the spectral method, most of the computation time is associated with evaluating the alias-free nonlinear terms, and the three-dimensional FFTs (3D-FFTs) are used in the evaluation, which requires data transfer among all parallel processes in the parallel computation.

We have developed a parallel DNS code (referred to as **base code**) in Fortran 90 using the Message Passing Interface (MPI) and OpenMP for parallel processing, and we optimized this code for the K computer. Two-directional domain decomposition or pencil decomposition for MPI parallelization is applied for the data distribution and the FFTW library [6] is used in the implementation.

Figure 2 shows a parallel 3D-FFT procedure used in the code. The complex valued data in the wavenumber space are partitioned among MPI processes in y- and z-directions. After executing the one-dimensional FFTs (1D-FFTs) in the x-direction, the domain decomposition changes to the z- and x-directions by performing all-to-all communications for the x-y plane processes. After executing the 1D-FFTs in the y-direction, the domain decomposition changes to the x- and y-directions by performing all-to-all communications for the y-z plane processes in the same manner. Finally, the 1D-FFTs are executed in the z-direction. We created sub-communicators, each of which contains several x-y or y-z planes, using the MPI function `MPI_COMM_SPLIT`, so that all-to-all communication can be confined in each two-dimensional plane (or slab) contained in each sub-communicator.

2.2 Two-Path Calculation of Nonlinear Terms

The evaluation of nonlinear terms represents a time-consuming part in the code. Considering that two paths of normal and phase shift series of transform methods can be calculated concurrently as shown in Fig. 3, the calculation time can be shortened by using twice number of parallel processes, if available.

One set of parallel processes computes the nonlinear terms of the normal series and the other set of parallel processes computes those of the phase shift series. In this case, the number of parallel processes are equal in both sets. Computation parts of the base code other than the computation of nonlinear terms is executed redundantly in both sets to avoid the additional data transfer in the wavenumber space. The computation results of nonlinear terms in each series should be exchanged between both sets to calculate the sum of them, and therefore additional send/receive operations are incorporated in this implementation. Thus computation time of nonlinear terms is expected to decrease to half. We also implemented and evaluated this revised phase shift method in the code (referred to as **revised code**).

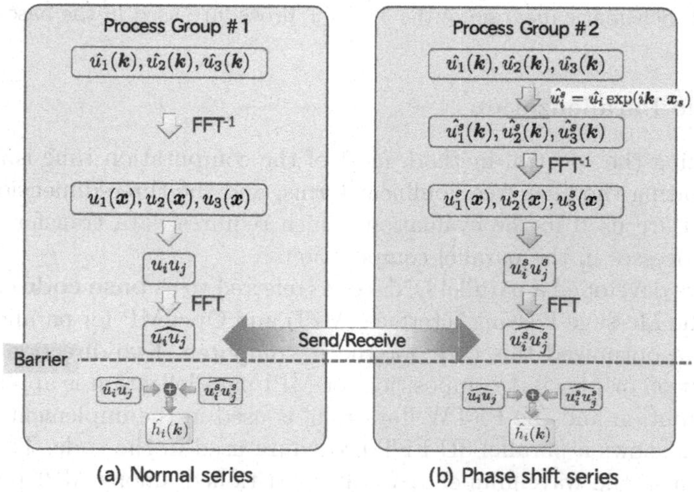

Fig. 3. Two-path phase shift method

3 Simulation Results and Code Performance

We performed DNSs of incompressible turbulence in the periodic boundary box with grid points $N^3 = 4096^3$, 6144^3, 8192^3, and 12288^3 on the K computer. In this section, brief results were followed by performance evaluations of the newly developed code. An extensive and detailed analysis of the DNSs results from the perspective of turbulence were reported in [5].

3.1 Simulation Results

Calculated velocities in the box were used to analyze the vortex dynamics in high Reynolds (*Re*) turbulent flows. Figure 4 depicts an iso-surface of large vorticity area $|\omega| = 6\langle\omega^2\rangle^{1/2}$ in the DNS with grid points $N^3 = 12288^3$ at an approximately statistical steady state. Similar figures are seen in Ref. [7]. This figure was drawn by an originally developed visualization code based on the marching cubes algorithm [8]. Visualization data can be generated in part with a few layers of grid points in the box and a visualization processing of a batch of layers can be applied independently to the other batch. Therefore the visualization can be carried out in parallel and on computational resources which are not much in performance.

The figure shows that there are tube-like vortex-clusters at various sizes from some integral length scale L down to several dozen Taylor micro scale λ. The lengths L and λ are shown in Fig. 4. It also shows that distinct vortex-cluster areas and void (vorticity-free) areas coexist in high *Re* turbulence.

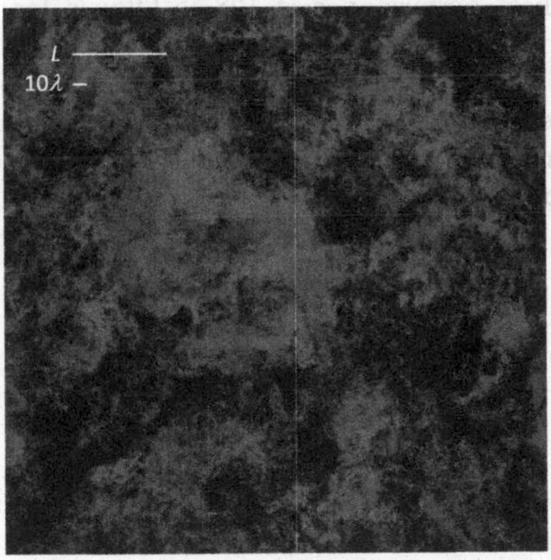

Fig. 4. Isosurface of vorticity $|\omega| = 6\langle\omega^2\rangle^{1/2}$.

Table 1. Sustained performance of the base code on the K computer.

N^3	Number of compute nodes	Performance (Tflop/s)	Efficiency (%)
4096^3	2,048 (64 × 32)	10.80	4.12
	4,096 (64 × 64)	12.48	2.38
6144^3	6,144 (96 × 64)	30.16	3.84
	12,288 (96 × 128)	44.03	2.80
8192^3	8,192 (128 × 64)	32.76	3.14
	16,384 (128 × 128)	42.01	2.00
12288^3	24,576 (192 × 128)	70.52	2.24
	49,152 (384 × 128)	128.45	2.04

3.2 Computational Performance

We measured the sustained performance of the base code using the hardware counter on the K computer. An MPI process is assigned to each compute node, and therefore the number of MPI processes equals to the number of compute nodes.

Table 1 shows the sustained performance of the code calculated in double precision with $N^3 = 4096^3$, 6144^3, 8192^3, and 12288^3 grid points. Here, note that the interconnect network of the K computer has a Tofu basic unit consisting of 12 compute nodes and 1D-FFT for grid points with a multiple of 3 is efficiently carried out. Thus, we took 6144^3 and 12288^3 grid points which are multiples of 3 into consideration.

We observe that performance (Tflop/s) increases with the number of compute nodes, and the highest performance is 128.45 Tflop/s for $N^3 = 12288^3$ with 49,152 compute nodes. The maximum efficiency is 4.12%, which was achieved for $N^3 = 4096^3$ using the smallest number of compute nodes (2,048 compute nodes). Higher efficiency is obtained with fewer compute nodes for the same value of N^3, because the ratio of communication time to calculation time increases as the number of compute nodes increases.

3.3 Computation Time of the Revised Code

Computation time of the base and revised codes were measured for the DNS with 768^3 grid points and 100 time steps. Three cases were considered, as described below.

Case 1 Computation time of the base code was measured with 96 MPI processes on 96 compute nodes of the K computer.

Case 2 Computation time of the revised code with 192 MPI processes on 192 compute nodes. These MPI processes were divided into two sets and the computation of normal series and shifted series is assigned to each set. The message size of each all-to-all communication in 3D-FFT is the same as that

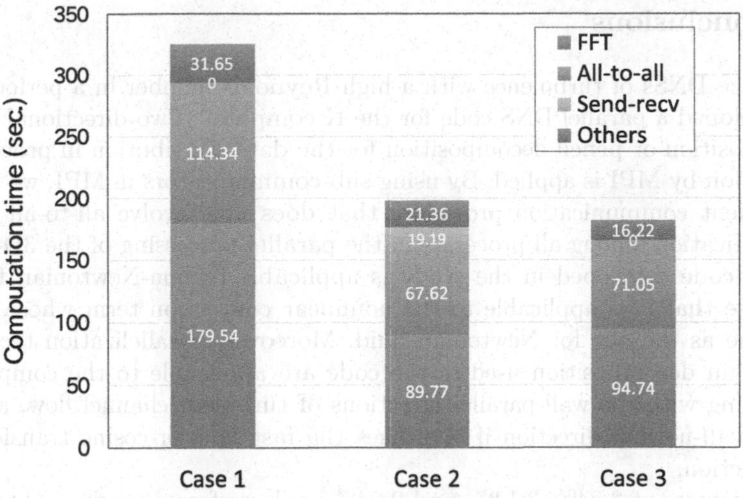

Fig. 5. Computation time of the three cases

in case 1 but additional send/receive operation for the exchange of nonlinear term values occurs between the two sets.

Case3 Computation time of the base code with 192 MPI processes. The message size of each all-to-all communication is halved but the number of counter nodes to be communicated in all-to-all communications is doubled.

Figure 5 shows the computation time for all cases. The computation time associated with case 2 is approximately half that of case 1 and additional send/receive time (approximately 10% of the total time) is observed. The revised phase shift method appears to work well. Further, computation time associated with case 3 (i.e. base code with double amount of MPI processes compared to case 1) is also half of that of case 1 and less than that of case 2. No time is spent on send/receive operations in case 3. To some extent the base code exhibits strong scaling as the number of MPI processes increases, and therefore the computation time of case 3 is half of that associated with case 1.

If a sufficient number of compute nodes can be used, and if an appropriate number of DNS parallelisms corresponding to the number of compute nodes for a given number of grid points can be found, the base code exhibits better performance than the revised code. However, where there are a large number of compute nodes available and no more parallelisms can be extracted from a DNS for a given number of grid points, which is substantially less than the number of compute nodes, the revised phase shift method appears to be useful. For example, DNS with 128^3 grid points normally has up to 128^2 parallelisms where pencil decomposition is carried out on a parallel system with greater than 256^2 nodes, and the base code cannot be assigned properly in this case but the revised two-path phase shift method can work efficiently.

4 Conclusions

To realize DNSs of turbulence with a high Reynolds number in a periodic box, we developed a parallel DNS code for the K computer. Two-directional domain decomposition or pencil decomposition for the data distribution in process parallelization by MPI is applied. By using sub-communicators in MPI, we realized an efficient communication procedure that does not involve all-to-all process communication among all processes in the parallel processing of the 3D-FFT.

The code developed in the study is applicable to non-Newtonian fluid, in the sense that it is applicable to the nonlinear convection term whose form is the same as the one for Newtonian fluid. Moreover, parallelization techniques by domain decomposition used in the code are applicable to the computation concerning with the wall-parallel directions of turbulent channel flow, and also to the wall-normal direction if one takes the fast Fourier-cosine transforms in the direction.

Efficiencies of 3.84%, 3.14%, and 2.24% peak performance were obtained in double precision calculations of DNSs with 6144^3, 8192^3, and 12288^3 grid points, respectively, on the K computer. Using the new code, we performed DNSs of turbulence on a larger scale than hitherto reported in the literature, specifically, with respect to the Reynolds number. This led to attaining a higher Reynolds number ($R_\lambda \sim 2300$, $Re \sim 1.5 \times 10^5$) than has thus far been reported at an approximately statistical steady state (see [5]).

A two-path de-aliasing calculation technique of the phase shift method for the alias-free calculation of nonlinear terms was also proposed, and its efficiency was explored for the case where there are many compute nodes relative to the number of DNS parallelisms with less grid points than parallel processes. Next-generation supercomputers will have many more cores compared to those of contemporary high-performance systems, thus computation time can be reduced by two-path de-aliasing technique.

Acknowledgments. This work was supported in part by JSPS KAKENHI Grant Number (C) 26390130 and obtained by using computational resources of the K computer provided by the RIKEN Advanced Institute for Computational Science through the HPCI System Research projects (Project IDs: hp140135, hp150174, hp160102, hp170087, hp180109, and hp190076). This work is also supported partially by MEXT Next Generation High-Performance Computing Infrastructures and Applications R&D Program, entitled R&D of a Quantum-Annealing-Assisted Next Generation HPC Infrastructures and Its Applications.

References

1. Yokokawa, M., Itakura, K., Uno, A., Ishihara, T., Kaneda, Y.: 16.4-Tflops direct numerical simulation of turbulence by a Fourier spectral method on the Earth Simulator. In: Proceedings of ACM/IEEE Conference on Supercomputing, pp. 1–17. IEEE, Baltimore (2002). https://doi.org/10.1109/SC.2002.10052

2. Kaneda, Y., Ishihara, T., Yokokawa, M., Itakura, K., Uno, A.: Energy Dissipation rate and energy spectrum in high resolution direct numerical simulations of turbulence in a periodic box. Phys. Fluids **15**(2), L21–L24 (2003). https://doi.org/10.1063/1.1539855
3. Ishihara, T., Kaneda, Y., Yokokawa, M., Itakura, K., Uno, A.: Small-scale statistics in high-resolution direct numerical simulation of turbulence: Reynolds number dependence of one-point velocity gradient statistics. J. Fluid Mech. **592**, 335–366 (2007). https://doi.org/10.1017/S0022112007008531
4. Yokokawa, M., Shoji, F., Uno, A., Kurokawa, M., Watanabe, T.: The K computer: Japanese next-generation supercomputer development project. In: IEEE/ACM International Symposium on Low Power Electronics and Design, pp. 371–372. IEEE, Fukuoka (2011). https://doi.org/10.1109/ISLPED.2011.5993668
5. Ishihara, T., Morishita, M., Yokokawa, M., Uno, A., Kaneda, Y.: Energy spectrum in high-resolution direct numerical simulations of turbulence. Phys. Rev. Fluids **1**(8), 082403 (2016). https://doi.org/10.1103/PhysRevFluids.1.082403
6. Frigo, M., Johnson, S.G.: The Design and Implementation of FFTW3. Proc. IEEE **93**(2), 216–231 (2005). https://doi.org/10.1109/JPROC.2004.840301
7. Ishihara, T., Uno, A., Morishita, M., Yokokawa, M., Kaneda, Y.: Vortex clusters and their time evolution in high-Reynolds-number turbulence. NAGARE **35**(2), 109–113 (2016). (in Japanese)
8. Lorensen, W.E., Cline, H.E.: Marching cubues: a high resolution 3D surface construction algorithm. Comput. Graph. **21**(4), 163–169 (1987). https://doi.org/10.1145/37402.37422

DNS of Mass Transfer from Bubbles Rising in a Vertical Channel

Néstor Balcázar-Arciniega[1,2(✉)], Joaquim Rigola[1], and Assensi Oliva[1]

[1] Heat and Mass Transfer Technological Center (CTTC), Universitat Politècnica de Catalunya-BarcelonaTech (UPC). ETSEIAT, Colom 11, 08222 Barcelona, Spain
nestorbalcazar@yahoo.es, {nestor,cttc}@cttc.upc.edu

[2] Termo Fluids S.L., Avda Jacquard 97 1-E, 08222 Terrassa (Barcelona), Spain
http://www.cttc.upc.edu/, http://www.termofluids.com/

Abstract. This work presents Direct Numerical Simulation of mass transfer from buoyancy-driven bubbles rising in a wall-confined vertical channel, through a multiple markers level-set method. The Navier-Stokes equations and mass transfer equation are discretized using a finite volume method on a collocated unstructured mesh, whereas a multiple markers approach is used to avoid the numerical coalescence of bubbles. This approach is based on a mass conservative level-set method. Furthermore, unstructured flux-limiter schemes are used to discretize the convective term of momentum equation, level-set advection equations, and mass transfer equation, to improve the stability of the solver in bubbly flows with high Reynolds number and high-density ratio. The level-set model is used to research the effect of bubble-bubble and bubble-wall interactions on the mass transfer from a bubble swarm rising in a vertical channel with a circular cross-section.

Keywords: Mass transfer · Bubbly flow · Vertical channel · Flux-limiters · Unstructured meshes · Level-set method · Finite volume method · High-Performance Computing

---

This work has been financially supported by the *Ministerio de Economía y Competitividad, Secretaría de Estado de Investigación, Desarrollo e Innovación* (MINECO), Spain (ENE2015-70672-P), and by Termo Fluids S.L. Néstor Balcázar acknowledges financial support of the *Programa Torres Quevedo, Ministerio de Economía y Competitividad, Secretaría de Estado de Investigación, Desarrollo e Innovación* (PTQ-14-07186), Spain. Three-dimensional simulations were carried out using computing time awarded by PRACE 14th-Call (project 2016153612), and RES (project $FI-2018-1-0025$), on the supercomputer MareNostrum IV based in Barcelona, Spain. Computational resources provided by RES (project $FI - 2018 - 3 - 0037$) on the supercomputer Altamira, supported by Santander Supercomputing group at the University of Cantabria (IFCA-CSIC), are acknowledged.

J. M. F. Rodrigues et al. (Eds.): ICCS 2019, LNCS 11539, pp. 596–610, 2019.
https://doi.org/10.1007/978-3-030-22747-0_45

1 Introduction

Mass transfer in bubbly flows is a ubiquitous phenomenon in natural and industrial applications. For example, bubble columns are used in chemical engineering to promote chemical reactions, as well as to improve heat and mass transfer rates. Therefore, understanding this phenomenon has both practical and scientific motivations. As a complement to theoretical and experimental approaches, the development of supercomputers has promoted High-Performance computing (HPC) and Direct Numerical Simulation (DNS) of Navier-Stokes equations, as another methodology to design non-invasive and controlled numerical experiments of bubbly flows. Indeed, during the last decades multiple numerical methods have been introduced for DNS of two-phase flows: volume-of-fluid (VOF) methods [26], level-set (LS) methods [33,36], conservative level-set (CLS) methods [4,32], front tracking (FT) methods [42], and hybrid VOF/LS methods [7,37,39]. Furthermore, some of these numerical approaches have been extended to include heat transfer or mass transfer phenomenon in two-phase flows [3,12,14,15,20,21]. On the other hand, few works have reported DNS of mass transfer in bubble swarms [2,12,29,35]. Although previous publications have researched mass transfer from bubbles rising on unconfined domains by using VOF, LS, VOF/LS, and FT methods, there are no previous studies in the context of wall-confined vertical columns and CLS method. Therefore, this work aims to present a numerical study of mass transfer from bubbles rising in a vertical pipe, in the framework of a multiple-marker CLS methodology introduced by [5,9,12]. As further advantages, the CLS method [4,12] was designed for three-dimensional collocated unstructured meshes, whereas the accumulation of mass conservation error inherent to standard level-set methods is circumvented. Moreover, unstructured flux-limiters schemes as first introduced in [4,8,12], are used to discretize convective terms of transport equations, in order to avoid numerical oscillations around discontinuities and to minimize the so-called numerical diffusion. This numerical approach has demonstrated to improve the numerical stability of the unstructured multiphase solver [4–8,12] for bubbly flows with high Reynolds number and high-density ratio.

This paper is organized as follows: The mathematical model and numerical methods are reviewed in Sect. 2. Numerical experiments are presented in Sect. 3. Concluding remarks and future work are discussed in Sect. 4.

2 Mathematical Model and Numerical Methods

2.1 Incompressible Two-Phase Flow

The Navier-Stokes equations for the dispersed fluid (Ω_d) and continuous fluid (Ω_c) are introduced in the framework of the so-called one-fluid formulation [42], which includes a singular source term for the surface tension force at the interface Γ [4,12,42]:

$$\frac{\partial}{\partial t}(\rho \mathbf{v}) + \nabla \cdot (\rho \mathbf{v}\mathbf{v}) = -\nabla p + \nabla \cdot \mu (\nabla \mathbf{v}) + \nabla \cdot \mu (\nabla \mathbf{v})^T + (\rho - \rho_0)\mathbf{g} + \mathbf{f}_\sigma, \quad (1)$$

$$\nabla \cdot \mathbf{v} = 0, \tag{2}$$

where $\mathbf{v}$ is the fluid velocity, p denotes the pressure field, $\mathbf{g}$ is the gravitational acceleration, ρ is the fluid density, μ is the dynamic viscosity, $\mathbf{f}_\sigma$ is the surface tension force per unit volume concentrated at the interface, subscripts d and c denote the dispersed phase and continuous phase respectively. Physical properties are constant at each fluid-phase with a jump discontinuity at Γ:

$$\rho = \rho_d H_d + \rho_c H_c, \; \mu = \mu_d H_d + \mu_c H_c. \tag{3}$$

Here H_c is the Heaviside step function that is one at fluid c (Ω_c) and zero elsewhere, whereas $H_d = 1 - H_c$. At discretized level a continuous treatment of physical properties is adopted in order to avoid numerical instabilities around Γ. The force $-\rho_0 \mathbf{g}$ included in Eq. (1), with $\rho_0 = V_\Omega^{-1} \int_\Omega (\rho_d H_d + \rho_c H_c)\, dV$, avoids the acceleration of the flow field in the downward vertical direction, when periodic boundary conditions are applied on the y–axis (aligned to $\mathbf{g}$) [5,9,12,22].

2.2 Multiple Marker Level-Set Method and Surface Tension

The conservative level-set method (CLS) introduced by [4,8,12] for interface capturing on three-dimensional unstructured meshes, is used in this work. Furthermore, the multiple markers approach [5,19] as introduced in [5,8,12] for the CLS method, is employed to avoid the so-called numerical coalescence inherent to standard interface capturing methods. In this context, each bubble is represented by a CLS function [5,8,9,12], whereas the interface of the ith fluid particle is defined as the 0.5 iso-surface of the CLS function ϕ_i, with $i = 1, 2, ..., n_d$ and n_d defined as the total number of bubbles in Ω_d. Since the incompressibility constraint (Eq. 2), the ith interface transport equation can be written in conservative form as follows:

$$\frac{\partial \phi_i}{\partial t} + \nabla \cdot \phi_i \mathbf{v} = 0, \; i = 1, .., n_d. \tag{4}$$

Furthermore, a re-initialization equation is introduced to keep a sharp and constant CLS profile on the interface:

$$\frac{\partial \phi_i}{\partial \tau} + \nabla \cdot \phi_i (1 - \phi_i) \mathbf{n}_i^0 = \nabla \cdot \varepsilon \nabla \phi_i, \; i = 1, .., n_d. \tag{5}$$

where $\mathbf{n}_i^0$ denotes $\mathbf{n}_i$ at $\tau = 0$. This equation is advanced in pseudo-time τ up to achieve the steady state. It consists of a compressive term, $\phi_i(1 - \phi_i)\mathbf{n}_i^0$, which forces the CLS function to be compressed onto the interface along the normal vector $\mathbf{n}_i$. Furthermore, the diffusive term, $\nabla \cdot \varepsilon \nabla \phi_i$, keeps the level-set profiles with characteristic thickness $\varepsilon = 0.5h^{0.9}$, where h is the grid-size [4,8,12]. Geometrical information at the interface, such as normal vectors $\mathbf{n}_i$ and curvatures κ_i, are computed from the CLS function:

$$\mathbf{n}_i(\phi_i) = \frac{\nabla \phi_i}{\|\nabla \phi_i\|}, \; \kappa_i(\phi_i) = -\nabla \cdot \mathbf{n}_i, \; i = 1, .., n_d. \tag{6}$$

Surface tension forces are calculated by the continuous surface force model [16], extended to the multiple marker CLS method in [5,8,9,12]:

$$\mathbf{f}_\sigma = \sum_{i=1}^{n_d} \sigma \kappa_i(\phi_i) \nabla \phi_i. \tag{7}$$

where σ is the surface tension coefficient. Finally, in order to avoid numerical instabilities at the interface, fluid properties in Eq. (3) are regularized by using a global level-set function ϕ [5,8,12], defined as follows:

$$\phi = min\{\phi_1, ..., \phi_{n_d}\}. \tag{8}$$

Thus, Heaviside functions presented in Eq. (3) are regularized as $H_d = 1-\phi$ and $H_c = \phi$. In this work $0 < \phi \leq 0.5$ for Ω_d, and $0.5 < \phi \leq 1$ for Ω_c. On the other hand, if $0.5 < \phi \leq 1$ for Ω_d and $0 < \phi \leq 0.5$ for Ω_c, then $H_d = \phi$, $H_c = 1-\phi$, and $\phi = max\{\phi_1, ..., \phi_{n_d}\}$ [12]. Further discussions on the regularization of Heaviside step function and Dirac delta function, as used in the context of the CLS method, are presented in [12].

2.3 Mass Transfer

This research focuses on the simulation of external mass transfer from bubbles rising in a vertical channel. Therefore, a convection-diffusion-reaction equation is used as a mathematical model for the mass transfer of a chemical species in Ω_c, as first introduced in [12]:

$$\frac{\partial C}{\partial t} + \nabla \cdot (\mathbf{v}C) = \nabla \cdot (\mathcal{D} \nabla C) + \dot{r}(C), \tag{9}$$

where C is the chemical species concentration, $\mathcal{D}$ is the diffusion coefficient or diffusivity which is equal to $\mathcal{D}_c$ in Ω_c and $\mathcal{D}_d$ elsewhere, $\dot{r}(C) = -k_1 C$ denotes the overall chemical reaction rate with first-order kinetics, and k_1 is the reaction rate constant. In the present model, the concentration inside the bubbles is kept constant [2,12,20,35], whereas convection, diffusion and reaction of the mass dissolved from Ω_d exists only in Ω_c.

As introduced by [12], the concentration (C_P) at the interface cells is computed by linear interpolation, using information of the concentration field from Ω_c (excluding interface cells), and taking into account that the concentration at the interface C_Γ is constant. As a consequence, the concentration at the interface is imposed like a Dirichlet boundary condition, whereas Eq. (9) is computed in Ω_c.

2.4 Numerical Methods

The transport equations are solved with a finite-volume discretization on a collocated unstructured mesh, as introduced in [4,8,12]. For the sake of completeness, some points are reviewed in this manuscript. The convective term of momentum equation (Eq. (1)), CLS advection equation (Eq. (4)), and mass transfer equation for chemical species (Eq. (9)), is explicitly computed approximating the fluxes at cell faces with a Total Variation Diminishing (TVD) Superbee flux-limiter scheme proposed in [4,12]. Diffusive terms of transport equations are centrally differenced [12], whereas a distance-weighted linear interpolation is used to find the cell face values of physical properties and interface normals, unless otherwise stated. Gradients are computed at cell centroids by means of the least-squares method using information of the neighbor cells around the vertexes of the current cell (see Fig. 2 of [4]). For instance at the cell Ω_P, the gradient of the variable $\psi = \{v_j, C, \phi_i, \phi_i(1-\phi_i), ...\}$ is calculated as follows:

$$(\nabla\psi)_P = (\mathbf{M}^T\mathbf{W}\mathbf{M})^{-1}\mathbf{M}^T\mathbf{W}\mathbf{Y}, \tag{10}$$

$\mathbf{M}$ and $\mathbf{Y}$ are defined as introduced in [4], $\mathbf{W} = \text{diag}(w_{P\to1}, .., w_{P\to n})$ is the weighting matrix [28,31], defined as the diagonal matrix with elements $w_{P\to k} = \{1, ||\mathbf{x}_P - \mathbf{x}_k||^{-1}\}$, $k = \{1, .., n\}$, and subindex n is the number of neighbor cells. The impact of the selected weighting coefficient ($w_{P\to k}$) on the simulations is evaluated in Sect. 3.1. The compressive term of the re-initialization equation (Eq. (5)) is discretized by a central-difference scheme [12]. The resolution of the velocity and pressure fields is achieved by using a fractional-step projection method [18]. In the first step a predictor velocity ($\mathbf{v}^*$) is computed at cell-centroids, as follows:

$$\frac{\rho\mathbf{v}^* - \rho^n\mathbf{v}^n}{\Delta t} = \mathbf{C}_\mathbf{v}^n + \mathbf{D}_\mathbf{v}^n + (\rho - \rho_0)\mathbf{g} + \sum_{i=1}^{n_d}\sigma\kappa_i(\phi_i)\nabla_h\phi_i, \tag{11}$$

where super-index n denotes the previous time step, $\mathbf{D}_\mathbf{v}(\mathbf{v}) = \nabla_h \cdot \mu\nabla_h\mathbf{v} + \nabla_h \cdot \mu(\nabla_h\mathbf{v})^T$, and $\mathbf{C}_\mathbf{v}(\rho\mathbf{v}) = -\nabla_h \cdot (\rho\mathbf{v}\mathbf{v})$. In a second step a corrected velocity ($\mathbf{v}$) is computed at cell-centroids:

$$\frac{\rho\mathbf{v} - \rho\mathbf{v}^*}{\Delta t} = -\nabla_h(p), \tag{12}$$

Imposing the incompressibility constraint ($\nabla_h \cdot \mathbf{v} = 0$) to Eq. (12) leads to a Poisson equation for the pressure field at cells, which is computed by using a preconditioned conjugate gradient method:

$$\nabla_h \cdot \left(\frac{1}{\rho}\nabla_h p\right) = \frac{1}{\Delta t}\nabla_h \cdot (\mathbf{v}^*), \quad \mathbf{e}_{\partial\Omega} \cdot \nabla_h p|_{\partial\Omega} = 0. \tag{13}$$

Here, $\partial\Omega$ denotes the boundary of Ω, excluding the periodic boundaries, where information of the corresponding periodic nodes is employed. Finally, to fullfill the incompressibility constraint, and to avoid the pressure-velocity decoupling on collocated meshes [34], a cell-face velocity $\mathbf{v}_f$ [4,8] is interpolated to advect momentum (Eq. (1)), CLS functions (Eq. (4)), and concentration (Eq. (9)), as explained in Appendix B of [8]. Temporal discretization of advection equation (Eq. (4)) and re-initialization equation (Eq. (5)) is done by using a TVD Runge-Kutta method [23]. Reinitialization equation (Eq. (5)), is solved for the steady state, using two iterations per physical time step [4,7,12].

Special attention is given to the discretization of convective (or compressive) term of transport equations. The convective term is approximated at Ω_P by $(\nabla_h \cdot \beta\psi\mathbf{c})_P = \frac{1}{V_P}\sum_f \beta_f \psi_f \mathbf{c}_f \cdot \mathbf{A}_f$, where V_P is the volume of the current cell Ω_P, subindex f denotes the cell-faces, $\mathbf{A}_f = ||\mathbf{A}_f||\hat{\mathbf{e}}_f$ is the area vector, $\mathbf{c} = \{\mathbf{v}, \mathbf{n}_i^o\}$, as introduced in [4,12]. Indeed, computation of variables $\psi = \{\phi_i, \phi_i(1-\phi_i), v_j, C, ...\}$ at the cell faces (ψ_f) is performed as the sum of a diffusive upwind part (ψ_{C_p}) plus an anti-diffusive term [4,8,12]:

$$\psi_f = \psi_{C_p} + \frac{1}{2} L(\theta_f)(\psi_{D_p} - \psi_{C_p}). \tag{14}$$

where $L(\theta_f)$ is the flux limiter, $\theta_f = (\psi_{C_p} - \psi_{U_p})/(\psi_{D_p} - \psi_{C_p})$, C_p is the upwind point, U_p is the far-upwind point, and D_p is the downwind point [12]. Some of the flux-limiters implemented on the unstructured multiphase solver [4–9,12], have the forms [40]:

$$\begin{cases} max\{0, min\{2\theta_f, 1\}, min\{2, \theta_f\}\} & \text{Superbee,} \\ 1 & \text{CD,} \\ 0 & \text{Upwind.} \end{cases} \tag{15}$$

Using TVD Superbee flux-limiter in the convective term of momentum equation benefits the numerical stability of the unstructured multiphase solver [4–9,12], especially for bubbly flows with high-density ratio and high Reynolds numbers, as demonstrated in our previous works [5,9]. Furthermore, $(\phi_i)_f$ in the convective term of Eq. (4) is computed using a Superbee flux-limiter (Eq. (15)). Nevertheless, other flux-limiters, e.g., TVD Van-Leer flux limiter, can be also employed as demonstrated in [12]. Regarding the variable $(\phi_i(1-\phi_i))_f$ of the compressive term in Eq.(5), it can be computed by a central-difference limiter (CD in Eq. 15), or equivalently by linear interpolation as detailed in Appendix A of [12]. The last approach is used in present simulations. The reader is referred to [4–6,8,9,12] for additional technical details on the finite-volume discretization of transport equations on collocated unstructured grids, which are beyond the scope of the present paper. Numerical methods are implemented in the framework of the parallel C++/MPI code TermoFluids [41]; whereas the parallel scalability of the multiple marker level-set solver is presented in [9,12].

3 Numerical Experiments

Validations, applications and extensions of the unstructured CLS method [4] are reported in our previous works, for instance: dam-break problem [4], buoyancy-driven motion of single bubbles on unconfined domains [4,6,7], binary droplet collision with bouncing outcome [5], drop collision against a fluid interface without coalescence [5], bubbly flows in vertical channels [9,11], falling droplets [10], energy budgets on the binary droplet collision with topological changes [1], Taylor bubbles [24,25], gas-liquid jets [38], thermocapillary migration of deformable droplets [7,13], and mass transfer from bubbles rising on unconfined domains

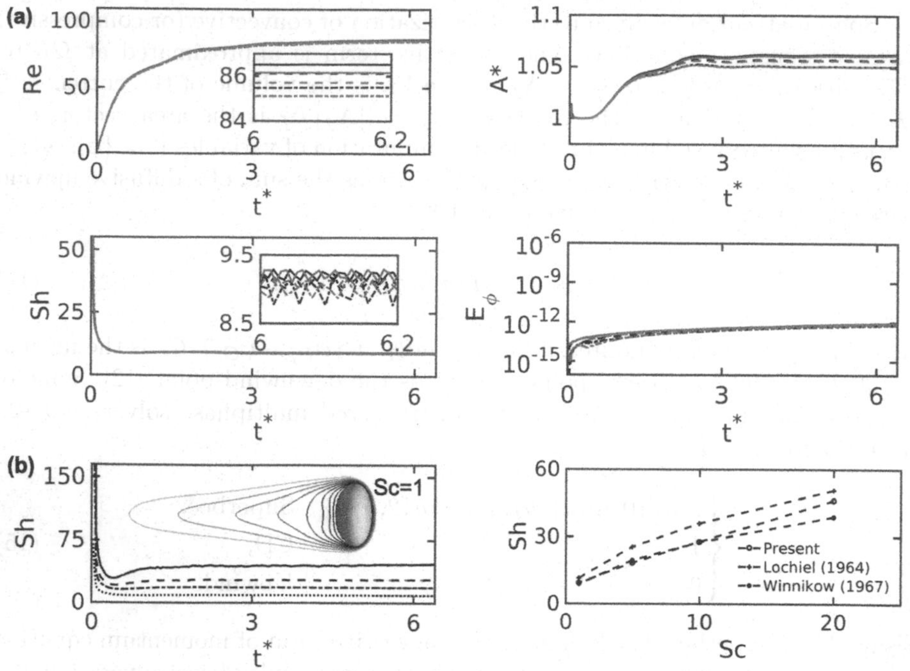

Fig. 1. Mass transfer from a single bubble, $Eo = 3.125$, $Mo = 1 \times 10^{-6}$, $\eta_\rho = \eta_\mu = 100$, $Sc = 20$, $Da = 0$ and $\alpha \approx 0\%$. Grid size $h = \{d/35(-), d/30(--), d/25(-\cdot)\}$. (a) Time evolution of Reynolds number (Re), normalized surface of the bubble ($A^*(t)$), Sherwood number ($Sh(t)$), and mass conservation error ($E_\phi = \int_\Omega(\phi(\mathbf{x},t) - \phi(\mathbf{x},0))dV / \int_\Omega \phi(\mathbf{x},0)dV$). Gradients evaluation (Eq.(10)) with $w_{P\to k} = 1$ (red lines) and $w_{P\to k} = ||\mathbf{x}_P - \mathbf{x}_k||^{-1}$ (black lines). (b) Sherwood number for $Sc = \{20(-), 10(--), 5(-\cdot), 1(\cdot\cdot)\}$ with a figure of mass concentration contours for $Sc = 1$, and comparison of present results against correlations [30,43]. (Color figure online)

[12]. Furthermore, a comparison of the unstructured CLS method [4] and coupled volume-of-fluid/level-set method [7] is reported in [10]. Therefore, this research can be considered as a further step for simulating mass transfer from buoyancy-driven bubbly flows in a wall confined vertical channel.

The hydrodynamics of bubbly flows in a vertical channel can be characterized by the following dimensionless numbers [17]:

$$Mo = \frac{g\mu_c^4 \Delta\rho}{\rho_c^2 \sigma^3}, \quad Eo = \frac{gd^2 \Delta\rho}{\sigma}, \quad Re_i = \frac{\rho_c U_{T i} d}{\mu_c},$$
$$\eta_\rho = \frac{\rho_c}{\rho_d}, \quad \eta_\mu = \frac{\mu_c}{\mu_d}, \quad C_r = \frac{D_\Omega}{d}, \quad \alpha = \frac{V_d}{V_\Omega}, \tag{16}$$

where, η_ρ is the density ratio, η_μ is the viscosity ratio, Mo is the Morton number, Eo is the Eötvös number, Re is the Reynolds number, d is the initial bubble diameter, $\Delta\rho = |\rho_c - \rho_d|$ is the density difference between the fluid phases, subscript d denotes the dispersed fluid phase, subscript c denotes the continuous fluid phase, α is the bubble volume fraction, C_r is the confinement ratio, D_Ω is the diameter of the circular channel, V_d is the volume of bubbles (Ω_d), V_Ω is the volume of Ω, and $t^* = t\sqrt{g/d}$ is the dimensionless time. Numerical results will be reported in terms of the so-called drift velocity [12,22], $U_{Ti}(t) = (\mathbf{v}_i(t) - \mathbf{v}_\Omega(t)) \cdot \hat{\mathbf{e}}_y$, which can be interpreted as the bubble velocity with respect to a stationary container, $\mathbf{v}_i(t)$ is the velocity of the ith bubble, $\mathbf{v}_\Omega(t)$ is the spatial averaged velocity in Ω.

Mass transfer with chemical reaction (first-order kinetics $\dot{r}(C) = -k_1 C$) can be characterized by the Sherwood number (Sh), Schmidt number (Sc) or Peclet number (Pe), and the Damköler (Da) number, defined in Ω_c as follows:

$$Sh = \frac{k_c d}{\mathcal{D}_c},\ Sc = \frac{\mu_c}{\rho_c \mathcal{D}_c},\ Pe = \frac{U_T d}{\mathcal{D}_c} = ReSc,\ Da = \frac{k_1 d^2}{\mathcal{D}_c}. \tag{17}$$

where k_c is the mass transfer coefficient at the continuous fluid side.

3.1 Validation and Sensitivity to Gradients Evaluation

In our previous work [12], extensive validation of the level-set model for mass transfer in bubbly flows has been presented. Here, the sensitivity of numerical simulations respect to gradients evaluation is researched, by simulating the mass transfer from a single buoyancy-driven bubble on an unconfined domain. Ω is a cylinder with height $H_\Omega = 10d$ and diameter $D_\Omega = 8d$, where d is the initial bubble diameter. Ω is discretized by three unstructured meshes with $\{4.33 \times 10^6(M_1), 3.65 \times 10^6(M_2), 1.5 \times 10^6(M_3)\}$ triangular-prisms control volumes, distributed on 192 CPU-cores. Meshes are concentrated around the symmetry axis y, in order to maximize the bubble resolution, whereas the grid size in this region corresponds to $h = \{d/35(M_1), d/30(M_2), d/25(M_3)\}$. Neumann boundary-condition is applied at lateral, top and bottom walls. The initial bubble position is $(x, y, z) = (0, 1.5d, 0)$, on the symmetry axis y, whereas both fluids are initially quiescent.

Mass transfer coefficient (k_c) in single rising bubbles is calculated from a mass-balance for the chemical species in Ω_c, as follows [12]:

$$k_c(t) = \frac{V_c}{A_d(C_{\Gamma,c} - C_\infty)} \frac{dC_c}{dt}, \tag{18}$$

where $C_c = V_c^{-1} \int_{\Omega_c} C(\mathbf{x}, t) dV$, $A_d = \int_\Omega ||\nabla \phi|| dV$ is the interfacial surface of the bubble, V_c is the volume of Ω_c, $C_{\Gamma,c}$ is the constant concentration on the bubble interface from the side of Ω_c, and $C_\infty = 0$ is the reference concentration. Dimensionless parameters are $Eo = 3.125$, $Mo = 1 \times 10^{-6}$, $Da = 0$, $Sc = \{1, 5, 10, 20\}$, $\eta_\rho = 100$ and $\eta_\mu = 100$.

Figure 1a shows the time evolution of Reynolds number (Re), normalized interfacial surface ($A^*(t)$), Sherwood number ($Sh(t)$), and mass conservation error (E_ϕ), The grid-independence study shows that $h = d/35$ is enough to

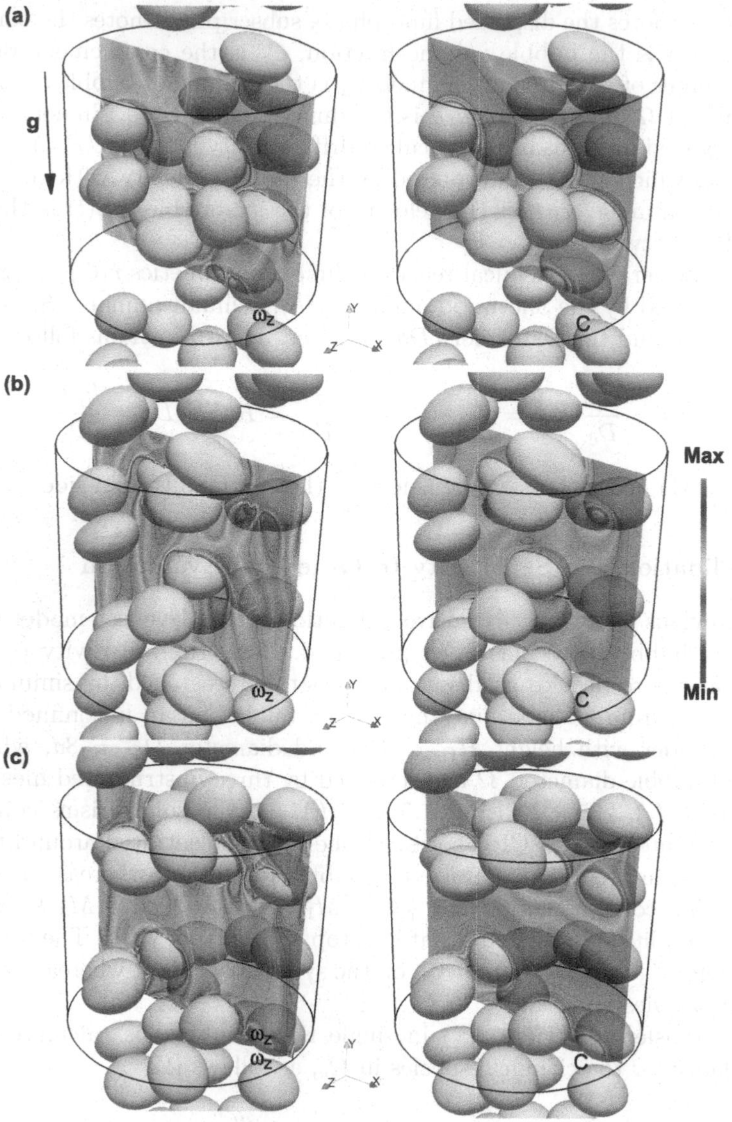

Fig. 2. Mass transfer from a bubble swarm (16 bubbles) in a periodic channel with circular cross-section, $Eo = 3.125$, $Mo = 5 \times 10^{-6}$, $\eta_\rho = \eta_\mu = 100$, $Sc = 1$, $Da = 7.97$, $\alpha = 13.4\%$. Vorticity ($\omega_z = \mathbf{e}_z \cdot \nabla \times \mathbf{v}$) and concentration ($C$) on the plane x–y at (a) $t^* = tg^{1/2}d^{-1/2} = 6.3$, (b) $t^* = 12.5$, (c) $t^* = 37.6$.

perform accurate predictions of hydrodynamics and mass transfer from single bubbles. Furthermore, the effect of gradients evaluation (Eq. (10)) on the simulations, is depicted for weighting factors $w_{P \to k} = 1$ (red lines) and $w_{P \to k} = ||\mathbf{x}_P - \mathbf{x}_k||^{-1}$ (black lines). It is observed that numerical results are

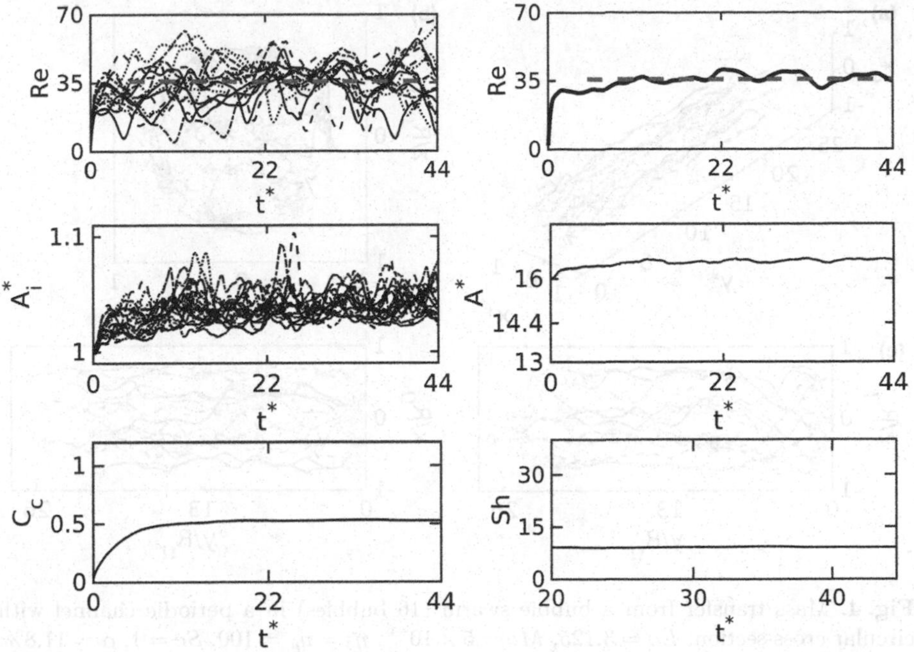

Fig. 3. Mass transfer from a bubble swarm (16 bubbles) in a periodic channel with circular cross-section, $Eo = 3.125$, $Mo = 5 \times 10^{-6}$, $\eta_\rho = \eta_\mu = 100$, $Sc = 1$, $\alpha = 11.8\%$. Time evolution of Reynolds number (Re) for each bubble (black lines), averaged Reynolds number (bold continuous line), time-averaged Reynolds number (red discontinuous line), normalized bubble surface $A_i^*(t)$, total interfacial surface of bubbles $A^*(t) = \sum_{i=1}^{n_d} A_i^*(t)$, spatial averaged concentration $C_c = V_c^{-1} \int_{\Omega_c} C dV$, and Sherwood number $Sh(t)$. (Color figure online)

very close, whereas the numerical stability is maintained independently of the selected weighting factor. In what follows $w_{P \to k} = ||\mathbf{x}_P - \mathbf{x}_k||^{-1}$ will be employed. Figure 1b depicts the effect of Schmidt number on the Sherwood number, as well as a comparison of present results against empirical correlations from literature [30,43]. These results also give a further validation of the model for mass transfer coupled to hydrodynamics in buoyancy-driven bubbles.

3.2 Mass Transfer from a Bubble Swarm Rising in a Vertical Channel

As a further step and with the confidence that the CLS model has been validated [12], the mass transfer from a bubble swarm rising in a vertical pipe, is computed. The saturation of concentration in Ω_c is avoided by the chemical reaction term in Eq. (9) [12,35]. On the other hand, the mass transfer coefficient (k_c) in Ω_c is computed by using a mass balance of the chemical species at steady state ($dC_c/dt = 0$), as follows [12]:

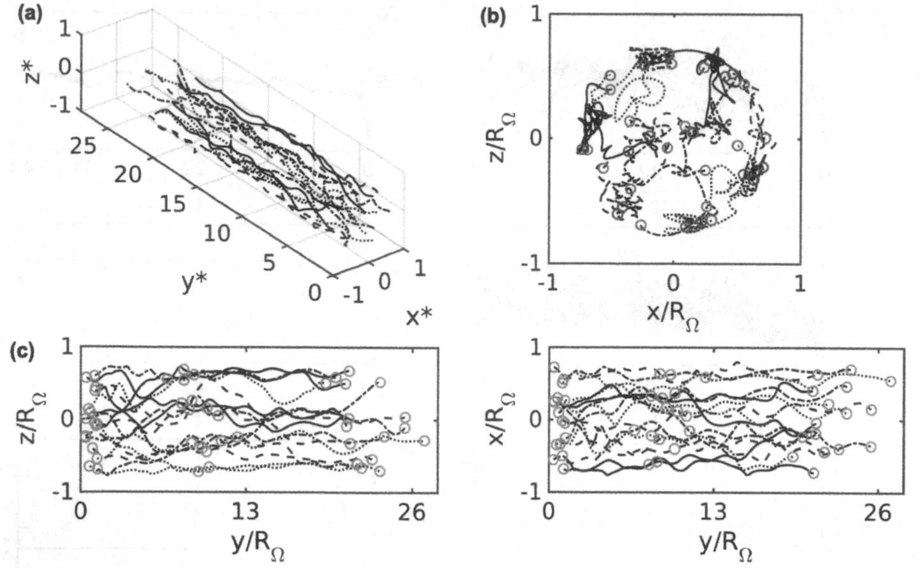

Fig. 4. Mass transfer from a bubble swarm (16 bubbles) in a periodic channel with circular cross-section, $Eo = 3.125$, $Mo = 5 \times 10^{-6}$, $\eta_\rho = \eta_\mu = 100$, $Sc = 1$, $\alpha = 11.8\%$. (a) 3D bubble trajectories. (b) Projection of bubble trajectories on the plane x–z. (c) Projection of bubble trajectories on the plane x–y and z–y. Here $R_\Omega = 0.5D_\Omega$ is the radius of the cylindrical channel, $(x^*, y^*, z^*) = (x/R_\Omega, y/R_\Omega, z/R_\Omega)$.

$$k_c = \frac{V_c k_1 C_c}{(C_{\Gamma,c} - C_c)\sum_{i=1}^{nd} A_i} \tag{19}$$

where $A_i = \int_\Omega ||\nabla \phi_i|| dV$ is the surface of the ith bubble, and $C_c = V_c^{-1} \int_{\Omega_c} C dV$. Ω is a periodic cylindrical channel, with height $H_\Omega = 4d$ and diameter $D_\Omega = 4.45d$, as depicted in Fig. 2. Ω is discretized by 9.3×10^6 triangular-prisms control volumes, with grid size $h = d/40$, distributed on 960 CPU-cores. Periodic boundary conditions are used on the $x - z$ boundary planes. On the wall, no-slip boundary condition for velocity, Dirichlet boundary condition for CLS markers ($\phi_i = 1$), and Neumann boundary condition for C. Bubbles are initially distributed in Ω following a random pattern, whereas fluids are quiescent. Since fluids are incompressible and bubble coalescence is not allowed, the void fraction ($\alpha = V_d/V_\Omega$) and number of bubbles are constant throughout the simulation.

Dimensionless parameters are $Eo = 3.125$, $Mo = 5 \times 10^{-6}$, $Sc = 1$, $Da = 7.97$, $\eta_\rho = 100$, $\eta_\mu = 100$, $\alpha = 13.4\%$ and $C_r = 4.45$, which corresponds to a bubbly flow with 16 bubbles distributed in Ω. Figure 2 illustrates the mass transfer from a swarm of 16 bubbles at $t^* = \{6.26, 37.6\}$. Concentration contours (C), and vorticity contours ($\omega_z = \hat{\mathbf{e}}_z \cdot \nabla \times \mathbf{v}$) are shown on the plane $x - y$. Figure 3 depicts the time evolution of Reynolds number for each bubble and the time-averaged Reynolds number (discontinuous line), normalized bubble surface $A_i^*(t)$, total surface of bubbles $A^*(t) = \sum_{i=1}^{n_d} A_i^*(t)$, spatial averaged chemical species concentration (C_c) in Ω_c, and Sherwood number $Sh(t)$ at steady state $(dCc/dt = 0)$. Figure 3 shows that $Re_i(t)$ presents fluctuations, due to oscillations in the bubble shapes (see $A_i(t)$), and bubble-bubble interactions such as bouncing interaction, and the so called drafting, kissing and tumbling processes [5,9,10]. On the other hand, the Reynold number of the bubble swarm, $\bar{Re} = n_d^{-1} \sum_{i=1}^{n_d} Re_i(t)$, achieves the steady-state. The spatial averaged concentration (C_c) tend to the steady-state after a short transient, indicating an equilibrium between mass transfer from the bubbles and chemical reaction in Ω_c. Furthermore, the mass transfer coefficient (Sh) achieves the steady-state, once $dC_c/dt = 0$. Finally, Fig. 4 depict bubble trajectories, which indicate a bubble-wall repulsion effect.

4 Conclusions

DNS of mass transfer from buoyancy-driven bubbles rising in a vertical channel has been performed using a parallel multiple marker CLS method [5,9,12]. These numerical experiments demonstrate the capabilities of the present approach, as a reliable method for simulating bubbly flows with mass transfer and chemical reaction in vertical channels, taking into account bubble-bubble and bubble-wall interactions in long time simulation of bubbly flows. The method avoids the numerical merging of bubble interfaces, which is an issue inherent to standard interface capturing methods. Interactions of bubbles include a repulsion effect when these are horizontally aligned or when bubbles interact with the wall, whereas two bubbles vertically aligned tend to follow the so-called drafting-kissing-tumbling mechanism observed also in solid particles. These bubble-bubble and bubble-wall interactions lead to a fluctuating velocity field, analogous to that observed in turbulence. Nevertheless, the time averaged Reynolds number $(\bar{Re})$ and mass transfer coefficient (Sh) tend to the steady-state. Turbulence induced by agitation of bubbles promote the mixing of chemical species in the continuous phase, whereas the spatial averaged chemical species concentration tends to the steady-state, indicating a balance between chemical reaction in Ω_c and mass transfer from bubbles. These results demonstrate that the multiple marker CLS approach [12] is a predictive method to compute $Sh = Sh(Eo, Re, Da, \alpha, C_r)$ in bubbly flows rising in a vertical channel. Future work includes the extension of this model to multicomponent mass transfer and complex chemical reaction kinetics, as well as parametric studies of $Sh = Sh(Eo, Re, Da, \alpha, C_r)$ to develop closure relations for models based on the averaged flow (e.g., two-fluid models [27]).

References

1. Amani, A., Balcázar, N., Gutiérrez, E., Oliva, A.: Numerical study of binary droplets collision in the main collision regimes. Chem. Eng. J. **370**, 477–498 (2019)
2. Aboulhasanzadeh, B., Thomas, S., Taeibi-Rahni, M., Tryggvason, G.: Multiscale computations of mass transfer from buoyant bubbles. Chem. Eng. Sci. **75**, 456–467 (2012)
3. Alke, A., Bothe, D., Kroeger, M., Warnecke, H.J.: VOF-based simulation of conjugate mass transfer from freely moving fluid particles. In: Mammoli, A.A., Brebbia, C.A. (eds.) Computational Methods in Multiphase Flow V, pp. 157–168 (2009). WIT Transactions on Engineering Sciences
4. Balcázar, N., Jofre, L., Lehmkhul, O., Castro, J., Rigola, J.: A finite-volume/level-set method for simulating two-phase flows on unstructured grids. Int. J. Multiphase Flow **64**, 55–72 (2014)
5. Balcázar, N., Lehmkhul, O., Rigola, J., Oliva, A.: A multiple marker level-set method for simulation of deformable fluid particles. Int. J. Multiphase Flow **74**, 125–142 (2015)
6. Balcázar, N., Lemhkuhl, O., Jofre, L., Oliva, A.: Level-set simulations of buoyancy-driven motion of single and multiple bubbles. Int. J. Heat Fluid Flow **56**, 91–107 (2015)
7. Balcázar, N., Lehmkhul, O., Jofre, L., Rigola, J., Oliva, A.: A coupled volume-of-fluid/level-set method for simulation of two-phase flows on unstructured meshes. Comput. Fluids **124**, 12–29 (2016)
8. Balcázar, N., Rigola, J., Castro, J., Oliva, A.: A level-set model for thermocapillary motion of deformable fluid particles. Int. J. Heat Fluid Flow **62**(Part B), 324–343 (2016)
9. Balcázar, N., Castro, J., Rigola, J., Oliva, A.: DNS of the wall effect on the motion of bubble swarms. Procedia Comput. Sci. **108**, 2008–2017 (2017)
10. Balcázar, N., Castro, J., Chiva, J., Oliva, A.: DNS of falling droplets in a vertical channel. Int. J. Comput. Methods Exp. Measur. **6**(2), 398–410 (2018)
11. Balcázar, N., Lehmkuhl, O., Castro, J., Oliva, A.: DNS of the rising motion of a swarm of bubbles in a confined vertical channel. In: Grigoriadis, D.G.E., Geurts, B.J., Kuerten, H., Fröhlich, J., Armenio, V. (eds.) Direct and Large-Eddy Simulation X. ES, vol. 24, pp. 125–131. Springer, Cham (2018). https://doi.org/10.1007/978-3-319-63212-4_15
12. Balcázar-Arciniega, N., Antepara, O., Rigola, J., Oliva, A.: A level-set model for mass transfer in bubbly flows. Int. J. Heat Mass Transf. (Accepted 2 Apr 2019). https://doi.org/10.1016/j.ijheatmasstransfer.2019.04.008
13. Balcázar, N., Antepara, O., Rigola, J., Oliva, A.: DNS of thermocapillary migration of deformable droplets. In: Salvetti, M.V., Armenio, V., Fröhlich, J., Geurts, B.J., Kuerten, H. (eds.) Direct and Large-Eddy Simulation XI. ES, vol. 25, pp. 207–213. Springer, Cham (2019). https://doi.org/10.1007/978-3-030-04915-7_28
14. Bothe, D., Koebe, M., Wielage, K., Warnecke, H.J.: VOF simulations of mass transfer from single bubbles and bubble chains rising in the aqueous solutions. In: Proceedings of FEDSM03: Fourth ASME-JSME Joint Fluids Engineering Conference, 6–11 July, Honolulu, HI, USA (2003)
15. Bothe, D., Fleckenstein, S.: Modeling and VOF-based numerical simulation of mass transfer processes at fluidic particles. Chem. Eng. Sci. **101**, 283–302 (2013)
16. Brackbill, J.U., Kothe, D.B., Zemach, C.: A continuum method for modeling surface tension. J. Comput. Phys. **100**, 335–354 (1992)

17. Clift, R., Grace, J.R., Weber, M.E.: Bubbles, Drops and Particles. Academic Press, New York (1978)
18. Chorin, A.J.: Numerical solution of the Navier-Stokes equations. Math. Comput. **22**, 745–762 (1968)
19. Coyajee, E., Boersma, B.J.: Numerical simulation of drop impact on a liquid-liquid interface with a multiple marker front-capturing method. J. Comput. Phys. **228**(12), 4444–4467 (2009)
20. Darmana, D., Deen, N.G., Kuipers, J.A.M.: Detailed 3D modeling of mass transfer processes in two-phase flows with dynamic interfaces. Chem. Eng. Technol. **29**(9), 1027–1033 (2006)
21. Davidson, M.R., Rudman, M.: Volume-of-fluid calculation of heat or mass transfer across deforming interfaces in two-fluid flow. Numer. Heat Transf. Part B: Fundam. **41**, 291–308 (2002)
22. Esmaeeli, A., Tryggvason, G.: Direct numerical simulations of bubbly flows Part 2. Moderate Reynolds number arrays. J. Fluid Mech. **385**, 325–358 (1999)
23. Gottlieb, S., Shu, C.W.: Total variation dimishing Runge-Kutta schemes. Math. Comput. **67**, 73–85 (1998)
24. Gutiérrez, E., Balcázar, N., Bartrons, E., Rigola, J.: Numerical study of Taylor bubbles rising in a stagnant liquid using a level-set/moving-mesh method. Chem. Eng. Sci. **164**, 102–117 (2017)
25. Gutiérrez, E., Favre, F., Balcázar, N., Amani, A., Rigola, J.: Numerical approach to study bubbles and drops evolving through complex geometries by using a level set - Moving mesh - Immersed boundary method. Chem. Eng. J. **349**, 662–682 (2018)
26. Hirt, C., Nichols, B.: Volume of fluid (VOF) method for the dynamics of free boundary. J. Comput. Phys. **39**, 201–225 (1981)
27. Ishii, M., Hibiki, T.: Thermo-Fluid Dynamics of Two-Phase Flow, 2nd edn. Springer, New-York (2010). https://doi.org/10.1007/978-1-4419-7985-8
28. Jasak, H., Weller, H.G.: Application of the finite volume method and unstructured meshes to linear elasticity. Int. J. Numer. Meth. Eng. **48**, 267–287 (2000)
29. Koynov, A., Khinast, J.G., Tryggvason, G.: Mass transfer and chemical reactions in bubble swarms with dynamic interfaces. AIChE J. **51**(10), 2786–2800 (2005)
30. Lochiel, A., Calderbank, P.: Mass transfer in the continuous phase around axisymmetric bodies of revolution. Chem. Eng. Sci. **19**, 471–484 (1964)
31. Mavriplis, D.J.: Unstructured mesh discretizations and solvers for computational aerodynamics. In: 18th Computational Fluid Dynamics Conference, AIAA Paper 2007–3955, Miami, FL (2007). https://doi.org/10.2514/6.2007-3955
32. Olsson, E., Kreiss, G.: A conservative level set method for two phase flow. J. Comput. Phys. **210**, 225–246 (2005)
33. Osher, S., Sethian, J.A.: Fronts propagating with curvature-dependent speed: algorithms based on Hamilton-Jacobi formulations. J. Comput. Phys. **79**, 175–210 (1988)
34. Rhie, C.M., Chow, W.L.: Numerical study of the turbulent flow past an airfoil with trailing edge separation. AIAA J. **21**, 1525–1532 (1983)
35. Roghair, I., Van Sint Annaland, M., Kuipers, J.A.M.: An improved front-tracking technique for the simulation of mass transfer in dense bubbly flows. Chem. Eng. Sci. **152**, 351–369 (2016)
36. Sussman, M., Smereka, P., Osher, S.: A level set approach for computing solutions to incompressible two-phase flow. J. Comput. Phys. **144**, 146–159 (1994)

37. Sussman, M., Puckett, E.G.: A coupled level set and volume-of-fluid method for computing 3D and axisymmetric incompressible two-phase flows. J. Comput. Phys. **162**, 301–337 (2000)
38. Schillaci, E., Antepara, O., Balcázar, N., Rigola, J., Oliva, A.: A numerical study of liquid atomization regimes by means of conservative level-set simulations. Comput. Fluids **179**, 137–149 (2019)
39. Sun, D.L., Tao, W.Q.: A coupled volume-of-fluid and level-set (VOSET) method for computing incompressible two-phase flows. Int. J. Heat Mass Transf. **53**, 645–655 (2010)
40. Sweby, P.K.: High resolution using flux limiters for hyperbolic conservation laws. SIAM J. Numer. Anal. **21**, 995–1011 (1984)
41. Termo Fluids S.L. http://www.termofluids.com/. Accessed 29 Jan 2019
42. Tryggvason, G., et al.: A front-tracking method for the computations of multiphase flow. J. Comput. Phys. **169**, 708–759 (2001)
43. Winnikow, S.: Letter to the editors. Chem. Eng. Sci. **22**(3), 477 (1967)

A Hybrid Vortex Method for the Simulation of 3D Incompressible Flows

Chloe Mimeau[1(✉)], Georges-Henri Cottet[2], and Iraj Mortazavi[1]

[1] Conservatoire National des Arts et Métiers, M2N Lab, 7340 Paris, France
{chloe.mimeau,iraj.mortazavi}@cnam.fr

[2] Grenoble-Alpes University, LJK Laboratory, 38041 Grenoble, France
georges-henri.cottet@univ-grenoble-alpes.fr

Abstract. A hybrid particle/mesh Vortex Method, called remeshed vortex method, is proposed in this work to simulate three-dimensional incompressible flows. After a validation study of the present method in the context of Direct Numerical Simulations, an anisotropic artificial viscosity model is proposed in this paper in order to handle multi-resolutions simulations in the context of vortex methods.

Keywords: Vortex method · Semi-Lagrangian · Anisotropic artificial viscosity model

1 Introduction

Among the numerous numerical approaches used in CFD, Lagrangian methods, also called particle methods, occupy an important place thanks to their intuitive and natural description of the flow as well as their low numerical diffusion and their stability. Indeed, in Lagrangian approaches, the physical quantities involved in the simulated problem are discretized onto a set of particles evolving spatially in the domain according to the problem dynamics. The particles are therefore characterized by their position in the computational domain and the value of the physical quantity they are carrying. Vortex methods [4] belong to this class of Lagrangian approaches and will constitute the key point of the present work. In Vortex methods, the particles discretize the Navier-Stokes equations in their velocity ($\mathbf{u}$) - vorticity ($\boldsymbol{\omega}$) formulation. This formulation allows to directly point to the essence of vorticity dynamics in incompressible flows, which is characterized by advection and diffusion as well as stretching and change of orientation.

However, Vortex methods exhibit difficulties inherent to particle methods and related to the particle distortion phenomenon, which manifests itself by the clustering or spreading of the flow elements in high strain regions, thus implying the loss of convergence of the method. The remeshing technique [7] may be considered as one of the most efficient and popular method to bypass the inherent problem of particle distortion. It consists in periodically redistributing the

J. M. F. Rodrigues et al. (Eds.): ICCS 2019, LNCS 11539, pp. 611–622, 2019.
https://doi.org/10.1007/978-3-030-22747-0_46

particles onto an underlying Cartesian grid in order to ensure their overlapping and thus the convergence of the solution. These hybrid Lagrangian/Eulerian approaches are characterized by the fact that the vorticity and the velocity variables are both resolved on the particles field and on a Cartesian grid.

In this work the remeshing procedure is performed in a directional way [8]. This approach transforms the usual tensorial computations (based on 3D-stencils) into 1D advection/remeshing problems in each direction, thus decreasing substantially the computational cost of this procedure. As the Cartesian grid used in the present work is uniform and fixed in time, the simulations of flows at high Reynolds numbers involve prohibitive computational efforts. To encounter this problem, we propose in the present paper *bi-level* simulations. The bi-level approach may be considered as a hybrid procedure since it relies on a resolved vorticity field while the related velocity field is filtered. The artificial viscosity model derived here for this purpose is directly based on Vortex Method framework, according to [2].

This paper is organized as follows. We will first describe the remeshed Vortex Method, giving the governing equations and the fractional step algorithm used to discretize them. Then we will expose the artificial viscosity model proposed here to perform bi-level simulations. The last section will be dedicated to the numerical results: both direct numerical simulations and bi-level simulations will be validated in the context of a Taylor-Green Vortex at $Re = 1600$.

2 Remeshed Vortex Method

2.1 Governing Equations

This study is based on the vorticity formulation of the incompressible Navier-Stokes equations, called the Vorticity Transport Equations. In a domain D, these equations read:

$$\frac{\partial \boldsymbol{\omega}}{\partial t} + (\mathbf{u} \cdot \nabla)\boldsymbol{\omega} - (\boldsymbol{\omega} \cdot \nabla)\mathbf{u} = \frac{1}{Re}\Delta\boldsymbol{\omega} \tag{1}$$

$$\Delta\mathbf{u} = -\nabla \times \boldsymbol{\omega}, \tag{2}$$

where $\boldsymbol{\omega}$, $\mathbf{u}$ and Re respectively denote the vorticity, the velocity and the Reynolds number. One can distinguish in Eq. 1 the advection term $(\mathbf{u} \cdot \nabla)\boldsymbol{\omega}$, the stretching term $(\boldsymbol{\omega} \cdot \nabla)\mathbf{u}$ (which vanishes in 2D) and the diffusion term $\Delta\boldsymbol{\omega}/Re$. The Poisson Eq. 2 is derived from the incompressibility condition $\nabla \cdot \mathbf{u} = 0$ and allows to recover the velocity field $\mathbf{u}$ from the vorticity field $\boldsymbol{\omega}$. This system of equations has to be complemented by appropriate conditions at the boundaries of computational domain D.

2.2 Fractional Step Algorithm

To solve the vorticity transport Eqs. 1–2, the flow is discretized onto particles that carry the vorticity field $\boldsymbol{\omega}$ transported at the velocity $\mathbf{u}$ and the resolution of the governing equations is based on a splitting algorithm, which consists at each time step in successively solving the following equations:

$$\Delta \mathbf{u} = -\nabla \times \boldsymbol{\omega} \tag{3}$$

$$\frac{\partial \boldsymbol{\omega}}{\partial t} = \boldsymbol{\nabla} \cdot (\boldsymbol{\omega} : \mathbf{u}) \tag{4}$$

$$\frac{\partial \boldsymbol{\omega}}{\partial t} = \frac{1}{Re} \Delta \boldsymbol{\omega} \tag{5}$$

$$\frac{\partial \boldsymbol{\omega}}{\partial t} + (\mathbf{u} \cdot \nabla)\, \boldsymbol{\omega} = 0 \tag{6}$$

$$\Delta t_{adapt} = \frac{\text{LCFL}}{\|\nabla \mathbf{u}\|_\infty} \tag{7}$$

Table 1. Time and space discretization methods used for the resolution of the viscous splitting algorithm (Eqs. 3–7).

Equation	Time discr. method	Space discr. method
Poisson Eq. (3)	-	Spectral method
Stretching (4)	RK3 scheme	4th order centered FD
Diffusion (5)	Implicit Euler scheme	Spectral method
Advection (6)	RK2 scheme	$\Lambda_{4,2}$ remeshed vortex method
Adaptive time step (7)	-	4th order centered FD (LCFL < 1)

The discretization of each equation of the fractional step algorithm is realized in this study by using a semi-Lagrangian Vortex method, called the remeshed Vortex method. Table 1 gives the time and space discretization schemes used in this work to solve them. The advection of vorticity field (Eq. 6) is performed in a Lagrangian way using a Vortex method. This Lagrangian approach provides a natural and efficient way to solve the non-linear convection term, with low numerical diffusion. Once the particles carrying the vorticity field have been transported, they are redistributed on an underlying Cartesian grid using a remeshing kernel of type $\Lambda_{4,2}$ [3]. The $\Lambda_{p,r}$ remeshing kernels are piecewise polynomial functions of regularity $\mathcal{C}^r$, satisfying the conservation of the first p moments. The $\Lambda_{4,2}$ kernel therefore satisfies the first 4 moments and is of regularity $\mathcal{C}^2$. It contains 6 points in its 1D-support, which means that each particle is redistributed onto 6 points in each direction.

In this work, the particle advection and the remeshing procedure are performed using a directional splitting approach [8]. It consists in solving the advection and remeshing problems direction by direction. As a consequence, if the chosen kernel contains S points in its 1D-support, the number of operations with the directional splitting method compared to the tensorial approach goes from $\mathcal{O}(S^2)$ to $\mathcal{O}(2S)$ in 2D and from $\mathcal{O}(S^3)$ to $\mathcal{O}(3S)$ in 3D (see Fig. 1). If we consider the $\Lambda_{4,2}$ kernel ($S = 6$) used in the present work, the directional splitting method thus allows to divide the number of operations by 12 for each particle. This directional splitting consequently allows for a drastic reduction of the computational cost in terms of regridding operations.

The systematic remeshing of particles onto an Eulerian grid at each time step enables to ensure the overlapping of particles required for the convergence of the method. Moreover the presence of the grid allows to discretize the other equations using efficient and/or fast grid methods (finite differences and spectral method based on FFT evaluations). In the present algorithm, Eqs. 3–5 are solved on the grid. Note that the stretching problem 4 is considered here in its conservative formulation, $\partial_t \boldsymbol{\omega} = \boldsymbol{\nabla} \cdot (\boldsymbol{\omega} : \mathbf{u})$.

Finally, the value of the adaptive time step is evaluated (on the grid) at the end of the fractional step algorithm according to the infinite norm of the velocity gradient (cf Eq. 7), which provides a more relaxed condition compared to classical CFL conditions. The Lagrangian CFL number, called LCFL, must be taken lower or equal to 1. In this work we set LCFL $= 1/8$.

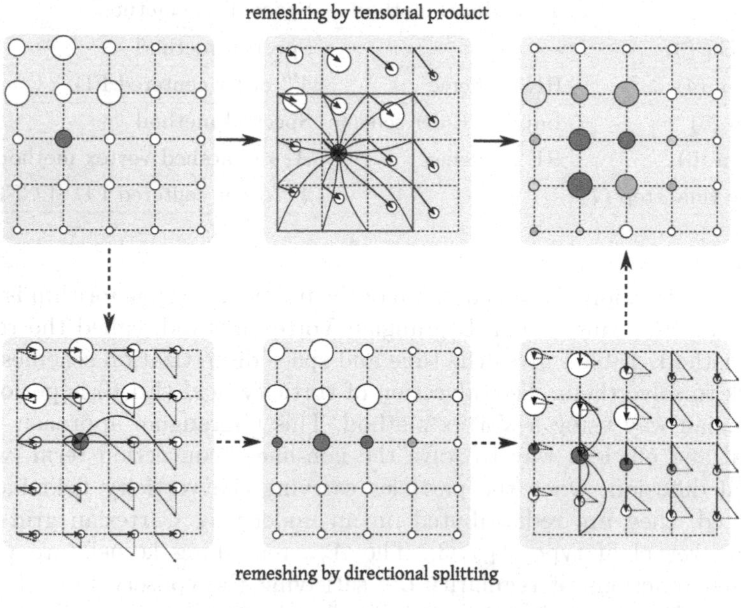

Fig. 1. 2D schematic representation of a remeshing procedure using a tensorial product (on top, depicted by plain arrows) and a directional splitting (on bottom, depicted by dashed arrows). The red lines indicate the support of the remeshing kernel. In this example, the kernel has a 1D-support of 4 points. (Color figure online)

3 Bi-Level Approach

3.1 Context and Motivation

One of the main limitation experienced by the present hybrid vortex method is related to the computational resources needed to deal with large problems. Indeed, the Eulerian grid used in this work to solve Eqs. 3–5 in the splitting algorithm is a uniform Cartesian grid. Therefore, when increasing the Reynolds number in the present direct numerical simulations, the handling of a uniform Cartesian grid requires to consider prohibitive mesh sizes in order to correctly capture the boundary layer, which makes the computation unaffordable. In order to overcome this problem, a good solution relies in multiresolution [1,9] and grid adaptivity [11]. However all these approaches require major modifications in the computational solver. A more straightforward technique may rely in a LES-like model, that is to say in the derivation of a eddy viscosity model.

The aim of LES models based on the filtered Navier-Stokes equations is to estimate the subfilter scale stress tensor τ_{ij}, which is usually modeled by a dissipation term, allowing the energy to cascade from the large scales to the smallest one.

In this work we propose a *bi-level* approach. It consists in solving the vorticity field $\boldsymbol{\omega}$ on a fine grid while the small scales of the related velocity field are filtered so that we only consider the velocity values on a coarse grid, $\overline{\mathbf{u}}$. The main goal of such approach is to reduce the overall computational cost, more specifically the cost dedicated to the resolution of the Poisson equation, and to afford higher Reynolds numbers. In terms of flow physical description on one side and computational cost on the other side, the bi-level simulations can therefore be considered in between DNS and LES.

Previous studies dealing with multi-resolution simulations in the context of remeshed Vortex Methods have already been performed in [5]. However, in that study, the multi-resolution simulations were dedicated to the transport of a passive scalar. In other words, a coarse grid was used to compute the flow quantities, namely the vorticity and velocity fields, while a fine grid was considered to compute the convected passive scalar. An interpolation was performed to exchange the informations between the coarse and the fine grid quantities, but no eddy viscosity model was needed since the different scales concerned non-coupled variables.

In the work proposed here, the bi-level approach is applied to the $\boldsymbol{\omega}$-$\mathbf{u}$ coupled variables involved in the incompressible Navier-Stokes equations, and relies on the use of a eddy viscosity model. In the context of vortex methods, the derivation of such eddy viscosity model can be done in a different way than the LES subgrid-scale models. We give hereafter the main steps of this model derivation, based on the former studies of [2], and we extend it to our 3D algorithm.

3.2 An Anisotropic Artificial Viscosity Model

When considering purely Lagrangian vortex methods, one obtains an exact weak solution of convection equations. Therefore, the truncation error of these methods only comes from the regularization used to compute the velocity of the particles from the vorticity field [4]. The artificial viscosity model proposed in this work to handle a *bi-level* approach is based on the dissipative mechanisms embedded in this truncation error. In [2], such artificial viscosity model has been derived in the 2D case by cancelling the positive enstrophy budget embedded in the truncation error of the regularized vorticity transport equation. This artificial diffusion model is anisotropic and is given in 2D by:

$$\frac{d\omega_p}{dt} = -C \sum_{q \sim p} v_q(\omega_p - \omega_q)\Big\{[\mathbf{u}(\mathbf{x}_p) - \mathbf{u}(\mathbf{x}_q)] \cdot (\mathbf{x}_p - \mathbf{x}_q) g(|\mathbf{x}_p - \mathbf{x}_q|)\Big\}_- \tag{8}$$

where the index $-$ denotes the negative part of the quantity.

In the present paper, the model is extended to the 3D case:

$$\frac{d\boldsymbol{\omega}_p}{dt} = \boldsymbol{\nabla}\cdot(\boldsymbol{\omega}_p \mathbf{u}_p) - C\Delta^{-4} \sum_{q \sim p} v_q(\boldsymbol{\omega}_p - \boldsymbol{\omega}_q)\Big\{[\mathbf{u}(\mathbf{x}_p) - \mathbf{u}(\mathbf{x}_q)] \cdot (\mathbf{x}_p - \mathbf{x}_q) g(|\mathbf{x}_p - \mathbf{x}_q|)\Big\}_- \tag{9}$$

where Δ is the regularization size (or filter size in a LES point of view) and where C is a coefficient depending on the nature and the state of the flow. Note that Eq. 9 allows to cancel the enstrophy production only in directions of antidiffusion, which provides an anisotropic artificial viscosity model.

In a procedural and algorithmic point of view, the use of such anisotropic artificial viscosity model in the present remeshed vortex method implies the replacement of the stretching Eq. 4 by Eq. 9, which now gives the following algorithm:

$$\begin{aligned}
\Delta \mathbf{u} &= -\nabla \times \boldsymbol{\omega} \\
\frac{d\boldsymbol{\omega}_p}{dt} &= \boldsymbol{\nabla} \cdot (\boldsymbol{\omega}_p \mathbf{u}_p) \\
&\quad - C\Delta^{-4} \sum_{q \sim p} v_q(\boldsymbol{\omega}_p - \boldsymbol{\omega}_q)\Big\{[\mathbf{u}(\mathbf{x}_p) - \mathbf{u}(\mathbf{x}_q)] \cdot (\mathbf{x}_p - \mathbf{x}_q) g(|\mathbf{x}_p - \mathbf{x}_q|)\Big\}_- \\
\frac{\partial \boldsymbol{\omega}}{\partial t} &= \frac{1}{Re} \Delta \boldsymbol{\omega} \\
\frac{\partial \boldsymbol{\omega}}{\partial t} + (\mathbf{u} \cdot \nabla)\, \boldsymbol{\omega} &= 0 \\
\Delta t_{adapt} &= \frac{\text{LCFL}}{\|\nabla \mathbf{u}\|_\infty}
\end{aligned} \tag{10}$$

4 Numerical Results

In order to validate the anisotropic artificial viscosity model proposed in this paper, we consider the Taylor-Green vortex benchmark, which is an unbounded periodic flow commonly used to study the capability of a numerical method to handle transition to turbulence.

The Taylor-Green vortex is an analytical periodic solution of incompressible Navier-Stokes equations. It describes the non-linear interaction of multiscales eddies under the influence of vortex stretching and their final decay. It is a classical benchmark used as an initial condition for numerical methods to study flow problems related to transition to turbulence. This benchmark has already been tested with success in the context of a remeshed vortex method by van Rees et. al [10]. Since this method was different from the present one in the sense of the remeshing procedure (tensorial versus directional approach in our case), we are eager to test the validity of the method proposed in this work.

We consider the flow that evolves in a periodic cubic box of side length $L = 2\pi$ and develops from the following initial condition, which satisfies the divergence-free constraint:

$$\begin{aligned} u_x(\mathbf{x}, t=0) &= \sin(x)\cos(y)\cos(z) \\ u_y(\mathbf{x}, t=0) &= -\cos(x)\sin(y)\cos(z) \\ u_z(\mathbf{x}, t=0) &= 0 \end{aligned} \tag{11}$$

The Reynolds number of the flow is defined by $Re = 1/\nu$. In the present study it is set to $Re = 1600$. At such regime, the minimum number of grid cells per direction is approximately given by:

$$n_x \approx \frac{l_0}{\eta} = Re^{3/4} \sim 253 \tag{12}$$

where $l_0 = 1$ denotes the integral length scale, that is to say the scale of the largest eddies, and where $\eta = \left(\frac{\nu^3 l_0}{u_0^3}\right)^{\frac{1}{4}}$ corresponds to the Kolmogorov length scale, that is to say the scale of the smallest eddies, with u_0 the characteristic velocity set to 1. Therefore, according to this estimation, we expect reliable results from a 253^3 total grid resolution.

4.1 DNS Results

First of all, we validate the present remeshed vortex method in the case of Direct Numerical Simulations. The results presented in this subsection are obtained from the splitting algorithm described in the first part (Eqs. 3–7).

Grid Convergence Study. In the present grid convergence study the simulations are performed on the following uniform Cartesian grids:

$$n_x \times n_y \times n_z = 64^3,\ 128^3,\ 256^3,\ 512^3 \tag{13}$$

The results are analyzed in terms of enstrophy evolution, where the enstrophy is the integral quantity defined as:

$$Z = \frac{1}{L^3}\int_D \omega^2 \, d\mathbf{x} = \nu^{-1}\varepsilon. \tag{14}$$

They are compared in Fig. 2 to the convergence study performed by Jammy et. al using an explicit finite difference solver [6]. We can notice that both methods converge with a 256^3 resolution, which corresponds to the minimum number of cells required in the domain to correctly solve the smallest scales, as explained previously. One can also emphasize an interesting feature of the Vortex Methods which relies on the fact that even with unconverged grids (e.g. 64^3), the correct maximum value of enstrophy at $T \approx 9$ is captured by the present method (Fig. 2a), which is not the case with a finite difference based method (Fig. 2b). This result highlights the low numerical diffusion produced by the present method due to the Lagrangian treatment of advection.

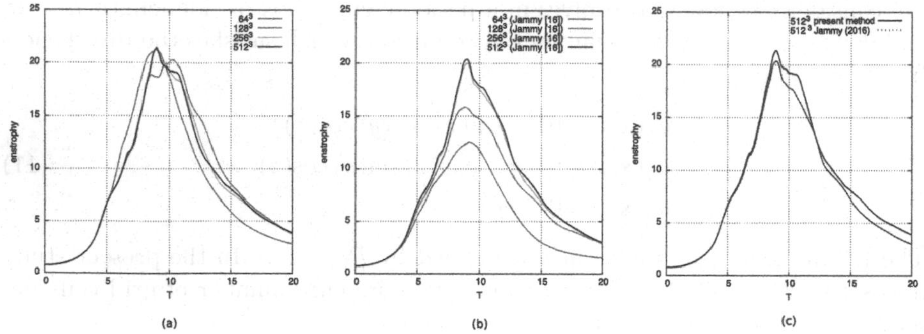

Fig. 2. Grid convergence study in terms of enstrophy evolution. (a) Present method. (b) Results obtained by Jammy et. al [6] with the OpenSBLI solver, based on a finite difference algorithm. (c) Superimposition of the solutions obtained by [6] and the present method with the converged 512^3 resolution.

Validation. Based on the results of the previous section, we consider the converged grid $n_x \times n_y \times n_z = 512^3$ for the simulations performed in this validation study. As can be seen on Fig. 2(c), the present solution is in good agreement with the one of [6], especially until $T \approx 9$, when the peak of energy dissipation is reached. A discrepancy between the two results is then observed during the flow mixing stage, showing a slightly more dissipative behavior provided by the present vortex method compared to the finite difference method of [6].

Our results are now qualitatively analyzed in terms of vortical structures in the flow. Figure 3 shows the norm of vorticity field $|\boldsymbol{\omega}|$ obtained with the present method using a 512^3 resolution in the $x = 0$ plane at $T = 9$, when the maximum of energy dissipation occurs. The vorticity isocontours of the eddy depicted on the close-up view are given below, in Fig. 4, and are compared to the one found in [10] using the same resolution of 512^3. As can be noticed, our results and the one of [10] coincide rather well. A noticeable discrepancy is noticed concerning the shape of the "eye" of the vortical structure. However, the size of the eddy and the thin elongated parts are almost similar for both methods, without any noisy regions.

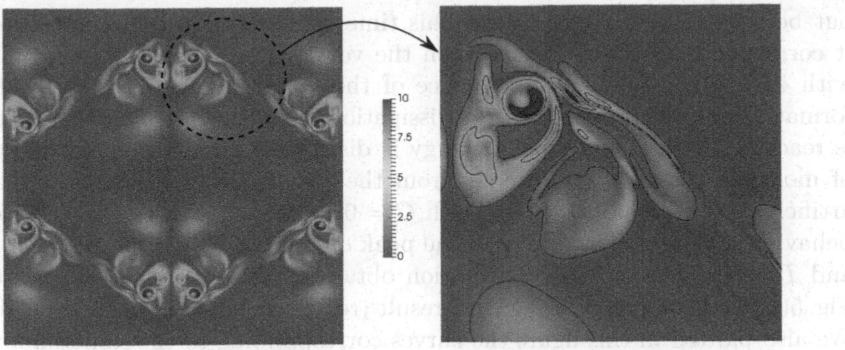

Fig. 3. T = 9. Instantaneous magnitude of vorticity field $|\omega|$ at $T = 9$ in the $x = 0$ plane with a 512^3 resolution. Global view (left) and close up view (right).

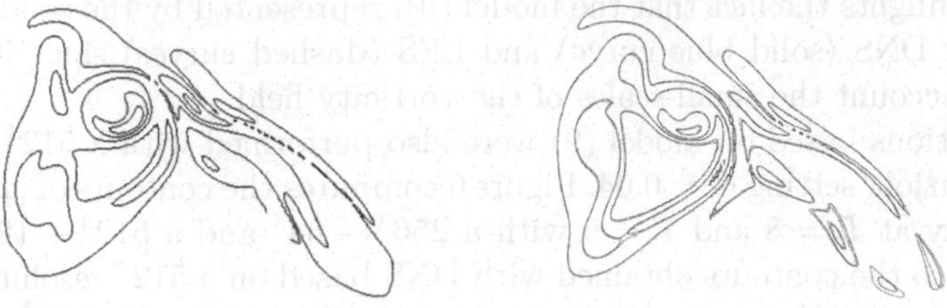

Fig. 4. T = 9. Instantaneous isocontours of $|\omega|$ for values 1, 5, 10, 20, 30 at $T = 9$ in the $x = 0$ plane with a 512^3 resolution. (Left) Present vortex method. (Right) Results obtained by [10].

4.2 Bi-level Results

This section is dedicated to the validation of the anisotropic viscosity model (9) in the context of the Taylor-Green vortex benchmark. The fractional step algorithm used here is the one given in Subsect. 3.2. As explained previously, the bi-level approach consists in using this artificial viscosity model within the incompressible $(\boldsymbol{\omega}, \mathbf{u})$ Navier-Stokes equations where the vorticity field $\boldsymbol{\omega}$ is solved on a fine grid and where the related velocity field $\mathbf{u}$ is filtered so that we only consider the velocity values on a coarse grid. The velocity filtering is performed through the following cutoff filter, defined in the Fourier space:

$$f_{k_\Delta}(|\mathbf{k}|) = 1 \quad \text{if } |\mathbf{k}| \leq k_\Delta = \frac{\pi}{\Delta} \tag{15}$$

$$0 \quad \text{otherwise} \tag{16}$$

where Δ is the filter size. In the simulations performed below we consider a 64^3 coarse mesh size for the velocity calculations and a 256^3 fine mesh size for the vorticity field. In this case the filter size is therefore equal to $\Delta = 4h$ where h, the fine grid step, is equal to $L/256$.

Figure 5(a) shows enstrophy evolution curves obtained with model (9) for different values of the constant C. This parametric study on constant C is carried out between $T = 0$ and $T = 10$. This time range is of great importance since it corresponds to the moments when the vortices roll-up and start to interact with each others under the influence of the vortex stretching, leading to the formation of regions of high energy dissipation, until the maximum of dissipation is reached ($T \approx 9$) and kinetic energy is dissipated into heat under the action of molecular viscosity. It emerges from the results given in Fig. 5(a) that the artificial viscosity model (9) set with $C = 0.04$ manages to capture the correct behavior of the flow, especially at the peak of energy dissipation between $T = 8$ and $T = 10$. The enstrophy evolution obtained with $C = 0.04$ is reported in Fig. 5(b) and compared to the DNS result (red and blue solid lines respectively). We also plotted in this figure the curves corresponding to the same simulations where the small scales of the vorticity field have also been filtered only for the evaluation of the enstrophy (denoted "Z" in the figure caption), so that Z is defined on the coarse 64^3 grid (red and blue dashed lines). Figure 5(b) thus clearly highlights the fact that the model (9), represented by the solid red curve, is between DNS (solid blue curve) and LES (dashed curves) since it allows to take into account the small-scales of the vorticity field.

Simulations based on model (9) were also performed with a $512^3 - 128^3$ bi-level resolution, setting $C = 0.04$. Figure 6 compares the contours of $|\boldsymbol{\omega}|$ obtained respectively at $T = 8$ and $T = 9$ with a $256^3 - 64^3$ and a $512^3 - 128^3$ bi-level resolution to the contours obtained with DNS based on a 512^3 resolution. These figures show that the contours obtained with model (9) and a $512^3 - 128^3$ resolution are very close in a qualitative point of view to the one given by the Direct Numerical Simulations obtained at 512^3, without spurious vortex structures. These results confirm the capability of the proposed anisotropic artificial viscosity model to adequately resolve the large scales of the flow.

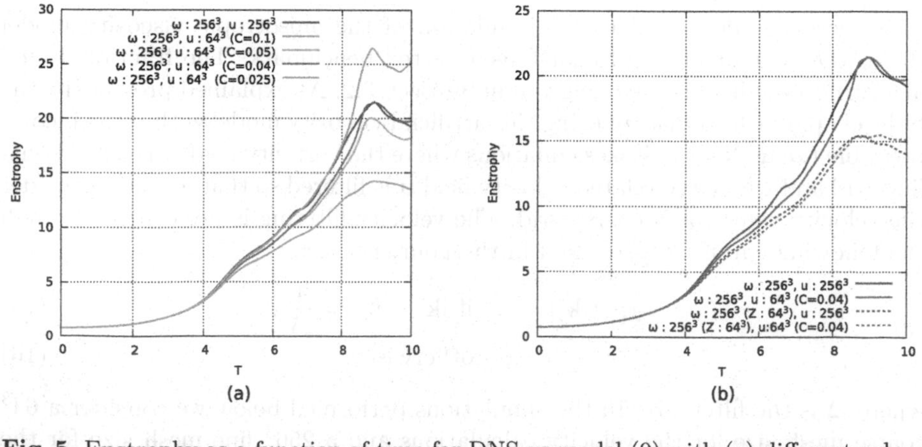

Fig. 5. Enstrophy as a function of time for DNS vs model (9) with (a) different values of C, (b) filtered vorticity values for enstrophy Z evaluation. (Color figure online)

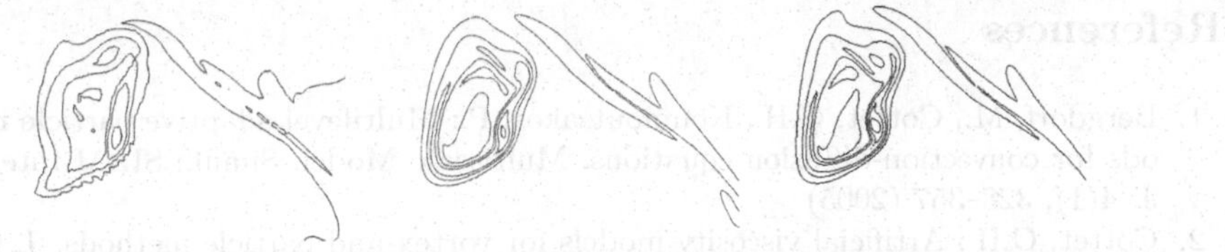

T=8

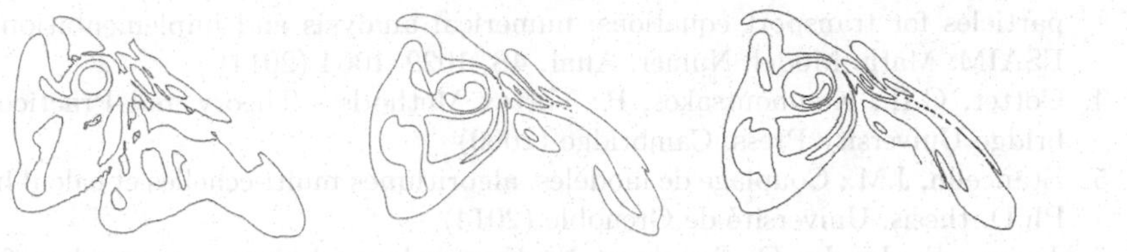

T=9

Fig. 6. Contours of $|\omega|$ in the YZ plane at $x = 0$ obtained through the artificial viscosity model (9) ($C = 0.04$) at $T = 8$ and $T = 9$ with a $256^3 - 64^3$ bi-level resolution (left) and a $512^3 - 128^3$ resolution (center). They are compared to the DNS results obtained with a 512^3 resolution (right).

5 Conclusion

In this work, a hybrid vortex method has been employed to simulate a three-dimensional flow. One original feature of this vortex method relies in the remeshing process. The particle advection-redistribution is indeed performed direction by direction, which allows significant computational savings in 3D compared to classical tensorial approaches. The present method has been first validated in the context of Direct Numerical Simulations, showing low numerical diffusion and good agreements with purely grid based high order methods. An anisotropic eddy viscosity model has also been presented in the context of this remeshed vortex method. The results obtained with such model for bi-level simulations constitute an encouraging preliminary study, which has to be tested on other 3D flow types.

Among the next tasks to consider, the main one would consist in dealing with multi-scale problems on hybrid CPU-GPU architecture in order to significantly enhance the computational performances. In practice, we aim to dedicate the high resolution vorticity transport/stretching sub-problems on multi-GPU's while solving the diffusion and Poisson equations on multi-CPU's with a velocity field defined on a coarse grid. This strategy would therefore require interpolation operations between the two coupled flow quantities.

References

1. Bergdorf, M., Cottet, G.H., Koumoutsakos, P.: Multilevel adaptive particle methods for convection-diffusion equations. Multiscale Model. Simul.: SIAM Interdisc. J. **4**(1), 328–357 (2005)
2. Cottet, G.H.: Artificial viscosity models for vortex and particle methods. J. Comput. Phys. **127**, 199–208 (1996)
3. Cottet, G.H., Etancelin, J.M., Perignon, F., Picard, C.: High order semi-Lagrangian particles for transport equations: numerical analysis and implementation issues. ESAIM: Math. Model. Numer. Anal. **48**, 1029–1064 (2014)
4. Cottet, G.H., Koumoutsakos, P.: Vortex Methods - Theory and Practice. Cambridge University Press, Cambridge (2000)
5. Etancelin, J.M.: Couplage de modèles, algorithmes multi-échelles et calcul hybride. Ph.D. thesis, Université de Grenoble (2014)
6. Jammy, S., Jacobs, C., Sandham, N.: Enstrophy and kinetic energy data from 3D Taylor-Green vortex simulations (2016). https://eprints.soton.ac.uk/401892/
7. Koumoutsakos, P., Leonard, A.: High-resolution simulations of the flow around an impulsively started cylinder using vortex methods. J. Fluid Mech. **296**, 1–38 (1995)
8. Magni, A., Cottet, G.H.: Accurate, non-oscillatory remeshing schemes for particle methods. J. Comput. Phys. **231**(1), 152–172 (2012)
9. Rasmussen, J.T., Cottet, G.H., Walther, J.H.: A multiresolution remeshed vortex-in-cell algorithm using patches. J. Comput. Phys. **230**, 6742–6755 (2011)
10. van Rees, W.M., Leonard, A., Pullin, D., Koumoutsakos, P.: A comparison of vortex and pseudo-spectral methods for the simulation of periodic vortical flows at high Reynolds numbers. J. Comput. Phys. **230**(8), 2794–2805 (2011)
11. Rossinelli, D., Hejazialhosseini, B., van Rees, W.M., Gazzola, M., Bergdorf, M., Koumoutsakos, P.: MRAG-I2D: multi-resolution adapted grids for remeshed vortex methods on multicore architectures. J. Comput. Phys. **288**, 1–18 (2015)

Accelerated Phase Equilibrium Predictions for Subsurface Reservoirs Using Deep Learning Methods

Tao Zhang(✉), Yiteng Li, and Shuyu Sun

Computational Transport Phenomena Laboratory (CTPL),
King Abdullah University of Science and Technology (KAUST),
Thuwal 23955-6900, Kingdom of Saudi Arabia
{tao.zhang.1,shuyu.sun}@kaust.edu.sa

Abstract. Multiphase fluid flow with complex compositions is an increasingly attractive research topic with more and more attentions paid on related engineering problems, including global warming and green house effect, oil recovery enhancement and subsurface water pollution treatment. Prior to study the flow behaviors and phase transitions in multi-component multiphase flow, the first effort should be focused on the accurate prediction of the total phase numbers existing in the fluid mixture, and then the phase equilibrium status can be determined. In this paper, a novel and fast prediction technique is proposed based on deep learning method. The training data is generated using a selected VT dynamic flash calculation scheme and the network constructions are deeply optimized on the activation functions. Compared to previous machine learning techniques proposed in literatures to accelerate vapor liquid phase equilibrium calculation, the total number of phases existing in the mixture is determined first and other phase equilibrium properteis will be estimated then, so that we do not need to ensure that the mixture is in two phase conditions any more. Our method could handle fluid mixtures with complex compositions, with 8 different components in our example and the original data is in a large amount. The analysis on prediction performance of different deep learning models with various neural networks using different activation functions can help future researches selecting the features to construct the neural network for similar engineering problems. Some conclusions and remarks are presented at the end to help readers catch our main contributions and insight the future related researches.

Keywords: Deep learning · Phase equilibrium · Multi-component multiphase flow

1 Introduction

Subsurface multiphase fluid low has attracted increasing attentions from researchers all over world, due to its various applications such as energy industry,

J. M. F. Rodrigues et al. (Eds.): ICCS 2019, LNCS 11539, pp. 623–632, 2019.
https://doi.org/10.1007/978-3-030-22747-0_47

including enhanced oil recovery and unconventional oil and gas development, and environmental production, including greenhouse effect and subsurface water pollution control. In petroleum industry, the study of multi-component multiphase flow is needed in all the expoitation and transportation stages to optimize the oil field development for the maximization of oil recovery [1,2,9–13]. Meanwhile, the simulation of subsurface multi-component multiphase fluid flow is critical to handle a large range of environmental concerned issues, for example, greenhouse gas emission and pollutant disposal. Especially, for the rapid development of $\mathbf{CO_2}$ sequestration technique, phase compositions and phase transition behaviors are urgently needed for the plan designing and optimization to better reduce the $\mathbf{CO_2}$ emission or controllably dispose the environmentally hazardous wastes [3,4]. In addition, the production of unconventional oil and gas resources has experienced a significant growth in all over the world and successfully becomes an important energy supply, due to the rapid production decline of conventional reservoirs [8]. In order to maximize the oil production from conventional and unconventional reservoir or resolve the aforementioned environment issues, it is great demand of the accurate numerical model of subsurface multiphase fluid systems, as well as the robust and efficient computational algorithm.

A number of studies have investigated the phase equilibrium problems in subsurface reservoirs, which often depend on different sets of given conditions [5,7]. One conventional phase equilibrium calculation approach is performed under constant chemical compositions, pressure and temperature, which is known as "NPT" flash. The long history of NPT flash has brought a wide range of applications related with phase splitting calculation and stability test, but at the same time some limitations have been found regarding this method. In some specified conditions, the system equilibrium cannot be determined uniquely, or the solution from the flash calculation fails to identify a clear and determined state. For the cubic equation of state, two separate solutions need to be conducted to determine the phase molar volumes of each phase. Furthermore, the root selection procedure has to be considered in the presence of multiple roots, although the middle root is often ignored because which corresponds to a physically meaningless solution. The remaining roots need to be paired to minimize the Gibbs free energy, and the root pairing could be challengeable as there might be two roots for each phase. If the selection of pairing is wrong, the whole procedure will fail with an unstable or metastable solution. Another limitation is that pressure is not always a priori, which makes it inefficient to iteratively solve phase equilibria problems of a different variable specification by the NPT flash. Another approach with priori conditions including constant chemical compositions, molar volume and temperature, namely the NVT flash, has been proposed to handle the above challenges faced by NPT flash. As an alternative, it has shown great potential in compositional multiphase flow simulation in subsurface porous media. However, the problem existing in the application of both the two types is the much CPU time cost in engineering scale. Generally, the temperature-pressure range or temperature-volume range, corresponding to NPT flash and NVT flash respectively, can be so wide that the iterations in each environmental condition should be reused for many times. Especially for the complex fluid mixture with a large number of components, both the two types of NPT and NVT flash calculations have problems in the quick but reliable phase equilibrium prediction.

To speed up flash calculation, different approaches have been proposed aiming at finding a path faster to estimate the phase equilibrium conditions. In this paper, we will review the general deep learning method and the detailed process, to help readers new in this area get a basic understanding of how this method can be used in the prediction. Comments on recent techniques proposed in deep learning methods have been involved with authors' own opinions, to show the current trend for a better prediction accuracy in engineering need. The performance of neural network models with different activation methods are tested from both the convergence analysis of total loss function and mean estimation errors. Furthermore, the determination of total phase numbers existing in the mixture, which is the key step in the numerical study of compositional multi-phase fluid flow in subsurface reservoirs is used as the main test target of our prediction. The performance of different activation functions, on the loss function convergence rate as well as mean absolute and mean relative errors, are compared and suggestions are made on future studies regarding similar problems.

The remainder of this paper is organized as follows. In Sect. 2, the deep learning method is explained in details, with clear procedures and instructions with author's own comments. Prediction examples are presented in Sect. 3 to show the robustness and efficiency of the concluded scheme. At the end, we make some conclusions in Sect. 4.

2 Deep Learning Method

Artificial neural networks (ANNs) are computational models designed to incorporate and extract key features of the original inputs and process data in a manner analogous to neurons in animals' central nervous systems (in particular the brain), which are capable of both machine learning, as well as pattern recognition, the former sometimes being called supervised and the latter unsupervised machine learning. The naming convention stems from the fact that in supervised machine learning, unlike the unsupervised fashion, the ANN is presented with the target variable(s) and seeks to find a functional relationship that can be used to predict the target variable(s) from input variable(s) with a desired degree of precision. Deep neural networks usually refer to those artificial neural networks which consist of multiple hidden layers. In this paper a deep fully connected neural network is applied to model the phase equilibrium calculation with data sourced from VT flash calculation are used as input. Following the input layer, a number of fully connected hidden layers, with a certain number of nodes, stack over the other, whose final output is fed into another fully connected layer, which is the final output layer. Since we are fitting the compositions of vapor phase and liquid phase in our model, the final output layer contains several nodes, which could be divided into three sets: one for total phase numbers existing in the mixture, one for vapor phase compositions and the final one for liquid phase compositions. The activation function of this layer is fixed as linear. Naturally the proposed ANN input variables include critical pressure (Pc), critical temperature (Tc), acentric factor (ω) and z of the components comprising the mixture, as well as the temperature T and mole volume c as the environmental condition. Different from previous studies, a much larger range of T and c can be included in our environmental conditions, as it is not a priori in our approach that the mixture should be ensured in a two phase area. On the contrast, our algorithm can decide whether the mixture is in single phase or two phase area, and then process the phase equilibrium estimation accordingly.

The whole package is developed using TFlearn. Trained on a Mac Laptop, which is a common equipment, the training iterations will converge in less than 10 min if the source data is with size 151 × 151. Formally, for the i-th hidden layer, let $\mathbf{a}_i$ denote the input of the layer, and $\mathbf{y}_i$ to denote the output of the layer. Then we have:

$$\mathbf{y}_i = f_i(\mathbf{W}_i * \mathbf{a}_i + \mathbf{b}_i), \tag{1}$$

where $\mathbf{W}_i$ is the weight; $\mathbf{b}_i$ is the bias; and f_i is the activation functions of the i-th layer. For a network with multiple layers, the output of one hidden layer is the input of the next layer. For example, we can replace the general neural network as shown in [14] as:

$$\mathbf{o} = f_3(\mathbf{W}_3 * f_2(\mathbf{W}_2 * f_1(\mathbf{W}_1 * \mathbf{x}_1 + \mathbf{b}_1) + \mathbf{b}_2) + \mathbf{b}_3), \tag{2}$$

where $\mathbf{o} = (X, T)$; f_1, f_2, f_3 are the activation functions; $\mathbf{W}_1, \mathbf{W}_2, \mathbf{W}_3$ are the weights for each layer; $\mathbf{b}_1, \mathbf{b}_2, \mathbf{b}_3$ are the bias terms of each layer.

As shown in Eq. (2), the activation function is where the non-linearity and the expressiveness power of deep neural network models comes from. There are numerous activation functions: Rectified linear unit (ReLU), Parametric rectified linear unit (PReLU), TanH, Sigmoid, Softplus, Softsign, Leaky rectified linear unit (Leaky ReLU), Exponential linear unit (ELU), and Scaled exponential linear unit (SELU). Here, we present the formula for two common used activation functions, ReLU and Sigmoid, as an example.

$$f(x) = \Big\{0, if x < 0; x, if\, x \geq 0, \tag{3}$$

$$\sigma(x) = \frac{1}{1 + \exp(-x)}. \tag{4}$$

3 Examples

In this section, the predictions of phase total phase numbers existing in a fluid mixture in various complex reservoir environmental conditions are performed based on the 8 components mixture detected in EagleFord Oilfield. Detailed component compositions and the parameters effecting phase equilibrium results using VT flash calculations based on Peng-Robinson Equation of State are presented in Table 1. The VT flash calculation approach is selected as described in [6,8], which is energy stable so as to accelerate the flash calculation for the large scale data. Besides, this scheme is consistent with the first and second thermodynamical laws, which makes the phase equilibrium results more reliable. Unlike machine learning methods reported in previous literatures, the environmental conditions need no more to ensure the mixtures to be in a two phase area, so that the applications of our method can be larger. The number of total phases is the result predicted using our method, and other phase equilibrium properties can be estimated correspondingly.

Table 1. Molar composition and compositional properties for the EagleFord2 oil.

Component	z_i	$T_{c,i}$ (K)	$P_{c,i}$ (MPa)	$M_{w,i}$ (mol/m^3)	ω_i
C_1	0.31231	190.72	4.6409	16.04	0.0130
N_2	0.00073	126.22	3.3943	28.01	0.0400
C_2	0.04314	305.44	4.8842	30.07	0.0986
C_3	0.04148	369.89	4.2568	44.10	0.1524
CO_2	0.01282	304.22	7.3864	44.01	0.2250
iC_4	0.01350	408.11	3.6480	58.12	0.1848
nC_4	0.03382	425.22	3.7969	58.12	0.2010
iC_5	0.01805	460.39	3.3336	72.15	0.2223
nC_5	0.02141	469.78	3.3750	72.15	0.2539
nC_6	0.04623	507.89	3.0316	86.18	0.3007
C_{7+}	0.16297	589.17	2.7772	114.40	0.3739
C_{11+}	0.12004	679.78	2.1215	166.60	0.5260
C_{15+}	0.10044	760.22	1.6644	230.10	0.6979
C_{20+}	0.07306	896.78	1.0418	409.20	1.0456

Using the deep learning approach described in Sect. 2, a neural network is designed with 5 hidden layers, 100 nodes in each layer, totally 4000 iterations. The performance of using different activation functions are tested in this section, to help provide a suggestion for future researchers. A 101 × 101 VT flash data source is generated for this mixture, so the total original data for training is 9180, and the tested data is 1021. Namely, the tested proportion in this paper is selected as 10 per cent. Especially, the key effort in phase equilibrium calculation, e.g. the decision of total phase numbers in a mixture, is also processed using our deep learning approach to test its capability to handle this problem.

3.1 Deep Learning Model Training

As explained in Sect. 2, the model takes the parameters of each components, as well as temperature (T) and mole volume (c) as the input and outputs the predicted value of X and Y in each phase. The key parameters of the model are the weights of each layer, which control what the model outputs given the input. At first, those weights are initialized randomly, which means that the model will output useless values given the inputs. To make the model useful for this problem, we need to train those weight parameters to fit our problem. The difference between the model's output and the ground truth is referred as loss. Here, for this regression problem, we use mean square error as the loss function. 10% of the data sources are selected as the test data, where the remaining data are input into the network for the training. The mean absolute error and mean relative error of the tested results from different networks with various activation functions used are listed in Table 2. As indicated from the results in [14], four activation functions are selected from the low error group: 'tanh', 'relu', 'sigmoid' and 'softsign' and one activation function selected from the high error group 'softplus' to test whether their performance will show similar results. It is glad to see from Table 2 that generally the four activations in low error group in [14] will show better estimation errors compared with the high error activation function 'softplus'. It is verified the statement resulted from the binary components cases can be extended to the complex multi-component cases.

Table 2. Estimation error of deep learning model trained with different activation functions

Activation functions	Mean absolute error	Mean relative error
softplus	0.00679	81540
tanh	0.01342	0.01455
relu	0.01378	0.01723
sigmoid	0.01900	0.02066
softsign	0.01398	0.01499

The loss function curve decreasing with iterations are also presented to show how the trained outputs are approaching the true value. Results from iteration 200 to iteration 400 are drawn as the loss function in this period represent a clear approaching trend. From Fig. 1, it can be referred that the total loss function using activation function 'softplus' is much larger than that of the other four neural networks. The loss function of 'sigmoid' and 'softsign' show similar lowest value, but they remain almost steady compared to the constant decrease of the network generated using 'tanh' and 'relu'.

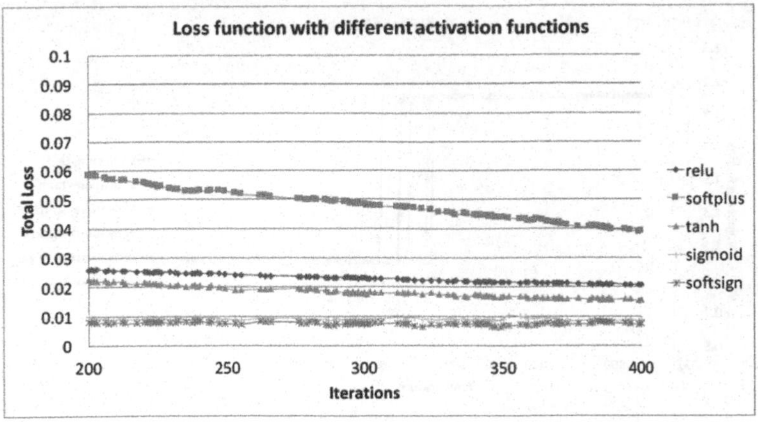

Fig. 1. The neural network to model VLE.

3.2 Phase Number Characterization

The key effort of phase equilibrium calculation is to decide the total number of phases existing in the mixture at certain conditions, which is priori to further prediction of phase behaviors. To test the capability of the deep learning approach to capture this value, a mixture of 8 components detected in EagleFord Oilfield is selected to generate the data source using our proposed VT flash calculation algorithm, with the data size of 101×101 and at the constant $c = 129.9$. As shown in Fig. 2, it can be referred that with the temperature increasing from 250 K to 850 K, this mixture will change from a two phases mixture to a single phase mixture. For all the five activation functions used in constructing the neural network, generally the deep learning result meet well with the original data generated from VT flash calculation, with only a small difference at the phase transition point. It is obvious to see that the results predicted using activation function 'sigmoid' and 'softplus' have the maximum prediction errors, with a larger phase change temperature point and a smaller temperature respectively. Meanwhile, the result of deep learning neural networks generated using activation function 'softsign' meets well with the original data resulted from VT flash data, with only a small difference on the prediction of phase change temperature point. Thus, it can be concluded that the activation function 'softsign' best fits the vapor liquid equilibrium problems, especially on the total phase numbers determination.

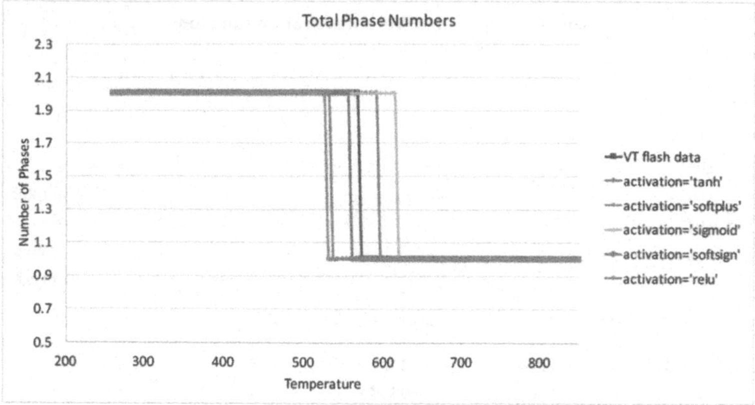

Fig. 2. Number of phases predicted in a mixture with different activation functions used in the neural network at $c = 129.9$

Combined with the error analysis in Sect. 3.1, it could be referred that activation function 'relu' and 'softsign' can be suggested for future deep learning methods considering to construct the neural networks. To prove this statement, another phase equilibrium prediction is performed with our trained model for constant mole volume set as $c = 249.8$. As shown in Fig. 3, our statement is verified with the fact the results using activation function 'sigmoid' and 'softplus' deviate the most with the original flash calculation data, while the result of activation function selected as 'softsign' totally agree with the original data. The result of activation function 'relu' also meets well with the VT flash result, while the 'tanh' activation function performs not as good in both the two cases.

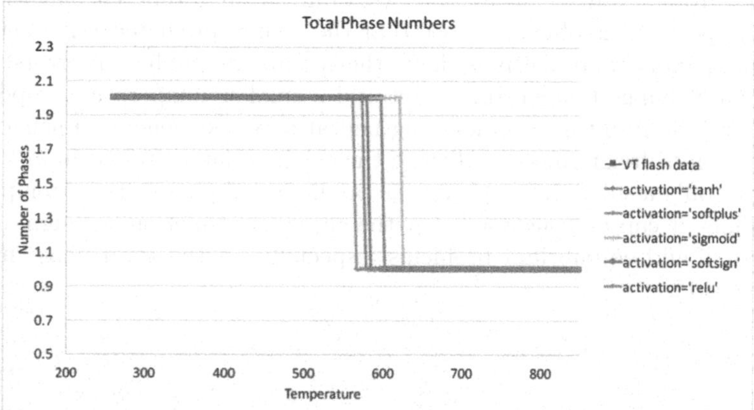

Fig. 3. Number of phases predicted in a mixture with different activation functions used in the neural network at $c = 249.8$

4 Conclusion

The concept and detailed procedures of deep learning method is described in this paper, especially instructions for the construction of deep neural networks used to predict vapor liquid phase equilibrium. Previous methods for phase equilibrium calculations are introduced and analyzed first, and this explains why we use VT flash algorithm for complex multicomponent mixtures to generate the original phase equilibrium data used for the training in our deep learning method. The robustness and efficiency of our designed neural network is verified by the accurate predictions for the complex mixture selected from a realistic oil field, and the comparison of different activation functions is presented from three different viewpoints: loss function convergence rate, mean absolute and relative error and total phase number prediction. Based on our results, it could be concluded that different activations perform with different accuracy in this type of problems, where 'relu' and 'softsign' are suggested to construct the deep neural networks. This statement is similar with the comparison of the performance of activation functions in [14] for binary components. It indicates that this optimized configurations set is capable to handle a large range of phase equilibrium problems, from components, and data sources. As the total phase number can be decided using our method, the priori that the mixture should stay in a two phase area is not necessary any more, which greatly increases the capability and application ranges of deep learning methods in the phase equilibrium predictions. On future studies, a potential direction is to take into consider more mechanisms in reservoirs, like capillary pressure, to see the capability of this network and approach.

References

1. Aziz, K.: Petroleum Reservoir Simulation. Applied Science, London (1979)
2. Dawson, C., Sun, S., Wheeler, M.F.: Compatible algorithms for coupled flow and transport. Comput. Methods Appl. Mech. Eng. **193**(23–26), 2565–2580 (2004)
3. El-Amin, M., Sun, S., Salama, A.: Modeling and simulation of nanoparticle transport in multiphase flows in porous media: CO_2 sequestration. Math. Methods Fluid Dyn. Simul. Giant Oil Gas Reserv. (2012). https://www.onepetro.org/conference-paper/SPE-163089-MS
4. Espinoza, D.N., Santamarina, J.C.: Water-CO_2-mineral systems: interfacial tension, contact angle, and diffusion-implications to CO_2 geological storage. Water Resour. Res. **46**(7), W07537(1–10) (2010)
5. Firoozabadi, A.: Thermodynamics of Hydrocarbon Reservoirs. McGraw-Hill, New York (1999)
6. Kou, J., Sun, S., Wang, X.: Linearly decoupled energy-stable numerical methods for multicomponent two-phase compressible flow. SIAM J. Numer. Anal. **56**, 3219–3248 (2018)
7. Jindrová, T., Mikyška, J.: General algorithm for multiphase equilibria calculation at given volume, temperature, and moles. Fluid Phase Equilib. **393**, 7–25 (2015)
8. Li, Y., Kou, J., Sun, S.: Thermodynamically stable two-phase equilibrium calculation of hydrocarbon mixtures with capillary pressure. Ind. Eng. Chem. Res. **57**(50), 17276–17288 (2018)

9. Mikyška, J., Firoozabadi, A.: A new thermodynamic function for phase-splitting at constant temperature, moles, and volume. AIChE J. **57**(7), 1897–1904 (2011)
10. Moortgat, J., Sun, S., Firoozabadi, A.: Compositional modeling of three-phase flow with gravity using higher-order finite element methods. Water Resour. Res. **47**(5), W05511(1–26) (2011)
11. Sun, S., Liu, J.: A locally conservative finite element method based on piecewise constant enrichment of the continuous Galerkin method. SIAM J. Sci. Comput. **31**(4), 2528–2548 (2009)
12. Wu, Y.S., Qin, G.: A generalized numerical approach for modeling multiphase flow and transport in fractured porous media. Commun. Comput. Phys. **6**(1), 85–108 (2009)
13. Zhang, T., Kou, J., Sun, S.: Review on dynamic Van der Waals theory in two-phase flow. Adv. Geo-Energy Res. **1**(2), 124–134 (2017)
14. Li, Y., Zhang, T., Sun, S., Gao, X.: Accelerating Flash Calculation through Deep Learning Methods. arXiv preprint arXiv:1809.07311 (2018)

Study on the Thermal-Hydraulic Coupling Model for the Enhanced Geothermal Systems

Tingyu Li[1], Dongxu Han[2(✉)], Fusheng Yang[1(✉)], Bo Yu[2], Daobing Wang[2], and Dongliang Sun[2]

[1] School of Chemical Engineering and Technology, Xi'an Jiaotong University, Xi'an 710049, China
yang.fs@xjtu.edu.cn

[2] School of Mechanical Engineering Key Laboratory of Pipeline Critical Technology and Equipment for Deepwater Oil and Gas Development, Beijing Institute of Petrochemical Technology Beijing, Beijing 102617, China
handongxubox@bipt.edu.cn

Abstract. Enhanced geothermal systems (EGS) are the major way of the hot dry rock (HDR) exploitation. At present, the finite element method (FEM) is often used to simulate the thermal energy extraction process of the EGS. Satisfactory results can be obtained by this method to a certain extent. However, when many discrete fractures exist in the computational domain, a large number of unstructured grids must be used, which seriously affects the computational efficiency. To solve this challenge, based on the embedded discrete fracture model (EDFM), two sets of seepage and energy conservation equations are respectively used to describe the flow and heat transfer processes of the matrix and the fracture media. The main advantages of the proposed model are that the structured grids can be used to mesh the matrix, and there is no need to refine the mesh near the fracture. Compared with commercial software, COMSOL Multiphysics, the accuracy of the proposed model is verified. Subsequently, a specific example of geothermal exploitation is designed, and the spatial-temporal evolutions of pressure and temperature fields are analyzed.

Keywords: Enhanced geothermal systems · Embedded discrete fracture model · Thermal-hydraulic coupling · Numerical simulation

1 Introduction

Hot dry rock (HDR), as one of geothermal energy, has high exploitation valuableness and prospect due to its high quality of thermal storage. Enhanced geothermal systems or engineering geothermal systems (EGS) are the exploitation technology of HDR (shown in Fig. 1). Its fundament thought is to obtain high-temperature water or steam by injecting cold water into the artificial heat storage system formed by the hydraulic fracturing technology. The thermal extraction processes mainly include the following four steps. ① The low-temperature working fluid is injected into the deep storage system through the injection well; ② After fully exchanging heat between the working fluid and high-temperature rock, the temperature of working fluid increases and flows

J. M. F. Rodrigues et al. (Eds.): ICCS 2019, LNCS 11539, pp. 633–646, 2019.
https://doi.org/10.1007/978-3-030-22747-0_48

to the bottom of production well; ③ The high-temperature working fluid is extracted through the production well; ④ The working fluid is transported to the ground power generation system. Due to the complexity of heat recovery process, the exploitation of EGS involves the complex spatial-temporal evolution processes of multi-physics fields such as fluid flow, convective heat transfer and mechanical deformation in fractured rock mass [1, 2]. Numerical simulation is considered one of the effective ways to study this process [3–11]. Because the artificial thermal reservoir contains the fracture network caused by the hydraulic fracturing, the difficulty of numerical simulation lies in how to characterize these fractures and establish efficient and reliable multi-physics field coupling model.

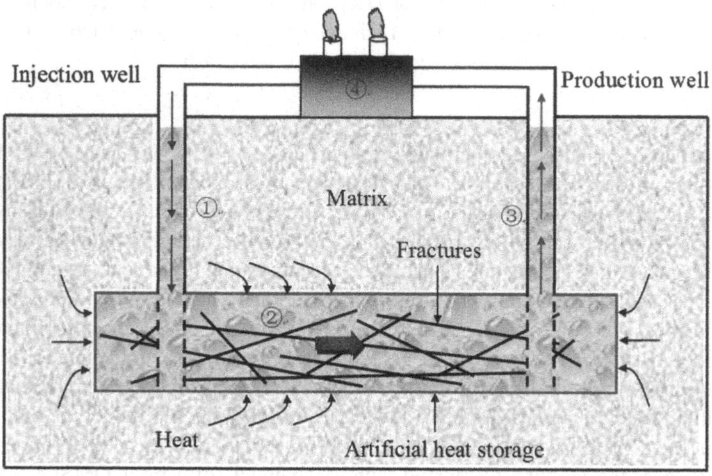

Fig. 1. Schematic of the HDR exploitation process

The characterization of fracture network has made a lot of progress in the simulation of petroleum exploitation. A widely used fracture model is the discrete fracture model (DFM) [12, 13]. Based on the DFM, researchers have established some geothermal exploitation models. Sun et al. [14] and Yao et al. [15] used COMSOL Multiphysics to establish a two-dimensional (2D) and three-dimensional (3D) thermal-hydraulic-mechanical (THM) coupling models of the HDR exploitation process, respectively. However, the DFM has a significant disadvantage when meshing. The fracture must be aligned with the matrix grid, and a large number of unstructured grids must be used for the simulation of complex fracture distribution, which seriously affects the efficiency of numerical calculation. This is the reason why the DFM cannot be directly applied to the actual engineering applications. To overcome the shortcomings of DFM, Lee et al. [16] developed an embedded discrete fracture model (EDFM) (shown in Fig. 2), which divides the computational domain into two kinds of media, termed matrix and fracture. The matrix can be divided by the structured grids, and the fractures are directly embedded in the grid system of the matrix, thus avoiding the complicated unstructured grids of the traditional DFM. Based on the EDFM,

Gunnar et al. [2] established an efficient method to investigate the role of thermal stress during hydraulic simulation over short and long periods. Yan et al. [17] developed a hydro-mechanical coupling model of the fractured rock mass. The results of the numerical calculation are not much different from those of COMSOL Multiphysics, and the computational speed is faster. Karvounis [18] improved the EDFM and proposed an adaptive hierarchical fracture model to simulate the operational process of EGS. In EDFM, the dimension of fracture is reduced. For the 2D problem, the fracture is regarded as a 1D line segment, while for the 3D problem, the fracture is considered as a 2D plane.

Therefore, in view of the advantages of the EDFM, the exploitation model of HDR is established by using this fracture model in this paper. Two sets of flow and energy conservation equations are used to describe the flow and heat transfer processes in the matrix and the fracture, respectively. The accuracy of the proposed model is verified by comparing with COMSOL Multiphysics. Finally, combined with fractal theory, an artificial thermal storage system with a fractal tree is generated to characterize the artificial fracture network in the exploitation process of HDR. The spatial-temporal evolution processes of pressure and temperature fields are analyzed.

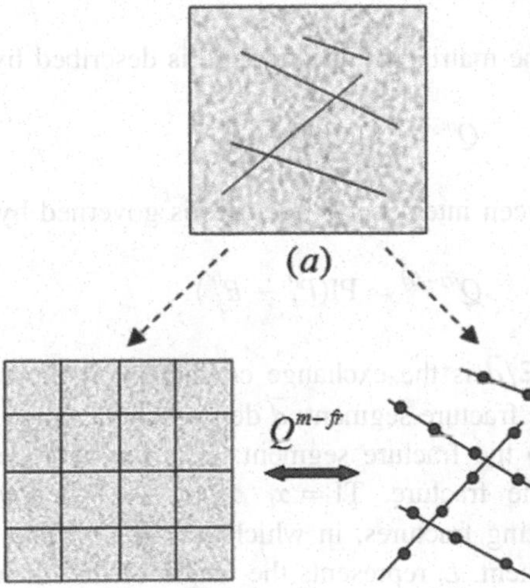

Fig. 2. Schematic of the embedded discrete fracture model (EDFM)

2 Governing Equations

2.1 Seepage Equations

As shown in Fig. 2, the EDFM regards the fractured rock mass as two sets of independent media. The coupling between the matrix and the fracture is implemented through the transfer flow function. The mass conservation equation of the matrix media is as follow.

$$\rho_f c_t \phi^m \frac{\partial p^m}{\partial t} + \rho_f \nabla \cdot (\mathbf{v}^m) = Q^{fr-m} + Q_w \tag{1}$$

For the fracture media, we have:

$$\rho_f c_t \phi^{fr} \frac{\partial p^{fr}}{\partial t} + \rho_f \nabla \cdot (\mathbf{v}^{fr}) = Q^{m-fr} + Q^{fri-frj} \tag{2}$$

In Eqs. (1)–(2), ρ_f is the fluid density; p and T represent the pressure and temperature field, respectively; t is the time; ϕ is the porosity; c_t is the coefficient of compressibility; Q_w is the source or sink term, described by the Peaceman formula [19, 20].

Velocity fields in Eqs. (1)–(2) are calculated by Darcy's equation.

$$v^i = -\frac{\mathbf{k}^i}{\mu_f}(\nabla p^i - \rho_f \mathbf{g}),\ i = m \,\text{or}\, fr \tag{3}$$

where μ_f is the dynamic viscosity; $\mathbf{k}$ is the permeability tensor; $\mathbf{g}$ is the gravity acceleration.

The coupling between the matrix and the fracture is described by the Eq. (4).

$$Q^{m-fr} = \mathrm{CI}(P^m - P^{fr}) \tag{4}$$

The exchange flow between intersecting fractures is governed by the Eq. (5).

$$Q^{fri-frj} = \mathrm{PI}(P_i^{fr} - P_j^{fr}) \tag{5}$$

In Eqs. (4)–(5), CI = $A\Xi/\bar{d}$ is the exchange coefficient; A represents the area of matrix grids occupied by the fracture segment; $\bar{d}$ denotes the average distance from all points in the matrix grid to the fracture segment; Ξ is the average flow coefficient between the matrix and the fracture. $\mathrm{TI} = \alpha_i \cdot \alpha_j/(\alpha_i + \alpha_j)$ denotes the exchange coefficient between intersecting fractures, in which $\alpha_i = 2b_i k_i/(\mu L_i)$, b_i indicates the aperture of the fracture segment. L_i represents the length of the ith fracture segment.

2.2 Energy Conservation Equations

According to the idea of the EDFM, two sets of energy equations are still used to describe the matrix and the fracture media, respectively, and the exchange energy between them should be considered. For the matrix media:

$$(\rho c_p)_{eff} \frac{\partial T^m}{\partial t} + (\rho c_p)_f \mathbf{v}^m \cdot \nabla T^m = \nabla \cdot (k_{eff} \nabla T^m) + E^{fr-m} + Q_E \tag{6}$$

where $E^{fr-m} = \mathrm{H}(\mathbf{V}^{fr-m})\mathbf{V}^{fr-m}h^{fr} + \mathrm{H}(-\mathbf{V}^{fr-m})\mathbf{V}^{fr-m}h^{m} + k_{eff}\mathrm{CI}(T^{fr} - T^{m})$ represents the energy exchange between the matrix and the fracture; $\mathbf{V}^{fr-m}$ represents the relative velocity between the matrix and the fracture; $H(\cdot)$ is the Heaviside function; Q_E is the energy term of well, including the injected and produced energy.

The energy conservation equation of the fracture media is described by the Eq. (7).

$$(\rho c_p)_{eff}\frac{\partial T^{fr}}{\partial t} + (\rho c_p)_f \mathbf{v}^{fr} \cdot \nabla T^{fr} = \nabla \cdot (k_{eff} \nabla T^{fr}) + E^{m-fr} + E^{fri-frj} \tag{7}$$

where $E^{fri-frj} = (\mathrm{H}(\mathbf{V}^{fri-frj})\mathbf{V}^{fri-frj}h^{fri}) + \mathrm{H}(-\mathbf{V}^{fri-frj})\mathbf{V}^{fri-frj}h^{frj} + k_{eff}\mathrm{TI}(T^{fri} - T^{frj})$ is the energy exchange between different fractures; $\mathbf{V}^{fri-frj}$ represents the relative velocity between the ith fracture and the jth fracture; $(\rho c_p)_{eff} = \phi^i(\rho c_p)_f + (1-\phi^i)(\rho c_p)_s$ denotes the effective physical parameters; $k_{eff} = \phi^i k_f + (1-\phi^i)k_s$ represents the effective thermal conductivity;

2.3 Evolutions of the Fluid Properties

In geothermal exploitation, the working fluid is usually injected into the target position at a low temperature. Due to the high temperature and pressure in the thermal reservoir, the thermal properties of the working fluid will be greatly affected [21]. The equation proposed by Wagner and Kurse [22] is used to describe the changing relationship of density and viscosity with pressure and temperature. For the change of density, we have:

$$\rho(\psi, \tau) = p/(RT\psi\lambda_{psi}) \tag{8}$$

where $\psi = p/p^*$; the reference pressure $p^* = 16.53\mathrm{Mpa}$; $\tau = T^*/T$; the reference temperature $T^* = 1386K$; the specific gas constant of water $R = 461.526\mathrm{J}/(kg \cdot \mathrm{K})$; $\lambda_{psi} = \sum_{i=1}^{34} -n_i l_i (7.1-\psi)^{l_i-1}(\tau - 1.222)^{J_i}$.

For the viscosity, we have:

$$\mu = \mu^*(\tau^{0.5}\sum_{i=0}^{3}\tau^i)^{-1}\exp(\delta\sum_{i=1}^{19}n_i(\delta-1)^{l_i}(\tau-1)^{J_i}) \tag{9}$$

where $\mu^* = 5.5071 \times 10^{-5}\mathrm{Pa \cdot s}$ is the reference viscosity; $\tau = 647.226/T$; $\delta = \rho/317.763\mathrm{kg/m^3}$.

Figure 3 shows the variations of density and viscosity with pressure and temperature. It can be seen from Fig. 3b, the change of viscosity with temperature is significant. From Eq. (3), it is obvious that the change of viscosity will affect the calculation of Darcy's velocity. Therefore, it is not accurate to set the viscosity as a fixed value in the actual simulation.

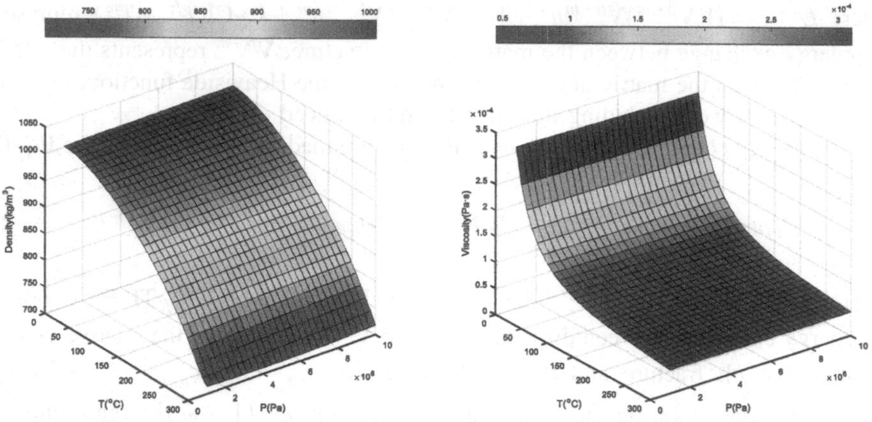

(a) Density (b) Viscosity

Fig. 3. Variations of the fluid property with pressure and temperature

3 Model Solution and Verification

3.1 Model Solution

In this paper, the finite volume method (FVM) is used to discretize the flow and heat transfer equations. The seepage equations are discretized by the two-point flux approximation scheme (TPFA). For the energy equations, the convection terms are discretized by the QUICK scheme, the diffusion terms are discretized by the second order central difference scheme, and the unsteady term is discretized by the first-order backward Euler scheme.

3.2 Verification-Five Discrete Fractures

Consider the 2D square matrix with a side length of 100 m as shown in Fig. 2a. Five discrete fractures are predefined in the computational domain. The left boundary simulates the injection well with a pressure of 10^6 Pa. The right boundary simulates the production well with a pressure of 10^5 Pa. The initial pressure is 10^5 Pa. The impermeable boundary conditions are applied on the top and bottom boundaries. The injection temperature at the left boundary is 20 °C and the initial temperature is 180 °C. The remaining boundaries are adiabatic. Other main parameters are shown in Table 1. The time step is one year and the total simulation time is 40 years.

To verify the correctness of the model, the results are compared with the commercial software-COMSOL. Figure 4a shows the grid used in COMSOL. The total number of free triangular grids is 7,019. Figure 4b is the computational grid adopted in the proposed model. The number of matrix grids is 2,500. The number of fracture grids is 121 and the total computational grid is 2,621. The computing time of COMSOL is 17 s and that of the proposed model is 8 s. Figure 5 shows the changes in pressure and temperature fields after 40 years. The results of COMSOL are displayed in the left and the right is the results of the proposed model. There is no apparent difference between these two groups of results. To quantitatively verify the computational accuracy of the pressure and temperature fields, the horizontal fracture in Fig. 4 is selected. With COMSOL as the reference solutions, the computational results are shown in Fig. 6. It can be seen that both the pressure and temperature field are in good agreement, which verifies the correctness of the model and algorithm in this paper.

Table 1. Main model parameters

Parameters	Values
Size (m)	100 × 100
Permeability of matrix and fractures (m^2)	10^{-16}, 10^{-11}
Porosity of matrix and fractures	0.2, 0.3
Fluid viscosity (kg/m·s)	10^{-3}
Fluid density (kg/m^3)	1000
Matrix density (kg/m^3)	2500
Thermal conductivity of fluid and matrix (W/m °C)	2.0, 0.5
Thermal capacity of fluid and matrix (J/kg K)	1000, 4000
Fluid compressibility (Pa^{-1})	4.4×10^{-10}
Initial pressure (Pa)	10^5
Initial temperature (°C)	180
Fracture aperture (m)	10^{-3}

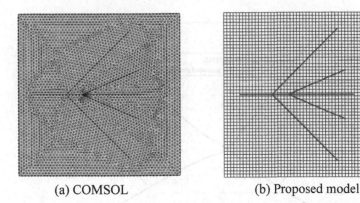

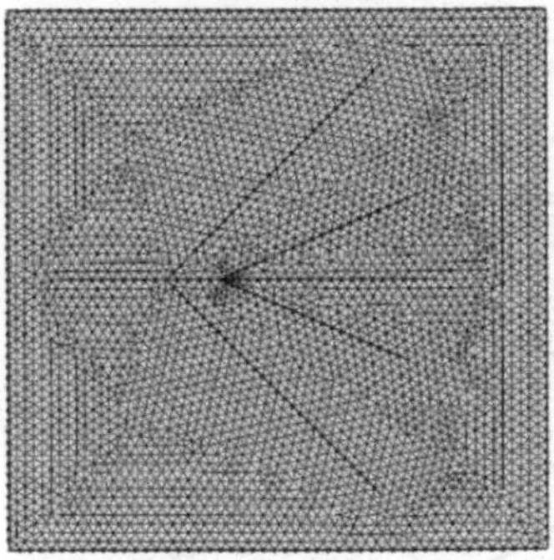

(a) COMSOL

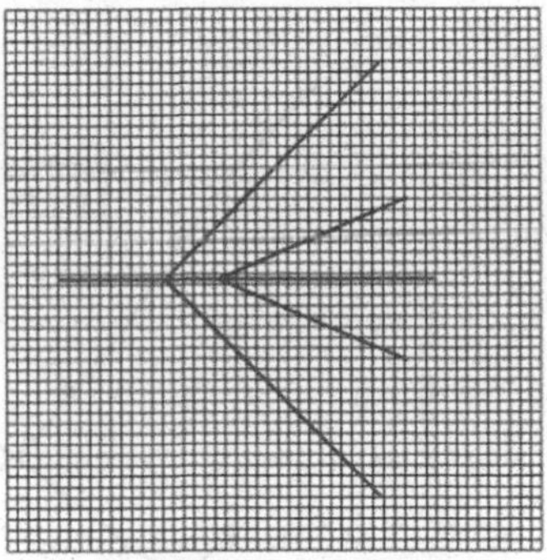

(b) Proposed model

Fig. 4. Computational grids for the COMSOL and the proposed model.

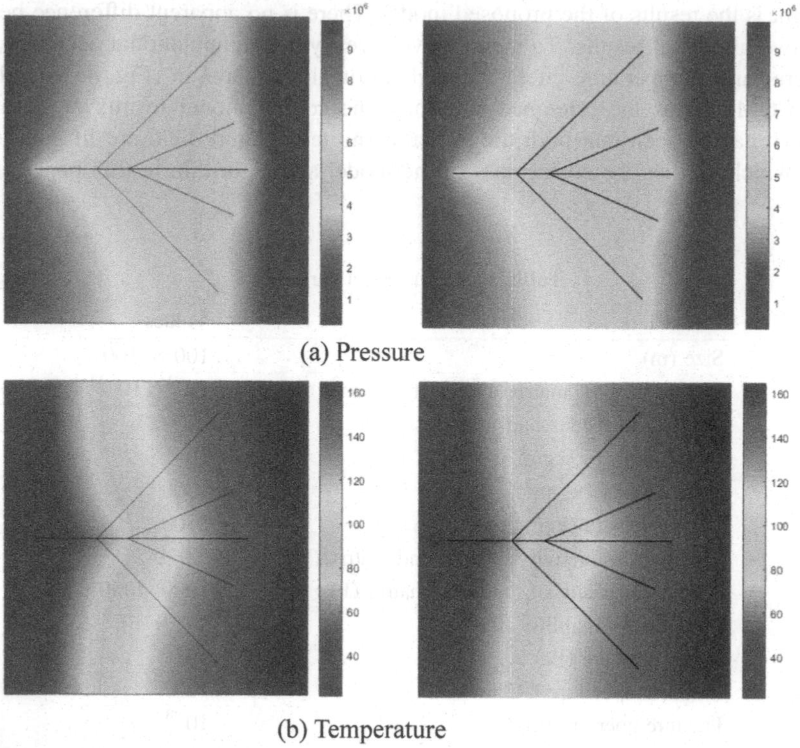

Fig. 5. Results comparison of COMSOL (left) and the proposed model (right).

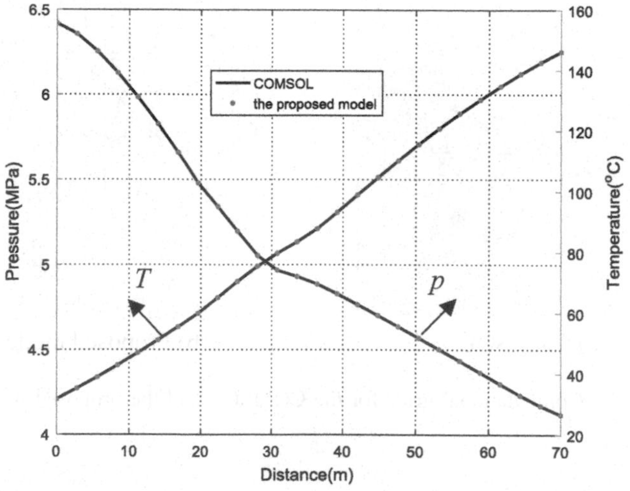

Fig. 6. Comparisons of pressure and temperature in the horizontal fracture with COMSOL.

4 Model Application

The EGS can be exploited after artificial hydraulic fracturing. The fracture network will affect the whole thermal extraction process. However, due to the complex structure of deep thermal reservoirs, it is difficult to get all the information of fracture distributions. Some studies have shown that the fracture network after artificial fracturing has fractal characteristics to a certain extent [23]. Therefore, based on fractal theory, fractal trees with 63 fractures are generated to represent the fracture network of deep thermal reservoir. As shown in Fig. 7b, the start-point and end-point coordinates of the first trunk are (20 m, 20 m) and (40 m, 40 m). The angle between the branch and the horizontal and vertical directions is 30°, and the number of iterations of the fractal tree is 4 times. The computational grids of fractal fracture tree are shown in Fig. 8. Each matrix grid contains only one fracture segment. The number of matrix grids is 10,000. The number of fracture grids is 409 and the total number of grids is 10,409. The time step is one year and the total simulation time is 40 years. The specific model parameters are shown in Table 2. The boundary conditions of the model are as follows.

(1) Pressure boundary: all boundaries are subjected to an impermeable boundary condition. The injection well is located at the bottom left corner, and the injected flow rate is 130 m^3/day. The produced pressure of the production well at the top right corner is 10^5 Pa.
(2) Temperature boundary: The matrix is surrounded by adiabatic boundary conditions. The initial reservoir temperature is 180 °C and the injected temperature is 20 °C.

Table 2. Parameters of the fractal tree model

Parameters	Values
Size (m)	100 × 100
Permeability of matrix and fractures (m^2)	10^{-16}, 10^{-10}
Porosity of matrix and fractures	0.2, 0.3
Fluid viscosity (kg/m s)	10^{-3}
Fluid density (kg/m^3)	1000
Matrix density (kg/m^3)	2500
Thermal conductivity of fluid and matrix (W/m·°C)	2.0, 0.5
Thermal capacity of fluid and matrix (J/kg·K)	1000, 4000
Fluid compressibility (Pa^{-1})	4.4×10^{-10}
Initial pressure (Pa)	10^5
Well diameter (m)	0.1
Fracture aperture (m)	10^{-3}

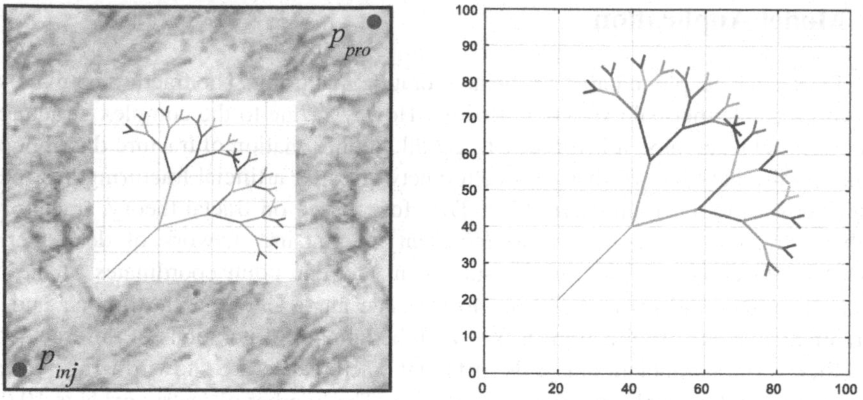

(a) Computational domain (b) Fractal fracture tree

Fig. 7. Fractal fracture tree including 63 fractures

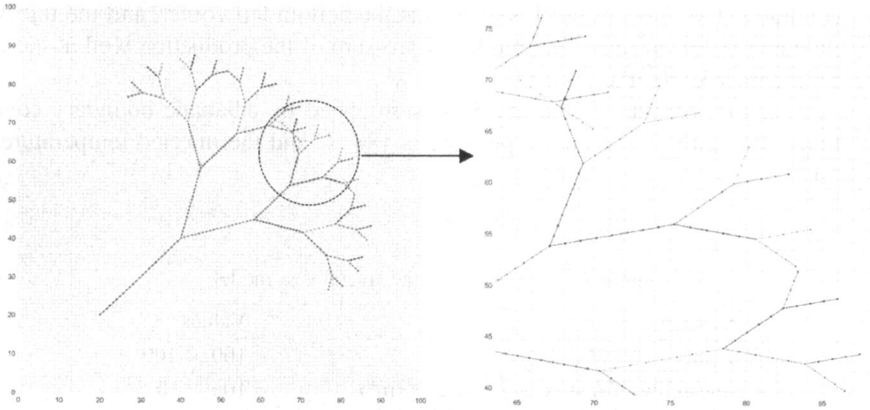

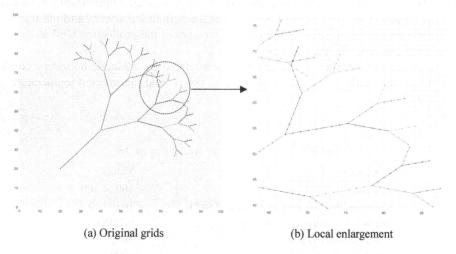

(a) Original grids (b) Local enlargement

Fig. 8. Computational grids of fractal fracture tree

4.1 Temporal and Spatial Evolutions of Pressure Field

Figure 9a, b and c show the spatial distributions of pressure field in thermal reservoirs at different years. The existence of fractures makes the seepage field exhibits anisotropy. Because of the high permeability of fractured media, it provides the main pathways of fluid. Therefore, the pressure in fractured media decreases rapidly. Due to an artificial fracture system existing between the injection well and the production well, the hydraulic exchange between the matrix and the fracture is significant in this area, which leads to a rapid rise in pressure of the surrounding matrix.

As the time of exploitation goes on, the pressure near the injection well increases gradually, while the pressure near the production well decreases gradually. This is mainly due to the obvious temperature difference between wells. In the early stage of exploitation, with injecting the cold water, the temperature near the injection well decreases rapidly. The density and viscosity become large, which causes the high flow resistance in the area, whereas the trend near production well is the opposite. However, after 20 years of exploitation, the temperature near the production well began to decrease. Therefore, the density and viscosity of fluid increase, which leads to the increasing pressure gradient around the production well, while that around the injection well is the opposite. If we do not consider the changes in density and viscosity with pressure and temperature in the simulation, the evolutions of the pressure field around the wells cannot be observed. This shows that the influences of fluid properties on the pressure field cannot be ignored.

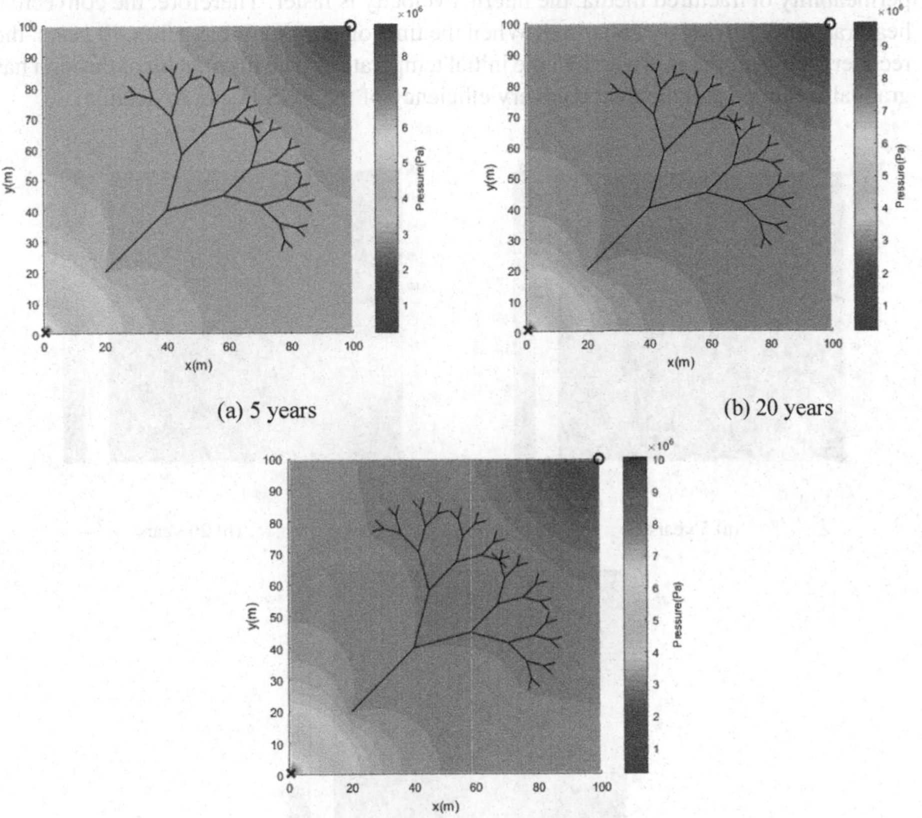

(a) 5 years

(b) 20 years

(c) 40 years

Fig. 9. Pressure field distributions at various time

4.2 Temporal and Spatial Evolutions of Temperature Field

Figure 10a, b and c show the spatial distributions of temperature field in thermal reservoirs for 5, 20 and 40 years. It can be seen that the range of temperature decrease gradually expands to the production well with the growing time of exploitation. After 5 years of exploitation, the cold front has advanced to the root zone of fracture tree. In the early stage of exploitation, the heat exchange between low-temperature fluid and high-temperature rock mass is mainly through the heat conduction. The temperature change near the injection well is roughly an arc-shaped distribution.

The leading edge of temperature after 5 years extends to a distance of about 56 m from the injection well and the trailing edge, which is consistent with the bottom hole temperature, is about 10 m. When the exploitation time is 20 years, the cold front has been pushed into most areas of the fracture trees. At this time, the effect of fracture conductivity become prominent and the low-temperature zone in the reservoir is also enlarged. Meanwhile, the temperature distributions show anisotropy. Because of the high permeability of fractured media, the internal velocity is faster. Therefore, the convective heat transfer is obviously enhanced. When the time of exploitation reaches 40 years, the recovery temperature is lower than the initial temperature. The thermal breakthrough has gradually formed and the heat recovery efficiency of the EGS began to decline.

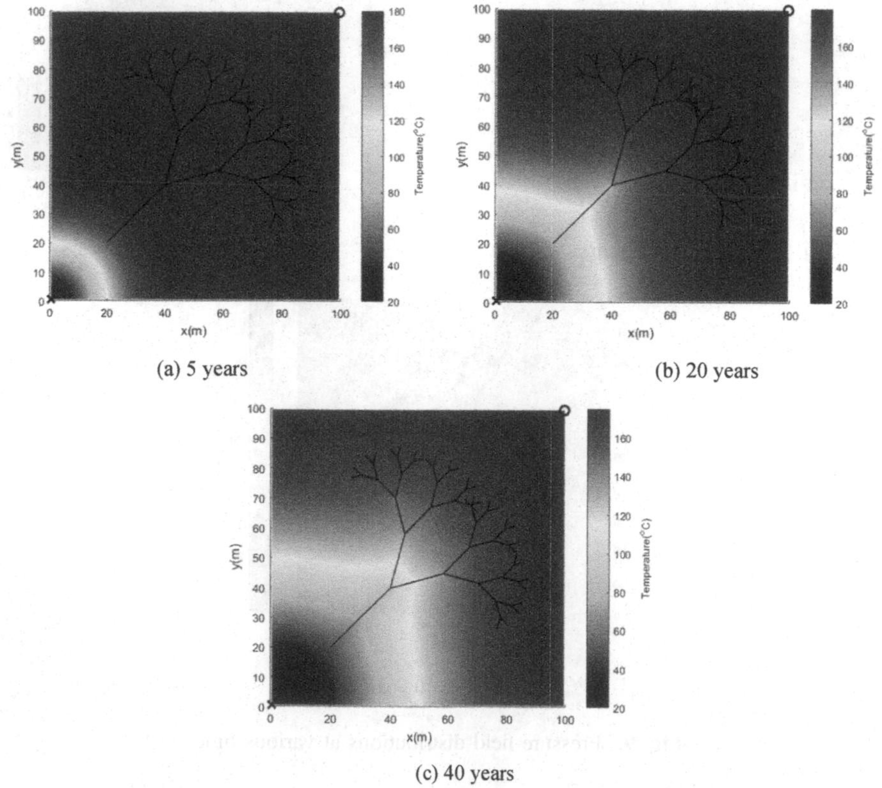

Fig. 10. Temperature field distributions at various time

5 Conclusion

This paper provides a way to simulate the process of geothermal energy exploitation. According to the idea of differentiating matrix and fracture in the EDFM, two sets of governing equations are adopted to describe the fluid flow and heat transfer processes of matrix and fracture, respectively. The proposed model also considers the influence of fluid properties. The spatial-temporal evolutions of pressure and temperature fields are analyzed for a hypothetical EGS case. The main conclusions are as follows.

(1) The accuracy of the proposed model and in-house code is proved by comparing with commercial software, COMSOL Multiphysics. For numerical simulation of the flow and heat transfer in EGS with fractures, the finite volume method based on the embedded discrete fracture model can be a good alternative. The matrix can be divided by the structured grids, which is not restricted by the fractures and avoids using the complicated unstructured grids.

(2) The distributions of the pressure field are mainly affected by the fracture distributions. The pressure of the matrix will increase rapidly in areas with dense fractures. The interwell pressure gradient is mainly influenced by the thermal reservoir permeability and the fluid properties. The distribution of leading edge of the temperature field is obviously affected by the fracture conductivity, while the trailing edge is mainly affected by the injected water temperature. The interwell temperature distributions are greatly influenced by the through fractures.

Acknowledgments. The study is supported by the National Key R&D Program of China (Grant No. 2016YFE0204200), the Project of Construction of Innovative Teams and Teacher Career Development for Universities and Colleges under Beijing Municipality (No. IDHT20170507), the Program of Great Wall Scholar (No. CIT&TCD20180313), the Beijing Youth Talent Support Program (CIT&TCD201804037), and the National Natural Science Foundation of China (No. 51706021, No. 51804033).

References

1. Pandey, S.N., Chaudhuri, A., Kelkar, S.A.: Coupled thermo-hydro-mechanical modeling of fracture aperture alteration and reservoir deformation during heat extraction from a geothermal reservoir. Geothermics **65**, 17–31 (2017)
2. Gunnar, J., Miller, S.A.: On the role of thermal stresses during hydraulic stimulation of geothermal reservoirs. Geofluids **2017**, 1–15 (2017)
3. Shaik, A.R., Rahman, S.S., Tran, N.H., et al.: Numerical simulation of fluid-rock coupling heat transfer in naturally fractured geothermal system. Appl. Therm. Eng. **31**(10), 1600–1606 (2011)
4. Zeng, Y.C., Su, Z., Wu, N.Y.: Numerical simulation of heat production potential from hot dry rock by water circulating through two horizontal wells at Desert Peak geothermal field. Energy **56**, 92–107 (2013)
5. Zeng, Y.C., Zhan, J.M., Wu, N.Y., et al.: Numerical simulation of electricity generation potential from fractured granite reservoir through vertical wells at Yangbajing geothermal field. Energy **103**(1), 290–304 (2016)

6. Zeng, Y.C., Tang, L., Wu, N., et al.: Numerical simulation of electricity generation potential from fractured granite reservoir using the MINC method at the Yangbajing geothermal field. Geothermics **75**, 122–136 (2018)
7. Hadgu, T., Kalinina, E., Lowry, T.S.: Modeling of heat extraction from variably fractured porous media in Enhanced Geothermal Systems. Geothermics **61**, 75–85 (2016)
8. Vik, H.S., Salimzadeh, S., Nick, H.M.: Heat recovery from multiple-fracture enhanced geothermal systems: the effect of thermoelastic fracture interactions. Renew. Energy **121**, 606–622 (2018)
9. Cheng, W.L., Wang, C.L., Nian, Y.L., et al.: Analysis of influencing factors of heat extraction from enhanced geothermal systems considering water losses. Energy **115**, 274–288 (2016)
10. Huang, W.B., Cao, W.J., Jiang, F.M.: Heat extraction performance of EGS with heterogeneous reservoir: a numerical evaluation. Int. J. Heat Mass Transf. **108**, 645–657 (2017)
11. Jiang, F.M., Chen, J.L., Huang, W.B., et al.: A three-dimensional transient model for EGS subsurface thermo-hydraulic process. Energy **72**, 300–310 (2014)
12. Noorishad, J., Mehran, M.: An upstream finite element method for solution of transient transport equation in fractured porous media. Water Resour. Res. **18**(3), 588–596 (1982)
13. Karimi-Fard, M., Durlofsky, L.J., Aziz, K.: An efficient discrete-fracture model applicable for general-purpose reservoir simulators. SPE J. **9**(2), 227–236 (2004)
14. Sun, Z.X., Zhang, X., Xu, Y., et al.: Numerical simulation of the heat extraction in EGS with thermal-hydraulic-mechanical coupling method based on discrete fractures model. Energy **120**, 20–33 (2017)
15. Yao, J., Zhang, X., Sun, Z.X., et al.: Numerical simulation of the heat extraction in 3D-EGS with thermal-hydraulic-mechanical coupling method based on discrete fractures model. Geothermics **74**, 19–34 (2018)
16. Lee, S.H., Lough, M.F., Jensen, C.L.: Hierarchical modeling of flow in naturally fractured formations with multiple length scales. Water Resour. Res. **37**(3), 443–455 (2001)
17. Yan, X., Huang, Z.Q., Yao, J., et al.: An efficient hydro-mechanical model for coupled multi-porosity and discrete fracture porous media. Comput. Mech. **62**(5), 943–962 (2018)
18. Karvounis, D.C., Jenny, P.: Adaptive hierarchical fracture model for enhanced geothermal systems. SIAM J. Multiscale Model. Simul. **14**(1), 207–231 (2016)
19. Peaceman, D.W.: Interpretation of well-block pressures in numerical reservoir simulation. Soc. Petrol. Eng. J. **18**(3), 183–194 (1978)
20. Peaceman, D.W.: Interpretation of well-block pressures in numerical reservoir simulation with nonsquare gridblocks and anisotropic permeability. Soc. Petrol. Eng. J. **8**(3), 183–194 (1983)
21. Cao, W.J., Huang, W.B., Jiang, F.M.: Numerical study on variable thermophysical properties of heat transfer fluid affecting EGS heat extraction. Int. J. Heat Mass Transf. **92**, 1205–1217 (2016)
22. Wagner, W., Kruse, A.: Properties of water and steam, the industrial standard IAPWS-IF97 for the thermodynamic properties and supplementary equations for other properties. J. Eur. Environ. Plann. Law **12**(1), 78–92 (1998)
23. Leveinen, J.: Composite model with fractional flow dimensions for well test analysis in fractured rocks. J. Hydrol. **234**(3), 116–141 (2000)

Modelling of Thermal Transport in Wire + Arc Additive Manufacturing Process

Edison A. Bonifaz(✉)

Mechanical Engineering Department,
Universidad San Francisco de Quito, Cumbayá, Ecuador
ebonifaz@usfq.edu.ec

Abstract. Due to the simultaneous effects of different physical phenomena that occur on different length and time scales, modelling the fusion and heat affected microstructure of an Additive Manufacturing (AM) process requires more than intelligent meshing schemes to make simulations feasible. The purpose of this research was to develop an efficient high quality and high precision thermal model in wire + arc additive manufacturing process. To quantify the influence of the process parameters and materials on the entire welding process, a 3D transient non-linear finite element model to simulate multi-layer deposition of cast IN-738LC alloy onto SAE-AISI 1524 Carbon Steel Substrates was developed. Temperature-dependent physical properties and the effect of natural and forced convection were included in the model. A moving heat source was applied over the top surface of the specimen during a period of time that depends on the welding speed. The effect of multi-layer deposition on the prediction and validation of melting pool shape and thermal cycles was also investigated. The heat loss produced by convection and radiation in the AM layers surfaces were included into the finite element analysis. As the AM layers itself act as extended surfaces (fins), it was found that the heat extraction is quite significant. The developed thermal model is quite accurate to predict thermal cycles and weld zones profiles. A firm foundation for modelling thermal transport in wire + arc additive manufacturing process it was established.

Keywords: Multi-layer deposition · Weld pool · Thermal cycles · Layer build-up process · Heat source

1 Introduction

Additive Manufacturing (AM) involves a comprehensive integration of materials science, mechanical engineering and joining technologies. Related literature [1–6] contains plenty information on energy insertion, dynamics of the molten pool, and mechanical properties generated by the evolution of thermal (residual) stresses. Directed Energy Deposition (DED) process predominantly used with metals in the form of either powder or wire, is often referred to as "metal deposition" technology [7]. DED encompasses all processes that use a laser, arc or e-beam heat source to generate a melt pool into which feedstock (powder or wire) is deposited. Wire + arc additive manufacturing (WAAM) process is essentially an extension of welding process. The origins of this category can be traced to welding technology, where material can be

J. M. F. Rodrigues et al. (Eds.): ICCS 2019, LNCS 11539, pp. 647–659, 2019.
https://doi.org/10.1007/978-3-030-22747-0_49

deposited outside a build environment by flowing a shield gas over the melt pool [8]. WAAM is a layer-by-layer build-up procedure that can be categorized as a form of multipass arc-welding for additive manufacturing (AM) purposes. In multipass welding, multiple welds interact in a complex thermomechanical manner. In the AM welding procedure, the current metal layers are deposited on top of previously solidified weld beads. Complex geometries and assemblies, weight reduction, rapid design iterations, creation of new graded materials with improved material properties, minimization of lead time and material waste, elimination of tooling, and component repair, are some of the numerous advantages to AM over traditional subtractive manufacturing. A component manufactured by the AM "welding" procedure can achieve equivalent mechanical properties to a forged or cast complex component. However, the surface finish, the accuracy to ensure a constant quality during the build-process, resulting in e.g. varying porosity throughout the part, and the high residual stresses that arise during manufacturing, are questionable. For instance, residual stresses may limit the load resistance and contribute to the formation of thermal cracks [4]. Therefore, to produce a high-quality part or product, it is important to establish a methodology for characterizing direct energy deposited metals by linking processing variables to the resulting microstructure and subsequent material properties.

Temperature history and thermal cycles during DED directly dictate the weld pool dimensions and resulting microstructure. However, the temperature magnitude and the molten pool size are very challenging to measure and control due to the transient nature and small size of the molten pool. Of this manner, modelling is needed to enable development of predictive relationships between material microstructure, material properties, and the resulting material lifecycle performance. The prediction of material lifecycle performance as a function of composition and processing parameters for DED-AM methods represents a grand challenge and consequently research in the following topics are necessary: (1) Prediction of microstructure evolution during single and multiple laser, arc, or e-beam passes (2) Prediction of the size, shape, and distribution of voids, inclusions and other defects during single and multiple laser, arc, or e-beam passes, and (3) Model validation focused on demonstrating a functional relationship between both the material composition and the AM processing parameters and resulting microstructure.

Material properties are dependent upon the microstructural characteristics of the part, while microstructure depends on the thermal gradients and cooling rates produced in the metal AM process. It is recognized that the material properties for AM parts are a strong function of the welding processing parameters. Thus, a fundamental understanding of how AM components behave in load-bearing applications depends critically on understanding the evolution of thermal (residual) stresses during component fabrication. The interaction of the residual stresses with localized stress concentrations and crack-like defects must also be taken into account to predict component reliability in load bearing and thermomechanical applications. In spite of solidification and melt pool dynamics influence the microstructure and defect development, the application of fundamental solidification theories to simulate the AM process has not been fully explored [8, 9]. Of the same manner, a strategy to the validation and assessment through the experimentation to demonstrate that the microstructure can be consistently and accurately predicted as a function of the processing parameters, is still required.

Several modelling techniques have been proposed in the literature [7–12]. Most of them are made to optimize the welding process input parameters by means of the finite element method (FEM). However, to control the formation of thermal cracks, distortion and porosity, a lot of hypothesis are introduced into the numerical analysis. In addition, due to the extreme cooling conditions, the events that follow welding are far from equilibrium and the formation of non-equilibrium phases cannot be avoided [13, 14]. On the other hand, the fine discretization required to capture field variable details in the vicinity of the moving heat source spot, increases the cost of the simulations. Moreover, the transient nature of the heat transfer problem considering the effect of multi-layer deposition on the prediction and validation of melting pool shape and thermal cycles, also increases the computational requirements. Currently, microstructure evolution models exist only for one melting and solidification step, such as would be encountered during an idealized (e.g., single pass) AM process. These simulations do not address the formation of non-equilibrium phases or the effect of multiple heating and cooling cycles, such as those encountered in production AM processes. In the present research, the mentioned effect and the heat loss (convection and radiation) between the weldment (layers) surfaces and the surroundings were included into the FE analysis.

Finite element thermo-plasticity multi-scale procedures in wire and arc AM process need to be developed. The thermo-mechanical analysis using non-standard domain decomposition methods based on the concept of Representative Volume Elements (RVEs) also needs to be addressed. The establishment of a methodology for characterizing direct energy deposited metals by linking processing variables to the resulting microstructure and subsequent material properties, is a pendent research goal. The present work aims to build a firm foundation for modeling thermal transport in wire + arc additive manufacturing process. The increasing use of state-of-the-art experimental techniques and also the increasing emphasis on the computation of unsteady flow phenomena are generating many highly informative moving flow sequences.

Industrially focused projects look for fundamental studies of modeling microstructure and stress formation across multiple scales. That is, thermal cycles at the macro, meso and micro-scale levels should be calculated to predict macro, meso and micro residual stresses. Although significant advances have been made, a multi-pass arc welding process has not yet been incorporated into a commercially available additive manufacturing system. It means that a fully automated system using wire + arc welding to additively manufacture metal components still need to be developed. In particular, the development of an automation software required to produce CAD-to-part capability is necessary. The focus of future work is also to integrate thermodynamics and materials thermo-mechanical modeling to provide a multidisciplinary solution to the control of microstructure, residual stress and surface finish in the DED-AM deposition process.

2 The Thermal Model

The field variable temperature T(x; y; z; t) at any location (x; y; z) and time (t) with respect to the moving heat source is calculated by solving the heat diffusion equation:

$$\frac{\partial}{\partial x}\left(k\frac{\partial T}{\partial x}\right)+\frac{\partial}{\partial y}\left(k\frac{\partial T}{\partial y}\right)+\frac{\partial}{\partial z}\left(k\frac{\partial T}{\partial z}\right)+\dot{Q}=\rho c_p\frac{\partial T}{\partial t} \tag{1}$$

Here, ρ is the density, c_p is the specific heat, k is the thermal conductivity, T is temperature, t is time, and $\dot{Q}$ the internal heat source term. In the present research, $\dot{Q}$ is zero and the latent heat was not considered. The convective and radiative energy outflow is calculated with Eq. (2)

$$-k\frac{\partial T}{\partial y}|_{top}+q(r)=h_t(T-T_s)+\sigma\varepsilon(T^4-T_s^4) \tag{2}$$

Here, $h_t = 10\ \mathrm{W\ m^{-2}\ K^{-1}}$ is the convection coefficient at the workpiece surfaces, T_s is the surrounding temperature, ε is the emissivity and σ is the Stefan-Boltzman constant. Due to the flow of the shielding gas, the region directly beneath the nozzle of the torch experiences forced convection. In this area, a value of $h_t = 242\ \mathrm{W\ m^{-2}\ K^{-1}}$ it was used. The ABAQUS [15] user subroutine FILM was constructed to account the convection and radiation effect, and the user subroutine DFLUX was written to account for the heat input from the moving heat source to the workpiece. The Gaussian power density distribution (Eq. 3) was used to represent the moving heat source. It was applied during a period of time that depends on the welding speed (v).

$$q(x,z,t)=\frac{3Q}{\pi C^2}exp^{\{-3[(z-vt)^2+x^2]/C^2\}} \tag{3}$$

Here, $Q = \eta Vi$, η is the process efficiency, V is voltage, i is electric current, and C is the heat distribution parameter.

Table 1. Other data used in the simulations

Property	Value
Thermal efficiency (η)	0.7
T_o (room temperature)	20 °C
Density steel (ρ_w) at T_o	7820 kg/m^3
Density alloy IN-738 at T_o	8110 kg/m^3
Surface emissivity	0.7

3 The Material Model

It has been proved that thermal conductivity and specific heat change significantly when material is heated up to liquid phase from solid phase [16]. These physical properties for added material and solid substrate are also different. Therefore, the inclusion of these property changes in DED simulations is extremely important. As the velocity of the molten metal in the weld pool increases with temperature, an effective high thermal conductivity in the molten region was used to represent the weld pool stirring [1]. The physical properties of the SAE-AISI 1524 carbon steel used in reference [17] are plotted in Fig. 1.

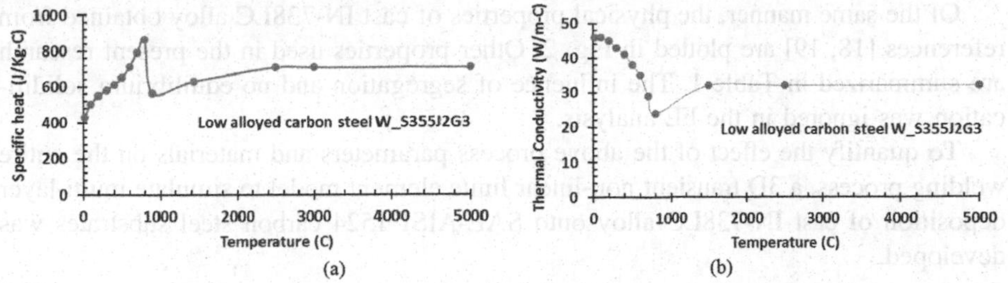

Fig. 1. Physical properties of the SAE-AISI 1524 carbon steel (a) specific heat (b) thermal conductivity.

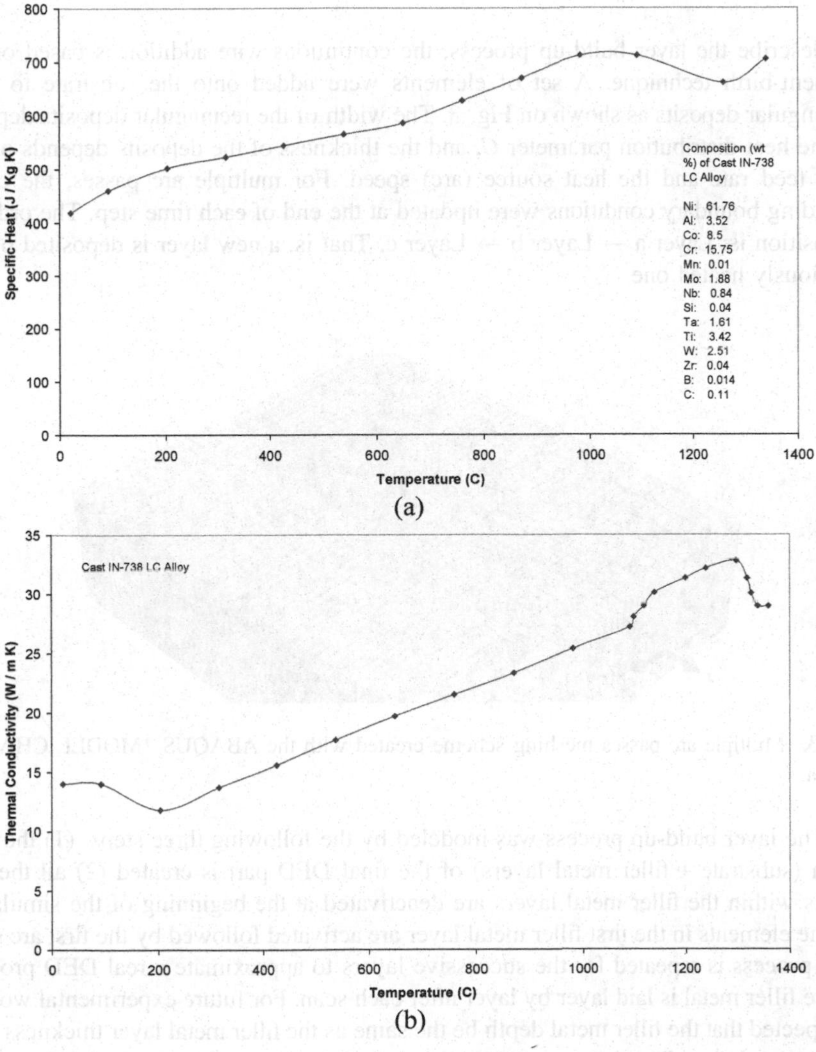

Fig. 2. Physical properties of cast IN-738LC alloy (a) specific heat (b) thermal conductivity.

Of the same manner, the physical properties of cast IN-738LC alloy obtained from references [18, 19] are plotted in Fig. 2. Other properties used in the present research are summarized in Table 1. The influence of segregation and no equilibrium solidification was ignored in the FE analysis.

To quantify the effect of the above process parameters and materials on the entire welding process, a 3D transient non-linear finite element model to simulate multi-layer deposition of cast IN-738LC alloy onto SAE-AISI 1524 carbon steel substrates was developed.

4 A Physics-Based Method to Describe the Layer Build-Up Process

To describe the layer build-up process, the continuous wire addition is based on the element-birth technique. A set of elements were added onto the substrate to form rectangular deposits as shown on Fig. 3. The width of the rectangular deposits depends on the heat distribution parameter *C*, and the thickness of the deposits depends on the wire feed rate and the heat source (arc) speed. For multiple arc passes, the corresponding boundary conditions were updated at the end of each time step. The order of deposition is: Layer a → Layer b → Layer c. That is, a new layer is deposited over a previously heated one.

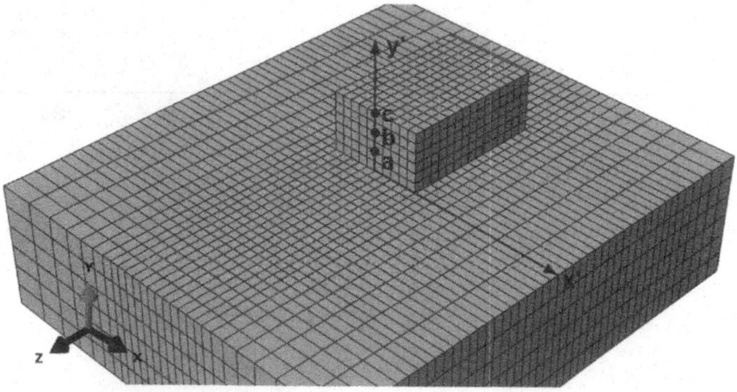

Fig. 3. Multiple arc passes meshing scheme created with the ABAQUS *MODEL CHANGE option.

The layer build-up process was modeled by the following three steps: (1) the total mesh (substrate + filler metal layers) of the final DED part is created (2) all the elements within the filler metal layers are deactivated at the beginning of the simulation (3) the elements in the first filler metal layer are activated followed by the first arc scan. This process is repeated for the successive layers to approximate a real DED process, where filler metal is laid layer by layer after each scan. For future experimental work, it is expected that the filler metal depth be the same as the filler metal layer thickness used in the simulation. The sequence of the layer build-up process is shown in Fig. 4.

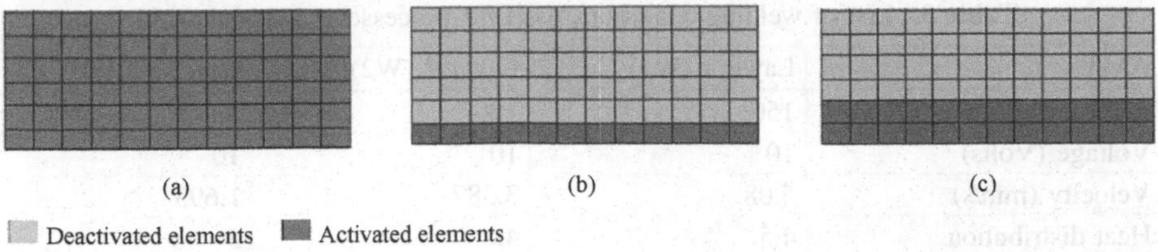

Fig. 4. (a) Filler metal mesh (b) Scan of 1st layer (c) Scan of 2nd layer

5 The Heat Source Material Interaction

The calculations conducted in the present research are divided in two parts: In the *first part*, two AM tests (AM1 and AM2) were made using the welding parameters documented in Table 2. Two layers of filler metal (alloy IN-738) were deposited in a SAE-AISI 1524 carbon steel substrate. The heat distribution parameter, *C*, was determined based on the experimental measurement of the fusion zone width reported in Fig. 3 of reference [2]. The shape and distribution of the heat input greatly depends on its value. It is important to note that in this first part, convection and radiation heat loss from the layers surfaces *were not* included into the FE analysis.

Table 2. List of welding conditions used for processes AM1 and AM2

AM1	Layer 1 (Weld1)	Layer 2 (Weld2)
Voltage (Volts)	9.5	9.5
Intensity (Amps)	150	150
Heat distribution parameter *C* (mm)	3.65	3.65
Weld velocity (mm/s)	2	2
AM2	*Layer 1 (Weld1)*	*Layer 2 (Weld2)*
Voltage (Volts)	10.5	10.5
Intensity (Amps)	200	200
Heat distribution parameter *C* (mm)	5.3	5.3
Weld velocity (mm/s)	2	2

In the *second part,* two additional AM tests (AM3 and AM4) were made using the welding parameters documented in Table 3. Three layers of alloy IN-738 were deposited in a SAE-AISI 1524 carbon steel substrate. The effect of convection and radiation heat loss from the layers surfaces *were* included into the FE analysis.

Table 3. List of welding conditions used for processes AM3 and AM4

AM3	Layer 1 (W1)	Layer 2 (W2)	Layer 3 (W3)
Intensity (Amps)	150	100	50
Voltage (Volts)	10	10	10
Velocity (mm/s)	5.08	3.387	1.693
Heat distribution parameter C (mm)	4.5	4.5	3
AM4	*Layer 1 (W1)*	*Layer 2 (W2)*	*Layer 3 (W3)*
Intensity (Amps)	120	120	120
Voltage (Volts)	10	10	10
Velocity (mm/s)	3.125, 6.25 and 12.5 in turn	3.125, 6.25 and 12.5 in turn	3.125, 6.25 and 12.5 in turn
Heat distribution parameter C (mm)	4.5	4.5	4.5

To predict the evolution of temperature distribution in the entire weldment (substrate, two and three cast IN-738LC alloy layers) for the entire welding and cooling cycle of the process, a 3D transient nonlinear heat flow analysis was performed. To observe the heat transference among layers, all welding layers had gluing contacts. Figure 5 shows the experimental set-up, specimen dimensions, and x, y and z directions for AM weld tests.

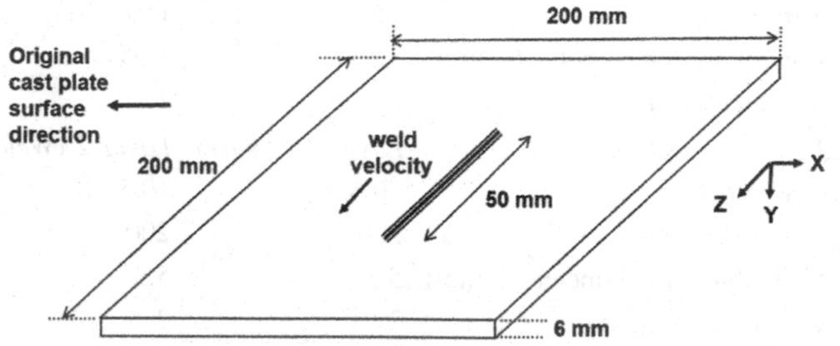

Fig. 5. Schematic of the experimental set-up for AM weld tests.

The initial temperature T_0 was set to 20 °C. The designed mesh is shown in Fig. 6a for processes AM1 and AM2, and in Fig. 6b for processes AM3 and AM4.

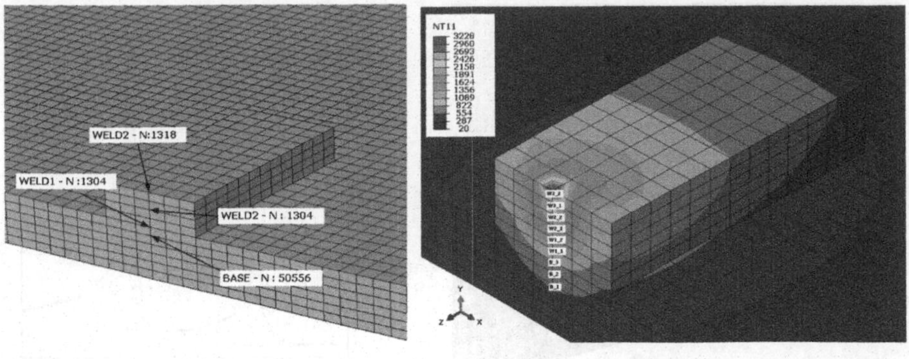

(a) (b)

Fig. 6. Selected nodes in a mesh composed of SAE-AISI 1524 Carbon Steel Substrate (200 × 6 × 200 mm) and layers of alloy IN-738 (12 × 2 × 50 mm each) (a) Mesh used for processes AM1 and AM2 (b) Mesh used for processes AM3 and AM4. The selected nodes are located in the central XY cross-sectional plane (i.e., from the top of cross section along Y axis). Element size 2 × 1 × 2 mm. Element type DC3D8.

6 Results and Discussions

Numerically predicted thermal gradients, isotherms and thermal cycles produced by the AM process were calculated with the developed thermal model. For the *first part* of this work, the following results corresponding to the processes AM1 and AM2 were obtained in a mesh representing two wire layers (composed of four finite element rows) plus the substrate (Fig. 6a). The welding conditions used in the simulations are shown on Table 2. Figure 7 shows thermal contourns and thermal cycles calculated at the reported nodes.

Compared with the AM2 process, higher peak temperatures are observed in the AM1 process. The reason is because in the AM1 process, a higher heat input is deposited to the workpiece.

The following results are obtained for the *second part* of this work. For instance, Fig. 8 shows thermal cycles calculated at the locations specified in Fig. 6b for process AM3. Even though the heat input is keep constant, it is apparent in Fig. 8 that as the weld velocity and heat distribution parameter decreases, the peak temperature increases. For these particular welding conditions, the effect of the heat distribution parameter is more representative.

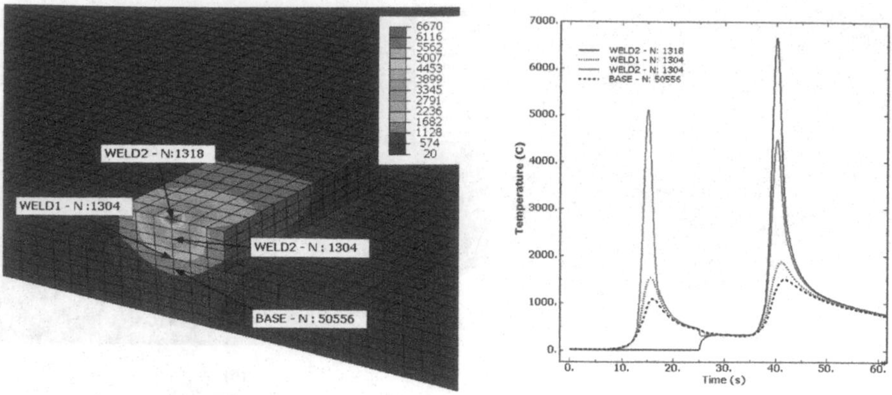

(a)

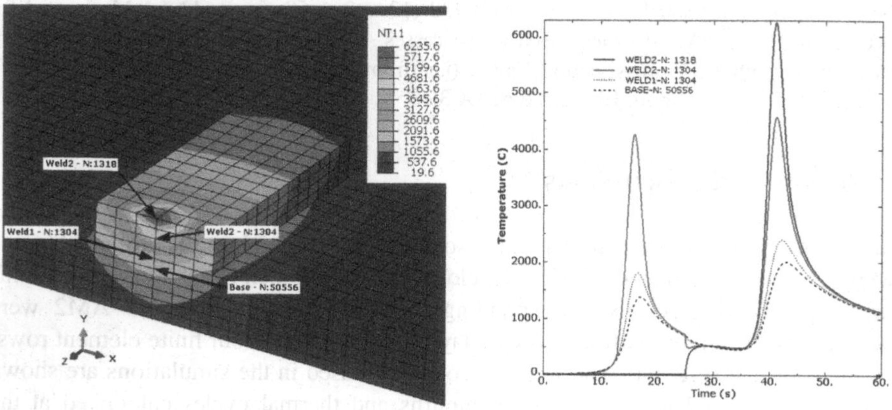

(b)

Fig. 7. Thermal contourns and thermal cycles for the AM process (a) For the AM1 process (b) For the AM2 process. Welding parameters documented in Table 2.

Figure 9 shows thermal cycles calculated at the reported nodes located in the substrate and in the three layers for the process AM4 at weld velocities of 12.5, 6.25 and 3.125 mm/s. The heat source is applied over the top surface of each layer (50 mm length) during a lapse of time that depends on the welding speed. The welding parameters are documented in Table 3.

It is observed in Fig. 9 that as the weld velocity decreases, higher peak temperature arise. This is because the heat input (*HI*) described by Eq. (4), is inversely proportional to the weld velocity

$$HI = \frac{\eta Vi}{v} \tag{4}$$

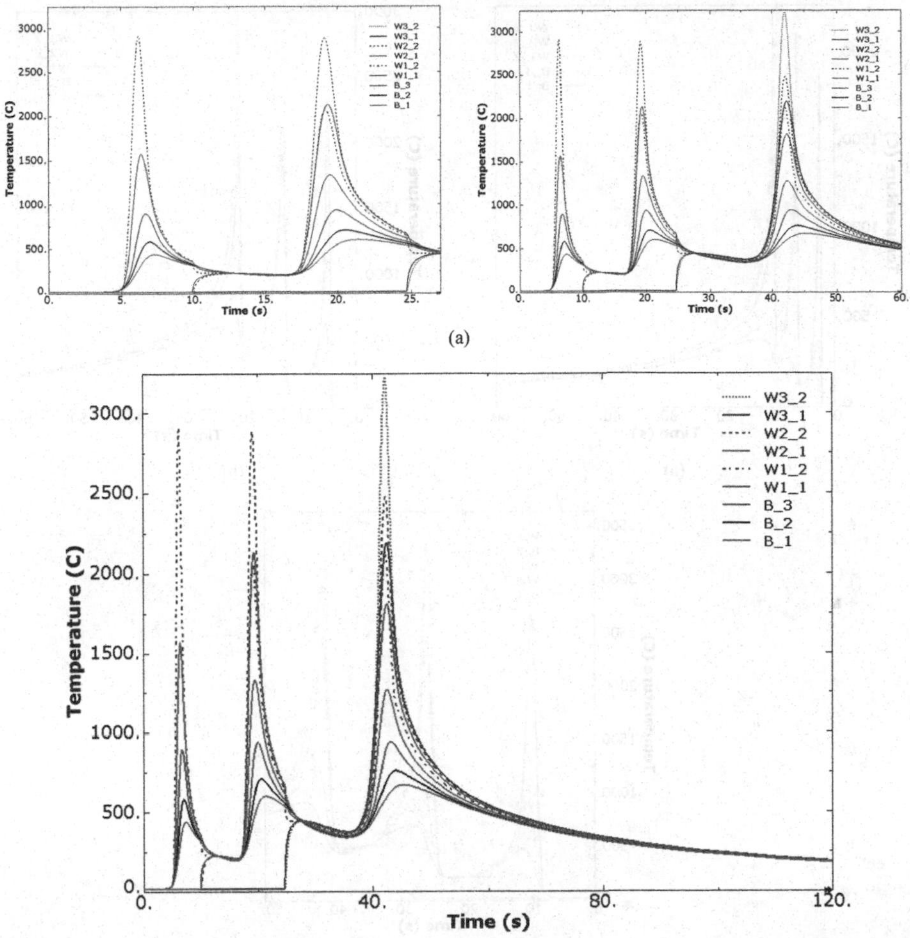

(b)

Fig. 8. Thermal cycles for the AM3 process (a) Cooling time limited to 30 and 60 s (b) Cooling time limited to 120 s. Welding parameters documented in Table 3.

As stated above, η is the process efficiency, V is voltage, i is electric current, and v is the weld velocity. Of the same manner, as the heating and cooling curves are different in a same analyzed location, the corresponding mechanical properties will be different as different the AM process. It is necessary to mention that the thermal model is quite accurate to predict thermal cycles and weld zones profiles. It is also important to note that the peak temperatures calculated in the processes AM3 and AM4 considered in this *second part*, are smaller than those obtained in the *first part*, that is, in the processes AM1 and AM2. The reason is because the heat extraction by convection and radiation through the surfaces of the AM layers included into the FE analysis in the second part, is quite significant. The AM layers itself act as extended surfaces (fins).

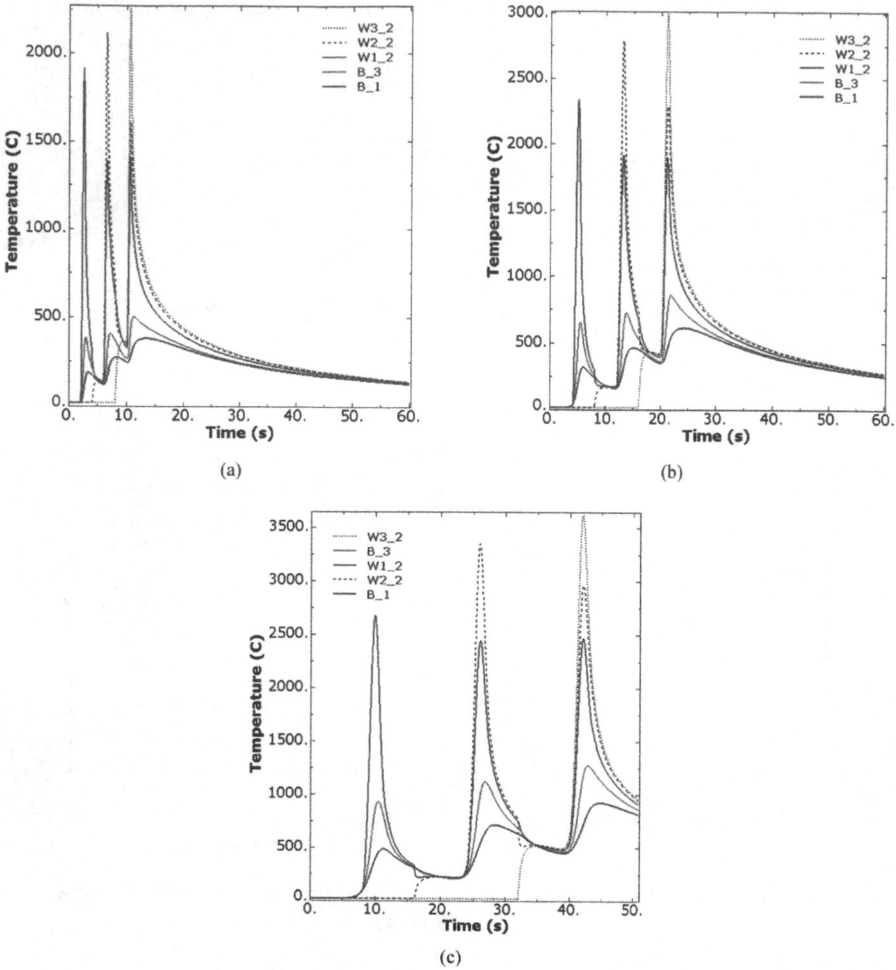

Fig. 9. Thermal cycles calculated at the reported nodes located in the substrate and in the three layers for different weld velocities (a) For a weld velocity of 12.5 mm/s (b) For a weld velocity of 6.25 mm/s (c) For a weld velocity of 3.125 mm/s.

7 Conclusions

1. A firm foundation for modeling thermal transport in wire + arc additive manufacturing process it was established.
2. A three dimensional transient non-linear finite element thermal model has been developed to generate weld profiles, thermal gradients, and thermal cycles in multi-layer deposition of cast IN-738LC alloy onto SAE-AISI 1524 Carbon Steel Substrates.
3. The effect of multi-layer deposition on the prediction and validation of melting pool shape and thermal cycles was also investigated.

4. The effect of convection and radiation heat loss from the layers surfaces were included into the FE analysis. As the AM layers itself act as extended surfaces (fins), it was found that the heat extraction is quite significant.
5. The developed thermal model is quite accurate to predict thermal cycles and weld zones profiles.

Acknowledgment. E. A. Bonifaz acknowledges to the Centro de Estudios e Investigaciones Técnicas de Guipuzcoa (CEIT) in Spain for allowing the use of its facilities during a short stay of scientific cooperation.

References

1. Bonifaz, E.A.: Finite element analysis of heat flow in single/pass arc welds. Welding J. **79** (5), 121_s–125_s (2000)
2. Bonifaz, E.A.: Thermo-mechanical analysis in SAE-AISI 1524 carbon steel gas tungsten arc welds. Int. J. Comput. Mater. Sci. Surface Eng. **7**(3/4), 269–287 (2018)
3. Bonifaz, E.A., Richards, N.L.: Modeling cast In-738 superalloy gas-tungsten-arc- welds. Acta Mater. **57**, 1785–1794 (2009)
4. Kou, S.: Welding Metallurgy, 2nd edn. Wiley, New York (2003)
5. Bonifaz, E.A., Richards, N.L.: Stress-strain evolution in cast IN-738 superalloy single fusion welds. Int. J. Appl. Mech. **2**(4), 807–826 (2010)
6. Brown, S.B., Song, H.: Implications of three-dimensional numerical simulations of welding of large structures. Welding J. **71**(2), 55-s–62-s (1992)
7. Gibson, I., Rosen, D., Stucker, B.: Additive Manufacturing Technologies. Springer Science, Heidelberg (2015)
8. Sames, W.J., List, F.A., Pannala, S., Dehoff, R., Babu, S.S.: The metallurgy and processing science of metal additive manufacturing. Int. Mater. Rev. **61**(5), 315–360 (2016)
9. Gu, D., Meiners, W., Wissenbach, K., Poprawe, R.: Laser additive manufacturing of metallic components: materials, processes and mechanisms. J. Int. Mater. Rev. **57**(3), 133–164 (2012)
10. http://www.farinia.com/additivemanufacturing/3dtechnique/modelingsimulationinmetaladditivemanufacturing
11. http://www.farinia.com/additivemanufacturing/3dtechnique/additivelayermanufacturing
12. Lindgren, L., Lundbäck, A., Fisk, M., Pederson, R., Andersson, J.: Simulation of additive manufacturing using coupled constitutive and microstructure models. Addit. Manuf. **12**, 144–158 (2016)
13. Debroy, T., David, S.A.: Physical processes in fusion welding. Rev. Modern Phys. **67**(1), 85–112 (1995)
14. Thiessen, R.G., Richardson, I.M.: A physically based model for microstructure development in a macroscopic heat-affected zone: grain growth and recrystallization. Metall. Mater. Trans. B **37**(4), 655–663 (2006)
15. ABAQUS documentation manual, V. 6.12. Dassault Systèmes Simulia Corp., Providence, RI, USA
16. Mudge, R.P., Wald, N.R.: Laser engineered net shaping advances additive manufacturing and repair. Weld. J. **86**(1), 44–48 (2007)
17. SYSWELD-Visual Weld material data base. Material Database Manager (2011)
18. Alloy IN-738 Technical Data. INCO, the International Nickel Company, Inc. One New York Plaza, New York, N.Y. 10004
19. J MatPro 4.1, Thermotech Sente Software © 2007. http://www.thermotech.co.uk/

Author Index

Printed in the United States
By Bookmasters